D0077103

Handbook for Electronics Engineering Technicians

OTHER McGRAW-HILL HANDBOOKS OF INTEREST

Handbook for Electronics Engineering Technicians

Editors

MILTON KAUFMAN
President
Electronic Writers and Editors, Inc.

ARTHUR H. SEIDMAN
Professor of Electrical Engineering
Acting Dean
Pratt Institute, School of Engineering

PERRY J. SHENEMAN, Associate Editor

McGRAW-HILL BOOK COMPANY
New York St. Louis San Francisco Auckland Bogotá
Düsseldorf Johannesburg London Madrid Mexico
Montreal New Delhi Panama Paris São Paulo
Singapore Sydney Tokyo Toronto

Library of Congress Cataloging in Publication Data

Main entry under title:

Handbook for electronics engineering technicians.
Includes index.
1. Electronics – Handbooks, manuals, etc. I. Kaufman, Milton. II. Seidman, Arthur H. III. Sheneman, Perry.
TK7825.H37 621.38 75-40457
ISBN 0-07-033401-3

4567890 DODO 7987

The editors for this book were Harold B. Crawford, Ross J. Kepler, and Betty Gatewood, the designer was Naomi Auerbach, and the production supervisor was Teresa F. Leaden.

To the fond memory of
David M. Kaufman
and
Rose and Isidore Sieser

Contents

3. COILS

4. MAGNETIC CIRCUITS

5. TRANSFORMERS

6. PRACTICAL CIRCUIT ANALYSIS

7. METERS AND MEASUREMENTS

8. SEMICONDUCTOR DEVICES AND TRANSISTORS

9. INTEGRATED CIRCUIT TECHNOLOGY

10. TUNED CIRCUITS

11. FILTERS

12. TRANSISTOR AMPLIFIERS AND OSCILLATORS

13. OPERATIONAL AMPLIFIERS

14. DIGITAL CIRCUIT FUNDAMENTALS

15. DIGITAL INTEGRATED CIRCUITS

16. POWER SUPPLIES

17. BATTERIES

Index follows the Appendix.

Preface

This is the first fully comprehensive handbook designed to meet the day-to-day needs of electronics technicians. While useful to engineers as well, it does not require an extensive background in high-level engineering principles and techniques. All that is required is a background no more extensive than an education in an electronics technical school or a two-year community college. And in many cases, because of the handbook's simplified approach, even a lesser background would be sufficient.

The handbook treats fundamental topics in discrete circuits, and also in analog and digital integrated circuits, from the point of view of practical applications. Each topic is illustrated with practical, numerical, worked-out examples that can be applied to the reader's own particular problems. Numerous practical examples illustrate, for example, how to choose the proper resistor, capacitor, transistor, integrated circuit, or operational amplifier and how to find current in, or voltage across, an element in dc and ac circuits.

Each one of the eighteen in-depth sections follows the same practical, concise format that has been developed to help readers find the information they need quickly and easily. The general approach to each topic in a section is as follows:

1. Definition of terms and parameters
2. Breakdown of the types and characteristics of components
3. Analysis of the basic and special functions
4. Detailed practical problems and clearly worked-out solutions
5. Clarifying charts, tables, nomographs, and illustrations

With this easy-to-follow format, every effort has been made to ensure that this handbook will have the greatest usefulness to its readers.

All mathematics, both in theory sections and in worked-out problems, has been kept to the level of relatively simple algebra or arithmetic. No calculus is employed.

The editors wish to gratefully acknowledge the cooperation and assistance of various individuals and companies in providing manuscript material, photographs, tables, graphs, and other technical data. Due credit is given within the book as applicable. We wish particularly to thank the following persons for their invaluable contributions: Donald R. Phillips of RCL Electronics, Inc; Dr. Howard J. Strauss of Gould, Inc.; J. A. "Sam" Wilson of Bank Wilson Services; Joshua A. Hauser of Lambda Electronics, Inc.; Raymond C. Miles; Bill Stutz; and the late Louis Toth. In addition, we wish to thank Sharon Wilson for her considerable efforts in typing and editorial services.

Milton Kaufman

Arthur H. Seidman

Handbook for Electronics Engineering Technicians

Chapter 1

Characteristics of Resistors

1.1 INTRODUCTION

Every physical material impedes the flow of electric current to some degree. Materials such as copper offer hardly any resistance to current flow; copper, therefore, is called a *conductor*, or a material having negligible resistance. Other materials, such as ceramic, which offer extremely high resistance to current flow are referred to as *insulators*.

In electric and electronic circuits, there is a need for materials with specific values of resistance in the range between that of a conductor and an insulator. These materials are called *resistors* and their values of resistance are expressed in ohms (represented by the Greek letter omega, Ω). The large variety of types and forms of discrete resistors available are indicated in Fig. 1.1.

Resistors may be classified as being *fixed* or *variable* in their value. Variable resistors are commonly referred to as *potentiometers*, or *pots*. Electrical symbols for fixed resistors and pots are illustrated in Fig. 1.2.

Resistors may also be classified as *linear* and *nonlinear*. For a linear resistor, as the voltage across it varies, the current flowing varies by a proportionate amount. The behavior of a nonlinear resistor is such that as the voltage varies, the current change is not proportional to the voltage change. Resistors generally used in circuits are linear. For special applications, to be described later, nonlinear resistors are available.

1.2 GENERAL DESCRIPTION

The resistance of any material is given by the following expression:

$$R = \frac{\rho L}{A} \tag{1.1}$$

where R = resistance, Ω
ρ = resistivity of the material, Ω-cm
L = length of material, cm
A = cross-sectional area of material, cm^2

Resistivity ρ (Greek letter rho) is an inherent property of materials. Values of ρ for some commonly used materials are summarized in Table 1.1.

Equation (1.1) illustrates two important facts. For a material of a given resistivity, the resistance varies *directly* with length L and *inversely* with cross-sectional area A. For example, a wire of long length has greater resistance than a short-length wire. Also, a wire of large diameter (large cross-sectional area) has less resistance than a wire with a small diameter.

Fig. 1.1 A sampling of some discrete resistors available to the user. *(Courtesy Victoreen Instrument Division)*

The voltage and current in a resistor are related by Ohm's law:

$$I = \frac{E}{R} \tag{1.2a}$$

$$E = IR \tag{1.2b}$$

$$R = \frac{E}{I} \tag{1.2c}$$

where E = voltage across resistor, V

I = current flowing in resistor, A

Power P (in watts, W) dissipated in a resistor may be expressed by any of the following expressions:

$$P = EI \tag{1.3a}$$

$$= I^2R \tag{1.3b}$$

$$= \frac{E^2}{R} \tag{1.3c}$$

The application of the preceding equations is illustrated in the following examples.

example 1.1 Calculate the resistance of a wire whose length is 10 m and cross-section area is 0.1 cm² if the material of the wire is (*a*) copper and (*b*) Nichrome.

solution From Table 1.1, for copper $\rho = 1.7 \times 10^{-6}$ Ω-cm and for Nichrome, $\rho = 100 \times 10^{-6}$. Because 1 m = 100 cm (units in an equation must always be consistent), $L = 10 \times 100$ cm = 1000

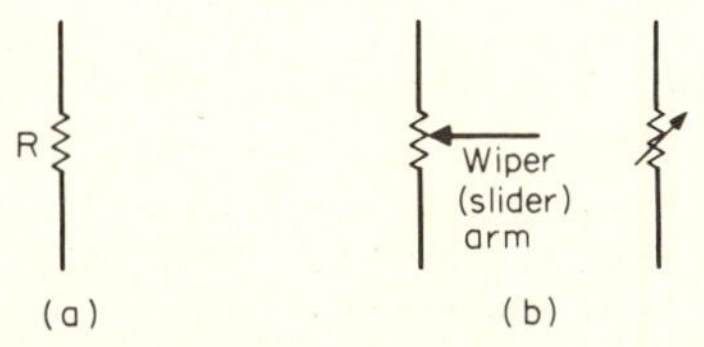

Fig. 1.2 Electrical symbols for resistors: (*a*) Fixed. (*b*) Variable (pot).

TABLE 1.1 Resistivities of Some Commonly Used Materials

Material	ρ, Ω-cm
Silver	1.5×10^{-6}
Copper	1.7×10^{-6}
Aluminum	2.6×10^{-6}
Carbon (graphite)	30×10^{-6} to 190×10^{-6}
Nichrome	100×10^{-6}
Glass	10^{10}–10^{14}

cm. Substitution of the given values in Eq. (1.1), $R = \rho L/A$, yields

(*a*) $R_{\text{copper}} = 1.7 \times 10^{-6}\ \Omega\ \text{cm} \times (1000\ \text{cm})/(0.1\ \text{cm}^2)$
$= 1.7 \times 10^{-2}\ \Omega$

(*b*) $R_{\text{Nichrome}} = 100 \times 10^{-6} \times 1000/0.1 = 1\ \Omega$

example 1.2 If the current flowing in the wires of Example 1.1 is 3 A, calculate the dissipated power in each wire.

solution Using Eq. (1.3*b*) $P = I^2R$, $P_{\text{copper}} = 3^2 \times 1.76 \times 10^{-2} = 9 \times 1.76 \times 10^{-2} = 0.158$ W. $P_{\text{Nichrome}} = 3^2 \times 1 = 9 \times 1 = 9$ W.

1.3 RESISTOR TERMS AND PARAMETERS

In this section, commonly used terms and parameters for characterizing fixed and variable resistors are defined. Where appropriate, typical curves showing the variation in resistance with temperature and other quantities of interest are supplied.

RESISTANCE The unit of resistance is the ohm (Ω). Generally, resistance values in thousands of ohms is expressed in kilohms (kΩ) and millions of ohms in megohms (MΩ). Nominal values of resistors are generally based on 25°C (room temperature) operation.

TOLERANCE Tolerance expresses the maximum deviation in resistance from its nominal value. For example, if the tolerance of a 1000-Ω resistor is ±10 percent, this denotes that the actual value of resistance is in the range of $1000 - 0.1 \times 1000 = 900\ \Omega$ to $1000 + 0.1 \times 1000 = 1100\ \Omega$.

TEMPERATURE COEFFICIENT OF RESISTANCE The temperature coefficient of resistance (abbreviated TCR, or *Tempco*) indicates how resistance changes with temperature. It is expressed as a percentage change in the nominal value at 25°C for each degree Celsius or in parts per million per degree Celsius (ppm/°C) of the resistor. The TCR may be positive or negative. Typical curves showing the variation in resistance with temperature are illustrated in Fig. 1.3. In this figure, a *logarithmic* (log) *scale* is used for plotting nominal resistance (ohms). A log scale permits a large range of values (in this case, 10^1 to 10^8 Ω) to be compressed in a convenient scale.

example 1.3 A 1000-Ω resistor has a TCR of +1000 ppm/°C. Calculate its resistance at 125°C.

solution The quantity 1000 ppm is equal to $10^3/10^6 = 0.001$. The change in resistance from room temperature (25°C) is equal to the product of the difference in temperature (125 − 25°), the TCR, and the nominal value of resistance at 25°C (1000 Ω). Hence, the change in resistance $= (125 - 25) \times 0.001 \times 1000 = 100\ \Omega$. The resistance value at 125°C is, therefore, $1000 + 100 = 1100\ \Omega$.

POWER RATING Power rating is the maximum continuous power, in watts, that a resistor can dissipate at a temperature as high as 70°C. At temperatures beyond 70°C, the power rating of the resistor is reduced, or *derated*. A typical derating curve is given in Fig. 1.4.

example 1.4 A particular resistor is rated at 60 W. Using the derating curve of Fig. 1.4, determine the power rating of the resistor when operating at 130°C.

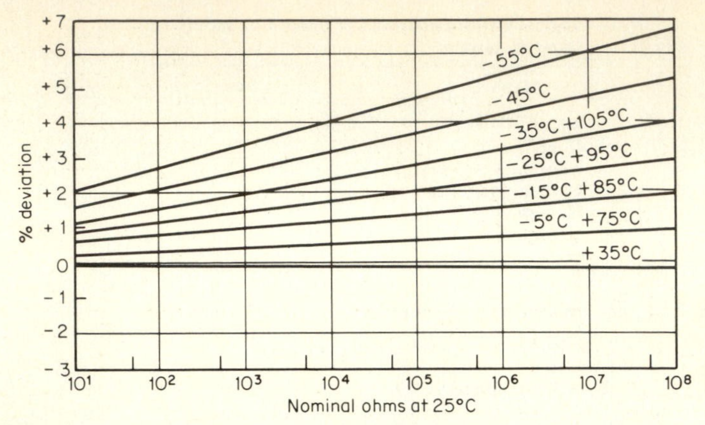

Fig. 1.3 Typical curves showing the variation in nominal resistance with temperature. (*Courtesy Allen-Bradley*)

solution From Fig. 1.4, at 130°C, the rated load is 30 percent. Expressing 30 percent by the decimal 0.3, the rating of the resistor at 130°C is $0.3 \times 60 = 18$ W.

In specifying the power rating of a resistor, for conservative operation it is usual practice to increase the actual dissipated power by a factor of 2. For example, if the calculated dissipation is 0.5 W, specify a $0.5 \times 2 =$ 1-W resistor.

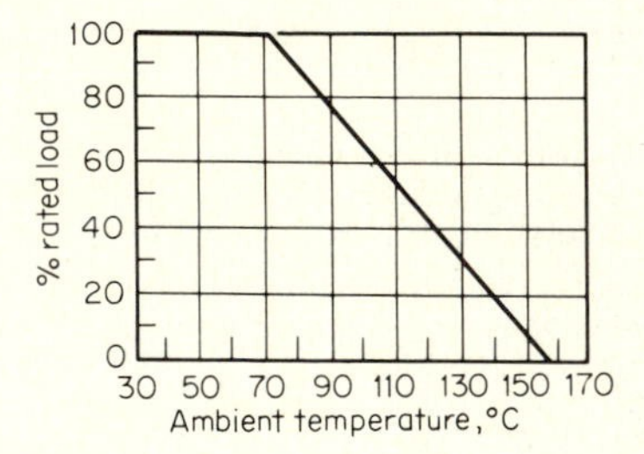

Fig. 1.4 Typical derating curve for a resistor. Ambient temperature refers to the temperature of the environment in which the resistor is operating.

RATED CONTINUOUS WORKING VOLTAGE The rated continuous working voltage (RCWV) is the maximum voltage that can be safely applied to a resistor. Typical curves for RCWV versus resistance are shown in Fig. 1.5. For example, the maximum rated continuous working voltage for a 50-Ω 2-W resistor is 10 V.

CRITICAL RESISTANCE The critical resistance R_c is the value of resistance where the maximum voltage and power rating occur simultaneously. For example, for a resistor rated at 500 V and 1 W, from Eq. (1.3*c*), $R_c = E^2/P = 500^2/1 = 250\,000\ \Omega$.

NOISE Because of the molecular structure of matter, electrons in a material exhibit a random motion. As a result, random voltages are produced which are referred to as *noise*. The noise increases with increasing resistance values, operating temperature, and the bandwidth of the circuit to which the resistor is connected. Typical curves of noise voltage (microvolts) at 25°C versus bandwidth (hertz) are illustrated in Fig. 1.6. It is seen that as the resistance value or bandwidth is increased, the noise voltage becomes greater.

FREQUENCY EFFECTS A *model* (equivalent circuit) of a resistor operating at high frequencies is illustrated in Fig. 1.7. In series with the desired resistance R is induct-

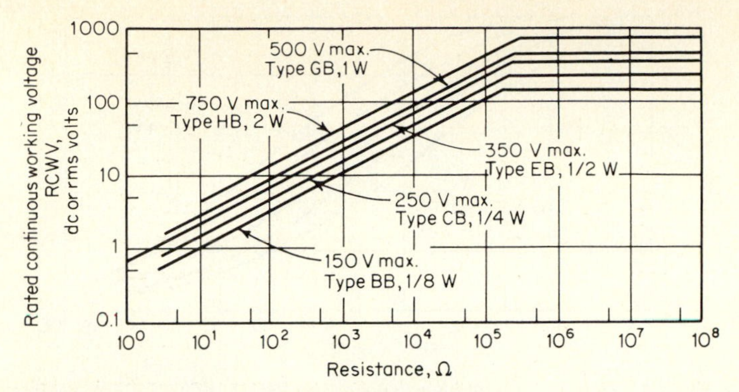

Fig. 1.5 Typical curves of RCWV for different power ratings as a function of resistance. (*Courtesy Allen-Bradley*)

ance L; shunted across R and L is capacitance C. Inductance L arises from the type of construction and connecting leads of the resistor. Capacitance C is present also because of the resistor's construction and the capacitance of the connecting leads. These undesired elements, L and C, are referred to as *parasitics*.

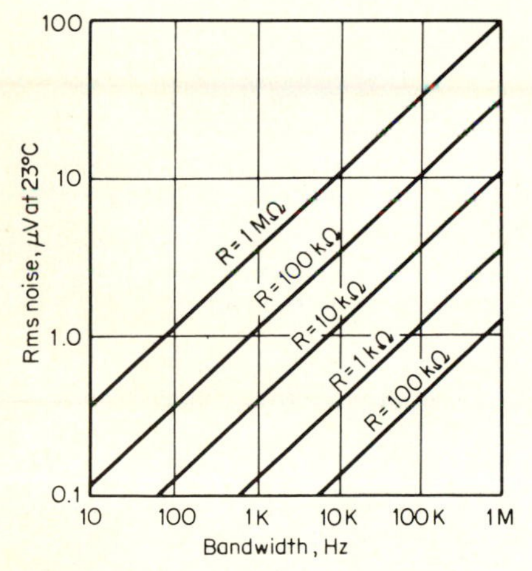

Fig. 1.6 Typical curves of noise for different values of resistance as a function of bandwidth. (*Courtesy MEPCO/ELECTRA, Inc.*)

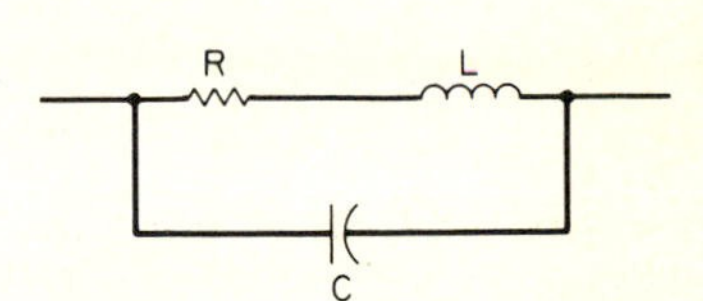

Fig. 1.7 A model of a resistor operating at high frequencies.

Owing to the parasitics in a resistor, the useful frequency range is limited. Typical high-frequency characteristics of a resistor are illustrated in Fig. 1.8. Expression $R_{h\text{-}f}/R_{dc}$ is the ratio of the resistance at high frequencies, $R_{h\text{-}f}$ to the resistance at direct current and low frequencies, R_{dc}. It is seen that as frequency is increased, $R_{h\text{-}f}/R_{dc}$ is reduced.

DRIFT Drift, or *time stability*, refers to the change in resistance value over a time interval of use, such as 1000 h. High-value resistors exhibit more drift than low-value units. Short-term drift is generally more pronounced than long-time drift.

TAPER The taper of a pot refers to the variation of its resistance as a function of the rotation of the wiper, or slider, arm. Examples of taper curves are illustrated in Fig. 1.9.

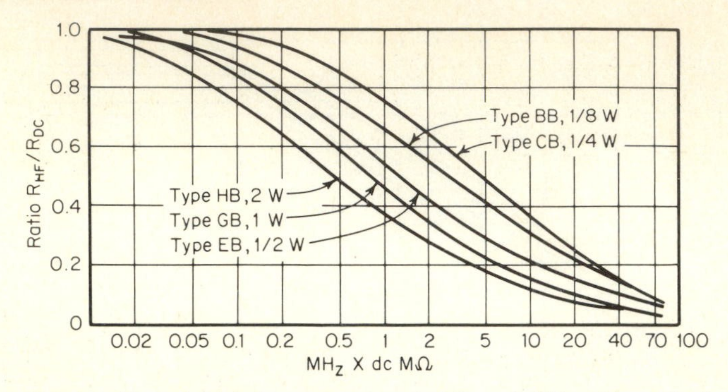

Fig. 1.8 Typical high-frequency characteristics of different-type fixed resistors. (*Courtesy Allen-Bradley*)

A *linear taper* is exhibited by curve *a*. For this taper, the variation in resistance is directly proportional to the displacement of the wiper. *Nonlinear tapers* are indicated by curves *b* and *c*. Curve *b* represents a logarithmic taper, and curve *c* a special nonlinear taper.

RESOLUTION The resolution of a pot is the smallest change in resistance that can be realized as the wiper arm is rotated.

END RESISTANCE End resistance is the resistance between the wiper and end terminals with the wiper positioned at the corresponding end point.

CONTACT RESISTANCE Contact resistance is the resistance between the wiper terminal and the resistance element in contact with the wiper.

WIPER (SLIDER) CURRENT Wiper current is the maximum current that may flow in or out of the wiper terminal.

SETTING STABILITY This parameter indicates the *repeatability* of a pot setting to a given resistance value.

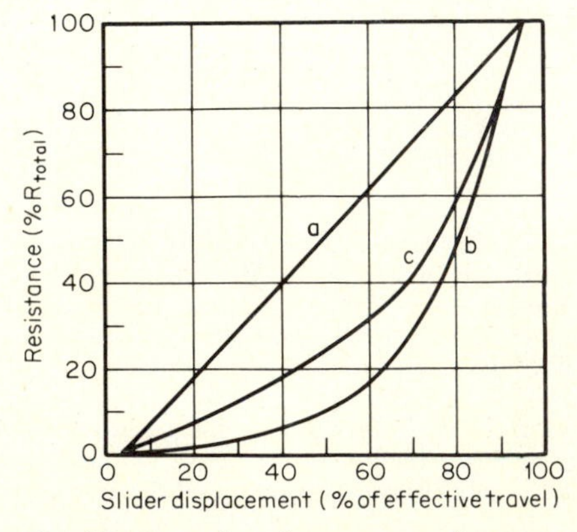

Fig. 1.9 Examples of taper curves for potentiometers. (*Courtesy MEPCO/ELECTRA, Inc.*)

TABLE 1.2 Summary of Fixed and Variable Resistor Types

Category	Type	Key property	Power	Temperature coefficient, °C	Resistance range
General purpose	Carbon composition	Cost	$\frac{1}{8}$–2 W	> 500	1 Ω ≥ 100 MΩ
≥ 5% tolerance	Molded wire-wound	Tempco	$\frac{1}{2}$–2 W	≥ 200	0.1 Ω–2.4 kΩ
≥ 200 ppm	Ceramic wire-wound	Low voltage W	2–50 W	≥ 200	0.1 Ω–30 kΩ
	Metal glaze	Flexibility	$\frac{1}{8}$–5 W	200	4.3 Ω–1.5 MΩ
	Tin oxide	Reliability	$\frac{1}{8}$–20 W	200	4.3 Ω–2.5 MΩ
	Carbon film (import)	Cost	$\frac{1}{4}$–2 W	> 200	10 Ω ≥ 1 MΩ
	Cermet film	Stability	$\frac{1}{4}$–3 W	150	10 Ω–10 MΩ
Semiprecision	Metal glaze	Flexibility	$\frac{1}{8}$–2 W	≤ 200	1 Ω–1.5 MΩ
> 1 < 5% ≤ 200 ppm	Tin oxide	Stability	$\frac{1}{8}$–2 W	≤ 200	4.3 Ω–1.5 MΩ
Power	Ceramic wire-wound	Cost	2–50 W	≥ 200	0.1 Ω–30 kΩ
≥ 2 W	Axial lead coated WW	Auto insertion	$\frac{1}{2}$–15 W	≤ 50	0.1 Ω–175 kΩ
	Tubular and flat WW		4–250 W	≤ 100	0.1 Ω ≥ 1 mΩ
Precision	Metal film	Tolerance	$\frac{1}{10}$–1 W	20	0.1 Ω–1 MΩ
≤ 1%	Metal glaze	Environment	$\frac{1}{10}$–1 W	≤ 100	1 Ω–1 MΩ
≤ 100 ppm	Tin oxide	Power	$\frac{1}{10}$–$\frac{1}{2}$ W	≤ 100	10 Ω–1 MΩ
	Thin film	Size, networks	$\frac{1}{20}$–5 W	≤ 100	10 Ω–100 MΩ
	Encaps. wire-wound	Power, Tempco	$\frac{1}{20}$–1 W	≤ 20	0.1 Ω ≥ 1 MΩ
Ultraprecision	Thin film	Flexibility	$\frac{1}{20}$–$\frac{1}{2}$ W	≤ 25	20 Ω–1 MΩ
≤ 0.5%	Encaps. wire-wound	Noise	$\frac{1}{20}$–1 W	≤ 20	0.1 Ω ≥ 1 MΩ
≤ 25 ppm					
Variable devices	Wire-wound	Tempco	5 at 70°C	± 20	10 Ω–100 kΩ
(pots, trimmers)	Conductive plastic	Rotational life	2 at 70°C	± 250–500	1 kΩ–100 kΩ
	Cermet	Environmental	12 at 70°C	± 250–500	500 Ω–2 MΩ
	Carbon	Cost	5 at 70°C	± 300–2000	100 Ω–2 MΩ
Networks	Thick film	Cost	≤ 2 W/pkg.	≤ 200	10 Ω–10 MΩ
	Thin film	Performance	≤ 2 W/pkg.	≤ 100	10 Ω–1 MΩ

Reprinted from *Electronics*, May 30, 1974; copyright McGraw-Hill, Inc., 1974.

1.4 FIXED RESISTORS

Fixed resistors are manufactured in four basic types:

1. Carbon composition
2. Metal film
3. Carbon film
4. Wire-wound

Resistors can also be classified in terms of their tolerance:

1. General purpose: tolerance 5 percent or greater
2. Semiprecision: tolerance between 1 and 5 percent
3. Precision: tolerance between 0.5 and 1 percent
4. Ultraprecision: tolerance better than 0.5 percent

A summary of different resistor types and their key properties appears in Table 1.2.

CARBON COMPOSITION The carbon-composition resistor is perhaps the most widely used fixed resistor in discrete circuits. Composition resistors are available in resistance values from 1 Ω to 100 MΩ and typical power ratings of $^1/_8$ to 2 W. Their temperature coefficient is high (greater than 500 ppm/°C), and their cost is low.

A cutaway view of a carbon resistor is shown in Fig. 1.10. The resistance material is a form of carbon, such as graphite, embedded in a binder. The resistance and in-

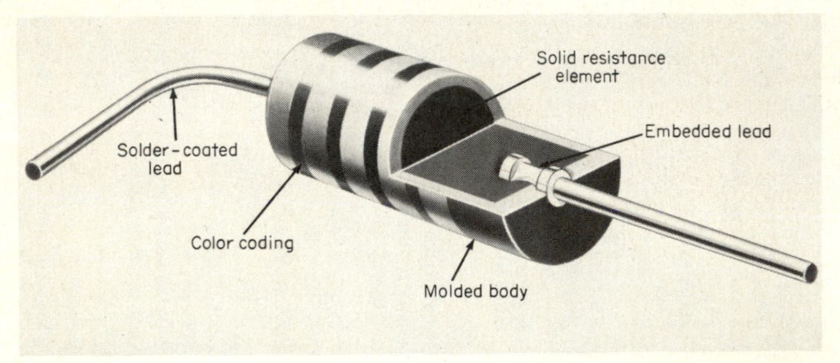

Fig. 1.10 Cutaway view of a carbon-composition resistor. (*Courtesy Allen-Bradley*)

sulation material, and the wire leads, are molded simultaneously under high temperature and pressure. The result is referred to as a slug-type structure. Different resistance values are obtained by varying the carbon and filler content.

METAL FILM Metal-film resistors are available as thin- and thick-film-type components. In the thick-film category are the tin-oxide, metal-glaze, cermet, and bulk-film resistors. Each type will be considered in this section.

Thin film The resistance element in a thin-film resistor is a film having a thickness in the order of one-millionth of an inch. (A thick film has a thickness greater than one-millionth of an inch.) Typically, the thin film is deposited on a ceramic substrate under a high vacuum; the technique is referred to as *vacuum deposition* (see Chap. 9). Metals used for deposition include nickel and chromium. Some characteristics of the thin-film resistor are: resistance range of 10 Ω to 1 MΩ; tolerance better than 0.5 percent, TCR less than 25 ppm/°C, power rating up to 5 W; low noise.

Tin oxide Tin oxide, in vapor form, is usually deposited on a ceramic substrate under high temperature. The vapor reacting with the substrate, which is heated, results in a tightly formed resistance film. Some characteristics of the tin-oxide resistor are: resistance range of a few ohms to 2.5 MΩ, tolerance better than 1 percent, TCR less than 200 ppm/°C, power rating up to 2 W, good stability.

Metal glaze A powdered glass and fine metal particle (palladium and silver) mixture is deposited on a ceramic substrate. The combination is then fired at a high temperature (typically 800°C). This results in a fusion of metal particles to the substrate.

Some characteristics of the metal-glaze resistor are resistance range of a few ohms to 1.5 MΩ, tolerance better than 1 percent, TCR as low as 20 ppm/°C, power rating up to 5 W, good stability.

Cermet A cermet-film resistor is made by screening (see Chap. 9) a mixture of precious metals and binder material on a ceramic substrate. Similar to the manufacture of a metal-glaze resistor, the combination is then fired at a high temperature. An example of a cermet film resistor is illustrated in Fig. 1.11. Some characteristics of this type of resistor are: resistance range of 10 Ω to 10 MΩ, tolerance as low as 1 percent, TCR in the order of 100 ppm/°C, power rating up to 3 W, good stability.

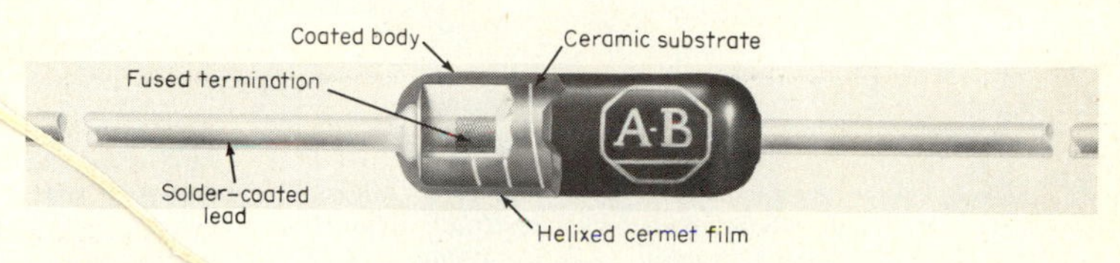

Fig. 1.11 Cutaway view of a cermet-film resistor. (*Courtesy Allen-Bradley*)

Bulk film In this resistor, the metal film is etched on a glass substrate (Fig. 1.12). Because of their unequal coefficients of expansion, the metal film is compressed slightly by the glass substrate. The compressed film has a negative temperature coefficient, which cancels out the inherent positive temperature coefficient of the film. As a result, the bulk-film resistor has a TCR close to zero. Some characteristics of the bulk-film resistor are resistance range of 30 Ω to 600 kΩ, tolerance as low as 0.005 percent, TCR in the order of 1 ppm/°C, power rating up to 1 W, very low noise, good stability.

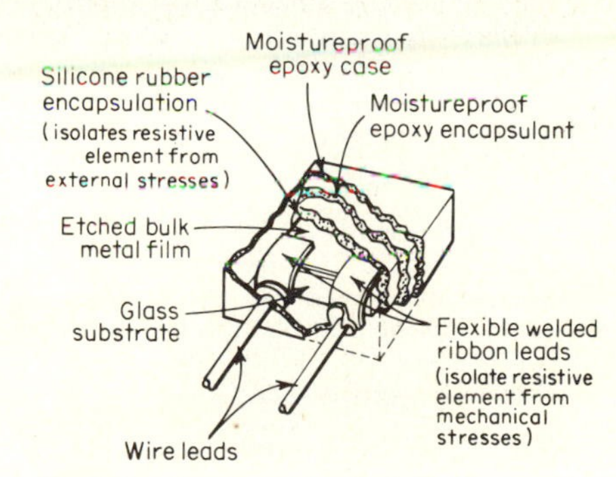

Fig. 1.12 Cutaway view of a bulk-film resistor. (*Courtesy Vishay Resistor Products*)

CARBON FILM This resistor is manufactured by depositing a carbon film on a ceramic substrate (Fig. 1.13). Some characteristics of carbon-film resistors are resistance range of 10 Ω to 10 MΩ, tolerance 5 percent or greater, TCR in the order of 150 ppm/°C, power rating up to 2 W, generally less noisy than the carbon-composition resistor, low cost.

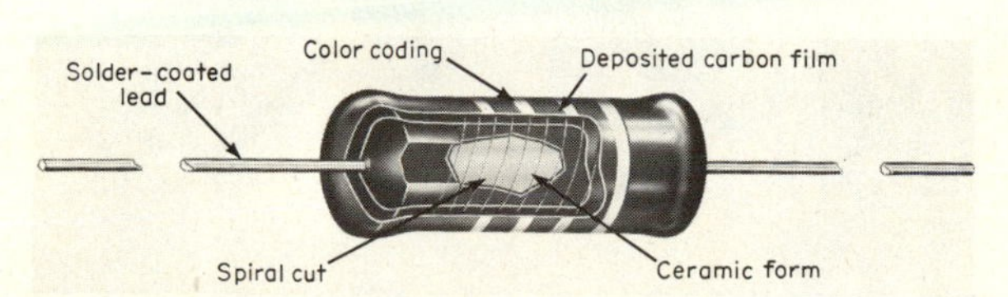

Fig. 1.13 Cutaway view of a carbon-film resistor. (*Courtesy Piher International Corporation*)

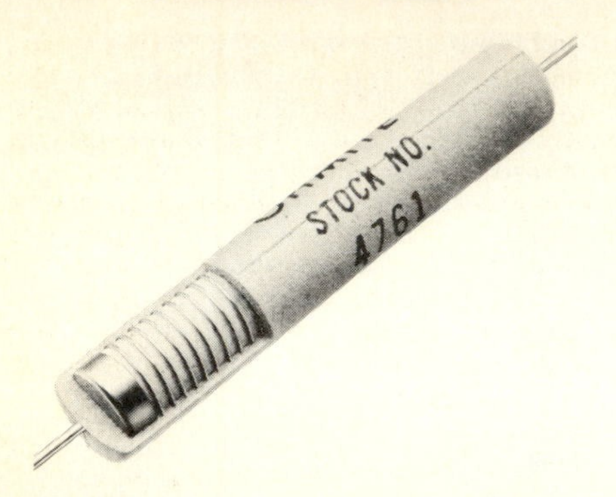

Fig. 1.14 Cutaway view of a power-style molded vitreous enamel wire-wound resistor. (*Courtesy Ohmite Manufacturing Company*)

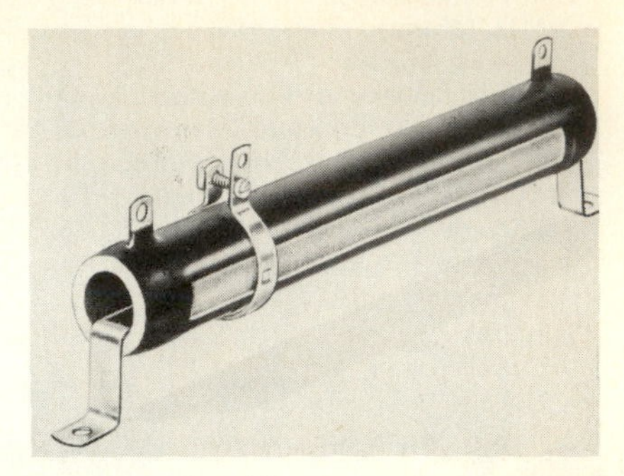

Fig. 1.15 Adjustable power-style wire-wound resistor. (*Courtesy Ohmite Manufacturing Company*)

WIRE-WOUND The wire-wound resistor enjoys a broad spectrum of applications. Made in many shapes and sizes, it is used as an ultraprecision resistor in instrumentation and as a power resistor in industrial applications. Examples of its construction are illustrated in Figs. 1.14 to 1.16.

The *power-style* wire-wound resistor is made by winding a single-layer length of special alloy wire, in the form of a coil, around an insulating core. The unit is then covered with a coating, such as vitreous enamel (an inorganic glasslike mixture) or silicone. The coating protects the winding against moisture and breakage.

The resistance wire used must have carefully controlled resistance per unit length of wire and low-temperature coefficient, and be able to operate at high temperatures. Alloys used include nickel-chromium-aluminum (800 alloy) and nickel-chromium-iron (Nichrome). The cylindrical core is ceramic, steatite, or a vitreous material. In the *precision-style* wire-wound resistor, typically a multilayer coil is wound on an epoxy form, or *bobbin*.

Because the wire-wound resistor acts like a coil, its inductance (as well as the capacitance between coil turns) is a problem at high-frequency operation. A number of techniques are utilized to minimize inductance. In one method, one-half the resistance wire is wound in one direction and the other half in the opposite direction. This type of winding is referred to as *bifilar*.

In another method, used in precision-style resistors, a thick-film *serpentine pattern* is deposited on a ceramic core (Fig. 1.17). Similar in behavior to the bifilar winding,

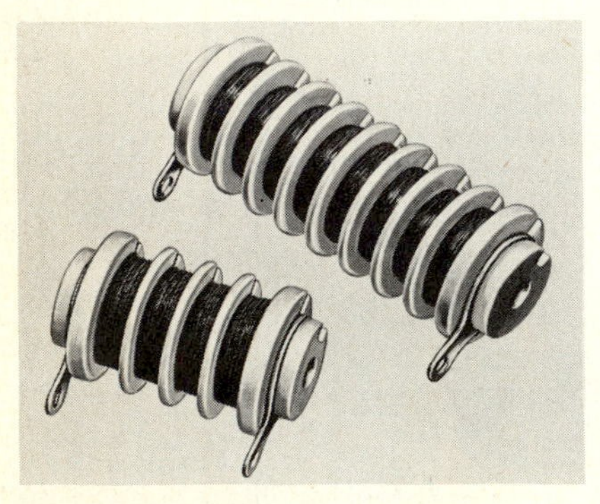

Fig. 1.16 Examples of precision-style wire-wound resistors. (*Courtesy Shallcross*)

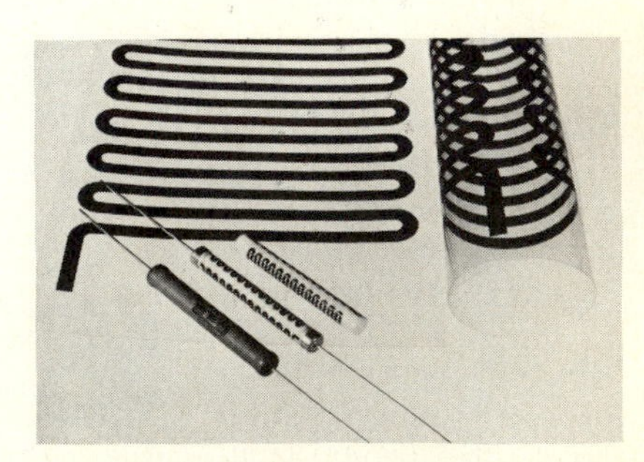

Fig. 1.17 A serpentine resistor pattern used to reduce inductance. (*Courtesy Caddock Electronics*)

opposite magnetic fields produced by current flowing in adjacent resistance paths cancel each other. The result is a resistor with virtually no inductance.

Precision-style wire-wound resistors are available having the following characteristics: resistance range of a fraction of an ohm to 10 MΩ, tolerance better than 0.5 percent, TCR less than 20 ppm/°C, power rating in the order of 2 W.

Some characteristics of power-style wire-wound resistors are: resistance range of less than an ohm to greater than a megohm; tolerance in the range of ±5 to ±20 percent; TCR as low as 5 ppm/°C; power rating as high as 1500 W. To increase its power rating, the resistor is sometimes placed in a metal housing, such as aluminum, illustrated in Fig. 1.18. For some types of resistors, the power rating of a resistor in a metal housing is increased by a factor of 2. For the realization of the increased power rating, however, the housing must be mounted on a metal chassis or suitable heat sink.

1.5 COLOR CODING OF RESISTORS

Figure 1.19 illustrates the standard color code for composition and some axial-type resistors. Different color bands are used to designate the resistance value and tolerance. The first two bands denote the first and second digits of the resistance value and the third band indicates how many zeros follow the first two digits. Tolerance is given by the fourth band. For example, a resistor with the following colored bands, yellow-violet-orange-silver, denotes a 47 000-Ω ±10 percent tolerance resistor.

Fig. 1.18 A power resistor mounted in an aluminum housing. *(Courtesy RCL Electronics, Inc.)*

First band – 1st digit
Second band – 2nd digit

Color	Digit	Multiplier	Tolerance
Black	0	1	–
Brown	1	10	±1%
Red	2	100	±2%
Orange	3	1000	–
Yellow	4	10 000	–
Green	5	100 000	–
Blue	6	1 000 000	–
Violet	7	10 000 000	–
Gray	8	–	–
White	9	–	–
Gold	–	0.1	± 5%
Silver	–	–	±10%
No color	–	–	±20%

Fig. 1.19 Standard color code for composition and some axial-type resistors.

For film and wire-wound resistors, in general, the resistance value and tolerance are stamped on the body of the resistor. Occasionally, a manufacturer may use his own code. For this reason, it is good practice to consult the manufacturer's catalog or data sheet.

1.6 CONNECTING RESISTORS

In this section we examine how fixed resistors are connected in series and parallel.

RESISTORS IN SERIES Figure 1.20 shows n resistors connected in series. In a series circuit, the current I flowing in each resistor is the same. The total equivalent resistance of the series circuit R_{TS} is equal to the sum of the individual resistors:

$$R_{TS} = R_1 + R_2 + \cdots + R_n \tag{1.4}$$

If n equal resistors are connected in series, the total equivalent resistance is equal to the product of the value of an individual resistor and n:

$$R_{TS} = nR \tag{1.5}$$

The maximum voltage E_{max} that can be applied across a series circuit of resistors is equal to the sum of the rated continuous working voltage RCWV of each resistor:

$$E_{max} = (\text{RCWV})_1 + (\text{RCWV})_2 + \cdots + (\text{RCWV})_n \tag{1.6}$$

example 1.5 Three resistors of 10, 20, and 30 Ω are connected in series. The current flowing in the circuit is 2 A, and the RCWV for each resistor is 500 V. Determine (*a*) the total equivalent resistance of the circuit, (*b*) the power dissipated in each resistor, (*c*) the maximum voltage that can be impressed across the series circuit. If the three resistors are replaced by a single equivalent resistor, determine (*d*) its power and maximum voltage ratings.

solution (*a*) By Eq. (1.4) $R_{TS} = 10 + 20 + 30 = 60\ \Omega$.
(*b*) By Eq. (1.3*b*), $P = I^2R$, we have

$$P_{10} = 2^2 \times 10 = 4 \times 10 = 40\ \text{W}$$
$$P_{20} = 2^2 \times 20 = 4 \times 20 = 80\ \text{W}$$
$$P_{30} = 2^2 \times 30 = 4 \times 30 = 120\ \text{W}$$

(*c*) By Eq. (1.6), $E_{max} = 500 + 500 + 500 = 1500$ V.
(*d*) In part *a* we found that the equivalent resistance is 60 Ω. The power rating, therefore, is $2^2 \times 60 = 4 \times 60 = 240$ W. (*Note:* This value must equal the sum of the dissipated power in each individual resistor found in part *b*.) From part *c*, the maximum voltage rating is 1500 V.

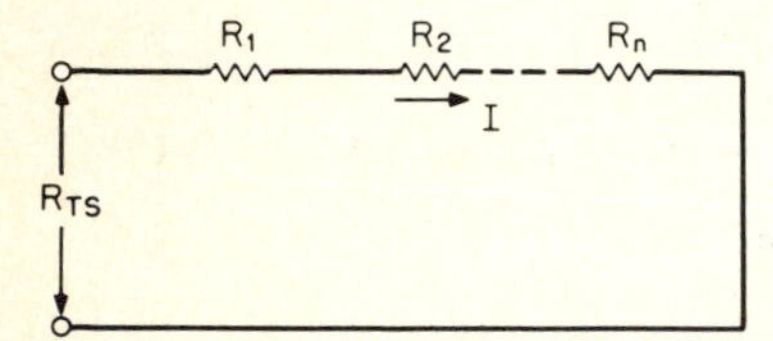

Fig. 1.20 Resistors connected in series.

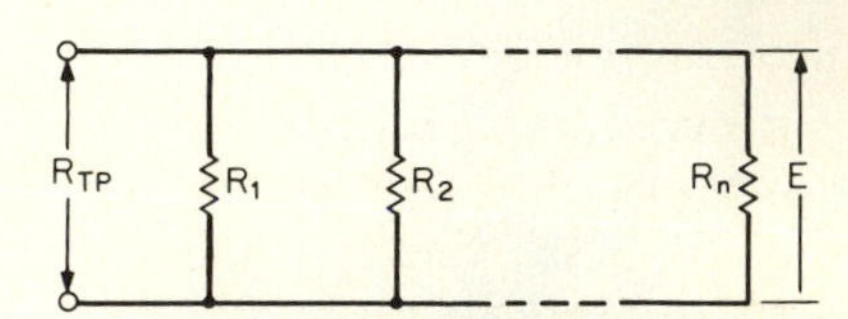

Fig. 1.21 Resistors connected in parallel (shunt).

example 1.6 A 90- and a 10-Ω resistor are connected in series. The impressed voltage across the series combination is 1000 V. Determine the voltage across each resistor.

solution By voltage division, the voltage across each resistor is proportional to its resistance value. For the 10-Ω resistor, $E_{10} = 1000 \times 10/(10 + 90) = 100$ V; for the 90-Ω resistor, $E_{90} = 1000 \times 90/(10 + 90) = 900$ V.
What is important in this example is that in a series circuit the voltage across the larger resistor (90 Ω) is much greater than that across the smaller (10 Ω) resistor. The RCWV of the 90-Ω resistor, therefore, must be at least 900 V. For the 10-Ω resistor, an RCWV of 100 V is adequate.

RESISTORS IN PARALLEL Figure 1.21 shows n resistors connected in parallel. In a parallel circuit, the voltage E across each resistor is the same. It is convenient to consider two resistors at a time when calculating the equivalent resistance R_{TP} of a parallel circuit. The equivalent resistance R of two resistors R_1 and R_2 in parallel (denoted as $R_1 \| R_2$) is equal to their product divided by their sum:

$$R = \frac{R_1R_2}{R_1 + R_2} \tag{1.7}$$

For n equal resistors in parallel, the equivalent resistance is equal to the value of an individual resistor R_1 divided by n:

$$R_{TP} = \frac{R_1}{n} \tag{1.8}$$

The maximum voltage that can be applied across resistors in parallel is limited by the smallest value of RCWV for a resistor in the circuit.

example 1.7 Three resistors of 120, 120, and 12 Ω are connected in parallel. The voltage across the parallel circuit is 240 V, and the RCWV of the 120-Ω resistors is 1000 V and 750 V for the 12-Ω resistor. Determine (*a*) the total equivalent resistance of the circuit, (*b*) the power dissipated in each resistor, (*c*) the maximum voltage that can be impressed across the parallel circuit. If the three resistors are replaced by a single equivalent resistor, determine (*d*) its power rating.

solution (*a*) Considering the two 120-Ω resistors, by Eq. (1.8), $R_{TP} = {}^{120}\!/_2 = 60\ \Omega$. By Eq. (1.7), the equivalent parallel resistance of the circuit is $60 \| 12 = (60 \times 12)/(60 + 12) = 10\ \Omega$.

(*b*) By Eq. (1.3*c*), $P = E^2/R$, we have

$$P_{120} = \frac{(240)^2}{120} = 480 \text{ W}$$

$$P_{12} = \frac{(240)^2}{12} = 4800 \text{ W}$$

(*c*) The maximum voltage, limited by the smaller value of RCWV, is 750 V.

(*d*) In *a* we found that the equivalent resistance is 10 Ω. The power rating, therefore, is $(240)^2/10 = 5760$ W.

Examples 1.5 and 1.7 illustrate that a high power rating can be realized by connecting lower power-rated resistors in series or in parallel.

1.7 SPECIAL RESISTORS

A number of special resistors and resistive networks are available. Some of these components are considered in this section.

HIGH-VOLTAGE RESISTORS Resistors that operate at voltages as high as 40 000 V are available. In one type of construction, a carbon-film resistance element is vacuum-sealed in a glass envelope.

HIGH-MEGOHM RESISTORS Using a semiconductor glass for the resistance element, resistance values as high as 1 000 000 MΩ (a million, million ohms) are obtainable. Measuring in length in the order of 0.25 in. and about 0.05 in. in diameter, the application for high-megohm resistors includes radiation detectors, electrometers, and use in FET circuits.

DIP NETWORKS Resistors, in various configurations packaged in the *dual-in-line package* (DIP), used for integrated circuits, constitute a DIP network. The DIP contains typically 14 or 16 pins. An example of a DIP ladder network is illustrated in Fig. 1.22. Innumerable resistor configurations are available. The resistors are generally thin or thick film.

1.8 VARIABLE RESISTORS (POTS)

A variable resistor, commonly referred to as a potentiometer or pot, converts the rotation of a shaft to an output voltage. There are basically three types of pots:

1. Single turn
2. Multiple turn
3. Trimmer

Examples of the three types of pots are illustrated in Figs. 1.23 to 1.25.

The single-turn pot may be regarded as the "workhorse" of variable resistors. Available in resistance ranges from 50 Ω to 5 MΩ and higher, in tolerances of ±10 and ±20 percent, and in power ratings of 2 and 3 W, the single-turn pot enjoys a wide spectrum of applications. It is used, for example, as a gain, treble, or base control in an amplifier and as the brightness and contrast controls in a TV receiver. In some applications, two or more pots sharing a common rotating shaft are required. Such a combination, referred to as a *ganged pot*, is illustrated in Fig. 1.26.

Multiple-turn pots are used in applications that require the precise setting of a resistance value. An example is the setting of coefficient pots in an analog computer.

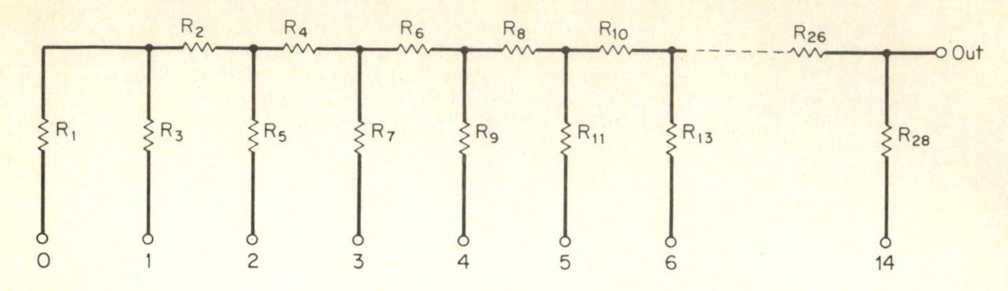

Resistors: R_1, R_3, R_5, R_7, R_9, R_{11}, R_{13}, R_{15}, R_{17}, R_{19}, R_{21}, R_{23}, R_{25}, R_{27}, R_{28}; each = 2R

Resistors: R_2, R_4, R_6, R_8, R_{10}, R_{12}, R_{14}, R_{16}, R_{18}, R_{20}, R_{22}, R_{24}, R_{26}; each = R

(a)

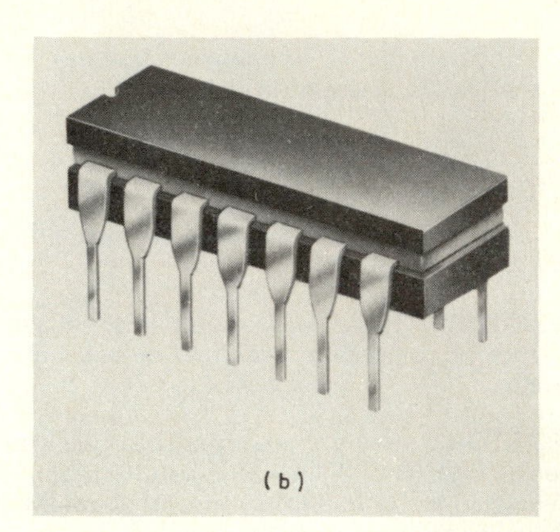

Fig. 1.22 An example of a DIP network: (*a*) Ladder network. (*b*) DIP with 14 pins. (*Courtesy RCL Electronics, Inc.*)

These pots generally have 10 turns and are available in typical resistance ranges of 50 Ω to 250 kΩ, in tolerances of ±3 percent, and in power ratings up to 5 W.

The trimmer pot, which may be viewed as a "set and forget me" pot, is used generally for a one-time adjustment of resistance. Its rotational life, compared to that of other pots, is therefore limited. Trimmer pots are available as single- and multiple-turn units. Typically, their resistance range is from a few ohms to 5 MΩ, tolerance is ±10 percent, and power rating is 1 W.

Fig. 1.23 An example of a single-turn pot that enjoys a wide variety of application. (*Courtesy Ohmite Manufacturing Company.*)

Fig. 1.24 An example of a multiple-turn pot. (*Courtesy Beckman Instruments, Inc., Helipot Division*)

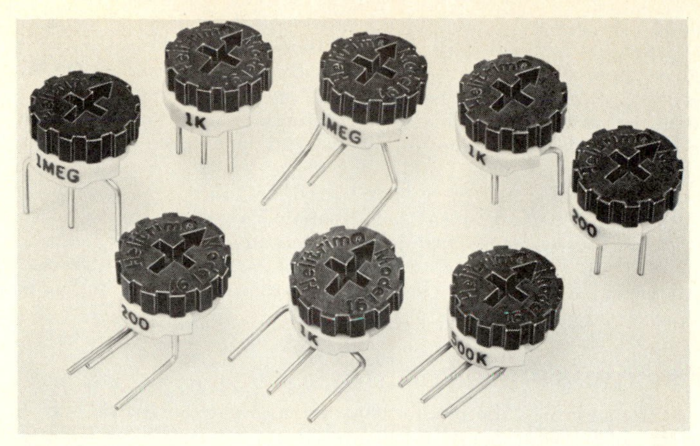

Fig. 1.25 Examples of trimmer pots. *(Courtesy Beckman Instruments, Inc., Helipot Division)*

TYPES OF MATERIALS Four basic types of materials are used in the construction of pots:

1. Carbon
2. Cermet
3. Conductive plastic
4. Wire-wound

Pots other than wire-wound are referred to as non-wire-wound. A comparison of different types of pots is provided in Table 1.2.

Carbon The resistance element is made by spraying or brushing a carbon-resistance compound on an insulating material, such as laminated plastic. Its chief merit is its low cost. The carbon pot exhibits a high TCR and is susceptible to moisture, and its maximum operating temperature is generally limited to 85°C.

Variations of the carbon pot that exhibit somewhat improved characteristics are the *carbon-ceramic* and *molded-carbon* types. In the carbon-ceramic pot, the carbon compound is screened on a ceramic substrate. Molded-carbon pots are made by molding, simultaneously, the carbon-resistance element and other parts of the pot.

Cermet In the cermet pot, a mixture of precious metals and glass or ceramic powder is screened on a ceramic substrate. The TCR of a cermet pot may be as low as 50 ppm/°C and its operating temperature as high as 150°C. The cermet pot is very stable and can withstand overloads.

Conductive plastic One method of making a conductive plastic pot is by spraying, under pressure, a special liquid suspension on a plastic material, such as Mylar. The suspension contains carbon, a solvent, and a filler. The resistance value depends on the mix of carbon and filler. Conductive plastic pots exhibit good linearity and long rotational life. They are, however, susceptible to humidity.

Fig. 1.26 A dual-ganged pot. *(Courtesy Ohmite Manufacturing Company)*

Wire-wound In the wire-wound pot, the resistance element is wound on a phenolic, plastic, or glass-filled laminate card. Low-resistance wire-wound pots are generally wound with a copper alloy wire and high-resistance pots with a nickel-chromium wire. The wire-wound pot exhibits the lowest TCR (20 ppm/°C), operates at temperatures as high as 150°C, and is available in high-power ratings. Because of its construction, the wire-wound pot has appreciable stray inductance and capacitance which may be a problem at high-frequency operation.

RHEOSTAT A wire-wound pot that can dissipate 5 and more watts is often referred to as a *rheostat*. The resistance wire is wound on an open ring of ceramic which is covered with vitreous enamel, except for the track of the wiper arm. Rheostats are used to control motor speed, x-ray tube voltages, welding current, ovens, and in many other high-power applications.

1.9 CHIP RESISTORS

A variety of thin- and thick-film resistor chips are available for hybrid microelectronic circuits (see Chap. 9). They measure as little as 30 by 30 mils (0.03 by 0.03 in.) and have resistance values from a few ohms to 1000 MΩ. Typical power rating is 0.25 W, and TCR is in the order of 20 to 200 ppm/°C.

1.10 THERMISTORS

A thermistor is a *nonlinear resistor*, made of semiconductor material, that is extremely sensitive to changes in temperature. For a small change in body temperature of a

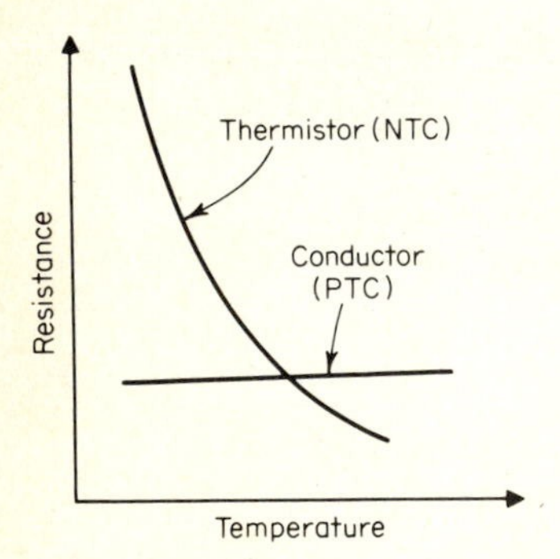

Fig. 1.27 The variation in resistance of a conductor (PTC) and a thermistor (NTC) with temperature.

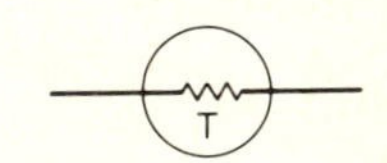

Fig. 1.28 Electrical symbol for a thermistor.

thermistor, there is an appreciable change in its resistance. Whereas most conductors have a positive temperature coefficient (PTC), the thermistor can exhibit a positive or negative temperature coefficient (NTC). Of particular interest is thermistors having a negative temperature coefficient.

A comparison of a conductor and an NTC thermistor is illustrated in Fig. 1.27. For the thermistor, the resistance decreases rapidly for elevated temperatures. The conductor having a positive temperature coefficient, however, exhibits a small increase in resistance with rising temperatures.

Thermistors are used in a wide variety of applications. These include the measurement and control of temperature, time delay, temperature compensation, and as liquid-level indicators. Thermistors are available as disk, washer, bead, and bolted assembly packages. The electrical symbol for a thermistor is illustrated in Fig. 1.28.

CHARACTERISTIC CURVES AND PARAMETERS In this section, typical characteristic curves and parameters for the thermistor are considered.

Resistance-temperature characteristic A typical resistance-temperature characteristic of a thermistor is illustrated in Fig. 1.29. Plotted is resistance as a function of temperature. The resistance, for zero power dissipation, decreases rapidly with elevated temperature.

Three useful parameters for characterizing the thermistor are the time constant, dissipation constant, and resistance ratio. The time constant is the time for a thermistor to change its resistance by 63 percent of its initial value, for zero-power dissipation. Typical values of time constant range from 1 to 50 s.

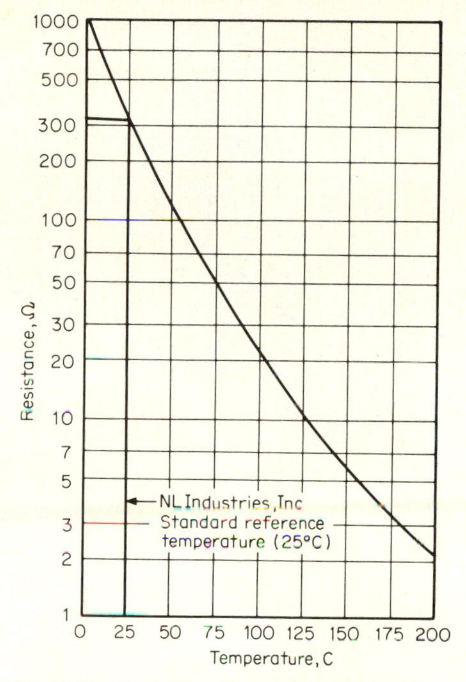

Fig. 1.29 Typical resistance-temperature characteristic of a thermistor. (*Courtesy N L Industries, Inc., Electronics Department*)

The dissipation factor is the power necessary to increase the temperature of a thermistor by 1°C. Expressed in milliwatts per degree Celsius, typical values of dissipation factor are 1 to 10 mW/°C.

Resistance ratio is the ratio of the resistance at 25°C to that at 125°C. Its range is approximately 3 to 60.

Static voltampere characteristics A typical voltampere characteristic is shown in Fig. 1.30. Plotted is the voltage across the thermistor as a function of current flow. For

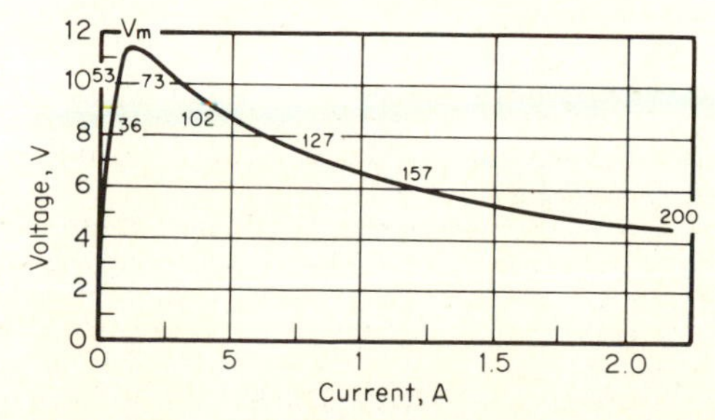

Fig. 1.30 Typical static voltampere characteristic of a thermistor. (*Courtesy N L Industries, Inc., Electronics Department*)

minute currents, the dissipated power is too small to heat the thermistor. In this region Ohm's law is obeyed. As the current is increased, the thermistor temperature rises and the resistance begins to decrease. At the *self-heating voltage* V_m, the characteristic begins to exhibit a negative slope. The temperature of the thermistor (in degrees Celsius) at various values of voltage and current is denoted by the numbers plotted along the curve.

Current-time characteristic Because of its mass, it takes a finite time for a current in a thermistor to reach its final value. This is illustrated in the current-time characteristic curves of Fig. 1.31*a* for three different thermistors.

The test circuit of Fig. 1.31*b* contains a 100-Ω resistor in series with a thermistor

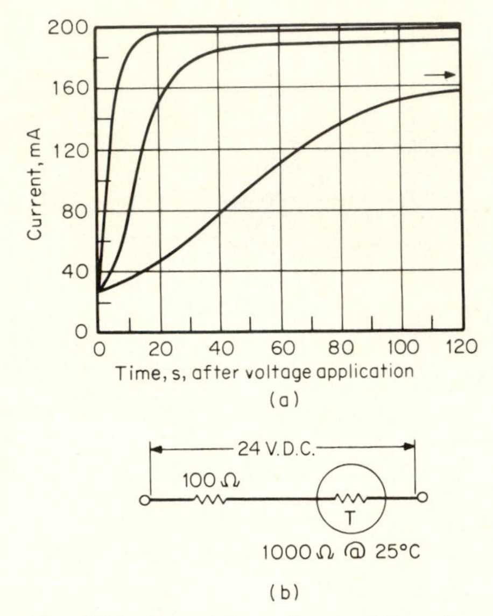

Fig. 1.31 Current-time characteristics of a thermistor: (*a*) Typical family of curves. (*b*) Test circuit. (*Courtesy N L Industries, Inc., Electronics Department*)

whose resistance is 1000 Ω at 25°C. A voltage of 24 V dc is impressed across the circuit. For example, the thermistor having the characteristic of the top curve reaches its final value of approximately 200 mA in 20 s. The thermistor having the bottom characteristic, however, takes 120 s to approach its final value of 160 mA.

example 1.8 Show how a thermistor may be used as a liquid-level indicator.

solution The circuit for a simple liquid-level indicator employing a thermistor is given in Fig. 1.32. Suspended above the liquid is a thermistor in series with an indicator, such as a lamp, and battery E. In air, the resistance of the thermistor is relatively low, thereby allowing sufficient current I to flow and light the lamp. When the liquid level reaches the thermistor, it

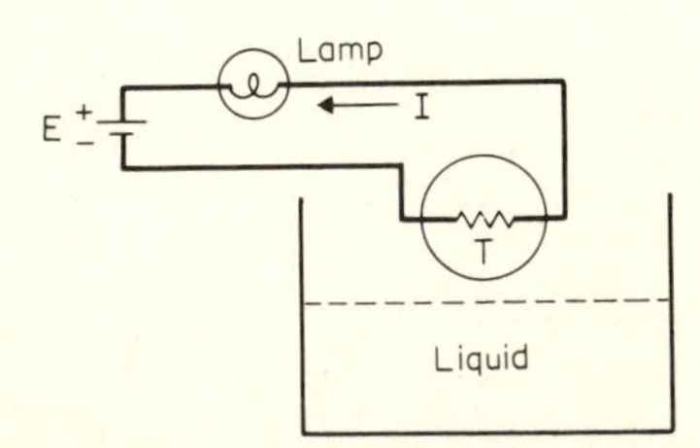

Fig. 1.32 A simple liquid-level indicator using a thermistor.

cools the device and its resistance is increased. Current I is reduced and the lamp is extinguished.

1.11 VARISTORS

The varistor, another example of a nonlinear resistor, is a device in which the current varies as a power of the impressed voltage. The resistance, therefore, is also reduced. In an ordinary resistor, the current is directly proportional to the impressed voltage and Ohm's law is obeyed. In the varistor, however, the current is proportional to the power of the impressed voltage E^n, where n is in the range of 2 to 6. A typical voltage-current characteristic curve and electrical symbol for the varistor are shown in Fig. 1.33.

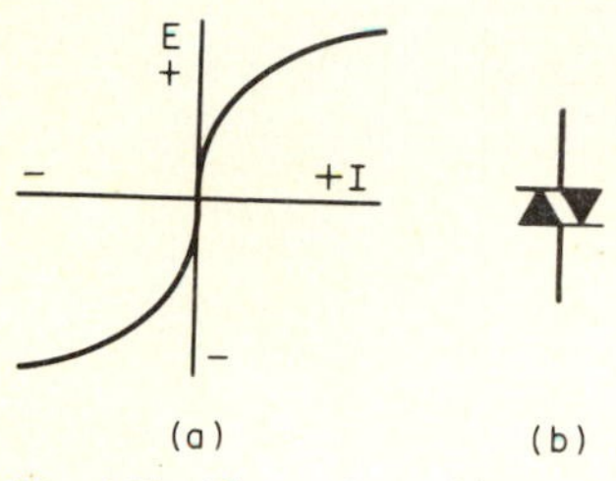

Fig. 1.33 The varistor: (*a*) Typical voltampere characteristic. (*b*) Electrical symbol. (*Courtesy N L Industries, Inc., Electronics Department*)

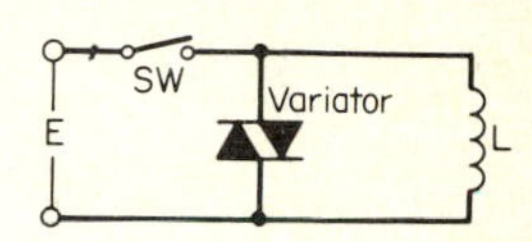

Fig. 1.34 A simple surge-protection circuit using a varistor.

Applications for the varistor include voltage surge and protective circuits and the generation of nonsinusoidal waveforms. The varistor is made of silicon carbide and is available in disk, rod, and washer forms. It can withstand dc voltages as high as 10 000 V.

example 1.9 Explain the operation of the surge protection circuit of Fig. 1.34.

solution Without the varistor connected across coil L, when the switch is opened, the high energy in the magnetic field develops a dangerously high voltage across L. This can result in the puncturing of the coil winding insulation. With a varistor in the circuit, the high voltage across it causes appreciable current to flow. The resistance of the device is thereby reduced and the magnetic field energy is dissipated in the varistor, instead of in the coil.

Chapter 2

Characteristics of Capacitors

2.1 INTRODUCTION

Next to resistors, capacitors are the most widely used passive element in circuits. They are available as fixed and variable units having capacitance values from a few picofarads (pF) to thousands of microfarads (μF). For achieving unique characteristics, a large variety of materials are used in their construction. The abundance of different fixed capacitors available to the user is vividly portrayed in the photograph of Fig. 2.1. Because of such a plethora of units, the technician and engineer are often in a dilemma in selecting a capacitor for a given application.

APPLICATIONS The applications for capacitors may be broadly categorized as follows:

1. Blocking direct current. A capacitor cannot conduct direct current.
2. Coupling a signal from one circuit, or system, to another.
3. Bypassing a resistor to permit the easy flow of alternating current.
4. Filtering.
5. Tuning.
6. Generation of nonsinusoidal waveforms, such as a sawtooth wave.
7. Energy storage. Capacitors are used to build up sufficient electric charge to fire, for example, a flash tube or a laser.

2.2 GENERAL DESCRIPTION

The basic fixed capacitor consists of two metal plates, called *electrodes*, separated by an insulator, called a *dielectric*, as shown in Fig. 2.2*a*. In a variable capacitor, the relative position of one or more plates is changed with respect to a set of fixed plates, thereby varying the capacitance. Air is generally used as the dielectric for variable capacitors. Symbols for fixed and variable capacitors are shown in Fig. 2.2*b* and *c*, respectively.

The capacitance of parallel-plate capacitors is directly proportional to the relative dielectric constant of the insulator and to the area of the plates. Further, the capacitance is greater if the separation between plates is small, and vice versa. These facts are conveniently expressed by the following equation:

$$C = \frac{k\epsilon_0 A}{d} \tag{2.1}$$

where C = capacitance in farads, F

ϵ_0 = a constant, called the *permittivity of free space* and is equal to 8.85×10^{-12} F/m

k = relative dielectric constant of insulator

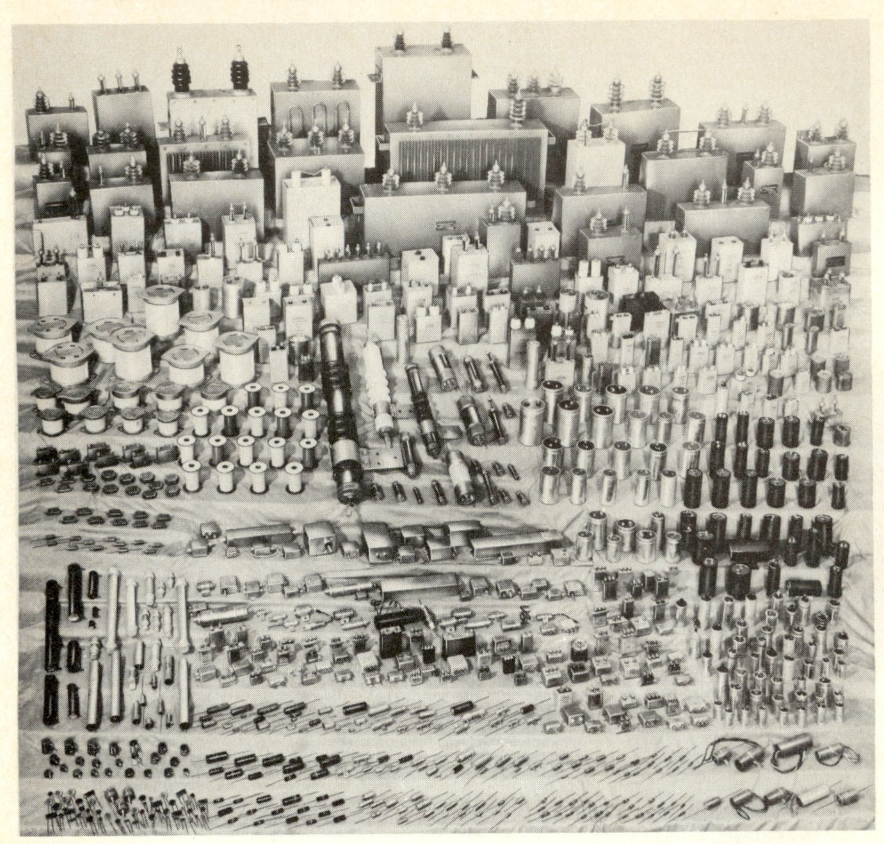

Fig. 2.1 A wide variety of types and styles of fixed capacitors are available to the technician and engineer. (*Courtesy Sprague Electric Company*)

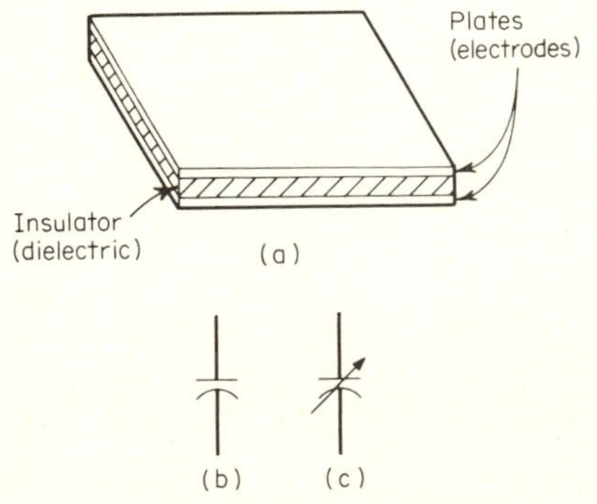

Fig. 2.2 The basic parallel-plate capacitor: (*a*) Construction. (*b*) Symbol for fixed capacitor. (*c*) Symbol for variable capacitor.

A = area of plates, m^2
d = separation between plates, m

From Eq. (2.1) it is seen that capacitance may be increased by increasing the area of the plates or the dielectric constant, and by decreasing the separation between plates.

RELATIVE DIELECTRIC CONSTANT The relative dielectric constant k compares different dielectrics with that of a vacuum for which $k = 1$. The dielectric constants for some commonly used materials are given in Table 2.1. Note that the dielectric constant of air is approximately equal to that of a vacuum.

TABLE 2.1 Dielectric Constants of Some Commonly Used Materials

Dielectric	k
Vacuum	1
Air	1.0006
Teflon	2
Polystyrene	2.5
Mylar	3
Paper, paraffin	4
Mica	5
Aluminum oxide	7
Tantalum oxide	25
Ceramic (low k)	10
Ceramic (high k)	100–10 000

example 2.1 A parallel-plate capacitor has the following dimensions: $A = 100\ \text{cm}^2$ and $d = 1$ mm. Determine the values of capacitance if the dielectric used is: (*a*) air; (*b*) paper, paraffin; and (*c*) ceramic ($k = 200$).

solution The given dimensions are first converted to meters. One square centimeter equals $10^{-4}\ \text{m}^2$; hence, $100\ \text{cm}^2 = 100 \times 10^{-4} = 10^{-2}\ \text{m}^2$. One millimeter equals 10^{-3} m. Substitution of these values in Eq. (2.1) yields

$$C = k \times 8.5 \times 10^{-12} \times \frac{10^{-2}}{10^{-3}} = k \times 85 \times 10^{-12}\ \text{F}$$

(*a*) The dielectric constant for air is approximately one. Hence, $C_{air} = 85 \times 10^{-12} \times 1 = 85 \times 10^{-12}\ \text{F} = 85\ \text{pF}$, since 10^{-12} F is equal to 1 pF.

(*b*) For paper coated with paraffin, from Table 2.1, $k = 4$. Therefore, $C_{paper} = 85 \times 10^{-12} \times 4 = 350 \times 10^{-12}\ \text{F} = 350\ \text{pF}$.

(*c*) $C_{ceramic} = 85 \times 10^{-12} \times 200 = 17 \times 10^{-9}\ \text{F} = 0.017\ \mu\text{F}$, since 10^{-6} F is equal to 1 μF.

The particular dielectric used exerts a strong influence on the value of a capacitor. As we shall see, capacitors are designated by the dielectric used in their construction.

CHARGE AND ENERGY The charge Q (coulombs) stored in a capacitor equals the product of the capacitance C (farads) and the voltage E (volts) across the capacitor:

$$Q = CE \tag{2.2a}$$

Solving Eq. (2.2*a*) for E and C,

$$E = \frac{Q}{C} \tag{2.2b}$$

$$C = \frac{Q}{E} \tag{2.2c}$$

From the preceding equations it is seen that to store a given charge, more voltage is needed for a small capacitor and less voltage for a large capacitor.

The maximum voltage that can be impressed across a capacitor, without causing irreparable damage, depends on the dielectric used and the separation d of the plates. For a given dielectric, the allowable maximum voltage increases as the separation between plates increases. From Eq. (2.1), as d grows large for a given plate area and dielectric the capacitance is reduced. One may conclude that for a given dielectric and physical size, a capacitor with a high-voltage rating has less capacitance than one with a low-voltage rating.

A charged capacitor has *energy stored* in the electric field existing between its plates. The expression for stored energy is

$$w = \frac{Ce^2}{2} \tag{2.3}$$

where w = stored energy in Joules, J (1 J = 1 W-s)
C = capacitance, F
e = voltage across capacitor terminals, V

example 2.2 Determine the energy stored in a 100-μF capacitor if the impressed voltage is 1 kV.

solution From Eq. (2.3) $w = 100 \times 10^{-6} \times (1000)^2/2 = 50$ J.

2.3 CAPACITOR TERMS AND PARAMETERS

In this section commonly used terms and parameters for characterizing capacitors are defined. Where appropriate, typical curves showing the variation of parameters with temperature and frequency are included.

CAPACITANCE The basic unit for capacitance is the farad. Because practical capacitors have a capacitance much less than a farad, typical units employed are the microfarad and picofarad.

AMBIENT TEMPERATURE The actual value of capacitance depends on the temperature of the medium surrounding the capacitor. This is referred to as the *ambient temperature*. A typical curve showing the variation in capacitance as a function of temperature is given in Fig. 2.3. Note that *percent capacitance change* is plotted as a

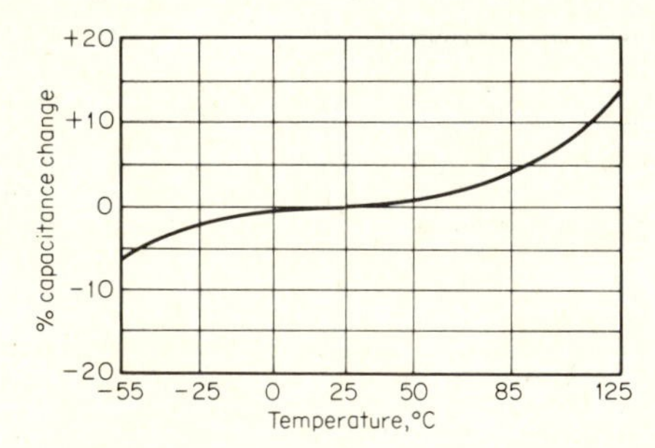

Fig. 2.3 A typical curve showing how capacitance varies with temperature. (*Courtesy Electrocube*)

function of temperature. At 25°C, which corresponds to room temperature, the percentage change is zero. This indicates that the changes in capacitance values are taken with respect to room temperature.

example 2.3 A capacitor is rated at 100 μF at 25°C. Using the curve of Fig. 2.3, determine the value of capacitance at −55 and +125°C.

solution From Fig. 2.3, at −55°C the change is approximately −6 percent. Hence, the value of C is $100 - 100 \times 0.06 = 94$ μF. At +125°C, the change is approximately +13 percent; the value of C at 125°C then is $100 + 100 \times 0.13 = 113$ μF.

TOLERANCE Tolerance is the variation in capacitance expressed as a percentage of its specified value at 25°C. The specified value at 25°C is referred to as the *nominal value*. For example, 20 μF ± 10% means that the actual value is between $20 - 20 \times 0.1 = 18$ μF and $20 + 20 \times 0.1 = 22$ μF.

TEMPERATURE COEFFICIENT The temperature coefficient TC is the change in capacitance per degree change in temperature. It is generally expressed in parts per million per degree Celsius (ppm/°C). The temperature coefficient may be positive, negative, or zero. If the TC is positive, the letter *P* precedes the coefficient; if negative, *N* precedes the coefficient. The designation NPO is used for a zero-temperature coefficient capacitor.

WORKING VOLTAGE The working voltage is the maximum voltage that can be impressed across a capacitor for continuous operation. This rating must indicate whether the voltage is dc or ac; in general, the rating is not the same.

BREAKDOWN VOLTAGE The breakdown voltage is the maximum capacitor voltage that causes the dielectric to become damaged. It is also referred to as the *surge* and *test* voltage.

DC LEAKAGE Dc leakage refers to a minute direct current that flows in a capacitor at a specified direct voltage. Leakage is due to the presence of a few free carriers of charge in the dielectric. For this reason, a charge in a capacitor cannot be stored indefinitely, and it ultimately leaks off.

INSULATION RESISTANCE The insulation resistance *IR* is the resistance of the dielectric. The greater is the resistance, the less is the leakage current. Insulation resistance decreases with increasing temperature, as illustrated in Fig. 2.4.

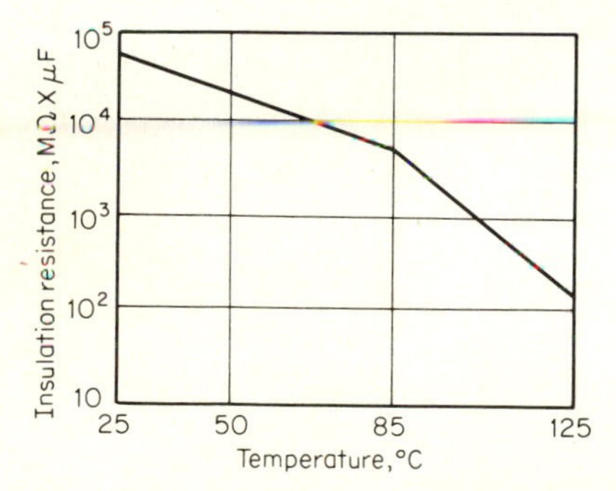

Fig. 2.4 The variation of insulation resistance with temperature. (*Courtesy Electrocube*)

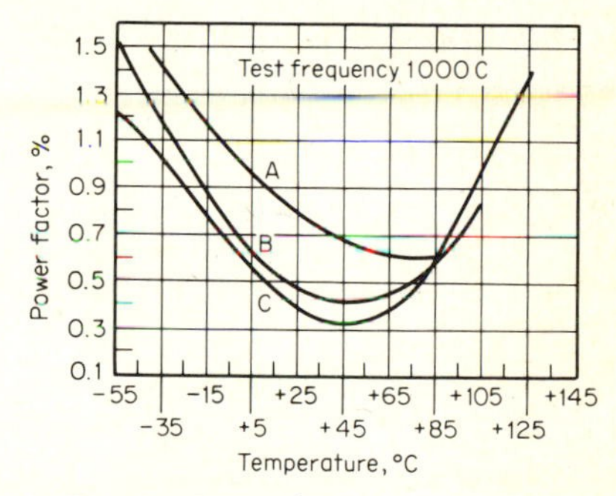

Fig. 2.5 Power factor as a function of temperature for three types of fixed capacitors. (*a*) Paper-dipped type (Type PD). (*b*) Mylar-paper-dipped (Type MPD). (*c*) Mylar. (*Courtesy The Electro Motive Manufacturing Company, Inc.*)

CAPACITIVE REACTANCE The capacitive reactance X_C (ohms) of a capacitor is equal to 0.159 divided by the product of the frequency (hertz) and capacitance (farads):

$$X_C = \frac{0.159}{fC} \tag{2.4}$$

The capacitive reactance increases as the frequency or capacitance decreases, and vice versa.

POWER FACTOR The power factor PF expresses the ratio of energy wasted to the energy stored in a capacitor. If, for example, PF = 2 percent, then 98 percent of the applied energy can serve a useful purpose and the remaining 2 percent is lost as heat in the capacitor. The variation of PF as a function of temperature is illustrated in Fig. 2.5. The power factor also increases with rising frequency.

EQUIVALENT SERIES RESISTANCE Owing to the resistance of the capacitor plates and leads, the equivalent series resistance, ESR, is the net series resistance of a capacitor. An example of variation in ESR for capacitors designed to operate at UHF and microwave frequencies is shown in Fig. 2.6. The ESR increases with increasing frequency.

IMPEDANCE The impedance Z (ohms) of a capacitor is the total opposition to the flow of alternating current. It is equal to the square root of the sum of the capacitive reactance squared and the equivalent series resistance squared:

$$Z = \sqrt{X_C^2 + (\mathrm{ESR})^2} \qquad (2.5)$$

Typical curves of impedance as a function of frequency and temperature are illustrated in Fig. 2.7. Impedance of a capacitor decreases with increasing frequency and temperature. For a high-quality capacitor, the impedance is approximately equal to the capacitive reactance.

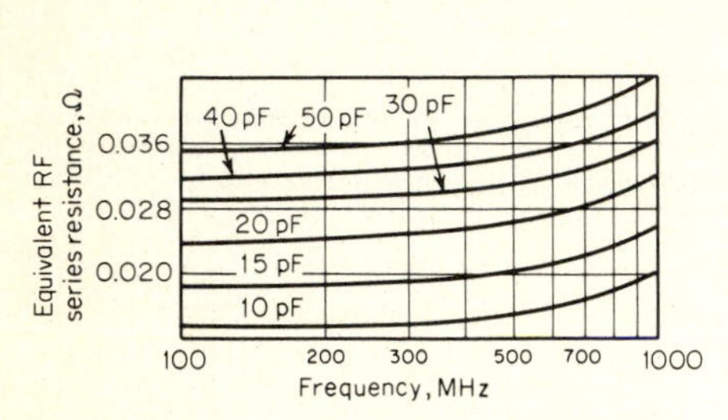

Fig. 2-6 Variation in ESR with frequency for a porcelain capacitor designed to operate at ultrahigh and microwave frequencies. (*Courtesy American Technical Ceramics*)

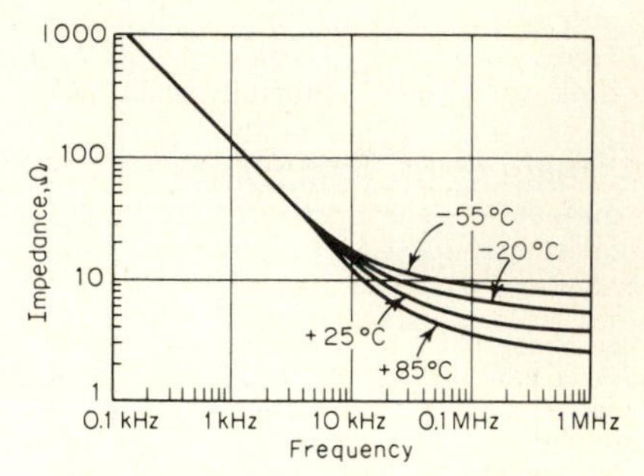

Fig. 2.7 The impedance of a fixed capacitor decreases with increasing frequency and temperature. (*Courtesy Sprague Electric Company*)

DISSIPATION FACTOR The dissipation factor DF is another parameter used to describe the quality of a capacitor. The DF is expressed as a percentage and defined as the ratio of ESR to X_C:

$$\mathrm{DF} = \frac{\mathrm{ESR}}{X_C} \times 100\% \qquad (2.6)$$

The lower the value of DF is, the better is the capacitor. A typical curve showing how DF varies with temperature is given in Fig. 2.8.

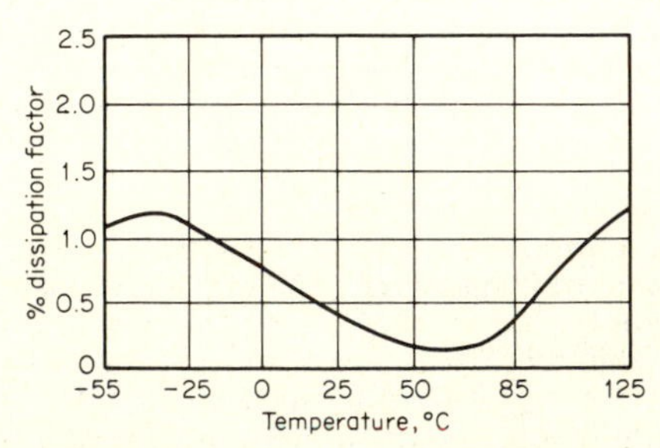

Fig. 2.8 The variation of dissipation factor DF with temperature. (*Courtesy Electrocube*)

QUALITY FACTOR (Q) The quality factor is one over the dissipation factor:

$$Q = \frac{1}{DF} = \frac{X_C}{ESR} \tag{2.7}$$

The greater the value of Q is, the better is the capacitor. Parameters DF and Q are relatively simple to determine on an impedance bridge.

RIPPLE CURRENT AND VOLTAGE Ripple current (voltage) is the ac component of a unidirectional current (voltage). Ripple is present, for example, in the output of a dc power supply. If the ripple component in a capacitor is less than 5 percent, it may be considered negligible. For values greater than 5 percent, the ripple contributes to the heating of the capacitor.

2.4 FIXED CAPACITORS

Although the types, shapes, and sizes of capacitors are nearly countless (see Fig. 2.1), the basic structure of Fig. 2.2, namely two plates separated by a dielectric, is common to all fixed capacitors. The plates and dielectric may be rolled in a tube, placed in a

TABLE 2.2 Typical Characteristics of Commonly Used Fixed Capacitors

Type	Capacitance range	Maximum working voltage, V	Maximum operating temperature, °C	Tolerance, %	Insulation resistance, MΩ
Mica	1 pF – 0.1 μF	50 000	150	±0.25 to ±5	>100 000
Silvered mica	1 pF – 0.1 μF	75 000	125	±1 to ±20	1000
Paper	500 pF – 50 μF	100 000	125	±10 to ±20	100
Polystyrene	500 pF – 10 μF	1000	85	±0.5	10 000
Polycarbonate	0.001 – 1 μF	600	140	±1	10 000
Polyester	5000 pF – 10 μF	600	125	±10	10 000
Ceramic:					
Low *k*	1 pF – 0.001 μF	6000	125	±5 to ±20	1000
High *k*	100 pF – 2.2 μF	100	85	+100 to −20	100
Glass	10 pF – 0.15 μF	6000	125	±1 to ±20	>100 000
Vacuum	1 – 5000 pF	60 000	85	±5	>100 000
Energy storage	0.5 – 250 μF	50 000	100	±10 to ±20	100
Electrolytic:					
Aluminum	1 μF – 1 F	700	85	+100 to −20	<1
Tantalum	0.001 – 1000 μF	100	125	±5 to ±20	>1

disk or rectangular package, or otherwise to reduce the overall size. There are nearly 200 capacitor manufacturers in the United States. Their general thrust has been to develop larger capacitance values in smaller packages. The ratio of the capacitance value to the volume of the package is the *volumetric efficiency* of the capacitor.

Capacitors may be grouped into four broad categories:

1. Fixed
2. Variable
3. Chip
4. Voltage-variable (varactor)

Because the dielectric is a prime influence on capacitance size, a host of dielectrics are used in the construction of capacitors. Consequently, capacitors are classified according to their dielectrics. The characteristics of the most commonly used capacitors are summarized in Table 2.2.

MICA One of the earliest made, and still used, capacitors employs mica for the dielectric. Mica, a natural mineral, is a chemically inert and highly stable dielectric. The mica capacitor generally has a "sandwich" structure (Fig. 2.9), consisting of interleaving layers of tin-lead foil and mica. Mica capacitors are used over a wide temperature range (−55 to +150°C), and they have a high insulation resistance. Their capacitance values range from approximately 1 pF to 0.1 μF.

SILVERED MICA In the silvered mica a thin layer of silver, which is screened and fired onto the surface of the mica, is the conducting electrode. Among the advantages of a silvered mica over a mica capacitor are greater mechanical stability and more uniform characteristics.

PAPER The dielectric in paper capacitors is kraft paper impregnated with a wax or resin. It is generally packaged as a "rolled sandwich," shown in Fig. 2.10. Paper capacitors come in a large variety of values (500 pF to 50 μF), can operate in ambient temperatures as high as 125°C, are low cost, and can withstand high voltages. They are, however, temperature-sensitive. A 10°C rise in ambient temperature may reduce the life of the capacitor by as much as 50 percent.

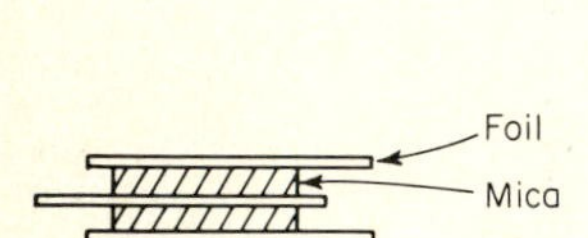

Fig. 2.9 The "sandwich" structure mica capacitor.

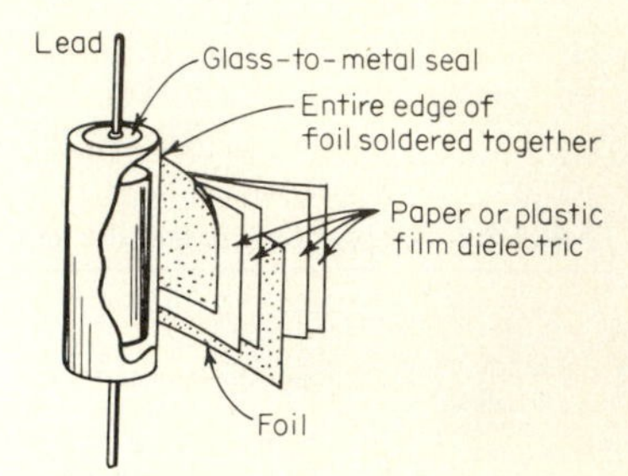

Fig. 2.10 The "rolled-sandwich" structure is commonly used for paper and plastic film capacitors.

The black band, usually printed on the package, is near the lead that is connected to the outer metal foil of the capacitor. To minimize noise, this lead should be connected to the lowest potential point in a circuit.

PLASTIC FILM Various plastics are used as the dielectrics of capacitors. Plastic capacitors are available in typical ranges of 500 pF to 10 μF. The plastics used include *polystyrene*, *polycarbonate*, and *polyester* (Mylar).

Polystyrene capacitors exhibit a low dissipation factor, small capacitance change with temperature, and very good stability. They tend to be large in size, and their maximum operating temperature is 85°C.

Polycarbonate capacitors have a dissipation factor and stability which approach those of polystyrene capacitors. They can operate at temperatures as high as 140°C.

Polyester capacitors, also referred to as Mylar capacitors, are the most widely used plastic capacitors. They are highly stable and exhibit a greater resistance to moisture than kraft paper. The maximum operating temperature of polyester capacitors is 125°C.

CERAMIC There are two basic types of ceramic dielectrics: low k and high k. Low-k ceramic capacitors can be made to exhibit zero temperature coefficient, and they find wide use in temperature-compensation networks. The low-k capacitor operates at voltages as high as 6 kV and up to 125°C. Its capacitance value is generally limited to 0.001 μF.

The high-k ceramic capacitor has capacitance values up to 2.2 μF and maximum working voltage of 100 V. These capacitors, however, change their value appreciably with temperature, dc voltage, and frequency.

GLASS This capacitor is made by stacking alternate layers of aluminum foil and glass ribbon in a sandwich structure which is similar to the mica capacitor. These capacitors are extremely reliable and stable. It is claimed that any two glass capacitors, regardless of their value or size, exhibit a temperature coefficient within 10 ppm/°C of each other. Glass capacitors can operate at voltages as high as 6000 V and up to 125°C. The range of capacitor values is 10 pF to 0.15 μF.

VACUUM For very high voltages, a vacuum is often used as the dielectric. Vacuum capacitors range from 1 to 5000 pF and can operate at voltages as high as 60 kV.

ENERGY STORAGE Energy-storage capacitors store energy which can be discharged in a time interval from a fraction of a microsecond to several hundred microseconds. The dielectric used for energy storage capacitors often consists of a specially prepared and impregnated kraft paper. Capacitance values of 0.5 to 250 μF and voltage ratings up to 50 kV are available.

METALLIZED DIELECTRICS If the dielectric, such as paper or plastic, is deposited with a thin metallic film instead of separate metal foils, volume savings up to 75 percent may be realized. Because of their low insulation resistance, however, these capacitors are not recommended for coupling and logic circuits.

ELECTROLYTIC In applications where a large value of capacitance in a small volume (high volumetric efficiency) is required, such as a filter in a power supply, the electrolytic capacitor is the choice. There are two basic types, *aluminum* and *tantalum* electrolytics, and are either *polarized* or *nonpolarized.*

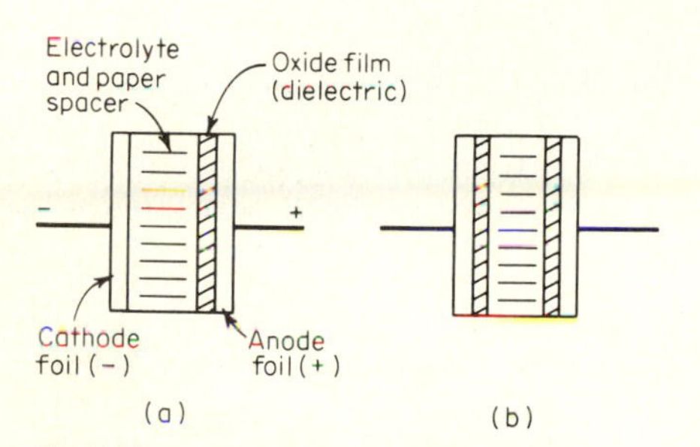

Fig. 2.11 An elementary electrolytic capacitor: (*a*) Polarized. (*b*) Nonpolarized.

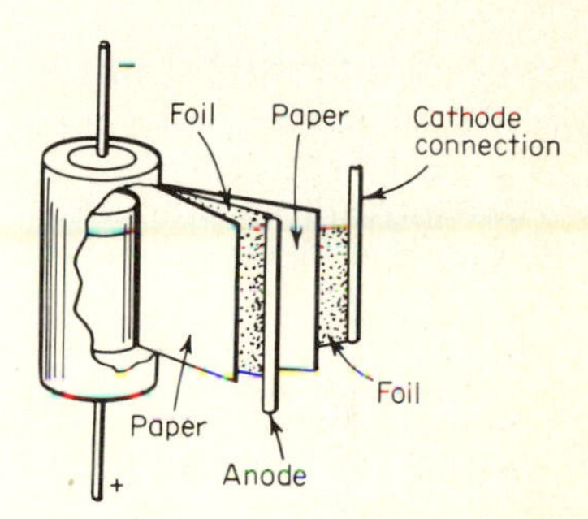

Fig. 2.12 The rolled-sandwich structure of an electrolytic foil capacitor.

In the polarized electrolytic capacitor, a plus sign is printed on the package near one of the two leads. When the capacitor is put in a circuit, the lead near the plus sign *must* be connected to a higher dc potential than the unmarked lead. If this is not done, the capacitor may be short-circuited and permanently damaged. No such restraint exists for connecting nonpolarized electrolytics in a circuit.

The basic structure of an electrolytic foil capacitor is illustrated in Fig. 2.11. For the polarized type of Fig. 2.11*a*, to increase its surface area, the aluminum (or tantalum) foil is etched. Then, a very thin layer of aluminum (or tantalum) oxide is electrochemically formed on the anode foil. The oxide layer becomes the dielectric for the capacitor. Because the oxide film is exceedingly thin and the surface area of the etched foil is large, the volumetric efficiency of electrolytic capacitors is high.

Separating the cathode and oxide-coated anode is a paper spacer which is soaked in an electrolyte solution. The spacer is required to prevent short circuiting between the cathode and anode foils. For high-voltage ratings, the spacer is thicker than for low-voltage units. The most common construction for electrolytic capacitors is the rolled sandwich, shown in Fig. 2.12.

The nonpolarized electrolytic capacitor in Fig. 2.11*b* has two oxide-coated anodes. For an equal-size polarized capacitor, the nonpolarized type has one-half the capacitance for the same voltage rating. Nonpolarized electrolytics should be used for such applications as ac motor starting, crossover networks, and large-pulse signals.

Aluminum electrolytic capacitors have high dc leakage and low insulation resistance.

Their shelf life is limited and their capacitance deteriorates with time and use. They enjoy, however, a high volumetric efficiency and are low in cost.

In addition to being very stable, tantalum oxide has nearly twice the dielectric constant of aluminum oxide. Tantalum electrolytic capacitors have a long shelf life, stable operating characteristics, increased operating temperature range, and a greater volumetric efficiency. The chief disadvantages of the tantalum, in comparison to the aluminum electrolytic, are its greater cost and lower voltage rating (see Table 2.2).

There are three types of tantalum electrolytics: *foil*, *wet anode*, and *solid anode*. The tantalum foil type is similar in construction to the aluminum foil electrolytic. The wet-anode capacitor is made by first molding, usually into the shape of a pellet, a mixture of tantalum powder and a binder. Under high temperature and in a vacuum, the pellet mixture is welded together (*sintered*), and the binder and impurities are driven off. The result is a porous pellet on which a layer of tantalum oxide is electrochemically formed. The volumetric efficiency of a wet-anode tantalum is about three times as great as the foil type.

The solid-anode tantalum is made by sintering an anode pellet on which is a layer of tantalum oxide. The pellet is then covered with a layer of manganese dioxide which serves as a solid electrolyte and cathode of the capacitor. This is followed by a layer of carbon and of silver paint, completing the cathode connection. The solid-anode construction is most widely used. It has the longest life and lowest leakage current of the three types of tantalum capacitors.

2.5 CONNECTING CAPACITORS

In this section we examine how fixed capacitors are connected in parallel and in series.

CAPACITORS IN PARALLEL One may think of connecting capacitors in parallel (Fig. 2.13) as effectively increasing the area of the plates. Because capacitance in-

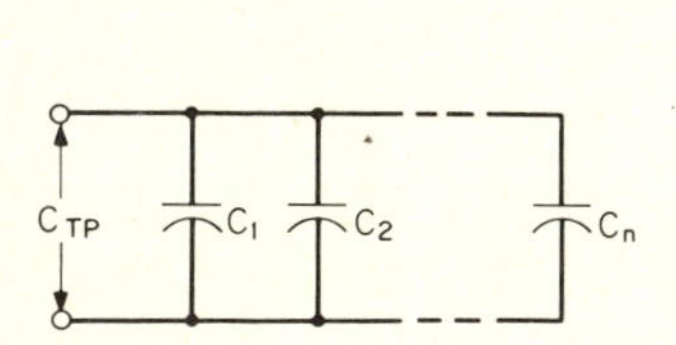

Fig. 2.13 Capacitors in parallel.

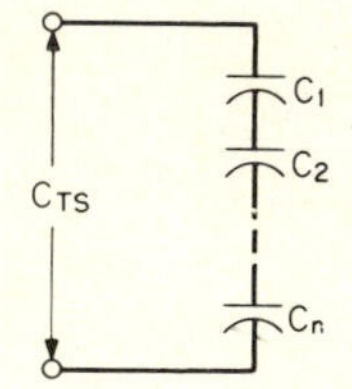

Fig. 2.14 Capacitors in series.

creases with plate area (more charge is stored in the capacitor), the total capacitance of capacitors connected in parallel, C_{TP}, is equal to the sum of the individual capacitances:

$$C_{TP} = C_1 + C_2 + \cdots + C_n \tag{2.8}$$

The working voltage of the parallel combination is limited by the smallest working voltage of the individual capacitors.

example 2.4 In Fig. 2.13 assume that three capacitors are connected in parallel ($n = 3$): $C_1 = 40$ μF, 300 V dc; $C_2 = 25$ μF, 450 V dc; $C_3 = 80$ μF, 200 V dc. Determine the capacitance and dc working voltage of the parallel combination.

solution The total capacitance C_{TP} is equal to the sum of C_1, C_2, and C_3. Hence, $C_{TP} = 40 + 25 + 80 = 145$ μF. The dc working voltage is equal to the smallest rating of the individual capacitors. In this example, it is equal to the voltage rating of the 80-μF capacitor, 200 V dc.

CAPACITORS IN SERIES Connecting capacitors in series may be thought of as increasing the separation of the outer plates of the combination (Fig. 2.14). The result is that the total capacitance C_{TS} is less than the smallest capacitance of the individual capacitors. The relationship for C_{TS} is similar to that for resistors in parallel. For two capacitors in series, C_1 and C_2, we have

$$C_{TS} = \frac{C_1 C_2}{C_1 + C_2} \tag{2.9}$$

If equal-valued capacitors are connected in series, the net value of capacitance equals the value of an individual capacitor divided by the number connected in series. For example, if three 150-pF capacitors are in series, the net capacitance is $^{150}/_{3} = 50$ pF.

The total dc working voltage rating of capacitors in series is equal to the sum of the dc ratings of the capacitors. This fact is sometimes used, for example, in the design of filters for high-voltage power supplies. For this purpose, electrolytics are generally chosen and connected in series to increase the dc working voltage. Because the insulation resistance of electrolytics is relatively low and variable with age and use, it is possible that one of the series capacitors will see a much greater voltage than it could withstand.

To prevent this from happening, *equalizing resistors,* of values between 20 and 50 kΩ, are placed across the capacitors (Fig. 2.15). The resistors form a voltage divider.

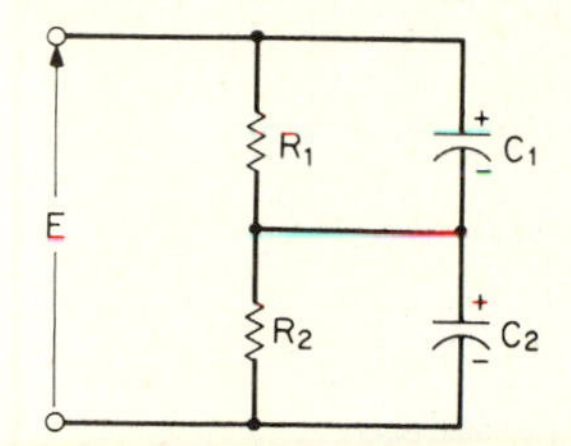

Fig. 2.15 Equalizing resistors R_1 and R_2 across electrolytic capacitors.

If they are equal, the voltage across each capacitor is $E/2$ V. If they are unequal, the voltage across a capacitor is proportional to the resistance value.

2.6 COLOR CODING OF CAPACITORS

For large-size capacitors, the values of capacitance, working voltage, and tolerance are usually stamped on the packages. Color coding, based on the resistor code, is used for small capacitors. Figure 2.16 illustrates the coding schemes for various shapes of ceramic and mica capacitors.

2.7 VARIABLE CAPACITORS

The need for a variable capacitor arises in tuning circuits found in receivers, transmitters, and oscillators. A commonly used type is the *air-variable capacitor.* The capacitor of Fig. 2.17*a* is a *single-gang* type; the *dual-gang capacitor* of Fig. 2.17*b* is used in tuning, for example, a superheterodyne receiver.

Capacitance values for air-variable capacitors range from a few picofarads up to 500 pF; their maximum voltage rating is approximately 9 kV. For higher operating voltages (up to 60 kV), a *vacuum-variable capacitor* is used. A cross-sectional view of such a capacitor is illustrated in Fig. 2.18.

The air-variable capacitor is composed of two sets of plates, usually made of aluminum. One set, referred to as the *rotor,* is mounted on a shaft and meshes with a set of fixed metal plates, referred to as the *stator.* The stator and rotor plates are never permitted to touch each other.

Upon rotation of the shaft, either more or less area exists between the rotor and stator plates. Because capacitance is directly proportional to the area of the plates, the capacitance is thereby varied. The shape of the plates determines the manner in

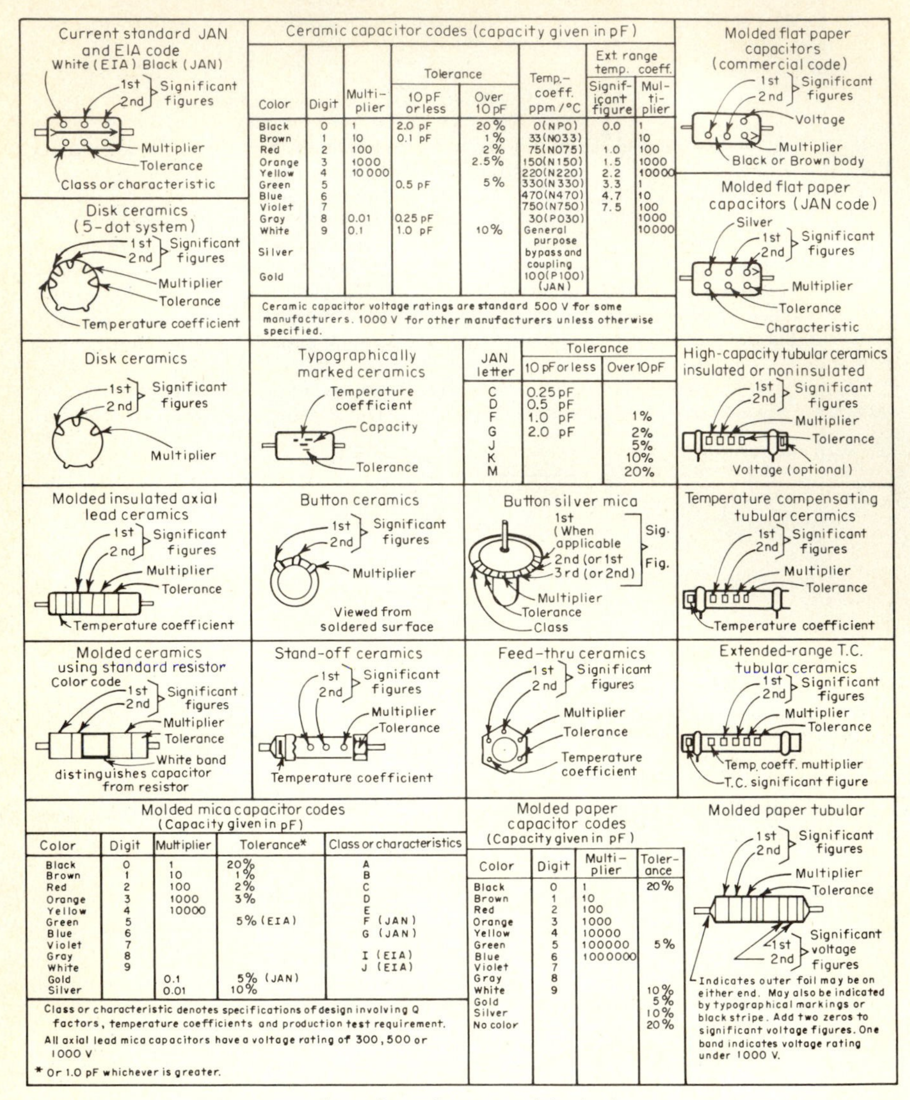

Ceramic capacitor codes (capacity given in pF)

Color	Digit	Multiplier	Tolerance: 10 pF or less	Tolerance: Over 10 pF	Temp.-coeff. ppm/°C	Ext. range temp. coeff.: Significant figure	Ext. range temp. coeff.: Multiplier
Black	0	1	2.0 pF	20%	0(NPO)	0.0	1
Brown	1	10	0.1 pF	1%	33(N033)		10
Red	2	100		2%	75(N075)	1.0	100
Orange	3	1000		2.5%	150(N150)	1.5	1000
Yellow	4	10 000			220(N220)	2.2	10000
Green	5		0.5 pF	5%	330(N330)	3.3	1
Blue	6				470(N470)	4.7	10
Violet	7				750(N750)	7.5	100
Gray	8	0.01	0.25 pF		30(P030)		1000
White	9	0.1	1.0 pF	10%	General purpose bypass and coupling		10000
Silver							
Gold					100(P100) (JAN)		

Ceramic capacitor voltage ratings are standard 500 V for some manufacturers. 1000 V for other manufacturers unless otherwise specified.

JAN letter	Tolerance: 10 pF or less	Tolerance: Over 10 pF
C	0.25 pF	
D	0.5 pF	
F	1.0 pF	1%
G	2.0 pF	2%
J		5%
K		10%
M		20%

Molded mica capacitor codes (Capacity given in pF)

Color	Digit	Multiplier	Tolerance*	Class or characteristics
Black	0	1	20%	A
Brown	1	10	1%	B
Red	2	100	2%	C
Orange	3	1000	3%	D
Yellow	4	10000		E
Green	5		5% (EIA)	F (JAN)
Blue	6			G (JAN)
Violet	7			
Gray	8			I (EIA)
White	9			J (EIA)
Gold		0.1	5% (JAN)	
Silver		0.01	10%	

Class or characteristic denotes specifications of design involving Q factors, temperature coefficients and production test requirement.

All axial lead mica capacitors have a voltage rating of 300, 500 or 1000 V

* Or 1.0 pF whichever is greater.

Molded paper capacitor codes (Capacity given in pF)

Color	Digit	Multiplier	Tolerance
Black	0	1	20%
Brown	1	10	
Red	2	100	
Orange	3	1000	
Yellow	4	10000	
Green	5	100000	5%
Blue	6	1000000	
Violet	7		
Gray	8		
White	9		10%
Gold			5%
Silver			10%
No color			20%

Fig. 2.16 Color-coding schemes used for fixed capacitors.

which capacitance varies with shaft rotation. For example, a linear or a nonlinear variation in capacitance with shaft position can be obtained.

TRIMMERS A variety of small variable capacitors using air, ceramic, mica, quartz, and other dielectrics are available for circuit applications. Referred to as *trimmer capacitors*, they are utilized for fine tuning and in hybrid microelectronic circuits. Their capacitance values range from a few picofarads to about 100 pF.

The capacitor of Fig. 2.19 is designed for hybrid circuits. It has a working voltage of 200 V dc, and in the 15- to 50-pF range, a Q of greater than 300 at 100 MHz. The ceramic trimmer of Fig. 2.20 has a working voltage of 250 V dc, and for the 4.5- to 50-pF range, a minimum Q of 100 at 100 MHz.

At high voltages and frequencies, the piston trimmer, with quartz or glass as the

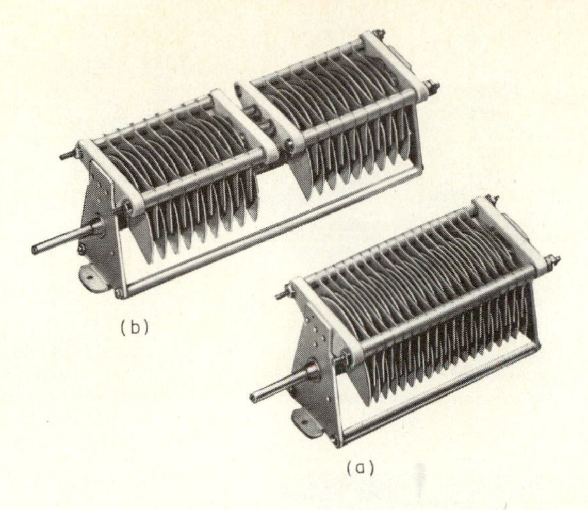

Fig. 2.17 Examples of air-variable capacitors: (*a*) Single-gang. (*b*) Dual-gang. (*Courtesy E. F. Johnson Company*)

dielectric, is suitable (Fig. 2.21). A 1.5- to 38-pF piston trimmer with a quartz dielectric has a Q greater than 750 at 100 MHz and a working voltage of 1000 V dc.

2.8 CHIP CAPACITORS

Often no larger than a match head, chip capacitors ranging in values from a few picofarads to 100 μF are available for hybrid microcircuits. Because of their high volu-

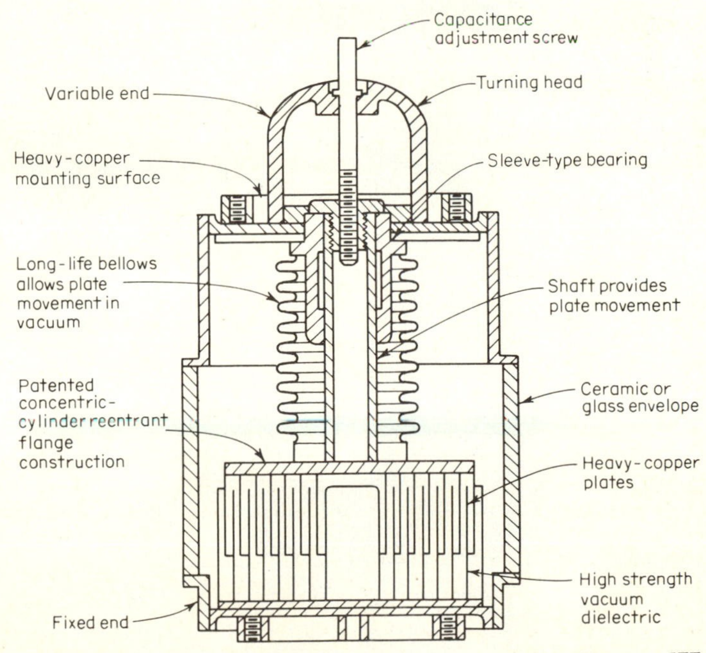

Fig. 2.18 Construction details of a vacuum-variable capacitor. (*Courtesy ITT Jennings*)

Fig. 2.19 A high-Q ceramic tuning capacitor. It is used, for example, in electronic wristwatches. *(Courtesy Johanson Manufacturing Corp.)*

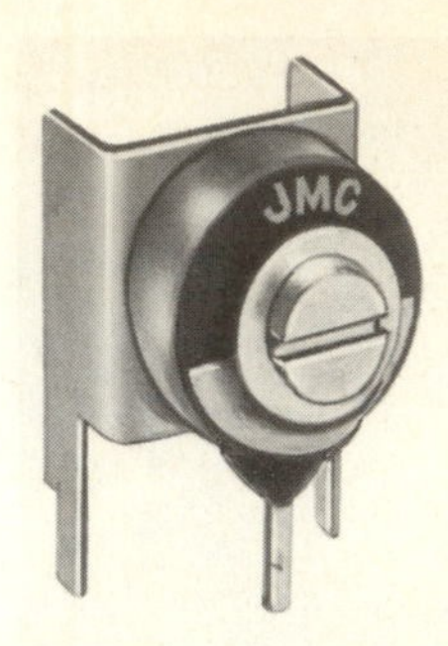

Fig. 2.20 Side-tuned ceramic disk trimmer capacitor. *(Courtesy Johanson Manufacturing Corp.)*

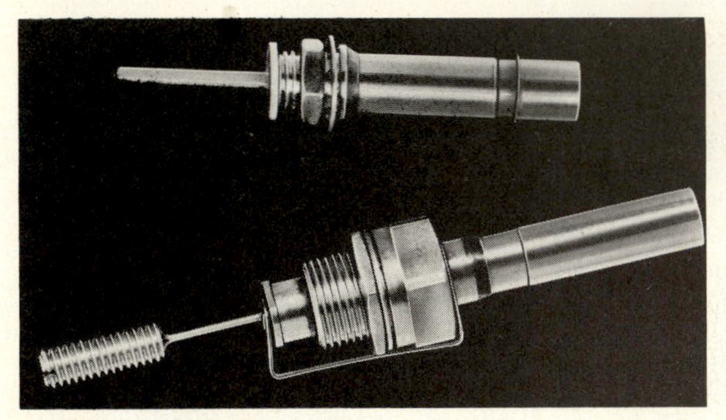

Fig. 2.21 Examples of piston tuning capacitors. *(Courtesy Johanson Manufacturing Corp.)*

Fig. 2.22 Ceramic chip capacitors. *(Courtesy Sprague Electric Company)*

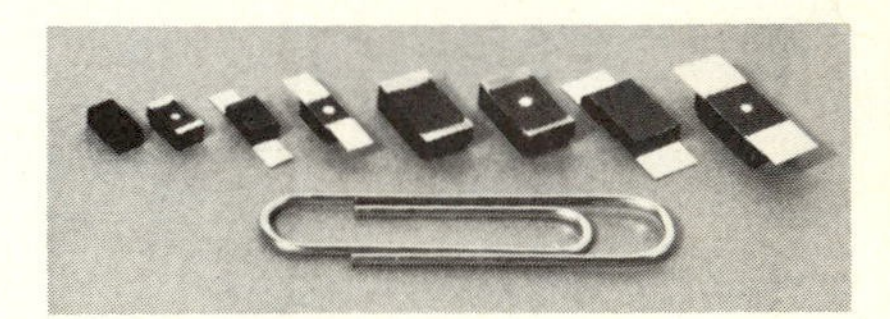

Fig. 2.23 Tantalum chip capacitors. *(Courtesy Sprague Electric Company)*

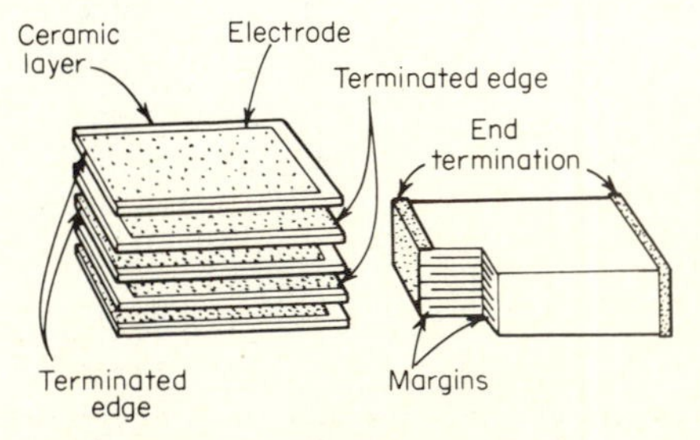

Fig. 2.24 Construction features of a ceramic chip capacitor. *(Courtesy West-Cap Division, San Fernando Electric Manufacturing Company)*

metric efficiency, *ceramic* (Fig. 2.22) and *tantalum* (Fig. 2.23) chips are commonly used. Porcelain chip capacitors are employed in microwave circuits.

The ceramic chip capacitor is usually constructed as a multilayered sandwich of ceramic and screened-on conductor, illustrated in Fig. 2.24. In the porcelain chip capacitor, porcelain is used instead of ceramic for low loss at microwave frequencies. The tantalum chip capacitor is constructed in essentially the same manner as the solid-anode electrolytic.

A comparison of chip capacitors is provided in Table 2.3. They are all capable of operating at temperatures from −55 to +125°C. Using low-*k* ceramic, chip capacitors up to approximately 0.047 μF with zero temperature coefficient are available. Although the maximum voltage rating of tantalum and porcelain chip capacitors is in the order of 50 V, this is no serious disadvantage. Transistors in monolithic and hybrid circuits generally operate at voltages that are well below 50 V. In terms of cost, tantalum chip capacitors tend to be the least expensive and porcelain chips the most expensive.

TABLE 2.3 Typical Characteristics of Chip Capacitors

Type	Capacitance range	Tolerance range, %	Maximum working voltage, V	Insulation resistance, MΩ
Ceramic	10 pF–3.5 μF	±1 to ±20	200	1000
Tantalum	100 pF–100 μF	±5 to ±20	50	1500
Porcelain	1–330 pF	±1 to ±10	50	100 000

2.9 VOLTAGE-VARIABLE CAPACITORS (VARACTORS)

Consider the pn junction diode of Fig. 2.25*a*. Separating the p and n regions is the depletion layer. When the pn junction is reverse-biased (the p region is negative with respect to the n region), the depletion region becomes devoid of free electrons and holes. One may therefore view the depletion layer of a reverse-biased diode as the dielectric, and the p and n regions as the plates of a capacitor.

As the reverse bias is increased, the depletion layer widens and the capacitance decreases, because the separation of the p and n regions has increased. If the reverse bias is reduced, the depletion layer contracts and the capacitance is increased. We thus have a diode whose capacitance varies inversely with the impressed reverse bias. The greater the magnitude of the reverse bias, the less the capacitance, and vice versa. Such a device is called a *voltage-variable* capacitor, or *varactor*. The symbol for this device is given in Fig. 2.25*b*.

Curves showing the variation in capacitance with reverse voltage for the varactor are illustrated in Fig. 2.26. Variation in capacitance is nonlinear. Over a range of

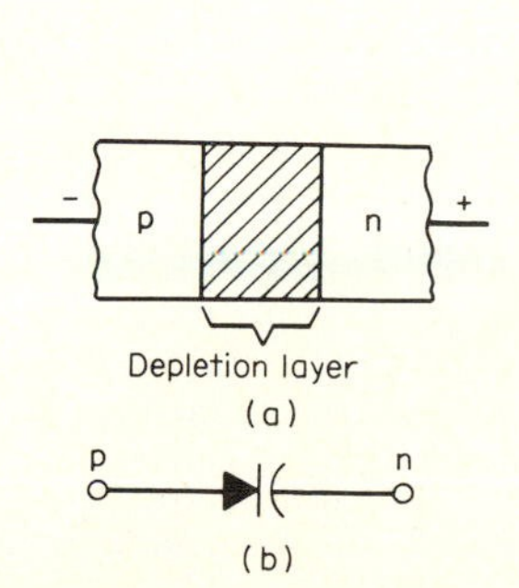

Fig. 2.25 Voltage-variable capacitor (varactor): (*a*) Reverse-biased p-n junction diode. (*b*) Symbol for voltage-variable capacitor.

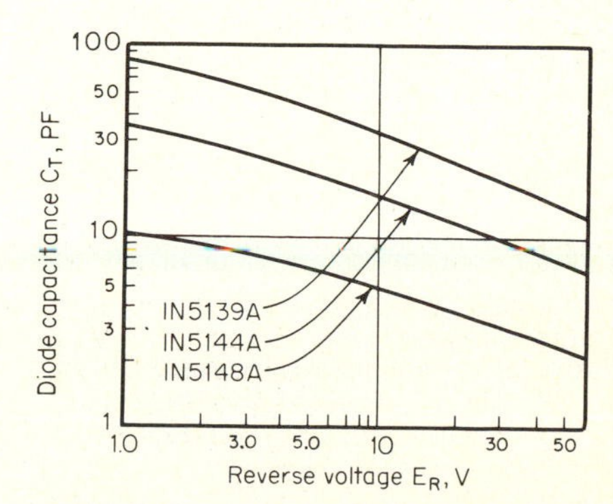

Fig. 2.26 Variations of varactor capacitance with reverse voltage. (*Courtesy Codi Semiconductor*)

reverse voltage from 1 to 60 V, for example, the variation in capacitance for the 1N5139A is approximately 80 to 14 pF. Diode capacitances as high as 2000 pF are possible for low-frequency applications. Microwave varactors may have a maximum capacitance of only 0.4 pF.

Applications for varactors are numerous and include TV tuners, high-power frequency multipliers, and frequency control. They can handle from 100 mW to hundreds of watts of r-f power. Their breakdown voltage ranges from approximately −6 to −300 V. Examples of various case styles used for varactors are illustrated in Fig. 2.27.

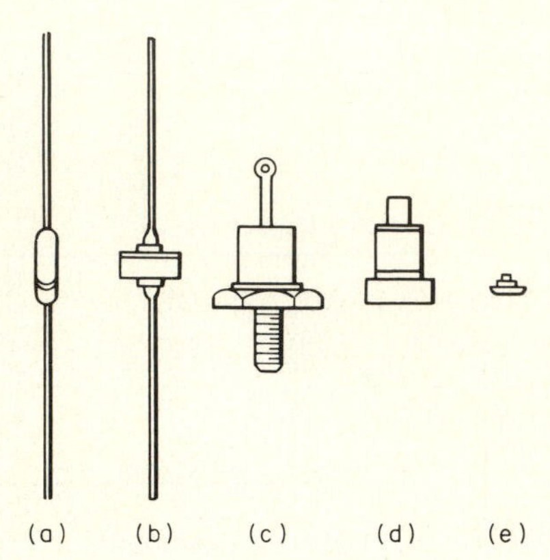

Fig. 2.27 Examples of case styles for varactors: (*a*) General purpose. (*b*) High capacitance. (*c*) Power. (*d*) UHF power. (*e*) Microwave.

PARAMETERS Some of the significant parameters for characterizing the varactor will now be defined.

Total diode capacitance (C_t) The total diode capacitance C_t is the sum of the junction and case capacitances. It is generally specified at a reverse voltage E_R of −4 or −6 V. This parameter is important because it indicates whether the diode can be used at a particular frequency of interest. Variation in C_t with E_R is illustrated in Fig. 2.26.

Series resistance (R_S) Resistance R_S is the resistance in series with the junction of the diode. It varies with the reverse voltage, decreasing in value as the reverse voltage is increased.

Figure of merit (Q) The Q of a varactor is equal to

$$Q = \frac{0.159}{fR_S C_t} \tag{2.10}$$

where R_S is in ohms, f in hertz, and C_t in farads. The variation in Q with reverse voltage and frequency is illustrated in Fig. 2.28. As the reverse voltage increases, the values of R_S and C_t decrease, and Q rises. For rising frequencies, the losses in the diode increase and Q falls.

Cutoff frequency (f_{co}) Cutoff frequency f_{co} is the frequency where the Q of the device is unity.

Variation of capacitance with temperature The variation of C_t with junction temperature T_j is often presented as a normalized plot shown in Fig. 2.29. At a junction temperature corresponding to room temperature (25°C), the normalized value of C_t is one. As the temperature decreases, the capacitance falls and increases with increasing temperature.

C_t ratio The ratio of capacitance variation at a reverse voltage of −4 or −6 V to the capacitance at approximately 80 percent of the breakdown voltage is referred to as the C_t ratio.

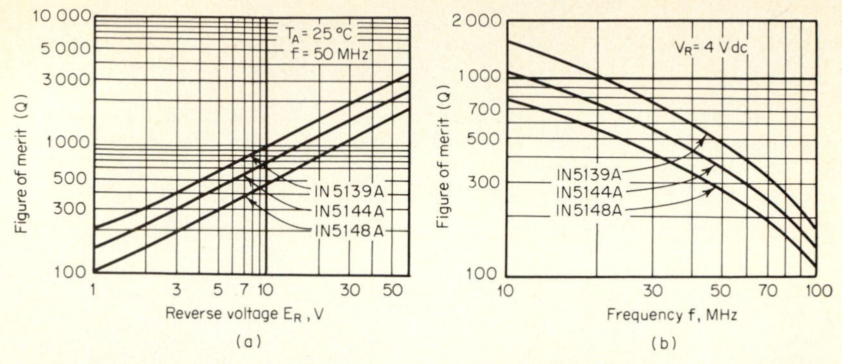

Fig. 2.28 Variation in Q of a varactor with (*a*) reverse voltage and (*b*) frequency. (*Courtesy Codi Semiconductor*)

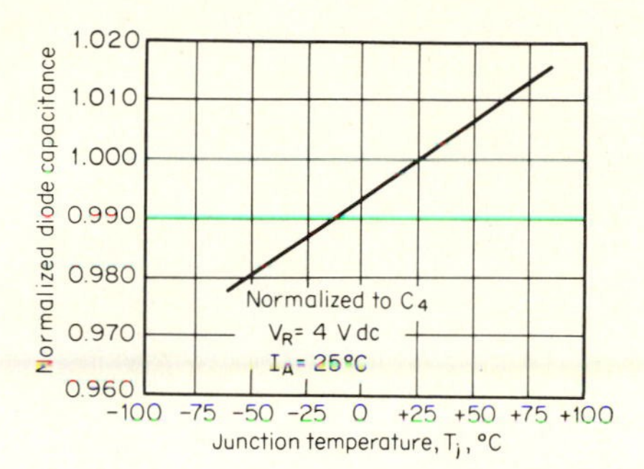

Fig. 2.29 A normalized plot of varactor capacitance as a function of junction temperature. (*Courtesy Codi Semiconductor*)

Conversion efficiency (η) The conversion efficiency defines the performance of a varactor used as a frequency multiplier. This parameter is defined as the ratio of the output power P_o to the input power P_i and is expressed as a percentage:

$$\eta = \frac{P_o}{P_i} \times 100\% \qquad (2.11)$$

example 2.5 A tripler using a varactor multiplies an input frequency of 50 MHz to an output frequency of 150 MHz. If the r-f input power is 52.5 W and the output r-f power is 40 W, what is conversion efficiency of the tripler?

solution From Eq. (2.11) $\eta = (40/52.5) \times 100\% = 76\%$.

Chapter 3

Coils

3.1 INTRODUCTION

Capacitors and resistors are available in a wide variety of sizes and component values. It is possible to purchase almost any value, tolerance, or other rating in these components without special orders.

Coils, on the other hand, may not be as readily available. Technicians may find it necessary to wind their own coils when they want a specific value of inductance. This chapter gives the equations and characteristics of the most frequently used coils. A nomogram is included for simplifying the design of coils for specific applications.

3.2 TYPES OF FERROUS CORES AND THEIR EFFECTS IN COILS

When a core of magnetic material is inserted into an air-core coil, there is a great increase in coil inductance. A magnetic material is any kind of substance that is attracted by a magnet, and may be classified as being either hard or soft. The hard magnetic materials retain their magnetism after the magnetizing source has been removed, but the soft magnetic materials immediately lose magnetism upon removal of the magnetizing source. Saying that a magnetic material is soft does not mean that it is physically soft. A magnetic material may be physically extremely hard and brittle and yet be magnetically very soft. In other words, it very quickly loses all traces of magnetism as soon as the magnetizing source is removed. Soft magnetic materials are used as core material in radio-frequency and in audio-frequency coils and transformers.

IRON-POWDER CORES Iron-powder cores are composed of finely powdered iron or iron alloys mixed with a plastic binder. The cores are formed and pressed into final shapes at pressures ranging from 10 to 50 tons/in^2, and then baked to set the binder.

Iron-powder cores are available in a large variety of different shapes and sizes and with different magnetic properties to meet different requirements.

FERRITE CORES Ferrites are also magnetic substances. They are unlike iron and steel and their various alloys in that ferrites are insulators. There are many different types of ferrites, and the exact chemical composition of each different type depends on its use; some ferrites that are useful at low frequencies cannot be employed at high frequencies and vice versa.

Table 3.1 gives the magnetic and physical properties of a number of different types of ferrites that are used as core material.

LAMINATED CORES The cores utilized in transformers and chokes that operate at power-line and audio frequencies are made of sheet steel laminations of various grades and thicknesses. These are treated in considerable detail in Chap. 5.

TABLE 3.1 Magnetic and Physical Properties of Ferrites Used as Core Material*

Material	Initial permeability, μ_o	Maximum permeability, μ_{max}	Flux density @ 14, B@H	Residual magnetism, Br	Coercive force, Hc	Frequency range, mHz	Loss factor, $\frac{1}{\mu_o Q}$	Curie temperature, °C	Volume resistivity, Ω-cm	Specific gravity, g/cm^3
AM04	40	160	2200	400	4.0	10–80	@50 mHz 1.8×10^{-4}	500	10^9	4.3
AM12	125	340	2300	1000	1.9	0.2–20	@10 mHz 1.7×10^{-4}	300	10^8	4.4
AM20	200	640	2700	1800	1.5	0.05–8	@2.5 mHz 5×10^{-5}	280	10^8	4.4
AK04	400	1300	3500	1500	0.75	0.01–2	@100 kHz 2.5×10^{-5}	230	10^3	4.6
AK08	800	1500	3400	1500	0.45	0.01–1	@100 kHz 2.0×10^{-5}	200	10^2	4.6
AK16	1600	2800	3900	1000	0.20	1–400 kHz	@100 kHz 8.0×10^{-6}	140	10^2	4.5
AK20	2000	3400	3900	1800	0.16	1–250 kHz	@100 kHz 7.0×10^{-6}	135	10^1	4.6
AK30	3000	4500	4200	1500	0.18	1–250 kHz	@100 kHz 5.0×10^{-6}	200	10^1	4.6

* Typical values as measured on toroidal cores.
Courtesy The Arnold Engineering Company.

Fig. 3.1 Ferrite tuning cores, toroids, special forms, and slugs. (*Courtesy The Arnold Engineering Company*)

TUNING CORES, SLUGS, AND TOROIDS When the core of a coil is adjustable within the coil, it is called a tuning core, or tuning slug. The nonadjustable cores are simply slugs or toroids. Ferrite tuning cores, toroids, special forms, and slugs of various sizes are shown in Fig. 3.1.

CORE PERMEABILITY Some idea of the increase in inductance achieved by the use of a core of magnetic material may be learned from a consideration of the permeability of the magnetic material—iron-powder, ferrite, or other. The permeability of air is always one, but the permeability of an iron-powder or ferrite core may be hundreds and even thousands of times greater than this. If the permeability of a magnetic core material is 100, its use increases the magnetic flux 100 times; this means that the inductance of the coil will also be increased 100 times.

Computations involving iron-powder and ferrite-core coils usually require technical data from core manufacturers. This will be discussed in a later paragraph.

3.3 TYPES OF NONFERROUS CORES

Besides cores of magnetic materials, cores of nonmagnetic metals (brass, copper, silver) are also sometimes employed at frequencies of about 50 mHz and above. Nonmagnetic cores have an effect on coil inductance just the opposite of that produced by a magnetic core; that is, they reduce, instead of increase, the inductance of a coil. When nonmagnetic cores are used, measures must be taken to reduce r-f core losses as much as possible. Since high-frequency currents flow mainly near the surface of a conductor, a thin metal sleeve or a silver-plated plastic core is sometimes utilized in the wire center. Solid nonmagnetic cores may be used when high coil efficiency is not a circuit requirement.

The effect of nonferrous cores on coil inductance is much less than that produced by cores of magnetic material. Nonferrous tuning cores are useful in equipment that requires a final adjustment of coil inductance before being put into service.

3.4 COMPUTING THE INDUCTANCE OF AIR-CORE COILS

A number of different formulas may be used for computing the inductance of air-core coils. Although these formulas take different forms, since they do give the same results, they may all be reduced to certain basic forms. The ones for the coil forms of Fig. 3.2*a* will give results sufficiently accurate for all practical purposes. The equations are listed here for convenience. Typical transmitter-type air-core coils are shown in Fig. 3.2*b*.

$$L = \frac{r^2N^2}{9r + 10l} \tag{3.1}$$

$$L = \frac{0.8r^2N^2}{6r + 9l + 10d} \tag{3.2}$$

$$L = \frac{r^2N^2}{8r + 11d} \tag{3.3}$$

If coil dimensions are given in centimeters, instead of in inches, use the following:

$$L = \frac{0.394r^2N^2}{9r + 10l} \tag{3.4}$$

$$L = \frac{0.315r^2N^2}{6r + 9l + 10d} \tag{3.5}$$

$$L = \frac{0.394r^2N^2}{8r + 11d} \tag{3.6}$$

The inductance of single-layer coils may also be determined by means of nomograms and by the use of special slide rules designed for this purpose.

The formulas of Fig. 3.2*a* all show that inductance is directly proportional to the square of the number of turns. Thus, if the number of turns in a coil is doubled, and

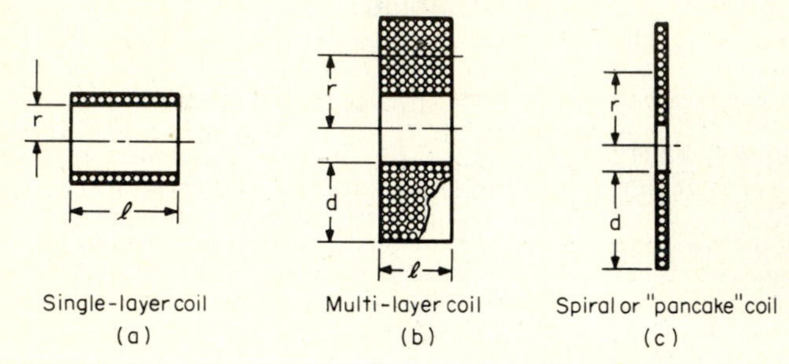

Fig. 3.2 Computing the self-inductance of some typical air-core coils:

(*a*) $L = \dfrac{r^2N^2}{9r + 10l}$ (*b*) $L = \dfrac{0.8r^2N^2}{6r + 9l + 10d}$ (*c*) $L = \dfrac{r^2N^2}{8r + 11d}$

where L = coil inductance, μH
N = number of turns
l = length of coil, inches
d = depth of coil, inches
r = mean radius of coil, inches

If coil dimensions are given in centimeters instead of inches, use the following:

(*a*) $L = \dfrac{0.394r^2N^2}{9r + 10l}$ (*b*) $L = \dfrac{0.315r^2N^2}{6r + 9l + 10d}$ (*c*) $L = \dfrac{0.394r^2N^2}{8r + 11d}$

coil dimensions are kept unchanged, inductance will be increased four times. This is strictly true only if there is very close coupling between turns, as there is between the turns and between the windings of iron-core transformers and chokes. However, for air-core coils, it is only approximately true.

example 3.1 A single-layer coil like the one shown in Fig. 3.2*a* is to be wound in a space 2 in. long. How many turns must be wound in the 2-in. space if coil diameter is 1 in., and the required coil inductance is 250 μH?

solution

$$L = \frac{r^2N^2}{9r + 10l} = 250$$

$$= \frac{(0.5)^2N^2}{9(0.5) + 10(2)}$$

Solving for the number of turns,

$$N = \frac{250[9(0.5) + 10(2)]}{(0.5)^2}$$

$$= 156.5 \text{ turns}$$

example 3.2 A multilayer coil, such as that shown in Fig. 3.2*b*, is wound in a slot 1 in. square. Mean radius *r* of the coil is 2.25 in., and $d = 1$. If the coil contains 900 turns, what is the inductance of the coil?

solution

$$L = \frac{0.8r^2N^2}{6r + 9l + 10d}$$

$$= \frac{(0.8)(2.25)^2(900)^2}{6(2.25) + 9(1) + 10(1)}$$

$$= 100\,938\ \mu\text{H}$$

$$= 100.938\ \text{mH}$$

3.5 MUTUAL INDUCTANCE

Mutual induction, or the mutual inductive effect between two magnetically coupled coils, is due to the magnetic flux of the first coil acting on the turns of the second coil, as at the same time, the magnetic flux of the second coil acts on the turns of the first coil. To put it another way, there is a mutual interaction between the magnetic fields of the two coils.

Mutual inductance is a measure of the amount of mutual induction that exists between two magnetically coupled coils. Figure 3.3 illustrates the meaning of mutual induction. If the two coils are wound on separate cores and placed some distance from each other, as shown in Fig. 3.3*a*, there will be no magnetic coupling between the mag-

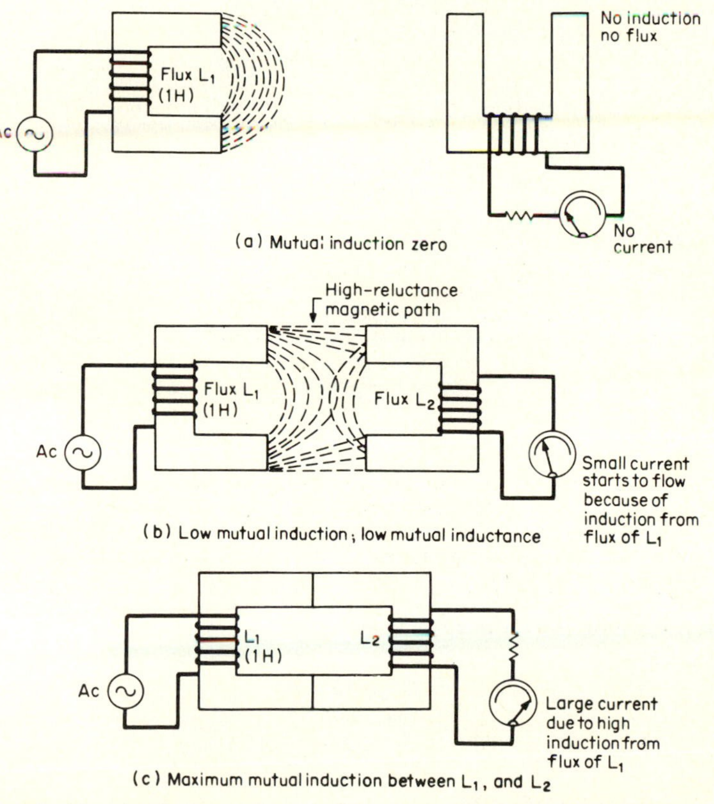

Fig. 3.3 Mutual induction is the mutual interaction between the magnetic fluxes of L_1 and L_2. The amount of mutual inductance depends on the coupling between L_1 and L_2, which, in turn, depends on the reluctance of the magnetic circuit between L_1 and L_2 and on the self-inductances of the coils.

netic fields. Hence, the two coils will not have any mutually inductive effect on each other, and their mutual inductance will be zero.

If the two coils are brought nearer to each other, as in Fig. 3.3*b*, the magnetic field of coil L_1 will start inducing a voltage in coil L_2, and a small current will flow in coil L_2 which will then generate a magnetic field of its own because of the current. However, since the reluctance of the air path between the two coils will still be very high, the magnetic coupling between them will be very low, and they will have only a very low mutual inductance.

If the coils are next brought in close contact with each other, as in Fig. 3.3*c*, the magnetic coupling between the two coils will be a maximum; thus, mutual induction and mutual inductance between the two coils will also be a maximum.

It may be seen, therefore, that mutual inductance depends on the distance between two magnetically coupled coils. It also depends on the self-inductances of the coils and on the reluctance of the magnetic path between the coils. All these factors have an effect on the amount of coupling between the coils.

If we again consider the very closely coupled coils of Fig. 3.3*c*, and denote the inductance of coil L_1 by L_1 and that of coil L_2 by L_2, then the combined inductance of the two coils connected in series aiding equals $L_1 + L_2$ plus the mutual inductance due to the magnetic lines of force of L_1 acting on the turns of L_2, plus the mutual inductance due to the magnetic lines of force of L_2 acting on the turns of L_1. Since L_1 and L_2 are identical, their mutual induction effects are equal. Therefore, the total inductance is determined by the equation

$$L = L_1 + L_2 + 2M \tag{3.7}$$

where M = mutual inductance. Since $L_1 = L_2$,

$$L = 2L_1 + 2M = 4L_1 = 4 \text{ H}$$

as determined by an actual measurement of the combined inductance of the two coils. Then, since $L_1 = 1$ H,

$$M = \tfrac{1}{2}(L - 2L_1)$$
$$M = \tfrac{1}{2}(4 - 2 \times 1) = 1 \text{ H}$$

In any other case in which two identical coils are as closely coupled as the two coils of Fig. 3.3*c*, mutual inductance will always be approximately the same as the inductance of either coil.

It should be noted that even if L_1 and L_2 are not equal to each other, it cannot be said that one of the coils has a mutual inductance that differs from the mutual inductance of the other. In fact, neither coil by itself has any mutual inductance. It is only when the two coils are magnetically coupled to each other that a mutual inductive effect exists between them, and it is never correct to say that one of the coils has a mutual inductance.

COUPLED COILS IN SERIES AND IN PARALLEL Not all coils will be as closely coupled magnetically as the two coils of Fig. 3.3*c*. For some applications this would not even be a desirable condition. In any case, it would be impossible to achieve with air-core coils or with any coils in which the coupling medium has a low permeability or high reluctance. Moreover, two coupled coils may have entirely different values of self-inductance. Hence in these cases the combined inductance of two coils connected in series aiding will be less than four times the inductance of one coil, even if the coupled coils are identical. The combined inductance of any two coils connected in series aiding and having self-inductances of L_1 and L_2, respectively, will be given by

$$L = L_1 + L_2 + 2M \tag{3.8}$$

where L = combined inductance, and M = mutual inductance.

But if the coils are connected in series opposing, their combined inductance will be

$$L = L_1 + L_2 - 2M \tag{3.9}$$

And if two coils are connected in parallel with fields aiding,

$$L = \frac{1}{1/(L_1 + M) + 1/(L_2 + M)} \tag{3.10}$$

And for two coils in parallel with fields opposing,

$$L = \frac{1}{1/(L_1 - M) + 1/(L_2 - M)} \tag{3.11}$$

COMPUTING THE MUTUAL INDUCTANCE OF SINGLE-LAYER AIR-CORE COILS

An important case for consideration is the mutual inductance between coils having air cores. Figure 3.4 shows two important examples, one with no spacing between the coils and the other with a space between them.

CASE 1: If there is no separation between the windings of two coupled coils wound on the same form, as shown in Fig. 3.4*a*, their combined inductance with the coils connected in series aiding is given by Eq. (3.8), and also by the equation for computing

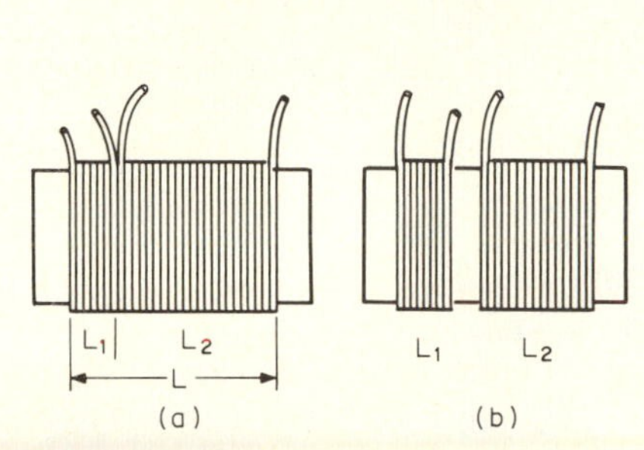

Fig. 3.4 Typical single-layer air-core coils: (*a*) L_1 and L_2 on the same form and touching; mutual inductance, $M = \frac{1}{2}(L - L_1 - L_2)$. (*b*) With L_1 and L_2 on the same form but separated by an air space, computing mutual inductance is more difficult.

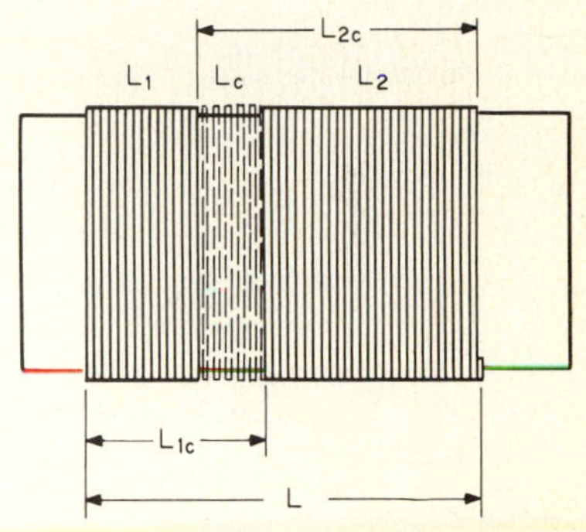

Fig. 3.5 L_1 and L_2 are single-layer coils wound on the same form. L_c is an assumed winding in the space between L_1 and L_2. Mutual inductance between L_1 and L_2 is computed from the following formula: $M = \frac{1}{2}(L + L_c - L_{1c} - L_{2c})$.

the inductance of a single-layer coil [Eq. (3.1)]. Since the separate inductances of the two coils may also each be computed by using Eq. (3.1), the only unknown in Eq. (3.8) will be M. This can be determined from the equation

$$M = \tfrac{1}{2}(L - L_1 - L_2) \tag{3.12}$$

CASE 2: When the windings of two coupled coils are separated by an air space, as in Fig. 3.4*b*, imagine that the space between the coils is also wound with wire that has the same size as the windings of L_1 and L_2. The fictitious center winding is called L_c. Also, assume that L_1, L_c, and L_2 are connected in series aiding, as shown in Fig. 3.5, and their combined inductance is found by using Eq. (3.1). Using the same equation, the inductance of L_1 in series with the imaginary center winding L_c is found, then of L_2 in series with L_c, and of L_c separately. L, the combined inductance of L_1, L_c, and L_2, is then given by

$$L = L_{1c} + L_{2c} - L_c + 2M$$

Then
$$2M = L + L_c - L_{1c} - L_{2c}$$

and
$$M = \tfrac{1}{2}(L + L_c - L_{1c} - L_{2c}) \tag{3.13}$$

example 3.3 Two single-layer coils, L_1 and L_2, are wound side by side on the same form, as in Fig. 3.4*a*. Coil L_1 has 40 turns and L_2 has 60. The total winding space is 2 in long, and coil radius r is 0.5 in. Compute:

(*a*) Total inductance with the coils connected in series aiding
(*b*) Mutual inductance
(*c*) Total inductance with the coils connected series opposing
(*d*) Total inductance with L_1 and L_2 connected in parallel and fields *aiding*
(*e*) Total inductance of L_1 and L_2 in parallel with fields opposing

solution (*a*) Using Eq. (3.1),

$$L = \frac{(0.5)^2 100^2}{9(0.5) + 10(2)}$$
$$= \frac{0.25(10\,000)}{4.5 + 20} = 102.04\ \mu\text{H}$$

(*b*) Since L_1 has 40 turns, and the total number of turns is 100 in a winding space 2 in long, the length of L_1 is 0.4 of 2, or 0.8 in. Therefore,

$$L_1 = \frac{0.5^2(40)^2}{9(0.5) + 10(0.8)} = 32\ \mu\text{H}$$

and

$$L_2 = \frac{0.5^2(60)^2}{9(0.5) + 10(1.2)} = 54.54\ \mu\text{H}$$

Since the series-aiding inductance,

$$L = 102.08 = 32 + 54.54 + 2M$$

it follows that

$$M = \tfrac{1}{2}(102.08 - 32 - 54.54) = 7.77\ \mu\text{H}$$

(*c*) With the coils connected in series opposing,

$$L = 32 + 54.54 - 2M$$
$$= 86.54 - 2(7.77) = 71\ \mu\text{H}$$

(*d*) With the coils connected in parallel and fields *aiding*,

$$L = \frac{1}{1/(L_1 + M) + 1/(L_2 + M)}$$
$$= \frac{1}{1/(32 + 7.77) + 1/(54.54 + 7.77)} \qquad (3.10)$$
$$= 24.28\ \mu\text{H}$$

(*e*) When the coils are connected in parallel with fields *opposing*,

$$L = \frac{1}{1/(L_1 - M) + 1/(L_2 - M)}$$
$$= \frac{1}{1/(32 - 7.77) + 1/(54.54 - 7.77)} \qquad (3.11)$$
$$= 15.96\ \mu\text{H}$$

example 3.4 Two single-layer coils, L_1 and L_2, are wound on the same form, as in Fig. 3.4*b*. The coils are separated by an air space of $\frac{1}{4}$ in, and coil radius, *r* is 1 in. Coil L_1 contains 24 turns and is 1 in long; L_2 contains 48 turns and is 2 in long. Compute:
(*a*) Mutual inductance
(*b*) Total inductance of L_1 and L_2 connected in series aiding
(*c*) Total inductance of L_1 and L_2 connected in parallel with fields *aiding*.

solution To determine the mutual inductance, the four inductances, L, L_{1c}, L_{2c}, and L_c must first be computed. Since L_1 contains 24 turns and is 1 in long, the $\frac{1}{4}$-in air space would contain six turns; total number of turns in the $3\frac{1}{4}$-in winding space would then be $24+6+48=78$ turns, and for 78 turns,

$$L = \frac{1(78)}{9(1)10(3\tfrac{1}{4})} = 146.6\ \mu\text{H}$$
$$L_{1c} = \frac{1^2(30)^2}{9(1) + 10(1\tfrac{1}{4})} = 41.86\ \mu\text{H}$$
$$L_{2c} = \frac{1^2(54)^2}{9(1) + 10(2\tfrac{1}{4})} = 92.57\ \mu\text{H}$$
$$L_c = \frac{1^2(6)^2}{9(1) + 10(\tfrac{1}{4})} = 3.13\ \mu\text{H}$$

(*a*) The mutual inductance,

$$M = \tfrac{1}{2}(L + L_c - L_{1c} - L_{2c}) = \tfrac{1}{2}(146.6 + 3.13 - 41.86 - 92.57) = 7.65\ \mu\text{H}$$

(*b*)

$$L_1 = \frac{1^2(24)^2}{9(1) + 10(1)} = 30.32\ \mu\text{H}$$

$$L_2 = \frac{1^2(48)^2}{9(1) + 10(2)} = 79.45\ \mu\text{H}$$

$$L = L_1 + L_2 + 2M$$
$$= 30.32 + 79.45 + 2(7.65) = 125.07\ \mu\text{H}$$

(*c*)

$$L = \frac{1}{1/(L_1 + M) + 1/(L_2 + M)}$$
$$= \frac{1}{1/(30.32 + 7.65) + 1/(79.45 + 7.65)}$$
$$= 26.44\ \mu\text{H}$$

3.6 COMPUTING THE VALUE OF INDUCTANCE FROM MEASUREMENTS

In some cases it is very difficult to determine mutual inductance by calculation. Two examples are shown in Fig. 3.6.

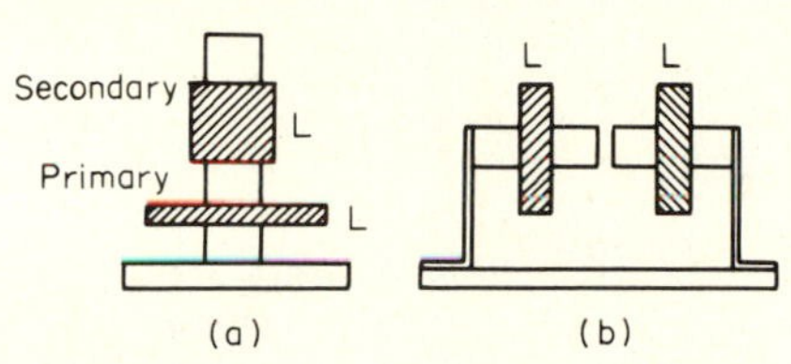

Fig. 3.6 (*a*) Two multilayer coils on the same form. (*b*) Ferrite core r-f chokes separated by an air gap. Mutual inductance is computed from $M = \frac{1}{4}(L_A - L_o)$, where M = mutual inductance, L_A = total inductance of L_1 and L_2 in series aiding, L_o = total inductance of L_1 and L_2 in series opposing.

The two r-f coils of Fig. 3.6*a* are wound on the same iron-powder core but separated from each other. Figure 3.6*b* shows two r-f chokes wound on separate ferrite cores and separated from each other by a small air gap. In both of these, and in many similar instances, it would be extremely difficult to calculate mutual inductance from such known factors as the self-inductance of the coils. It is possible, however, to measure the combined inductance of the coils connected in series aiding and then in series opposing, denoting the series-aiding inductance by L_A and the series-opposing inductance by L_o. Two simultaneous equations are then formed.

$$L_A = L_1 + L_2 + 2M$$
$$L_o = L_1 + L_2 - 2M$$

Subtracting the second equation from the first,

$$L_A - L_o = 4M$$

from which

$$M = \tfrac{1}{4}(L_A - L_o) \tag{3.14}$$

where M = mutual inductance
L_A = combined inductance of L_1 and L_2 connected in series aiding
L_o = combined inductance of L_1 and L_2 connected in series opposing

example 3.5 Two radio-frequency coils are separated by a small air gap, as shown in Fig. 3.6*b*. When the coils are connected in series aiding, their combined inductance, as measured on an inductance bridge, is 17 mH, and when the coils are connected in series opposing, combined inductance is 3 mH. What is the mutual inductance?

solution

$$M = \frac{17 - 3}{4} = 3.5\text{ mH}$$

3.7 THE COEFFICIENT OF COUPLING

The coefficient of coupling, or coupling coefficient, is a measure of the amount of coupling between two coils; it shows whether the coupling is "tight" or "loose." The greatest possible value of the coefficient of coupling is one, in which case the *coupling* may be said to be 100 percent. This is very tight coupling. Coupling coefficient may be computed from the equation

$$k = \frac{M}{\sqrt{L_1 L_2}} \tag{3.15}$$

where k = coefficient of coupling (There are no units for k)
M = mutual inductance in the same units as L_1 and L_2
L_1, L_2 = self-inductances of the separate coils

A coupling coefficient of 1, or 100 percent coupling, is approached only in well-designed iron-core transformers and chokes. In air-core transformers and coils it is very much less, but increases in transformers with iron-powder and ferrite cores.

example 3.6 When the primary of an r-f antenna coil, such as that shown in Fig. 3.6*a*, is connected to the secondary in series aiding, the measured inductance is 3085 μH; when the primary and secondary are in series opposing, the measured inductance is 1405 μH. If primary and secondary self-inductances are 2000 and 245 μH, respectively, what is the coefficient of coupling between the primary and secondary?

solution

$$M = \tfrac{1}{4}(L_A - L_o)$$
$$= \tfrac{1}{4}(3085 - 1405) = 420\ \mu\text{H}$$

and the coefficient of coupling,

$$k = \frac{M}{\sqrt{L_1 L_2}} = \frac{420}{\sqrt{2000(245)}} = 0.6$$

3.8 COIL AND CONDUCTOR RESISTANCE AT RADIO FREQUENCIES

The resistance that a conductor offers to the flow of high-frequency alternating current through it is greater than the resistance of the same conductor to direct current. This is due to a characteristic of coils and conductors that is known as the skin effect. It is a characteristic of conductors that carry alternating current, but since it is a magnetic effect, it is much more pronounced in coils than in straight conductors.

Skin effect causes the current in a conductor to be much denser near the surface of the conductor than at its center, and is a result of the magnetic lines of force associated with any current-carrying conductor. These magnetic lines of force, or flux lines, not only surround the conductor but also exist in the interior of the conductor and are densest at its center. Consequently, as the instantaneous value of the alternating current changes, the inductance of a conductor in its interior is much greater than the inductance near its surface. The result is that more current flows near the surface, or "skin," of the conductor than in its interior. Since inductive effects increase with increasing frequency, skin effect also increases as frequency is increased.

EDDY CURRENTS IN CONDUCTORS Besides the skin effect, every conductor that carries alternating current also has eddy currents induced in it by its own changing magnetic field—just as a changing magnetic field induces eddy currents in the core material of a transformer or choke. These eddy currents constitute an additional loss which causes the effective ac or r-f resistance of a conductor (or coil) to be increased.

At very high frequencies, because of the skin effect and induced eddy currents in a conductor, a hollow, thin-walled tube is really a very much better conductor than a solid conductor of the same diameter. Some short-wave (high-frequency) coils are actually made with copper tubing, instead of with solid wire.

Since the resistance of a coil at ac frequencies depends on the frequency, in all ac problems this resistance must be considered at the operating frequency of the coil.

DIELECTRIC LOSSES Another source of loss in coils at high frequencies is that which occurs in the forms on which the coils are wound and in the insulating covering of the wire. Since both of these are dielectric materials, any loss of this sort is called a dielectric loss. Even though a coil form may be made of the highest quality dielectric material, such as molded mica, steatite, resinite, or polystyrene, there is no perfect insulator, and dielectric losses cannot be entirely avoided at very high frequencies.

Dielectric loss, like skin effect and the eddy current loss in a conductor, also tends to increase the effective resistance of a coil at high frequencies.

DISTRIBUTED CAPACITANCE Since every turn of a coil is separated from its adjacent turns by an insulating material—which may be air in certain cases—and a difference of potential always exists between turns when the coil is conducting current, there is always a capacitive effect between the turns. This is shown in Fig. 3.7. This

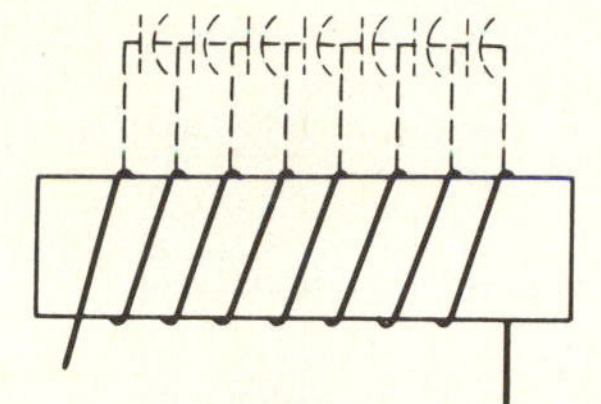

Fig. 3.7 Capacitive effects between the turns of a coil (distributed capacitance).

capacitive effect is known as the distributed capacitance of the coil, because it is distributed throughout the entire length of the coil.

Distributed capacitance causes a decrease in the effective inductance of a coil and an increase in its effective resistance. Hence, it is another undesirable effect, and measures are usually taken to decrease it as much as possible.

3.9 METHODS USED TO DECREASE SKIN EFFECT, EDDY CURRENT LOSS, DIELECTRIC LOSS, AND DISTRIBUTED CAPACITANCE IN COILS

LITZENDRAHT, OR LITZ, WIRE One of the earliest methods for reducing skin effect in conductors and coils at high frequencies was the use of litzendraht, commonly known as litz wire. Litz is a braided conductor made up of a large number of very thin strands of enamel insulated wire. The individual strands are all connected in parallel at the start and finish of a conductor or coil, but because of the enamel insulation, there is no connection between the individual strands of the conductor at any other points along the conductor. The braided conductor is also twisted at regular intervals so that a particular insulated strand will at some points be at the exact center of the conductor, and at the surface of the conductor at other points.

One of the reasons why skin effect is reduced with litz wire is that skin effect is not as pronounced in thin wires as it is in thick ones. Another reason is the twisting of the conductor, which tends to keep the inductance of a litz conductor at its center the same as it is at its surface.

A very large number of individual insulated strands in a litz conductor are most advantageous at frequencies below 600 kHz; but for frequencies above this, more than 25 strands are unfavorable. According to some authorities, there is a certain critical frequency above which a solid conductor of the same diameter again has less high-frequency resistance than litz wire.

FLAT COPPER STRIP A flat copper strip conductor, instead of a round one, is also sometimes used to reduce skin-effect losses. Although skin effect still exists in a flat strip conductor, it is much less than in a round conductor of the same cross-sectional area.

Inductors wound with flat copper strip are used in many high-power radio transmitters; the strip is wound edgewise, and the turns are widely spaced so that the distributed capacitance between turns is not excessive.

TUBULAR CONDUCTORS A thin-walled hollow tube actually offers less resistance to high-frequency currents than a solid conductor of the same diameter. Tubular conductors may be used most advantageously in coils designed for high-frequency operation. An added advantage is that coils wound with tubular conductors generally require no supporting forms, because the tubing is rigid enough to make such coils self-supporting.

SPECIAL WINDING METHODS USED TO REDUCE DISTRIBUTED CAPACITANCE Figure 3.8 shows a number of different ways of winding coils so as to reduce the distributed capacitance. In any ordinary multilayer coil, such as that shown in Fig. 3.8*a*, since the first and last turns are adjacent, and the second and next to the last turn, etc., are all adjacent, the distributed capacitance tends to be high. If the same number of turns is wound in a narrow slot, as in Fig. 3.8*b*, the first and last turns will not be adjacent, and the coil will have a lower distributed capacitance. Other methods of winding coils to reduce capacitance are the bank windings shown in Fig. 3.8*c* and *d*, and the universal or honeycomb winding of Fig. 3.9.

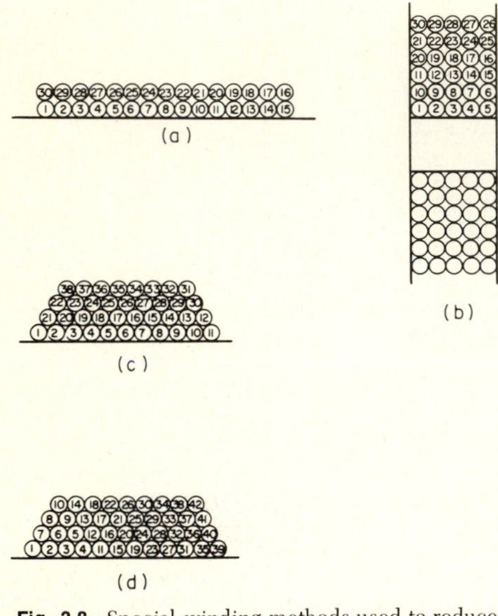

Fig. 3.8 Special winding methods used to reduce distributed capacitance: (*a*) Ordinary multilayer coil has high distributed capacitance. (*b*) Distributed capacitance may be reduced by winding the coil in a narrow slot. (*c*) and (*d*) Bank windings used to reduce distributed capacitance. Note difference between the windings of (*c*) and (*d*).

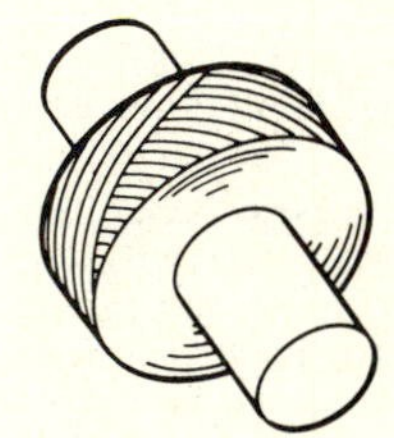

Fig. 3.9 The "universal" or "honeycomb" winding, also known as a pie winding.

THE USE OF POWDERED IRON AND FERRITE CORES Since all the undesirable effects, skin effect, eddy currents, dielectric losses, and distributed capacitance in a coil are proportional to the amount, or length, of wire in a coil, all these unwanted effects may be greatly reduced by the use of powdered iron and ferrite cores. The permeability of powdered iron and ferrite-core materials is very much greater than the permeability of air. Thus, coils that are wound over such cores have a much higher inductance than air-core coils of the same diameter and the same number of turns; accordingly, much less wire may be used in a powdered iron or ferrite-core coil than in an air-core coil which has the same inductance, and all the aforementioned undesirable effects and losses in a coil may be reduced to a minimum.

At frequencies above those where core losses with powdered iron outweigh its advantages, ferrite cores may be most advantageously substituted for the powdered iron ones.

3.10 THE Q OR FIGURE OF MERIT OF A COIL

If a curve is plotted showing how the current through a series-tuned coil varies as the coil is tuned through resonance while the impressed voltage is kept constant, it will take a form similar to the curves of Fig. 3.10. If the ratio of coil reactance to effective coil resistance is high, the curve will also be high and steep, like curve *a*. However, if the coil has more losses, such as core, copper, dielectric, skin effect, its effective re-

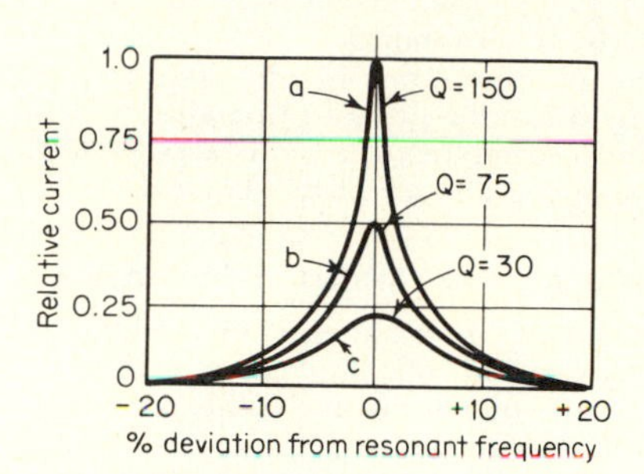

Fig. 3.10 Selectivity curves of tuned circuits which show how selectivity depends on coil Q.

sistance will be higher, and the ratio of coil reactance will be lower. The current at resonance will also be lower because of the higher effective resistance, and the curve will then take a form similar to curve *b*, which is less steep than curve *a*. If a coil has still more losses, and consequently an even lower reactance to resistance ratio, the resonance curve will take form *c*.

Since increased losses in a coil will always yield a lower reactance to resistance ratio, and the efficiency of a coil—just as the efficiency of any device—decreases as losses increase, a coil with a high reactance to resistance ratio will always be more efficient than one with a low reactance to resistance ratio. Hence, the reactance to resistance ratio of a coil may be used as a figure of the merit, or Q of a coil.

In the form of an equation;

$$Q = \frac{X}{R} \tag{3.16}$$

where X = reactance of the coil or capacitor at resonance
R = effective resistance of the coil

Figure 3.10 shows how the selectivity of a tuned circuit, or the sharpness of resonance, depends on Q.

example 3.7 In a series-tuned circuit the current at resonance with an impressed voltage of one volt is 50 mA. Circuit capacitance is 160 pF, and inductance is 250 μH. What is the Q of this circuit?

solution Effective resistance,

$$R = \frac{E}{I} = \frac{1}{0.05} = 20\ \Omega$$

Resonant frequency,

$$f_r = \frac{1}{2\pi\sqrt{LC}}$$

$$= \frac{1}{2\pi\sqrt{(250 \times 10^{-6} \times 160 \times 10^{-12})}}$$

$$= \frac{10^8}{2\pi\sqrt{16 \times 25}} = \frac{10^8}{2\pi(20)} = \frac{10^7}{2\pi(2)}$$

At resonance,

$$X_L = 2\pi f_r L$$
$$= 2\pi \left(\frac{10^7}{4\pi}\right)(250 \times 10^{-6}) = 1250\ \Omega$$
$$Q = \frac{X}{R} = \frac{1250}{20} = 62.5$$

The letter Q is sometimes the Quality factor, which implies that the quality of a tuned circuit is dependent only upon this value. However, there are many applications where a high-Q coil or circuit is very undesirable. For example, if a circuit is to pass a band of frequencies, then the coil with curve c of Fig. 3.10 would be better than the coil with curve a. Even though the current is smaller at resonance, the falloff in response at the skirts of curve c is not as rapid.

In tuned circuits where a lower Q (that is, a broader frequency response) is desired, the coil of the tuned circuit may be placed in parallel with a resistor. This resistor, which is sometimes called a *swamping resistor*, is used to deliberately introduce resistive losses into the tuned circuit and lower its Q.

3.11 TRUE INDUCTANCE AND APPARENT INDUCTANCE

When inductance measurements are made on a coil, and the test frequency is at or near the self-resonant frequency of the coil, the results may be misleading by very wide margins. The self-resonant frequency of a coil is the frequency at which the inductance of the coil and its distributed capacitance are in resonance. Since coil impedance rapidly increases as the self-resonant frequency of a coil is approached, and since impedance also increases as inductance increases, the result, when measurements are made at or near the self-resonant frequency of a coil will provide an inaccurate inductance reading.

Any measured inductance at a frequency at or near the self-resonant frequency of a coil is called the *apparent inductance*, to distinguish it from *true inductance*, which should always be measured at a frequency far below the self-resonant frequency of a coil.

A formula for determining true inductance from an apparent measured inductance is the following:

$$L_{\text{appar}} = \frac{L_{\text{true}}}{1 - (f_{\text{test}}/f_o)^2} \tag{3.17}$$

where L_{appar} = apparent inductance of the coil, μH
L_{true} = true inductance of the coil, μH
f_{test} = frequency at which the apparent inductance measurement is made, MHz
f_o = self-resonant frequency of the coil, MHz

Equation (3.17) shows that at test frequencies above the self-resonant frequency of a coil the computed (not measured) apparent inductance would be negative. Since there is no such thing as a negative inductance, a negative result merely shows that the reactance of the coil has become negative; in other words, above its self-resonant frequency, the coil acts as a capacitor, and the current leads the impressed voltage.

example 3.8 The self-resonant frequency of a 100-μH choke is 5 MHz. What would be the apparent inductance of this choke if tested on an inductance bridge at a test frequency of 4.5 MHz?

solution Using (3.17),

$$L_{\text{appar}} = \frac{100}{1 - (4.5/5)^2}$$
$$= \frac{100}{1 - 0.81} = \frac{100}{0.19} = 526\ \mu\text{H}$$

3.12 DETERMINING THE SELF-RESONANT FREQUENCY OF AN INDUCTOR

Figure 3.11 shows the test setup for measuring the self-resonant frequency of an inductor. A resistor is placed in series with the coil and its distributed capacitance C_d.

The combination is placed across a variable-frequency signal generator. At resonance, the parallel-tuned circuit will have maximum impedance. Thus, the current through R will be minimum, and the voltage across R will also be minimum. This is indicated by a dip in the VTVM reading. The response curve in Fig. 3.11 shows the voltage across R as the signal generator is tuned through resonance.

For low-Q coils the exact resonance point may be hard to determine, because the impedance curve of a low-Q coil is rather broad and not very steep.

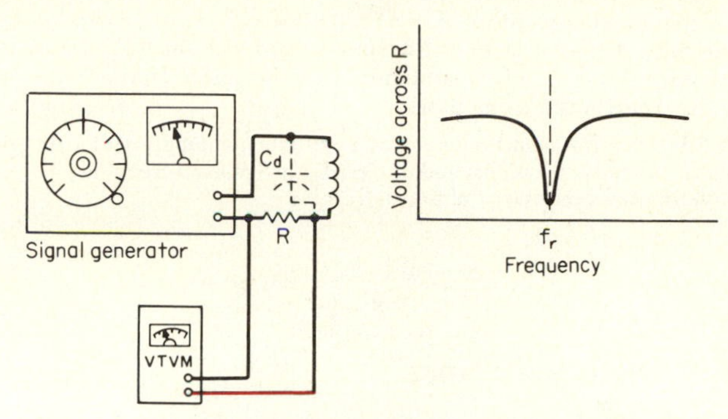

Fig. 3.11 The self-resonant frequency of a coil. Any coil, in conjunction with its distributed capacitance, forms a parallel-resonant circuit.

It is also necessary to watch for resonance points at frequencies that are harmonics of the fundamental. All these will cause some dip in the voltmeter reading across the resistor, but the dip caused by the fundamental frequency will be very much greater than any dip resulting from a harmonic of the fundamental. Harmonics are always either an even or an odd number times the frequency of the fundamental.

If the coil being measured has a core that changes permeability considerably with changes in applied voltage, the signal generator output voltage should be adjusted for a voltage across the coil that will be the same as the voltage across the coil under actual operating conditions.

3.13 DETERMINING THE DISTRIBUTED CAPACITANCE OF A COIL BY MEASUREMENT

For this measurement a grid-dip oscillator and two precise, calibrated capacitors are required—or a standard variable calibrated capacitor may be used. The capacitors should be high-quality low-loss units such as silvered mica or polystyrene types. For coils in the standard broadcast range, capacitors having values of 100 and 150 pF will be suitable.

One of the capacitors, C_1, is connected across the coil, which is loosely coupled to the pickup loop of the grid-dip oscillator, as shown in Fig. 3.12. The oscillator frequency is adjusted for resonance, which will be indicated by the grid-dip oscillator meter, and the oscillator frequency is recorded as f_1. The second capacitor, C_2, is then substituted

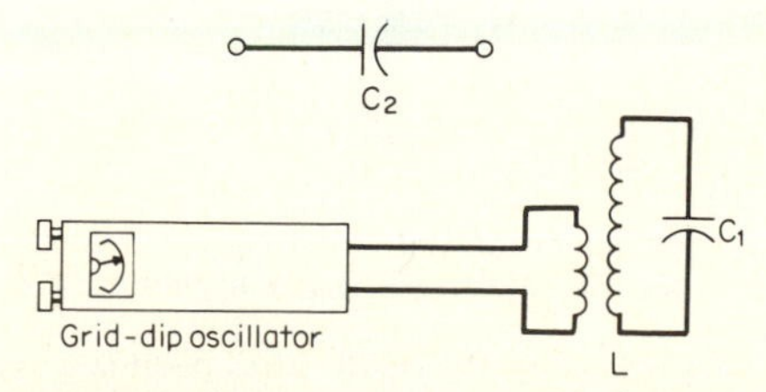

Fig. 3.12 Equipment and circuit used to determine the distributed capacitance of a coil.

for C_1, and the coil is again resonated at the new frequency, f_2. The distributed capacitance of the coil is then computed from the equation

$$C_d = \frac{(f_1/f_2)^2 C_1 - C_2}{1 - (f_1/f_2)^2} \tag{3.18}$$

where C_d = distributed capacitance of coil, pF
f_1, f_2 = two resonant frequencies, kHz
C_1, C_2 = two capacitor values, pF

If, instead of the fixed capacitors, a calibrated variable capacitor is used, it is only necessary to resonate the coil at two different settings of the variable capacitor; substituting the observed values of capacitance and oscillator frequencies in Eq. (3.18) will then give the distributed capacitance.

example 3.9 The resonant frequency of a coil when shunted with a 100-pF capacitor is 1200 kHz, but if the capacitor is changed to 155 pF, the resonant frequency becomes 996 kHz. What is the distributed capacitance of the coil?

solution Using Eq. (3.18),

$$C_d = \frac{(1200/996)^2\ 100 - 155}{1 - (1200/996)^2} = 21.79 \text{ pF}$$

3.14 MEASURING COIL INDUCTANCE

Figure 3.13 shows how a grid-dip meter can be utilized for measuring inductance. A high-quality (high-Q) capacitor, silvered mica or polystyrene, is connected across

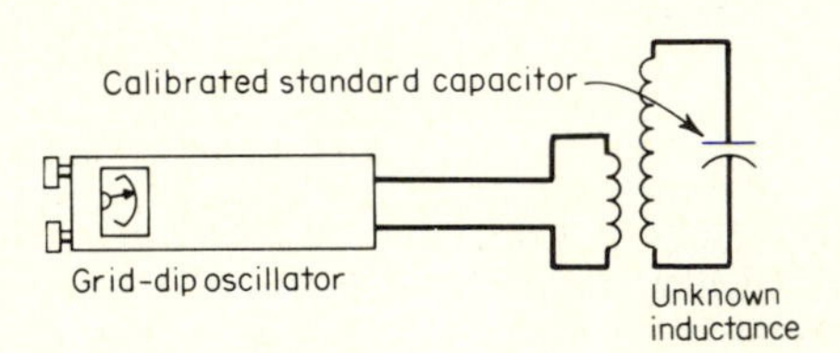

Fig. 3.13 Method used to measure coil inductance.

the coil whose inductance is to be measured. For coils that are to be used at standard broadcast frequencies—500 to 1600 kHz—any capacitance value from about 100 to 200 pF will be suitable. For coils to be used at higher frequencies capacitance values should be reduced in proportion, and should be increased for coils to be used at frequencies below 500 kHz. For approximate inductance measurements, capacitors having a tolerance rating of ±5 percent will do, but for more accurate closer tolerance, calibrated capacitors should be used.

The coil is loosely coupled to the pickup loop of the grid-dip oscillator, as shown in Fig. 3.13. Only the minimum amount of coupling that will give an indication on the grid-dip meter should be used. The oscillator frequency is adjusted for resonance which will be indicated on the grid-dip meter, and the coil inductance is computed from

$$L = \frac{25\ 330}{Cf^2} \tag{3.19}$$

where L = inductance of coil, μH
C = tuning capacitance, pF
f = frequency, MHz

Since the distributed capacitance of the coil is ignored in the above method, the inductance, as given by Eq. (3.19), will be slightly higher than the true inductance of the coil.

For more precise results, the capacitance of the capacitor used to resonate the coil for the inductance test should also be more precisely known; a calibrated capacitor should be used, and the distributed capacitance of the coil should be determined.

The distributed capacitance of the coil is added to the capacitance of the calibrated capacitor, and their sum is then substituted for C in Eq. (3.19).

MEASURING HIGH INDUCTANCE R-F CHOKES For high-inductance r-f chokes out of the frequency range of grid-dip oscillators, the circuit of Fig. 3.14 may be employed. For some tests it may be necessary to use an audio frequency instead of an r-f

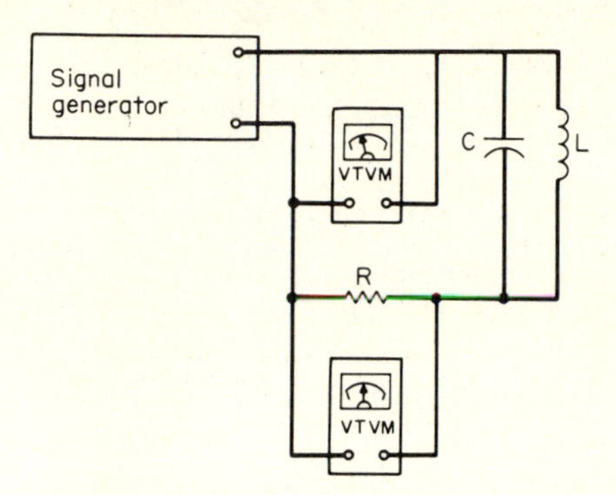

Fig. 3.14 Method used to measure the inductance of chokes having high inductive values—beyond the measuring range of grid-dip oscillators.

generator. Shunting capacitor values should have reactance values in proportion to the inductive reactance values.

Resonance will be indicated by a very pronounced dip in the voltage across the series resistor R. The inductance may then be computed by using Eq. (3.19) and solving for L.

example 3.10 When a capacitor that has a capacitance of 198 pF is shunted across a coil, the resonant frequency of the combination is 800 kHz. The distributed capacitance of the coil is 14 pF. Compute:

(*a*) The approximate inductance of the coil by neglecting the distributed capacitance of the coil

(*b*) The inductance more precisely by taking the distributed capacitance into consideration

solution (*a*) Using Eq. (3.19), and without taking distributed capacitance into consideration:

$$L = \frac{25\ 330}{198 \times 0.8^2} = 199.89\ \mu\text{H}$$

(*b*) Taking the distributed capacitance into consideration:

$$L = \frac{25\ 330}{(198 + 14) \times 0.8^2} = 186.68\ \mu\text{H}$$

3.15 DETERMINING THE Q OF A COIL

A circuit for determining coil Q is shown in Fig. 3.15. The capacitance of capacitor C need not be precisely known, but it should have a value sufficient to resonate the coil, and it should be of the best quality low-loss type—silvered mica or polystyrene. The voltmeters should have high input impedance. If low-impedance meters are employed, they should be used with high impedance probes. The input capacitance of V_2 should be very much lower than the capacitance C.

When resonance occurs, the voltage indicated by V_1 will dip to a minimum, and V_2 will peak. The Q of the coil at resonance is given by

At resonance

$$Q = \frac{V_2}{V_1} \tag{3.20}$$

where V_1 = voltage across capacitor and coil in series
V_2 = voltage across capacitor

example 3.11 If the voltmeter across the capacitor of Fig. 3.15 indicates 144 V while V_1 indicates 0.6 V at the resonant frequency of the coil and capacitor, what is the Q of the coil?

solution Using Eq. (3.20), the Q at resonance is equal to

$$\frac{144}{0.6} = 240$$

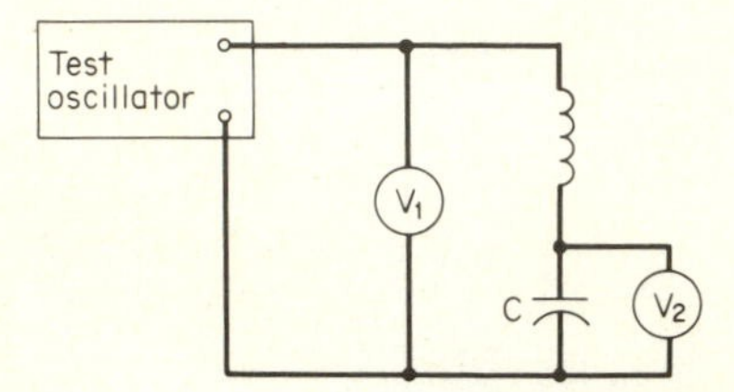

Fig. 3.15 A very convenient method used to measure the Q of a coil. At the resonant frequency, Q equals V_2/V_1.

example 3.12 If the inductance of the coil of Example 3.11 is 300 μH, what is the effective resistance of the coil if the resonant frequency is 750 kHz?

solution The inductive reactance of the coil is obtained by the equation $X_L = 2\pi fL$. At 750 kHz,

$$2\pi fL = 6.2832 \times 750 \times 10^3 \times 300 \times 10^{-6}$$
$$= 1413.72\ \Omega$$

Then

$$Q = 240 = \frac{X}{R} = \frac{1413.72}{R}$$

Solving for R, the result is

$$R = \frac{1413.72}{240} = 5.89\ \Omega \qquad \textit{Answer}$$

3.16 COILS WITH POWDERED IRON OR FERRITE CORES

Unlike air, which always has a permeability of 1, the permeability of iron-powder and ferrite-core materials may have values ranging from just a few times to many thousands of times the permeability of air. Even though manufacturers do produce iron-powder and ferrite cores in certain standard permeabilities and sizes, there are so many of these that it would be extremely difficult to develop formulas for computing coil characteristics that would be applicable to all the different types of iron-powder and ferrite cores. Even if such were available, they would be extremely voluminous—and difficult to apply, except to single-layer coils, such as those used in the tuning circuits of radio receivers.

In any case, when computations involving iron-powder or ferrite-core coils are to be made, it is best to obtain the required computational data directly from the manufacturer of the particular core material that is to be used.

Iron-powder and ferrite-core materials are constantly being improved, and some of the core material produced only a few years ago is already obsolete. Newest types of core materials are more stable against environmental changes—changes of coil inductance with changes in temperature, induction, direct current and frequency.

3.17 COMPUTING IRON-POWDER AND FERRITE-CORE COILS

Typical of the technical data available from manufacturers are those given on pages 3-19 to 3-22 taken from a manufacturer's* technical bulletin which contains more than

* The Arnold Engineering Company.

PHYSICAL SPECIFICATIONS

CORE DIMENSIONS	INCHES			MILLIMETERS		
	O.D.	I.D.	HT.	O.D.	I.D.	HT.
NOMINAL	2.000	1.250	0.530	50.8	32.0	13.46
AFTER FINISH	2.035 MAX.	1.218 MIN.	0.565 MAX.	51.7 MAX.	30.9 MIN.	14.35 MAX.

MEAN LENGTH OF MAGNETIC PATH		CROSS SECTIONAL AREA OF MAGNETIC PATH A		APPROXIMATE WEIGHT OF FINISHED CORE		MINIMUM WINDOW AREA		
in.	cm.	in.^{2}	cm.^{2}	LBS.	GRAMS	CIRCULAR MILS	in.^{2}	cm.^{2}
5.13	13.03	0.1922	1.24	0.298	135	1,480,000	1.165	7.52

	CORE HEIGHT	in.	cm.
MEAN LENGTH OF TURN FOR FULL WINDING	0.550	2.01	5.11

ELECTRICAL SPECIFICATIONS

CORE SPECIFICATIONS				WINDING DATA FOR FULL-WOUND CORE		
PART NUMBER	NOM. PERM. μ	INDUCTANCE mh FOR 1000 TURNS ±8%	NOMINAL DC RESISTANCE OHMS PER mh	AWG WIRE SIZE	TURNS	DC RESISTANCE OHMS
A-424426-2	350	426	0.0060	12	113	0.0324
A-404365-2	300	365	0.0070	13	141	0.0505
A-382304-2	250	304	0.0084	14	177	0.0788
A-217249-2	205	249	0.0103	15	221	0.1227
D-217249-4	205	249	0.0103	16	276	0.1922
W-217249-4	205	249	0.0103	17	344	0.298
A-181210-2	173	210	0.0122	18	430	0.467
D-181210-4	173	210	0.0122	19	537	0.728
W-181210-4	173	210	0.0122	20	668	1.133
A-327195-2	160	195	0.0131	21	832	1.768
D-327195-4	160	195	0.0131	22	1042	2.79
W-327195-4	160	195	0.0131	23	1289	4.31
A-154179-2	147	179	0.0142	24	1608	6.76
D-154179-4	147	179	0.0142	25	1999	10.55
W-154179-4	147	179	0.0142	26	2495	16.62
A-715152-2	125	152	0.0169	27	3084	25.7
D-163152-4	125	152	0.0169	28	3855	40.6
W-715152-4	125	152	0.0169	29	4730	61.8
A-106073-2	60	73	0.035	30	5941	98.8
D-164073-4	60	73	0.035	31	7391	154.8
W-106073-4	60	73	0.035	32	9048	234.
A-348032-2	26	32	0.080	33	11330	371.
A-349017-2	14	17	0.151			

(Reproduced by permission from the catalogue of The Arnold Engineering Company)

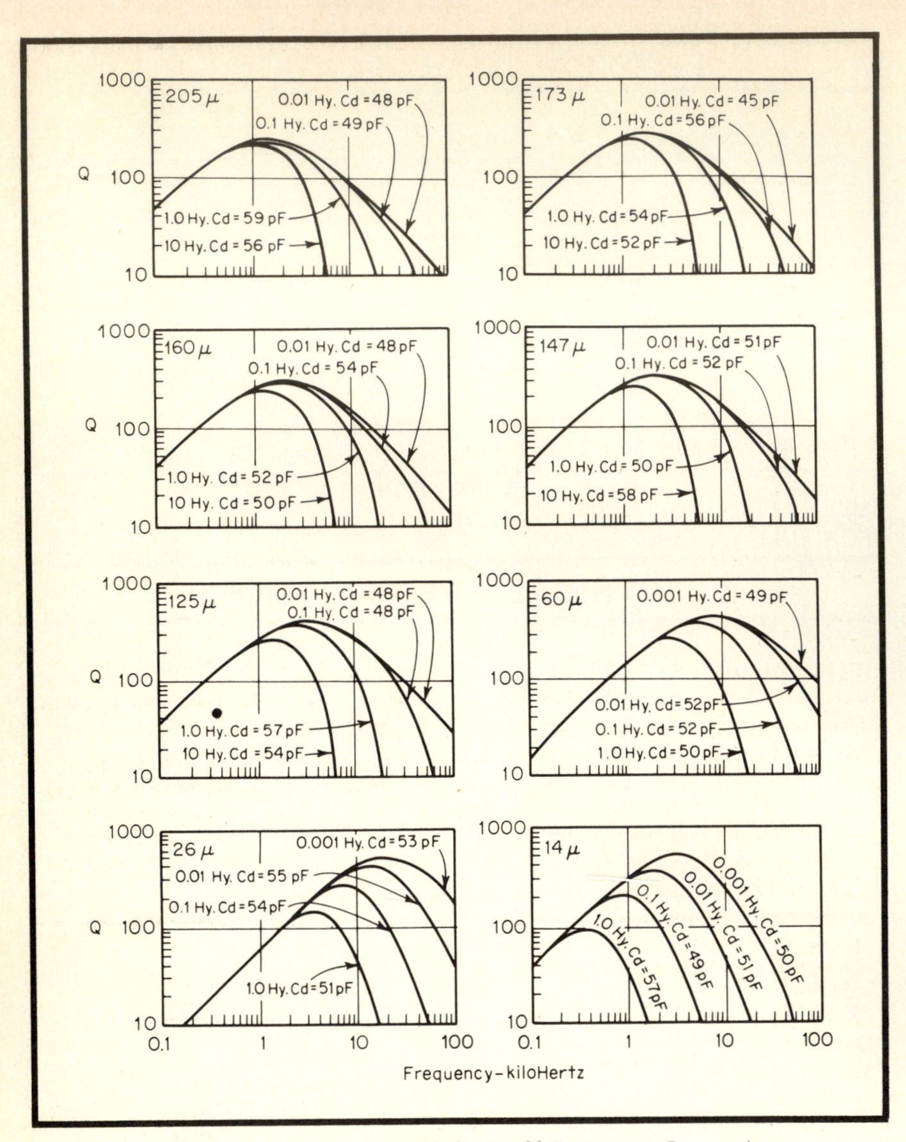

(Reproduced by permission of The Arnold Engineering Company)

Formulae and Mathematics of MAGNETICS

PERMEABILITY

Permeability of a test core can be calculated by either formula below (inches or millimeters). Inductance is measured on a sensitive bridge with a uniformly distributed winding on the core.

H = core height (in. or mm.)
r = corner radius (in. or mm.)
H_e = effective height before finish, corrected for corner radii (in. or mm.)
L = measured inductance (henries)
OD = mean outside diameter before finish (in. or mm.)
ID = mean inside diameter before finish (in. or mm.)
N = number of turns in test winding.

INCHES

$$\mu = \frac{L \times 10^9}{11.7 \left[\log_{10} \frac{OD}{ID}\right] H_e N^2}$$

$$H_e = H - \frac{1.717\ r^2}{OD - ID}$$

MILLIMETERS

$$\mu = \frac{L \times 10^9}{.4606 \left[\log_{10} \frac{OD}{ID}\right] H_e N^2}$$

$$H_e = H - \frac{.0675\ r^2}{OD - ID}$$

TURNS VS INDUCTANCE

The number of turns (N) necessary to obtain a specific inductance (L) can be calculated by:

$$N = 1000\sqrt{\frac{L}{L_{1000}}}$$

L = desired inductance (millihenries)
L_{1000} = core inductance (millihenries for 1,000 turns)

"Q" FORMULA

The Q formula calculates the ratio of reactance to effective resistance for an inductor and thus indicates its quality. For electrical wave filters, an increase in Q provides sharper cut-off, higher attenuation ratios, and better defined resonance. Q is affected by the distributed capacitance of an inductor's winding.

Neglecting the effects of self-resonance caused by the distributed capacitance, Q can be calculated, when designing inductors, by this formula:*

$$Q = \frac{\omega L}{R_{dc} + R_{ac} + R_{cd}}$$

Q = quality factor
L = inductance (henries)
ω = $2\pi \times$ frequency (frequency in hertz)
R_{dc} = dc winding resistance (ohms)
R_{ac} = resistance due to core losses (ohms)
R_{cd} = resistance due to dielectric losses in winding (ohms)

DC WINDING RESISTANCE

DC winding resistance for an average winding can be calculated by:

$$R_{dc} = \frac{\ell_w N r}{12000}$$

ℓ_w = mean length of turn (in.)
N = number of turns
r = resistance of wire in ohms per 1000 feet

(Reproduced by permission from the catalogue of The Arnold Engineering Company)

AC CORE LOSS RESISTANCE

AC core loss resistance is calculated from:

$$R_{ac} = \mu Lf\,(a\,B_{max} + c + ef),$$

which is Legg's equation.

DIELECTRIC LOSS RESISTANCE

Dielectric loss resistance is significant at higher frequencies and can be calculated from the equation found in Terman's Handbook:**

$$R_{cd} = d\omega^3\,L^2C_d$$

d = power factor of distributed capacitance

$\omega = 2\pi \times$ frequency (frequency in hertz)

L = inductance (henries)

*C_d = distributed capacitance.

NOTES ON "Q" CURVES

The Q curves published in this manual have been developed through the use of digital computer techniques. They have been checked with various core sizes and inductances to assure reasonable correspondence to the real world of wire, insulation, and winding. The user's ability to get equivalent results depends in part upon his ability to duplicate the assumed conditions. These are:

1. A "full-wound core" is defined to be one in which the minimum winding ID or residual hole left after winding is one-half of the inside diameter of the core.

2. This leaves a useful winding area which is three-quarters of the available window area. It was assumed that 70% of this space would be filled with copper wire including heavy synthetic film insulation.

3. The dc resistance of a full wound core varies as the square of the number of turns in the same manner that the resultant inductance varies as the square of the turns. Therefore, each core size has a table of calculated ohms per millihenry based on the "full-wound core" definition above. This resistance determines the positive slope of the low frequency portion of the Q curve and is assumed to be independent of inductance.

4. Three factors affect the high frequency performance of an inductor.

a. The most fundamental is the eddy current loss of the core material which is mostly responsible for the negative slope of the low inductance curves at frequencies above the frequency of maximum Q. This is calculated from Legg's equation.

b. The second factor is caused by dielectric loss and is calculated as R_{cd} in the formula above.

c. The most dramatic factor is the effect of self resonance of the distributed capacitance and the inductance. For small inductances, such as the 0.001 henry or the 0.01 henry curve for each core, the self resonant frequency f_o is well above the normal useful frequency range of the component. Therefore, these curves tend to indicate the component performance with a negligible effect of self resonance. The distributed capacitance and the self inductance determine a self resonant frequency according to:

$$f_o = \frac{1}{2\pi\sqrt{LC_d}}\ \text{hertz.}$$

At some lower frequency, f, the value Q_f can be calculated from:

$$Q_f = Q\left[1 - \left(\frac{f}{f_o}\right)^2\right]$$

where Q is calculated from determined values of loss resistances as indicated above, and Q_f is the apparent Q taking into account the effect of the distributed capacitance. It should be noted that when f is 20% of f_o, Q_f is 96% of its original value. However, when f is 70% of f_o, Q_f drops to 51% of its original value. The apparent value of the inductance, L_a, is also affected as follows:

$$L_a = \frac{L}{1 - \left(\frac{f}{f_o}\right)^2}$$

5. Because the distributed capacitance is determined by the winding method, the user can obtain different results from those plotted, depending on this value of the capacitance.

* This analysis follows Herman Blinchikoff, "Toroidal Inductor Design," *Electro-Technology*, November, 1964

** *Radio Engineer's Handbook*, F. E. Terman, McGraw-Hill, Inc. New York (1943), p. 84

70 pages of data pertaining to the toroidal, MPP iron-power cores manufactured by this company.

MPP is a special magnetic alloy composed of approximately 2 percent molybdenum, 81 percent nickel, and 17 percent iron. After being pulverized, the resulting powder is pressed into cores of various sizes. The cores are available in standard permeabilities ranging from 14 to 350, and in toroidal core sizes having an outer diameter of less than 0.2 to greater than 5 in. The data on page 3–19 are for a core having an outer diameter of 2 in. These data show at once how difficult it might be to compute the characteristics of iron-powder and ferrite-core coils without such data, and how easy the computations are when such data are available.

Similar data are available for other types of iron-powder and ferrite cores.

The Q curves show that beyond a certain frequency, which is different for cores having different permeabilities, the Q rapidly decreases, owing to the increased core and other losses in a coil. Below the frequency at which Q is a maximum, although core losses are not excessive, the ratio of coil reactance to resistance is lower, and the result is a lower Q.

example 3.13 Use the manufacturer's data on page 3–19 and determine the number of turns required to produce an inductance of 100 mH in an MPP OD 2,000 toroidal core that has a nominal permeability of 147.

solution From the turns versus inductance formula on page 3–21 and the data for a 147-μ core on page 3–19,

$$N = 1000\sqrt{\frac{100}{179}} = 748 \text{ turns}$$

example 3.14 What size wire should be used to wind the coil of Example 3.13 for highest Q?

solution The dc resistance of the winding should be kept as low as possible. From the winding data for a full-wound core, we see that if we used number 20 wire, a full-wound core would contain only 668 turns. Therefore, we must use the next smaller size—higher gage number—wire, number 21, to get the 748 turns into the available winding space.

3.18 CHANGE OF COIL INDUCTANCE WITH DIRECT CURRENT

Anyone familiar with TV receivers and circuits may have noted that most of the coils used in TV receivers have adjustable tuning slugs. One reason for this is that if a coil's inductance is above or below its rated or computed value, a slight adjustment of the tuning slug quickly brings the inductance to the required value; but another reason is that the permeability, and hence the inductance, of an iron-powder or ferrite-core coil may change appreciably with changes of direct current in the coil. Since many of these coils are used in the plate circuits of vacuum tubes, and plate currents for the same type of tubes may differ widely from tube to tube, a slug-tuned coil allows immediate readjustment of a coil's inductance to correspond to the dc plate current in the coil winding.

This same sort of inductance change with current is also one of the reasons why a coil in a TV circuit may have to be readjusted when a tube is changed.

3.19 TESTING AND MEASURING IRON-POWDER AND FERRITE-CORE COILS

Since the inductance of some iron-powder and ferrite-core coils is likely to change considerably with changes in coil current, either ac or dc, it is best to make inductance measurements under actual operating conditions for the coil. Special measuring equipment is now available which will measure iron-core and ferrite-core inductors under almost any combination of ac and dc voltages. In the absence of such equipment, a combination of other equipment may be set up which will do the job just as well—but not as conveniently.

When coils are merely to be tested, rather than measured, it is usually not necessary to test them under actual operating conditions. For instance, a test instrument may be designed merely to indicate whether or not the inductance of a coil falls within certain limits. In this case, it is not necessary to know the actual inductance of the coil being tested.

3.20 MEASURING METHOD FOR COILS CARRYING BOTH R-F and DC

A convenient method that may be used to measure both the Q and the inductance of a coil that carries direct current combined with a certain amount of r-f is shown in Fig. 3.16. The inductance of the choke coil L_2 should be very much greater than the inductance of L_1, which is to be measured. A convenient value for L_2 when L_1 is about 100 mH would be 10 H. This would be 100 times the inductance of L_1, but the greater

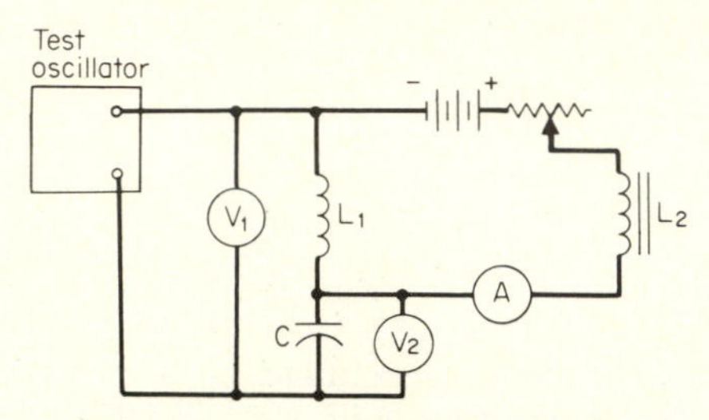

Fig. 3.16 Measuring the inductance of a coil that carries both r-f and direct currents. This method should be used if coil inductance is likely to change considerably when carrying current.

the inductance of L_2, the less will it load the ac impedance of L_1. The current through the inductors is adjusted to the level at which the inductance of L_1 is to be measured. The rheostat R, in series with the battery and coils, is used to adjust the current.

At resonance $Q = E_2/E_1$, in which E_1 is the voltage across the coil and capacitor in series (the generator output voltage), and E_2 is the voltage across the capacitor.

The inductance,

$$L = \frac{1}{(2\pi f_r)^2 C} \tag{3.21}$$

where L = coil inductance, H
f_r = frequency at resonance, Hz
C = capacitance, F

example 3.15 What will be the inductance of the coil L_1 in the test circuit of Fig. 3.16 if a 0.05-μf capacitor is needed to resonate the coil at a frequency of 2 kHz?

solution Using Eq. (3.21),

$$L = \frac{1}{(6.28 \times 2000)^2 \times 0.05 \times 10^{-6}}$$
$$= 0.126\,78 \text{ H} = 126.78 \text{ mH}$$

3.21 NOMOGRAMS

Nomograms provide a very convenient method of quickly designing a coil to meet some specified design requirement, for solving coil design problems, or for checking the results of pencil and paper computations.

The use of most types of nomograms may be mastered in just a few minutes; nevertheless, the results obtained are remarkably accurate—accurate enough for most practical purposes.

A very useful coil design nomogram is shown in Fig. 3.17. This is a dual-purpose nomogram which makes it doubly useful. To use the nomogram in coil design problems, three values must be known. A straightedge is placed across the nomogram to connect two of the known values on the scale according to the key. A light pencil line is drawn at the point where the straightedge intersects the index line. The straightedge is then run from the third unknown value across the intersecting point to the fourth scale which will give the quantity to be determined at its point of intersection with the straightedge.

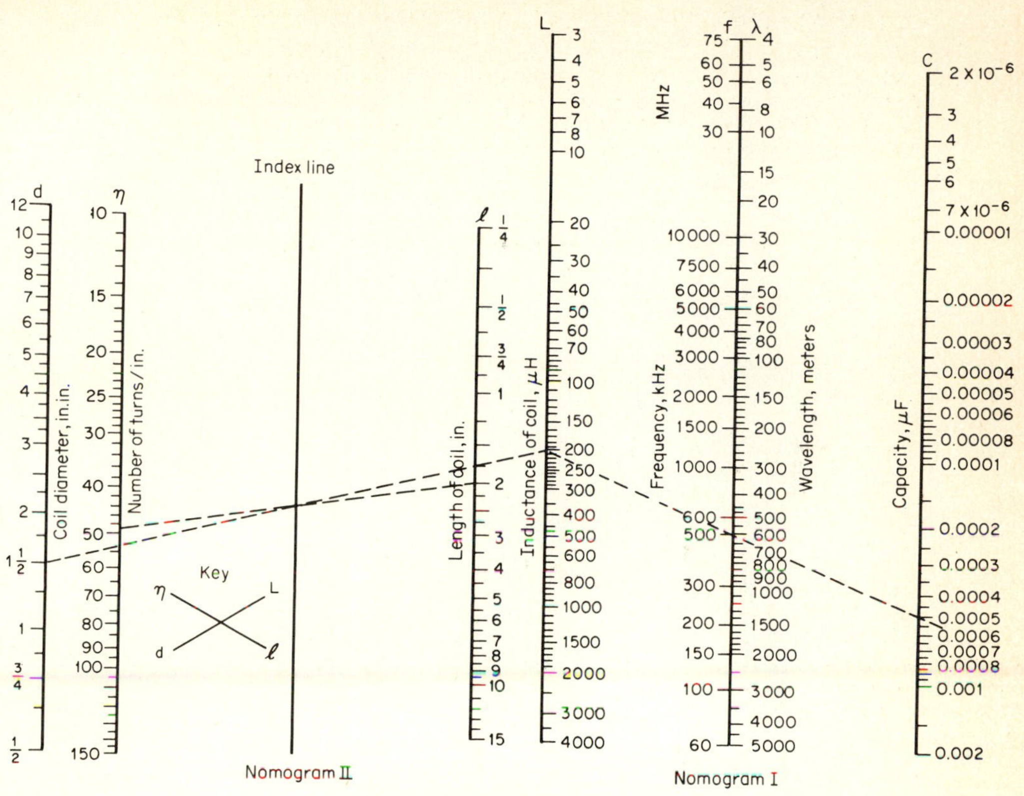

Fig. 3.17 Coil design nomogram.

Nomogram I is used to determine the inductance required to tune to a certain frequency with a known value of capacitance or vice versa or to find the frequency to which a given inductance will tune with a given capacitance. Here it is merely necessary to run a straightedge from the capacitance scale across the frequency, or wavelength, scale at the given frequency value. The required inductance value will then be found on the inductance scale at the point where the straightedge intersects it.

example 3.16 (*a*) Determine the required value of inductance to tune to a frequency of 500 kHz with a 500-pF capacitor.

(*b*) How many turns must a coil contain for the required inductance if the coil is to be $1\frac{1}{2}$ in in diameter and 2 in long?

solution (*a*) A straightedge from the given capacitor value (500 pF) on scale *C* through the frequency scale (*f*) at 500 kHz and over to the inductance scale (*L*) shows *L* to be 203 μH.

(*b*) The straightedge is placed on the nomogram to intersect the known values of coil inductance (203 μH) and coil diameter ($1\frac{1}{2}$ in) on scales *L* and *d*, and a thin penciled line is drawn at the point where the straightedge crosses the index line; the straightedge is then run from 2 in (the coil length) on scale (*l*) through the marked line of intersection on the index line and over to the scale *n*, which shows turns per inch to be 49. Total number of coil turns will therefore be

$$2 \times 49 = 98 \text{ turns}$$

Chapter 4

Magnetic Circuits

4.1 INTRODUCTION

Today it is hardly possible to be concerned with electronics without at the same time being concerned with magnetism or electromagnetism in one form or another. Consider, for instance, the numerous applications of electromagnetics in the modern television receiver with its loudspeaker, centering magnet, purity and focus magnets, deflection yoke, and various types of transformers and chokes. In other branches of electronics such as computers or navigational instruments—to mention only two—the importance of magnetism and electromagnetism is well established. The electronics industry is dependent (at least in part) on the supporting roles played by companies that specialize in the manufacture of magnetic and electromagnetic materials and components. Figure 4.1 shows a comprehensive selection of various magnetic materials including magnets, cores, bobbins, and iron-powder cores.

4.2 MAGNETS, NATURAL AND ARTIFICIAL

Natural magnets are found in the form of ore deposits known as *magnetite,* or more commonly, *lodestone.* This form of magnetic substance was known as early as 600 B.C. All other magnets are artificial. The first artificial magnets were made by contacting pieces of iron with magnetite. Even today, for some very minor applications, we may occasionally make a magnet by stroking a piece of iron or steel with any conveniently available magnet.

If a piece of soft iron is magnetized by induction from any magnetic source, the soft iron very quickly loses most of its induced magnetism after the magnetizing source is removed. However, if a piece of hardened steel or cast iron is magnetized, the induced magnetism will be retained for extremely long periods of time. Magnetized materials that very quickly lose most of their induced magnetism are called *temporary* magnets. Those that retain their magnetism are *permanent* magnets. Permanent magnets are made of wrought iron, hardened steel, or the various kinds of alloyed steels used for permanent magnets in loudspeakers and electric meters.

Figure 4.2 shows the difference between temporary and permanent magnets. Figure 4.3 illustrates a variety of permanent magnets which are used for making electric motors and other electric and electronic components.

4.3 MAGNETIC MATERIALS

When we speak of magnetic materials, we usually mean substances that are very strongly attracted by magnets or magnetic fields. Substances of this type are said to be either *ferromagnetic* or *ferrimagnetic.* Ferromagnetic substances are the various kinds

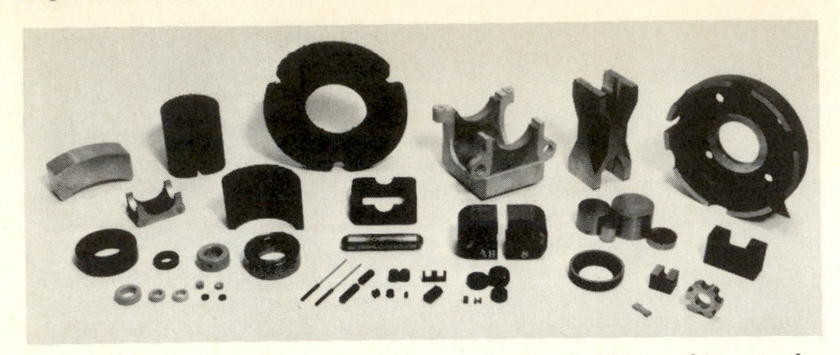

Fig. 4.1 A selection of various types of magnets, cores, bobbins, and iron-powder cores. (*Courtesy The Arnold Engineering Company*)

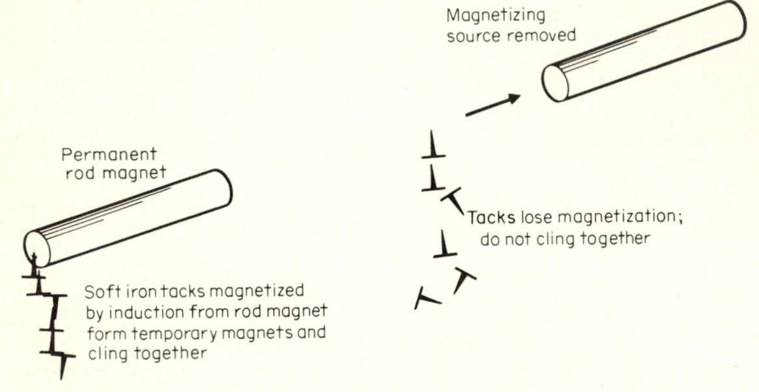

Fig. 4.2 Comparison of permanent and temporary magnetism.

Fig. 4.3 Permanent-magnet devices, used for motors and for other applications.

of iron and steel, including the powdered iron core materials that are used in some radio-frequency coils. Ferrimagnetic substances are the ferrites and other magnetic oxides that are used as core materials in coils operating at microwave frequencies, in high-frequency pulsing transformers such as the TV flyback transformer, and in the memory and switching units of high-speed computers.

Magnetic materials may also be classified as being either magnetically *soft* or magnetically *hard*. The magnetically soft materials are the kinds used in the cores of power and audio-frequency transformers and chokes. Magnetically hard materials are used for making permanent magnets.

Substances, such as aluminum, chromium, manganese, and air on which even very intense magnetic fields have only a very mild, barely detectable, attractive effect are called *paramagnetic*. Some substances, such as bismuth, antimony, copper, silver, and a few others on which very intense magnetic fields have a scarcely perceptible repellent effect, are said to be *diamagnetic*. Most substances, with the exception of those that are either ferromagnetic or ferrimagnetic, are commonly said to be *nonmagnetic*. Nonmagnetic substances do allow magnetism to pass through them, but they never become magnetized to any noticeable degree. This is illustrated in Fig. 4.4.

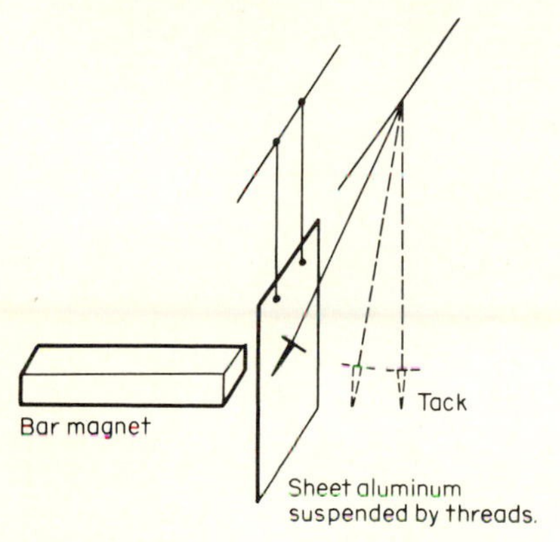

Fig. 4.4 Magnetic lines of force from the bar magnet pass through the nonmagnetic material (sheet aluminum) and attract the magnetic material (iron tack).

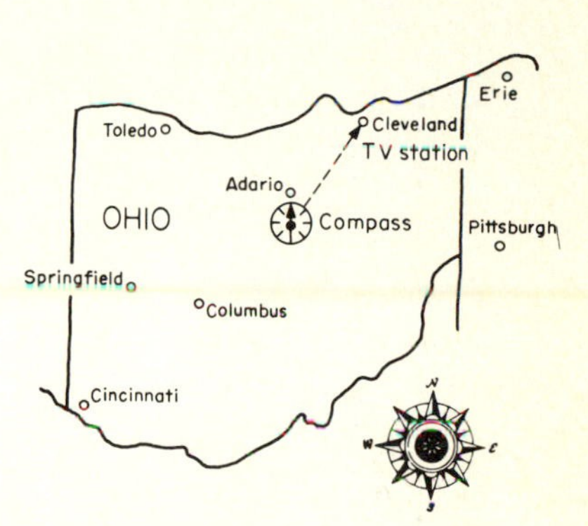

Fig. 4.5 A compass and a map can be used for orienting a TV antenna.

4.4 THE MAGNETIC COMPASS

If a piece of magnetite is suspended by a thread, it will align itself so that one part of it will always point in the general direction of the Earth's north magnetic pole. This was the very earliest form of compass, to be followed later by an artificial magnet suspended by a string. The modern magnetic compass takes the form of a magnetized steel needle balanced on a jeweled pivot and free to rotate horizontally. The end of the compass needle which points toward the north is called the *north seeking pole*. The other end, of course, points toward the south pole.

The compass, in conjunction with a map, is sometimes used to orient a TV antenna for reception from some distant station. The technique is illustrated in Fig. 4.5. The antenna being oriented is located at the point where the compass is lying on the map, and the desired station is in Cleveland. The dotted line shows the direction in which the antenna must be pointed to receive the signal directly from the station.

A compass may also be used in this manner to beam a radio signal toward some specific location.

4.5 MAGNETIC POLES

When the north pole of one magnet is placed near to the north pole of another magnet, there is a repelling force between them. Likewise, there is a repelling force between two south magnetic poles, but a north pole and a south pole are strongly attracted. This characteristic of magnetism is expressed in the law of magnetic attraction and repulsion which states:

RULE: *Magnetic poles that are alike will repel each other, while those that are unlike will attract each other.*

COULOMB'S LAW Coulomb's law, which defines the *amount* of attraction or repulsion, may be stated as follows:

RULE: *The force of attraction or repulsion between two magnetic poles is inversely proportional to the square of the distance between the poles and directly proportional to the product of the pole strengths.*

THE UNIT MAGNETIC POLE To use Coulomb's law in computations, we must have some method of precisely designating the strength of a particular magnetic pole in terms of some unit. This unit is the unit pole, derived from Coulomb's law and based on the concept that two magnetic poles of equal strength could be so selected that they would repel each other with a force of one dyne when the distance between the poles is exactly one centimeter. It is from this concept that the following definition of the unit magnetic pole is obtained:

The unit magnetic pole is of such strength that it will repel an exactly similar pole with a force of one dyne when the distance between the poles is one centimeter.

MAGNETIC POLE COMPUTATIONS From Coulomb's law we get the following equation for computing the force of attraction or repulsion between two magnetic poles:

$$F = \frac{M_1 M_2}{d^2} \tag{4.1}$$

where F = force between poles, dyn
M_1 = strength of first pole, unit poles
M_2 = strength of second pole, unit poles
d = distance between poles, cm

example 4.1 The force of repulsion between the north poles of two rod magnets is 100 dyn when the distance between them is 2 cm. What is the force of repulsion between the rod magnets if the distance between them is decreased to 1 cm?

solution In order to solve this type of problem, it is necessary to assume that the south magnetic poles of the rods are at a sufficient distance from the north poles so that their effects are negligible. In other words, the south pole of one of the rods is not noticeably attracted by the north pole of the other. Denoting the quantity to be determined by X, and using Coulomb's law, an inverse proportion can be written as

$$100:X = 1^2:2^2$$

or

$$\frac{100}{X} = \frac{1^2}{2^2}$$

Solving for X, the result is

$$\frac{(100)(2^2)}{1^2} = \frac{(100)(4)}{1} = 400 \text{ dyn}$$

example 4.2 The force of repulsion between the north poles of two bar magnets is 75 dyn when the distance between them is 10 cm. If the north pole of one of the bar magnets has a strength of 100 unit poles, what is the strength of the other bar magnet?

solution Again, it must be assumed that only the north poles of the magnets are exerting a measurable force, and the attraction or repulsion of the south poles can be ignored. Substituting into Eq. (4.1),

$$75 = \frac{100 \times M_2}{10^2}$$

and solving this for M_2, the result is

$$\frac{10^2 \times 75}{100} = \frac{100 \times 75}{100} = 75 \text{ unit poles}$$

4.6 THE EARTH'S MAGNETIC FIELD

From what is known about magnetic poles, it is evident that the only reason why a magnetic compass functions as it does is that the Earth itself is a huge magnet, with one of its poles in the north and the other just opposite, in the south. The Earth's magnetic pole that is located in the *geographical* north is actually a south *magnetic* pole. This is evident because it attracts the north pole of the magnetic compass.

Because the magnetic poles of the Earth are some distance from its geographic poles, the magnetic compass does not point precisely north from all points of the Earth's surface. In the eastern parts of the United States the compass will point west of geographic or true north; and in the western parts of the United States, it will point east of true north. The angle by which a compass points away from geographic or true north is known as *magnetic declination.* Tables and charts showing the magnetic declination in degrees at various points of the Earth's surface may be found in handbooks on surveying and navigation.

Some knowledge of the Earth's magnetic field and the compass is very often of importance to those working with or on electronic navigational equipment, radar or other locating equipment, missile tracking, and radio astronomy.

example 4.3 In making tests on a gyrocompass, it is necessary to orient it so that it points to geographic or true north. A precise magnetic compass is available. What procedure should be followed?

solution The gyrocompass should first be oriented to point to magnetic north, using the magnetic compass. The necessary correction should then be made by consulting a table of magnetic declinations.

4.7 ELECTROMAGNETISM

If a compass is brought near a conductor carrying current, as shown in Fig. 4.6, the compass needle will point in the direction shown. If the current through the conductor

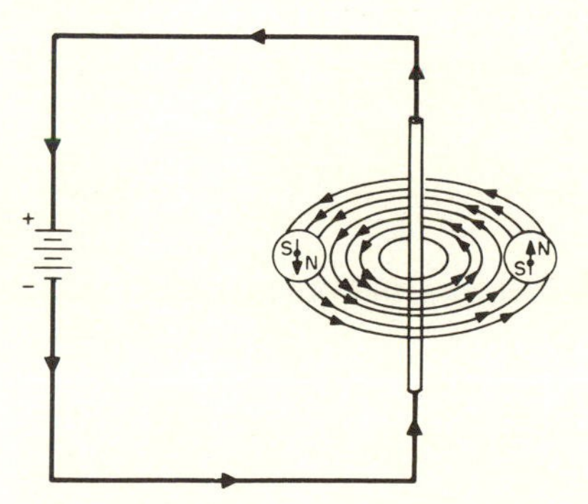

Fig. 4.6 The magnetic field encircling a current-carrying conductor. (Conventional current shown by arrows in circuit.)

is reversed, the compass needle will turn and point in the opposite direction. From this it may be seen that a conductor carrying electric current is surrounded by a magnetic field, and this field exhibits bipolar effects. In other words, the field has direction, just as the magnetic field of a permanent magnet.

The magnetic field is said to come out of the magnet at its north pole and enter at its south pole. Actually the field lines do not move. Instead, their direction is defined as the direction that a unit north pole would move if placed in the field. There is no such thing in real life as a unit north pole, but it is a useful imaginary concept for defining the direction of magnetic fields.

To summarize, the magnetic field around a current-carrying wire has a north-to-south direction just as the field around a permanent magnet has direction. This direction is defined as the direction that a unit north pole will move, and is indicated by the direction a compass points when placed in the field.

THE RIGHT-HAND RULE AND THE LEFT-HAND RULE FOR MAGNETIC FIELDS AROUND WIRES To determine the direction of the magnetic field surrounding any current-carrying conductor, use the following rule which is illustrated in Fig. 4.7.

RULE: *Grasp the wire in the right hand, and let the thumb point in the direction in which conventional current (that is, current from positive to negative) is flowing through the wire. The fingers will then point in the same direction as the direction of the magnetic field, or lines of force, that encircle the wire.*

If electron current is assumed, the left-hand rule is used.

RULE: *Grasp the wire with the left hand so that the thumb points in the direction of electron current flow and the fingers will encircle the wire in the direction of the magnetic field.*

4.8 THE RIGHT-HAND RULE AND THE LEFT-HAND RULE FOR COILS

Any coil or solenoid that carries current exhibits magnetic effects, and magnetic effects are always of a bipolar nature. Any current-carrying coil must always have both a north and a south pole. To determine the magnetic polarity of a current-carrying coil, use the right-hand rule for coils (see Fig. 4.8):

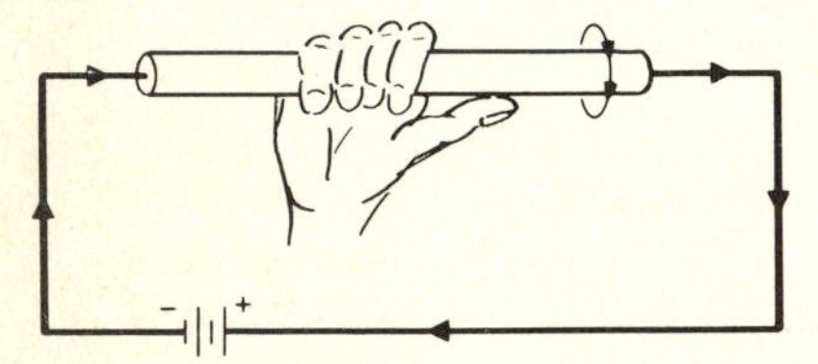

Fig. 4.7 The right-hand rule for determining the direction of the magnetic field around a current-carrying conductor. (Conventional current shown.)

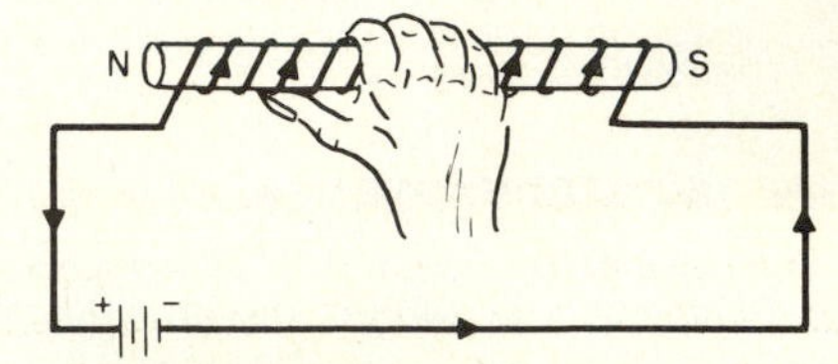

Fig. 4.8 The right-hand rule for determining the magnetic polarity of a coil.

RULE: *Grasp the coil in the right hand, and let the fingers point in the direction of the conventional current around the coil; the thumb will then point in the direction of the north pole of the coil.*

If the coil is grasped with the left hand so that the fingers encircle the coil in the same direction that *electron* current is flowing, then the thumb of the left hand will point in the direction of the north pole.

example 4.4 Two identical coils, *A* and *B*, are wound on an iron core of rectangular cross section, as shown in Fig. 4.9. The coils are wound in opposite directions, and the start of coil *A* is connected to the positive terminal of a battery. How should the other terminals of the two coils be connected together in series and to the negative terminal of the battery so that their magnetic fields will aid each other?

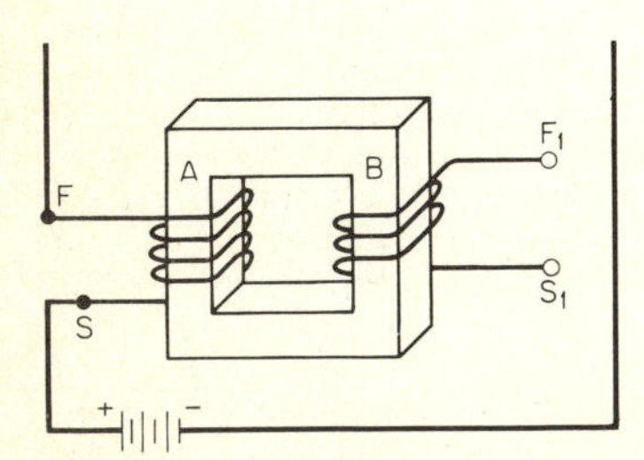

Fig. 4.9 Connect the coils to the battery so that their magnetic fields combine.

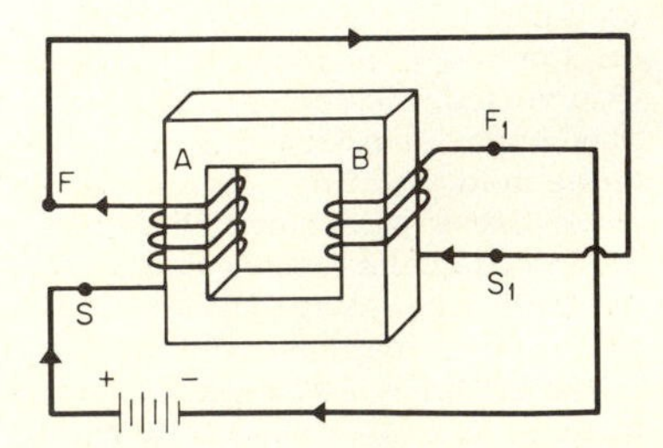

Fig. 4.10 Coils *A* and *B* are connected so that their fields are aiding.

solution Terminal F of coil A should go to terminal S_1 of coil B, and F_1 of coil B should be connected to the negative terminal of the battery, as shown in Fig. 4.10. Check this illustration with the right-hand (or left-hand) rule to make sure that the fields of the coils are aiding.

example 4.5 What would be the effect if the two coils, A and B, were to be connected in parallel and to a battery, as in Fig. 4.11?

solution In this case also, the magnetic fields of the two coils would aid each other. Application of either the right-hand rule or the left-hand rule for coils will prove this.

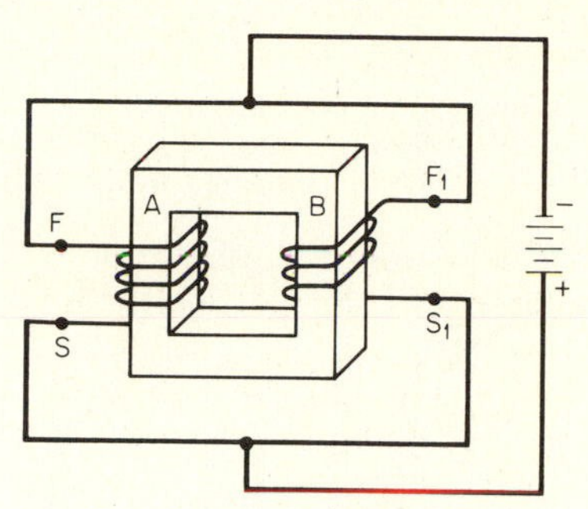

Fig. 4.11 Coils A and B are connected in parallel. Do the magnetic fields of the coils aid or oppose?

4.9 UNITS OF MEASUREMENT IN MAGNETICS

The basic units of measurement in electricity are amperes, volts, and ohms. It would be convenient if there were also only three basic units of magnetism, but unfortunately such is not the case. Instead, there are several different systems of measurement, and there is no universally used system. (The mks system has been suggested as a logical universal system of measurement in magnetism, but at the time of this writing the literature from American manufacturers does not define their products in terms of the mks system exclusively.)

One way to understand the myriad of units for magnetic measurements would be to describe all the units in a particular system. For example, the method of measuring flux, flux density, magnetomotive force, etc., in the cgs system could be described carefully, and then a conversion table could be included at the end of the discussion for changing from cgs to any other system.

Another method—the one used in this book—is to describe each term and its unit of measurement in the system that describes that particular term most simply. Thus, we will define the unit of *flux* in terms of the cgs system, and the unit of magnetomotive force in terms of the mks system. A table will then be included for converting from one system to another.

MAGNETIC FLUX Magnetic fields exist in lines of force, called flux. In the cgs system, each individual line of flux is called a *maxwell.* If there are three lines of flux, the amount of flux is said to be three maxwells in the cgs system, whereas in the British system it is simply called three *lines.* The greater the number of lines of flux, the stronger the magnetic field. The lines are imaginary, and the concept of lines of flux probably comes from the popular experiment in which iron filings are sprinkled on a paper that is placed over a magnet. The iron filings tend to arrange themselves in a group of lines, as shown in Fig. 4.12. These are actually lines of equal magnetic intensity.

UNIT POLES All magnets contain a north and a south pole. Some magnets can be made to have more than one north and south pole, but there is always a south pole for each north pole. For the purpose of discussion, however, let us suppose that it is possible to make an extremely small north pole with no associated south pole. This very small imaginary north pole is called a unit north pole in the cgs system. This unit north pole is helpful in describing some of the terms used in magnetism, and some of the units of measurement. However, it must always be remembered that it does not actually exist anywhere in the world physically.

How small should this unit north pole be? It is not enough to just say small, or very small; it is necessary to say exactly how small if it is going to be useful for making measurements, and for describing units of measurement. Figure 4.13 shows how the unit north pole is defined. If two unit north poles are placed in a vacuum one centimeter apart, they will *repel* each other with a force of one dyne. This defines, then, how small a unit north pole is.

If a unit north pole was placed on the piece of paper in Fig. 4.12, it would move away from the north pole of the magnet and toward the south pole. Its path will be along one of the flux lines shown in the illustration. For this reason, you will often see arrows along the lines of flux. These arrows are supposed to show the direction of a unit north

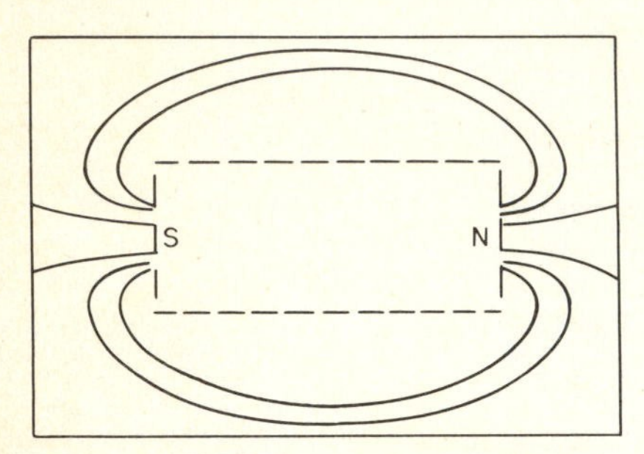

Fig. 4.12 Arrangement of iron filings on a piece of paper covering a bar magnet.

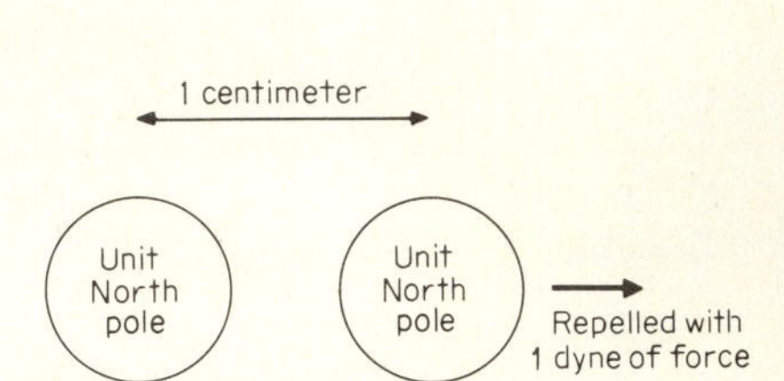

Fig. 4.13 Definition of unit north pole.

pole along the flux line, but often give the mistaken idea that the flux line itself is supposed to be moving. The direction of a line of flux is defined as the direction that a unit north pole will move if placed on the flux line. One line of flux, or one maxwell, will act upon a unit north pole with a force of one dyne. Thus, the unit north pole is helpful in defining the magnetic strength of a line of flux.

FLUX DENSITY If one line of flux passes perpendicularly through a square centimeter of this page, then there is a certain amount of magnetic field strength present. If *two* lines of flux pass through that same square centimeter, then the field strength will be twice as great as when one flux line was present. If three lines of flux pass through that square centimeter, the field strength will be three times as great, and so on. It should be easy to visualize that two lines of flux passing through one square centimeter will produce a much greater concentration of magnetic field strength than would be produced if the two lines of flux pass through one square centimeter. An important way of defining the strength of a magnetic field, then, is in terms of the number of flux lines that pass through a square centimeter of area. Obviously, a very strong magnet will cause more flux lines to pass through a square centimeter of area than a weak magnet will. This is illustrated in Fig. 4.14.

The number of lines of flux that pass through one square centimeter of area is called the flux density. In the cgs system, the unit of flux density is the *gauss*. When one maxwell (that is, one line of flux) passes through one square centimeter of cross-sectional area, the flux density is said to be one gauss. Figure 4.15 illustrates one gauss of flux density. The gauss is a unit of measurement in the cgs system.

MAGNETOMOTIVE FORCE Flux lines can be established in a number of ways. One of the easiest ways is to use an electric current. For every electric current there is an associated magnetic field. The force that causes a magnetic flux to be established is called the magnetomotive force. It can be compared to the voltage in electricity which is sometimes considered to be the "force" that produces an electric current. (Actually, of course, voltage is a measure of work, *not* a unit of force.)

The easiest unit of magnetomotive force to understand is the *ampere-turn*. If one ampere of current flows through one turn of wire, then the magnetomotive force is said to be one ampere-turn.

Figure 4.16 illustrates the ampere-turn. The magnetomotive force is always equal to the number of amperes of current multiplied by the number of turns of wire. The ampere-turn is a mks unit of measurement. (In the cgs system, a gilbert is used as the unit of mmf. One gilbert = 1.257 × ampere-turns.)

RELUCTANCE When a magnetomotive force is used to establish flux in a material, there is always some opposition to the flux. This opposition is called *reluctance*. There

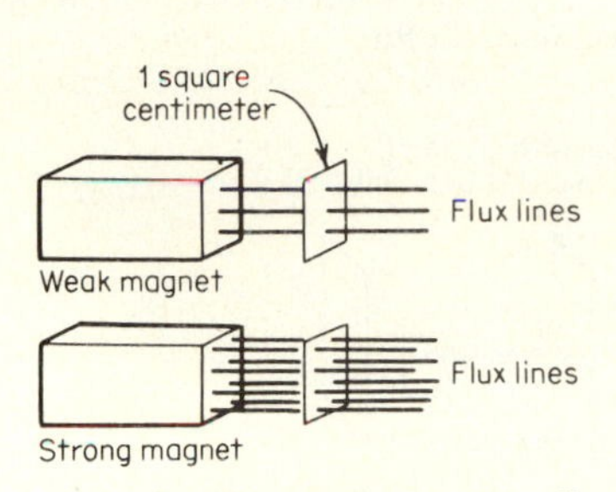

Fig. 4.14 The strong magnet will cause more lines of flux per square centimeter than the weak magnet.

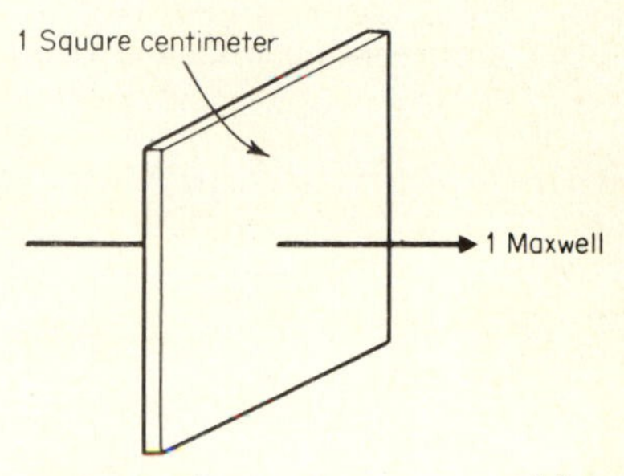

Fig. 4.15 Illustrating one gauss of flux density.

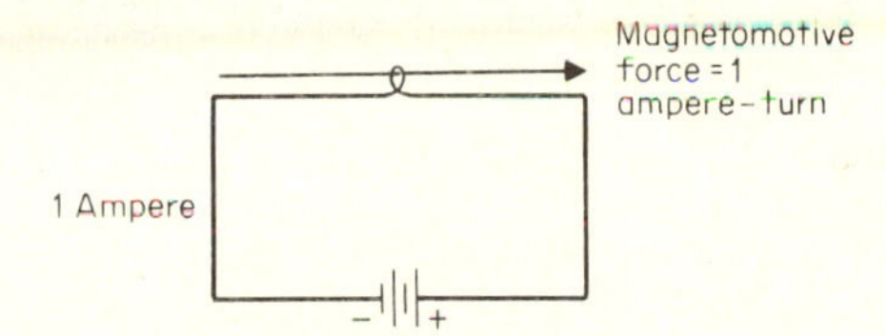

Fig. 4.16 The ampere-turn is a unit of magnetomotive force.

are no *English* units for measuring reluctance. The term *rel* has been proposed but is not in common usage. In the cgs system reluctance is measured in *gilberts per maxwell*, but there is no common name for the unit of reluctance, and there is no name for it in the mks system.

In magnetic circuits, it is sometimes more convenient to know the ease with which flux lines can be established in a material, rather than the opposition that the material offers to establishment of flux. The ease with which flux lines can be established in a material is called the *permeance* of that material, and it is the reciprocal of reluctance.

COMPARISON OF ELECTRIC AND MAGNETIC CIRCUITS There is a similarity between electric and magnetic units. Table 4.1 lists the more important electrical terms and the comparable magnetic terms.

TABLE 4.1 Comparison of Electric and Magnetic Terms

Electric units	Magnetic units
Electric current I	Magnetic flux ϕ
Electromotive force E	Magnetomotive force, mmf
Resistance R	Reluctance R
Conductance G	Permeance P

There are certain things about electric circuits that are quite different from magnetic circuits, and should always be kept in mind. The more important differences will now be discussed.

In an electric current, all the current is normally confined to the conductors carrying it. In a magnetic circuit, some of the flux lines actually leave the circuit along the route, so that it is somewhat more difficult to calculate the amount of flux at a certain point in comparison to calculating current in a circuit.

The resistance of a wire or resistor is a fixed value for a particular circuit over the range of currents for which it is intended to be used. In a magnetic circuit, the amount of reluctance is dependent upon the amount of flux already present. The greater the amount of flux already present in a piece of iron, the greater the reluctance becomes—that is, the greater the opposition to a further increase in flux.

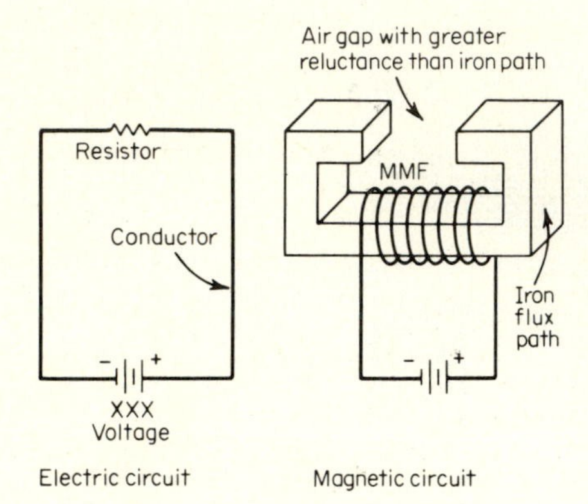

Fig. 4.17 Comparison of an electric and a magnetic circuit.

In electric circuits, there are insulating materials that offer great opposition to current flow. In magnetic circuits there are no materials that can prevent flux from being established. In order to prevent magnetic flux lines from entering a particular area, the only course that can be taken is to employ a very low reluctance path around that area. Nearly all the flux lines will follow the low-reluctance path in such cases, and therefore they will not enter the region being shielded.

Figure 4.17 compares an electric and a magnetic circuit. A battery supplies the power to the electric circuit. In the magnetic circuit the coil of current-carrying wire provides the magnetomotive force that establishes flux in the circuit.

Electric current flows through the conductor in the electric circuit. In the magnetic circuit a low-reluctance iron provides a path for the flux. As mentioned before, some of the flux may leave this low-reluctance path and return to the source of mmf by another path. If the circuit is properly designed, the flux leakage can be kept to a minimum.

A resistor in the electric circuit offers opposition to the flow of current. In the magnetic circuit, the air gap offers a greater opposition to establishment of flux than the iron path. Therefore, the air gap may be thought of as being the resistor in the magnetic circuit.

Table 4.2 lists some of the units of magnetism, and conversion factors between units.

MAGNETIZING FORCE One ampere of current flowing through one turn of wire will give a magnetomotive force of one ampere-turn. (Mmf in ampere-turns = number of amperes × number of turns of wire.) Likewise, one ampere of current flowing through ten turns of wire will give a magnetomotive force of ten ampere-turns. Now, suppose you wind the ten turns very closely together so that each turn is tight against the adjacent turns. The magnetomotive force will then be concentrated along a short length of the coil form. On the other hand, if you spread the turns out so that they are evenly distributed along a length of one foot, then the magnetomotive force will be

TABLE 4.2 Magnetic Units

Unit	cgs	mks	British
Flux	Maxwells	Webers	Lines
Magnetomotive force	Gilberts	Ampere-turns	Ampere-turns
Magnetizing force	Oersteds	Ampere-turns per meter	Ampere-turns per foot
Reluctance	Gilberts per maxwell	Ampere-turns per weber	No unit (rel suggested)
Flux density (also called magnetic induction)	Maxwells per square centimeter°	Webers per square meter	Lines per square inch

CONVERSION FACTORS

Inches × 2.54 = centimeters Gilberts × $10/4\pi$ = ampere-turns
Maxwells × 10^{-8} = webers Oersteds × $10^3/\pi$ = ampere-turns per meter
Gauss (maxwells per square centimeter) × 10^{-4} = webers per square meter
1 gamma = 10^{-5} gauss

° One maxwell per square centimeter is called a gauss; so flux density in the cgs system can be given in either the number of maxwells per square centimeter, or gauss.

distributed along a path one foot long. In both cases *the magnetomotive force* is ten ampere-turns, but in one the mmf is concentrated in a short length, whereas in the other it is spread out along a length of one foot.

It is important to know the amount of magnetomotive force, but it is also essential to know how it is distributed. The distribution of mmf along a length is called the *magnetic gradient*. It is also known as the *magnetizing force*. In the above example, ten turns per inch will produce a greater magnetizing force than ten turns per foot. In the mks system, magnetizing force is measured in ampere-turns per meter, while in the British system either ampere-turns per foot or ampere-turns per inch may be used. In the cgs system, the oersted is used. One gilbert per centimeter gives a magnetizing force of one oersted.

RELATIVE PERMEABILITY Relative permeability, which is often simply called permeability, is the ratio of the flux density in a material for a given magnetizing force to the flux density that would be produced in a vacuum for the same amount of magnetizing force. Mathematically, the relative permeability in the cgs system is expressed by the equation

$$\mu = \frac{B}{H} \tag{4.2}$$

where μ = relative permeability of the material
B = flux density
H = magnetizing force

There are no units for relative permeability. Relative permeability depends on the type of material. In other words, every different type of material will have a different value of relative permeability. For example, the relative permeability of air, like that of a vacuum, is one. For all paramagnetic materials, the relative permeability is slightly greater than one. The relative permeability of diamagnetic materials is less than one, while the relative permeability of ferromagnetic materials is much greater than one (and also much greater than the relative permeability of paramagnetic materials).

A relative permeability greater than one simply means that it is easier to establish flux lines in that material than it is to establish them in a vacuum. A material having a relative permeability less than one indicates that it is more difficult to establish flux lines in that material than it is to establish them in a vacuum.

4.10 ROWLAND'S LAW

Rowland's law expresses the relationship among magnetomotive force, flux, and reluctance. It is sometimes called *Ohm's law for magnetic circuits*.

The three fundamental quantities, the maxwell (ϕ), the gilbert (F), and the unit of reluctance (R), are used in Rowland's law as follows:

$$F = \phi R \tag{4.3}$$

where F = magnetomotive force, Gb
ϕ = flux, Mx
R = reluctance, Gb/Mx

If any two factors are known, the third can be obtained by solving Eq. (4.3). Thus,

$$\phi = \frac{F}{R} \tag{4.4}$$

$$R = \frac{F}{\phi} \tag{4.5}$$

example 4.6 What current flows through the coils of an electromagnet that has 400 turns, develops a flux of 250 000 lines and has a core reluctance of 0.04 unit?

solution The magnetomotive force developed by the coils is

$$1.257 \times I \times N = 1.257 \times I \times 400$$

And since the flux $\phi = F/R$,

$$250\,000 = \frac{1.257 \times I \times 400}{0.04}$$

Solving for I, the result is

$$\frac{0.04 \times 250\,000}{400 \times 1.257}$$

or 19.89 A.

example 4.7 The magnetizing coil of a magnetic circuit has 300 turns and carries a current of 0.5 A. The flux in the magnetic circuit is 100 000 Mx. What is the reluctance of the circuit?

solution The magnetomotive force

$$F = 1.257 \times 0.5 \times 300 = 188.55 \text{ Gb}$$

The reluctance is

$$R = \frac{F}{\phi} = \frac{188.55}{100\,000}$$
$$= 0.001\,885\,5 \text{ unit or Gb/Mx}$$

Owing to the fact that the reluctance in a magnetic circuit is not linear, Rowland's law is not as versatile as Ohm's law for electric circuits. The reluctance in the circuit is only a given value when a certain amount of flux is present. If you change the flux, you change the reluctance. The situation is very similar to that of an electric circuit with a thermistor. The resistance of the thermistor depends on its temperature which, in turn, depends on the amount of current through it. You can use Ohm's law only when you know the resistance of the thermistor for some given amount of current. If you change the current (or temperature), the resistance changes, and a new set of conditions exist.

To summarize, magnetic circuits (like thermistor circuits) are nonlinear. You cannot apply a linear equation, such as $I = E/R$ or $\phi = \text{mmf}/R$, to nonlinear circuits for all possible conditions in the circuit.

4.11 RELUCTANCES IN SERIES

To find the total reluctance of any number of reluctances in series, we merely add them together, just as we would add resistances to find the total resistance of a number of resistors in series.

example 4.8 Figure 4.18 shows a magnetic structure that is designed for a special-purpose application. The total flux in the core material, as determined by means of a flux-meter, is 50 000 lines with an applied magnetomotive force of 1600 Gb. The reluctance of section A is 0.001 25 unit, and the reluctance of section B is 0.000 75 unit. One of the air gaps has a length of 0.01 in, while the other is 0.02 in long. Compute (a) the combined reluctance of the air gaps, and (b) the reluctance of each air gap taken separately.

solution (*a*) We denote the unknown reluctance of the air gaps by Z. Since this is a series magnetic circuit, the combined reluctance of the magnetic circuit is

$$0.001\,25 + 0.000\,75 + Z$$

Total flux is

$$\phi = \frac{F}{R} = 50\,000$$
$$= \frac{1600}{0.001\,25 + 0.000\,75 + Z}$$
$$= \frac{1600}{0.002 + Z}$$

Solving for Z, the combined reluctance of the two air gaps is

$$\frac{1500}{50\,000} = 0.03 \text{ unit}$$

(*b*) Since one air gap is twice as long as the other, its reluctance will also be twice as great as the reluctance of the other air gap, just as the resistance of a wire that is 2 ft long is twice as

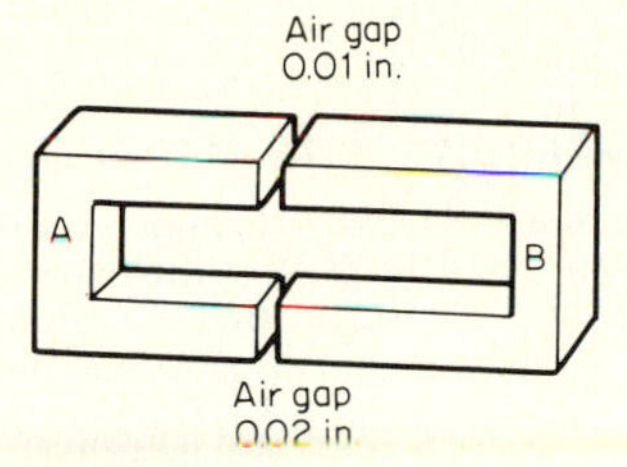

Fig. 4.18 Special-purpose magnetic structure.

great as the resistance of a wire that is only 1 ft long. Denoting the reluctance of one air gap by Y, the reluctance of the other will be $2Y$; then $Y + 2Y = 0.03$ unit. Solving this for Y, the reluctance of one air gap is $0.03/3 = 0.01$ unit; and the reluctance of the other air gap is $2Y$, or 0.02 unit.

4.12 RELUCTANCES IN PARALLEL

The permeance of any magnetic circuit is equal to 1 divided by the total reluctance of the circuit, or

$$\text{Permeance } P = \frac{1}{\text{reluctance}} = \frac{1}{R} \tag{4.6}$$

Permeance in magnetic circuits may be likened to conductance in electric circuits and is used to compute the combined reluctance of a number of reluctances in parallel, just as conductance is used in parallel resistance calculations. To find the combined reluctance of a number of reluctances in parallel, we take the reciprocal of the sum of the permeances of the separate paths. Thus,

$$R = \frac{1}{1/R_1 + 1/R_2 + 1/R_3 + \cdots + 1/R_N} \tag{4.7}$$

where R = combined reluctance

$R_1, R_2, R_3, \cdots, R_N$ = reluctances of the separate paths

For two reluctances in parallel, Eq. (4.7) becomes

$$R = \frac{1}{1/R + 1/R} = \frac{R_1 R_2}{R_1 + R_2} \tag{4.8}$$

and for three reluctances in parallel,

$$R = \frac{R_1 R_2 R_3}{R_1 R_2 + R_1 R_3 + R_2 R_3}$$

example 4.9 What current flows through the coils of an electromagnet that has 400 turns, develops a flux of 250 000 lines, and has a core reluctance of 0.04 unit?

solution From Eq. (4.4) and the equation for magnetomotive force

$$\phi = \frac{F}{R} = 250\ 000 = \frac{I \times 400 \times 1.257}{0.04}$$

Solving for I, the result is

$$\frac{0.04 \times 250\ 000}{400 \times 1.257}$$

or 19.89 A.

example 4.10 The core of a hastily improvised choke coil is made up of three different grades of steel laminations having reluctance values of 0.05, 0.08, and 0.12, respectively. What is the reluctance of the core?

solution Using Eq. (4.9) for computing the reluctance of three reluctances in parallel, we get

$$R = \frac{0.05 \times 0.08 \times 0.12}{(0.05 \times 0.08) + (0.05 \times 0.12) + (0.08 \times 0.12)}$$
$$= 0.003 \text{ unit}$$

4.13 VARIATION OF PERMEABILITY WITH MAGNETIZING FORCE

Since the reluctance of a magnetic substance is not constant but changes with changes in flux density, the permeability of a magnetic substance will also change with every change in flux density. The reason for this is that the reluctance of a magnetic material changes with every change in the flux. The reluctance of only a 1-cm cube of that material will, of course, also change at the same time. The reluctance of a 1-cm cube of any material is its *reluctivity,* and permeability is equal to one divided by reluctivity. It follows, then, that the permeability of a magnetic material will also change with every change in flux. Since the change in reluctance will *not* be *directly proportional* to the change in flux, the change in permeability will also not be directly proportional to changes in flux. These facts should always be taken into consideration when magnetic circuit calculations are to be made.

MAGNETIZATION CURVES Magnetization curves and tables showing the relations between B and H or among B, H, and μ at various stages of magnetization for

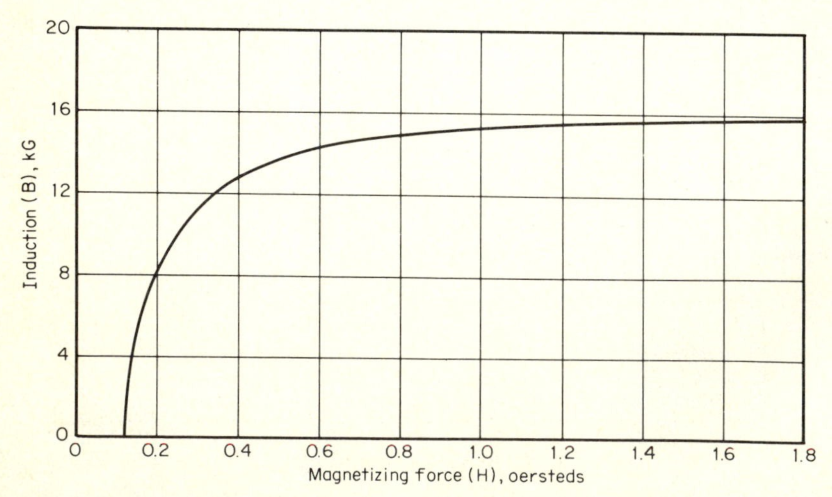

Fig. 4.19 Typical dc magnetization curve of 12-mil Silectron uncut core—not impregnated. *(Courtesy The Arnold Engineering Company)*

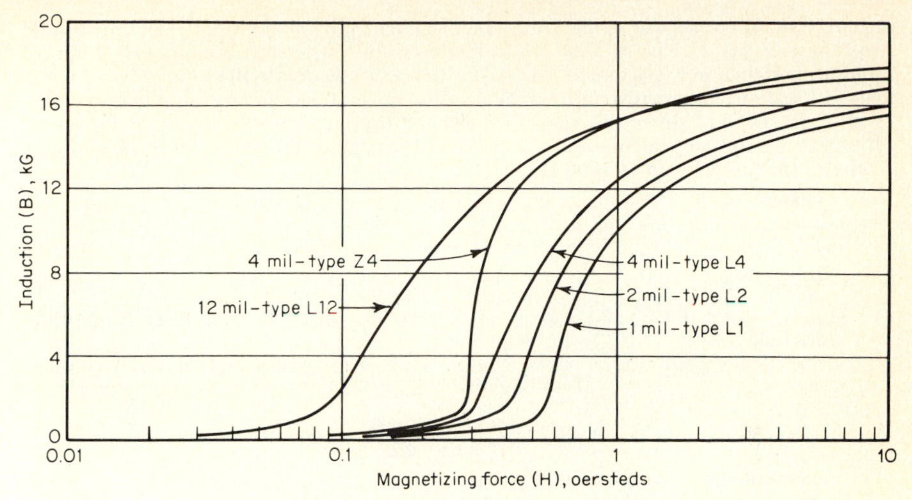

Fig. 4.20 Typical dc magnetization curves for 1-, 2-, 4-, and 12-mil Silectron toroids—not impregnated. (*Courtesy The Arnold Engineering Company*)

different grades of iron and steel and various other kinds of magnetic materials are available from manufacturers of transformer steel and other core materials.

Magnetization curves may take various forms. Some may be graphed on ordinary, linear graph paper, while others may be on logarithmic or on semilogarithmic paper. The use of magnetization curves greatly simplifies the solution of magnetic circuit problems. Typical magnetization curves are shown in Figs. 4.19 to 4.21. These

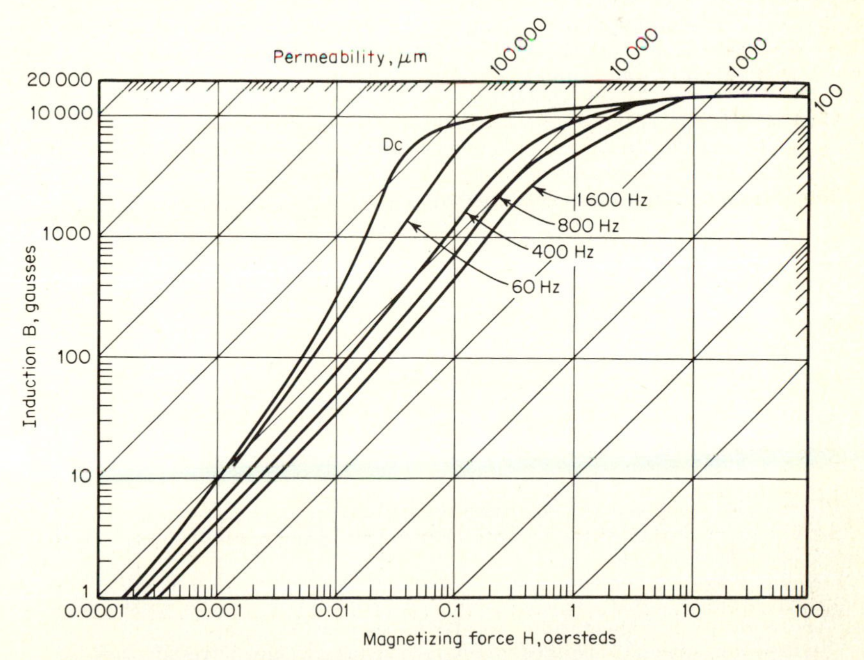

Fig. 4.21 Magnetization curves for 0.014-in 4750 induction and permeability versus magnetizing force: type AL 4750, gage 0.014 in, sample from rings 1.9 by 2.5 in. (*Courtesy The Arnold Engineering Company*)

magnetization curves are something like the characteristic curves of electron tubes, and they are just as easy to interpret. To determine permeability from a B–H curve, such as is shown in Figs. 4.19 and 4.20, it is necessary to divide the induction B on the vertical scale by the corresponding value of the magnetizing force H on the horizontal scale. Figure 4.21 gives the value of permeability for any corresponding values of B and H without any calculations. Figure 4.21 also gives the values of B, H, and μ at different frequencies up to 1600 Hz.

example 4.11 Using the B–H curve of Fig. 4.20, determine (*a*) the magnetizing force needed for an induction (flux density) of 12 kG (12 000 lines/cm^2) in a 12-mil-type L12 toroid, and (*b*) the permeability of the core material for this value of induction.

solution (*a*) Find the value of the induction, 12 kG, on the vertical scale, and from this point follow the horizontal line to its point of intersection with the curve for 12-mil-type 12 material; then find the corresponding value of the magnetizing force, 0.25 Oe, on the horizontal scale.

(*b*)

$$\text{Permeability } \mu = \frac{B}{H}$$

$$= \frac{12\,000}{0.25} = 48\,000$$

example 4.12 Using the B–H–μ curve of Fig. 4.21, determine (*a*) the dc induction in the core material when the magnetizing force is 0.008 Oe, and (*b*) the permeability of the core for the given value of magnetizing force.

solution (*a*) Find 0.008 on the horizontal scale, and follow the vertical line up from this point to its intersection with the dc curve; the corresponding value of the induction on the vertical scale is 200 G. (*b*) From the point on the dc curve corresponding to the determined values of B and H, place a straightedge parallel to and between the diagonal lines leading to the upper scale; and the corresponding permeability value of 25 000 will be found on the upper scale.

4.14 MAGNETIC CIRCUIT COMPUTATIONS

Just as the resistance of an electric conductor is equal to the specific resistance, or resistivity, of the conductor multiplied by its length and divided by its cross-sectional area, the reluctance of a magnetic material (or conductor of flux or lines of force) is similarly equal to the specific reluctance, or reluctivity, of the magnetic material multiplied by the length of the material and divided by its cross-sectional area. In other words, as the length of a magnetic circuit increases, its reluctance also increases in proportion; but as the cross-sectional area increases, the reluctance decreases in proportion. Then the reluctance of a magnetic material is directly proportional to its length and inversely proportional to its cross-sectional area. And since permeability is equal to 1 divided by reluctivity, the reluctivity of a magnetic material is equal to

$$\frac{1}{\text{Permeability}} = \frac{1}{\mu}$$

and the equation for finding reluctance may be written in the form

$$R = \frac{(1/\mu) \times L}{A} = \frac{L}{\mu A} \tag{4.9}$$

where R = reluctance in units
μ = permeability
L = length of magnetic circuit
A = cross-sectional area of magnetic circuit

If L is given in inches, A is in square inches; but if L is in centimeters, A will be in square centimeters.

Dividing 1 by the permeability of a magnetic substance to obtain its reluctivity is a procedure similar to that used to obtain the resistivity, or specific resistance, of an electric conductor by taking the reciprocal of its conductivity.

Our basic fundamental equation for finding total flux may now be restated as

$$\text{Total flux in maxwells } (\phi) = \frac{F}{R} = \frac{1.257 \times IN}{L/\mu A}$$

or

$$\phi = \frac{1.257 \times IN \times \mu A}{L} \tag{4.10}$$

where IN = ampere-turns
μ = permeability
A = cross-sectional area, cm^2
L = length, cm

4.15 ELECTROMAGNETS

The pulling power or tractive force of an electromagnet may be computed from an equation derived from the fundamental definition of a unit magnetic pole. A unit magnetic pole is a pole of such strength that it will repel an exactly similar pole with a force of one dyne when the distance between the poles is one centimeter.

The fundamental equation of electromagnet computations is

$$T = \frac{AB^2}{72\ 134\ 000} \tag{4.11}$$

where T = pull, lb
A = cross-sectional area of pole faces, in^2
B = flux density, lines/in^2

To find A, when T and B are known, the equation is

$$A = \frac{72\ 134\ 000T}{B^2} \tag{4.12}$$

and to find B, when T and A are known, we use

$$B = \sqrt{\frac{72\ 134\ 000T}{A}} \tag{4.13}$$

example 4.13 The current in the windings of an electromagnet induces a flux density of 110 000 lines/in^2 in the core. If the core and pole face have a cross-sectional area of 20 in^2, what weight will the electromagnet support?

solution

$$T = \frac{AB^2}{72\ 134\ 000} = \frac{20 \times (110\ 000)^2}{72\ 134\ 000}$$

$$T = \frac{242\ 000\ 000}{72\ 134} = 3355 \text{ lb}$$

4.16 SATURATION

Any magnetic material is said to be saturated when any further increase in the magnetizing force, or field intensity, no longer causes any appreciable increase in the flux density in the material.

The value of the flux density in the core material of any electromagnetic circuit is of considerable importance; this value will depend upon the use of the magnetic material. An iron-core choke in which the core material is highly saturated loses most of its inductive properties and becomes almost wholly resistive; for this reason, inductive chokes that carry direct current are usually designed with an air gap in the magnetic structure to keep the core from becoming magnetically saturated. If the core of a transformer in the plate circuit of a power output tube becomes magnetically saturated, the result will be severe distortion and poor transfer of energy from the primary to the secondary. Or if the biasing current in the recording head of a tape recorder is either too high or too low, any recorded material will be badly distorted. On the other hand, some electromagnetic devices will operate only very poorly, or not at all, if the initial flux density in the core material is not at its maximum value. The saturation point of any magnetic iron or steel may easily be determined from its B–H magnetization curve.

4.17 RETENTIVITY

A measure of retentivity is the amount of magnetism, or value of B, that remains in a magnetic material after the magnetizing force has been raised to a value more than sufficient to saturate the material and then reduced to zero; the value of B remaining in the material is called *residual magnetism* or *remanence*. If two different kinds of permanent magnet materials of the same size are both magnetized to saturation, the one that makes the stronger permanent will, of course, have the higher retentivity. However, this does not necessarily mean that the stronger magnet will be harder to demagnetize, since this depends upon other factors besides retentivity.

4.18 COERCIVITY OR COERCIVE FORCE

Coercivity is the value of the opposing magnetizing force that is needed to reduce the residual magnetism, or remanence, in a magnetic substance down to zero.

A clear concept of coercivity may be obtained by referring to Fig. 4.22. Assume that the iron core of the coil is not magnetized, and that the potentiometer setting is at

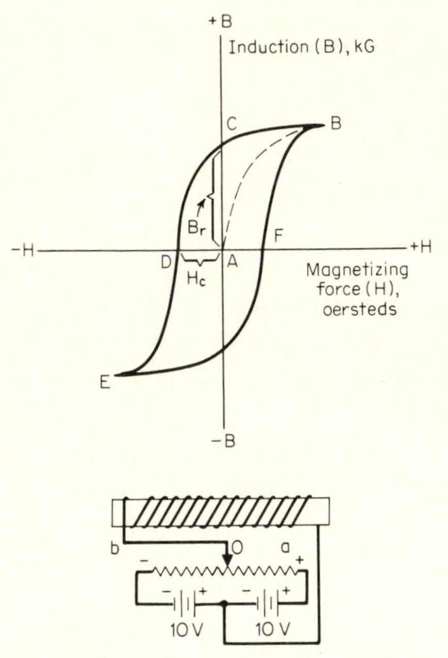

Fig. 4.22 Illustrating coercivity, or coercive force H_c, and residual magnetism, or remanence B_r.

its midpoint, or zero. As the potentiometer setting is shifted toward point a, the core material will become magnetized to saturation along the line A–B of the magnetization curve. If the potentiometer setting is then shifted back toward the center, or zero position, the flux density B in the core will not fall back to zero along the line A–B on the curve, but instead along the line B–C–D; and when the potentiometer setting is again at zero, the flux density will still be at point c. This will be the residual magnetism, or remanence, denoted by the symbol B_r. To reduce the remanence to zero, it is necessary to shift the potentiometer setting toward point b to produce a magnetizing force opposing the residual field in the core. This will cause the residual field to fall back to zero along the line from C to D. The negative value of H needed to reduce the residual magnetism in the core to zero is the coercivity of the core material. Coercivity is denoted by the symbol H_c.

If the potentiometer setting is shifted still further to the left, the core will again become magnetized, but the magnetization will now be in a negative direction. (This is based on the assumption that the original direction was positive.) Saturation will next occur at point E of the magnetization curve. If the negative value of H is now reduced to zero and then increased again to its highest positive value, by shifting the potentiometer setting toward the right, the flux density in the core will follow a new line from E to F, or zero, and then back up to B.

4.19 HYSTERESIS

The loop formed by the curves of Fig. 4.22 is known as a hysteresis loop, and the area enclosed by the loop is a measure of the energy used in reversing the magnetic molecules of the core. The energy thus lost is called the *hysteresis loss.* Hysteresis loss is of importance only when the core material is being subjected to pulsating dc or ac

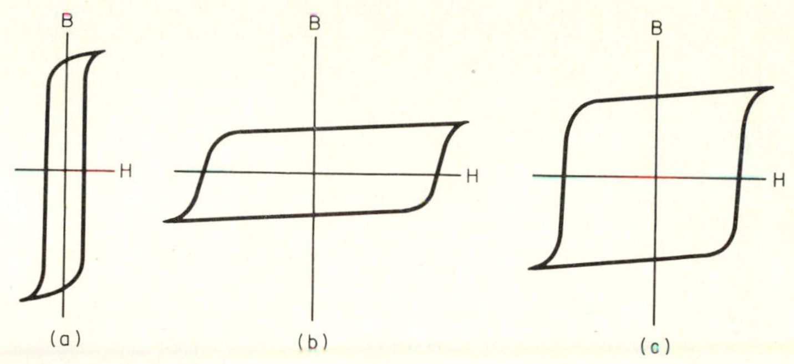

Fig. 4.23 Typical hysteresis loops: (*a*) Typical magnetic core material; small coercive force H_c required to eliminate large residual magnetism B_r. (*b*) Typical of a class of permanent-magnet material; large coercive force H_c required to eliminate small residual magnetism. (*c*) Typical of another class of permanent-magnet material; large coercive force H_c required to eliminate large residual magnetism B_r. (*Courtesy The Arnold Engineering Company*)

fields. Since the hysteresis loop is repeated over and over again for every cycle, core losses become greater and greater as the frequency is increased. For this reason, the core materials used for transformers should have very narrow hysteresis loops. Some typical hysteresis loops are shown in Fig. 4.23.

Core-loss data may be obtained in both tabular and graphical form from manufacturers of transformer core materials. An example of core-loss data in graphical form is shown in Fig. 4.24, which gives the total core loss (hysteresis loss plus eddy current loss) at various frequencies up to 10 kHz and inductions up to 18 kG. The form in which the data are presented may depend upon the engineering practices of the companies supplying the data. Some older publications give equations for computing hysteresis loss, but such equations would hardly be applicable to the large variety of core materials now available.

4.20 EDDY CURRENTS

Currents that are set up in magnetic core materials by induced voltages because of changing magnetic fields are known as eddy currents. The energy lost in the form of heat as a result of eddy current circulating in the core material is called the eddy-current loss. Since eddy-current loss is proportional to the square of the inducing frequency, such losses become very appreciable as the frequency is increased.

Eddy currents may be reduced by stacking the core laminations so that they are parallel to the flux. Natural oxidation of the laminate surfaces also helps greatly in reducing eddy currents; or the laminations may be coated with a thin layer of insulating varnish. Figure 4.25 shows how eddy currents are reduced by laminated cores. But even with laminated cores there is still some circulation of eddy currents in the in-

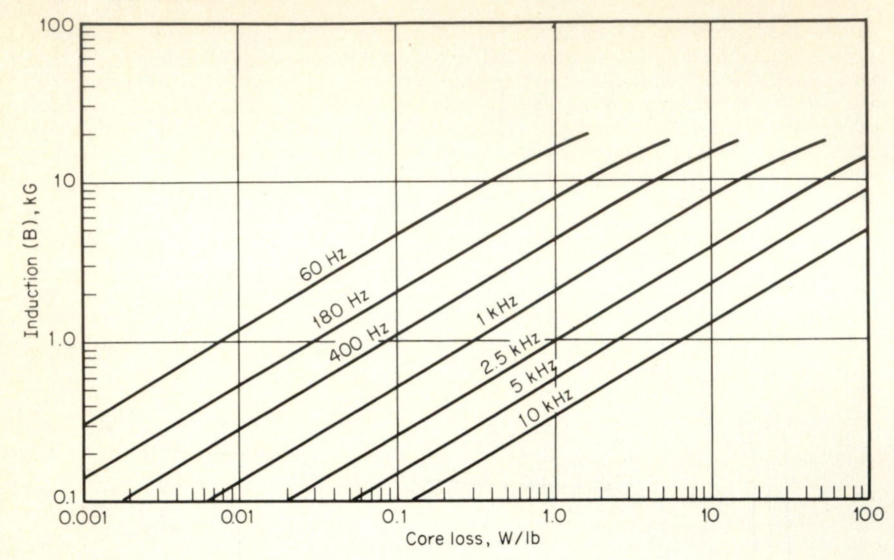

Fig. 4.24 Design curves showing maximum core loss for 4-mil Silectron C cores, type AH. Core loss guarantee 10.0 W/lb maximum at 400 Hz, 15 kg. (*Courtesy The Arnold Engineering Company*)

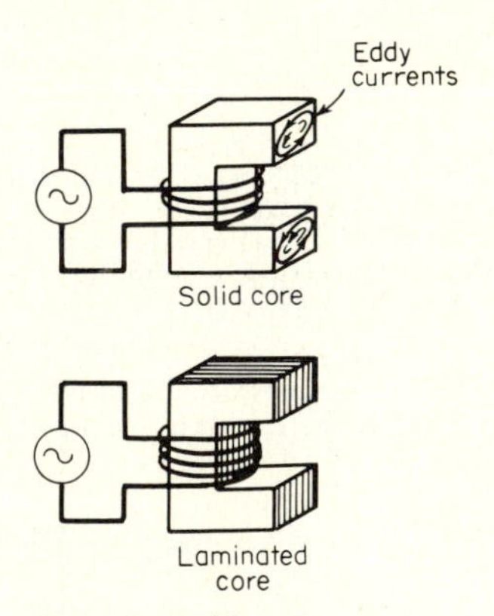

Fig. 4.25 Eddy currents circulating in solid core are stopped by laminated core.

dividual laminations. Therefore, at radio frequencies, the powdered iron-core materials are used. As the frequency is further increased, even the powdered iron cores are no longer effective in reducing eddy-current loss to a sufficient degree. Then the ferrite-core materials must be used. Ferrites, unlike iron or steel, are not electric conductors.

Some equations are available for computing eddy-current loss in ordinary transformer steel, but such equations could hardly be used for computing the eddy-current loss in the large variety of other core materials that have become available in recent years.

4.21 IRON-POWDER CORES

Iron-powder cores for radio coils first came into existence in the early 1930s, when they were used in i-f transformers. Since that time there has been an ever-increasing demand for iron-powder cores and great improvement in production techniques.

Fig. 4.26 Ferrite cores for TV horizontal output transformers, and some ring cores for coils.

Iron-powder cores are now available for a wide variety of applications and for use at frequencies as high as 250 mHz. Every TV receiver now contains a number of coils that have iron-powder cores. (See Fig. 4.26.) Some of these may be fixed-tuned, but others may be of the so-called slug-tuned variety, such as the horizontal width and the horizontal frequency-control coils.

The powdered iron used for iron-powder cores is first subjected to certain chemical processes which cause the individual tiny particles of iron to take on an insulating layer. Then the binder, or glue, used to bind the particles of iron together adds an additional insulating layer over the insulating layer formed by the chemical processing. Thus, in the finished core, the individual tiny particles of iron are effectively insulated from one another; and, as a result, eddy current cannot circulate in the core. Still there will be some eddy-current loss, because the individual particles of iron are electric conductors and not insulators; eddy current can still circulate within each individual particle.

The fineness of the individual iron particles, the amount of binder used in proportion to the amount of iron powder, will be different for different applications and frequency ranges. The iron powder may also be made up of two or three different grades of iron mixed in various proportions.

Some users of iron-powder cores mistakenly assume that all iron-powder cores have the same characteristics; for example, in some electronic construction article, it may be stated that a particular coil was wound on a one-half-inch-diameter form that had an iron-powder core. But then nothing is said about the type of core material. The result is that many of those following the data presented in the article are often disappointed by the results. They wonder what has gone wrong, since they, too, are unaware of the fact that iron-powder cores are not all alike.

Chapter 5

Transformers

5.1 INTRODUCTION

The function of every type of transformer is to transform or change electricity in some manner. This is true whether the transformer is only a tiny sub-ouncer unit or is one of the huge devices used at the substations and terminal points of high-voltage electric-power distribution systems. For example, the tiny output transformer of a transistor pocket-portable radio transforms the current in the output stage to a form that is suitable for actuating the loudspeaker. As another example, the large transformer at the power distribution substation transforms (steps down) the high voltage of the power lines so that it may safely be used in homes and industry.

Since the iron-cored transformer and choke came into use much earlier than other kinds of transformers and chokes, it seems only fitting to take up the study of such transformers and chokes (see Fig. 5.1) before taking up the study of other types. Figure 5.1*a* shows a variety of ferrite-core coils. An iron-core transformer is shown in Fig. 5.1*b*.

5.2 THE BASIC TRANSFORMER

The basic transformer is shown in Fig. 5.2. It consists of two coils wound upon the same iron core which forms a closed magnetic circuit. The flux path is shown by dotted lines around the core. The transformer will not work on pure direct current, but instead must be supplied with power from a source of alternating current or from a source of pulsating direct current.

The winding to which power is supplied is called the *primary*, and the winding from which power is taken is the *secondary*. Any change in the primary current results in a change in flux (ϕ). In accordance with Faraday's law, a change in the flux linking the secondary winding will induce a voltage across the secondary. If a path for current flow exists across the secondary winding, then in accordance with Lenz's law the direction of current flow will be such that its magnetic field opposes the changing flux that produced it.

If the secondary voltage is the same as the primary voltage, the transformer is called a *one-to-one transformer;* if the secondary voltage is greater than the primary, it is a *step-up transformer,* and when the secondary voltage is less than the primary, it is a *step-down transformer.* Power supplied to the primary at a specific frequency produces power at the same frequency in the secondary.

5.3 MUTUAL INDUCTION

If the transformer primary of Fig. 5.2 is connected to a source of alternating current, as in Fig. 5.3, the instantaneous values of the primary voltage and current and the resulting

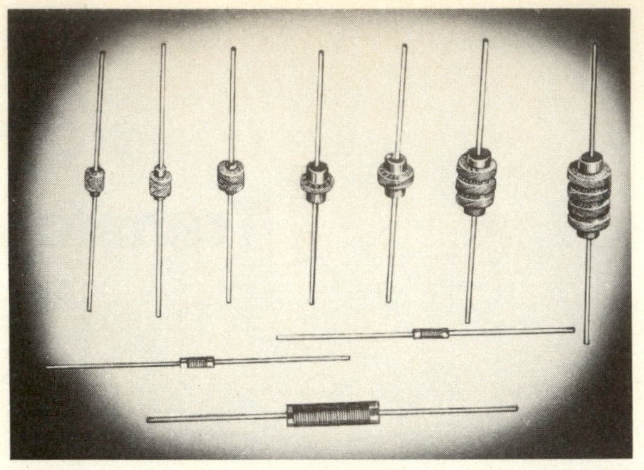

Fig. 5.1*a* A variety of ferrite-core coils and chokes. (*Courtesy James Millen Manufacturing Company, Inc.*)

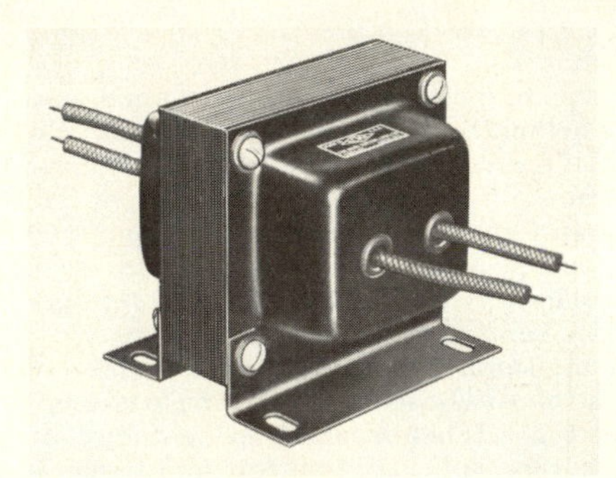

Fig. 5.1*b* Iron-core transformer with end bell construction. (*Courtesy Acme Electric Corporation.*)

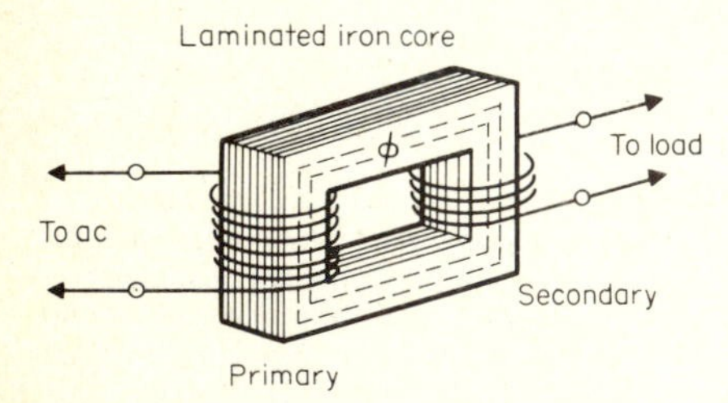

Fig. 5.2 Illustrating the basic transformer. Power is transferred from the primary to the secondary winding by means of the continuously changing flux (ϕ) in the core.

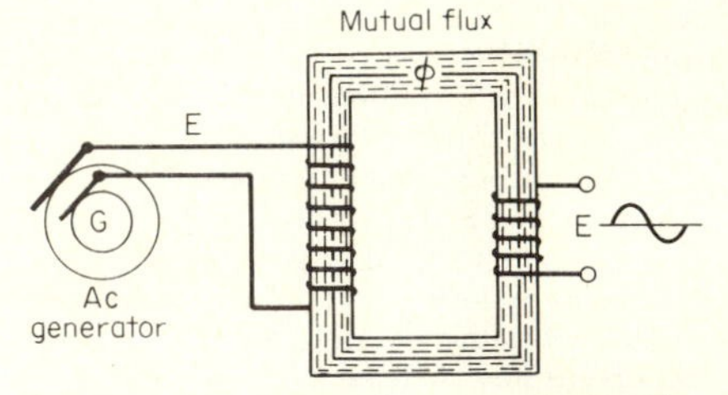

Fig. 5.3 When the transformer is connected to a source of alternating current, the continuously changing magnetic flux in the core induces a voltage in the secondary winding and at the same time also induces a back emf (or counter-emf) in the primary winding.

magnetic field will all be continuously changing and inducing a back voltage in the primary. The instantaneous value of the back voltage will be very nearly equal to the instantaneous value of the applied voltage, but it will always be just a bit lower than the applied voltage. If the applied and back voltages were exactly equal to each other, it would be impossible to get the current flowing in the circuit.

Since the core of the transformer forms a closed magnetic circuit, the primary and secondary windings are interlinked through the mutual magnetic flux in the core. The changing flux in the core induces a voltage in the secondary at the same time that it is inducing a back voltage in the primary, and the combined induction effect is called mutual induction.

The secondary induced voltage may be the same as the primary voltage, or it may be higher or lower than the primary voltage, depending on the ratio of the number of turns of wire between primary and secondary.

When the secondary is connected to a load, current flows from the secondary into the load just as it would from any other voltage source. According to Lenz's law, the current flow in the secondary must always be in such a direction that it will set up an opposition to any *change* in the flux of primary current. For example, if the secondary has a load connected to it, the load current in the secondary winding creates a magnetic field, and as this field changes, it induces a voltage in the primary in opposition to the applied voltage.

example 5.1 Figure 5.4 represents a transformer supplying power to a load connected to its secondary. A waveform of the current is also shown. Draw a diagram showing terminal polarities of both primary and secondary windings and the direction of current flow in each winding at the instant when the primary current is rising from zero to its maximum positive value and has reached the point x on the current curve. Also show the directions of the magnetic fields developed by each winding at this instant. Use the right-hand rule for coils and Lenz's law.

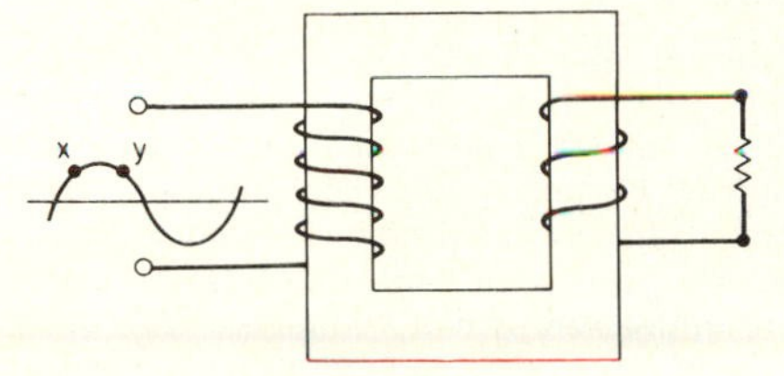

Fig. 5.4 Transformer with sine-wave current input supplying a load.

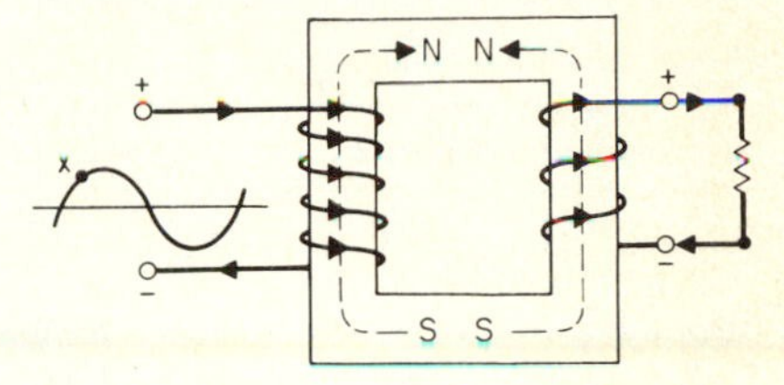

Fig. 5.5 Terminal polarities and the directions of the magnetic fields produced by currents in the primary and secondary windings at the instant when the primary current is rising and is at point x on the current curve.

solution Figure 5-5 shows that as the current is rising from zero to its maximum positive value, the upper terminal of the primary will be positive. Therefore, current flow through the primary winding will be in the direction shown by the arrows, and by the right-hand rule for coils, the direction of the resulting magnetic field will be clockwise in the core. Since the field is building up, the direction of the magnetic field developed in the secondary must, according to Lenz's law, oppose the building up of the field developed by the primary; hence, the magnetic field developed by the secondary must be in a counterclockwise direction in the core. The right-hand rule for coils will show that the magnetic field developed by the secondary will be in a counterclockwise direction in the core when current flows through the coil windings in the direction indicated by the arrows. The upper terminal of the secondary will then be positive.

example 5.2 Draw a diagram showing conditions in the transformer of Fig. 5.4 at the instant when the primary current is falling from its maximum positive value toward zero and has reached the point y on the current curve.

solution The upper terminal of the primary winding is still positive and the primary current will be in the direction indicated. Since the current is now decreasing, its associated magnetic field will be collapsing. According to Lenz's law, the magnetic field developed in the secondary must oppose any *decrease* of the primary field. Therefore, the magnetic field of the secondary winding at this instant will be in the same direction in the core as the primary field, or clockwise, as shown in Fig. 5.6. Application of the right-hand rule for coils will then show that current flow in the secondary will be in the direction indicated by the arrows, and the lower terminal of the secondary winding will be positive.

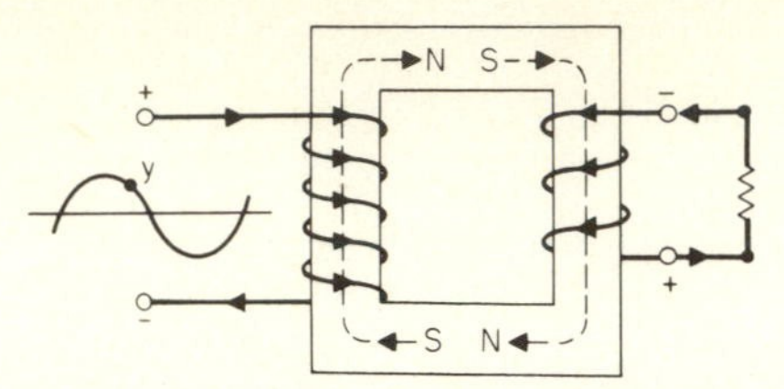

Fig. 5.6 Showing terminal polarities and the directions of the magnetic fields produced by currents in the primary and secondary windings of a transformer at the instant when the current is falling and has reached point y on the curve of current.

5.4 TRANSFORMER VOLTAGE RELATIONSHIPS

Since the mutual flux in the core of any transformer interlinks the turns of the primary and secondary windings, any change in this flux that induces a certain voltage in *each* turn of the primary winding must also induce the same voltage in *each* turn of the secondary winding. Then, if the secondary winding has only half as many turns as the primary, total secondary voltage will be only half the induced or impressed primary voltage. If the secondary has twice as many turns as the primary, the total induced secondary voltage will be twice as great as the impressed primary voltage.

It may be concluded, then, that the total voltage induced in each winding is always proportional to the number of turns in the winding:

$$\frac{E_p}{E_s} = \frac{N_p}{N_s} \tag{5.1}$$

where E_p = impressed primary voltage
E_s = secondary voltage
N_p = number of primary turns
N_s = number of secondary turns

When there is no load connected to the secondary, the induced primary voltage has practically the same value as the impressed voltage; and even when the transformer is supplying a load, if the transformer is properly designed, there will be only a very slight difference between the impressed and induced primary voltages. In an efficient, well-designed transformer the voltage drops due to resistance and reactance will be very low. Therefore, the ratio E_p/E_s or N_p/N_s may be taken as the voltage ratio of the transformer.

example 5.3 What is the secondary voltage of the transformer shown in Fig. 5.7?

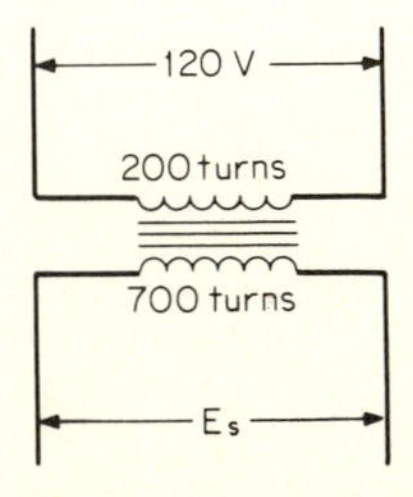

Fig. 5.7 What is the voltage across the secondary?

solution Substituting in Eq. (5.1),

$$\frac{E_p}{E_s} = \frac{N_p}{N_s}$$

$$\frac{120}{E_s} = \frac{200}{700}$$

Solving for the unknown,

$$E_s = \frac{120 \times 700}{200} = 420 \text{ V} \qquad \textit{Answer}$$

example 5.4 The primary winding of a transformer contains 100 turns. How many turns must the secondary contain to step down a voltage of 120 to 12.6 V?

solution By substitution in Eq. (5.1),

$$\frac{120}{12.6} = \frac{100}{N_s}$$

Solving for N_s,

$$N_s = \frac{100 \times 12.6}{120} = 10.5 \text{ turns}$$

5.5 TRANSFORMER CURRENT RELATIONSHIPS

The modern transformer is a highly efficient device. In fact, some large iron-core transformers have an efficiency exceeding 99 percent. This means that almost as much power may be taken out of the transformer as is put into it. Therefore, for practical purposes we may assume that the transformer has an efficiency of 100 percent. If the efficiency is 100 percent, primary and secondary power factors are equal. Power input to the transformer is given by $E_pI_p \times$ (power factor), and the output power is $E_sI_s \times$ (power factor). Assuming that the input and powers are equal,

$$E_pI_p \times \text{(power factor)} = E_sI_s \times \text{(power factor)}$$

Since power factor in the first member has the same value in the second member of the equation, it cancels out, leaving $E_pI_p = E_sI_s$ or

$$\frac{E_p}{E_s} = \frac{I_s}{I_p} \tag{5.2}$$

where E_p = primary impressed voltage
I_s = secondary current
E_s = secondary voltage
I_p = primary current

In a step-up transformer, if the secondary voltage is twice the primary voltage, secondary current is only one half the primary current.

Since

$$\frac{E_p}{E_s} = \frac{N_p}{N_s}$$

it follows that

$$\frac{N_p}{N_s} = \frac{I_s}{I_p} \tag{5.3}$$

Equation (5.3) shows that the currents in the primary and secondary windings are inversely proportional to the corresponding turns.

example 5.5 A transformer supplies 2.4 A at 6.3 V to a secondary load. What is the primary current if the primary is connected to a 120-V ac source?

solution By substitution in Eq. (5.2),

$$\frac{E_p}{E_s} = \frac{I_s}{I_p}$$

$$\frac{120}{6.3} = \frac{2.4}{I_p}$$

Solving for I_p,

$$I_p = \frac{6.3 \times 2.4}{120}$$
$$= 0.126 \text{ or } 126 \text{ mA} \qquad \textit{Answer}$$

example 5.6 A 120-V generator supplies a current of 20 A to a load some distance from the source. The total resistance of the transmission lines from the generator to the load is 2 Ω. The line drop, in this case, is therefore 2 × 20, or 40 V, and the voltage across the load is only 120 − 40, or 80 V. The power loss in the lines is $20^2 \times 2$, or 800 W. These conditions are shown in Fig. 5.8.

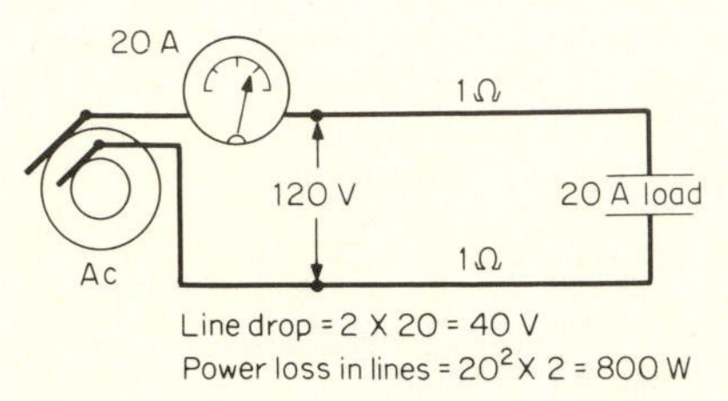

Fig. 5.8 Resistance of line wires between generator and load causes large power loss.

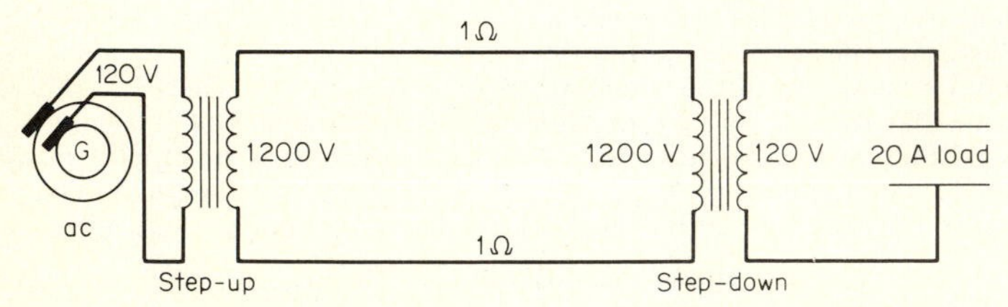

Fig. 5.9 Stepping up the line voltage at the generator results in reduction of line loss to insignificant value. Line current = 2 A; line drop = 2 × 2 = 4 V; power loss in lines = $2^2 \times 2 = 8$ W.

Using the same transmission lines, what will be the line drop and the line loss if a transformer is used to step up the voltage at the generator to 1200 V, and another transformer at the load is then used to step down the line voltage from 1200 to 120 V, as shown in Fig. 5.9.

solution Since the load takes 20 A at 120 V, and the primary of the load transformer operates at 1200 V,

$$\frac{E_p}{E_s} = \frac{I_s}{I_p} \tag{5.2}$$

$$\frac{120}{1200} = \frac{I_s}{20}$$

$$I_s = \frac{20 \times 120}{1200} = 2 \text{ A} \qquad \textit{Answer}$$

The line drop is found by Ohm's law:

$$E = IR = 2 \times 2 = 4 \text{ V} \qquad \textit{Answer}$$

The power loss is

$$P = I^2R = (2)^2 \times 2 = 8 \text{ W} \qquad \textit{Answer}$$

5.6 THE FUNDAMENTAL EQUATION OF THE TRANSFORMER

The average of any varying quantity is the average of all the instantaneous values that the variable will assume in changing from one value to another.

If the voltage impressed on the primary of a transformer has a sine waveform, as shown in Fig. 5.10, this will produce similar sine waves of both current and flux. (This assumes that the flux in the magnetic circuit is linearly related to the current in the winding. Although this is usually not true, it is a reasonable assumption when the flux does not approach the saturation point.) The average value of the induced voltage in either the primary or secondary of a transformer is given by

$$E_A = \frac{N\phi_m}{10^8 t} \tag{5.4}$$

where E_A = average induced voltage
N = number of turns
ϕ_m = maximum flux
t = time in seconds for the flux to rise from zero to its maximum value

Figure 5.10 shows that a sine wave rises from zero to maximum in one-fourth the time it takes to complete one cycle. Therefore, for a sine wave of frequency f

$$t = \frac{1}{4f}$$

Equation (5.4) may be written as

$$E_A = \frac{N\phi_m}{10^8(1/4f)} = \frac{4fN\phi_m}{10^8} \tag{5.5}$$

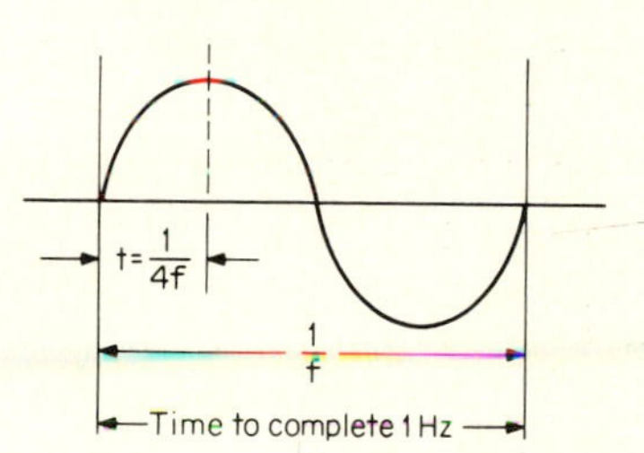

Fig. 5.10 A voltage or current having a sine-wave form rises from zero to maximum in one-fourth of the time it takes to complete 1 Hz.

To change Eq. (5.5) to a form that gives effective or root-mean-square voltage E, instead of average values, we multiply by

$$\frac{0.707}{0.636}$$

Thus,

$$E = \frac{0.707}{0.636} \times \frac{4fN\phi_m}{10^8}$$
$$= \frac{1.11 \times 4fN\phi_m}{10^8}$$
$$= \frac{4.44fN\phi_m}{10^8} \tag{5.6}$$

These equations show that the induced voltage is directly proportional to the maximum flux in the core, the frequency of the applied voltage, and the number of turns in the winding.

The permissible flux in a magnetic circuit is usually expressed in terms of the flux density B_m.

$$B_m = \frac{\phi_m}{A}$$

where B = lines per square inch if the area A is in square inches
B_m = gausses if the area is in square centimeters

Then Eq. (5.6) may be written as

$$E = \frac{4.44fNB_mA}{10^8} \tag{5.7}$$

Equation (5.7) is the fundamental equation of the transformer. In the transformer primary, the effective value of the induced voltage,

$$E_p = \frac{4.44fN_pB_mA}{10^8} \tag{5.8}$$

In the transformer secondary, the effective value of induced voltage,

$$E_s = \frac{4.44fN_sB_mA}{10^8} \tag{5.9}$$

Effective values are of interest since transformer ratings are in these values, and all meters, unless marked otherwise, indicate effective values.

example 5.7 The primary winding of a transformer contains 300 turns. Determine (*a*) at what maximum flux (ϕ) must the core operate if the impressed primary voltage is 120 V at 60 Hz? (*b*) How many turns must the secondary winding have to develop 12.6 V?

solution (*a*) The maximum flux $\phi = B \times A$; by substitution in Eq. (5.8),

$$E_p = \frac{4.44 \times 60 \times 300 \times \phi_m}{10^8} = 120$$

Solving for ϕ_m, we get

$$\phi_m = \frac{120 \times 10^8}{4.44 \times 60 \times 300} = 150\ 150 \text{ Mx}$$ *Answer*

(*b*) By substitution in Eq. (5.9),

$$E_s = 12.6 \frac{4.44 \times 60 \times N_s \times 150\ 150}{10^8}$$

Solving for N_s, the result is

$$\frac{12.6 \times 10^8}{4.44 \times 60 \times 150\ 150} = 31.5 \text{ turns}$$ *Answer*

example 5.8 A transformer primary winding has 450 turns, and its secondary has 2700. The cross-sectional area of the core is 1 in², and maximum flux density B_m in the core is 100 000 lines in². Line frequency is 60 Hz. Compute: (*a*) rated primary voltage E_p, (*b*) rated secondary voltage E_s, (*c*) effective voltage induced in each primary turn, (*d*) effective voltage induced in each secondary turn.

solution (*a*) Substituting in Eq. (5.8),

$$E_p = \frac{4.44 \times 60 \times 450 \times 100\ 000 \times 1}{10^8} = 119.88$$

(*b*) Substituting in Eq. (5.9),

$$E_s = \frac{4.44 \times 60 \times 1200 \times 100\ 000 \times 1}{10^8} = 719.28$$

(*c*) $\dfrac{119.88}{450} = 0.266 \text{ V}$

(*d*) $\dfrac{719.28}{2700} = 0.266 \text{ V}$

5.7 DISCUSSION OF NO-LOAD CONDITIONS

If the secondary winding of a transformer is left open-circuited, the primary current is very small, since the primary current is then opposed not only by the resistance of the winding but also by its inductive reactance which has a *choking* effect on the primary current. The no-load current produces the magnetic flux, and at the same time supplies the eddy current and hysteresis losses in the core. The no-load current, therefore, consists of two components: the magnetizing current component and a power-loss component. The magnetizing current component lags the applied primary voltage by 90°, but is always in phase with the flux. The core loss, or power component, is always in phase with the applied voltage.

Since the magnetizing current component lags the applied primary voltage by 90°, it must lead the induced primary voltage by 90°, because the primary applied and induced voltages are 180° out of phase with each other.

According to Lenz's law, the magnetizing current must lead the induced secondary voltage by 90°, since as the magnetizing current and flux decrease from their maximum positive values and pass through zero in a negative direction, both current and flux

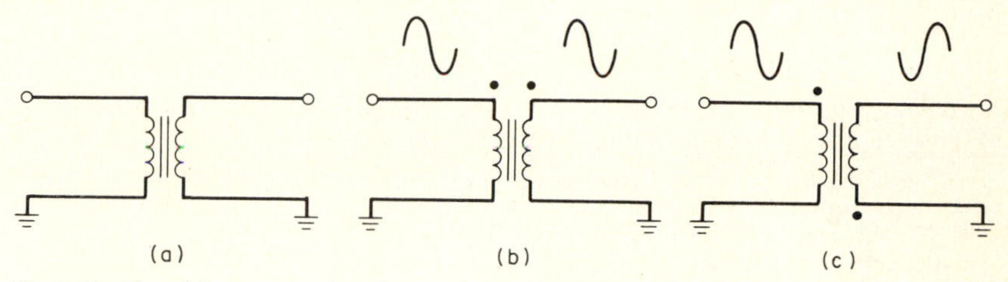

Fig. 5.11 Use of dot notation to indicate phase of input and output signals: (*a*) The transformer symbol does not indicate the phase of input and output signals. (*b*) The dots indicate leads having the same phase. (*c*) Connection of a transformer for phase reversal.

will be changing at their greatest rates at the exact instant when they are passing through their zero values. Therefore, the greatest rate of change of flux in the core occurs at this very instant, and the induced voltages in primary and secondary windings will both reach their maximum values at this instant. This should not be misconstrued to mean that the voltage across the secondary winding must always be 180° out of phase with the voltage across the primary.

Figure 5.11 shows how the phase of the secondary voltage is related to the phase of the primary voltage. The symbol for a transformer in Fig. 5.11*a* does not give any indication of the phase of voltage across the secondary, since the phase of that voltage actually depends on the direction of the winding around the core. To solve this problem, the *dot notation* is used when the phase of the secondary is important. This notation is shown in Fig. 5.11*b* where the primary and secondary currents are in phase, and Fig. 5.11*c* where the currents are 180° out of phase.

5.8 VECTOR REPRESENTATION OF NO-LOAD CONDITIONS

Figure 5.12 is a vector diagram of conditions in a typical, unloaded step-down transformer. The flux lags the applied primary voltage by 90° and leads the induced primary and secondary voltages by 90°. The total no-load current is represented by I_E, and resolved into its core loss and magnetizing components by I_H and I_M, respectively. Since the core-loss current I_H is small in comparison with I_M, the magnetizing current is very nearly equal to the total primary current. Thus, the total no-load current is often called the *magnetizing current* or the *exciting current*. The cosine of the angle θ by which the no-load current lags the applied voltage is the power factor of the transformer at no load. Power factor is given by

$$\text{Power factor} = \frac{I_H}{I_E} \tag{5.10a}$$

$$= \frac{I_H}{\sqrt{I_H^2 + I_M^2}} \tag{5.10b}$$

where I_H = core-loss current
I_E = total no-load current
I_M = magnetizing current

The total power expended in the transformer at no load is equal to $EI \times \cos\theta$. The copper, or I^2R, loss at no load is very small and need not be taken into account in most practical problems.

For the following problem draw a vector diagram approximating the conditions of the problem. Fill in the known values and then the unknown values as they are determined.

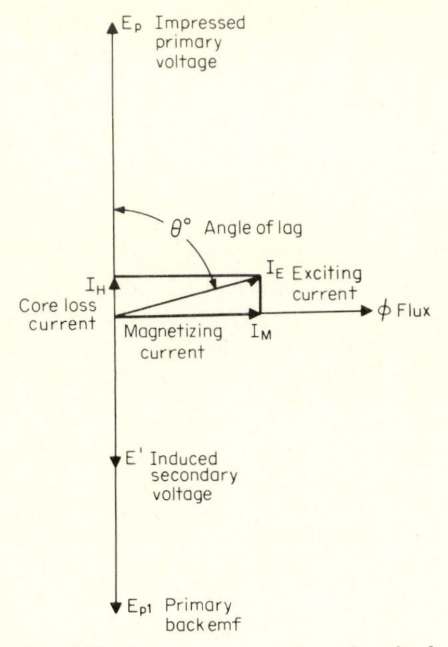

Fig. 5.12 Vector representation of no-load conditions.

example 5.9 When the secondary of a 120- to 240-V transformer is open, the primary current from a 120-V source is 0.24 A at a power factor of 0.16. The transformer is rated at 2 kVA. Calculate: (*a*) the primary no-load exciting current in percentage of full-load current; (*b*) the core-loss current in amperes and in percentage of full-load current; (*c*) the magnetizing current in amperes and in percentage of full-load current.

solution (*a*) The transformer being rated at 2000 kVA, full-load primary current is

$$\frac{2000}{120} = 16.666 \text{ A}$$

$$\frac{\text{No-load primary current} \times 100}{\text{Full-load primary current}} = \% \text{ of full-load current}$$

Then,

$$\frac{0.24 \times 100}{16.666} = 1.44\%$$

(*b*) Since the power factor is the cosine of the angle of lag between the exciting current and the core-loss current, core-loss current equals

$$I_E \cos\theta = I_E \times \text{PF}$$
$$= 0.24 \times 0.16 = 0.0384 \text{ A}$$

and

$$\frac{0.0384 \times 100}{16.666} = 0.23\%$$

(*c*) Reference to Fig. 5.12 will show that the magnetizing current I_m may be determined by use of the Pythagorean theorem. (The square of the hypotenuse of a right triangle equals the sum of the squares of the other two sides.) In this case,

$$I_E^2 = I_M^2 + I_H^2$$

or

$$0.24^2 = I_M^2 + 0.0384^2$$

Solving for I_M results in

$$I_M = \sqrt{0.24^2 - 0.0384^2} = 0.2369 \text{ A}$$

and
$$\frac{0.2369 \times 100}{16.666} = 1.4214\% \text{ of full-load current}$$

In these results, it should be noted that the magnetizing current has nearly the same value as the no-load primary current. A vector diagram showing the conditions of this problem is given in Fig. 5.13.

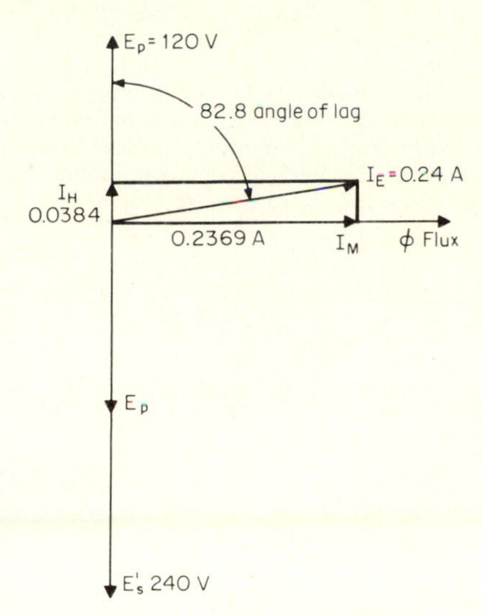

Fig. 5.13 Vector diagram of no-load conditions in the transformer of Example 5.9.

5.9 CONDITIONS UNDER LOAD

Figure 5.14 shows the transformer with a load. This illustration indicates the conditions existing at the moment when the terminal P_1 of the transformer is positive, and is increasing. Application of the right-hand rule for coils shows that the magnetic field produced by the primary is then in the direction shown in the illustration. By Lenz's law, the current in the secondary must be in such a direction that its associated magnetic field must oppose the building up of a magnetic field by the primary current. In other words, the magnetic field of the secondary will oppose the field of the primary when the field due to the primary current is building up.

Also, by Lenz's law, the induced secondary voltage must be in a direction that opposes the change that produced it. Hence, when a load is connected to the secondary

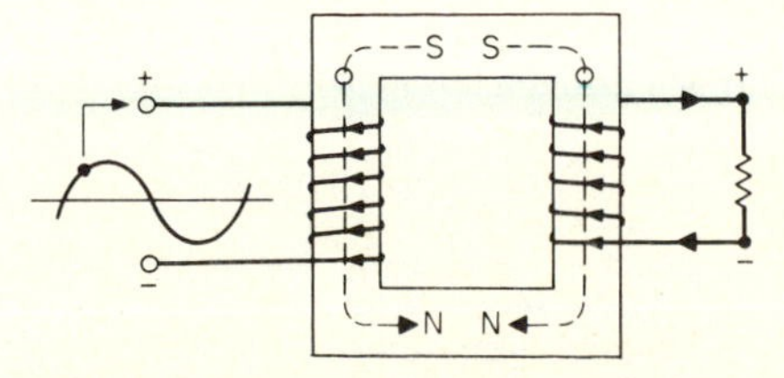

Fig. 5.14 As secondary load current increases, opposing magnetic field of secondary winding reduces inductance of primary, and more current flows into primary, thus maintaining core flux at its normal operating value.

winding, the current flowing in the secondary produces a magnetomotive force which opposes, or tends to oppose, the flux ϕ produced by the primary. Since an opposing flux, according to Eq. (5.6), results in a decrease in the back voltage of the primary, more current will flow into the primary as the secondary current increases! Therefore, any decrease of magnetization in the transformer core caused by current flowing in the secondary winding will immediately be counterbalanced by an increase in primary current. In this manner, the transformer automatically adjusts itself to load conditions, and the core flux is practically the same with full load as with no load.

Since the transformer is a highly efficient device, almost as much power is used up in the secondary as is fed to the primary. The power factor of the secondary, therefore, differs only slightly from that of the primary, and in most practical computations primary and secondary power factors may be considered equal to each other.

Because the core-loss and magnetizing current components of primary current are very much less than the load-current component, the core-loss and magnetizing current components may be added to the load-current component directly, instead of vectorially, except in those cases requiring the greatest accuracy. Such cases will be mostly of theoretical interest only.

example 5.10 The no-load current taken by a 120- to 240-V transformer is 0.6 A. The transformer is rated at 2.4 kVA. Assuming that the primary and secondary power factors are equal to each other, determine the primary current when the secondary is supplying its rated kilovoltamperes to a load that has a power factor of 0.9.

solution Full-load secondary current:

$$I_s = \frac{2400}{240} = 10 \text{ A}$$

Since it is assumed that full-load primary and secondary power factors are equal, power factor may be disregarded in the calculations.

The transformer voltage ratio being 1 to 2, the primary component of the load current will be

$$2 \times 10 = 20 \text{ A}$$

Adding this directly to the 0.6-A no-load, or core-loss and magnetizing current component, the result is 20.6 A.

example 5.11 Determine the ratio of full-load secondary power to full-load primary power for the transformer of Example 5.10.

solution

$$\begin{aligned}\text{Secondary power} &= 240 \times 10 \times 0.9 \\ &= 2160 \text{ W} \\ \text{Primary power} &= 120 \times 20.6 \times 0.9 \\ &= 2224.8 \text{ W}\end{aligned}$$

Ratio of full-load secondary power to full-load primary:

$$\frac{2400}{2472} = 0.9708$$

5.10 OPERATING A TRANSFORMER ON LOWER OR HIGHER THAN RATED FREQUENCY

Solving Eq. (5.8) for B_m (the maximum flux density in the core), we obtain

$$B_m = \frac{E_p \times 10^8}{4.44 \times f \times N_p \times A}$$

From this it may be seen that if we increase the voltage applied to the transformer primary, the flux will increase in proportion. The increased flux is, of course, due to the increased magnetizing ampere-turns, or magnetizing current.

For efficient operation, it is necessary to operate a transformer with a *maximum* flux density below the knee of the core magnetization curve, as shown in Fig. 5.15. Below the knee of the curve, an increase in magnetizing current will be out of proportion to the resulting increase in B_m. For example, with a magnetizing current of 0.5 A, the flux density will be 70 kG. If the applied primary voltage is increased to raise the flux density to 105 kG, the magnetizing current will rise to 2.0 A. Thus, as the flux density increased by 50 percent, the magnetizing current increased by 300 percent.

Equation (5.8) also shows that if the frequency f is decreased, B_m will again increase—and so will the primary current. Operating a transformer at considerably lower than rated frequency *could* cause the primary current to rise excessively, even to a value far in excess of the full-load primary current. The increased current would then burn out the winding.

From Eq. (5.8) it is evident that if the number of primary turns is increased, B_m decreases; but if the number of turns is decreased, B_m increases. Therefore, since B_m is directly proportional to the applied voltage and inversely proportional to the number of turns, changing the applied voltage and the number of turns in direct proportion would leave B_m and the primary current unchanged.

If we wanted to operate a transformer on a lower or higher than rated frequency, it would be impractical to change the number of its turns. However, since B_m is also directly proportional to the applied voltage and inversely proportional to the frequency,

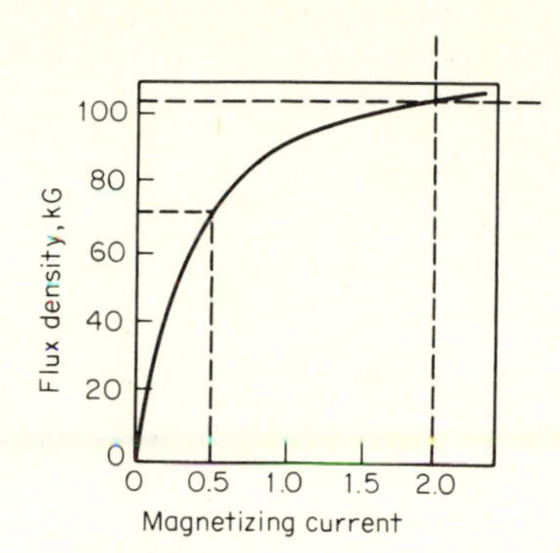

Fig. 5.15 Above the knee of the curve a large increase in magnetizing current results in a comparatively small increase in induction.

we could operate a transformer on lower than rated frequency, with normal flux, by decreasing the applied voltage in direct proportion to the decrease in frequency. For example, if we wished to operate a 120-V 60-Hz transformer at a frequency of only 25 Hz, it would be necessary to reduce the applied voltage to $^{25}/_{60}$ of 120, or to 50 V. The transformer would then still operate at its normal flux and safe current levels, but its power output—and input—capacity and its core losses would all be lower in almost direct proportion to the decrease in operating frequency. It follows that if a transformer is to be operated at normal flux and current levels but at a higher than rated frequency, its primary applied voltage must be increased. Its power output capacity and core losses would, consequently, also increase.

From the above relations between operating frequencies and voltages, we get the proportion

$$F_1 : F_2 = E_1 : E_2 \tag{5.11}$$

where F_1 = rated operating frequency of transformer
F_2 = changed operating frequency
E_1 = rated operating voltage of transformer
E_2 = changed operating voltage

A transformer may be safely operated at normal flux and current levels at any frequency if the ratio of applied voltage to frequency is kept at approximately its same, normal operating value.

example 5.12 A 120-V 60-Hz transformer is to be operated at a frequency of 400 Hz. What voltage should be applied to the primary if the transformer is to operate at normal flux and current levels?

solution By substitution in Eq. (5.11),

$$60{:}400 \simeq 120{:}E_2$$

and solving for E_2;

$$E_2 = \frac{120 \times 400}{60} = 800 \text{ V} \qquad \textit{Answer}$$

5.11 CONNECTING TRANSFORMERS IN SERIES AND PARALLEL

Occasions may often arise when it is necessary to connect two or more transformers in series or parallel, usually to obtain more current or more voltage than is available from only one transformer. It is essential that the transformers be properly phased when they are connected in series or parallel. Transformers may be damaged or ruined if they are connected together improperly.

SERIES OPERATION Connecting two transformers together to obtain a higher voltage is a very simple matter. It is only necessary to connect the primaries together in parallel, and connect their secondaries in series and to a voltmeter, as shown in Fig. 5.16. If the secondaries are properly phased, the meter will indicate the sum of

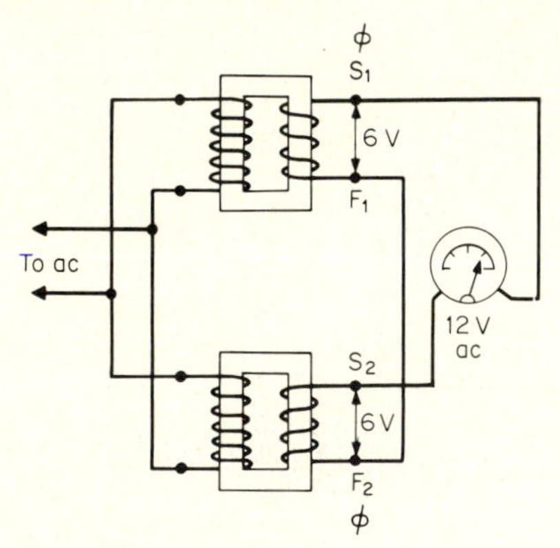

Fig. 5.16 Phasing two transformers for series or parallel operation.

the secondary voltages. If the meter indicates the difference of the secondary voltages, the connection to one of the secondaries may be reversed, or one of the primary windings may be reversed.

PARALLEL OPERATION If two transformers are to be connected in parallel, the transformers should be alike in every respect. They should have the same number of primary and secondary turns and the same secondary voltages, and the secondary voltages should remain the same under load. If the secondaries of two transformers having unequal voltages were to be connected in parallel, the transformer with the higher secondary voltage would, in addition to supplying its load, also send a current through the secondary winding of the other transformer. The result would be a large copper (I^2R) loss in the windings, which might produce enough heat to damage the winding insulation.

Even though two transformers may have the same primary and secondary voltages, they may differ in other respects. One winding may have a greater internal drop than the other when the transformers are supplying a load, and this would cause a difference in their terminal voltages under load.

The usual reason for connecting two transformers for parallel operation is that one does not have the necessary current-delivering capacity.

PHASING TRANSFORMERS FOR PARALLEL OPERATION If the secondaries of two transformers are not properly phased for parallel operation, the result is a short circuit that burns out the windings. Even if two transformers of the same make have the same numbered or lettered terminals, it should not be assumed that they will be properly phased for parallel operation.

To phase two transformers for parallel operation, we connect the primaries in parallel and the secondaries in series—just as we would in phasing the transformers for series operation—as shown in Fig. 5.16. The secondary terminals of one of the transformers are arbitrarily tagged S_1 and F_1 and those of the other transformer S_2 and F_2. If the voltmeter across the terminals, S_1 and S_2, reads twice the voltage of either transformer, we connect terminal S_1 to F_2 and terminal F_1 to S_2. However, if the voltmeter reading is zero, S_1 and F_2 have opposite polarities, and we connect S_1 to S_2 and F_1 to F_2.

The phasing of transformers with high-voltage secondaries or primaries should be done with lowered excitation voltages in order to reduce high-voltage shock hazard.

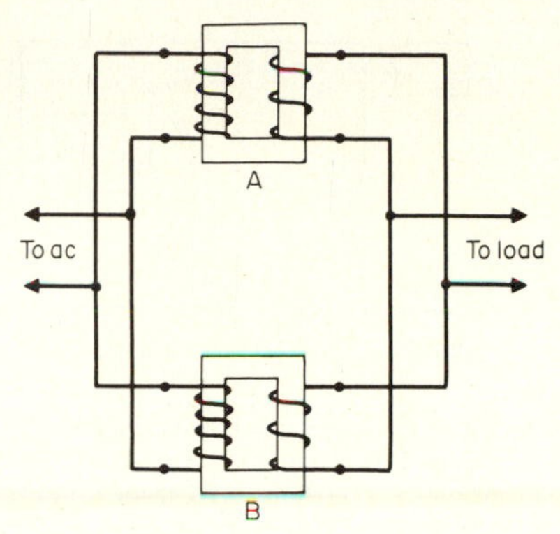

Fig. 5.17 What will be the result if connections to one of the primaries are reversed?

example 5.13 Two transformers are connected together for parallel operation, as in Fig. 5.17. If the connections to the primary of transformer *A* are accidentally reversed, what will be the result?

solution The polarity of the secondary winding of transformer *A* will also be reversed, and the result will be a short circuit of the secondary windings.

5.12 TRANSFORMER LOSSES

Power losses in a transformer occur in the transformer core material and in the transformer primary and secondary windings.

Eddy-current losses, as explained in Chap. 4, are the result of currents circulating in the core and producing heat. The circulating current in the core is set up by induced voltages in the core material. These voltages are the result of the changing magnetic flux.

Hysteresis loss is the energy lost by reversing the magnetic field in the core as the magnetizing alternating current rises and falls and reverses direction. Hysteresis loss was also described in Chap. 4.

COPPER LOSS Copper loss is the power lost in the primary and secondary windings of the transformer. It is due to the ohmic resistance of the windings. Copper loss is obtained by the equation

$$\text{Copper loss} = I_p^2 R_p + I_s^2 R_s$$

where R_p = resistance of primary
R_s = resistance of secondary
I_p, I_s = respective primary and secondary currents

DETERMINATION OF COPPER LOSS Copper loss may be determined by measuring the ohmic resistance of each winding and then computing the I^2R loss in both primary and secondary windings for any given secondary load current.

The copper loss in both windings may also be determined by means of a wattmeter, and a circuit diagram showing the proper connections for this method is given in Fig. 5.18. If a transformer has multiple primaries, secondaries, or both, all the coils must be connected into the circuit for this test.

At the start of the test, the variable-voltage autotransformer is set at zero; its setting is then gradually raised until the ammeter connected across the secondary indicates rated full-load secondary current (assuming that we wish to determine copper loss at full load). At this point the wattmeter will indicate the power loss in both windings and also a very small core loss. Usually the core loss is too low to be indicated on the wattmeter.

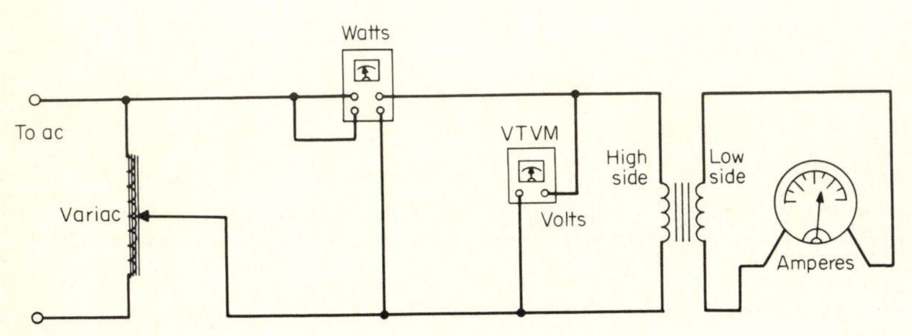

Fig. 5.18 Short-circuit test for copper loss.

Although rated primary and secondary currents flow through the windings when this method is used, core loss is negligible because the voltage impressed on the primary has only a very low value—perhaps less than 5 percent of normal. This low primary voltage is still enough to cause rated secondary current to flow in the secondary winding, which is practically short-circuited by the ammeter connected across its terminals. Since the transformer is operating at only about five percent of rated voltage, the maximum flux density B_m in the core is also only about five percent of normal. Since core loss is almost directly proportional to B_m^2, the core loss with five percent of normal impressed primary voltage will be 0.05×0.05 or 0.0025 ($^{25}/_{100}$ of 1 percent) of normal. As an example, if normal core loss in a transformer is 50 W, the loss on short circuit with five percent of rated primary voltage will be $0.05^2 \times 50$, or 0.125 W, a value too small to be noticeable on the wattmeter.

The above test for copper loss is known as the *short-circuit test* method. Copper loss by this method may be determined from either the primary or secondary side of the transformer; the high side will usually be more convenient because on short circuit it would take only about five percent of full-rated voltage to cause full-load current to flow through the windings, but it might not always be convenient to adjust the applied voltage to the proper low value needed for a test from the low side.

CORE-LOSS TEST Core loss in a transformer may be determined by applying the rated voltage to the transformer primary at its rated frequency and measuring the power input with the secondary open-circuited. Power input is measured by means of a wattmeter. Figure 5.19 shows a diagram of the necessary connections. The variable voltage autotransformer is adjusted to apply rated voltage to the transformer primary. This will be indicated by the VTVM (or other high-resistance meter) connected across the primary. (If a low-resistance voltmeter were to be used to measure the applied voltage, its power consumption would be added to the core-loss power, and the wattmeter would then give a false core-loss reading. Therefore, if a low-resistance voltmeter is used to measure the applied voltage, the wattmeter reading should be taken with the circuit to the voltmeter open. The input power measured by the wattmeter will include a small copper loss, but at no load this is of no consequence and may be ignored.

If the rated primary voltage of the transformer has a very high value, or whenever it is higher than the secondary voltage, core loss may be determined by applying rated secondary voltage to the transformer. This is called the *open-circuit test,* and it is

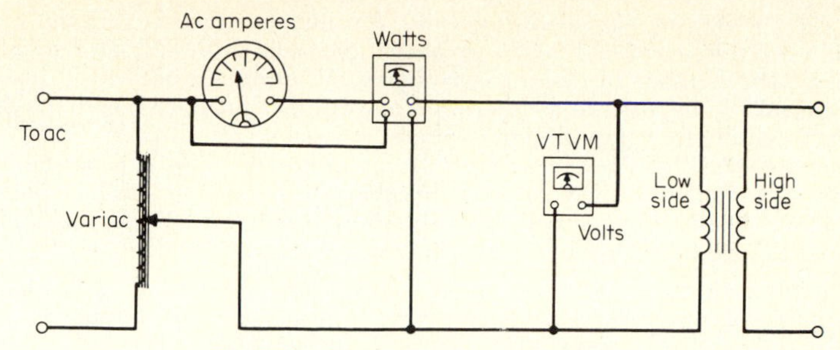

Fig. 5.19 Open-circuit test for core loss.

usually more convenient and safer for any step-down transformer. It gives the same results as a core-loss measurement taken on the high or primary side of the transformer.

example 5.14 A 240- to 24-V step-down transformer has a full-load secondary current rating of 50 A. A short-circuit test for copper loss at full load gives a wattmeter reading of 100 W. If the resistance of the primary is 0.8 Ω, what is the resistance of the secondary, and what is the secondary power loss? (Ignore the small core loss.)

solution Since full-load secondary current is 50 A, and the voltage ratio of the transformer is 240 to 24, or 10 to 1, the full-load primary current is $^{50}/_{10}$ or 5 A. Total copper loss equals

$$100 \text{ W} = I_p^2 R_p + I_s^2 R_s = 5^2$$

Then,

$$R_s = \frac{100 - 5^2(0.8)}{50^2} = \frac{100 - 20}{2500}$$

$$= \frac{80}{2500} = 0.032 \ \Omega$$

Secondary power loss is

$$50^2 \times 0.032 = 80 \text{ W}$$

example 5.15 On an open-circuit test for core loss, the transformer of Example 5.14 takes 1.2 A from a 240-V ac source. If the wattmeter reading is 80 W, determine (*a*) the copper loss at zero load; (*b*) the core loss.

solution (*a*) Since the primary resistance is 0.8 Ω, and the secondary is an open circuit, copper loss occurs only in the primary and is equal to

$$I_p^2 R_p = 1.2^2 \times 0.8 = 1.15 \text{ W}$$

(*b*) The wattmeter reading minus the copper loss equals the core loss, or

$$80 - 1.15 = 78.8 \text{ W}$$

5.13 EFFICIENCY

The efficiency of a transformer, like the efficiency of any machine, is the ratio of power output to power input, or in the form of an equation:

$$\text{Efficiency} = \frac{\text{power output}}{\text{power input}}$$

$$= \frac{\text{power output}}{\text{power output} + \text{copper loss} + \text{core loss}}$$

$$= \frac{E_s I_s \times \text{PF}}{(E_s I_s \times \text{PF}) + \text{copper loss} + \text{core loss}} \quad (5.12)$$

where PF = power factor of the load.

It might be assumed that transformer efficiency may be determined by connecting wattmeters in the primary and secondary of a transformer—that is, supplying a load, and then dividing the secondary wattmeter reading by that of the primary. However, this would be an incorrect assumption. Transformer efficiency is very high, and therefore the copper and core losses are very low. A wattmeter connected in the secondary circuit of a transformer would read very nearly the same as a wattmeter connected in the primary, and because of meter inaccuracies and errors introduced in reading the meters, it would be impossible to determine power input and output with the degree of precision necessary for computing transformer efficiency.

To compute efficiency, core loss must be determined. This can be done by the core-loss test method described earlier in this chapter. Copper loss may be determined by the measured resistance method, or by the short-circuit test. These core and copper loss factors are substituted in Eq. (5.12), which will then give the efficiency.

example 5.16 An open-circuit test for core loss in a 240- to 720-V 5-kVA transformer gives a value of 60 W. The measured resistance of the low side winding (primary) is 0.025 Ω, and that of the high side (secondary) is 1.25 Ω. If the power factor of the load is 0.84, what is the efficiency of the transformer at full load?

solution Full-load secondary current 5000/720; and full-load secondary copper loss,

$$\left(\frac{5000}{720}\right)^2 1.25 = 60.28 \text{ W}$$

Full-load primary current 5000/240; and full-load primary copper loss,

$$\left(\frac{5000}{240}\right)^2 \times 0.025 = 10.85 \text{ W}$$

Total copper loss,

$$60.28 + 10.85 = 71.1 \text{ W}$$

Efficiency,

$$\frac{5000 \times 0.84}{(5000 \times 0.84) + 60 + 71.1} = 0.971 \text{ or } 97.1\%$$

5.14 EQUIVALENT RESISTANCES

In certain computations it is sometimes useful to assume that the copper loss in both primary and secondary windings of a transformer actually occurs in only the primary; this is equivalent to assuming that the wattmeter in a short-circuit test for copper loss actually indicates power being expended in a purely resistive load. In the form of an equation this may be written as

$$W = I_p^2 R_{ep}$$

or

$$R_{ep} = \frac{W}{I_p^2} \tag{5.13}$$

where W = wattmeter reading on a short-circuit test
I_p^2 = primary current squared
R_{ep} = equivalent resistance of transformer

It is important to note that R_{ep} in this case is *not* the dc resistance of the primary, but merely an assumed resistance that would give the same IR drop as the *combined IR* drops of both primary and secondary windings.

It is also possible to express the equivalent primary resistance in terms of the resistances of both primary and secondary windings in the form

$$R_{ep} = R_p + R_s\left(\frac{N_p}{N_s}\right)^2 \tag{5.14}$$

And it is possible to express the *equivalent secondary resistance* R_{es} also in terms of both primary and secondary resistances, as follows:

$$R_{es} = R_s + R_p\left(\frac{N_s}{N_p}\right)^2 \tag{5.15}$$

From Eqs. (5.14) and (5.15) we also get

$$\frac{R_{ep}}{R_{es}} = \frac{N_p^2}{N_s^2} \tag{5.16}$$

example 5.17 A 1200- to 120-V transformer is rated at 5 kVA. The primary resistance is 1 Ω, and secondary resistance is 0.05 Ω. Determine (*a*) the equivalent primary resistance R_{ep}; (*b*) the equivalent secondary resistance R_{es}; (*c*) the total copper loss I^2R, using the determined equivalent secondary resistance; (*e*) the total copper loss, using the dc resistances of the primary and secondary windings.

solution (*a*) The equivalent primary resistance,

$$R_{ep} = 1 + 0.05 \left(\frac{N_p}{N_s}\right)^2$$

Since the voltage ratio of the transformer is 1200/120, or $^{10}/_1$, N_p/N_s also equals $^{10}/_1$, or 10. Therefore

$$R_{ep} = 1 + 0.05(10)^2$$
$$= 6\ \Omega$$

(*b*) The equivalent secondary resistance,

$$R_{es} = 0.05 + 1 \left(\frac{N_s}{N_p}\right)^2$$

$$= 0.05 + (^1/_{10})^2 = 0.06\ \Omega$$

(*c*) Primary current,

$$I_p = \frac{5000}{1200} = 4.175\ \text{A}$$

Copper loss computed from R_{ep} and I_p values equals

$$4.175^2 \times 6 = 104.58\ \text{W}$$

(*d*) Secondary current,

$$I_s = \frac{5000}{120} = 41.75\ \text{A}$$

Copper loss computed from R_{es} and I_s values equals

$$41.75^2 \times 0.06 = 104.58\ \text{W}$$

(*e*) Primary copper loss equals

$$(I_p^2R_p) = 4.175^2 \times 1 = 17.43\ \text{W}$$

Secondary copper loss equals

$$(I_s^2R_s) = 41.75^2 \times 0.05 = 87.15\ \text{W}$$

Total copper loss,

$$I_p^2R_p + I_s^2R_s = 17.43 + 87.15 = 104.58\ \text{W}$$

5.15 VOLTAGE REGULATION

The voltage regulation of a transformer in percent of the rated full-load secondary voltage is given by the equation

$$\text{Percent regulation} = \frac{\text{no-load secondary volts} - \text{full-load secondary volts}}{\text{full-load secondary volts}} \times 100 \quad (5.17)$$

The voltage regulation of small transformers may easily be determined by adjusting the primary voltage so that the transformer delivers its rated kilovoltamperes at its rated secondary voltage to a load at some specified power factor. Then, disconnecting the load but keeping the primary voltage constant will cause the secondary voltage to rise to its no-load value. Substituting the observed values of full-load and no-load voltages in Eq. (5.17) will give the percent regulation.

The above method, however, would not be very convenient for large heavy-duty transformers, especially when such transformers may also have high-voltage primary or secondary windings. For such transformers it is better to compute regulation from the short-circuit test data. The computation for a load having either a lagging or leading power factor is somewhat more complicated than for a load having unity power factor.

Regulation may be computed in terms of either the primary or secondary winding, and the first step, in either case, is to compute the equivalent IR and IX drops for the winding from the short-circuit test data. Determining regulation in terms of the secondary winding is equivalent to assuming that the secondary is really the primary; and regardless of whether we compute regulation in terms of either the primary or secondary winding, the no-load voltage will always be greater than the rated full-load voltage of the transformer.

Vector diagrams for computing regulation from known values of $\mathbf{E}_R$, I_p, and $\mathbf{IX}$ are shown in Fig. 5.20. A vector diagram for a transformer supplying a unity power factor load is shown in Fig. 5.20*a* and for a load with a lagging power-factor in Fig. 5.20*b*. The vector $\mathbf{OE}_R$ represents the rated full-load voltage (primary or secondary) of the transformer. Vectors **IR**, **IX**, and **IZ** are the equivalent (primary or secondary) resistance, reactance, and impedance drops. Vector $\mathbf{OE}_I$ is the vector sum of $\mathbf{E}_R$ and **IZ**, and it represents the no-load voltage of the transformer. In (*a*) the current I is in phase with $\mathbf{E}_R$, since the load has a power factor of unity. In (*b*) the current I lags $\mathbf{E}_R$ by the angle θ, corresponding to the power factor (cosine of the angle of lag) of the load. In

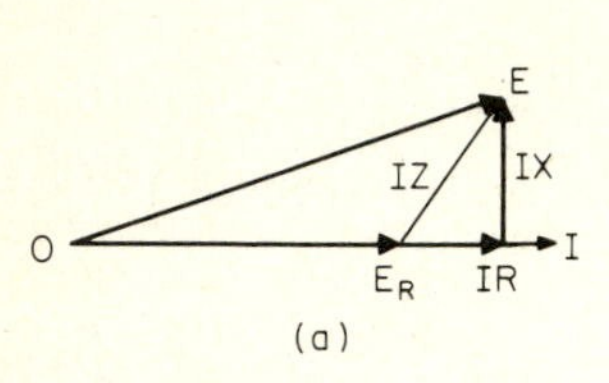

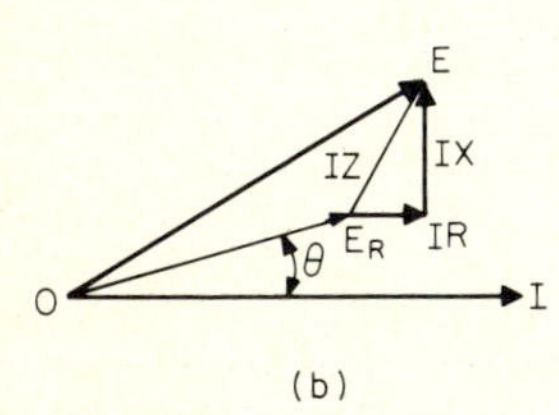

Fig. 5.20 Vector diagrams for computing regulation from known values of $\mathbf{E}_R$, $\mathbf{I}_p$, and **IX**. $\mathbf{E}_R$ equals the rated voltage of the transformer. **IR**, **IX**, and **IZ** are the equivalent values of resistance, reactance, and impedance. (*a*) Unity power-factor load. (*b*) Lagging power-factor load.

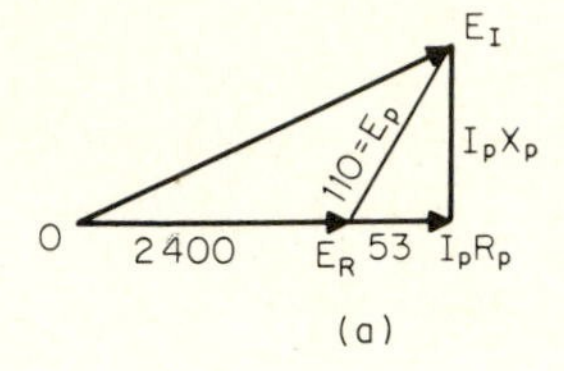

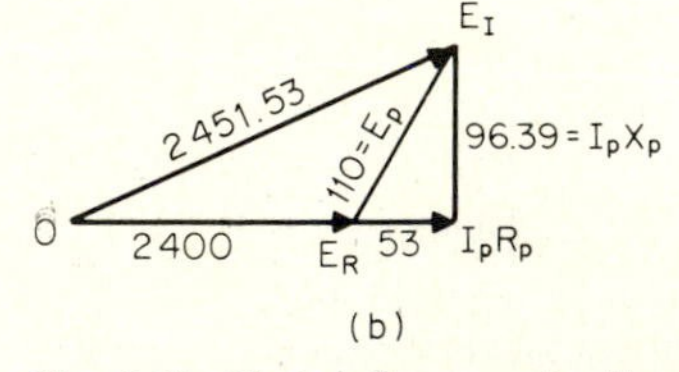

Fig. 5.21 Vector diagrams for Example 5.18. Computing $\mathbf{OE}_I$, the no-load voltage, is a problem in simple trigonometry. (*a*) Partially solved problem. (*b*) Completed problem.

either case, the problem is to find the value of the vector $\mathbf{OE}_I$, the no-load voltage of the transformer, from the known values of $\mathbf{E}_R$, I_p, and **IX**.

example 5.18 On a short-circuit test, a 2400- to 240-V, 10 kVA transformer takes 4.17 A at an applied primary voltage of 110 V. The wattmeter reading is 221 W. Assuming a unity power factor load, compute the regulation in terms of the primary. Draw a vector diagram illustrating the conditions of the problem, and fill in the unknown values as they are calculated.

solution Since the power factor of the load is unity, the applied voltage and the current will be in phase. A vector diagram of circuit conditions will, therefore, be similar to the one shown in Fig. 5.20*a*. The voltmeter reading is the equivalent primary impedance times the primary current:

$$I_p Z_{ep} = 110 \text{ V}$$

The equivalent primary resistance in volts is

$$I_p R_{ep} = \frac{W}{I_p} = \frac{221}{4.17} = 53 \text{ V}$$

Figure 5.21 shows the vector diagrams for this problem. Figure 5.21*a* illustrates the conditions of the problem at this stage. By simple trigonometry,

$$(I_pX_p)^2 = 110^2 - 53^2$$

Hence, the equivalent primary reactance in volts is

$$I_pX_p = \sqrt{110^2 - 53^2} = 96.39 \text{ V}$$

Since $\mathbf{E}_R$ is in phase with I_pR_p, we may add $\mathbf{E}_R$ to I_pR_p directly, or

$$\mathbf{E}_R + I_pR_p = 2400 + 53 = 2453$$

and again using simple trigonometry,

$$(\mathbf{OE}_l)^2 = 2453^2 + (I_pX_p)^2$$
$$= 2453^2 + 96.39^2$$

Solving for the no-load voltage

$$\mathbf{OE}_l = \sqrt{2453^2 + 96.39^2}$$
$$= 2451.53 \text{ V}$$
$$\text{Regulation} = \frac{2451.53 - 2400}{2400} \times 100 = 2.147\%$$

Figure 5.21*b* is a vector diagram of the completed problem.

5.16 THE ISOLATION TRANSFORMER

An isolation transformer is used whenever it is necessary to avoid a direct electric connection between a piece of electric equipment and the power lines or other source of power. This is useful and very often even necessary where equipment must be grounded for increased efficiency, to reduce noise, or for other reasons. Since one side of the ac power line is grounded, connecting a piece of grounded equipment to the lines could result in a short circuit unless precautions are taken to assure that the grounded side of the equipment is connected to the grounded side of the power lines. There are also certain types of equipment—such as those with ac-dc power supplies—in which a lethal shock *could* be received when the chassis is exposed. (It would be necessary to touch a ground point *and* the exposed chassis to receive such a shock.) By using an isolation transformer, the chassis becomes a floating ground, and it is nearly impossible to receive a deadly shock under such conditions.

Since equipment may frequently be used in different locations and because power outlets in the same building or even in the same room are not usually polarized, a transformer between the equipment and the power lines is a very convenient and practical method of avoiding the possibility of damaging a piece of electric equipment that must be grounded. Since the transformer serves to isolate the equipment from the power lines, it is called an isolation transformer; and it is evident that any transformer may be used as an isolation transformer if it has the necessary voltampere rating and the required primary to secondary turns or voltage ratio.

5.17 THE AUTOTRANSFORMER

In an ordinary transformer the primary and secondary windings are magnetically coupled, but there is no direct electric connection between the primary and secondary windings. In an autotransformer, the primary and secondary windings are not only magnetically coupled, but there is also a direct electric connection between the windings. In fact, the secondary winding may be merely a continuation of the primary.

Any ordinary transformer may be used as an autotransformer. For example, consider the 1-to-1 isolation transformer of Fig. 5.22 in which there is a direct series connection between the primary and secondary windings. With the transformer phased as shown, the impressed primary voltage is added to the induced secondary voltage, and the total voltage across the load, connected to terminals A and Y, is

$$120 + 120 = 240 \text{ V}$$

Thus we have a step-up ratio of 1 to 2 in a transformer with a primary to secondary turns ratio of only 1 to 1. Figure 5.23 shows some additional connections of autotransformers in which a transformer with a 1-to-1 ratio is used to step up the voltage (Fig. 5.23*a*), and to step down the voltage (Fig. 5.23*b*). In Fig. 5.23*a* the windings of the transformer are wound on a straight, open core, and this more clearly illustrates the conditions of the transformer of Fig. 5.22. This same transformer may also be used as a step-down autotransformer by connecting the load across terminals *A* and *B* and connecting terminals *A* and *Y* to a power source, as shown in Fig. 5.23*b* which merely interchanges the line and load connections of Fig. 5.23*a*. In other words, the transformer of Fig. 5.23*a* is operated in reverse—just as we might operate any transformer in reverse by using the secondary, with rated applied voltage, as the primary and then connecting the load to the normal primary.

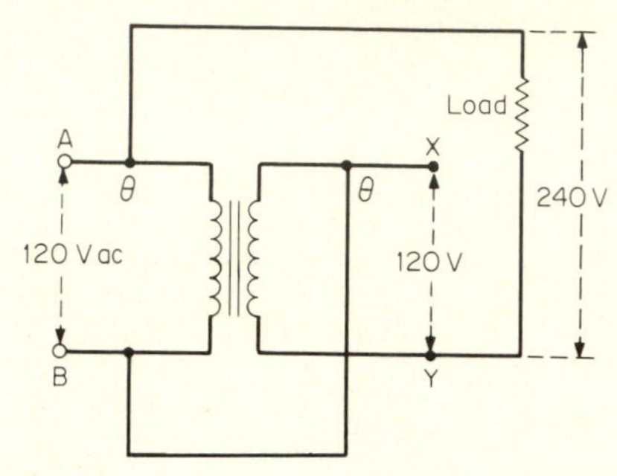

Fig. 5.22 A 1-to-1 isolation transformer, connected for use as an autotransformer, gives a stepped-up voltage ratio of 1 to 2.

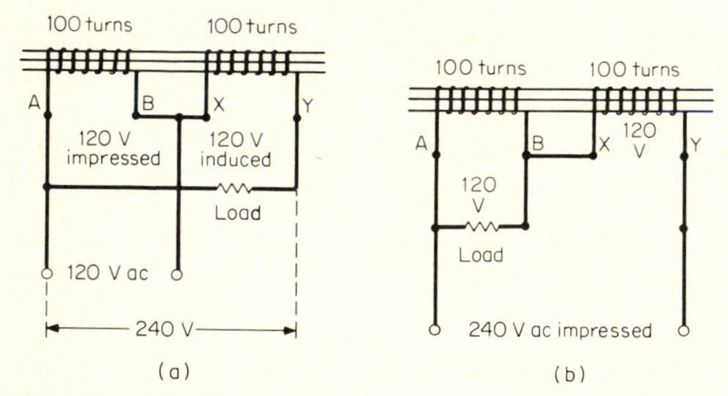

Fig. 5.23 Some additional connections of autotransformers: (*a*) Autotransformer connection steps up impressed primary voltage from 120 to 240 V. (*b*) Same 1:1 ratio transformer used in reverse steps down impressed line voltage of 240 to 120 V.

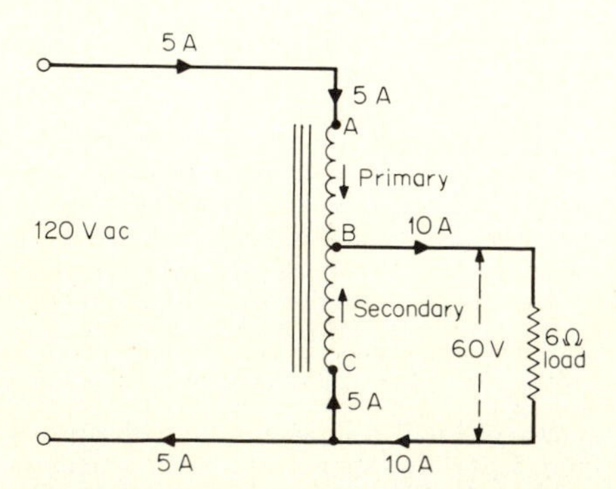

Fig. 5.24 Autotransformer supplying a load. Current *transformed* is 5 A, and in this case the current *transferred* is also 5 A.

A schematic diagram of an autotransformer supplying a load is shown in Fig. 5.24. In this case we have a 2-to-1 step-down transformer. The number of turns between A and B is the primary, and the turns between B and C the secondary. If we assume that the transformer is 100 percent efficient and supplies a load having unity power factor, power input to the transformer will be

$$120 \times 5 = 600 \text{ W}$$

Power output will be

$$60 \times 10$$

which also equals 600 W, but the transformer primary power is only 60 × 5, or 300 W. This is also the transformer secondary power. Hence, the power *transformed*, by transformer action, is only 300 W; and the power *transferred* directly to the secondary load, by means of the connection between the primary and secondary windings, is the remaining 300 W, making the total of 600 W supplied to the load.

Since, in this case, each half of the autotransformer supplies five amperes to the load, copper loss is only one-half the copper loss that would occur in a transformer with separate primary and secondary windings and supplying the same load with the same current.

As the primary to secondary voltage ratio of an autotransformer decreases, the copper loss decreases in proportion, and the amount of power *transferred* directly—instead of by transformer action—increases in proportion.

Autotransformers are seldom used when it is necessary to reduce a very high voltage to a much lower value. For example, a voltage ratio of 1200 to 120, or 10 to 1, would result in a small reduction of copper loss, and the power *transferred* directly would also be small. Furthermore, a dangerous shock hazard would exist in the secondary, since there would be no isolation between the high-voltage primary and the low-voltage secondary.

If an ordinary transformer having several windings is to be used as an autotransformer by connecting all the windings together, the current carried by any winding must not be greater than its normal rated current.

example 5.19 The windings of an ordinary 1200- to 120-V 5-kVA transformer are connected in series to form a step-up autotransformer, as shown in Fig. 5.25. Assuming that the

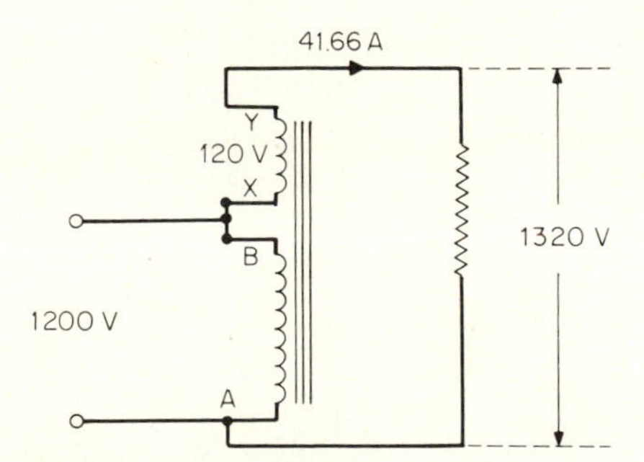

Fig. 5.25 Ordinary transformer connected as autotransformer; in this case power capacity is increased for 5000 to 55 000 W.

transformer is 100 percent efficient and operates at a power factor of unity with rated secondary current, determine (*a*) the power delivered to the load; (*b*) the total current supplied by the source; (*c*) the increase in the power rating of the transformer when it is used as an autotransformer; (*d*) the power transferred.

solution (*a*) Since rated secondary current is

$$\frac{5000}{120} = 41.66 \text{ A}$$

the current in the load may also be 41.66 without overloading the secondary winding, XY. Also, since the voltage across the load is 1200 + 120 or 1320 V, the power supplied to the load is

$$1320 \times 41.66 = 55\,000 \text{ W}$$

(*b*) Since we have assumed 100 percent efficiency, power supplied by the source is equal to the power delivered; current taken from the source is, therefore,

$$\frac{55\,000}{1200} = 45.83 \text{ A}$$

(*c*) Normal power rating of the transformer is 5000 W. Power rating of the transformer when used as an autotransformer is 55 000 W. Percentage increase in power rating,

$$\frac{55\,000}{5000} \times 100 = 1100\%$$

(*d*) Power transferred is

$$55\,000 - 5000 = 50\,000 \text{ W}$$

5.18 THE VARIABLE AUTOTRANSFORMER

The variable autotransformer is usually better known under one of its commercial trade names, such as Variac, Powerstat, Adjust-A-Volt. But regardless of what we call it, its basic principle of operation remains the same. Figure 5.26 is a photograph of a commonly used Variac. The exploded view clearly shows all the basic parts. Figure 5.27 shows several autotransformer circuits.

Figure 5.27*a* illustrates the most basic type of variable autotransformer. The coil is a single-layer winding over a toroidal core of high permeability. The sliding brush contact, which is controlled by its attached knob or dial, may be set to tap the winding at any point about its periphery. Thus, the output voltage across the load, connected to terminals 1 and 2, may be varied from zero to full line voltage. A slight modification of the design, as shown in Fig. 5.27*b*, permits the voltage to be varied from zero to slightly (approximately 17 percent) above line voltage. This feature is useful when-

Fig. 5.26 An exploded view of a Variac. The brush, shown at lower left, would normally contact the front outer perimeter of the winding. (*Courtesy General Radio Company.*)

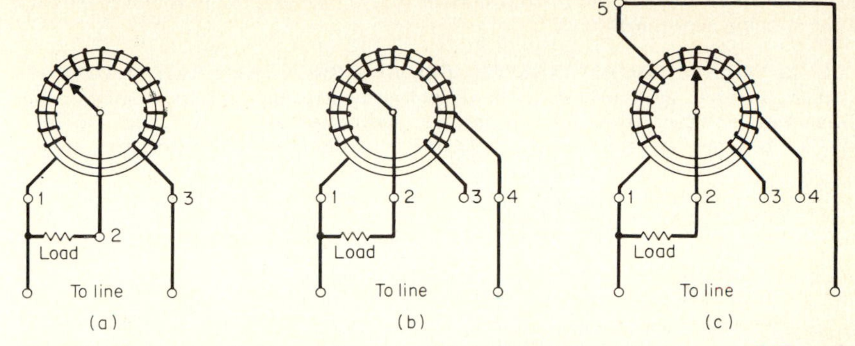

Fig. 5.27 Connections for autotransformers: (*a*) The basic variable autotransformer. (*b*) Tapped winding permits adjustment from zero to slightly above line voltage. (*c*) Tapping the winding at a lower point permits adjustment from zero to more than twice the line voltage.

ever it is necessary to compensate for the drop in line voltage that occurs when the power lines are heavily loaded. But if necessary, this design may also be connected for a range of zero to line voltage by joining terminals 1 and 3, instead of 1 and 4, to the lines.

An additional modification of the basic design, as shown in Fig. 5.27*c*, gives a voltage range adjustable from zero to slightly more than twice the line voltage (from 0 to 280 V for one model). Terminal 5 is connected to a tap on the winding at a point slightly less than midway between the total number of turns.

The rated current of a variable autotransformer should not be exceeded at any voltage setting of the transformer, and it should not be assumed that greater than rated current may safely be drawn at voltage settings below line voltage—as we might use an ordinary step-down transformer to supply a much greater current to a secondary load than the current supplied to the primary.

When the variable autotransformer is used to supply a fixed, known load, without using the overvoltage connection, maximum current may be supplied to the load at full line voltage. Then as the voltage setting is reduced, the decrease in current with decreasing output voltage will tend to keep the load current within safe limits for any voltage setting. Accidental overload damage may easily be avoided by fusing the variable autotransformer for rated current.

When it is necessary to supply a load at some reduced voltage but with a current considerably in excess of the normal current rating of a particular variable autotransformer, an auxiliary transformer, connected to the output terminals of the variable autotransformer, may be used, as shown in Fig. 5.28. For the transformer and circuit of Fig. 5.28, the primary-to-secondary voltage ratio of the transformer is 12 to 1, and normal rated current of the variable autotransformer is two amperes. Current taken by the load may, therefore, be as great as 24 A at voltages between 0 and 10 V. In this case, the use of an auxiliary transformer has increased current capacity 12 times. In any application of this

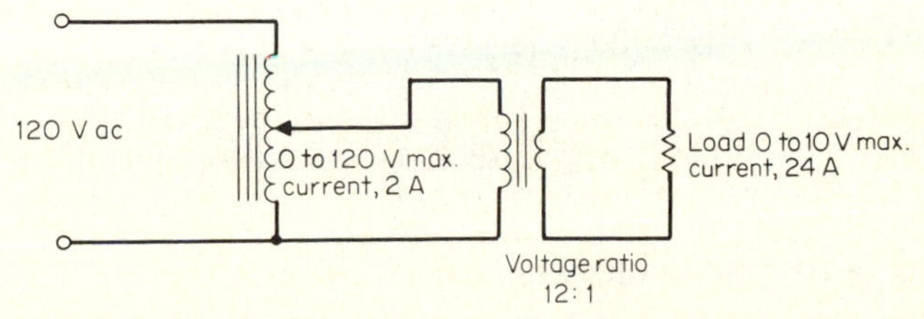

Fig. 5.28 The use of supplementary transformer in conjunction with low-current variable autotransformer increases current capacity 12 times.

sort, the voltampere rating of the transformer should be equal to the voltampere rating of the variable autotransformer.

USING VARIABLE AUTOTRANSFORMERS AT HIGHER THAN RATED VOLTAGES A single variable autotransformer should not, of course, be operated at higher than normal rated voltage. However, two variable autotransformers may be ganged for high-voltage operation, as shown in Fig. 5.29. Thus, if the normal voltage rating of each

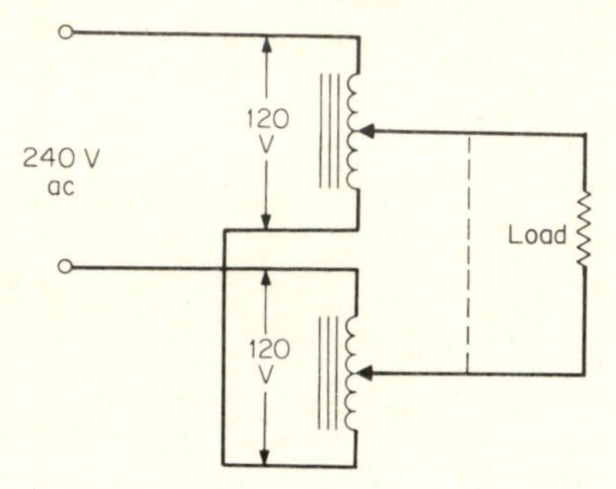

Fig. 5.29 Two 120-volt variable autotransformers ganged for operation on 240 V.

transformer is 120 V, the ganged transformers may be used at 240 V, or two 240-V models may be used at 480 V.

SWITCHING Before a variable autotransformer is switched into or out of a circuit, its setting should be reduced to zero to prevent excessive surge currents from damaging the windings. This is especially important when a variable autotransformer is used to supply a load that has a very high ratio of hot-to-cold resistance. For example, the hot-to-cold resistance ratio of the heaters or filaments of vacuum tubes or of incandescent lamps may be as great as 15 to 1.

example 5.20 A 120-V 1-to-1 isolation transformer is available. Draw a diagram showing how this transformer should be connected to a 120-V Variac and to a load that is to be operated at various voltages ranging from 0 to 240 V.

solution The primary winding of the transformer is connected to the Variac, and then the transformer itself is connected to form an autotransformer, as shown in Fig. 5.30. This gives a

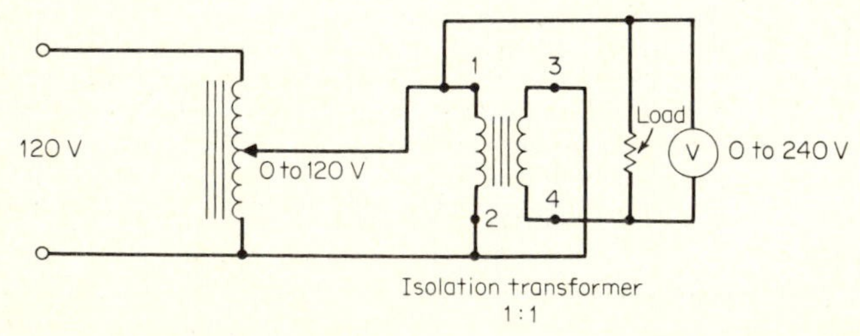

Fig. 5.30 Increasing the voltage range of a Variac.

step-up ratio of 1 to 2. If the voltmeter indicates zero as the Variac setting is increased, the connections to either terminals 1 and 2 or 3 and 4 should be reversed – for proper phasing.

5.19 INSTRUMENT TRANSFORMERS

Instrument transformers are designed for use with measuring instruments, relays, or control devices. They are especially designed to maintain a specific relationship in magnitude and phase between the primary and secondary currents or voltages.

THE POTENTIAL TRANSFORMER Since it would hardly be feasible to construct voltmeters to directly measure the extremely high voltages encountered in power distribution systems, such high voltages cannot be measured by connecting a voltmeter directly across the lines. Even if voltmeters having the necessary range and insulation resistance were available, it would be very dangerous to bring leads from the high-voltage lines directly into the instrument switchboards and to the meters. A dangerous high-voltage shock hazard would then be a constant threat to personnel at the switchboard. Therefore, instead of bringing the high-voltage leads directly to a meter, these leads are connected to one side of a small transformer, called a potential transformer, as shown in Fig. 5.31. In Fig. 5.31*a* the potential transformer steps down the high voltage to a very much lower value that may be safely brought to a low-voltage meter. Usually, the meter has its scale calibrated to indicate the high voltage across the transformer's primary terminals. Since the potential transformer has very high primary-to-secondary insulation resistance, there is no high-voltage shock hazard at the meter terminals.

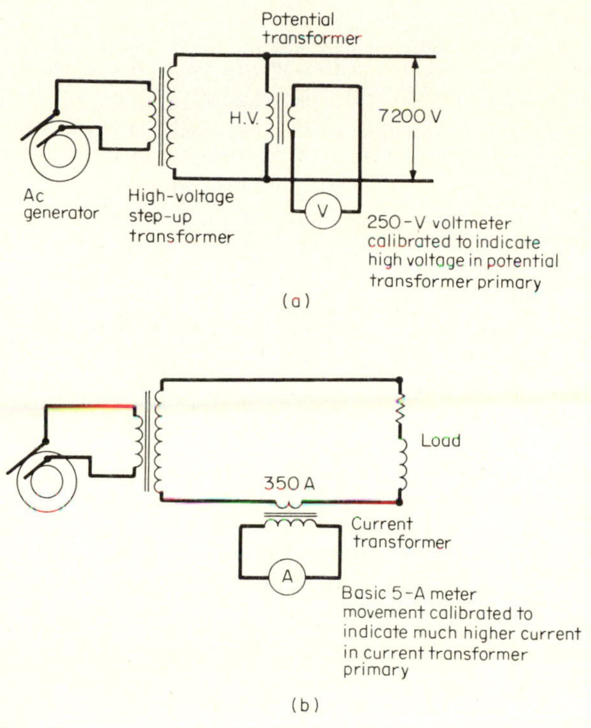

Fig. 5.31 Instrument transformers connected to measure high voltage and current: (*a*) Potential transformer. (*b*) Current transformer.

THE CURRENT TRANSFORMER A current transformer, instead of being used to step down a high voltage, is used to step down a high current to a much smaller and more conveniently measurable value.

The method of connecting a current transformer into the line is shown in Fig. 5.31*b*. Since a current transformer is always connected in series with the line, it is also known as a *series* transformer. The primary, or line side, of a current transformer usually has only a few turns—sometimes only one—since it must carry a very large current. The secondary, which is connected to a low-range ammeter, has a much larger number of turns. The meter scale is calibrated to indicate the line current in the primary of the current transformer. This line current is, of course, very much greater than the current in the secondary of the current transformer.

Although the use of a current transformer completely insulates the secondary and the ammeter from the high-voltage lines, the secondary of a current transformer should never be left open-circuited. To do so may result in a dangerous high voltage being induced in the secondary. Therefore, if on any occasion it should be necessary to disconnect the meter used with a current transformer, the transformer secondary terminals should first be short-circuited, and this can do no harm because the primary current will still be limited to the amount that the load will permit to flow.

example 5.21 The voltage across the secondary of a potential transformer is 180 V. The primary, or line side, of the transformer has 24 000 turns, and the secondary has 600 turns. What is the line voltage?

solution Dividing 24 000 by 600 shows that the turns or voltage ratio of the transformer is 40 to 1. Then, since the secondary voltage is 180, the primary or line voltage is

$$40 \times 180 = 7200 \text{ V}$$

example 5.22 The current ratio of a current transformer is 100 to 1. What is the line current when an ammeter connected to the secondary indicates 3.5 A?

solution The current ratio being 100 to 1, the line current is

$$100 \times 3.5 = 350 \text{ A}$$

5.20 TRANSFORMERS IN TWO- AND THREE-PHASE SYSTEMS

Power transformation in a three-phase system may be accomplished by using either three separate but exactly similar single-phase transformers or a single three-phase transformer. In either case, the transformers may be of the closed core type or of the shell type.

A three-phase system may also be changed to two phase, or a two phase to three phase. Secondary voltage of the changed system may be the same as its primary voltage, or it may be stepped up or stepped down as the phase is being changed.

CONNECTING THREE SEPARATE TRANSFORMERS FOR THREE-PHASE OPERATION Figure 5.32 shows separate transformers of both the closed core (Fig. 5.32*a*) and shell types stacked for three-phase operation (Fig. 5.32*b*). The three primary windings

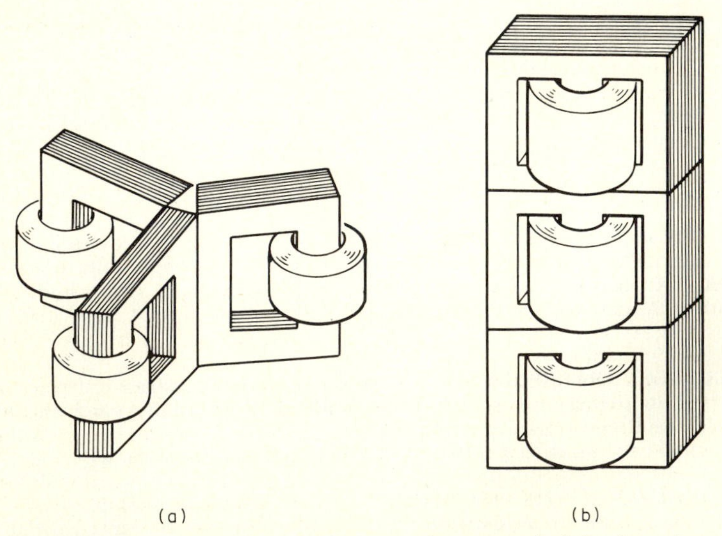

Fig. 5.32 Three separate single-phase transformers stacked for three-phase operation: (*a*) Core types. (*b*) Shell types.

and the three secondaries may be connected together in two different ways. Thus, there is a total of four different ways in which a bank of three transformers may be used for transforming three-phase power. These are the four ways: (*a*) wye (Y) primary to wye (Y) secondary; (*b*) delta (Δ) primary to delta (Δ) secondary; (*c*) wye (Y) primary to delta (Δ) secondary; (*d*) delta (Δ) primary to wye (Y) secondary.

In using the three separate transformers, the primaries may be connected together in either wye or delta in any arbitrary manner; but the secondaries will then have to be phased, and terminal polarities must be taken into consideration in connecting them together.

WYE-TO-WYE TRANSFORMATION Figure 5.33 illustrates the two different methods that may be used to show a Y-to-Y connected transformer bank (Fig. 5.33*a*). This

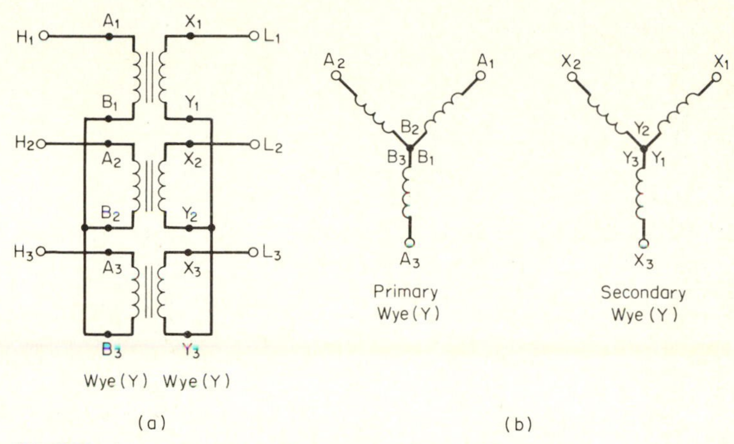

Fig. 5.33 A wye-to-wye transformer connection: (*a*) The wye-to-wye connection of three separate but identical transformers for three-phase to three-phase transformation. (*b*) Alternate method of representing a wye-to-wye-connected three-phase bank.

connection is called a *Y connection* because of its resemblance to the letter Y when the connections are drawn as in Fig. 5.33*b*. A vector diagram of conditions in a Y-connected primary or secondary also takes the form of a Y.

If we assume that the transformer bank of Fig. 5.33 is connected to a three-phase 2200 V source, as shown, the actual voltage across each transformer primary winding will be only

$$\frac{2200}{\sqrt{3}}$$

or

$$0.5773 \times 2200 = 1270 \text{ V}$$

For any Y-connected primary the voltages across the transformer windings will differ in phase by 120°, and the vector sum of any two of these voltages, or the line voltage, will be $\sqrt{3}$ times the voltage across any primary. The current in each primary winding will be equal to the line current.

These same relations will also hold true for the secondary windings. In other words, the voltage between any two secondary output leads, L_1–L_2, L_2–L_3, or L_1–L_3, will be $\sqrt{3}$ times the voltage across any secondary winding, and the current in each secondary winding will be the same as the secondary line current.

DELTA-TO-DELTA TRANSFORMATION Two methods of representing the transformer windings for a delta (Δ)-to-delta (Δ) transformation are presented in Fig. 5.34. Figure 5.34*a* shows the connections of the primary and secondary windings. Drawing the connected windings as in Fig. 5.34*b* shows that this connection resembles the

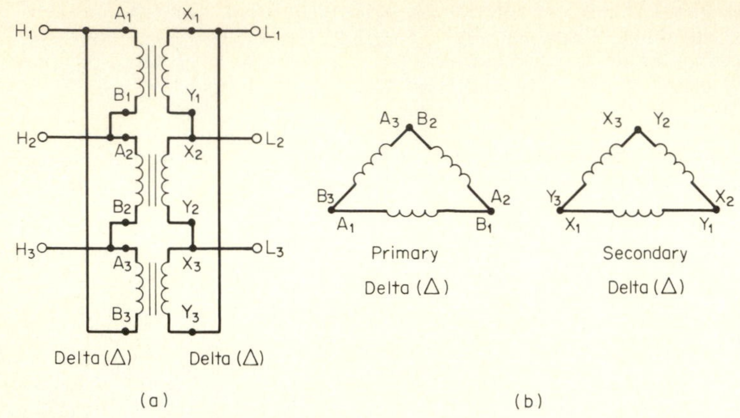

Fig. 5.34 A delta-to-delta transformer connection: (*a*) The delta-to-delta connection for three-phase transformation. The transformers must be identical. (*b*) Alternate method of representing a delta-to-delta three-phase bank.

Greek letter Δ. Hence, it is called a *delta connection*. Here, it is evident that the total voltage across any winding equals the line voltage because each winding is directly across the line. The secondary line voltages will be equal to the transformer secondary voltages. But the current in each winding, with a delta connection, will be

$$\frac{\text{Line current}}{\sqrt{3}}$$

The primaries in a delta-connected bank may be connected together in any random order, but the secondaries must be phased before they are connected together.

WYE-TO-DELTA TRANSFORMATION When the primaries are connected in wye and the secondaries in delta, as in Fig. 5.35, the primary-to-secondary voltage ratio of each transformer will not be the same as the ratio of primary-to-secondary line voltages. One method of illustrating this is shown in Fig. 5.35*a* and an alternative method is illustrated in Fig. 5.35*b*. In this case, the ratio of line voltages will be $\sqrt{3}$ times the voltage ratio of each transformer. For example, in transforming from a three-phase line

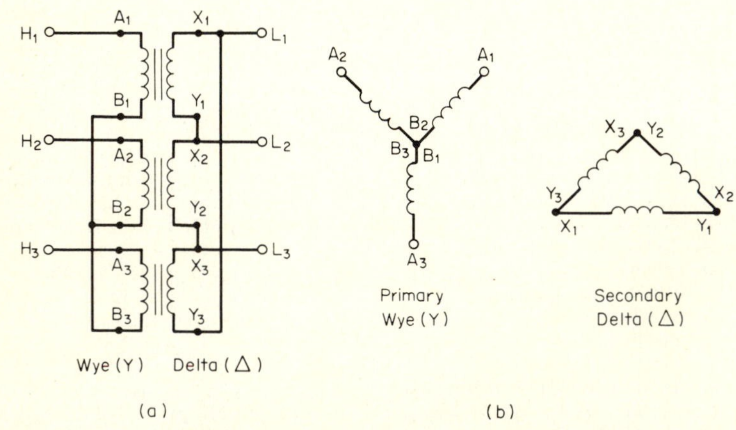

Fig. 5.35 A wye-to-delta transformer connection: (*a*) A wye-to-delta three-phase transformation. (*b*) Alternate method of representing a wye-to-delta transformation.

voltage of 210 V to a secondary line voltage of 105 V, the primary-to-secondary turns ratio of each transformer in a wye-to-delta transformer bank will be

$$\frac{210}{105\sqrt{3}} = \frac{2}{1\sqrt{3}} = \frac{1.156}{1} = 1.156{:}1$$

DELTA-TO-WYE TRANSFORMATION The delta-to-wye connection, shown in Fig. 5.36, is merely a wye-to-delta connection in reverse. Therefore, in this case, the ratio of line voltages will be equal to

$$\frac{\text{Voltage ratio of each transformer}}{\sqrt{3}}$$

One way to illustrate a delta-to-wye connection is shown in Fig. 5.36*a*, and the alternative method is shown in Fig. 5.36*b*.

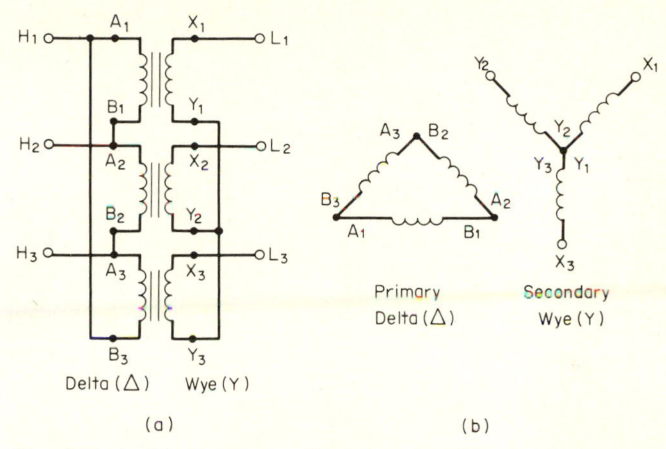

Fig. 5.36 A delta-to-wye transformer connection: (*a*) Delta-to-wye-connected transformer bank for three-phase operation. (*b*) Alternate method of representing the delta-to-wye connection.

Tables 5.1 and 5.2 give the voltage and current relations for both wye- and delta-connected transformers in a convenient form.

example 5.23 A bank of three transformers is connected for wye-to-delta transformation. Each transformer has a voltage ratio of 3 to 1, and the primary line voltage is 650 V. De-

TABLE 5.1 Current and Voltage Relationships in Δ and Y Transformers

	Wye (Y)	Delta (Δ)
Line voltage	Coil voltage × $\sqrt{3}$	Same as coil voltage
Line current	Same as coil current	Coil current/ $\sqrt{3}$

TABLE 5.2 Voltage Ratios for Three φ Transformers

	Transformer voltage ratios
Y to Y	Same as ratio of line voltages
Δ to Δ	Same as ratio of line voltages
Y to Δ	Ratio of line voltages/ $\sqrt{3}$
Δ to Y	Ratio of line voltages × $\sqrt{3}$

termine (*a*) the secondary line voltage; (*b*) the voltage across each primary winding; (*c*) the voltage across each secondary winding.

solution (*a*) Transformer voltage ratio is

$$\frac{3}{1} = \frac{\text{ratio of line voltages}}{\sqrt{3}}$$

The ratio of line voltage is

$$\frac{3\sqrt{3}}{1} = \frac{5.1963}{1}$$

$$= \frac{650}{\text{secondary line voltage}}$$

Solving for secondary line voltage, we obtain

$$\frac{650}{5.1963} = 125.08 \text{ V}$$

(*b*) Since the primary line voltage is equal to

$$650 = \text{coil voltage} \times \sqrt{3}$$

primary coil voltage equals

$$\frac{650}{\sqrt{2}}$$

or

$$650 \times 0.5773 = 375.25 \text{ V}$$

(*c*) Secondary line voltage equals 125.08 V, and since the secondaries are connected in delta, the voltage across each secondary winding is also 125.08 V.

example 5.24 The secondary line voltage of a delta-to-wye transformer bank is 416 V. (*a*) If the transformers have a primary-to-secondary ratio of 5 to 1, what is the primary line voltage? (*b*) If the current in each secondary line is 40 A, what is the current in each secondary coil or winding? (*c*) Assuming that power delivered to the load equals the power taken from the lines, what is the primary line current?

solution (*a*) Transformer ratio is 5 to 1:

$$\text{Ratio of line voltages} \times \sqrt{3} = \frac{\text{primary line voltage}}{\text{secondary line voltage}} \times \sqrt{3}$$

$$= \frac{\text{primary line voltage}}{416} \times \sqrt{3}$$

or

$$\text{Primary line voltage} = \frac{5 \times 416}{\sqrt{3}} = 1200.78 \text{ V}$$

(*b*) When the secondaries are connected in Y, the current in each secondary coil equals the line current, or 40 A.

(*c*) Since the transformer ratio is 5 to 1, and secondary current is 40 A, primary coil current is

$$\frac{40}{5} = 8 \text{ A}$$

The primary line current is

$$\frac{\text{primary coil current}}{\sqrt{3}} = \frac{8}{\sqrt{3}} = 4.618 \text{ A}$$

5.21 THE SCOTT OR T CONNECTION

For some industrial control or servosystem applications, it is at times necessary to convert from three-phase power to two phase. Use is then made of the Scott or T connection, which may not only transform from three phase to three phase but also from three phase to two phase or from two phase to three phase.

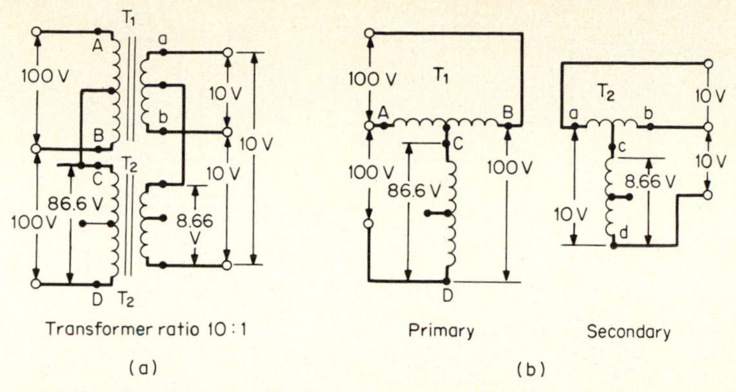

Fig. 5.37 Characteristics of the Scott connection: (*a*) The Scott or T connection for three-phase to three-phase transformation. (*b*) Voltage relations in the 10:1 ratio transformer. Drawing the connection here shows why this connection is also called the T connection.

TWO-PHASE TO THREE-PHASE TRANSFORMATION Two exactly similar transformers, having a primary-to-secondary voltage ratio of 10 to 1, and connected for three-phase to three-phase transformation, are shown in Fig. 5.37. Figure 5.37*a* shows a circuit configuration and Fig. 5.37*b* shows the voltage relationships. Transformer T_1 is called the main transformer, and the center taps of its primary and secondary windings are connected to terminals *C* and *c* of transformer T_2, called the *teaser*.

The connections for transforming from two-phase power to three phase are shown in Fig. 5.38. In this connection (Fig. 5.38*a*) also, the two transformers T_1 and T_2 should have the same number of primary and the same number of secondary turns—and also the same regulation. The secondary of T_1 is tapped at its midpoint and connects to terminal *c* of the other transformer, T_2, which is tapped at a point that gives only 86.6 percent of its rated secondary voltage; thus, only 86.6 percent of the secondary turns of transformer T_2 are used. (See Fig. 5.38*b*.)

THREE-PHASE TO TWO-PHASE TRANSFORMATION The connections for changing from three phase to two phase are merely the reverse of the connections for changing from two phase to three phase. For example, applying a three-phase voltage of 100

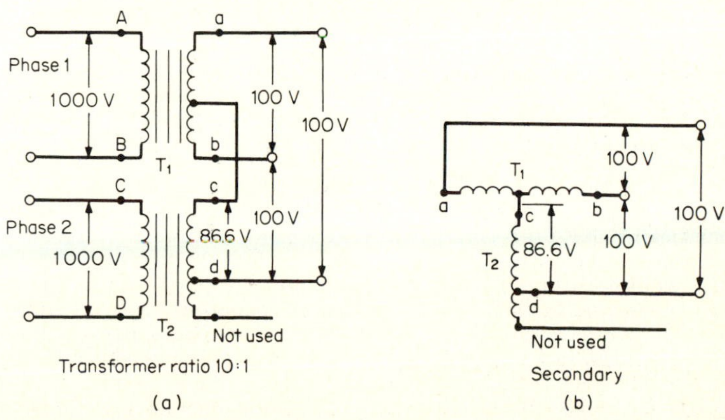

Fig. 5.38 The connections for transforming from two-phase power to three-phase: (*a*) Two Scott-connected transformers being used for two-phase to three-phase conversion. (*b*) Voltage relations in the secondary windings.

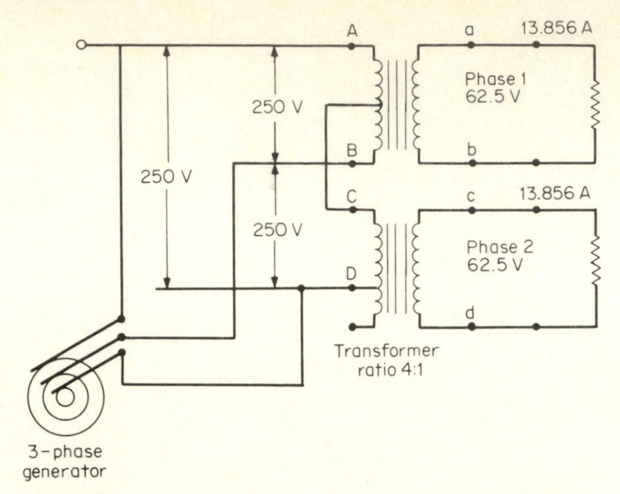

Fig. 5.39 Using two Scott-connected transformers to transform from three-phase power to two-phase reduces current capacity of each transformer to 86.6 percent of normal current rating—from 16 to 13.856 A in this case.

V to the secondaries of the transformers of Fig. 5.38 would result in a 1000-V two-phase voltage across the primary terminals.

For any Scott-connected transformation, the power capacity of the two transformers is only 86.6 percent of their normal single-phase rating.

example 5.25 Two similar transformers that have a normal 1-kVA rating are connected to change from three-phase power to two phase, as shown in Fig. 5.39. Assuming a unity power factor load, what current may be supplied by each of the two transformers, if the primary three-phase line voltage is 250 V, and the primary-to-secondary turns ratio of each transformer is 4 to 1?

solution Normal single-phase rated primary current of either transformer,

$$\frac{1000}{250} = 4 \text{ A}$$

Since the primary-to-secondary voltage ratio of the transformers is 4 to 1, normal secondary current of each transformer is

$$4 \times 4 = 16 \text{ A}$$

The reduced current rating for each of the Scott-connected secondaries is

$$0.866 \times 16 = 13.856 \text{ A}$$

5.22 FILTER CHOKES

An important application of chokes is in the filter sections of power supplies, where their function is to help smooth out the pulsating dc wave from the rectifier and change it to approximately pure direct current that is needed for the operation of radio, TV, and other electronic equipment.

Typical power-supply filter circuits are shown in Fig. 5.40. One of the functions of a choke in these circuits is to store energy as the filter input voltage and current are rising. Then, as the filter input decreases from maximum to zero, the stored energy is released to the load by inductive action. Thus, the output from the filter remains fairly constant. The inductance of the choke coil tends to keep the current flowing through it constant even though the dc input to the filter is a constantly varying quantity.

A filter choke is made with a tiny air gap in the core to keep the core from becoming magnetically saturated at high currents and thereby losing inductance to some extent

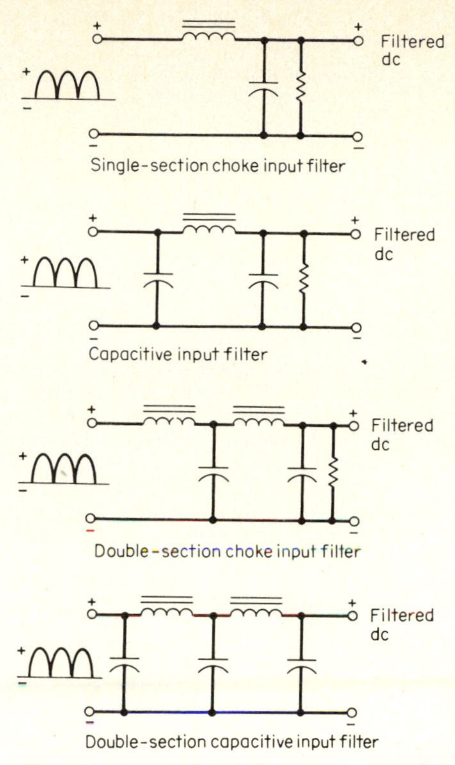

Fig. 5.40 Use of filter chokes in power-supply filter circuits.

or even entirely with excessively high currents. But, even with an air gap in the core, there will still be some change in inductance as current changes, and it is customary to specify choke inductance at the current which the choke will normally carry.

The dc output voltage of a choke input filter is about twenty percent less than the output of a capacitor input filter, and its hum level is greater. However, the choke input filter does have better regulation.

A small ac component always gets through to the output of any of the filters shown in Fig. 5.40. This ac component is called the *ripple*. For a full-wave rectifier ripple frequency is twice the line frequency, and for a half-wave rectifier it is the same as line frequency. When the ripple frequency is 60 Hz, ripple frequency with full-wave rectification will be 120 Hz.

A formula for determining the approximate ripple percentage for a single-section choke-input filter is for a choke-input single-section filter at a ripple frequency of 120 Hz.

$$\text{Percent ripple} = \frac{100}{LC} \tag{5.18}$$

where L = inductance, H
C = capacitance, μF

The lower the value of percent ripple, the more effective the filter. For a 60-Hz ripple, twice as much inductance and capacitance will be needed to reduce ripple percentage to the same value.

Adding more filter sections greatly improves filtering and reduces ripple to very low values. The amount of tolerable ripple will depend on the equipment to be supplied from the power supply. For a high-gain high-fidelity amplifier, it may be necessary to reduce ripple below 0.05 percent.

example 5.26 In a single-section choke-input filter, the capacitor has a capacitance of 32 μF. What should be the inductance of the input choke for an output ripple of 0.05 percent if the ripple frequency is 120 Hz?

solution

$$0.05 = \frac{100}{LC}$$

or

$$L = \frac{100}{0.05 \times 32} = \frac{2000}{32} = 62.5 \text{ H}$$

5.23 SWINGING CHOKES

A swinging choke, instead of being designed to maintain constant inductance with changes in current, is devised to change inductance with changes in current. Thus, its inductance will decrease with increases in current, and increase as the current decreases. For example, a choke may have an inductance of 5 H when choke current is 250 mA, but will increase to 25 H when choke current is only 25 mA.

Another function of chokes, besides filtering, is to keep the dc output voltage of a power supply below the average value of the *rectified* ac input to the filter. The critical value of inductance required for this purpose increases as load current decreases, and decreases as load current increases. Hence, if a choke is so designed that it will have its critical inductance with only the filter bleeder load, it will not be necessary to use a choke that will have this critical inductance at higher load currents. Therefore, a smaller, less expensive choke may be used.

An approximate formula for determining critical inductance is given by

$$L_{\text{crit}} = \frac{\text{load resistance, ohms} \times 120}{1000 \times f} \tag{5.19}$$

where f = ripple frequency

example 5.27 If the bleeder resistance in a power-supply filter is 25 000 Ω, and full-load resistance which includes the bleeder is 5000 Ω, between what values of inductance must a choke "swing" to maintain critical inductance at a ripple frequency of 120 Hz?

solution Critical inductance at 25 000 Ω is

$$\frac{25\,000 \times 120}{1000 \times 120} = 25 \text{ H}$$

Critical inductance at full load is

$$\frac{5000 \times 120}{1000 \times 120} = 5 \text{ H}$$

Therefore, a swinging choke that "swings" from 25 to 5 H and has the required current rating will meet requirements.

5.24 THE TUNED FILTER CHOKE

In a power supply that is designed to deliver a large but well-filtered current to a connected load, the filter chokes would have to be very large and bulky. For instance, assuming that a power supply is capable of supplying six amperes of smooth, low-hum current to a load, this same six amperes must also flow through the filter chokes. Hence, the filter chokes must be wound with wire large enough to carry the six amperes without overheating. Furthermore, in order to obtain sufficient inductance for smooth filtering, the chokes must have a large number of turns. To keep the cores from becoming magnetically saturated, they must have a large cross-sectional area. Thus, a large wire size, combined with a large number of turns and a bulky core, results in a large and costly choke.

By using a tuned filter choke, instead of an ordinary choke, the same amount of filtering may be obtained with much smaller and less expensive units. Figure 5.41 shows a pair of tuned filter chokes connected into the filter section of a high-current

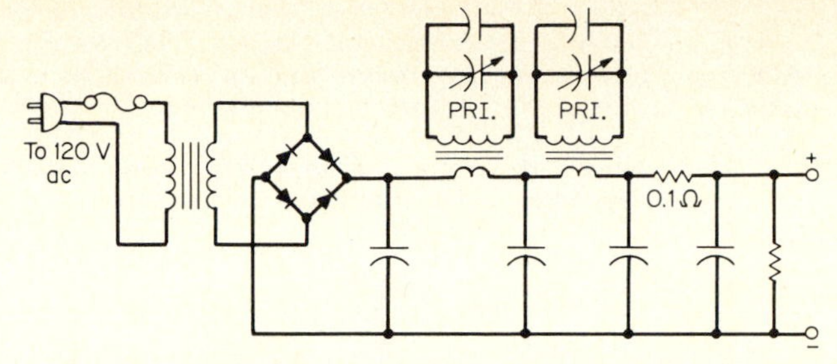

Fig. 5.41 By using tuned filter chokes in a high-current power supply, very effective filtering may be attained without the necessity of bulky, heavy-duty, expensive chokes. The tuned primaries reflect a high impedance back into the secondaries at the 120-Hz ripple frequency.

power supply. The principle of operation is very simple. The transformers that are used as tuned chokes may have a primary-to-secondary turns ratio as high as 20 to 1. Even ordinary filament transformers with a primary-to-secondary turns ratio of 117 to 6.3 will be suitable, if secondary current capacity is large enough to carry the load current. The primaries of the transformers are shunted with capacitors having the necessary capacitance values to obtain resonance at 120 Hz (the ripple frequency of the full-wave rectifier). It may be necessary to try several different combinations of smaller and larger capacitors before resonance occurs. Shunted capacitance will range from about 0.1 to 0.3 μF. Exact values, in any case, will have to be determined experimentally and will depend on transformer primary inductance. Hum output will be lowest at resonance.

When the transformer primaries are tuned to resonate at 120 Hz, a high impedance is reflected back into the secondaries, and the secondary windings then function as high-impedance chokes at the ripple frequency. Since the secondary windings do not have a large number of turns, they are well able to carry the high load current without saturating the core or overheating.

The condensers across the primaries should have a voltage rating great enough to safely withstand the high voltage developed across them at the 120-Hz ripple frequency.

5.25 TRANSFORMER CORES

The three basic types of transformer cores are the *open core*, the *closed core*, and the *shell core*. These three types are illustrated in Fig. 5.42. The open-core type is far

Open core

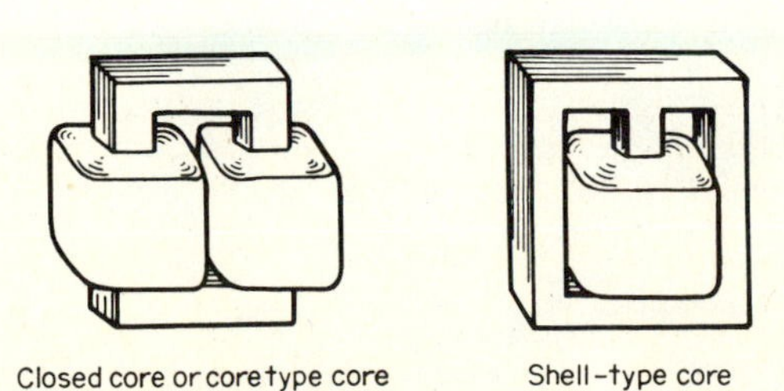

Closed core or core type core

Shell-type core

Fig. 5.42 Core types.

less efficient than the closed-core or shell-core types, but it is still used in some special applications in which efficiency is not an important factor. The closed-core type is commonly called the core type; it may have its windings on only one leg of the core or on both legs. In the shell type the windings are on the center leg.

LAMINATIONS Some of the various types of laminations used for stacking transformer cores are shown in Fig. 5.43. Laminations are available in various thicknesses

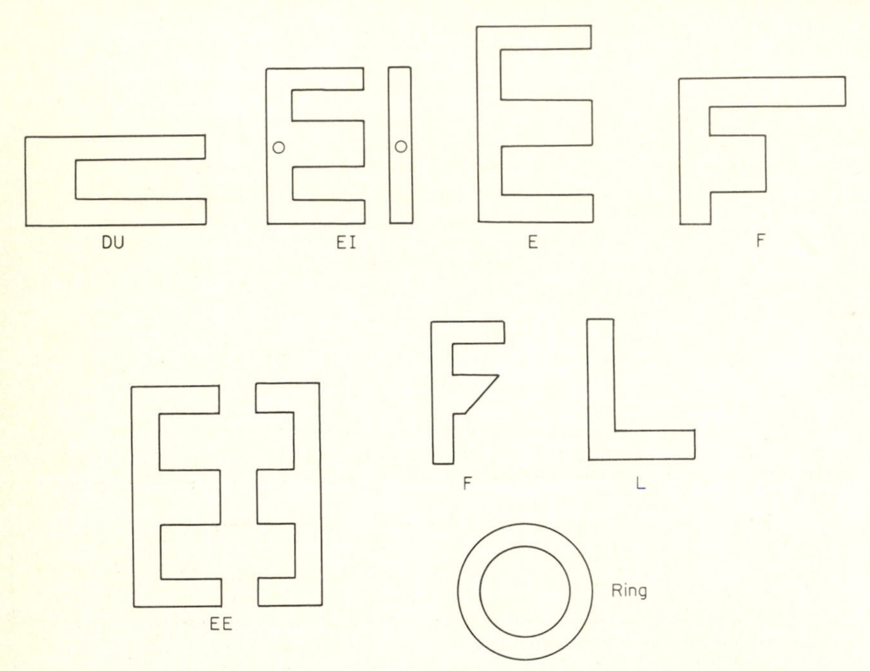

Fig. 5.43 Showing some of the various types of laminations used for the cores of transformers and chokes.

Fig. 5.44 This photo shows a variety of the laminations used to construct various types of transformers. (*Courtesy The Arnold Engineering Company.*)

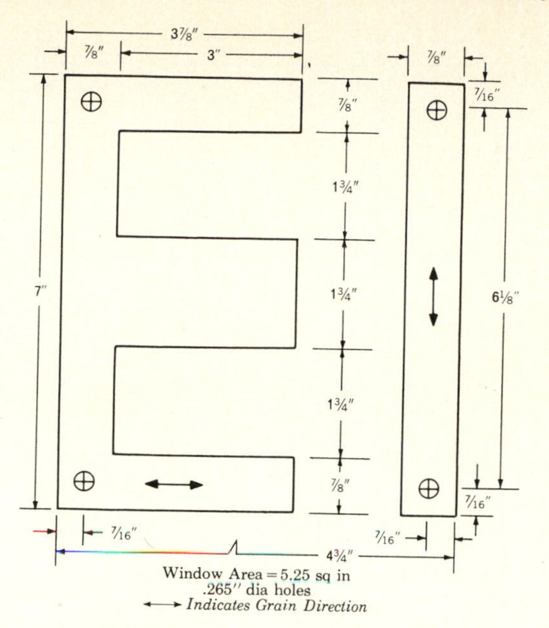

PROPERTIES OF SQUARE STACK

V = 646.5 cu cm = 39.43 cu in
A = 19.74 sq cm = 3.063 sq in
l = 33.0 cm = 13.0 in

$B_{max} = \frac{19.0 \times 10^3}{K_1 N}$ gausses per volt at 60 cycles (N is number of turns)

H_o = $(.038 \times 10^{-3})$ N oersteds per milliampere of direct current in winding

L = $(.750 \times 10^{-8})\ K_1 N^2 \mu_{ac}$ henries

Solid Core Weight

High Silicon = 4935 g = 10.86 lb
50% Nickel = 5325 g = 11.72 lb
80% Nickel = 5638 g = 12.40 lb

SETS PER INCH	HIGH SILICON WEIGHT			50% NICKEL WEIGHT			80% NICKEL WEIGHT		
	Ga	Lb/M Prs	Prs/Lb	Ga	Lb/M Prs	Prs/Lb	Ga	Lb/M Prs	Prs/Lb
71	.014″	86.8	11.5	.014″	93.6	10.7	.014″	99.2	10.1
54	.0185″	114.7	8.7						
40	.025″	155.0	6.4						

Fig. 5.45 Typical dimensions of EI laminations. (*Courtesy The Arnold Engineering Company.*)

ranging from about 0.004 to 0.025 in. The L and DU laminations are used in the core-type transformer; all others are used in stacking the cores of the shell type. Figure 5.44 shows some of the actual laminations used in transformer construction.

Figures 5.45 and 5.46 give actual dimensions of some typical laminations. The accompanying tables list the properties of the square stacks, and also give design data. The center leg of any square-stacked core is just as thick as it is wide. The core weights for the square stacks are given in both pounds and grams. This is useful in determining core loss from core-loss data curves. Core volume is in cubic centimeters and in cubic inches. Core cross-sectional area is given in square centimeters and in square inches. The length of the magnetic circuit l is in centimeters and in inches. The equation for finding B_{max} is similar to

$$B_m = \frac{E_p \times 10^8}{4.44 \times f \times N_p \times A}$$

given in Sec. 5.10, where B_m is the maximum flux density for the total applied voltage, instead of for only one volt. The *stacking factor K* may be taken equal to one for most

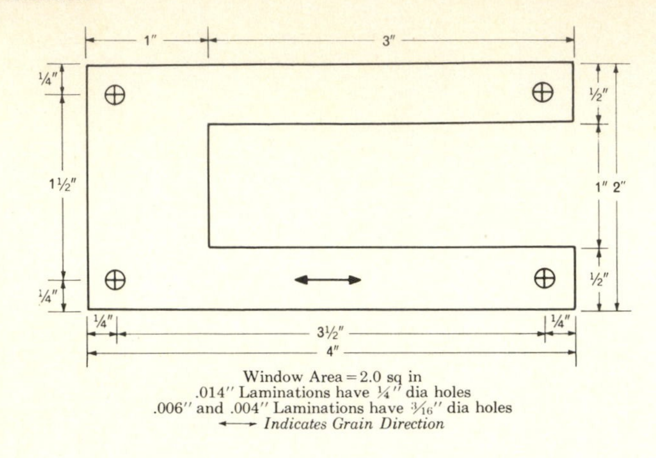

PROPERTIES OF SQUARE STACK

V = 40.15 cu cm = 2.45 cu in
A = 1.61 sq cm = .25 sq in
l = 22.9 cm = 9.0 in

$B_{max} = \frac{233 \times 10^3}{K_1 N}$ gausses per volt at 60 cycles (N is number of turns)

H_o = $(.055 \times 10^{-3})$ N oersteds per milliampere of direct current in winding

L = $(.089 \times 10^{-8})\ K_1 N^2 \mu_{ac}$ henries

Solid Core Weight

High Silicon = 307.0 g = .675 lb
50% Nickel = 331.0 g = .728 lb
80% Nickel = 351.0 g = .773 lb

SETS PER INCH	HIGH SILICON WEIGHT			50% NICKEL WEIGHT			80% NICKEL WEIGHT		
	Ga	Lb/M Pcs	Pcs/Lb	Ga	Lb/M Pcs	Pcs/Lb	Ga	Lb/M Pcs	Pcs/Lb
71	.014"	18.95	52.8	.014"	20.45	48.8	.014"	21.65	46.1
167	.006"	8.13	123.0	.006"	8.76	114.0	.006"	9.28	107.8
250	.004"	5.42	184.5	.004"	5.84	171.0	.004"	6.20	161.2

Fig. 5.46 Typical dimensions of DU laminations. (*Courtesy The Arnold Engineering Company.*)

practical applications, but for greater accuracy the following factors may be used for the corresponding lamination gages:

0.025 – 97%
0.0185 – 96%
0.014 – 95%
0.006 – 93%
0.004 – 88%

In the formula for L, the inductance of the core in henrys μ_{ac} is the ac permeability of the core.

GRAIN-ORIENTED LAMINATIONS The arrows on the laminations shown in Figs. 5.45 and 5.46 show the direction of grain orientation in the steel. Grain orientation is brought about by the hot- and cold-rolling processes used in the manufacture of sheet steel for transformers. These processes cause the individual crystals of silicon steel to become aligned in the direction of rolling. Grain-oriented steel is much easier to magnetize in the direction of grain orientation. Hence, grain-oriented laminations will have their maximum permeabilities when the magnetic fields induced in them are parallel to grain orientation.

A recent development has been the production of a magnetic steel that has two directions of easy magnetization. One is in the direction of rolling, and the other is perpendicular to it.

TAPE-WOUND CORES Tape-wound cores are wound from long strips of transformer sheet steel having the necessary width and thickness for a particular application. Since grain orientation is always in the direction in which the tape is wound, tape-wound cores may be made to have extremely high permeabilities. Insulating varnish between layers not only serves to insulate the layers from each other—to prevent the formation of transverse eddy currents—but also binds the layers together and adds rigidity to the core.

Uncut tape-wound cores are shown in Fig. 5.47. To use these cores in the construction of transformers, the cores are cut and then reassembled about the transformer

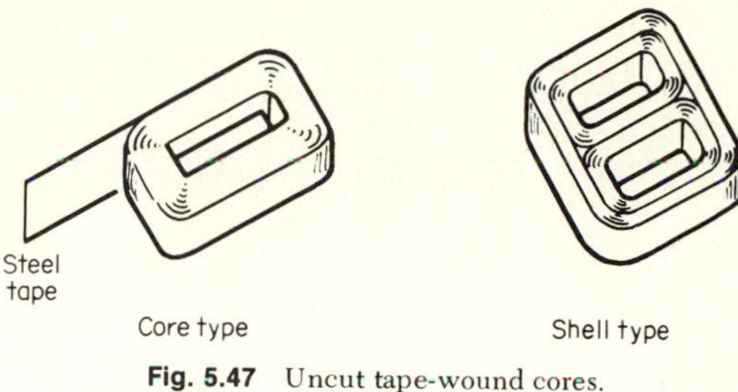

Fig. 5.47 Uncut tape-wound cores.

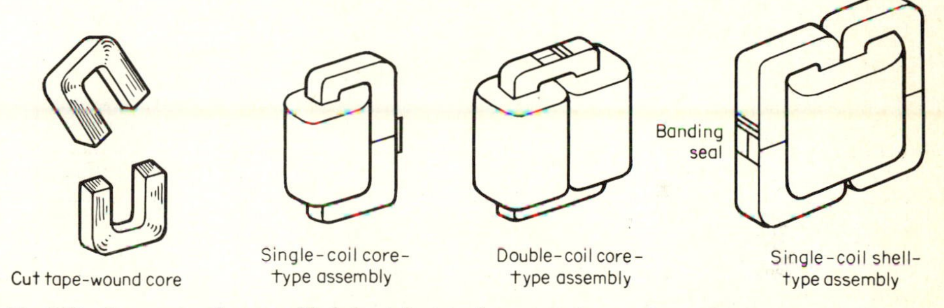

Fig. 5.48 Cores cut and reassembled about the transformer windings to form either a core or a shell-type core.

windings to form either a core or a shell-type core, as shown in Fig. 5.48. Besides improving performance and efficiency, the use of tape-wound cores eliminates much of the tedious and time-consuming labor involved in stacking ordinary laminated cores.

TAPE-WOUND BOBBIN CORES Not all tape-wound cores are cut; some small tape-wound cores are made of such extremely thin and narrow magnetic steel tape that cutting would ruin them. In fact, to maintain their mechanical rigidity, it is even necessary to wind them on bobbins. Magnetic steel tape for tape-wound bobbin cores is manufactured in standard gage sizes from $\frac{1}{8}$ to 1 mil (0.000 125 to 0.001 in), and tape widths may be as narrow as $\frac{1}{32}$ in. Core magnetization in tape-wound bobbin cores is usually produced by only a few turns of wire—about 1 to 10. Tape-wound bobbin cores operate at low power levels and are used in many computer circuits, shift registers, ring counters, high-frequency magnetic amplifiers, etc.

example 5.28 A 60-Hz transformer operates with an applied primary voltage of 120 V. The primary winding contains 152 turns. The core is square-stacked and is made up of the EI-19 laminations shown in Fig. 5.45. Use the appended design data and determine (*a*) the maximum flux density in the core in gauss and in lines per square inch; (*b*) the total maximum flux in the core. (Assume that stacking factor $K = 1$.)

solution Maximum flux density B_{max} in gauss per volt equals

$$\frac{19 \times 10^3}{1 \times 152} = 125$$

Then maximum flux density B_m with an applied primary voltage of

$$120\text{ V} = 120 \times 125 = 15\,000\text{ G}$$

In lines per square inch

$$15\,000\text{ G} = 6.45 \times 15\,000 = 96\,750$$

(*b*) Since the center leg of a lamination is $1^3/_4$ in. wide, the cross-sectional area of a square-stacked core is

$$(1^3/_4)^2 = {}^{49}/_{16}\text{ in}^2$$

Therefore, maximum core flux ϕ_m equals

$$^{49}/_{16} \times 96\,750 = 296\,297\text{ Mx}$$

example 5.29 Use the data supplied for the DU-50 laminations of Fig. 5.46 and determine the magnetizing force in oersteds developed in a square-stacked core wound with a 200-turn coil that carries a direct current of 100 mA.

solution From the data, the magnetizing force in oersteds for each milliampere of direct current in the coil is

$$(0.055 \times 10^{-3})200 = 0.011$$

Then, with a direct current of 100 mA, the magnetizing force will be

$$100 \times 0.011 = 1.1\text{ Oe}$$

Chapter 6

Practical Circuit Analysis

6.1 INTRODUCTION

Electronic technicians are often required to analyze simple circuits energized by dc or ac sources. Generally, only elementary algebra and trigonometry are required in their analysis. The object of this chapter is to review some basic circuit laws and their application in the analysis of dc and ac circuits. Topics discussed will be illustrated by pertinent examples and their complete solution.

6.2 NOTATION

By convention, for time-varying quantities, and often in general discussions, *lowercase* letters are used. Voltage is designated by e, current by i, energy by w, and power by p. *For specific dc and ac quantities, uppercase* letters are used: for voltage E, current I, energy W, and power P.

6.3 DEFINITIONS

We shall first define some basic terms used in the analysis of circuits.

PASSIVE AND ACTIVE ELEMENTS An element is considered *passive* if it is only capable of *absorbing* electric energy. A resistor, an inductor, and a capacitor are passive elements. Elements that *provide* electric energy are *active elements.* Examples of active elements include the battery (used to energize circuits) and the transistor (which may be used to energize a loudspeaker).

IDEAL ELEMENTS An ideal element *exhibits a specific property*, regardless of the frequency of operation or the impressed voltage and current flow through the element. For example, an ideal resistor of 50 Ω remains 50 Ω whether it operates at direct current or at hundreds of gigahertz (GHz), or has 1 or 10 000 V impressed across its terminals. Ideal elements, in terms of terminal voltage and current, are defined as follows:

a. Resistor (R) Current i (amperes) in an ideal resistor of R (ohms) is equal to the voltage e (volts) across R divided by R:

$$i = \frac{e}{R} \tag{6.1a}$$

Equation (6.1*a*) is the basic expression of Ohm's law. By algebraic manipulation, one may also write

$$e = iR \tag{6.1b}$$

$$R = \frac{e}{i} \tag{6.1c}$$

b. Inductor (L) The voltage e_L (volts) across an ideal inductor equals the value of the inductance L (henrys) multiplied by the rate of change of current with respect to time (amperes per second). A symbol used for the rate of change of current is $\Delta i/\Delta t$, where Δ (*delta*) denotes a *small change*. Hence,

$$e_L = L\left(\frac{\Delta i}{\Delta t}\right) \tag{6.2}$$

Assume, for example, that at time $t_1 = 6$ s, current i_1 in the inductor is 1 A. At time $t_2 = 8$ s, the current is $i_2 = 2$ A. Hence, $\Delta i = 2 - 1 = 1$ A, and $\Delta t = 8 - 6 = 2$ s. The rate of change of current is, therefore, $\Delta i/\Delta t = \frac{1}{2}$ A/s.

c. Capacitor (C) The current i_C (amperes) in an ideal capacitor equals the value of capacitance C (farads) multiplied by the rate of change of the impressed voltage with respect to time (volts per second). The symbol for the rate of change of the impressed voltage is $\Delta e/\Delta t$. Hence,

$$i_C = C\left(\frac{\Delta e}{\Delta t}\right) \tag{6.3}$$

d. Voltage source (e) An ideal voltage source is a source whose output voltage is constant, regardless of the current it supplies to a circuit. A practical voltage source, with a low output resistance, approximates an ideal voltage source.

e. Current source (i) An ideal current source is a source whose output current is constant, regardless of the resultant voltage appearing across the circuit to which the source is connected. Symbols used for ideal voltage and current sources are given in Fig. 6.1.

The symbol of Fig. 6.1*a* is general and can be used to denote a dc or time-varying voltage source. Figure 6.1*b* shows an ideal dc voltage source. The symbol for the ideal current source of Fig. 6.1*c* is used for dc and time-varying current sources.

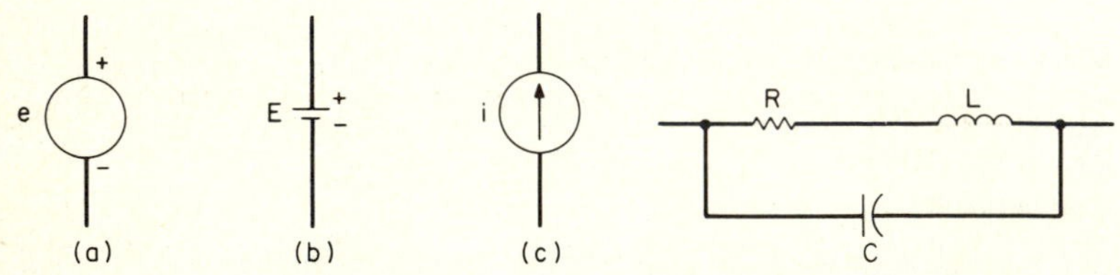

Fig. 6.1 Symbols for ideal sources: (*a*) Dc or time-varying voltage source. (*b*) Dc voltage source. (*c*) Dc or time-varying current source.

Fig. 6.2 A possible model of a resistor operating at very high frequencies.

MODEL A model is a representation of a *physical* device or circuit by *ideal elements*. Consider, for example, a resistor of R Ω. At very high frequencies, such as in the gigahertz range, the lead inductance and shunt capacitance enter into the picture. Hence, a possible model of a resistor operating at high frequencies may appear as in Fig. 6.2. Note that the model is composed of ideal elements R and L in series, shunted by an ideal capacitor C.

LINEAR ELEMENT A linear element is an element in which a change of its input (such as voltage) results in a proportionate change in its response, or output (such as current). If, for example, the voltage across an 8-Ω resistor is 16 V, $I_1 = \frac{16}{8} = 2$ A. If the voltage is *doubled* to 32 V, $I_2 = \frac{32}{8} = 4$ A, and the current is also *doubled.*

NONLINEAR ELEMENT An element that is not linear is said to be nonlinear. In a nonlinear element, its response is not proportional to a change in its input. An example of a nonlinear element is a *thermistor,* whose resistance depends on its current flow.

LINEAR CIRCUIT A circuit that contains only linear elements. (The term *network* is also used for circuit.)

NONLINEAR CIRCUIT A circuit that contains one or more nonlinear elements.

SERIES CIRCUIT A circuit in which the current in each element is the same.

PARALLEL CIRCUIT A circuit in which the voltage across each element is the same.

6.4 RC AND RL CIRCUITS

Consider the series RC circuit of Fig. 6.3. Initially, that is, at time equals zero ($t = 0$), capacitor C acts like a *short circuit*. Hence, the current I at $t = 0$, denoted by $I(0)$, is

$$I(0) = \frac{E}{R} \tag{6.4}$$

The product RC is the *time constant* T (seconds) of the circuit,

$$T = RC \tag{6.5}$$

where R is in ohms and C in farads. After an elapsed time of approximately 5 time constants ($5T$), the circuit is said to be in *steady state*, denoted by $t = \infty$. In steady state,

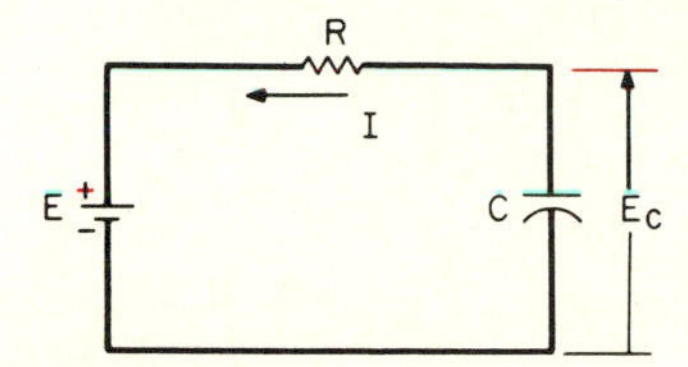

Fig. 6.3 Series RC circuit excited by a dc source.

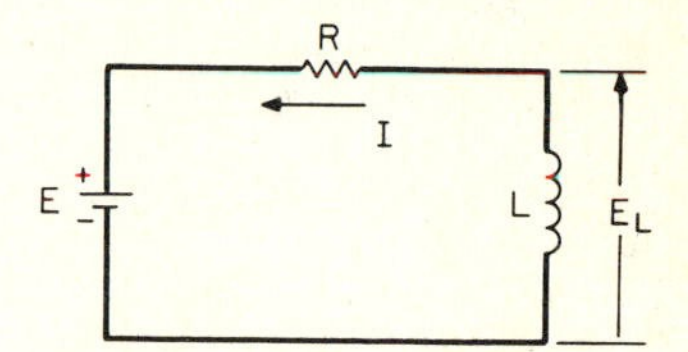

Fig. 6.4 Series RL circuit excited by a dc source.

the voltages and currents in the circuit are not changing and have settled down to a fixed value. For dc excitation in steady state, $\Delta e/\Delta t = 0$. From Eq. (6.3), if $\Delta e/\Delta t = 0$, the current in the capacitor is also zero ($i_C = 0$). Therefore, a capacitor acts like an *open circuit* to direct current in the steady state.

example 6.1 In Fig. 6.3 assume that $E = 100$ V, $R = 1$ kΩ, and $C = 1$ μF. Determine (*a*) time constant T, (*b*) initial and steady-state values of current, and (*c*) initial and steady-state values of voltage across the capacitor.

solution (*a*) From Eq. (6.5), $T = RC = 10^3 \times 10^{-6} = 10^{-3}$ s (one millisecond).

(*b*) Because at $t = 0$ the capacitor behaves like a short circuit to direct current, $I(0) = E/R = 100/1000 = 0.1$ A. In steady state the capacitor acts like an open circuit to direct current, and $I(\infty) = 0$.

(*c*) Because C acts like a short circuit at $t = 0$ (and has no charge), $E_C(0) = 0$ V. In steady state, no current flows in C, and the capacitor is fully charged by the source voltage; therefore, $E(\infty) = 100$ V.

Consider now the series RL circuit of Fig. 6.4. To direct current L acts as an open circuit at $t = 0$ and as a short circuit in steady state ($t = \infty$). The current in the circuit is therefore maximum in steady state for dc excitation.

The time constant T (seconds) is

$$T = \frac{L}{R} \tag{6.6}$$

where L is in henrys and R in ohms. At $t = 0$, the voltage across the inductor is $E_L(0)$ E volts. In steady state, $\Delta i/\Delta t = 0$; hence, from Eq. (6.2), $E_L(\infty) = 0$ V.

example 6.2 In Fig. 6.4, let $E = 25$ V, $R = 100$ Ω, and $L = 1$ H. Determine (*a*) T, (*b*) $I(0)$ and $I(\infty)$, (*c*) $E_L(0)$ and $E_L(\infty)$.

solution (*a*) From Eq. (6.6), $T = L/R = 1/100 = 0.01$ s.

(*b*) At $t = 0$, L acts like an open circuit to direct current; hence, $I(0) = 0$. In steady state, L acts like a short circuit to direct current, and $I(\infty) = E/R = 25/100 = 0.25$ A.

(*c*) $E_L(0) = 25$ V and $E_L(\infty) = 0$ V.

For times between approximately 0 and $5T$, the current i_C and voltage e_C as a function of time for a series RC circuit excited by a dc source (Fig. 6.3) are

$$i_C = \left(\frac{E}{R}\right)\epsilon^{-t/RC} \tag{6.7}$$

and

$$e_C = E(1 - \epsilon^{-t/RC}) \tag{6.8}$$

For the series RL circuit excited by a dc source (Fig. 6.4),

$$i_L = \left(\frac{E}{R}\right)(1 - \epsilon^{-tR/L}) \tag{6.9}$$

and

$$e_L = E\epsilon^{-tR/L} \tag{6.10}$$

for times between 0 and $5T$.

Equations (6.7), i_C, and (6.10), e_L, are of the same form and are therefore plotted in Fig. 6.5*a*. For convenience, the maximum value of either i_C or e_L is denoted by one. The unit for the x axis is t/T. Note that when t is equal to five time constants ($5T$),

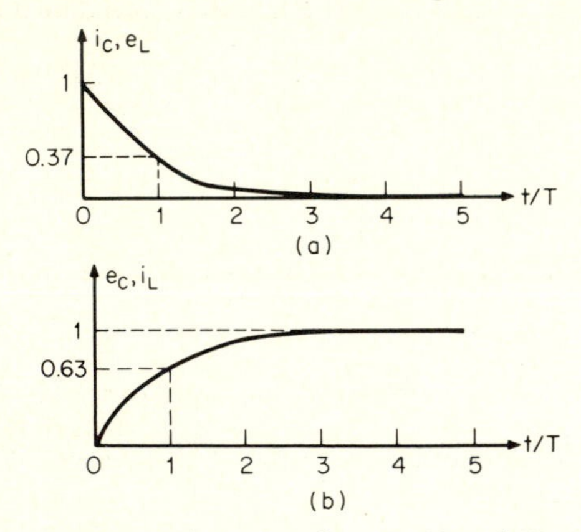

Fig. 6.5 Performance of series RC and RL circuits energized by a dc voltage source: (*a*) i_C and e_L as a function of time. (*b*) e_C and i_L as a function of time.

the current i_C in a capacitor and the voltage e_L across an inductor are approximately equal to zero. In a single time constant ($t = T$), i_C (or e_L) falls to approximately 0.37, or 37 percent of its maximum value.

Figure 6.5*b* is a plot of e_C and i_L [Eqs. (6.8) and (6.9)]. The maximum value is again denoted by one, and in $t = 5T$ steady state has been reached. For $t = T$, e_C (or i_L) have reached 63 percent of their maximum value.

6.5 ENERGY AND POWER

Energy is the ability to do work. In the meter-kilogram-second (mks) system of units, the unit of energy is the *joule;* 1 joule equals 1 *watt-second.* Power is the *rate of expending energy;* its unit is *joule per second* which equals a *watt.* Energy w equals the product of power p and time t:

$$w = pt \tag{6.11}$$

For electric circuits, power p may be expressed by

$$p = ei \tag{6.12a}$$

$$= \frac{e^2}{R} \tag{6.12b}$$

$$= i^2R \tag{6.12c}$$

where R is the resistance, e the impressed voltage across the resistance, and i the current flowing in the resistance.

Only resistors are capable of *dissipating energy;* capacitors and inductors only *store energy*. The energy stored in a capacitor w_C is

$$w_C = \frac{Ce^2}{2} \tag{6.13}$$

where e is the voltage across C. For an inductor the stored energy w_L is

$$w_L = \frac{Li^2}{2} \tag{6.14}$$

where i is the current flowing in the inductor.

example 6.3 Referring to Example 6.1, determine the energy stored in the capacitor in steady state.

solution In Example 6.1, $C = 10^{-6}$ F, and the voltage across C equals 100 V. From Equation 6.13, $w_C = Ce^2/2 = 10^{-6} \times 100^2/2 = 5 \times 10^{-3}$ J.

example 6.4 Referring to Example 6.2, determine (*a*) the power dissipated in R, and (*b*) the energy stored in L, in steady state.

solution (*a*) In Example 6.2, $I = 0.25$ A in steady state; $L = 1$ H; and $R = 100\ \Omega$. From Eq. (6.12*c*), letting $i = I$, $p = I^2R = (0.25)^2 \times 100 = 6.25$ W.
(*b*) From Eq. (6.14), $w_L = LI^2/2 = 1 \times (0.25)^2/2 = 31.75 \times 10^{-3}$ J.

6.6 CIRCUIT LAWS AND THEOREMS

A number of simple, yet powerful, laws and theorems for the analysis of circuits are available to the technician. Although the application of these laws and theorems will first be demonstrated in the analysis of dc circuits, they are also applicable in the analysis of ac circuits. For dc circuits the quantities of concern, such as voltage and current, are always expressed by real numbers. For ac circuits, however, these quantities are generally represented by complex numbers. A review of complex numbers, and the application of the circuit laws and theorems to ac circuits, will be covered later in the chapter.

KIRCHHOFF'S LAWS Kirchhoff's voltage law (KVL) and current law (KCL) provide the technician with two useful tools for the analysis of circuits, regardless of their complexity. The voltage law states that the algebraic sum of voltages in a closed-circuit path is always equal to zero. The current law states that the algebraic sum of currents leaving a circuit node is always equal to zero. A circuit node may be thought of as a connecting point that is common to two or more circuit elements.

In the application of KVL and KCL it is essential that we designate the polarities of voltages across elements and the direction of electron flow at a circuit node. In this chapter we adopt the following conventions:

1. In traversing a closed-circuit path, a *plus sign* precedes a voltage quantity, such as e or iR, in going *with* the direction of electron flow or from a *low* to a *high* potential. In going *against* the electron flow, or from a *high* to a *low* potential, a *minus sign* precedes the voltage quantity.

2. Current flowing *away* from a circuit node will have a *negative sign;* current flowing *toward* a circuit node will have a *positive sign.*

The application of KVL and KCL is illustrated in the next group of examples.

example 6.5 Find current I in Fig. 6.6.

solution Starting at point 2 and traversing the circuit in the direction (counterclockwise) of current I, we have, according to KVL and our conventions,

$$-100 + 30I + 20I + 50I = 0$$

From point 2 to point 1, we go from a *high* to a *low* potential; hence, the minus sign before 100. In traversing the remainder of the circuit, we go in the direction of current. The voltage across each resistor, therefore, has a positive sign.

Bringing the −100 term to the right side of the equation and simplifying, we have

$$(30 + 20 + 50)I = 100$$

or

$$100I = 100$$

Solving, $I = {}^{100}/_{100} = 1$ A.

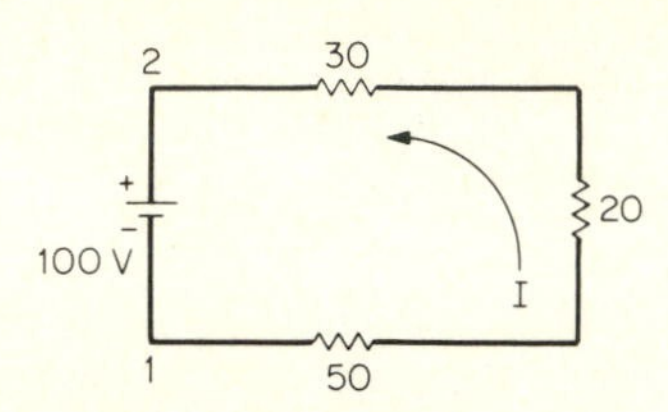

Fig. 6.6 Finding current I in a series circuit by KVL (Example 6.5).

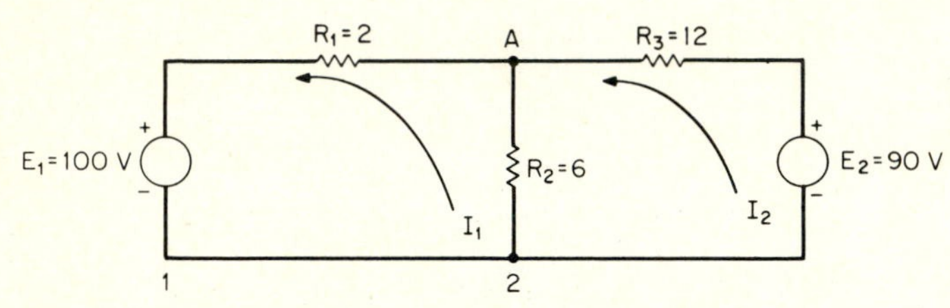

Fig. 6.7 Finding currents I_1 and I_2 in a two-mesh circuit by KVL (Example 6.6).

example 6.6 Figure 6.7 shows a circuit having *two* closed-circuit paths. This configuration is referred to as a *two-mesh* or *two-loop* circuit. Circuits containing two or more meshes are fairly common.

In Fig. 6.7 currents I_1 and I_2 are assumed to be flowing counterclockwise. (If it turns out that the actual current flows in a direction opposite to that assumed, the solved current will have a minus sign.) Applying KVL, write the necessary equations and solve for the currents.

solution Starting at point 1 for the first mesh, we obtain

$$-100 + (2 + 6)I_1 - 6I_2 = 0$$

or

$$100 = 8I_1 - 6I_2$$

Dividing by 2,

$$50 = 4I_1 - 3I_2 \tag{a}$$

Starting at point 2 in the second mesh, we have

$$-6I_1 + (12 + 6)I_2 + 90 = 0$$

or

$$-90 = -6I_1 + 18I_2$$

Dividing by 2,

$$-45 = -3I_1 + 9I_2 \tag{b}$$

Equations (*a*) and (*b*) constitute a pair of linear simultaneous equations. One may use determinants or substitution to find the currents. With the latter procedure and solving Eq. (*b*) for I_2,

$$I_2 = \frac{-45 + 3I_1}{9} = -5 + \frac{I_1}{3} \tag{c}$$

Substitution of Eq. (*c*) in Eq. (*a*) yields

$$50 = 4I_1 - 3\left(-5 + \frac{I_1}{3}\right)$$

$$= 4I_1 + 15 - I_1 = 15 + 3I_1$$

Solving, $I_1 = (50 - 15)/3 = {}^{35}\!/_3 = 11.67$ A. Substituting $I_1 = 11.67$ A in Eq. (*b*) gives

$$-45 = -3(11.67) + 9I_2 = -35 + 9I_2$$

Solving, $I_2 = -{}^{10}\!/_9 = -1.11$ A. The minus sign denotes that the actual direction of current flow in the second mesh is *clockwise*.

In the literature, the sum of resistances in a mesh is referred to as the *self-resistance* of the mesh. The resistance *common* to two meshes is called the *mutual resistance*. Returning to Fig. 6.7, the sum $R_1 + R_2$ is the self-resistance of mesh I and is denoted by $R_{11} = R_1 + R_2$. Resistors $R_2 + R_3$ constitute the self-resistance of mesh II ($R_{22} = R_2 + R_3$). Resistor R_2 is the mutual resistance linking meshes I and II ($R_{12} = R_{21} = R_2$).

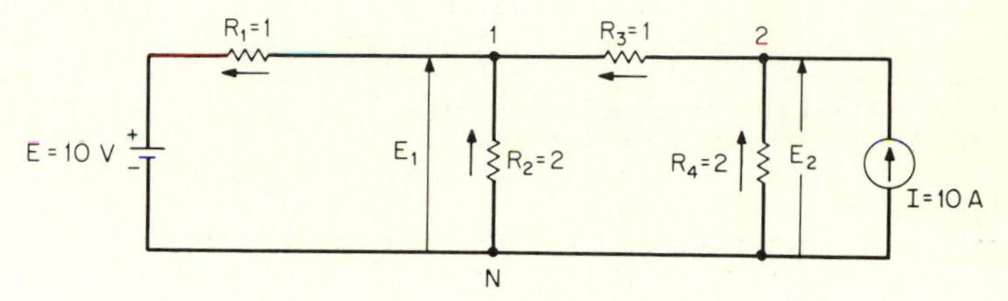

Fig. 6.8 Finding voltages E_1 and E_2 by KCL (Example 6.7).

example 6.7 Apply KCL, and solve for E_1 and E_2 in Fig. 6.8.

solution There are two independent nodes: 1 and 2. Node N is a *common*, or *reference*, node for E_1 and E_2. Taking the direction of currents as flowing from right to left and from bottom to top then, according to our conventions, at node 1 we have

$$-\frac{10 - E_1}{1} + \frac{E_1}{2} + \frac{E_1 - E_2}{1} = 0 \qquad (a)$$

The current flowing in a resistor is equal to the voltage difference across the resistor divided by its value. Thus, the current in R_1 is $(10 - E_1)/1$; in R_2, $E_1/2$; etc.

Simplifying Eq. (*a*),

$$10 = E_1(1 + {}^1\!/_2 + 1) - \frac{E_2}{1}$$

$$= 2.5E_1 - E_2 \qquad (b)$$

At node 2,

$$-\frac{E_1 - E_2}{1} + \frac{E_2}{2} + 10 = 0$$

or

$$-10 = -E_1 + 1.5E_2 \qquad (c)$$

Solving Eq. (*c*) for E_1,

$$E_1 = 10 + 1.5E_2 \qquad (d)$$

Substituting Eq. (*d*) in Eq. (*b*),

$$10 = 2.5(10 + 1.5E_2) - E_2$$
$$= 25 + 2.75E_2$$

Hence,

$$E_2 = \frac{10 - 25}{2.75} = -5.45 \text{ V}$$

and

$$E_1 = 10 + 1.5(-5.45) = 1.8 \text{ V}$$

The negative sign before E_2 denotes that node 2 is *actually negative* with respect to node N. In Fig. 6.8 it was assumed that node 2 was positive with respect to node N.

CONDUCTANCE Conductance G is equal to one divided by the resistance R:

$$G = \frac{1}{R} \qquad (6.15)$$

The unit for conductance is the *mho* (ohm spelled backward).

Letting $G_1 = 1/R_1$, $G_2 = 1/R_2$, $G_3 = 1/R_3$, and $G_4 = 1/R_4$, Eqs. (*b*) and (*c*) in Example 6.7 may be expressed by

$$10 = (G_1 + G_2 + G_3)E_1 - G_3E_2$$
$$-10 = -G_3E_1 + (G_3 + G_4)E_2$$

where $G_1 + G_2 + G_3 = G_{11} = 1 + 0.5 + 1 = 2.5$ mhos is the *self-conductance* of node 1; $G_3 = G_{12} = G_{21} = 1$ mho is the *mutual conductance* between nodes 1 and 2; and $G_3 + G_4 = G_{22} = 1 + 0.5 = 1.5$ mhos is the *self-conductance* of node 2.

6.7 INDEPENDENT AND DEPENDENT SOURCES

Ideal voltage and current sources are termed *independent*, because their values are independent of the circuit in which they are connected. Another type of source, called a *dependent*, or *controlled*, source arises in the representation of active devices, such as transistors, by their models. *A dependent source is one whose value depends on a specific current (or voltage) in a circuit.*

A simplified *hybrid* model of a bipolar junction transistor (BJT), useful under certain operating conditions, is shown in Fig. 6.9*a*. Note the presence of the *current-dependent source*, $h_{FE}I_B$. The value of this source *depends*, for a given value of h_{FE}, on base current I_B.

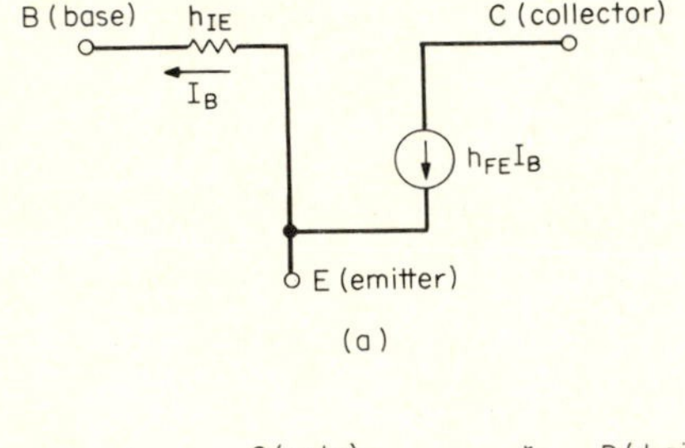

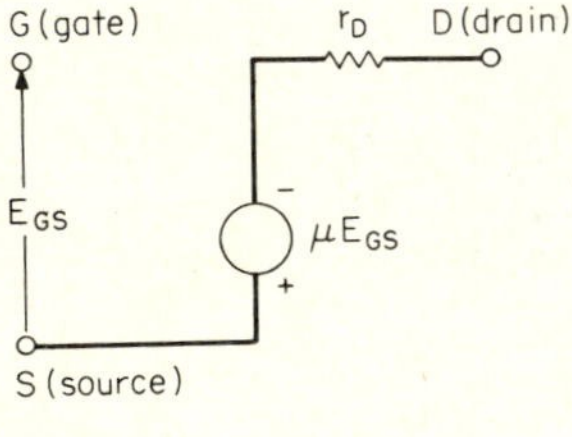

Fig. 6.9 Examples of circuits containing dependent (controlled) sources: (*a*) A current-dependent source, $h_{FE}I_B$. (*b*) A voltage-dependent source, μE_{GS}.

As another example, consider the simplified model of a field-effect transistor (FET) in Fig. 6.9*b*. Here, a *voltage-dependent source*, μE_{GS}, is part of the model. For a given value of μ the voltage *depends* on the gate-to-source voltage E_{GS}.

6.8 TOPOLOGY OF NETWORKS

In complex networks containing many nodes and meshes, it can be exceedingly difficult to choose mesh or nodal analysis for fewest equations. Furthermore, one may not always be certain whether all the nodes or meshes have been considered. Borrowing some results from a branch of mathematics called *topology*, it becomes relatively simple to remove these doubts. The procedure to use is as follows:

1. All sources, independent as well as dependent, in the circuit are set to zero.

In setting a voltage source to zero, it is *shorted out* of the circuit. In setting a current source to zero, it is *open-circuited*, or *removed*, from the circuit.

2. All circuit elements, such as a resistor, are replaced by line segments, called *branches*. The result is referred to as a *graph*.

The minimum number of nodal equations, referred to as *independent equations* N, is

$$N = n - 1 \tag{6.16}$$

where n is the number of nodes contained in the graph. The number of independent mesh (loop) equations M is

$$M = b - n + 1 \tag{6.17}$$

where b is the number of branches contained in the graph.

example 6.8 Draw the graphs and determine the number of independent nodal and mesh equations required for the circuits of (*a*) Fig. 6.7 and (*b*) Fig. 6.8.

solution (*a*) The graph of Fig. 6.7 is drawn in Fig. 6.10*a*. Note that in setting E_1 and E_2 to zero, resistors R_1 and R_2 become connected between nodes A and 1–2. Also, the resistors are represented by line segments (branches). The number of branches is $b = 3$; the number of nodes is $n = 2$. Hence, from Eqs. (6.17) and (6.16), $M = 3 - 2 + 1 = 2$ mesh equations, and $N = 2 - 1 = 1$ nodal equation.

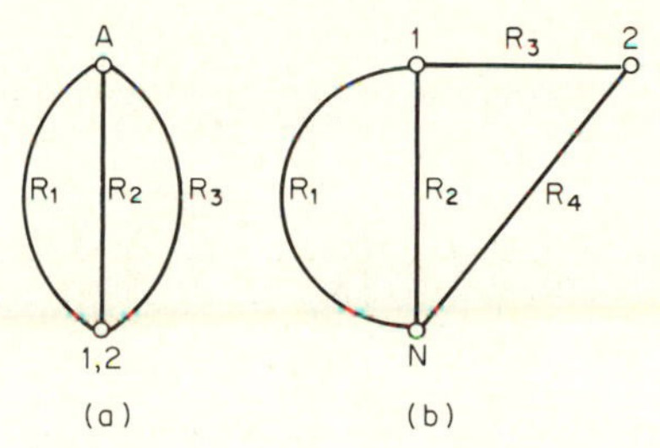

Fig. 6.10 Graphs for the circuits of (*a*) Fig. 6.7 and (*b*) Fig. 6.8 (Example 6.8).

In Example 6.6 we saw that two mesh equations employing KVL were required in the solution of Fig. 6.7. If KCL was used, only one nodal equation would have been required.

(*b*) The graph of Fig. 6.8 is given in Fig. 6.10*b*. Voltage source E is short-circuited out, thereby placing resistor R_1 between nodes 1 and N. Current source I is removed from the circuit. From the graph, $b = 4$ and $n = 3$. Hence, $N = 3 - 1 = 2$ nodal equations, and $M = 4 - 3 + 1 = 2$ mesh equations.

6.9 EQUIVALENT RESISTANCE

RESISTORS IN SERIES For n resistors in series (Fig. 6.11), current I flowing in each resistor is the same. Applying KVL to the circuit,

$$E = I(R_1 + R_2 + \cdots + R_n) = IR_{TS}$$

where the equivalent resistance for n resistors in series R_{TS} is

$$R_{TS} = R_1 + R_2 + \cdots + R_n \tag{6.18}$$

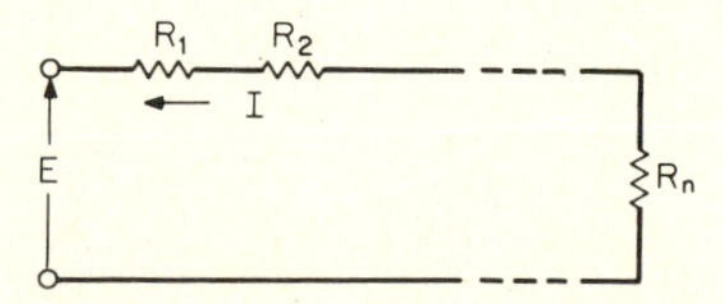

Fig. 6.11 A series circuit containing n resistors.

This result shows that the equivalent resistance of a number of resistors in series is equal to the *sum of the individual resistors.*

If the resistors are *equal,* the equivalent resistance is the product of the number of equal resistors n and the value of the resistor R:

$$R_{TS} = nR \tag{6.19}$$

example 6.9 In a particular prototype it is found that a 1.5-, 2.7-, and 5.8-kΩ resistor are connected in series. What single equivalent resistor can be used in their place?

solution From Eq. (6.18), $R_{TS} = 1.5 + 2.7 + 5.8 = 10$ kΩ.

RESISTORS IN PARALLEL For n resistors in parallel, or in *shunt* (Fig. 6.12), the voltage across each resistor is the same. Applying KCL:

$$I = E\left(\frac{1}{R_1} + \frac{1}{R_2} + \cdots + \frac{1}{R_n}\right)$$

Letting $G_1 = 1/R_1$, $G_2 = 1/R_2$, ... , $G_n = 1/R_n$, we obtain

$$I = E(G_1 + G_2 + \cdots + G_n) = EG_{TP}$$

where the equivalent conductance G_{TP} is

$$G_{TP} = G_1 + G_2 + \cdots + G_n \tag{6.20}$$

The equivalent conductance of a number of resistors in parallel is therefore equal to the sum of their conductances. It is usual, however, to think in terms of resistances.

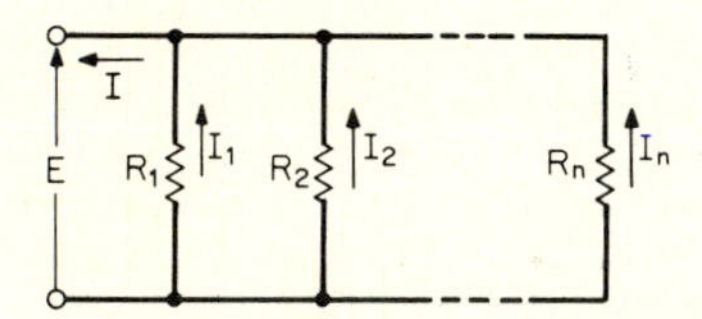

Fig. 6.12 A parallel circuit containing n resistors.

Letting $G_{TP} = 1/R_{TP}$, Eq. (6.20) may be expressed as

$$\frac{1}{R_{TP}} = \frac{1}{R_1} + \frac{1}{R_2} + \cdots + \frac{1}{R_n}$$

Solving for R_{TP},

$$R_{TP} = \frac{1}{1/R_1 + 1/R_2 + \cdots + 1/R_n} \tag{6.21}$$

For two resistors in parallel, $n = 2$, Eq. (6.21) simplifies to

$$R_{TP} = \frac{R_1 R_2}{R_1 + R_2} \tag{6.22}$$

Hence, for two resistors R_1 and R_2 in parallel (often denoted by $R_1 \| R_2$), their equivalent resistance is equal to their product divided by their sum. For more than two resistors in parallel, it is practical to determine their equivalent resistance by considering two resistors at a time.

For n *equal* resistors in parallel, the equivalent resistance is equal to one of the resistors R divided by n:

$$R_{TP} = \frac{R}{n} \tag{6.23}$$

example 6.10 A 60-, a 12-, and a 10-Ω resistor are in parallel. Determine their equivalent resistance.

solution Taking first 60‖12, from Eq. (6.22), $R_{TP1} = (60 \times 12)/(60 + 12) = 10$ Ω. Combining R_{TP1} (10 Ω) with the remaining 10-Ω resistor, from Eq. (6.23), $R_{TP} = {}^{10}\!/_{2} = 5$ Ω.

RESISTORS IN SERIES-PARALLEL Consider a circuit containing some resistors in parallel connected to other resistors in series. To find the equivalent resistance, first reduce the parallel combinations to their equivalent values, and add in series with the remaining resistors in the circuit.

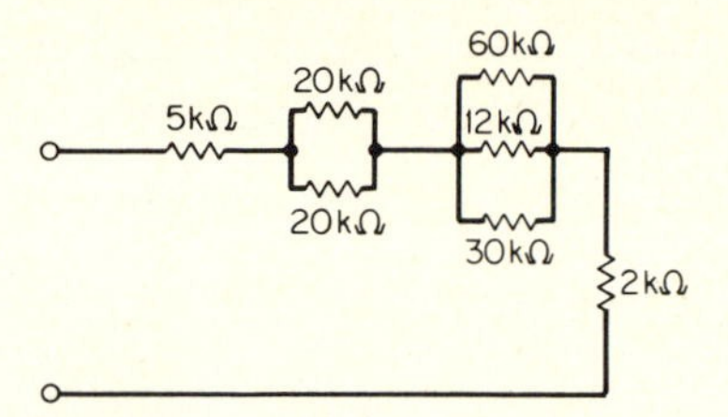

Fig. 6.13 A series-parallel circuit (Example 6.11).

example 6.11 Replace the resistors in the circuit of Fig. 6.13 with a single equivalent resistor.

solution First, we reduce the parallel combinations to their equivalent resistance values. From Eq. (6.23), $20\|20 = {}^{20}\!/_2 = 10$ kΩ. From Eq. (6.22), $60\|12 = (60 \times 12)/(60 + 12) = 10$ kΩ. Combining this with the remaining 30-kΩ resistor, $10\|30 = (10 \times 30)(10 + 30) = 7.5$ kΩ. Hence, $R = 5 + 10 + 7.5 + 2 = 24.5$ kΩ.

6.10 VOLTAGE DIVIDER

Voltage dividers are used to tap off a portion of an impressed voltage across a circuit. An example of a simple series-resistance voltage divider is illustrated in Fig. 6.14. Because the resistors are connected in series, the same current flows in each resistor. Voltage E_n across resistor R_n is therefore equal to

$$E_n = \frac{R_n E}{R_1 + R_2 + \cdots + R_n} \tag{6.24}$$

Voltage E_n is said to be *proportional* to R_n. It must be remembered that Eq. (6.24) is only valid if the *same current* flows in each resistor.

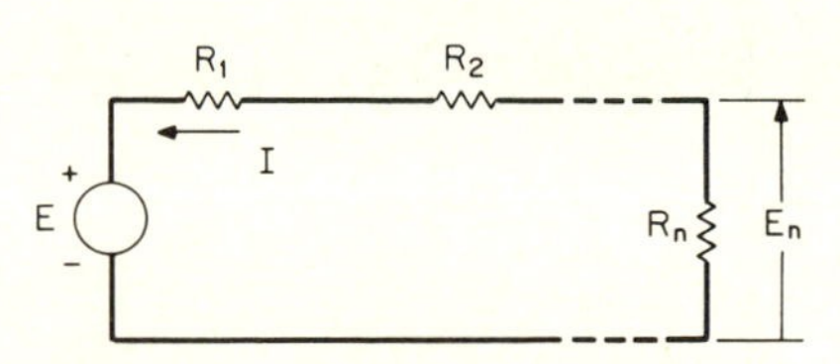

Fig. 6.14 A generalized series voltage divider.

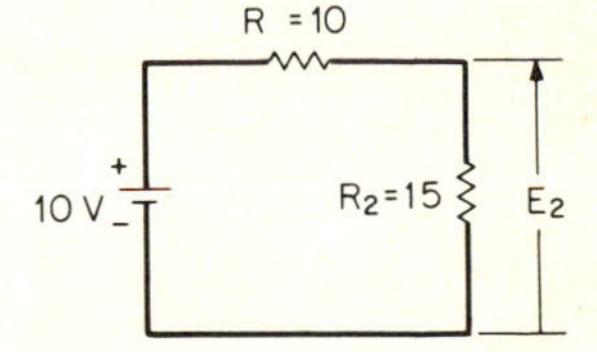

Fig. 6.15 A simple voltage divider (Example 6.12).

example 6.12 Determine E_2 across $R_2 = 15\ \Omega$ in Fig. 6.15.

solution Since the same current flows in each resistor, we can use Eq. (6.24) with $n = 2$. Hence, $E_2 = 15 \times 10/(10 + 15) = 6$ V.

6.11 CURRENT DIVIDER

Current division arises when a current is forced to divide among a number of resistors connected in parallel (Fig. 6.16). Because the voltage across each resistor is the same for n resistors in parallel, the current I_n in resistor R_n is

$$I_n = \frac{G_n I}{G_1 + G_2 + \cdots + G_n} \tag{6.25}$$

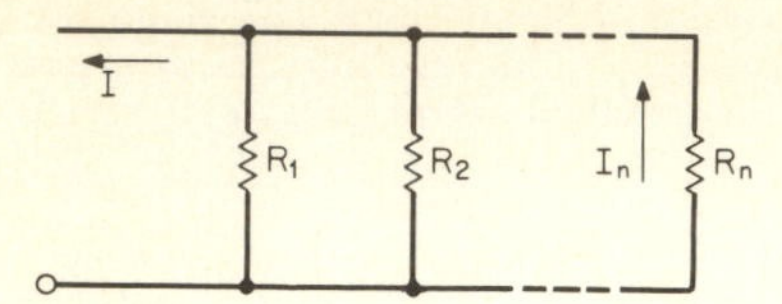

Fig. 6.16 A generalized parallel current divider.

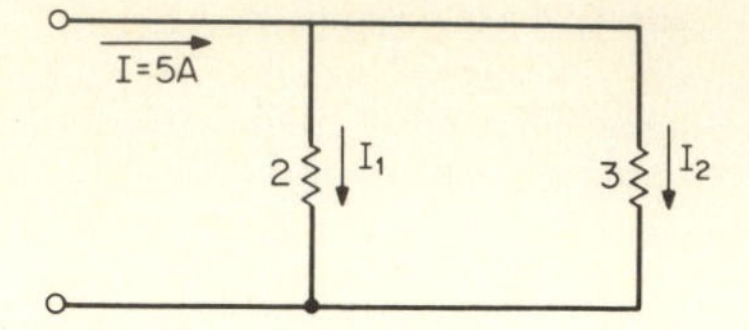

Fig. 6.17 A simple current divider (Example 6.13).

where $G_1 = 1/R_1$, etc. For $n = 2$, in terms of resistance values,

$$I_1 = \frac{R_2 I}{R_1 + R_2} \tag{6.26a}$$

$$I_2 = \frac{R_1 I}{R_1 + R_2} \tag{6.26b}$$

The current is said to divide *inversely* with the resistances in a parallel circuit. Most current flows in the smallest resistor and least current flows in the largest resistor.

example 6.13 Using current division, find I_1 and I_2 in Fig. 6.17.

solution From Eqs. (6.26*a*) and (6.26*b*), $I_1 = 3 \times 5/(2+3) = {}^{15}/_5 = 3$ A; $I_2 = 2 \times 5/(2+3) = {}^{10}/_5 = 2$ A. Note that the greater current, $I_1 = 3$ A, flows through the smaller resistor (2 Ω); and $I_2 = 2$ A flows through the larger resistor (3 Ω). Also, from KCL, $I = I_1 + I_2$ (5 A = 2 A + 3 A). In any problem, Kirchhoff's laws must always be satisfied.

6.12 SUPERPOSITION THEOREM

The superposition theorem, *which applies only to linear circuits*, provides a powerful tool for finding the response of a circuit that is energized by more than one source. In addition, the effect on a circuit to a particular source can be easily obtained.

The superposition theorem states that the response of a circuit energized by a number of independent sources is equal to the sum of the responses of each source with all remaining sources set to zero. Dependent sources are not set to zero—only independent sources. As mentioned earlier in the chapter, an independent voltage source is set to zero by short-circuiting it out; an independent current source is set to zero by removing it from the circuit.

example 6.14 Referring to Fig. 6.18, determine the current in the 60-Ω resistor, I_{60}, using superposition.

solution Setting the 72-V source to zero yields Fig. 6.19*a*. By current division, I_1, due to the 6-A source, is $6 \times 12/(60 + 12) = 1$ A.

Setting the 6-A source to zero yields Fig. 6.19*b*. Current I_2, due to the 72-V source, is $72/(12 + 60) = 1$ A. Hence, $I_{60} = I_1 + I_2 = 1 + 1 = 2$ A.

example 6.15 Using superposition, determine current I_5 flowing in the 5-Ω resistor in Fig. 6.20.

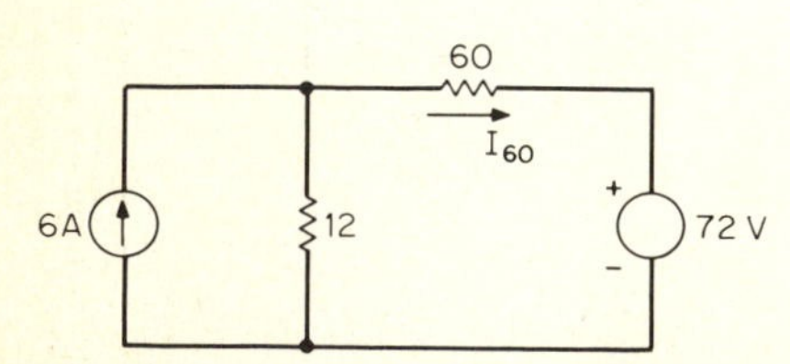

Fig. 6.18 Superposition used to find the current in the 60-Ω resistor, I_{60} (Example 6.14).

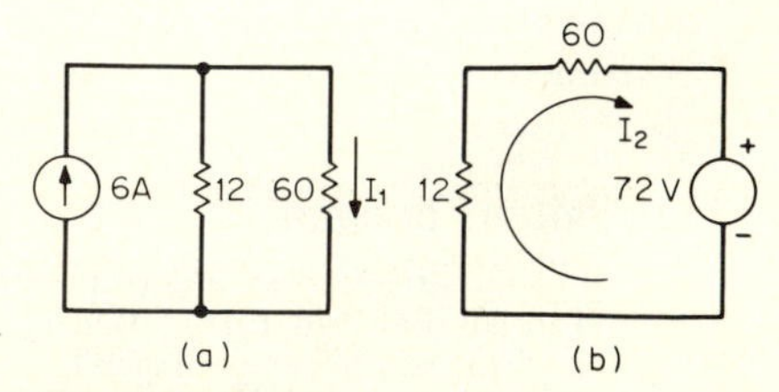

Fig. 6.19 Resultant circuits after setting sources in Fig. 6.18 to zero: (*a*) Voltage source set to zero. (*b*) Current source set to zero.

solution The circuit of Fig. 6.20 contains two independent sources, 20 V and 10 A, and one current-dependent source, $2I$. Current I is the current in the 10-Ω resistor. In the application of the superposition theorem, a dependent source is never set to zero.

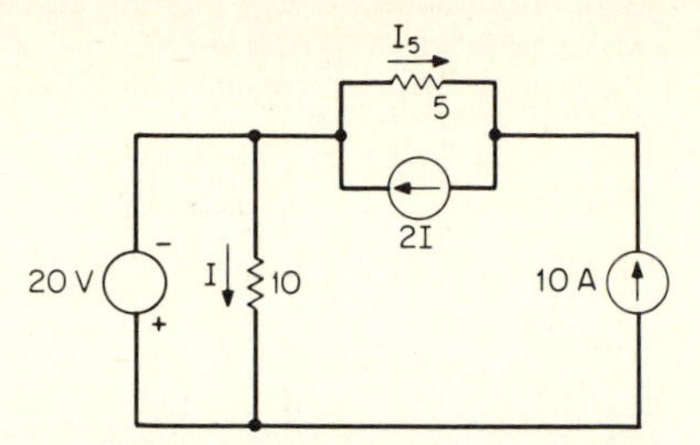

Fig. 6.20 Superposition used to find current I_5 in the 5-Ω resistor (Example 6.15).

Setting first the 10-A source to zero results in Fig. 6.21*a*. Current $I = {}^{20}/_{10} = 2$ A; hence, $2I = I_1 = 2 \times 2 = 4$ A.

Setting the 20-V source to zero yields Fig. 6.21*b*. Because the 10-Ω resistor is now short-circuited, $I = 0$, and $2I$ is therefore also equal to zero. Current $I_2 = -10$ A (the minus sign signifying that the 10 A are flowing in a direction opposite to that assumed for I_2). Hence, $I_5 = I_1 + I_2 = 4 - 10 = -6$ A.

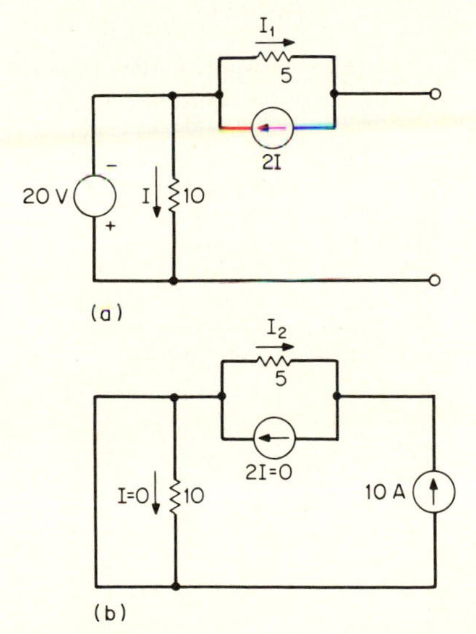

Fig. 6.21 Resultant circuits after setting independent sources in Fig. 6.20 to zero: (*a*) Current source set to zero. (*b*) Voltage source set to zero.

6.13 THÉVENIN'S THEOREM

Consider the linear *two-terminal* (also referred to as a *one-port*) network of Fig. 6.22*a*. The network may contain dependent as well as independent sources. According to Thévenin's theorem, network N, regardless of its complexity, may be replaced by a voltage source, E_{TH}, in series with a resistor, R_{TH} (assuming that the sources in N are direct current) as far as the external terminals a–b are concerned. The resulting circuit, shown in Fig. 6.22*b*, is the Thévenin equivalent circuit of N.

Voltage E_{TH} is called the Thévenin, or *open-circuit*, voltage across terminals *a–b*. One may perceive E_{TH} as the voltage a voltmeter reads when connected across *a–b*. Resistance R_{TH} is called the Thévenin, or *equivalent*, resistance of the network. If the network contains only independent sources, R_{TH} may be calculated as the resistance seen across *a–b* with all independent sources in *N* set to zero. If, however, the network also contains dependent sources, the procedure used is different. Terminals *a–b* are

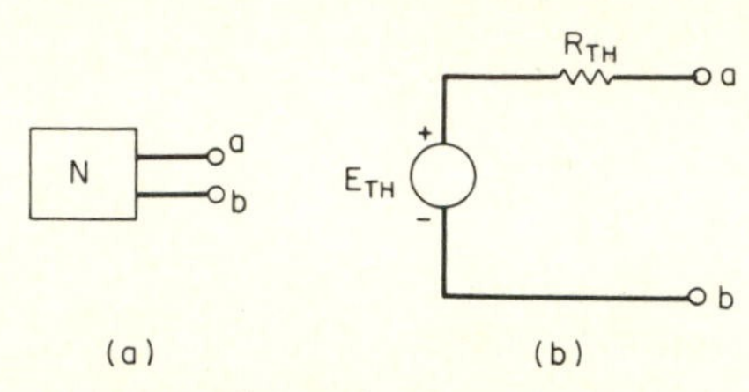

Fig. 6.22 Thévenin's theorem. For the linear two-terminal network *N* of (*a*), the Thévenin equivalent circuit is given in (*b*).

connected together (short-circuited) and the *short-circuit current* I_{sc} flowing in *a–b* is determined. Then,

$$R_{TH} = \frac{E_{TH}}{I_{sc}} \tag{6.27}$$

Equation (6.27) may also be used for networks containing only independent sources.

example 6.16 Determine the Thévenin equivalent circuit for the network of Fig. 6.23*a*.

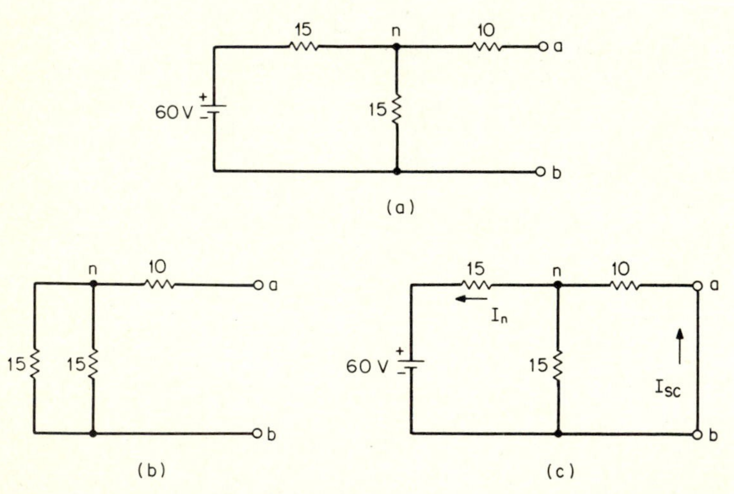

Fig. 6.23 Determining the Thévenin equivalent circuit of a network: (*a*) Given network. (*b*) Finding R_{TH}. (*c*) Finding I_{sc}. (Example 6.16.)

solution Because no current flows in the 10-Ω resistor, the open-circuit (Thévenin) voltage at *a* with respect to *b* is equal to the voltage at *n* with respect to *b*. Hence, by voltage division, $E_{TH} = 60 \times 15/(15 + 15) = 30$ V.

Setting the 60-V source to zero yields Fig. 6.23*b*. The equivalent resistance is $R_{TH} = 15 \| 15 + 10 = {}^{15}\!/_{2} + 10 = 17.5\ \Omega$. Figure 6.24 shows the complete Thévenin equivalent circuit of Fig. 6.23*a*.

We may also find R_{TH} by obtaining I_{sc}. Referring to Fig. 6.23*c*, terminals *a–b* are shown connected together. The current I_n flowing away from *n* is equal to 60/(15 +

$10\|15) = {}^{60}\!/_{21}$ A. By current division, $I_{sc} = ({}^{60}\!/_{21}) \times 15/(15 + 10) = {}^{12}\!/_{7} = 1.7$ A. By Eq. (6.27), $R_{TH} = E_{TH}/I_{sc} = 30/1.7 = 17.5\ \Omega$. This is identical with the value found by setting the 60-V source to zero.

Although it requires additional calculations to find R_{TH} in terms of the open-circuit voltage and short-circuit current, this is the only method that is permitted when dependent sources are present in a network.

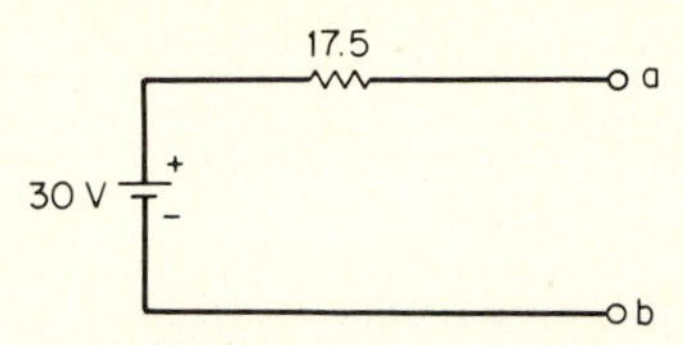

Fig. 6.24 Thévenin equivalent circuit of the network of Fig. 6.23*a*.

example 6.17 Determine the Thévenin equivalent circuit for the network of Fig. 6.25*a* with respect to terminals *S–N*.

solution Figure 6.25*a* is a simplified model of a FET amplifier. The dependent source is $50E_{GS}$, where E_{GS} is the voltage between *G* and *S*. Applying KVL to the mesh, we obtain $50E_{GS} = (1 + 9)I_D = 10I_D$. From the figure, $E_{GS} = 1 - 1(I_D) = 1 - I_D$. Hence, substituting this expression for E_{GS} in the previous equation, we obtain $50(1 - I_D) = 10I_D$. Solving, $I_D = {}^{5}\!/_{6}$ mA (the answer is in milliamperes because the resistance values are in kilohms). The open-circuit voltage is $E_{TH} = 1000 \times {}^{5}\!/_{6} \times 10^{-3} = {}^{5}\!/_{6}$ V.

Connecting terminal *S* to *N* results in Fig. 6.25*b*. For this condition, $E_{GS} = 1$ V and $I_{sc} = {}^{50}\!/_{9}$ mA. Hence, $R_{TH} = E_{TH}/I_{sc} = ({}^{5}\!/_{6})/({}^{50}\!/_{9} \times 10^{-3}) = 150\ \Omega$. The Thévenin equivalent circuit for the FET amplifier is given in Fig. 6.26.

NORTON'S THEOREM According to this theorem, the linear two-terminal network of Fig. 6.22*a* can be represented by the Norton equivalent circuit of Fig. 6.27, where current source I_{sc} is in parallel with the equivalent resistance R_{TH}. Regardless of the

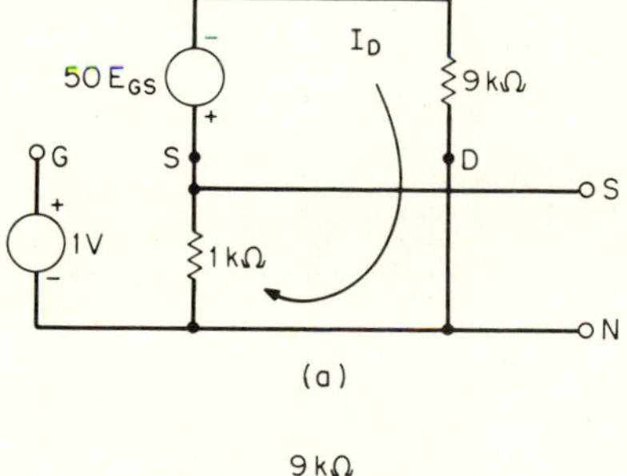

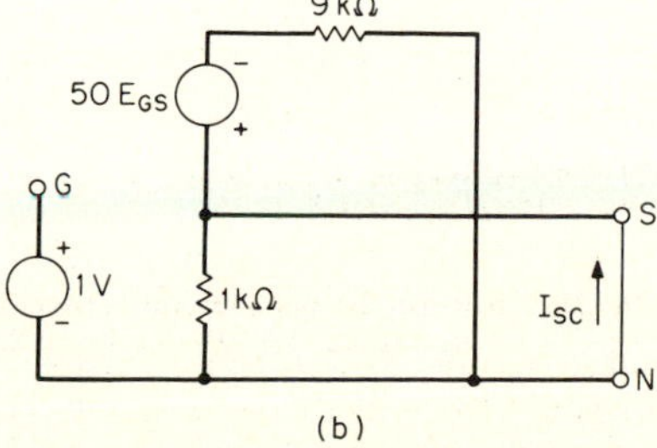

Fig. 6.25 Determining the Thévenin equivalent circuit of a network containing a dependent source: (*a*) Network. (*b*) Finding I_{sc}. (Example 6.17.)

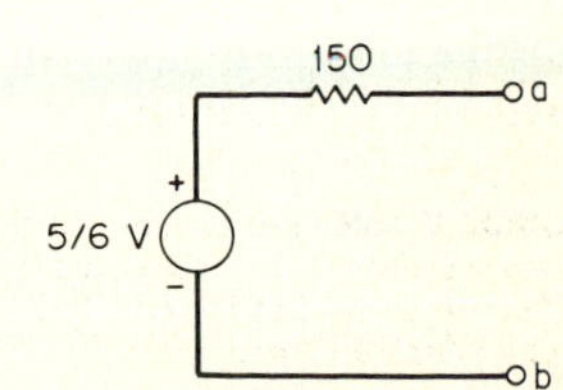

Fig. 6.26 Thévenin equivalent circuit for the network of Fig. 6.25*a*.

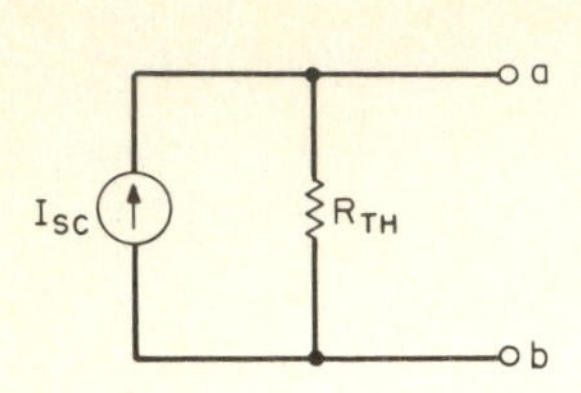

Fig. 6.27 Norton equivalent circuit of a network.

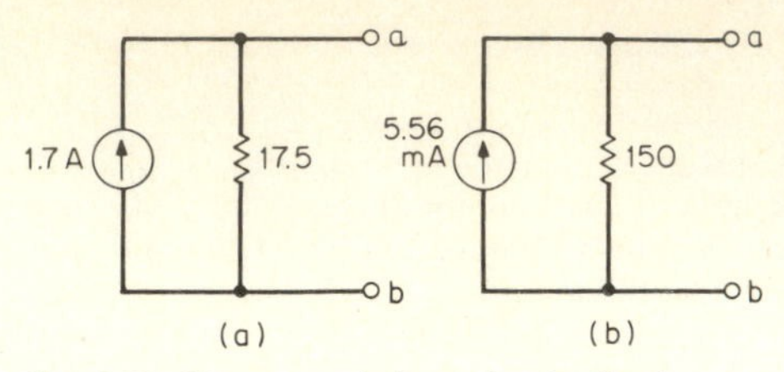

Fig. 6.28 Norton equivalent circuits for the networks of (*a*) Fig. 6.24 and (*b*) Fig. 6.26 (Example 6.18).

complexity of network N, the Norton equivalent circuit of Fig. 6.27 is identical to N with respect to terminals a–b. The procedures for determining I_{sc} and R_{TH} are the same as given for their determination in the Thévenin equivalent circuit.

example 6.18 Determine the Norton equivalent circuits for the networks of (*a*) Example 6.16 and (*b*) Example 6.17.

solution (*a*) From Fig. 6.24, $I_{sc} = 30/17.5 = 1.7$ A, and $R_{TH} = 17.5\ \Omega$. The Norton equivalent circuit is shown in Fig. 6.28*a*.

(*b*) From Fig. 6.26, $I_{sc} = (5/6)/150 = 5.56$ mA, and $R_{TH} = 150\ \Omega$. The Norton equivalent circuit is given in Fig. 6.28*b*.

It should be noted that the direction of the arrow in the current source I_{sc} always points to the plus terminal of the open-circuit voltage source E_{TH}.

6.14 PERIODIC WAVEFORMS

In addition to direct current, there are many waveforms generated and used for voltage and current sources in electronic circuits. Examples of some common waveforms are illustrated in Fig. 6.29. One feature in common with these examples is that their waveshape repeats itself with time. Waveforms having this property are called *periodic waves*. The *sine wave* of Fig. 6.29*a*, also referred to as a *sinusoid*, or *sinusoidal function*, is the most common periodic wave found in circuits. Alternating current (ac), for example, is a sine wave.

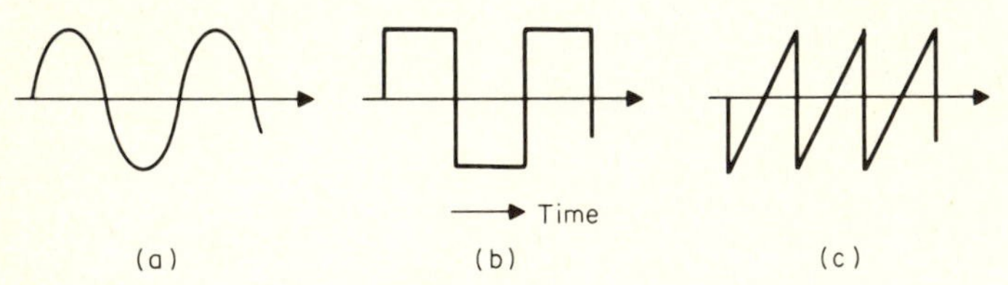

Fig. 6.29 Some commonly used periodic waveforms: (*a*) Sine wave. (*b*) Square wave. (*c*) Triangular wave.

A periodic wave which is not a sine wave is said to be *nonsinusoidal*. The square and triangular waves of Fig. 6.29*b* and *c*, respectively, are examples of nonsinusoidal waves. It is possible to express a periodic nonsinusoidal wave as a sum of sine waves. Such an expression is called a Fourier series.

6.15 SINE WAVE

Because of its importance in the analysis of ac circuits, terms used in characterizing the sine wave are defined in this section.

BASIC DEFINITION *One cycle* of a current and voltage sine wave is illustrated in Fig. 6.30. The current sine wave is expressed by

$$i = I_m \sin \alpha \qquad (6.28a)$$

and the voltage sine wave by

$$e = E_m \sin \alpha \qquad (6.28b)$$

Coefficient I_m is the *peak*, or *maximum amplitude*, of the current; similarly, E_m is the peak, or maximum amplitude, of the voltage. Angle α (*alpha*) is the *displacement angle* of the waveform. Values of sin α for 45° increments are given in Table 6.1. Note, for example, that when $\sin \alpha = 1$, the sine wave reaches its maximum *positive* amplitude; $i = I_m$ and $e = E_m$. When $\sin \alpha = -1$, the sine wave reaches its maximum *negative* amplitude; $i = -I_m$ and $e = -E_m$. These points are indicated in Fig. 6.30.

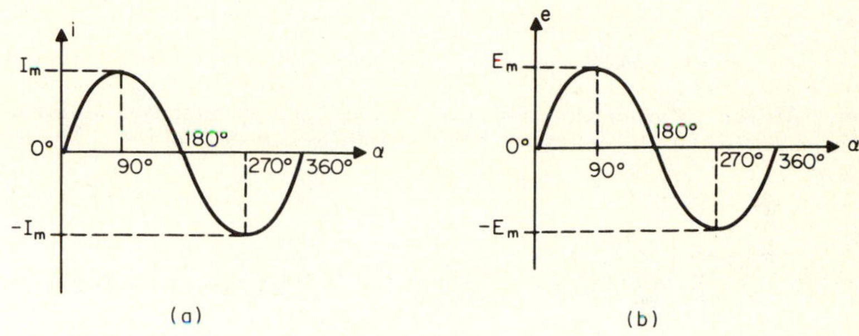

Fig. 6.30 One cycle of a sine wave: (*a*) Current sine wave. (*b*) Voltage sine wave.

TABLE 6.1 Some Values of Sin α

α, degrees	sin α
0	0
30	0.500
45	0.707
90	1
135	0.707
180	0
225	−0.707
270	−1
315	−0.707
360	0

FREQUENCY The frequency f of a sine wave is the number of times a cycle of the wave repeats itself in one second of time. The unit of frequency is the *hertz* (Hz) which equals *one cycle per second* (cps). Using scientific notation, $1000 = 10^3$, $10\,000 = 10^4$, etc., 10^3 Hz = 1 kilohertz (1 kHz); 10^6 Hz = 1 megahertz (1 MHz); and 10^9 Hz = 1 gigahertz (1 GHz). For example, household alternating current has a frequency of 60 Hz (60 cycles per second) in the United States.

PERIOD The period T is the time required for a periodic waveform to complete one cycle. The period is equal to one over the frequency

$$T = \frac{1}{f} \qquad (6.29)$$

If f is expressed in hertz, T is in seconds. For example, the period of 60 Hz is $1/60 = 0.0167$ s = 16.7 ms.

ANGULAR FREQUENCY It is common practice to express the displacement angle of a sine wave by ωt, where ω (*omega*) is the *angular frequency* and is expressed in *radians per second* (rad/s). One radian = 57.3° and 2π rad = 360°. Omega is related to frequency by

$$\omega = 2\pi f \qquad (6.30a)$$

and

$$f = \frac{\omega}{2\pi} = 0.159\omega \qquad (6.30b)$$

Equations for current and voltage written in terms of ω appear as

$$i = I_m \sin \omega t \qquad (6.31a)$$
$$e = E_m \sin \omega t \qquad (6.31b)$$

example 6.19 Given the following expression: $e = 141 \sin 377t$ volts. Find (*a*) its frequency f and (*b*) peak amplitude E_m.

solution (*a*) From the given expression, $\omega = 377$ rad/s. Applying Eq. (6.30*b*), $f = 0.159\omega = 0.159 \times 377 = 60$ Hz.

(*b*) The coefficient of the sine wave is the peak amplitude E_m. In this example, $E_m = 141$ V.

PHASE ANGLE Consider the sine wave of Fig. 6.31*a*. Comparison of this figure with Fig. 6.30*a* at $\alpha = 0$ ($\omega t = 0$) reveals that $i = -0.707I_m$ instead of zero. The reason

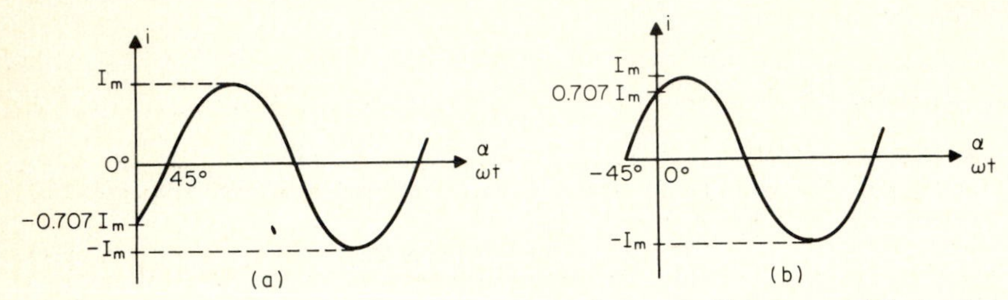

Fig. 6.31 Phase angle: (*a*) Sine wave shifted to the *right* by 45°. (*b*) Sine wave shifted to the *left* by 45°.

for this is that the sine wave of Fig. 6.31*a* has been *shifted to the right* by 45°. To express this mathematically, we write

$$i = I_m \sin (\omega t - 45°) \qquad (6.32a)$$

Similarly, for a voltage, we have

$$e = E_m \sin (\omega t - 45°) \qquad (6.32b)$$

We can verify Eqs. (6.32*a*) and (6.32*b*) with the aid of Table 6.1. If $\omega t = 0$, $i = I_m \sin (-45°) = -0.707I_m$, and $e = -0.707E_m$. For $\omega t = 45°$, $i = I_m \sin (45° - 45°) = I_m \sin (0°) = 0$. Likewise, $e = 0$.

The sine wave of Fig. 6.31*b* has been *shifted to the left* by 45°. This condition is represented by

$$i = I_m \sin (\omega t + 45°) \qquad (6.33a)$$

For a voltage,

$$e = E_m \sin (\omega t + 45°) \qquad (6.33b)$$

Referring again to Table 6.1, if $\omega t = -45°$, $i = I_m \sin (-45 + 45°) = I_m \sin (0°) = 0$. If $\omega t = 0°$, $i = I_m \sin (0 + 45°) = 0.707I_m$. Similar results are found for voltage sine waves.

The angle by which a sine wave has been shifted to the right or to the left is called the *phase angle*. Letting the phase angle be designated by θ (*theta*) in general terms, for current, we can write

$$i = I_m \sin (\omega t + \theta) \qquad (6.34a)$$
$$i = I_m \sin (\omega t - \theta) \qquad (6.34b)$$

Similar expressions can be written for voltage.

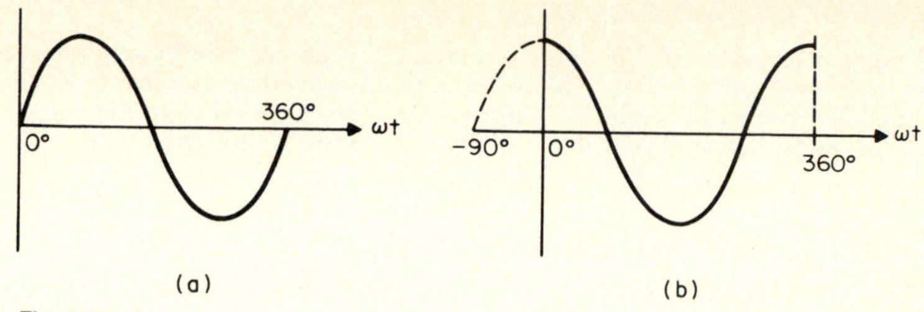

Fig. 6.32 A cosine wave viewed as a sine wave shifted 90° to the left: (*a*) Sine wave. (*b*) Cosine wave.

The relationship between sine and *cosine* waves is illustrated in Fig. 6.32. Figure 6.32*a* displays a sine wave. In Fig. 6.32*b* we have a cosine wave. The cosine wave can be viewed as a sine wave which has been shifted by 90° to the left. Mathematically, we may therefore write

$$\cos \omega t = \sin (\omega t + 90°) \tag{6.35}$$

It is common practice to refer to *both* sine and cosine waves as sinusoidal waves.

PHASE RELATIONSHIPS Consider the two sine waves of Fig. 6.33*a*. In this example, e is said to *lead* i by θ degrees, because the peak value of e is reached before i. An alternative statement is i *lags* e by θ degrees.

Because i reaches its peak value before e in Fig. 6.33*b*, i *leads* e by θ degrees (or e *lags* i by θ degrees). In Fig. 6.33*c* the phase angle between e and i is 0 degrees. In this case, e and i are said to be *in phase*.

In characterizing the phase angle between sinusoids, it is necessary that the waveforms in question *have the same frequency*. If the frequencies of the waveforms differ, one cannot speak of phase angle between waves.

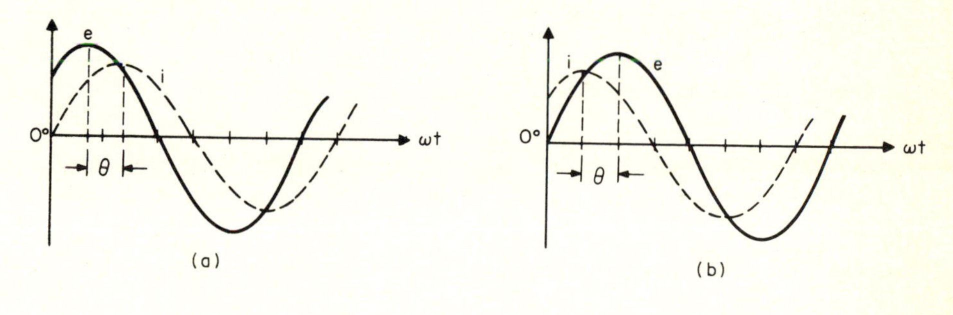

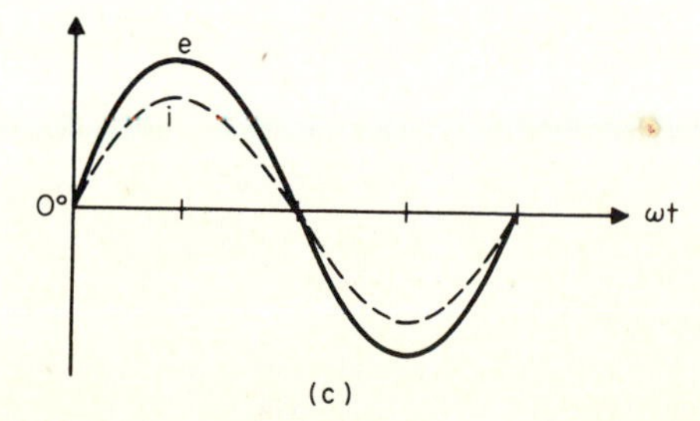

Fig. 6.33 Phase relationships between sine waves of the same frequency: (*a*) e leads i by θ degrees. (*b*) i leads e by θ degrees. (*c*) e and i are in phase.

6.16 AVERAGE (DC) VALUE OF WAVEFORMS

The average, or dc, value of a periodic waveform is what one would read on a dc ammeter or voltmeter. The average value of a periodic waveform is equal to its *net area* over one cycle divided by its period. For example, because the positive half-cycle is equal to the negative half-cycle, the average (dc) value of the sine, square, and triangular waveforms of Fig. 6.29 is equal to zero.

6.17 EFFECTIVE (RMS) VALUE OF WAVEFORMS

The effective, or root-mean-squared (rms), value of a periodic waveform is equal to the direct current (or voltage) which dissipates the same energy in a given resistor. To read the effective values of any current or voltage waveform, a rms meter is required. A multimeter, such as a VOM, is generally calibrated to read the rms value of only a sine wave. For a nonsinusoidal wave, a true rms meter must be used.

The effective value of a sinusoidal voltage E is equal to 0.707 times its peak value E_m:

$$E = 0.707E_m \tag{6.36a}$$

Also,

$$E_m = \frac{E}{0.707} = \sqrt{2}E \simeq 1.4E \tag{6.36b}$$

Similarly, for current,

$$I = 0.707I_m \tag{6.37a}$$

$$I_m = \sqrt{2}I \tag{6.37b}$$

In calculating the effective value of any periodic waveform, its peak value is squared. The squared area is determined and then divided by the period of the waveform. Finally, the square root is taken of the preceding result.

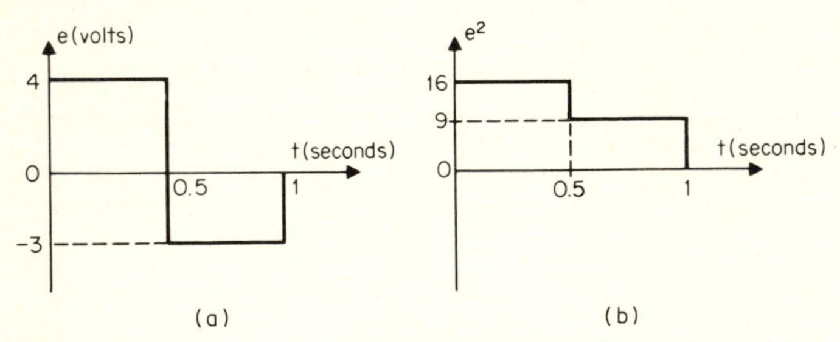

Fig. 6.34 Determining the dc and rms values of a periodic waveform: (*a*) One cycle of periodic waveform. (*b*) Squared area of waveform. (Example 6.20.)

example 6.20 Determine for the periodic waveform of Fig. 6.34*a* the (*a*) dc and (*b*) rms values.

solution (*a*) The area of the positive half-cycle is $4 \times 0.5 = 2$; the area of the negative half-cycle is $-3 \times 0.5 = -1.5$. Hence, the net area is $2 - 1.5 = 0.5$. Dividing by the period 1 s, the dc value is $0.5/1 = 0.5$ V.

(*b*) Squaring 4 yields $4^2 = 16$; also, $(-3)^2 = 9$. The squared waveform is illustrated in Fig. 6.34*b*. Its area is $16 \times 0.5 + 9 \times (1 - 0.5) = 8 + 4.5 = 12.5$. Dividing by the period 1 s, $12.5/1 = 12.5$. Taking the square root of 12.5, the rms voltage is $\sqrt{12.5} = 3.54$ V.

6.18 COMPLEX NUMBERS

Upon excitation of a circuit containing capacitors and/or inductors, in addition to resistors, changing waveforms of current and voltage, called ***transients***, are produced. After the decay of these transients (as mentioned earlier in the chapter, after approximately five time constants), the circuit is said to be in *steady state*. What is of significance is that in steady state, networks energized by sinusoidal sources may be analyzed, using nothing more complicated than simple algebra. In fact, this is exactly what was required in the analysis of dc circuits. In general, however, currents and voltages in ac circuits are represented by *complex numbers*, instead of real numbers as in dc circuits.

A complex number has two parts: a *real part* and an *imaginary part*. Mathematically, a complex number A is represented as $A = a + jb$, where a is the real part and b is the imaginary part of the number. (This representation is called the *rectangular form* of a complex number.) Coefficient j is an *operator* and is equal to the square root of -1: $j = \sqrt{-1}$. Hence, $j^2 = -1$, $j^3 = -j$, $j^4 = 1$, etc. Operator j and its raised powers are illustrated in Fig. 6.35. In plotting complex numbers, the j axis is referred to as the *imaginary axis*, and the x axis as the *real axis*. It is important to note that in multiplying a number by j, the number is located on the imaginary (or j) axis.

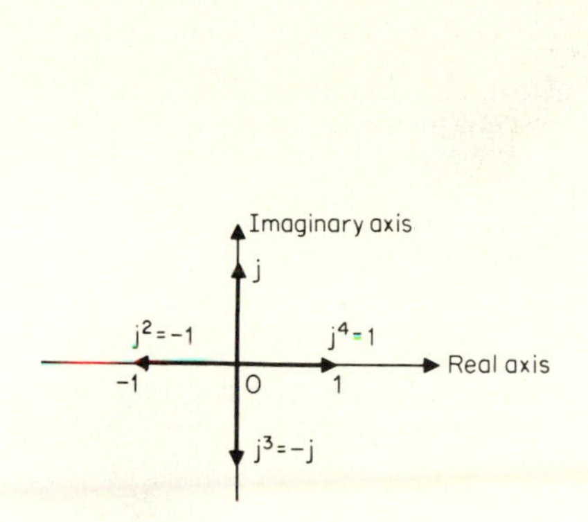

Fig. 6.35 Operator j and its raised powers.

Fig. 6.36 Examples of complex numbers.

A complex number may be thought of as being a *general* number. If the imaginary part is equal to zero, the complex number becomes *real;* if the real part is zero, the complex number becomes *imaginary*. Examples of complex numbers are plotted in Fig. 6.36. The roman numeral refers to the quadrant occupied by the complex number.

COMPLEX CONJUGATE The complex conjugate of a complex number is the complex number with *only* the sign of its imaginary part changed: plus to minus or minus to plus. Examples of complex numbers and their conjugates are listed in Table 6.2.

In each example, only the sign of the imaginary term was changed; the real part was untouched. In general terms, if $A = a + jb$, its complex conjugate, denoted by A^*, is $A^* = a - jb$. If a complex number is multiplied by its conjugate, a *real* number is obtained:

$$AA^* = (a + jb)(a - jb) = a^2 + b^2 \tag{6.38}$$

TABLE 6.2 Examples of Complex Numbers and Their Conjugates

Complex number	Complex conjugate
$3 - j4$	$3 + j4$
$3 + j$	$3 - j$
$-2 + j6$	$-2 - j6$
$-4 - j5$	$-4 + j5$
5	5
$j2$	$-j2$

ADDING AND SUBTRACTING COMPLEX NUMBERS Adding or subtracting complex numbers is easily accomplished. The real and imaginary parts of the numbers are added (or subtracted) *separately*.

example 6.21 Perform the following additions and subtractions: (*a*) $(2 + j6) + (3 - j2)$; (*b*) $(9 + j10) - (-3 + j4)$.

solution (*a*) The real and imaginary parts are added separately: $(2 + j6) + (3 - j2) = (2 + 3) + j(6 - 2) = 5 + j4$.
(*b*) The real and imaginary parts are subtracted separately: $(9 + j10) - (-3 + j4) = [9 - (-3)] + j(10 - 4) = 12 + j6$.

MULTIPLYING AND DIVIDING COMPLEX NUMBERS Although one can multiply or divide, as well as raise to a power or take a root of a complex number in rectangular form, it is easier to perform these operations when the complex number is expressed in *polar form*. In polar form, a complex number A is expressed as

$$A = |A|\underline{/\theta}$$

where $|A|$ is the *magnitude* of the complex number. The magnitude is equal to the square root of the sum of the real part squared and the imaginary part squared:

$$|A| = \sqrt{a^2 + b^2} \tag{6.39a}$$

and θ is the *angle measured from the positive real axis*. The angle is equal to the arc tangent of the imaginary part divided by the real part:

$$\theta = \tan^{-1} \frac{b}{a} \tag{6.39b}$$

A complex number in polar form is depicted graphically in Fig. 6.37. From the

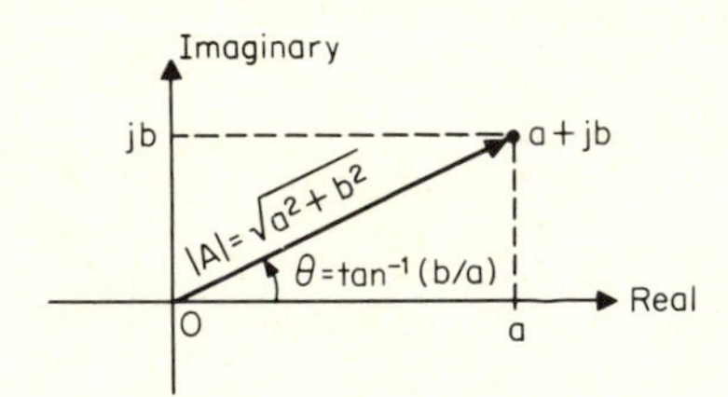

Fig. 6.37 A complex number represented in polar form.

figure, it is seen that the real part is equal to the magnitude multiplied by the cosine of the angle ($\cos \theta$):

$$a = |A| \cos \theta \tag{6.40a}$$

The imaginary part is equal to the magnitude multiplied by the sine of the angle:

$$b = |A| \sin \theta \tag{6.40b}$$

example 6.22 Express the complex numbers of Fig. 6.36 in polar form.

solution (*a*) $3 + j4 = \sqrt{3^2 + 4^2}\ \underline{/\tan^{-1}(4/3)} = 5\underline{/53.1^\circ}$.
(*b*) $-2 + j3 = \sqrt{2^2 + 3^2}\ \underline{/\tan^{-1}(3/-2)} = 3.61\underline{/123.7^\circ}$.
(*c*) $-3 - j3 = \sqrt{3^2 + 3^2}\ \underline{/\tan^{-1}(-3/-3)} = 4.25\underline{/225^\circ}$.
The sign of the angle is always positive when it is taken in a *counterclockwise* direction with respect to the positive real axis. If the angle is taken in a *clockwise* direction, the sign of the angle is *negative*. In Example (*c*), therefore, the positive angle 225° may be expressed as a negative angle: $-(360^\circ - 225^\circ) = -135^\circ$. Hence, one can also write $-3 - j3 = 4.25\underline{/-135^\circ}$.
(*d*) $4 - j2 = \sqrt{4^2 + 2^2}\ \underline{/\tan^{-1}(-2/4)} = 4.5\underline{/331.3^\circ} = 4.5\underline{/-28.7^\circ}$.

With a number expressed in polar form, it is a simple task to multiply complex numbers. The rule is: Multiply the magnitudes, and add algebraically the angles. For example, $5\underline{/30^\circ} \times 12\underline{/-12^\circ} = (5 \times 12)\underline{/30^\circ - 12^\circ} = 60\underline{/18^\circ}$.

For division, the rule is: Divide the magnitudes, and subtract algebraically the angle of the divisor from the angle of the dividend. For example, $24\underline{/-30^\circ} \div 4\underline{/-16^\circ} = (^{24}/_4)\underline{/-30^\circ - (-16^\circ)} = 6\underline{/-14^\circ}$.

For raising a complex number to the *n*th power, the rule is: Raise the magnitude to the *n*th power, and multiply the angle by *n*. For example, $(3\underline{/20^\circ})^3 = (3^3)\underline{/20^\circ \times 3} = 27\underline{/60^\circ}$.

For finding the *n*th root of a complex number, the rule is: Take the *n*th root of the magnitude, and divide the angle by *n*. For example, $(16\underline{/36^\circ})^{1/2} = (16^{1/2})\underline{/36^\circ/2} = 4\underline{/18^\circ}$.

Occasionally, a complex number is given in polar form, and it is required to convert it into the rectangle form. This is accomplished with the aid of Eqs. (6.40*a*) and (6.40*b*).

example 6.23 Convert the following complex numbers from polar to rectangular form: (*a*) $18\underline{/30^\circ}$, (*b*) $10\underline{/-45^\circ}$.

solution (*a*) From Eq. (6.40*a*) the real part $a = |A| \cos \theta$. Hence, $a = 18 \cos 30^\circ = 18 \times 0.866 = 15.6$. From Eq. (6.40*b*) the imaginary part $b = |A| \sin \theta$; therefore, $b = 18 \sin 30^\circ = 18 \times 0.5 = 9$. Finally, $18\underline{/30^\circ} = 15.6 + j9$.

(*b*) $10\underline{/-45^\circ} = 10 \cos(-45^\circ) + j \sin(-45^\circ)$. Because $\cos(-\theta) = \cos\theta$ and $\sin(-\theta) = -\sin\theta$, we can express $10\underline{/-45^\circ}$ by $10 \cos 45^\circ - j10 \sin 45^\circ = 10 \times 0.707 - j10 \times 0.707 = 7.07 - j7.07$.

6.19 REACTANCE

The resistance of an ideal resistor to alternating current is *constant* and *independent of frequency*. The resistance of an ideal inductor or capacitor to alternating current, referred to as *inductive reactance* and *capacitive reactance*, respectively, is *not constant* and *is dependent on frequency*.

Let L = henrys, C = farads, ω = radians per second, and f = hertz. The resistance of an ideal inductor to alternating current is $j\omega L$, where ωL is the inductive reactance X_L,

$$X_L = \omega L = 2\pi f L \tag{6.41}$$

The resistance of an ideal capacitor to alternating current is $-j/\omega C$, where $1/\omega C$ is the capacitive reactance X_C:

$$X_C = \frac{1}{\omega C} = \frac{0.159}{fC} \tag{6.42}$$

As for resistance, the unit for inductive and capacitive reactance is the *ohm*.

It should be noted that, from Eq. (6.41), the inductive reactance *increases* with frequency. At direct current, which can be taken as $f = 0$, the inductive reactance is zero. On the other hand, from Eq. (6.42), the capacitive reactance *decreases* with increasing frequency. At direct current, the capacitive reactance is infinite; a capacitor, therefore, *blocks* direct current.

EQUIVALENT REACTANCE The equivalent reactance of a number of inductors and capacitors in series, parallel, or series-parallel, is calculated in the same manner as for resistors. If magnetic coupling exists between inductors, however, it must be considered in the calculation of equivalent inductive reactance. In this chapter it will be assumed that no magnetic coupling is present.

example 6.24 Calculate the equivalent reactance for the circuit of Fig. 6.38*a*.

solution The first step is to reduce the parallel combinations to single reactances. For the inductors, $(j60)(j12)/j(60 + 12) = j10\ \Omega$. For the capacitors, because they are equal, the

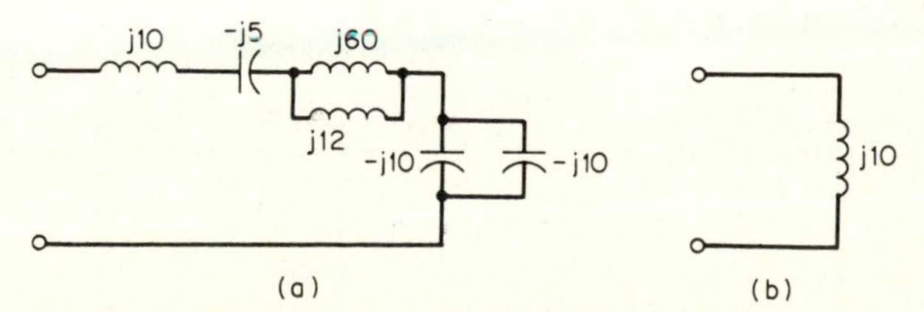

Fig. 6.38 Calculating the equivalent reactance of a circuit: (*a*) Circuit. (*b*) Equivalent reactance. (Example 6.24.)

equivalent capacitive reactance is $-j10/2 = -j5\ \Omega$. The net reactance for the network, therefore, is $j(10 - 5 + 10 - 5) = j10\ \Omega$, as illustrated in Fig. 6.38*b*.

6.20 IMPEDANCE AND ADMITTANCE

In general, ac circuits contain resistors, inductors, and capacitors. Letting the net resistance equal R and the net reactance $X_L - X_C = X$, the impedance Z in ohms is

$$Z = R + jX = |Z|\underline{/\theta}.$$

where

$$|Z| = \sqrt{R^2 + X^2} \tag{6.43a}$$

$$\theta = \tan^{-1}\frac{X}{R} \tag{6.43b}$$

The relationship of R, X, and $|Z|$ is illustrated by the *impedance triangle* of Fig. 6.39. It is seen from the triangle that when $X = 0$, the angle is zero and the impedance is

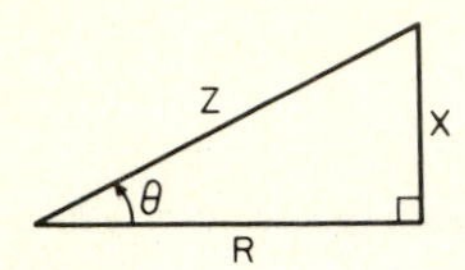

Fig. 6.39 Impedance triangle.

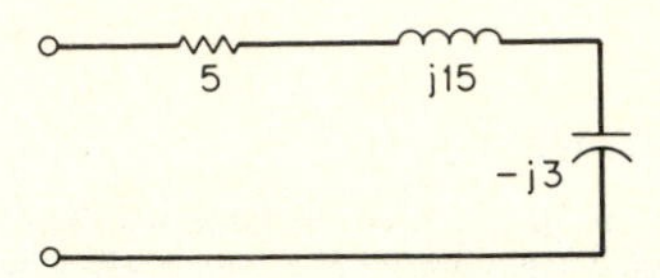

Fig. 6.40 Finding the impedance of a circuit (Example 6.25).

equal to R. For this condition, the circuit is purely resistive. As the value of X increases, the angle also increases. When $R = 0$, $\theta = 90°$, $|Z| = X$, and the network is purely reactive.

example 6.25 Find the impedance of Fig. 6.40.

solution The net resistance is $R = 5\ \Omega$, and the net reactance is $X = 15 - 3 = 12\ \Omega$. Hence, $Z = 5 + j12\ \Omega$.

Expressed in polar form, $|Z| = \sqrt{5^2 + 12^2} = \sqrt{169} = 13$; $\theta = \tan^{-1}(^{12}/_5) = 67.4°$. In polar form, $Z = 13\underline{/67.4°}$.

ADMITTANCE Admittance Y is defined as one over Z:

$$Y = \frac{1}{Z} \tag{6.44}$$

As for conductance, the unit of admittance is the *mho*. In general, Y may be expressed as

$$Y = G + jB = |Y|\underline{/\phi}$$

where G is the net conductance and B the net *susceptance* of the circuit. The conductance and susceptance, expressed in terms of R and X, are

$$G = \frac{R}{R^2 + X^2} \tag{6.45a}$$

$$B = -\frac{X}{R^2 + X^2} \tag{6.45b}$$

The magnitude of Y, $|Y|$, and angle ϕ are expressed by

$$|Y| = \sqrt{G^2 + B^2} \tag{6.46a}$$

$$\phi = \tan^{-1}\frac{B}{G} \tag{6.46b}$$

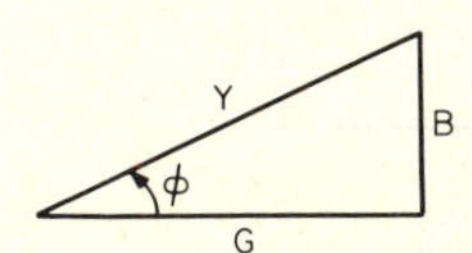

Fig. 6.41 Admittance triangle.

As for impedance, the *admittance triangle* of Fig. 6.41 can be drawn. If $B = 0$, then $Y = G = 1/R$, and the circuit is purely resistive. As B increases, ϕ increases; when $G = 0$, the angle is 90°, $Y = jB$, and the circuit is purely reactive.

example 6.26 Verify Eqs. (6.45*a*) and (6.45*b*).

solution By definition, $Y = 1/Z$. Substituting $Z = R + jX$, we obtain $Y = 1/(R + jX)$. The complex conjugate of the denominator is $R - jX$. Multiplying the numerator and denominator by the complex conjugate of the denominator yields $Y = (R - jX)/(R^2 + X^2) = R/(R^2 + X^2) - jX/(R^2 + X^2)$. Hence, $G = R/(R^2 + X^2)$ and $B = -X/(R^2 + X^2)$.

OHM'S LAW FOR AC CIRCUITS If Z is substituted for R, then Ohm's law is expressed by

$$I = \frac{E}{Z} \tag{6.47a}$$

If G is replaced by Y, we have

$$I = EY \tag{6.47b}$$

In general, current I and voltage E will be complex numbers.

6.21 PHASOR DIAGRAMS

A *phasor diagram* is a graphical description of the relationships of currents and voltages present in an ac circuit. Phasor diagrams provide insight into circuit performance and are useful in the graphical solution of some ac problems.

Figure 6.42*a* shows a voltage source of $E\angle 0°$ volts impressed across an ideal resistor

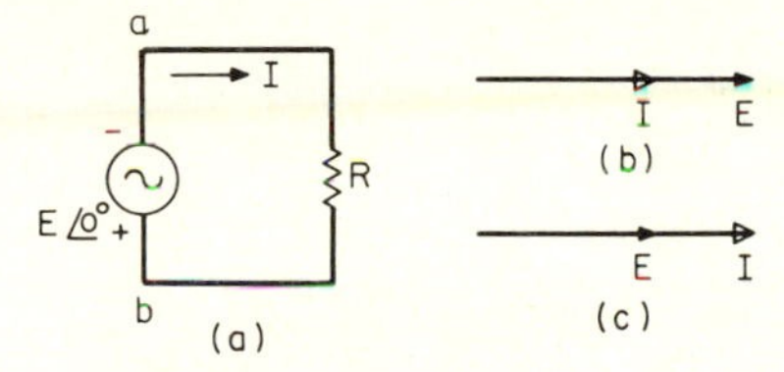

Fig. 6.42 Phasor diagrams for an ideal resistor: (*a*) Circuit. (*b*) Voltage used as the reference phasor. (*c*) Current used as the reference phasor.

of R ohms. Note that the impressed voltage is expressed in polar form. Often, a sine-wave symbol is drawn within the circle to indicate a sine-wave source. The plus sign indicates that at the time we are examining the circuit, point b is *positive* with respect to point a.

By Ohm's law, current I is

$$I = \frac{E\angle 0°}{R} = \left(\frac{E}{R}\right)\angle 0°$$

Because the angle of I is identical with that of E (zero degrees) the voltage and current are in phase.

The relationship between E and I is illustrated in the phasor diagram of Fig. 6.42*b*. A directed line, to a convenient scale, is drawn to represent $E\angle 0°$; this is called the *reference phasor*. Because the current and voltage are in phase, another directed line, to a convenient scale to represent the current phasor I, is superimposed on the voltage phasor. To distinguish between the voltage and current phasors, the current phasor is denoted by an open arrow.

Current may also be selected as the reference phasor. This is shown in Fig. 6.42*c*, where I and E are drawn to scales different from those of Fig. 6.42*b*.

In Fig. 6.43*a* a voltage source is impressed across an ideal inductor. By Ohm's law,

$$I = \frac{E}{jX_L}$$

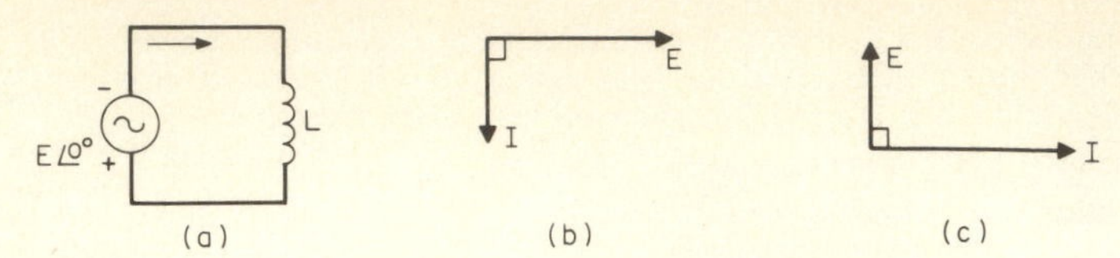

Fig. 6.43 Phasor diagrams for an ideal inductor: (*a*) Circuit. (*b*) Voltage used as the reference phasor. (*c*) Current used as the reference phasor.

Expressing jX_L in polar form, from Eqs. (6.43*a*) and (6.43*b*), $|Z| = X_L$, and $\theta = \tan^{-1}(X_L/0) = 90°$. Hence, $jX_L = X_L\angle 90°$. Substitution of this quantity in the expression for I yields

$$I = \frac{E}{X_L\angle 90°} = \frac{E}{X_L}\angle -90°$$

From the final equation it is seen that the current in an ideal inductor *lags* the impressed voltage by 90° (or the impressed voltage *leads* the current by 90°).

Phasor diagrams using voltage and current as the reference phasors are drawn in Fig. 6.43*b*, and *c*, respectively. One may think of a phasor diagram as a *rigid body*. By convention, positive rotation of a phasor is taken as being counterclockwise. It is seen, for example, that if the phasor diagram of Fig. 6.43*b* is rotated 90° counterclockwise, the resultant diagram appears as in Fig. 6.43*c*. The relation of I lagging E (or E leading I) by 90° is always maintained.

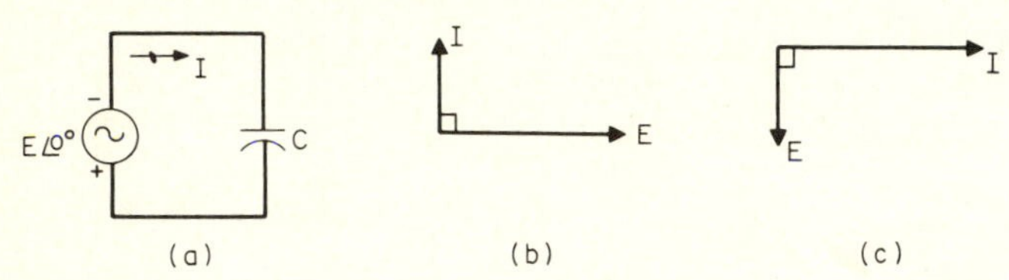

Fig. 6.44 Phasor diagrams for an ideal capacitor: (*a*) Circuit. (*b*) Voltage used as the reference phasor. (*c*) Current used as the reference phasor.

An ideal capacitor excited by a voltage source is illustrated in Fig. 6.44*a*. By Ohm's law,

$$I = \frac{E}{-jX_C} = \frac{E}{X_C}\angle 90°$$

since, in polar form, $-jX_C = X_C\angle -90°$. The current flowing in an ideal capacitor *leads* the impressed voltage by 90° (or the voltage *lags* the current by 90°). This is opposite to that of the ideal inductor. Phasor diagrams, with voltage and current used as reference phasors, are given in Fig. 6.44*b* and *c*, respectively.

6.22 AC CIRCUIT ANALYSIS

The techniques for analyzing ac circuits in steady state are identical with those used for the analysis of dc circuits. As mentioned earlier, the difference is that complex numbers are generally employed in ac analysis whereas real numbers are used in dc analysis. The following group of examples illustrate the methods of analyzing ac circuits.

example 6.27 Referring to Fig. 6.45: (*a*) find the rms current I; (*b*) draw the phasor diagrams with the impressed voltage and current as reference phasors; (*c*) write an expression for the current as a function of time.

solution (*a*) The impedance of the circuit is $Z = (2 + 2) + j(5 - 1) = 4 + j4\ \Omega$. In polar form, $|Z| = \sqrt{4^2 + 4^2} = \sqrt{32} \simeq 5.7$; $\theta = \tan^{-1}(4/4) = 45°$. Hence, $Z = 5.7\angle 45°$.

Current I is $100\angle 0°/5.7\angle 45° = 17.5\angle -45°$ A. Expressed in rectangular form, $I = 17.5 \cos(-45°) + j17.5 \sin(-45°) = 17.5 \cos 45° - j17.5 \sin 45° = 17.5(0.707 - j0.707) = 12.4 - j12.4$ A.

(*b*) Using the voltage as the reference phasor, the phasor diagram appears as in Fig. 6.46*a*. With current as the reference phasor, the phasor diagram is shown in Fig. 6.46*b*. In either representation, the current lags the voltage by 45°.

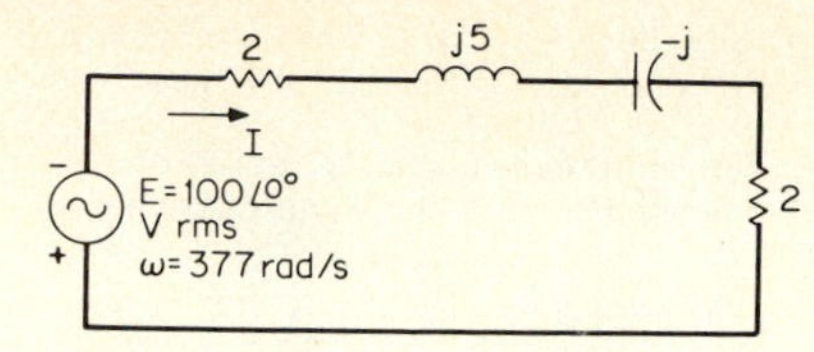

Fig. 6.45 Analyzing an ac circuit (Example 6.27).

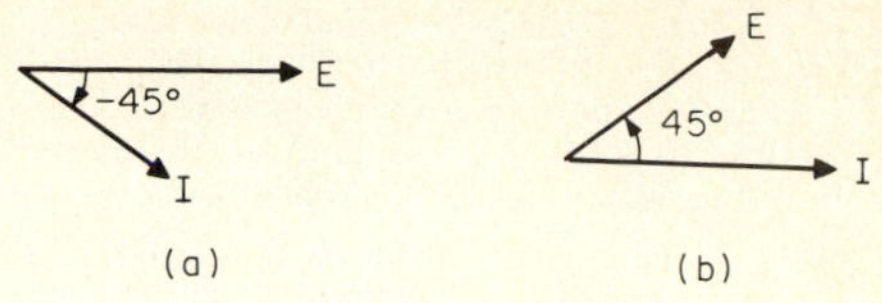

Fig. 6.46 Phasor diagrams for Fig. 6.45: (*a*) Voltage used as the reference phasor. (*b*) Current used as the reference phasor.

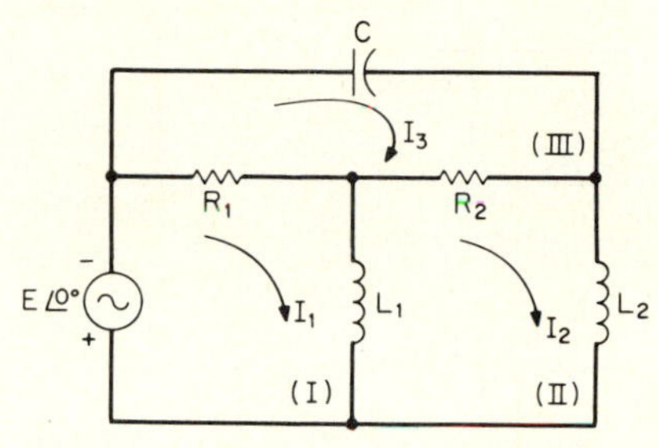

Fig. 6.47 Writing loop equations for a three-mesh circuit (Example 6.28).

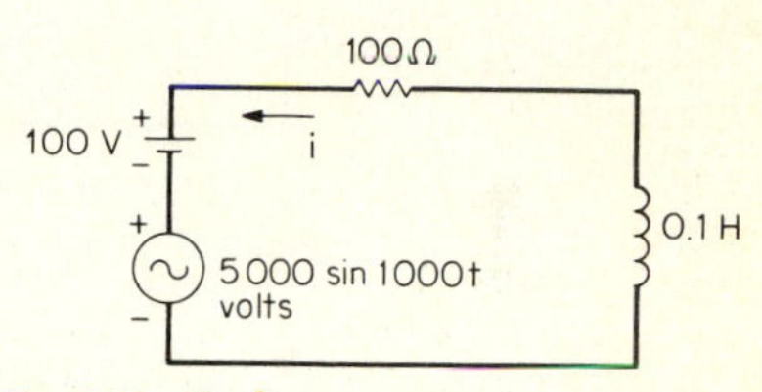

Fig. 6.48 Finding current i by superposition (Example 6.29).

(*c*) From Eq. (6.34*b*), current as a function of time for a lagging phase angle is $i = I_m \sin(\omega t - \theta)$. In (*a*) it was found that the magnitude of the rms current is $I = 17.5$ A. From Eq. (6.37*b*), $I_m = \sqrt{2}I = 1.4 \times 17.5 = 24$ A. The angular frequency is $\omega = 377$ rad/s, and $\theta = 45°$. Therefore, $i = 24(377t - 45°)$.

example 6.28 Write the required loop equations for the network of Fig. 6.47. Assume that no magnetic coupling exists between L_1 and L_2.

solution Based on the topology of the network, three independent equations are required. Using KVL, we obtain

For Mesh I: $E_1\underline{/0°} = (R_1 + j\omega L_1)I_1 - j\omega L_1 I_2 - R_1 I_3$
For Mesh II: $0 = -j\omega L_1 I_1 + [R_2 + j\omega(L_1 + L_2)]I_2 - R_2 I_3$
For Mesh III: $0 = -R_1 I_1 - R_2 I_2 + (R_1 + R_2 - j/\omega C)I_3$

example 6.29 Using superposition, find the steady-state current as a function of time for for the circuit of Fig. 6.48.

solution In steady state, the inductor acts like a short circuit to direct current. Setting the ac source to zero results in Fig. 6.49*a*. Current I_1, owing to the dc source, is $I_1 = {}^{100}/_{100} = 1$ A.
Setting the dc source to zero yields Fig. 6.49*b*. The angular frequency (from 5000 sin 1000*t*) is 1000 rad/s; hence, $X_L = \omega L = 1000 \times 0.1 = 100\ \Omega$. Note that the peak value of the ac source, $5000\underline{/0°}$ V, is shown instead of its rms value. The reason for this is that we are solving directly for the peak current.
The impedance of the circuit of Fig. 6.49*b* is $Z = 100 + j100 = 141\underline{/45°}$ in polar form. The peak value of i_2, I_{m2}, is $5000\underline{/0°}/141\underline{/45°} = 35.4\underline{/-45°}$. Therefore, $i_2 = 35.4(\sin 1000t - 45°)$. Finally, by superposition, $i = I_1 + i_2 = 1 + 35.4 \sin(1000t - 45°)$ A.

example 6.30 With respect to terminals *a–b*, determine the Thévenin and Norton equivalent circuits of Fig. 6.50.

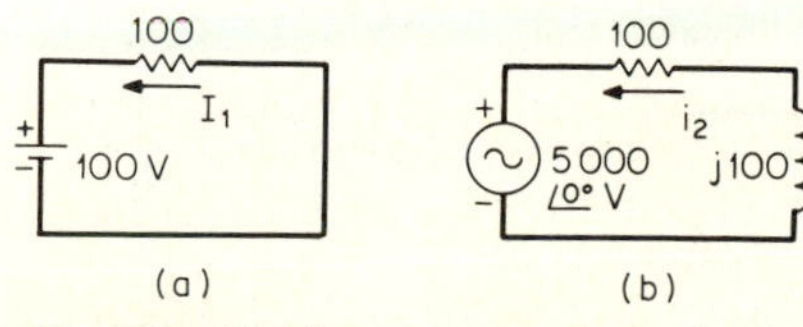

Fig. 6.49 Resultant circuit obtained after setting sources in Fig. 6.48 to zero: (*a*) Ac source set to zero. (*b*) Dc source set to zero.

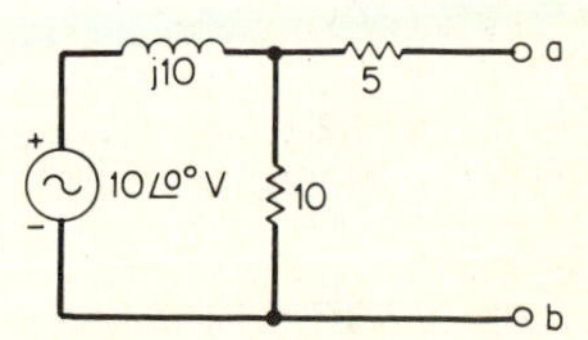

Fig. 6.50 Finding the Thévenin and Norton equivalent circuits of the network with respect to terminals *a–b* (Example 6.30).

solution Because no current flows in the 5-Ω resistor, the open-circuit (Thévenin) voltage across terminals a–b is equal to the voltage across the 10-Ω resistor. By voltage division, $E_{TH} = 10 \times 10\underline{/0^\circ}/(10 + j10) = 100\underline{/0^\circ}/14.1\underline{/45^\circ} = 7.07\underline{/-45^\circ}$ V.

Setting the voltage source to zero, the equivalent (Thévenin) *impedance*, Z_{TH}, is $10 \| j10 + 5 = j100/(10 + j10) + 5$. Multiplying the numerator and denominator by the complex conjugate of the denominator of the fractional term, we have

$$\frac{j100(10 - j10)}{(10 + j10)(10 - j10)} = \frac{j1000 + 1000}{100 + 100} = \frac{1000}{200} + \frac{j1000}{200}$$

$= 5 + j5$. Hence, $Z_{TH} = 5 + j5 + 5 = 10 + j5$ Ω. The Thévenin equivalent circuit is shown in Fig. 6.51*a*.

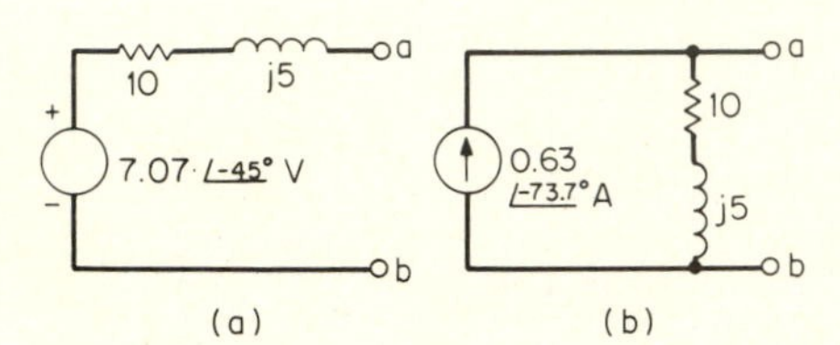

Fig. 6.51 Equivalent circuits for the network of Fig. 6.50: (*a*) Thévenin. (*b*) Norton.

To arrive at the Norton equivalent circuit, the short-circuit current I_{sc} is determined from the expression $I_{sc} = E_{TH}/Z_{TH}$. Expressing Z_{TH} in polar form, $|Z_{TH}| = \sqrt{10^2 + 5^2} = \sqrt{125} = 11.2$; $\theta = \tan^{-1} (5/10) = 28.7^\circ$. Hence, $Z_{TH} = 11.2\underline{/28.7^\circ}$. The short-circuit current is $I_{sc} = 7.07\underline{/-45^\circ}/11.2\underline{/28.7^\circ} = 0.63\underline{/-45^\circ - 28.7^\circ} = 0.63\underline{/-73.7^\circ}$ A. The Norton equivalent circuit is shown in Fig. 6.51*b*.

6.23 POWER IN AC CIRCUITS

The average (*real*) power P dissipated in an ac circuit is

$$P = EI \cos \theta \tag{6.48}$$

where E is the rms voltage across the circuit, I is the rms current flowing in the circuit, and θ is the phase angle between E and I. Term $\cos \theta$ is defined as the *power factor* PF of the circuit:

$$\text{PF} = \cos \theta \tag{6.49}$$

Maximum power occurs when the power factor is equal to *unity* ($\cos \theta = 1$). Referring to Table 6.3, it is seen that $\cos \theta = 1$ when θ is equal to 0°, 360° (or a multiple of 360°). This occurs only in a *purely resistive circuit* where the voltage and current are in phase. In a purely inductive circuit the current lags the voltage by 90° ($\theta = -90^\circ$) and $\cos(-90^\circ) = 0$. The average power dissipated in a purely inductive circuit is, therefore, equal to zero. Similarly, the current in a purely capacitive circuit leads the voltage by 90° ($\cos 90^\circ = 0$), and the average power dissipated is also zero.

In summary, only resistors and resistive circuits dissipate power and is equal to EI (the power factor is unity). For inductors, capacitors, or circuits containing only capacitors and inductors, no power is dissipated (the power factor is zero). For a circuit containing resistance *and* reactance, the phase angle is between 0 and 90°. If resistance

TABLE 6.3 Some Values of Cos θ

θ, degrees	cos θ
0	1
±30	0.866
±45	0.707
±60	0.5
±90	0
±360	1

predominates, the angle will be close to 0°; if reactance predominates, the angle will be close to 90°.

Average power dissipated in an ac circuit also equals I^2R and E^2/R, where I is the rms current flowing in the resistor, and E is the rms voltage across the resistor.

REACTIVE POWER A useful concept in ac circuit analysis is *reactive power* P_q. The unit of reactive power is the *var*. Reactive power is equal to

$$P_q = EI \sin \theta \tag{6.50}$$

where E, I, and θ have been previously defined. Reactive power is also equal to I^2X and E^2/X, where I is the rms current in the reactance, E is the voltage across the reactance, and X is the net reactance.

APPARENT POWER Apparent power P_a is the product of the voltage E impressed across the circuit and the current I flowing in the circuit:

$$P_a = EI \tag{6.51}$$

The unit of apparent power is the voltampere (VA).

The relationships among real, reactive, and apparent power may be displayed graphically by a *power triangle*. The power triangle of Fig. 6.52*a* is for a circuit with net inductive reactance. For a circuit with net capacitive reactance, the power triangle appears as in Fig. 6.52*b*.

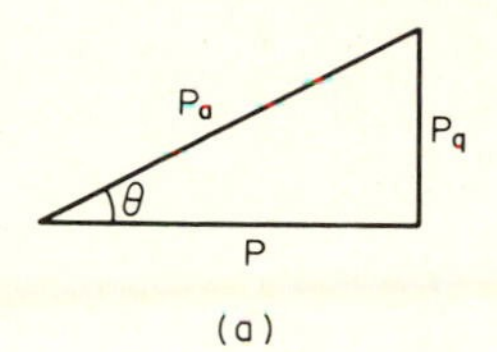

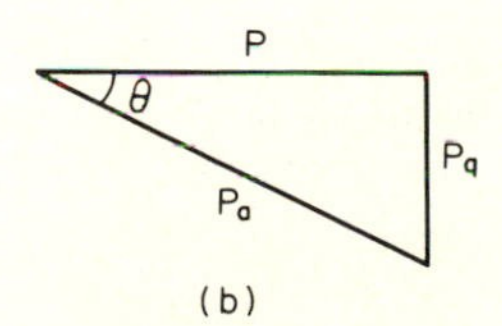

Fig. 6.52 Power triangles: (*a*) For a circuit with net inductive reactance. (*b*) For a circuit with net capacitive reactance.

By the Pythagorean theorem, the relationship of P, P_q, and P_a is expressed by

$$P_a^2 = P^2 + P_q^2 \tag{6.52}$$

example 6.31 For an ac circuit, $E = 100$ V rms, and $I = 10$ A rms with a lagging phase angle of 30°. Determine (*a*) power factor, (*b*) average power, (*c*) reactive power, and (*d*) apparent power.

solution (*a*) The power factor, pf = cos θ. From Table 6.3, cos 30° = 0.866. Hence, pf = 0.866 lagging.

(*b*) $P = EI \cos \theta = 100 \times 10 \times 0.866 = 866$ W.

(*c*) $P_q = EI \sin \theta$; from Table 6.1, sin 30° = 0.5. Hence, $P_q = 100 \times 10 \times 0.5 = 500$ vars.

(*d*) $P_a = EI = 100 \times 10 = 1000$ W = 1 kW.

The results may be verified by Eq. (6.52): $1000^2 = 866^2 + 500^2$.

Chapter 7

Meters and Measurements

7.1 INTRODUCTION

It is important for a technician to have an understanding of the difficulties involved in making exact measurements if he is to accurately evaluate the readings on his meter. The first part of this chapter is devoted to the problems of obtaining and interpreting accurate measurements. This information is also valuable for someone who is planning to purchase meters.

Measurements of current are made indirectly—either by the effect of their magnetic field or by the heat produced when the current flows through a resistance. A *meter movement* is the electromechanical part of a meter that operates in accordance with either the magnetic or heat effect of the current. A number of different meter movements are discussed in this chapter.

By using a sensitive meter movement for the indicator, it is possible to make ammeters, voltmeters, ohmmeters, and electronic multimeters for a wide variety of electrical measurements. The required circuitry is discussed in this chapter, including the procedure for calculating the shunt and series resistance values for the meters.

When accurate measurements of resistance, inductance, or capacitance are to be made, a bridge circuit is generally used. The theory of bridge circuits and the circuits for various bridges are included in this chapter.

7.2 PARALLAX AND OTHER PROBLEMS IN USING INSTRUMENTS

Consider the problem of reading an ammeter. If two different people sight along the needle of the meter movement, the slight difference in the position of their eyes while they are making the reading will make a difference in their interpretation of the current value. You may have encountered the same problem when reading a speedometer in an automobile. The driver, who sits *behind* the speedometer, will see the speed as one value, whereas the passenger looking at the speedometer from his position will interpret its reading as being entirely different. This problem is known as *parallax*. It can be reduced somewhat by placing a mirror behind the meter.

Figure 7.1 shows a meter with a mirrored scale. The proper way to use this meter is to sight along the needle with one eye in such a way that you cannot see the reflection of the needle in the mirror. When you do this, your eye is looking exactly perpendicular to the needle, and your measurements will be most accurate.

Even though you eliminate the problem of parallax, you still have the problem of interpreting the position of the meter when it falls between marks on the scale. As with the yardstick, one person will see a reading as six-tenths of the way between the marks whereas another person will see it as seven-tenths of the way between the marks. The interpretation of the reading is very important. To improve accuracy in

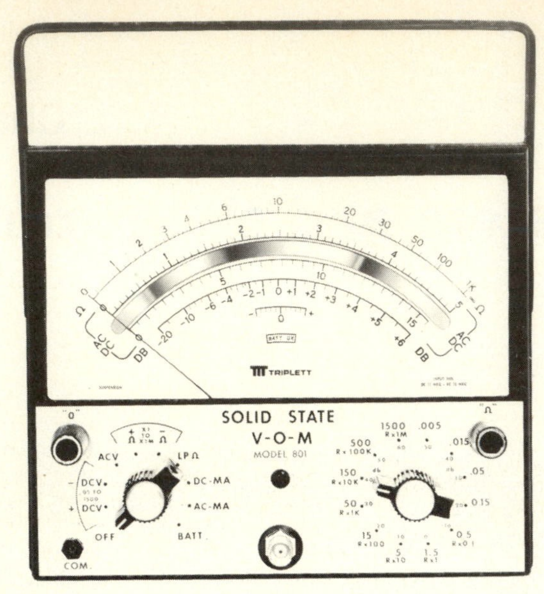

Fig. 7.1 A meter with a mirrored scale. (*Courtesy Triplett Electrical Instrument Company*)

this regard some practice will be required. This is one of the advantages of a *digital voltmeter* like the one shown in Fig. 7.2. Instead of using a meter movement with a needle, this type of meter actually shows the digits of the reading accurately to four or five decimal places. This eliminates the problem of parallax.

The most difficult part of making a reading with a meter is calibration. Here we encounter the same problem that we had with the yardstick. Two different voltmeters will take the same voltage measurements and give two different readings. Which one is correct? The answer is that probably neither one is 100 percent correct. Again, if we wanted a very, very, precise measurement, we would need to use a more expensive meter, and it would have to be accurately calibrated at the Bureau of Standards.

Another problem in making accurate instrument measurements is that when the instrument is inserted into the circuit, changes in circuit current, frequency, voltage, and resistance may occur.

When all these factors are taken into consideration, we are faced with a difficulty in making highly accurate measurements. If a vacuum tube is supposed to have 100 V on its plate, it will operate satisfactorily if the voltage is 101 or 99 V. In fact, if the voltmeter reading at the plate indicates that the voltage is within 5 or even 10 percent of the manufacturer's stated value, we can usually presume that the stage is operating satis-

Fig. 7.2 A digital, solid-state volt-ohm-milliammeter. (*Courtesy Triplett Electrical Instrument Company*)

factorily. The same is true of transistor circuits. When we make a voltage, current, or resistor check during a troubleshooting procedure, we must learn to interpret the reading we get in terms of the circuit complexity and decide whether the inaccuracy is sufficient to cause the trouble we are looking for. Learning to read the instruments is the easiest part. Learning to interpret the readings is more difficult.

7.3 PERCENT ACCURACY

When the percent accuracy of an instrument is stated by a manufacturer, the value given usually refers to the accuracy at the *full-scale reading.* As an example, suppose that the accuracy of a 0- to 10-mA meter movement is given as ±10 percent. This means that a full-scale reading of 10 mA may actually be due to a current that is

$$10 \text{ mA} + 10\% \text{ of } 10 \text{ mA} = 11 \text{ mA}$$

or

$$10 \text{ mA} - 10\% \text{ of } 10 \text{ mA} = 9 \text{ mA}$$

The manufacturer is saying, in so many words, that his instrument may be in error by as much as 1 mA for the full-scale reading. To show why this is misleading (unless we fully understand the rating), suppose that the meter in the above example is used to measure a current of 5 mA, and the reading is in error by 1 mA—that is, the meter indicates 6 mA when it is actually measuring a current of 5 mA. Now the percent error is much greater than the ±10 percent stated for the meter at a full-scale reading. The percent error is calculated as follows:

$$\% \text{ error} = \frac{\begin{pmatrix}\text{current indicated}\\ \text{by meter}\end{pmatrix} - \begin{pmatrix}\text{actual current}\\ \text{being measured}\end{pmatrix}}{\text{current indicated by meter}} \tag{7.1}$$

By using the *indicated* value of current of 6 mA for an *actual* current flow of 5 mA, the percent error is calculated as follows:

$$\% \text{ error} = \frac{\begin{pmatrix}\text{current indicated}\\ \text{by meter}\end{pmatrix} - \begin{pmatrix}\text{actual current}\\ \text{being measured}\end{pmatrix}}{\text{current indicated by meter}} \tag{7.1}$$

$$\% \text{ error} = \frac{6 - 5 \text{ mA}}{6 \text{ mA}} = 16.1\%$$

Therefore, *an error of 1 mA* will cause an error of 10 percent at full scale (which is the manufacturer's rating) but will cause an error of 16.6 percent at half scale. The manufacturer's rating, then, is for the *maximum possible accuracy* when the meter is deflected full scale.

It might be argued that an error of 1 mA at full scale is the maximum possible error of the instrument, and that the error when reading at half scale will be correspondingly less. However, this is not necessarily true. In fact, the reading error at half scale *may* actually be *greater* than at full scale. For this reason the calibration of an instrument should be checked (whenever possible) in the part of the scale where most of the readings are going to take place.

7.4 BASIC INSTRUMENTS AND THEIR USE FOR SERVICING ELECTRONICS EQUIPMENT

Ammeters, voltmeters, and ohmmeters are the three basic instruments used for measurements in electric and electronic circuits.

AMMETERS The instrument used for measuring electric current is an *ammeter.* See Fig. 7.3. (When designed for measuring very small currents, it is called a milliammeter or a microammeter.) In order to measure a current flow with an ordinary ammeter, it is necessary to have the current flow *through* the instrument, as shown in Fig. 7.3*a*. A special instrument, called a clamp-on ammeter, operates by measuring the

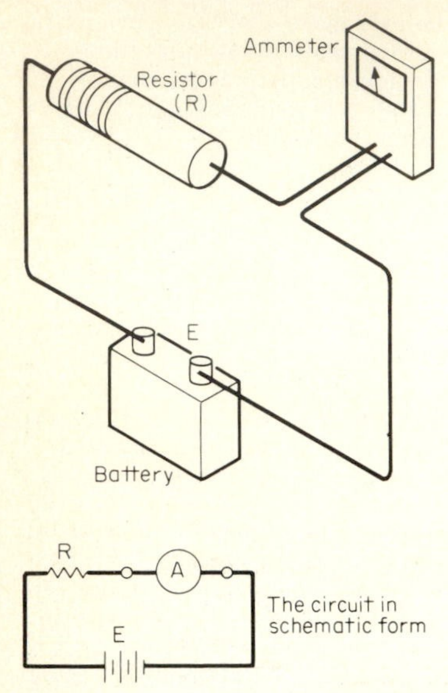

Fig. 7.3*a* To measure current with an ordinary ammeter it is necessary to open the circuit and insert the ammeter so that the current being measured flows *through* it.

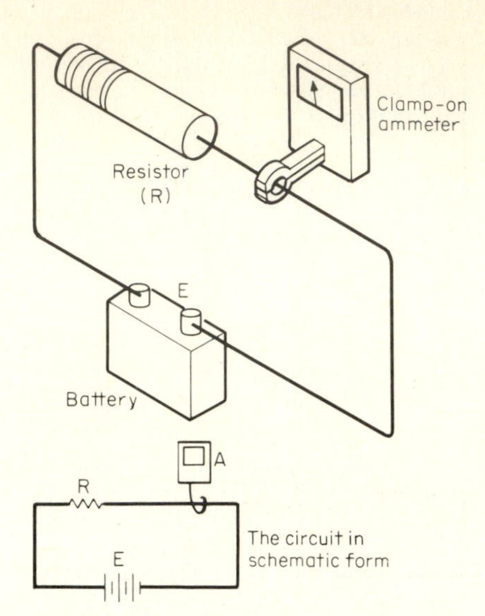

Fig. 7.3*b* The clamp-on ammeter can measure current without the need for opening the circuit to insert the instrument.

Fig. 7.3 Two instruments used for measuring current.

strength of the magnetic field around the current-carrying wire, as shown in Fig. 7.3*b*. The clamp-on ammeter is more convenient for general servicing because it does not require that the circuit be opened in order to insert the ammeter in series with the circuit components. Clamp-on ammeters are a relatively recent introduction in the instrument field, and they are becoming more and more popular with electronic servicemen.

It is seldom necessary to measure the current in vacuum-tube receivers. It is usually easier to measure the voltage across a resistor, and if it is desired to find the current through that resistor, it is a simple matter to apply Ohm's law ($I = E/R$). Vacuum tubes are voltage-operated devices, and therefore we are interested in the plate and grid voltages when troubleshooting this type of receiver. In transistor receivers, however, current measurements are sometimes important. Specifically, we may be concerned about the total amount of current drawn by a transistor radio because this is an indication of whether the transistor amplifiers are operating properly.

VOLTMETERS Manufacturer's literature often gives values of voltages that should be measured at various points in a receiver. In order to make these measurements, a voltmeter is used. To measure a voltage, the voltmeter must be placed *across* the circuit or component (not in series with it). Figure 7.4 shows how a voltmeter is employed to measure the voltage across a resistor.

Generally the voltage measurements in a receiver are taken with respect to a *common point*, often called the *ground point*. This means that one of the meter probes is connected to the ground point, and the other probe is connected to the point where the voltage is being measured. If the meter reads +15 V, it means that the point in question (that is, the point where the voltage is being measured) is 15 V positive with respect to ground. There are exceptions, however. If we want to measure the actual grid voltage of a vacuum tube, we must connect the voltmeter between the grid and cathode because, by definition, the grid voltage is the voltage on the grid *with respect to the cathode* (not with respect to ground).

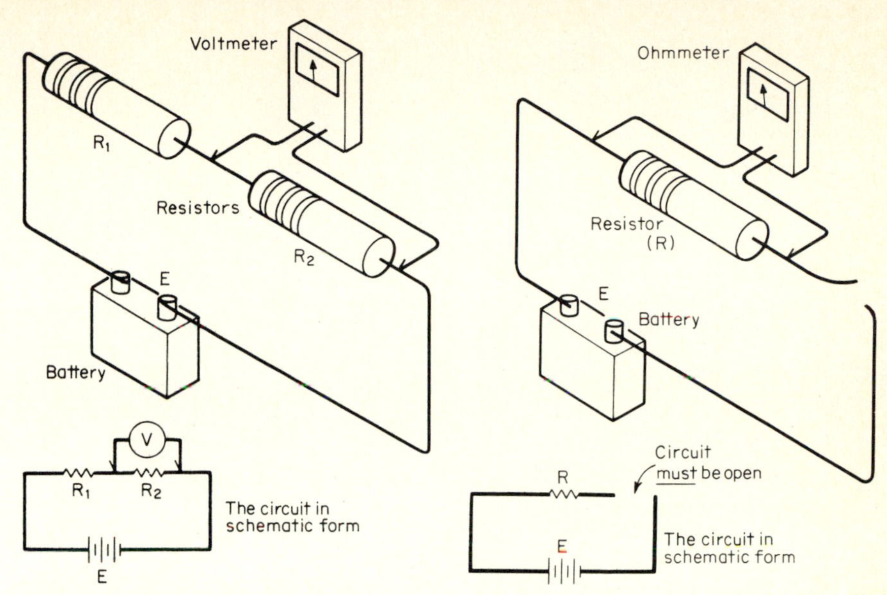

Fig. 7.4 To measure a voltage the meter probes are placed across the difference of potential.

Fig. 7.5 To measure a resistance the ohmmeter is placed across the resistor. The circuit must be opened so there will be no voltage across the resistor.

Many voltmeters are not sensitive enough to measure the difference in voltage between the emitter and the base of a properly operating transistor amplifier. However, the base or emitter voltage is sometimes measured with respect to ground for the purpose of troubleshooting.

OHMMETERS An ohmmeter is an instrument used to measure resistance. Most ohmmeters operate by sending a current through the unknown resistance and measuring the resulting voltage drop across that resistor. A voltmeter, having a scale that is actually marked in ohms, is used for the indicator. In using an ohmmeter it is very important that it be in a circuit that has *no voltage applied*—that is, no voltage other than the voltage source inside the meter. To make a resistance measurement, the ohmmeter is placed *across* the component to be measured, as shown in Fig. 7.5.

In using an ohmmeter it is always important to be sure that the component being measured is not in parallel with another component. If the components are in parallel, as shown in Fig. 7.6, the ohmmeter will measure the equivalent resistance of the parallel combination. If both resistors in this circuit are 1000 Ω in value, the equivalent resistance of the parallel combination is 500 Ω; and this is the resistance that will be indicated by the ohmmeter. It will appear, then, that the 1000-Ω resistor is defective because it measures only half of its rated value. Parallel branches of resistance are not always so obvious as the one shown in Fig. 7.6. It may be necessary to check the circuit schematic diagram to make sure that there is not a parallel path. If there is a parallel path, then it is necessary to disconnect the resistor before measuring its ohmic value.

Ohmmeters are also convenient instruments for use in measuring *continuity*. Continuity simply means that there is a continuous path for current to flow from one point to another. As an example of a continuity measurement, consider the multiwire cable of Fig. 7.7. There are five wires coming out of the cable at each end. It is desired to trace one wire through the cable. To do this, one probe of the meter is connected to one of the wires at one side of the cable. Then, the other probe is touched to the wires (one at a time) at the other side until the ohmmeter reads zero resistance. The wire that indicates zero resistance is obviously connected to the other ohmmeter probe.

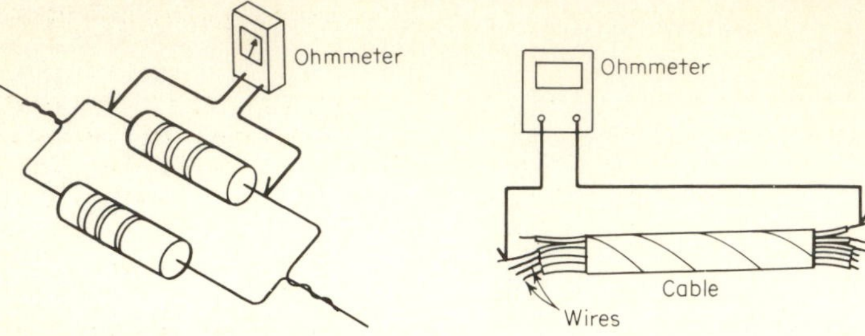

Fig. 7.6 When making a resistance measurement, the component being measured must not be in parallel with another resistance path.

Fig. 7.7 An ohmmeter can be used for continuity checks.

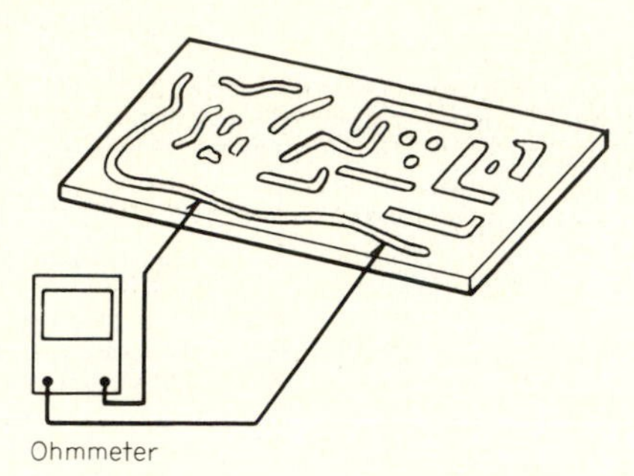

Fig. 7.8 Using an ohmmeter to check a printed-circuit board for cracks in the metal foils.

Continuity measurements are a valuable aid in servicing printed-circuit boards. If there is a tiny crack in a conductor on the board, it may not be visible to the naked eye. However, by placing an ohmmeter across the conductor and flexing the board, the ohmmeter will show that the path is being intermittently opened and closed. This type of test is shown in Fig. 7.8.

MULTIMETERS Instead of having separate instruments for measuring current, voltage, and resistance, all three may be included in a single instrument, as shown in Fig. 7.9*a*. Figure 7.9*a* presents an example of a volt-ohm-milliammeter (VOM). It can measure current, voltage, and resistance and has high input impedance.

The transistorized volt-ohmmeter of Fig. 7.9*b* has a very high input impedance (usually about 20 MΩ). The solid-state equivalent of a VTVM (vacuum-tube voltmeter), is the FET volt-meter shown in Fig. 7.9*c*. This instrument is battery-operated, and therefore it has the advantage of portability over the VTVM. Like the VTVM, it has the advantage of a very high input impedance. Both the VTVM and FET voltmeters have a higher input impedance than the VOM. They can measure voltages without noticeably changing the circuit voltage. Figure 7.10 shows why this is important. The 5-kΩ resistor of Fig. 7.10*a* has two volts across it. When the VTVM is placed across it, we have 5 kΩ and 20 MΩ in parallel. For all practical purposes, the voltmeter is an open circuit, and only 5 kΩ is in the circuit during measurement.

Figure 7.10*b* shows what happens when the voltage across the 5-kΩ resistor is measured with a voltmeter that has only 5000 Ω of resistance. The two resistances in parallel have a combined resistance of 2.5 kΩ. Since the circuit resistance is decreased by the presence of the voltmeter, both the voltage across the resistor and the circuit current are changed during measurement. In general, the higher the input resistance of the measuring device, the less its effect on the circuit being measured, and the more accurate the measurement.

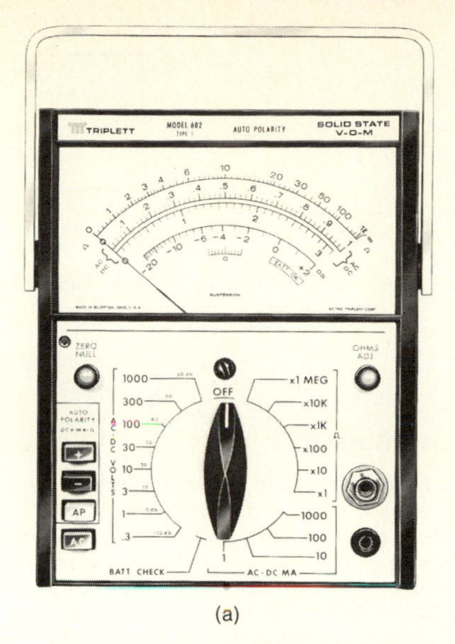

(a)

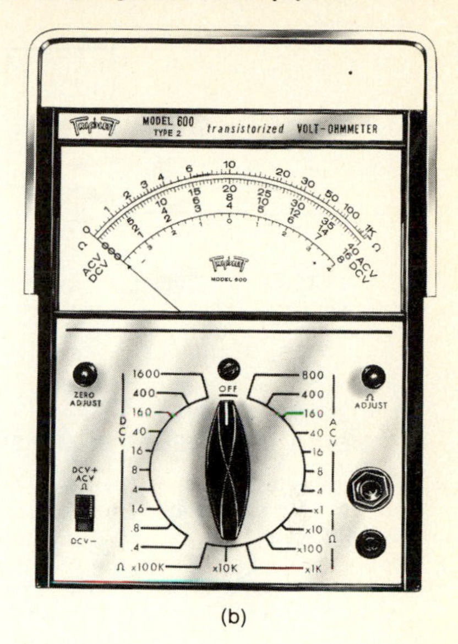

(b)

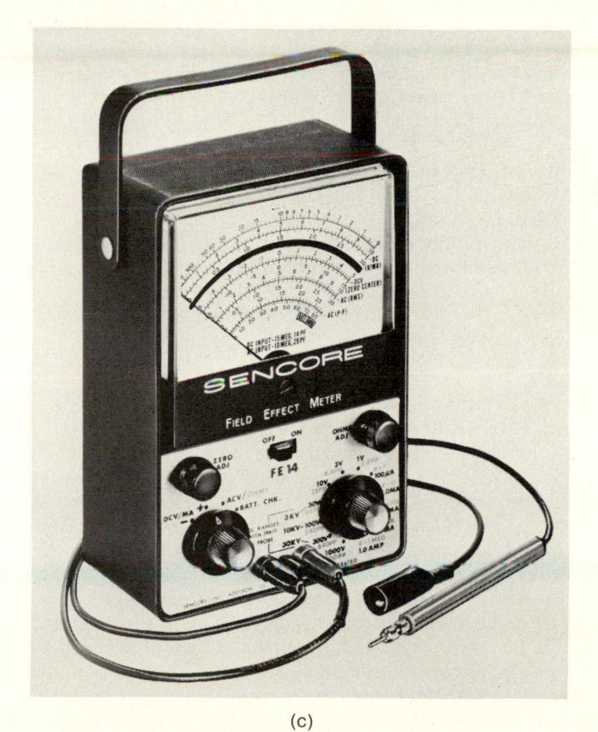

(c)

Fig. 7.9 Three examples of multimeters: (*a*) Volt-ohm-milliammeter. (*Courtesy Triplett Corporation*) (*b*) Transistorized volt-ohmmeter. (*Courtesy Triplett Corporation*) (*c*) FET meter. (*Courtesy Sencore, Inc.*)

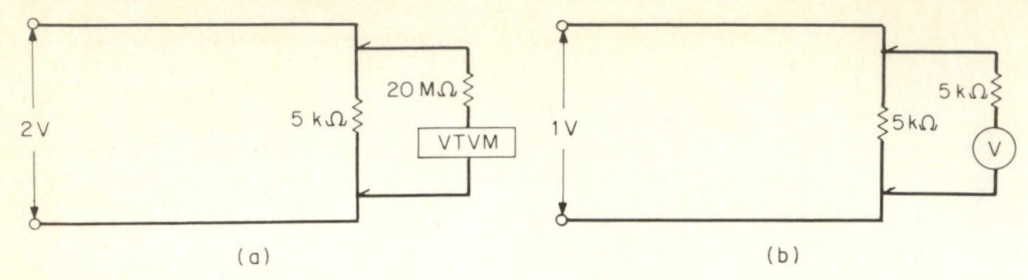

Fig. 7.10 This shows the importance of a high-impedance voltmeter: (*a*) The VTVM does not disturb the circuit when making measurements. (*b*) The low-impedance voltmeter reduces the circuit voltage.

7.5 METER MOVEMENTS IN MEASURING INSTRUMENTS

An ampere may be defined as a flow of electric current equal to one coulomb per second. In other words, an ampere of current is flowing whenever 6 250 000 000 000 000-000.0 electrons (that is: 6.25×10^{18} electrons) flow past a point on a conductor every second. Obviously, it would not be possible to count this number of electrons directly even if they could be seen. Therefore, some form of indirect measurement must be used to determine the amount of current flow. By "indirect method of measurement," we mean that the current is measured according to some effect that it produces in a circuit.

When an electric current flows through a circuit, there are several effects that are always produced. For example, heat is produced when current flows through a resistance. The amount of heat is directly related to: (*a*) the square of the amount of current flow, and (*b*) the amount of resistance in the circuit. This is evident from the equation for power dissipated by a resistor given in equation form as

$$P = I^2R \tag{7.2}$$

where P = power dissipated in the form of heat, W
I = current flow, A
R = resistance, Ω

If the amount of heat produced in a resistor can be accurately measured, and if, at the same time, the value of resistance is known, then the current can be determined by transposing the above equation:

$$I = \sqrt{\frac{P}{R}} \tag{7.2a}$$

We cannot directly substitute the temperature for P in this equation. Temperature and power dissipation are two different things, and they must be treated as such. Thus, if we measure the amount of temperature rise due to a current flowing in a resistor, this temperature rise must be converted to the power dissipated in watts. The method of measuring power dissipation by the amount of heat generated is called the *calorimetric method*. Once the power is known, the current can be obtained from Eq. (7.2*a*).

The calorimetric method may seem like a rather roundabout method of measuring current, but it is very useful in certain applications. For example, a high-powered transmitter may be radiating 50 000 or even 100 000 W of power. This amount of power is difficult to measure accurately with ordinary measuring instruments, but the calorimetric method can be used. The output of the transmitter is fed through a *dummy load* (an artificial load) which is immersed in water. The temperature rise of the wire is accurately measured, and from this information the output power can be calculated quite accurately. The manufacturer of equipment for making caloric-type measurements usually supplies charts for interpreting temperature rise into power.

The calorimetric method of measuring current can be used in almost any application where it is necessary to measure a current, specifically a large current, by precise measurements. However, it is much too inconvenient for such applications as servicing electronic equipment. (Later we will discuss another method of measuring current by utilizing the amount of heat produced by the current.)

In addition to the heat generated by a current, there is also a magnetic field produced around the current-carrying conductor. The strength of this magnetic field is related to the amount of current flow. One way to measure the strength of a magnetic field around a current-carrying conductor is to measure the effects of a force produced when that magnetic field is inserted into another magnetic field with a known intensity. The force produced on a current-carrying conductor when placed in a magnetic field is given by Ampere's law. Stated in words, Ampere's law tells us that when a conductor carries an electric current in a magnetic field, there is a force on the conductor that is directly proportional to the current, directly proportional to the length of the conductor in the magnetic field, and directly proportional to the magnetic flux density of the field. Mathematically this law is stated as

$$F = \frac{BIL \sin \Theta}{10} \tag{7.3}$$

where F = force on conductor, dyn
I = current flowing through conductor, A
L = length of conductor, cm
B = flux density of magnetic field into which conductor is inserted, G
Θ = angle between direction of current flow and direction of flux

Ampere's law is the principle upon which a number of measuring instruments are designed. In fact, most of the meter movements used in voltmeters, ammeters, ohmmeters, and multimeters are based on the principle stated by Ampere's law. The needle of the meter movement is moved upscale as a result of the magnetic field of a current flowing through a coil which reacts with either a permanent magnetic field or the field of another coil.

METERS THAT UTILIZE THE MAGNETIC FIELD OF A CURRENT FOR MEASUREMENT To summarize, instruments for measuring current employ indirect measurements based on either of two principles: the effect of heat produced by the current, or the effect of the magnetic field produced by the current. With the knowledge of these basic principles we can now study meter movements and see how they operate.

First, we will discuss measurements based on the use of a magnetic field accompanying a current; then we will discuss measurements based on the heat generated by a current.

MOVING-COIL METERS Instead of holding the coil stationary in a meter and having a piece of soft iron move in response to a magnetic field produced by a coil, some instruments are made in which a current-carrying coil itself is moved in a permanent-magnet field. Figure 7.11 shows an example of a moving-coil meter. In this case a movable coil is inserted into a permanent magnetic field. As current flows through the coil, the magnetic field surrounding the coil reacts with the permanent magnetic field and causes the coil to turn. The coil is connected in such a way that its magnetic field will always cause the pointer to move upscale. This is an important characteristic of this type of meter: in order to get an upscale movement of the pointer, it is necessary that the current flow through the coil in a specified direction! If the direction of current is reversed, the coil will try to move in the reverse direction. Very often this will destroy the meter movement. In order to measure an alternating current with this type of movement it is necessary to employ some type of a rectifying device to convert the alternating current to direct current flowing in the proper direction.

The name D'Arsonval meter movement is often used to describe the moving-coil meter movement of Fig. 7.11. In the strictest sense of the word, this is not a D'Arsonval type of movement. However, the terminology appears now to be firmly established. This type of meter movement is also referred to as a *jeweled-type* instrument because the pivot of the coil rests in tiny near-frictionless jewels. Instead of a jeweled pivot, many of the newer meters use what is known as a *taut-band movement*. Physically, the taut band is a flat metal band that twists as the pointer moves upscale.

Figure 7.12 shows the construction of a taut-band meter. The force that returns the

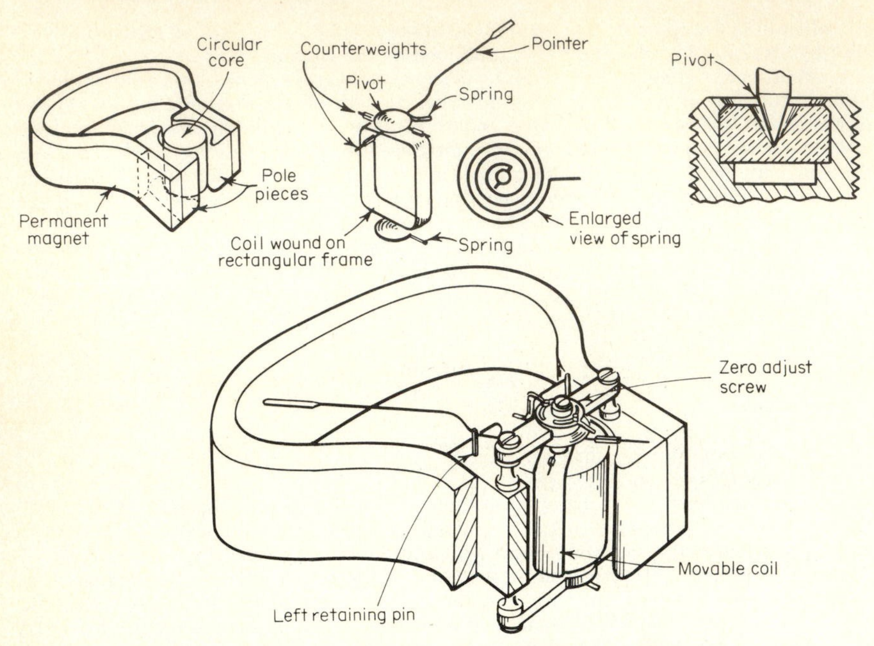

Fig. 7.11 A moving-coil meter and the parts that combine to make it.

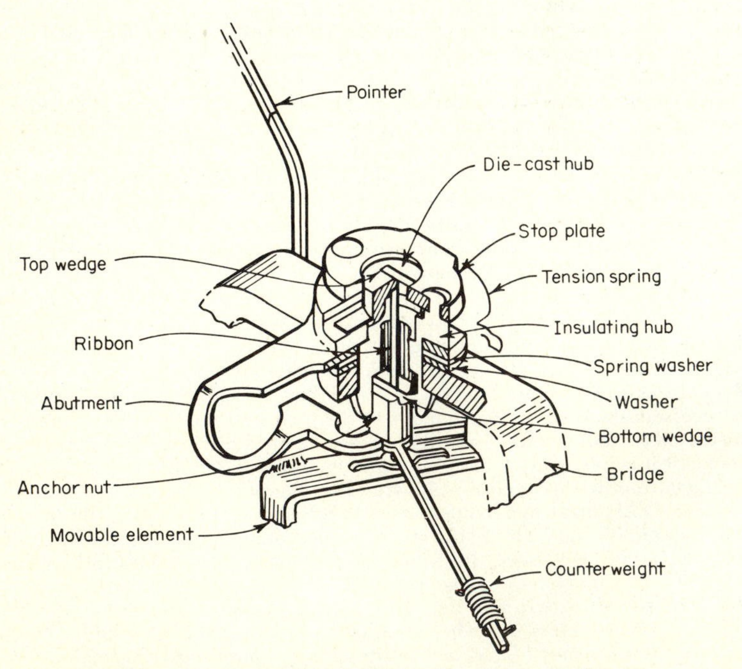

Fig. 7.12 Construction details of a taut-band moving-coil meter.

pointer to zero is due to the twist of the taut band. Taut-band meters are usually more accurate than jeweled types because of the reduction in the number of moving parts.

METERS THAT DEPEND ON THE HEATING EFFECT OF A CURRENT The meter movements that have been discussed so far depend for their operation on the interaction of magnetic fields. In general, the magnetic field surrounding a wire or a coil carrying an unknown current reacts with a magnetic material or a magnetic field in such a way that the attraction or repulsion moves a meter pointer. We will now discuss a type of meter that depends for its operation on the heat produced by a current flow.

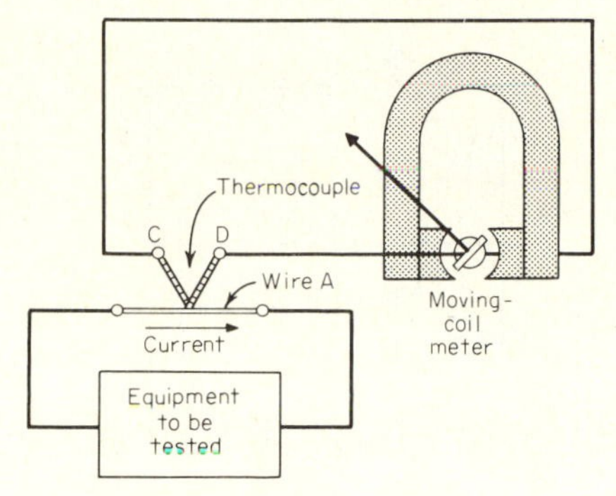

Fig. 7.13 A thermocouple ammeter.

The instrument of Fig. 7.13 is called a *thermocouple meter*. (A thermocouple is simply a junction of two dissimilar metals. When the junction is heated, a voltage is generated across the junction.) In the meter movement of Fig. 7.13, the junction is heated by current flowing in wire A, and the voltage appears across points C and D. The amount of voltage is proportional to the amount of heat at the junction of the thermocouple. This voltage generated by the thermocouple causes a current to flow through a moving-coil meter. A moving-coil meter is used here because it is one of the most sensitive available.

Wire A in the thermocouple meter movement will be heated regardless of whether a direct or an alternating current flows through it. An especially valuable application of this instrument is for measuring r-f current. R-f currents flowing through the wire cause it to heat, and therefore the current can be measured by the action of the thermocouple. However, since the r-f current flows through a straight piece of wire, there are no reactive effects that would change the magnitude of the current. A simple moving-coil meter cannot be used to measure radio frequency because the inductance in the moving coil would cause an inductive reactance in the circuit that would change the amount of r-f current flow.

7.6 METER CIRCUITS

The meter movements described in the previous section are designed for measuring very small currents. In order to measure larger currents, or to measure voltage and resistance, the meter movement must be placed in a circuit that will protect it from an *overload*—that is, from an excessive current which would destroy the meter movement. The most frequently encountered meter movement is the moving-coil permanent-magnet type. A typical movement of this type will deflect to its maximum full-scale reading when there are only 50 μA flowing through it. Obviously, it cannot be used directly for most of the measurements required for servicing electronic equipment.

CALCULATION OF METER SHUNTS Figure 7.14 shows the circuit for converting a sensitive microammeter to read larger current values, such as amperes and milliamperes. The meter M and its internal resistance R_M are paralleled with a shunt resistor R_{sh}. The purpose of the shunt resistor is to bypass excessive current around the meter movement. The arrows on the illustration show how the current divides. In the operation of the circuit, the portion of the current that flows through the meter does not exceed the value required for full-scale deflection. If the value of the shunt resistance was equal to the value of the meter resistance, then the current would divide evenly so that half the current being measured would flow through each branch. However, for measuring large currents, such as an ampere, it is necessary to divide the currents in such a way that the amount flowing through the shunt resistance is, by far, *greater* than the amount flowing through the meter. Thus, the shunt resistor has a very low resistance value—often as low as a fraction of an ohm. At first appearance, the shunt may often seem to be a short circuit across the meter movements.

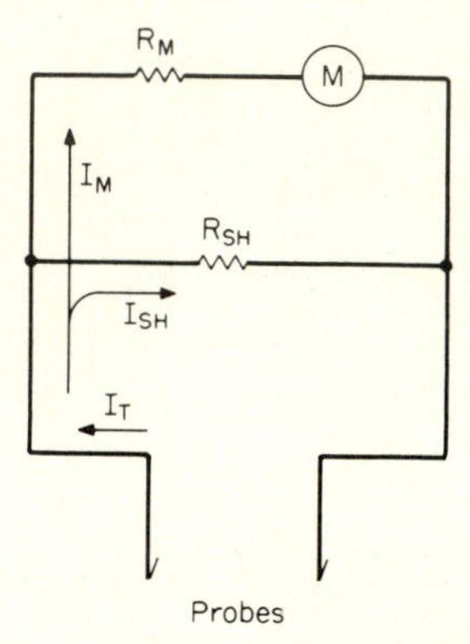

Fig. 7.14 A circuit for reading milliamperes or amperes.

The resistance value of R_{sh} (Fig. 7.14) can be calculated by application of Ohm's law and basic reasoning. The current being measured, I_T, will divide so that part of it flows through the meter, I_M, and the rest flows through the shunt I_{sh}. Mathematically, according to Kirchhoff's current law, it can be stated that

$$I_T = I_M + I_{sh} \tag{7.4}$$

where I_M = maximum current that can flow through the meter, A
I_{sh} = current that must flow through the shunt, A
I_T = sum of I_M and I_{sh}

The maximum value of I_M is fixed by the maximum reading of the meter M. Except for the small amount of current flowing through the meter, all the current being measured must flow through the shunt. By subtracting I_M from both sides of Eq. (7.4), we obtain

$$I_{sh} = I_T - I_M \tag{7.4a}$$

(The value of I_{sh} will be needed later.)

Our basic knowledge of electricity tells us that the voltage across all branches of a parallel circuit is the same. In the circuit of Fig. 7.14 the voltage across the meter branch is, by Ohm's law, the meter current times the meter resistance $I_M R_M$. This must be equal to the voltage across the shunt $I_{sh}R_{sh}$; so we can set them equal in equation form:

$$I_{sh}R_{sh} = I_M R_M \tag{7.5}$$

where I_M and I_{sh} = the same meaning as for Eq. (7.4)
R_{sh} = current that must flow through the shunt, A
R_M = sum of I_M and I_{sh}

Since we are interested in calculating the value of R_{sh}, we will isolate it by dividing both sides of the equation by I_{sh}:

$$R_{sh} = \frac{I_M R_M}{I_{sh}} \tag{7.5a}$$

When calculating the value of a meter shunt, we normally have the following information: the *meter resistance* R_M, the *meter current* I_M, and the *total* (*maximum value of*) *current to be measured* I_T. We usually do *not* know the value of shunt current I_{sh}; so Eq (7.5*a*) cannot be used directly. However, Eq. (7.4*a*) tells us that $I_T - I_M$ can be substituted for I_{sh}. Making this substitution into (7.5*a*) gives

$$R_{sh} = \frac{I_M R_M}{I_T - I_M} \text{ ohms} \tag{7.5b}$$

The currents must be in amperes, and the resistances in ohms. Equation (7.5*b*) can be used directly for calculating the required value of shunt resistance to convert a meter movement for measuring higher current values. However, instead of memorizing the equation, it is a better practice to make the calculations by the same reasoning that was used for the derivation.

example 7.1 A certain 50-μA meter movement has a resistance of 90 Ω. What value of shunt resistance is needed to convert the meter so that it can measure a (maximum) current of 0.5 mA?

solution Given: $I_M = 50\ \mu\text{A} = 0.000\,05$ A, $I_T = 0.5$ mA $= 0.0005$ A, and $R_M = 90\ \Omega$. The current through the shunt is

$$\begin{aligned} I_{sh} &= I_T - I_M \\ &= 0.0005 - 0.000\,05 = 0.000\,45 \text{ A} \end{aligned} \tag{7.4a}$$

$$\begin{aligned} R_{sh} &= \frac{I_M I_M}{I_{sh}} \\ &= \frac{0.000\,05 \times 90}{0.000\,45} = \frac{0.0045}{0.000\,45} = 10\ \Omega \textit{ Answer} \end{aligned} \tag{7.5a}$$

It is important to remember that this type of meter must be placed *in series* with the circuit current being measured. The very low value of shunt resistance—less than a fraction of an ohm in many instances—makes the meter circuit resistance negligible. Therefore, when the meter is placed in series with the current being measured, this does not noticeably affect its value.

CALIBRATION OF CURRENT-READING METERS Occasionally it becomes necessary to check the accuracy of an ammeter, a milliammeter, or a microammeter. The circuit of Fig. 7.15 shows how this can be done. It is based on the principle that the same current flows through all parts of a series circuit. Since meter M_1 and M_2 are in series, it follows that they should indicate the same amount of current flow. Meter M_1 is known to be accurate, whereas M_2 is being checked for accuracy. If there is a discrepancy between the two readings, then M_2 can (in many cases) be adjusted to read correctly, and this process is known as calibration. Resistor R_2 is a series-limiting resistor, to prevent the current through the meters from becoming excessive, as would happen if resistor R_1 were accidentally adjusted to zero ohms in the circuit. Resistor R_1 can be adjusted for an exact reading for meter M_1, and the reading of M_2 can be compared with this value. It is not necessary to know precise values of resistors R_1 and R_2 since we are only interested in the current value produced by the circuit.

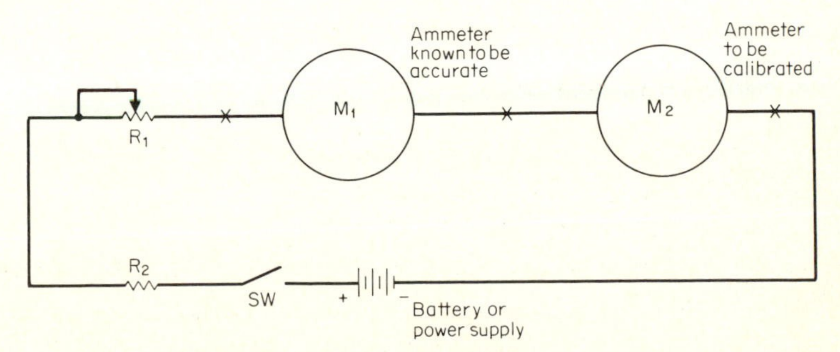

Fig. 7.15 Circuit used for calibrating ammeters.

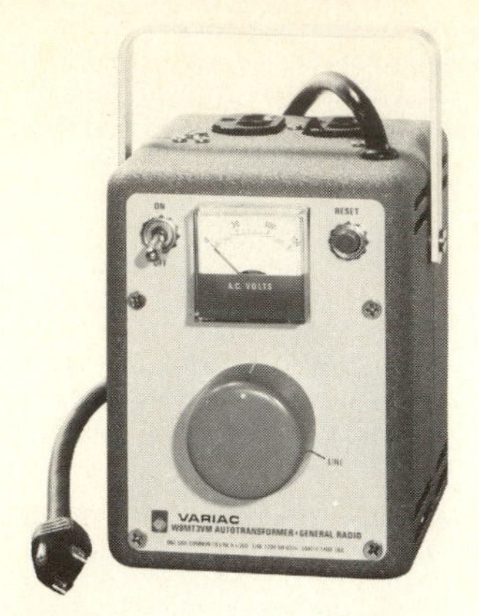

Fig. 7.16 A variable transformer that is useful for calibrating ac meters. (*Courtesy General Radio Company*)

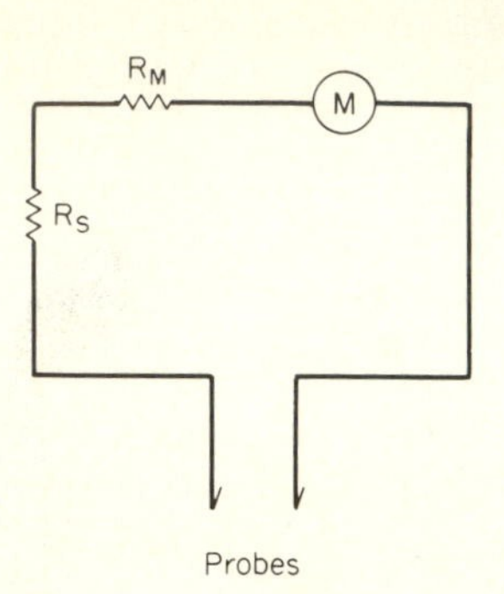

Fig. 7.17 A simple voltmeter circuit.

If the meters in the circuit of Fig. 7.15 are designed for measuring alternating current, then the battery should be replaced with an ac power supply. A variable power transformer, such as a *Variac* (see Fig. 7.16), is useful for this purpose. If a Variac is not available, a step-down filament transformer can be used.

CALCULATION OF METER MULTIPLIERS Though it is desirable for an ammeter to have an extremely low resistance so that it will not alter the circuit current value, a *voltmeter* must have a very high resistance. The voltmeter is placed across the voltage source, and ideally it will draw very little current. A typical moving-coil permanent-magnet meter movement may have a coil resistance less than 100 Ω. If such a meter movement were placed across a 20-V source, the amount of meter current flow would be so large that it would damage the meter movement. Therefore, a series resistor R_S is placed in the circuit to limit the current flow through the meter. The series resistor is sometimes called a multiplier. Figure 7.17 shows the basic voltmeter circuit. The resistance value of R_S may be quite high, and it is often a precision resistance value.

In the circuit of Fig. 7.17, a sensitive meter movement M with an internal resistance value of R_M is placed in series with the multiplier R_S. When the probes are placed across the voltage to be measured, V, current flows in the meter circuit. This current causes a voltage drop V_S across R_S and a voltage drop V_M across the meter movement resistance R_M.

According to Kirchhoff's voltage law, the sum of the voltage drops around a circuit, $V_M + V_S$, must be equal to the applied voltage V. Applying this law to the circuit of Fig. 7.17 results in

$$V = V_S + V_M \tag{7.6}$$

where V = voltage being measured, V
V_S = resistance of multiplier, Ω
V_M = resistance of meter movement, Ω

We are interested in the value of V_S at the moment. If we know the voltage across R_S (which is V_S), and we know the current through R_S (which is the meter current I_M), then we can calculate the resistance value of R_S.

Solving Eq. (7.6) for V_S gives

$$V_S = V - V_M \tag{7.6a}$$

Since $V_S = I_M R_S$, and $V_M = I_M R_M$, we can substitute these values into the equation to get

$$I_M R_S = V - I_M R_M \tag{7.6b}$$

where V = voltage being measured, V
I_M = maximum allowable meter current, A
R_S = multiplier resistance, Ω
R_M = resistance of meter movement, Ω

Dividing both sides of the equation by I_M to get an equation for R_S,

$$R_S = \frac{V - I_M R_M}{I_M} \text{ ohms} \tag{7.7}$$

All the values on the right side of Eq. (7.7) are normally known for determining the value of a meter multiplier.

example 7.2 A certain 50-μA meter movement has a resistance of 90 Ω. What value of series multiplier is needed to make an instrument that will read 5 V (maximum)?

solution Given: $I_M = 50\ \mu\text{A} = 0.000\,05$ A, $R_M = 90\ \Omega$, and $V = 5$ V. Therefore,

$$R_S = \frac{V - I_M R_M}{I_M} \tag{7.7}$$

$$= \frac{5 - (0.000\,05 \times 90)}{0.000\,05} = 99\,910\ \Omega \qquad \textit{Answer}$$

If a moving-coil permanent-magnet meter movement is used in an ac voltmeter circuit, provision must be made for converting ac to dc voltage. You will remember that the pointer of this type of meter deflects in a direction that depends upon the direction of current flow through it. The circuit of Fig. 7.18 shows an ac voltmeter arrangement

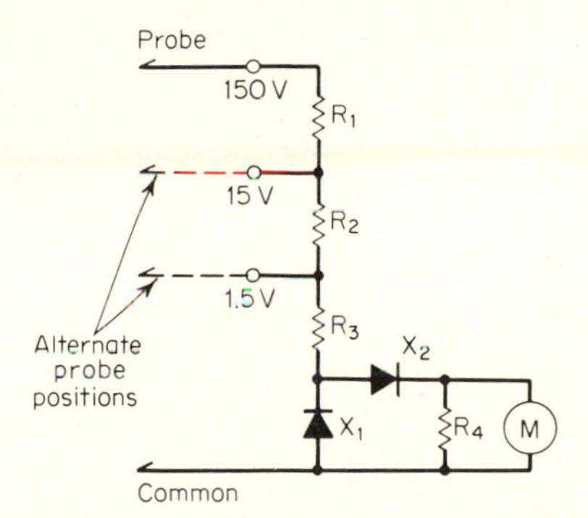

Fig. 7.18 A meter circuit for making ac voltage measurements.

for three different ranges of voltage values. In this arrangement, the meter range is determined by the point where the probe is inserted into the circuit. For example, when the probe is plugged into the 150-V terminal, then resistors R_1, R_2, and R_3 serve the same purpose as R_S in Fig. 7.17. When the probe is plugged into the 1.5-V terminal, R_3 alone serves as the series resistor.

On one half-cycle of input alternating current, diode X_1 of Fig. 7.25 conducts and places an effective short circuit across the meter. On the next half-cycle diode X_1 is cut off, and current flows through X_2, R_4, and the meter movement. Resistor R_4 serves as a meter shunt.

It is desirable to make the series resistance of the voltmeter as large as possible to prevent loading the circuit that is being measured. Consider, for example, the circuit of Fig. 7.17. The value of R_S cannot be so large that it prevents the meter from deflecting on full scale when the maximum voltage is being measured. The greater the current required for full-scale deflection of M, the smaller the required value of R_S, and the greater the amount of loading of the circuit by the voltmeter.

A commonly used method of rating a voltmeter according to its sensitivity is the ohms-per-volt rating. The higher the ohms-per-volt rating, the greater the sensitivity of the meter movement, and hence the lower the loading effect of the meter. The reciprocal of the ohms-per-volt rating gives the amount of current required for full-scale deflection of the meter movement.

$$\begin{pmatrix}\text{Amperes of current required}\\ \text{for full-scale deflection}\end{pmatrix} = \frac{1}{\text{ohms-per-volt rating}} \tag{7.8}$$

example 7.3 A certain meter is advertised as having a sensitivity of 50 000 Ω/V. How much current is required to deflect the meter movement to full scale?

solution

$$\begin{pmatrix}\text{Amperes of current required}\\ \text{for full-scale deflection}\end{pmatrix} = \frac{1}{\text{ohms-per-volt rating}} \tag{7.8}$$

$$= \frac{1}{50\ 000\ \Omega/\text{V}} = 0.000\ 02\ \text{A} = 20\ \mu\text{A} \qquad \textit{Answer}$$

7.7 VACUUM-TUBE VOLTMETERS

The vacuum-tube voltmeter arrangement of Fig. 7.19 causes very little loading of the circuit in which a voltage is being measured. The probe is connected directly into the grid circuit of a vacuum tube which has an extremely high input resistance. The voltage of E_1 is sufficient to hold the vacuum tube at cutoff when the probes are connected together. This means that the current flow through M will be zero when the probes are short-circuited together. Capacitor C_1 provides an ac bypass around the grid power supply, while capacitor C_2 provides a bypass around plate power supply E_2. When a positive voltage is applied to probe a, it partially cancels the negative voltage due to E_1, and the tube begins to conduct. The amount of plate current is proportional to the amount of positive voltage at a, and therefore the meter deflection is proportional to that voltage. Resistor R_L limits the plate-current flow through the meter movement.

The positive voltage applied to probe a must never exceed E_1 or the tube will conduct to saturation. Under the condition of saturation, the meter deflection will no longer be proportional to the input voltage. Therefore, when changing scales of the voltmeter, it will also be necessary to change the value of the grid voltage E.

The circuit of Fig. 7.19 can be used to measure either a dc or an ac voltage. When an ac voltage is applied, point a becomes alternately positive and negative. On one half-cycle, probe a will be negative with respect to probe b, and this voltage will be in series with that of E_1. Under this condition the tube is driven *beyond cutoff*. On the next half-cycle, probe a is positive with respect to probe b, producing a voltage that partially counteracts grid voltage E_1. This unblocks the tube and allows current to flow through the meter movement. Although the current that flows through the meter movement is flowing only on alternate half-cycles, the inertia of the needle prevents it from dropping back to zero during the half-cycles when no plate current flows. The

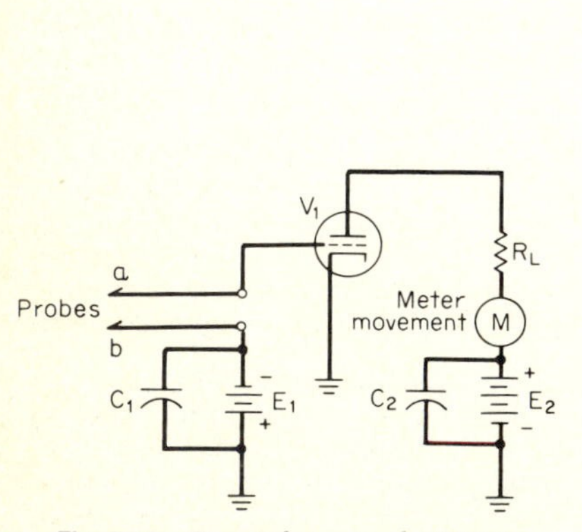

Fig. 7.19 Circuit for a simple VTVM.

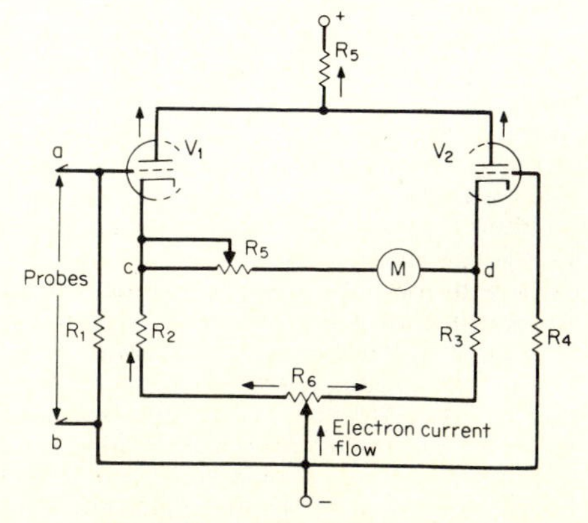

Fig. 7.20 A better VTVM circuit.

needle deflection is read on a scale that is related to the value of input ac voltage.

An alternative method of measuring ac voltage is to rectify the input voltage before it is applied to the probes. Full-wave rectification—such as obtained by a bridge rectifier—is preferable because plate current will then flow through the meter during both half-cycles of input voltage. Since the plate current does not drop to zero for a full half-cycle, it follows that, for a given ac input, the meter deflection of the meter movement will be greater for a given ac voltage than would be obtained without full-wave rectification.

In vacuum-tube voltmeters the circuit arrangement shown in Fig. 7.20 is often used. As in the case of the simple VTVM shown in Fig. 7.19, field-effect transistors may be used instead of tubes for this circuit. Tubes and FETs are interchangeable because their input impedances are quite high. Transistors, on the other hand, require current flow in their base circuit, and therefore cannot be directly interchanged with tubes or FETs.

The tube circuit of Fig. 7.20 is sometimes called a *difference amplifier;* it is also known by the names *bridge amplifier, differential amplifier,* and *dc amplifier.* Assuming that there is no input voltage to the probes, we see that there are two identical tube circuits, V_1 and V_2. The resistance R_1 and R_4 are equal, R_2 equals R_3, and V_1 and V_2 are presumed to be identically matched. If it should turn out that there are slight differences in the two tube circuits, they can be compensated for by adjusting variable resistor R_6 until conduction through the two tubes is identical. Under this condition, the two cathode voltages will be identical, and therefore no current can flow through R_5 and M. Assume that during measurement of a dc voltage, point a is made positive with respect to point b. This will cause V_1 to conduct harder than V_2, and a cathode current through R_2 will be greater than that through R_3. As a result, point c becomes more positive than point d, and current will flow through the meter. The amount of deflection of the meter needle is dependent on the amount of positive voltage applied through the probe.

If an ac voltage is applied between points a and b during the measurement, V_1 will conduct harder than V_2 on alternate half-cycles. Assuming that V_1 is biased near the cutoff point, the negative half-cycles will quickly drive it to cutoff. Thus, for all practical purposes there will be current flow only during positive half-cycles of input ac voltages. (A better arrangement is to full-wave-rectify the alternating current before it is applied to the input probe so that V_1 will conduct during both half-cycles of input ac voltage. This will increase the sensitivity of the meter.)

The circuit of Fig. 7.20 is often used in modern-day electronic systems. One very important advantage of the circuit is that it is not *tube-sensitive.* In other words, as the tubes age—assuming that the effects of their aging are identical—the circuit will still be balanced. If both tubes are placed in the same envelope, they must both be replaced when one fails, thus assuring that the tubes are matched. In the circuit of Fig. 7.19, as the tube ages, the cutoff voltage changes, and furthermore the total plate current tends to decrease with aging of the cathode. Operation of the simpler circuit, then, depends somewhat upon the age of the tubes.

Let's assume that the cathode emission of the two tubes in the difference amplifier decreases by one-third. This will still mean that point c will be identical with point d when no input voltage is applied to the probes. There will, however, be some effect on the meter reading with a voltage applied, because the amount by which V_1 conducts (compared to V_2 conduction) will decrease somewhat with tube aging. However, it is not nearly so critical as a single tube circuit.

Differential amplifiers—employing transistors instead of tubes—are used in many integrated-circuit applications. Transistors, like tubes, tend to change their characteristics with age. By using differential amplifiers, the problem of aging is reduced. Since both transistors are mounted on the same chip, they will age equally, and their characteristics change simultaneously.

The FET voltmeter circuit shown in Fig. 7.21 is similar to the simple tube circuit shown in Fig. 7.19. (It should be reemphasized that an FET meter may also employ a difference amplifier like the one shown in Fig. 7.27.) In the circuit of Fig. 7.21 an N-channel FET is used. When there is no input voltage to the circuit, conduction in the FET is held at approximately cutoff value. The positive voltage on the drain makes the transistor ready for conduction. The negative voltage, E_2, produces a small loop

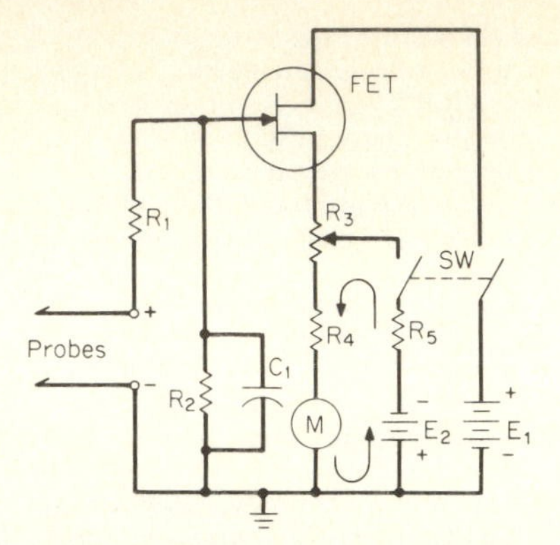

Fig. 7.21 Circuit for a FET voltmeter.

current, as shown by the arrows. This loop current exactly equals and cancels the current through the FET with no input voltage applied to the probes.

When a dc input voltage is applied to the probes in such a way that the positive terminal is at the gate, the FET will conduct. This conduction overrides the bucking current produced by E_2, and the resulting current flowing through meter M will cause the meter to deflect. The amount of deflection is proportional to the input voltage delivered to the gate. Resistor R_1 is part of a voltage divider; resistor R_2 forms the other part of the divider. Any input signal will appear as a voltage drop across the two resistors. The part of the voltage that appears across R_2 is delivered to the gate electrode. Capacitor C_1 serves as a noise signal bypass to ground.

When an ac voltage is applied to the probes, it will cause the FET to conduct on alternate half-cycles. In a better circuit arrangement, the ac voltage is full-wave-rectified with a bridge before being applied to the input probes. This would cause the FET to conduct during both half-cycles, resulting in a higher meter needle deflection for a given input ac voltage. As with other types of meters, full-wave rectification of the input ac signal being measured results in a more sensitive measuring instrument. The *sensitivity* of a measuring instrument refers to the ability of that instrument to measure very small values.

The ganged switch (SW) is used in the circuit to simultaneously remove the bucking voltage and drain voltage from the circuit when the meter is turned off, thus prolonging battery life.

An important advantage of the FET circuit over the simple circuit of Fig. 7.19 is that it does not require any filament voltage for its operation. This makes a completely portable high-impedance meter more feasible.

7.8 DIGITAL MULTIMETERS

All the meters discussed previously in this chapter may be referred to as "analog"-type meters. That is, they all require scales and a pointer to read a particular point on the appropriate scale. The scale readings are all continuous and change smoothly.

The analog meters all have several disadvantages in common, namely, difficulty in actual reading of the scale, the need to interpolate between printed numbers, the possibility of parallax errors, the need for multiplying factors on some ranges, limited accuracy, need for zero pointer setting, and the possibility of damage to the pointer or meter movement because of overload.

Digital multimeters overcome all the above-stated disadvantages. (A typical solid-state digital multimeter is shown in Fig. 7.2.) Such multimeters have large, easily read numbers and automatic placing of the decimal point. In some models a plus or minus sign appears at the left of the numbers, indicating the polarity of the voltage being measured. This feature is automatic and it is not necessary to switch leads or change a switch position when measuring either polarity or voltage. All measurements appear as numbers, and obviously there is no need for interpolation, no parallax error, no need for multiplying factors, and no such thing as zero pointer setting.

The accuracy of a digital multimeter on dc is commonly 0.1 to 1.0 percent, as compared with 2.0 percent for the usual good-quality vacuum-tube (or FET) voltmeter. Generally no damage can result to a digital meter from overload. Also, most of these meters have some sort of overload indication, ranging from a separate indicator to a scheme which causes all the digits to blink. This scheme operates in the case where the applied potential exceeds the selected scale. Some of the more expensive models also feature an "autoranger," which automatically selects the correct range required without the need to operate a range switch. Although the unit shown in Fig. 7.2 contains a read-out of four digits, other units are available with six or more digits. Obviously these latter units would be somewhat more expensive.

DIGITAL DISPLAYS An early type of digital display unit, which is still in fairly wide use, is the so-called "Nixie" tube. This is a small vacuum tube having a common anode and ten individual cathodes. The cathodes take the form of numbers from 0 through 9. These are stacked front to rear in the tube, which has a transparent front. The tube is a sort of glorified neon bulb. The separate cathodes are connected to the control circuitry and when a cathode is energized, its number is illuminated with an orange glow which is clearly visible. Obviously there will be as many tubes used in a particular display as the desired number of digits.

The Nixie tubes generally provide a display which is brighter and has larger numbers than the LED types (discussed below). However, they require a relatively high exciting voltage and draw too much current for practical battery operation. Thus these units are generally line-operated. The Nixie-tube display is not as visible under high ambient light conditions as the reflective liquid crystal type (discussed below).

A very popular digital display is the one which is illuminated by LEDs (light emitting diodes). This is generally found in one of two forms: the 7-segment or the 35-segment (5×7) display. These configurations are shown in Fig. 7.22. Although the number 8 is shown in both parts of this figure, any number between 0 and 9 could be shown. It is also possible to configure the displays so that a plus or minus sign may be produced. A plus sign is shown in Fig. 7.2.

The LED displays draw much less power than the Nixie types. The LEDs operate from a 1.5-V source and draw approximately 25 mA per segment. Many LED-type displays are "strobed" (intermittent, rapid lighting). This further reduces the power consumption, so that battery operation becomes very practical. The types of batteries in common use are the throwaway types, rechargeable types, and external battery packs. The readability of LED displays in high ambient light is not as good as that of the Nixie or the reflective liquid crystal types.

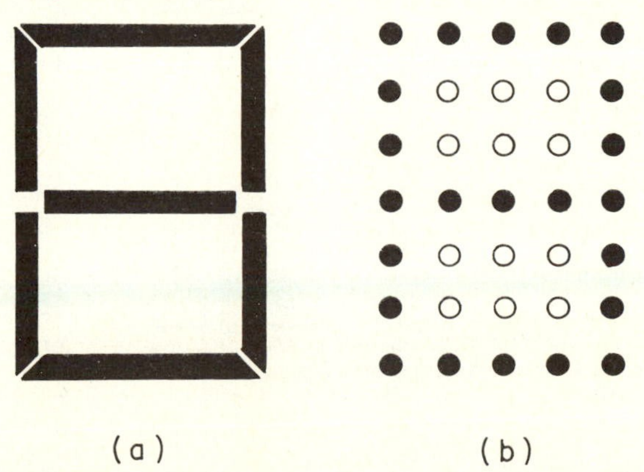

Fig. 7.22 Two common methods of producing digital segmented number displays. (*a*) Seven-segment display showing number 8. (*b*) A 5×7 (35-segment) display also showing number 8.

The liquid crystal displays draw by far the least current of the three types described here. Each liquid crystal segment draws only a few microwatts. Consequently this type is the most suitable for battery operation. This device is used in the seven-segment digit display shown in Fig. 7.22*a*. The liquid crystal displays come in two forms: the transmissive (back-lighted) type and the reflective (ambient-lighted) type. The transmissive type results in a greater power drain because of the back lighting and thus produces a shorter battery life. The reflective type has a very low power drain and is also used in devices such as watches and clocks, where maximum battery life is required.

Each segment of a liquid crystal display consists of two clear glass plates, with a thin layer of liquid crystal material between them. The glass plates each have conductive coatings which connect to the energizing source. When energized, the character of the liquid crystal material in the appropriate segments changes with regard to its light-transmission properties. In this manner the desired number appears, either from back lighting or from reflective lighting.

THEORY OF OPERATION The operation of a typical digital multimeter is explained here only in general terms. The reader should refer to Chap. 14, "Digital Circuit Fundamentals," and Chap. 15, "Digital Integrated Circuits," for the various details involved in such circuits.

The operation of a typical multimeter will be described with the aid of the simplified block diagram of Fig. 7.23. The quantity to be measured, ac or dc voltage, ac or dc current, or resistance, is fed into the input selector. The output of this selector is always a dc voltage, which is proportional to the quantity being measured. (The input selector also contains the range switching.) This proportional dc voltage is then fed to the ramp generator-comparator, which is the analog to digital converter. The output of this block will consist of digital pulses which are related to the proportional dc voltage, but only in terms of the time between them.

The ramp generator produces a linear, negative-going ramp, which begins at a positive value, goes through zero, and continues to an equally negative value. This is shown in Fig. 7.24. Note that the ramp (an upside-down sawtooth wave) voltage is proportional to time. Now assume that the voltage to be measured is 5 V. When the negative-going ramp reaches the same value as the measured voltage (5 V), the comparator puts out a pulse, often called a "start" pulse, which permits the LSI (large-scale integrated circuit) to begin counting pulses generated by the clock oscillator. The counting continues until the ramp decreases to 0 V. At this time the comparator puts out a "stop" pulse, which tells the LSI (logic circuits) to stop counting the clock oscillator pulses. Now note that the original voltage to be measured (5 V) has been translated from an anolog quantity to a time interval and then to a number of clock pulses, which is exactly proportional to the original 5 V. Obviously both the ramp repetition rate and the clock oscillator frequency are quite high, to minimize the time required to display readings (about 1 ms) and to permit "strobing" (or multiplexing) of the displays. This feature will now be explained.

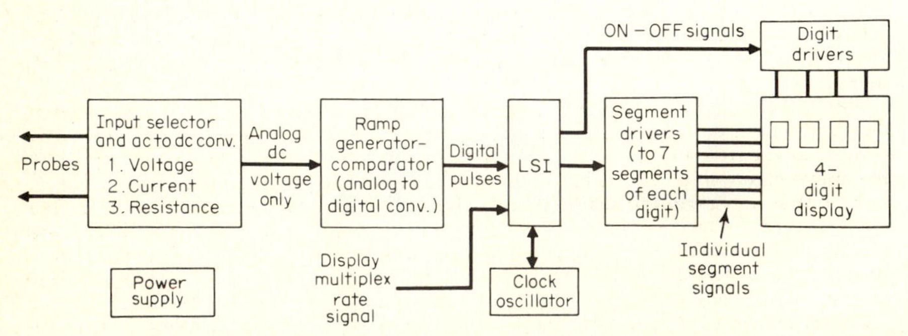

Fig. 7.23 Simplified block diagram of a digital multimeter. Corresponding segments of each digit are connected in parallel.

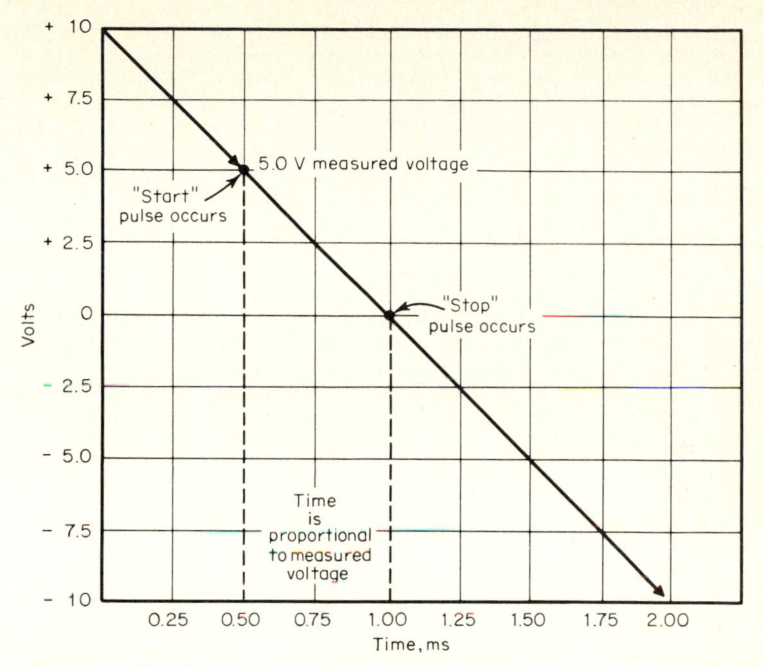

Fig. 7.24 When the ramp reaches the measured voltage (5 V), a "start" pulse is generated. When the ramp reaches zero volts, a "stop" pulse is generated (see text).

In the block diagram of Fig. 7.23 (and Fig. 7.2), the display is seen to consist of four digits. This means the maximum display number is 9999. However, an automatic decimal-point placement is included in the logic circuits, so that exact value is immediately evident. Note the placement of the decimal point in Fig. 7.2, where the reading is +17.76 V. When the strobing or multiplexing feature is present, as shown in Fig. 7.23, each of the four digits is lighted separately, in turn. However, this is done at such a high rate of speed that they all appear to be lighted simultaneously. Strobing saves an appreciable amount of battery power, as previously mentioned.

Now to tie it all together. The multiplexing (strobing) signal is fed to the LSI and affects both the operation of the digit drivers and the segment drivers. Thus, as one particular digit is caused to be energized, at that same exact time the segment drivers cause the proper combination of segments for that digit to appear. Since all corresponding segments of each digit are tied in parallel and connected to the segment drivers in this way, it can be seen that in order to have only one digit at a time light up, showing the correct number, the strobing or multiplexing scheme just described is necessary.

The ½ digit In Figs. 7.2 and 7.23, four-digit displays are shown, which means that the maximum number which can be displayed is 9999. However, for relatively little additional expense and circuitry it is possible to provide an additional digit in front of the above-mentioned four. This added digit can only be programmed (as in this case) to be either unlighted (no reading) or to show a 1. Thus, if energized, this so-called ½ digit can increase the above reading from 9999 to 19 999, or practically double the original reading. Thus a 1-V range could now read 1.9999, or a 1000-V range could now read 1 999. When the ½ digit is added to a four-digit display, it is called a 4½-digit display. When the ½ digit is switched on, the meter is said to be "overranging."

Digital meter specifications There are certain types of specifications which are unique to digital meters. Some of the more common ones are: sensitivity, resolution, and accuracy.

Sensitivity defines the capability of a digital voltmeter to respond to the smallest changes of measured voltage. This, of course, varies with different types. However, a rule of thumb is that the sensitivity is approximately equal to the least significant digit (extreme right-hand digit) of the reading. For example, if we are reading (as in Fig. 7.2), 17.76 V, the sensitivity would be 0.06 V.

Resolution is the inherent ability of a display to resolve one digit out of the total number capable of being displayed. For example, in a four-digit display the largest number which can be displayed is 9999 (for practical reasons, 10 000). Since the display can resolve 1 digit (least significant) out of 10 000, the resolution is 1 part in 10 000, or 0.01 percent.

The accuracy of the conventional D'Arsonal-type multimeter is expressed as a percentage of the full-scale reading. For example, a conventional meter with an accuracy rating of 2 percent, when used on the 10-V scale, would have an accuracy within plus or minus 0.2 V at full scale. However, the accuracy at lower readings is not usually known, although it must be not worse than 2 percent.

The accuracy of a digital multimeter is expressed in a somewhat more complex manner. It is given in two parts. The first part is a percentage of the actual reading; the second part (or modifier) is given as a percentage of the full scale reading, or a number of digits (least significant digit). This is best illustrated by the following three examples taken from commercial units.

Unit A:
- DCV: ±1% of reading, ±1 digit
- DCI and ACI and ACV (20 Hz–1 kHz): ±1.5%, ±1 digit
- Ohms: ±2%, ±1 digit

Unit B:
- DCV: ± 0.1% of reading, +.05% of full scale
- ACV (50–500 Hz): 0.5% of reading, +0.1% of full scale
- Ohms: ± 0.15% of reading, +0.1% of full scale

Unit C:
- DCV: ±0.1% of reading, +1 digit
- DCI: ±0.3% of reading, +1 digit
- ACV (45 Hz–10 kHz): 0.5% of reading, plus 2 digits
- ACI: ±1% of reading, plus 2 digits

7.9 CALIBRATION OF VOLTMETERS

The circuit of Fig. 7.25 shows how the accuracy of a voltmeter can be checked by comparing its reading with the reading of a meter known to be accurate. Meter M_1 is the one known to be accurate, and M_2 is to be calibrated. When the switch SW is closed, current flows through resistors R_1 and R_2 due to the applied voltage E. This may be either a dc voltage supplied by a battery or other power supply; or it may be an ac voltage if the voltmeters are being compared for ac voltage readings. Variable resistor R_1 is adjusted to vary the circuit current until the voltage drop across R_2 is some convenient value; then the meter readings are compared. In most VTVM's and FET voltmeters, provisions are made for adjusting the reading of M_2 until it is identical with that of M_1 during calibration.

Another method of calibrating a voltmeter (not illustrated) is to connect it across a voltage source of known value. An example of such a source is a dry cell that is manufactured for this purpose. Such batteries should never be employed as power sources in ordinary circuitry. They are to be utilized only for calibrating the voltage readings of voltmeters—a situation in which they are not required to supply any appreciable current. When properly used, calibration batteries may last a very long time.

OHMMETER CIRCUITS Most ohmmeters are actually voltmeters with a power supply included. One example is the circuit of Fig. 7.26. Note that the meter and its internal resistance combine with R_A and R_B to form a simple voltmeter circuit. When a voltage E is added to one leg of this circuit, it becomes a series ohmmeter.

When the ohmmeter probes are touched together—producing *zero* resistance between them—the battery will produce a full-scale deflection on meter M. Therefore zero ohms will be marked at full scale in this type of ohmmeter. Variable resistor R_B is used to adjust the full-scale deflection so that it reads exactly zero when the probes

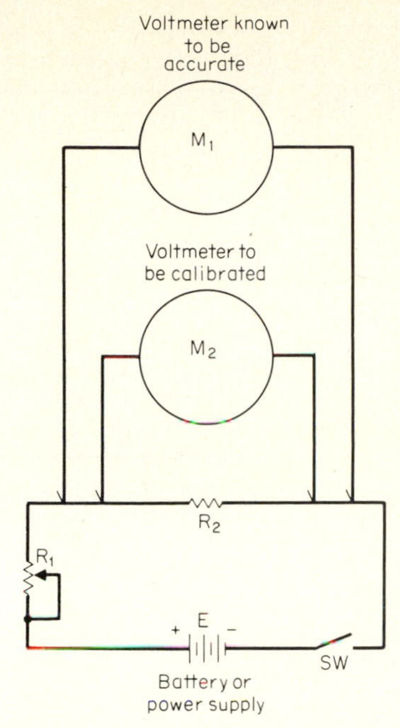

Fig. 7.25 Circuit for calibrating voltmeters.

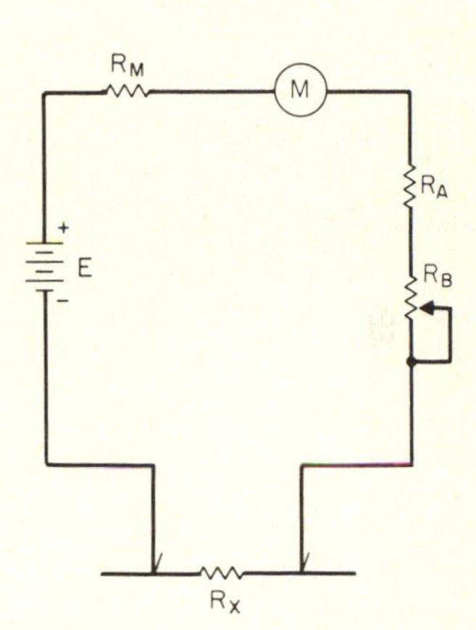

Fig. 7.26 Circuit for a series ohmmeter. The resistance being measured is R_X.

are touched together. This variable resistor is sometimes called "zero ohms." When the probes are apart, no current can flow through the circuit, and the meter needle does not deflect. Since there is nothing between the probes, we can assume that the resistance is infinitely large, and that is the marking on the left-hand side of the scale for ohmmeters of this type.

Figure 7.27 shows an ohmmeter scale with the zero value on the right side. Some beginning students have trouble with this type of scale because it is necessary to read it backward. When an unknown resistance (R_X in Fig. 7.26) is placed across the probes, the meter will deflect to some value between zero and infinity (that is, to some value between full-scale deflection and zero deflection). The ohmmeter scale is calibrated to read the resistance value of R_X.

Figure 7.28 shows a different kind of ohmmeter. It is called a *shunt ohmmeter* because the unknown resistance value R_X is placed in parallel with the meter movement. When the meter probes of this ohmmeter are touched together—producing *zero* resistance—the current of the circuit is shunted around the ohmmeter movement, and no deflection occurs. This means that at zero ohms, the needle of the scale is at the left side (zero deflection). When the probes are not touching, all the current from battery E will flow through the meter movement. With the probes open, we can assume that there is infinitely large resistance between them, and the right side of the scale will be marked infinity. Variable resistor R_B is adjusted to read full scale when the ohmmeter probes are not touching.

Of the instruments used by radio technicians, the ohmmeter is usually the most inaccurate. An accuracy of 10 to 20 percent of the reading is acceptable for most ohmmeter designs. The accuracy of the ohmmeter tends to deteriorate somewhat as the battery ages. This is because the internal resistance of the battery increases with an increase in battery age, and therefore the unknown resistance being measured represents a smaller percentage of the total resistance in the circuit. When the battery has deteriorated to the point where it can no longer produce full-scale deflection on the

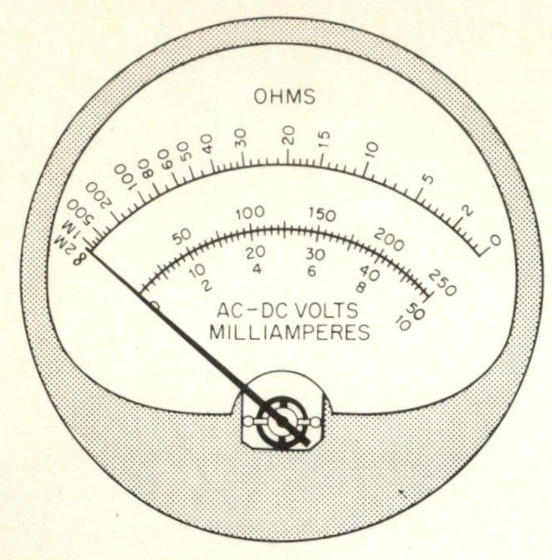

Fig. 7.27 Meter with ohmmeter scale that reads backward.

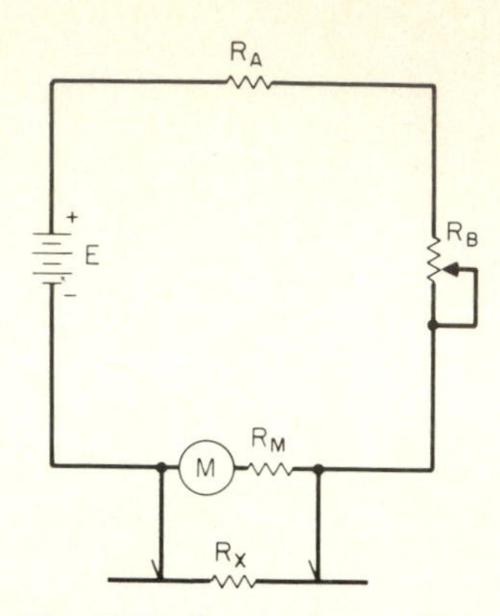

Fig. 7.28 Circuit for a shunt ohmmeter. The resistance being measured is R_X.

ohmmeter scale, it must be replaced. A VTVM, which operates on house current, uses batteries for the ohmmeter scale.

The best way to make a quick check of ohmmeter accuracy is to measure a precision resistance value with it.

7.10 DETERMINATION OF METER RESISTANCE

In order to calculate the value of the multiplier or shunt resistance needed to convert a sensitive meter movement into an ammeter or voltmeter, it is necessary to know the resistance of the meter movement. Before explaining how this can be done, it is very important to emphasize that an ohmmeter must *never* be used in an attempt to measure meter resistance. It has been shown that an ohmmeter contains a source of voltage. This voltage can send an excessive amount of current through a meter movement, thus causing damage to the coil or moving parts.

The circuit of Fig. 7.29 can be used for determining the resistance of meter M by the *half-deflection method.* With switch SW open, variable resistor R_1 is adjusted until meter M deflects to its full-scale reading. The reading of meter I is carefully noted. (Meter I should have a higher full-scale reading than meter M.)

Next, switch SW is closed. This will, of course, reduce the total resistance and increase the circuit current as measured by meter I. At the same time, it will reduce the amount of current flowing through meter M. Variable resistor R_X is adjusted until meter M reads its half-scale value. At the same time, variable resistor R_1 is adjusted to keep the total circuit current I equal to the value that it was with switch SW open. When the reading of M is equal to one-half its original value, then the resistance of R_X is equal

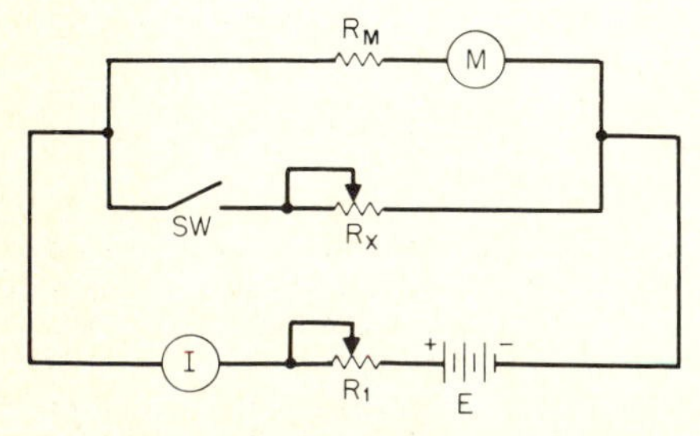

Fig. 7.29 Circuit for measuring the resistance of meter M.

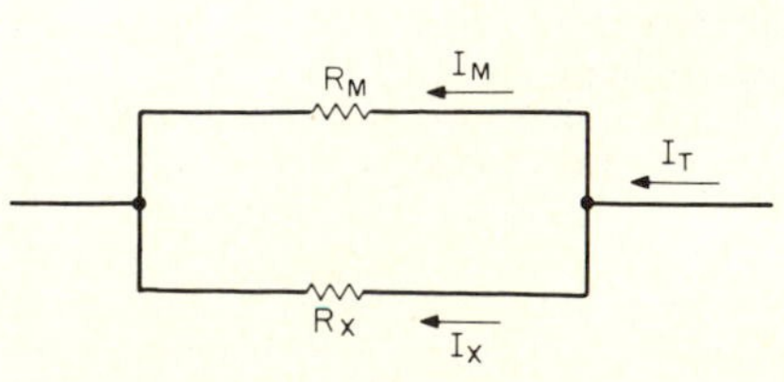

Fig. 7.30 A circuit used for explaining the proportional method of finding current.

to the meter resistance R_M. Without disturbing the setting of R_X when it has been adjusted for a half-scale reading of M, it is removed from the circuit and its resistance value is measured.

Another means of determining the value of meter resistance is based on the proportional method for finding current in a parallel branch. Figure 7.30 shows a parallel branch comprised of R_M and R_X, with a current I flowing into the junction of the branches. By the proportional method, the following equations are used for determining I_M and I_X:

$$I_M = I_T \frac{R_X}{R_M + R_X} \tag{7.9}$$

$$I_X = I_T \frac{R_M}{R_M + R_X} \tag{7.10}$$

Figure 7.31 shows a circuit that can be used for determining the meter resistance by using the proportional current equation. The current through meter M is

$$I_M = I_T \frac{R_X}{R_M + R_X} \tag{7.9}$$

Solving this equation for R_M gives

$$R_M = \frac{(I_T - I_M)R_X}{I_M} \tag{7.11}$$

In the circuit of Fig. 7.31 any precision resistance value that is comparable to the meter resistance can be used. Variable resistor R_Y is adjusted to provide a convenient reading on I_T and M.

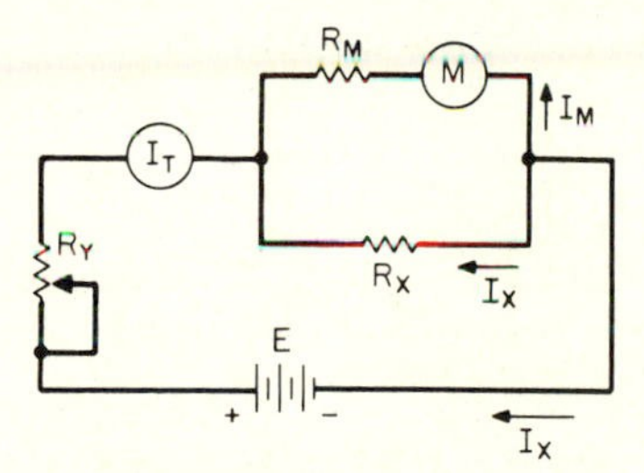

Fig. 7.31 An alternate method of finding the value of meter resistance.

7.11 THE WHEATSTONE BRIDGE

The ohmmeter is one of the least accurate of the instruments used by electronics technicians. If a more accurate method of making measurements is needed, a Wheatstone bridge may be used. These instruments may be made commercially, or they may be constructed with a few parts. In either case a highly accurate measurement is possible with a bridge circuit.

Before discussing the theory and use of the Wheatstone bridge, we will review the proportional method of calculating voltage drops. This will be useful for understanding the bridge circuit.

Figure 7.32 shows two resistors, R_1 and R_2, connected across a voltage source E. It does not matter whether this is a dc or an ac source; the relationships to be discussed hold true in either case. If the values of R_1, R_2, and E are known, then the circuit current I can be found by Ohm's law. By knowing the current through the resistors, the voltage drop across each resistor, V_1 and V_2, can be found.

Using the proportional method, the values of V_1 and V_2 can be found directly without the need for determining the current first. The derivation of the proportional voltage equations is quite simple, and will be given here.

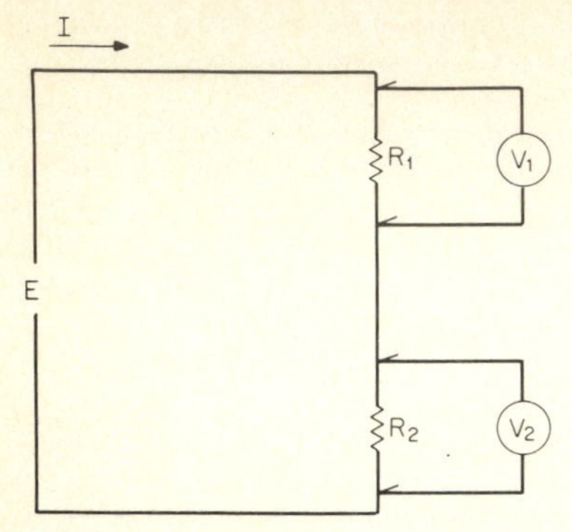

Fig. 7.32 Circuit used for obtaining a formula for V_1 and V_2.

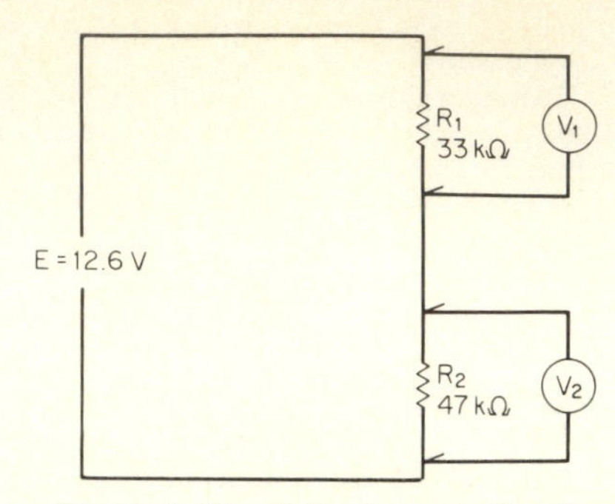

Fig. 7.33 Circuit for demonstrating the proportional method of calculating voltage drops.

Circuit total resistance:

$$R_T = R_1 + R_2$$

Circuit current:

$$I = \frac{E}{R_T} = \frac{E}{R_1 + R_2}$$

Voltage across R_1:

$$V_1 = IR_1 = \frac{E}{R_1 + R_2} \times R_1 = \frac{ER_1}{R_1 + R_2} \qquad (7.12)$$

Voltage across R_2:

$$V_2 = IR_2 = \frac{E}{R_1 \pm R_2} \times R_2 = \frac{ER_2}{R_1 \pm R_2} \qquad (7.13)$$

example 7.4 Calculate the values of V_1 and V_2 for the circuit of Fig. 7.33.

solution

$$V_1 = \frac{ER_1}{R_1 + R_2} = \frac{12.6 \times 33 \text{ k}\Omega}{33 + 47 \text{ k}\Omega} \cong 5.2 \text{ V}$$

$$V_2 = \frac{ER_2}{R_1 + R_2} = \frac{12.6 \times 47 \text{ k}\Omega}{33 + 47 \text{ k}\Omega} \cong 7.4 \text{ V}$$

Check: $E = V_1 + V_2$
$12.6 = 5.2 + 7.4$ V

It is obvious from the proportional method of calculating voltage that the amount of voltage across a resistor (such as R_1) in a branch (such as R_1 and R_2) depends upon the *proportion* of total resistance represented by the resistor R_1. In other words, $R_1/(R_1 + R_2)$ of the total applied voltage appears across R_1, and $R_2/(R_1 + R_2)$ of the applied voltage appears across R_2.

In the circuit of Fig. 7.34, the voltage at point a with respect to point c is

$$V_{a-c} = \frac{ER_2}{R_1 + R_2}$$

and the voltage at point b with respect to point c is

$$V_{b-c} = \frac{ER_4}{R_3 + R_4}$$

If $V_a = V_b$, then it follows that

$$\frac{\not{E}R_2}{R_1 + R_2} = \frac{\not{E}R_4}{R_3 + R_4}$$

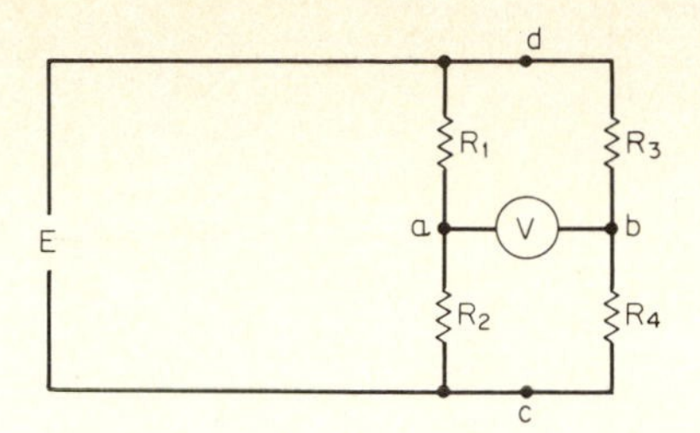

Fig. 7.34 The basic Wheatstone bridge circuit.

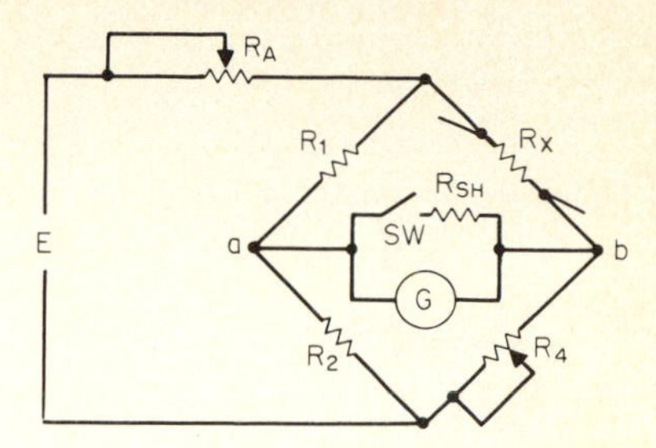

Fig. 7.35 A circuit for a commercial bridge.

This equation may be rewritten by a basic rule of algebra:

$$\frac{R_2}{R_4} = \frac{R_1 + R_2}{R_3 + R_4}$$

The voltage at point a with respect to point d is

$$V_{a-d} = \frac{R_1}{R_1 + R_2}$$

The voltage at point b with respect to point d is

$$V_{b-d} = \frac{R_3}{R_3 + R_4}$$

Under the condition that the voltage at point a is equal to the voltage at point b,

$$V_{a-d} = V_{b-d} \quad \text{and} \quad \frac{R_1}{R_1 + R_2} = \frac{R_3}{R_3 + R_4}$$

By using the same rule of algebra, this equation becomes

$$\frac{R_1}{R_3} = \frac{R_1 + R_2}{R_3 + R_4}$$

Since R_2/R_4 and R_1/R_3 are both equal to $(R_1 + R_2)/(R_3 + R_4)$, it follows that they are equal to each other.

$$\frac{R_1}{R_3} = \frac{R_2}{R_4} \tag{7.14}$$

This important equation holds true only when the voltage at point a is equal to the voltage at point b! This condition means that the bridge is *balanced*.

Figure 7.35 shows the Wheatstone bridge circuit used for measuring a resistance. The values of R_1 and R_2 are accurately known. Variable resistor R_4 has an accurately calibrated dial that makes it possible to know the exact resistance at any setting. The unknown resistance value—that is, the resistance being measured—is R_X. An ac or a dc voltage E is applied, and variable resistor R_A regulates the amount of current flowing to the bridge.

During a resistance measurement, the unknown resistance value is connected to terminals X and Y of Fig. 7.35. Then variable resistor R_4 is adjusted until no current flows through the galvanometer G—thus indicating that there is no voltage difference between points a and b. In other words, R_4 is adjusted *until the bridge is balanced.*

When the unknown resistance is first placed in the bridge circuit, there may be a considerable amount of unbalance, which would result in a large current flow through the galvanometer. To prevent damage to the meter movement, a shunt R_{SH} may be switched into the circuit. The switch is usually marked *sensitivity*. It is spring-loaded so that the shunt is always across the meter except when SW is depressed. This prevents the meter from being accidentally damaged during times when the bridge is greatly unbalanced.

When the bridge is balanced, the following equation holds true:

$$\frac{R_1}{R_X} = \frac{R_2}{R_4}$$

Solving this equation for R_X gives

$$R_X = \frac{R_1 R_4}{R_2} \tag{7.15}$$

example 7.5 The values of R_1 and R_2 in the circuit of Fig. 7.35 are 33 and 47 kΩ, respectively. The bridge is balanced when R_4 is set to 94 kΩ, with an unknown resistance R_X inserted into the bridge. What is the exact value of R_X?

solution

$$R_X = \frac{R_1 R_4}{R_2} = \frac{33\ \text{k}\Omega \times 94\ \text{k}\Omega}{47\ \text{k}\Omega} = 66\ \text{k}\Omega$$

7.12 AC BRIDGES

Variations of the Wheatstone bridge circuit can be used for measuring inductance, capacitance, impedance, and resonant frequency. In order to make these measurements, the input voltage to the bridge must be ac, and some minor changes must be made to the ratio arms. A few of the more important but commonly encountered ac bridge circuits will now be discussed.

RESONANCE AND FREQUENCY BRIDGE Figure 7.36 shows a bridge circuit that can be used for measuring resonance or frequency. The series-resonant circuit comprised of C_X and L_X is placed across one arm of the bridge. Resistance R_X may be the resistance of this series-resonance circuit, or it may be a resistance added to the bridge in order to make it possible to balance the bridge. An audio or r-f generator is applied across the bridge input terminals. The indicator in this case may be a pair of earphones, an ac voltmeter, or an oscilloscope. In some commercially manufactured bridges, tuning-eye tubes are employed as indicators.

For this discussion we will assume that an audio generator is used across the bridge terminals, and the indicator is a pair of earphones. As the generator frequency is varied so that its output frequency passes through the resonant frequency of $C_X - L_X$, the volume of sound heard will vary. The volume will be minimum when the audio generator is tuned to the resonant frequency. In other words, the signal generator is tuned for a null in the earphones. The resonant frequency of L_X and C_X is then read from the frequency dial of the audio generator. In order to get a sharper null, it may be necessary to adjust the resistance value of R_C so that the bridge is balanced.

The procedure just described is especially useful for finding the resonant frequency of an audio filter. The filter is connected into the bridge circuit in place of R_X, L_X, and C_X. The bridge will be balanced when the following conditions are met:

$$\frac{R_A}{R_B} = \frac{R_C}{R_X} \tag{7.16}$$

$$f_r = \frac{1}{2\pi\sqrt{L_X C_X}} \tag{7.17}$$

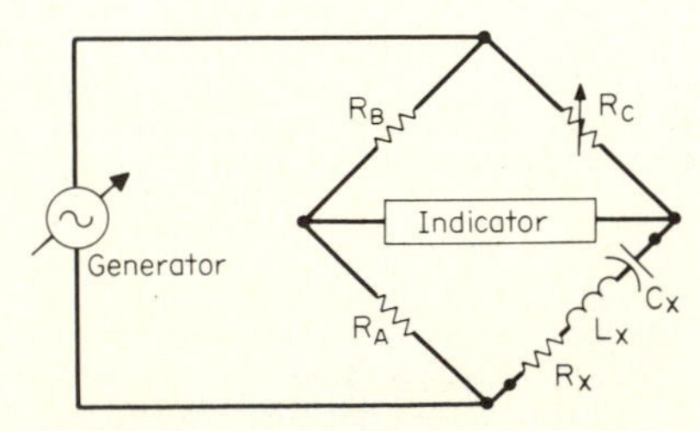

Fig. 7.36 Bridge circuit for measuring resonance and impedance.

where R_A, R_B, R_C, R_X = resistance values, Ω
f_r = frequency of audio generator, Hz
L_X = inductance of series-tuned circuit (or effective inductance of filter), H
C_X = capacitance of series-tuned circuit (or effective capacitance of filter), F

It is important to note that *both* conditions described by Eqs. (7.16) and (7.17) must be met simultaneously in order for a true null to be obtained.

If R_C is a calibrated resistor, the effective series resistance of the resonant circuit will be known by its setting when a true null is obtained. If R_X is in actuality a combination of the series resistance of the resonant circuit and a resistance value added to it to effect a balance, then the resistance of R_X must be subtracted from the calculated value to find the actual value of the resistance of the series-resonant circuit. In most cases R_X will be simply the resistance of the coil. If the resistance value of R_X is very small, then it might not be possible to obtain its value on this type of bridge. The reason is that the junction connections will affect the balance more than the small resistance.

If an r-f generator is employed in this circuit, the r-f signals must be modulated and a detector circuit used in conjunction with earphones (or other type of indicator).

Equation 7.17 shows that the frequency of the ac generator can be accurately determined if the values of C_X and L_X (as well as the resistance values) are accurately known. In this application the usual procedure is to vary C_X in series with an accurately known inductance L_X. The frequency is read directly from the dial of C_X.

INDUCTANCE BRIDGES The inductance of a coil can be accurately obtained by means of an ac bridge. Figure 7.37 shows a *resistance ratio bridge* that can be used

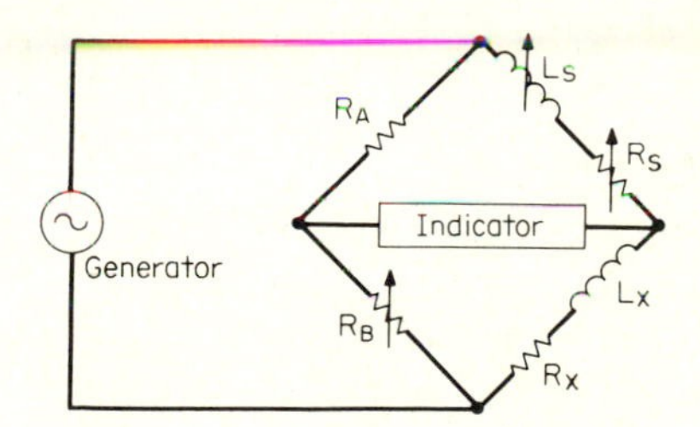

Fig. 7.37 Measurement of inductance can be accomplished with this resistance ratio bridge.

for this purpose. As in all the ac bridges discussed in this section, the indicator may be a set of earphones, an oscilloscope, or an ac meter, provided frequency response of the indicator is not exceeded by the frequency of the ac generator. In this simple circuit, R_X and L_X comprise the unknown inductor and its equivalent dc resistance value. These are represented schematically as a coil and a resistor in series, although it must be understood that the resistance is actually the resistance of the wire in the coil. Resistor R_S and inductor L_S represent the *Standard resistance* and *Standard inductance,* respectively. They must be accurately calibrated so that when they are adjusted for a balance, the value of R_X and L_X can be read directly from their dials. In order to effect a balance with this bridge, the following conditions must be met simultaneously:

$$L_X = \frac{R_B}{R_A} L_S \tag{7.18}$$

$$R_X = \frac{R_B}{R_A} R_S \tag{7.19}$$

where R_A, R_B, R_S, R_X = resistance values, Ω
L_S, L_X = inductance values, H

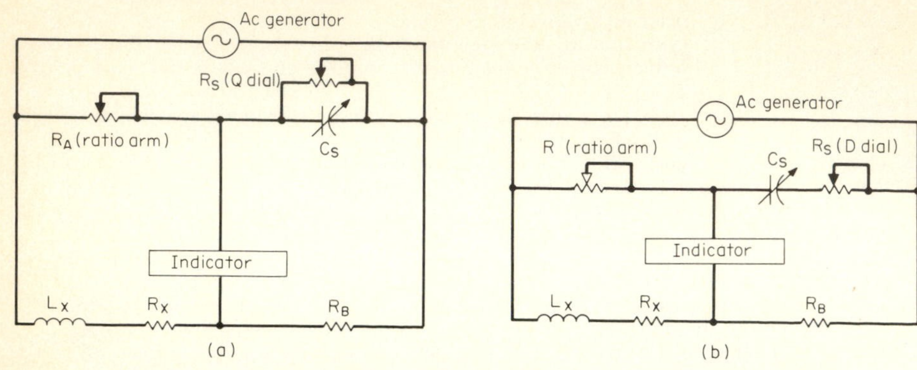

Fig. 7.38 Commonly used inductance bridges: (*a*) Maxwell bridge. (*b*) Hay bridge.

In the circuit of Fig. 7.37 resistor R_b is made variable to obtain the larger number of resistance ratios, and therefore increase the range of inductance values that can be measured with the bridge. Normally R_b is comprised of four or five precision resistors, any one of which can be connected into the circuit with a selector switch.

The most important disadvantage of the resistance ratio bridge of Fig. 7.34 is that it is difficult to obtain precision variable inductors. Such inductors are considerably more expensive than precision capacitance values. For this reason, the bridge circuits of Fig. 7.38 are more often encountered. In these circuits precision variable capacitors C_S are used for balancing the bridge (instead of precision inductors).

Although the two circuits of Fig. 7.38 are quite similar in construction, they actually have different applications. The Maxwell bridge (Fig. 7.38*a*) is used whenever the Q of the inductance being measured is relatively low—that is, when the dc resistance value of R_X is relatively high. When the Q of the inductor is high (indicating a low value of R_X), the Hay bridge (Fig. 7.38*b*) is employed for inductance measurement. The equations for L_X and R_X under the condition of a balanced bridge must be *simultaneously* met.

For the Maxwell bridge:

$$L_X = R_A R_B C_S \tag{7.20}$$

and

$$R_X = \frac{R_A R_B}{R_S} \tag{7.21}$$

The Q of the coil can be obtained from the equation

$$Q = \frac{\omega L_X}{R_X} = \omega C_S R_S \tag{7.22}$$

For the Hay bridge,

$$R_X = \frac{R_A R_B R_S(\omega C_S)}{1 + (R_S \omega C_S)^2} \tag{7.23}$$

and

$$L_X = \frac{R_A R_B C_S}{1 + (R_S \omega C_S)^2} \tag{7.24}$$

The Q of the coil can be obtained from the equation

$$Q = \frac{\omega L_X}{R_X} = \frac{1}{\omega C_S R_S} \tag{7.25}$$

where L_X = unknown inductance value, H
R_A, R_B, R_S, R_X = resistance values, Ω
$\omega = 2\pi f$
f = frequency of ac generator, Hz
C_S = capacitance value of precision variable capacitor, F

In the case of the Maxwell bridge, if the value of C_S is accurately known, then the value of Q will be related to the setting of R_S. In most commercially manufactured

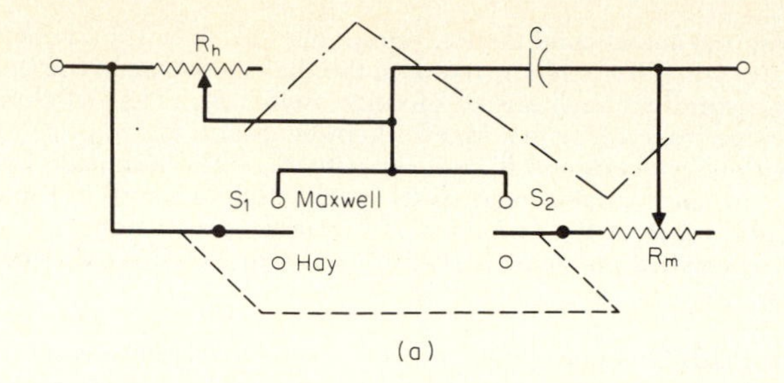

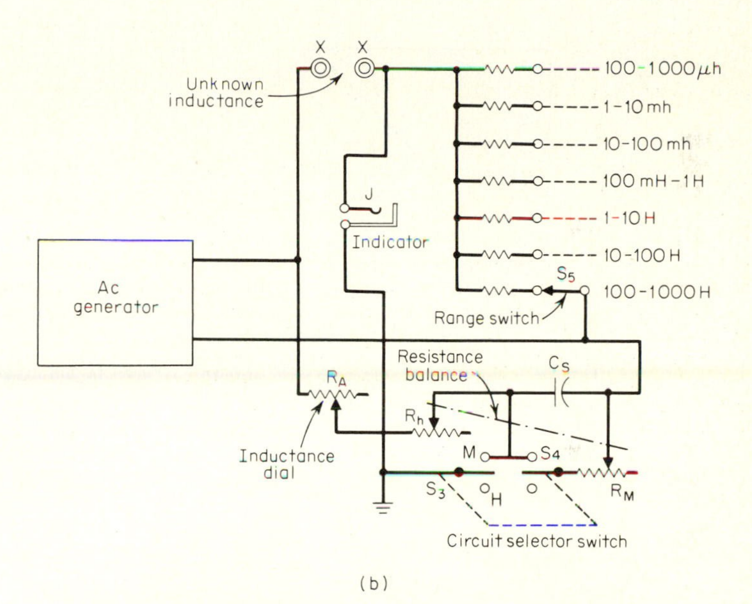

Fig. 7.39 Combination of Maxwell and Hay bridges. (*a*) Circuit for switching between series and parallel *RC* connections. (*b*) Circuit for combination bridge.

Maxwell bridge circuits, R_S is calibrated directly to read the value of Q. For the Hay bridge, the setting of R_S is proportional to the *reciprocal* of the Q of the coil, and it is normally marked D on the bridge dial. (The letter D stands for *dissipation factor*, which is defined as the reciprocal of the Q.) Once the value of D is obtained, the Q is easily determined by taking the reciprocal.

In a variation of the Hay bridge, resistor R_A is replaced with a variable capacitor in the ratio arm. The operation of the circuit, however, is the same. This version is called an *Owen bridge.*

It would be inconvenient to have two different conductance bridges—one for measuring the inductance of coils with high Q and one for measuring the inductance of coils with low Q. The only difference between Maxwell and Hay circuits is that the variable resistor is in parallel with the capacitor in one case, and in series with the capacitor in the other case. It is a simple matter to arrange a switching circuit for converting the same bridge circuit to either a Maxwell bridge or a Hay bridge. Figure 7.39*a* shows the switching arrangement. The dotted line between R_h and R_m (which stands for *Hay resistance* or *Maxwell resistance*) indicates that these two variable resistors are actually connected onto the same shaft. This simplifies the operation, but it requires that both the Q values and the D values be marked on the variable-resistor indicator.

Figure 7.39*b* shows the actual combination bridge circuit. It will be noted that the resistances in the ratio arm are actually marked on the dial according to the inductance range that can be measured for each setting of range switch S_5. The switching circuit that converts the bridge from a Hay to a Maxwell configuration is comprised of S_3 and S_4 in conjunction with resistors R_h and R_m and capacitor C_S. The unknown inductance is placed in the circuit, and variable resistor R_A is adjusted for a null in the detector. The actual inductance value of the unknown inductance is read from the calibrated dial of R_A. The Q or dissipation factor is read from the dial of R_m or R_h—depending on the type of bridge circuit in use at the time.

CAPACITANCE BRIDGES Figure 7.40 shows a simple *resistance ratio bridge* for measuring capacitance values. The unknown capacitance C_X and its resistance R_X

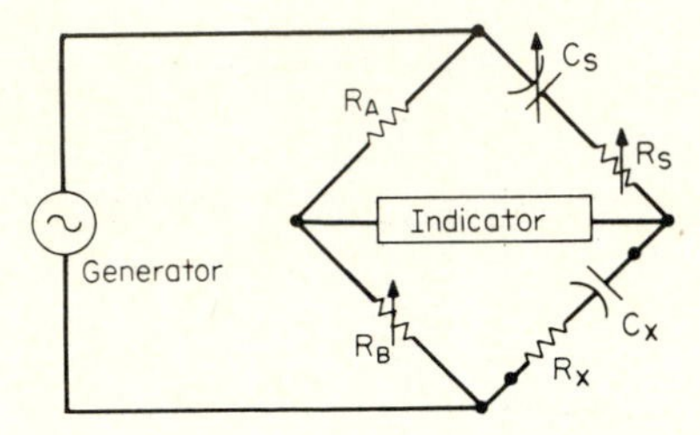

Fig. 7.40 Resistance ratio bridge for measuring capacitance.

are balanced by calibrated capacitors C_S and R_S. When the bridge is balanced, as indicated by a null, the following equations must simultaneously hold true:

$$C_X = \frac{R_A}{R_B} C_S \tag{7.26}$$

and

$$R_X = \frac{R_B}{R_A} R_S \tag{7.27}$$

where R_A, R_B, R_S, R_X = resistance values, Ω
C_S, C_X = capacitance values, F

The capacitance bridges shown in Fig. 7.41 are variations of the resistance ratio bridge. The Wien bridge of Fig. 7.41*a* accurately measures the capacitance in terms of resistance and frequency values. This circuit can be used for precision determination of the capacitance value if the frequency of the generator and the resistance values are accurately known. The equations for the bridge under the condition of balance are as follows:

$$C_X = \frac{1}{\omega^2 R_C R_X C_C} \tag{7.28}$$

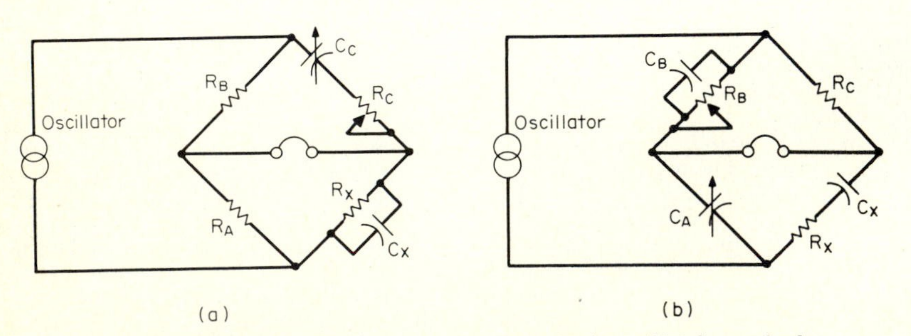

Fig. 7.41 Two popular bridge circuits: (*a*) Wien bridge. (*b*) Schering bridge.

The following alternate equation is sometimes more useful:

$$C_X = C_C\left(\frac{R_B}{R_A} - \frac{R_C}{R_X}\right) \tag{7.28a}$$

where R_A, R_B, R_C, and R_X are in ohms
C_X and C_C are in farads
$\omega = 2\pi f$ and f are in hertz

The Schering bridge of Fig. 7.41*b* is useful when both the capacitance and the power factor (or dissipation factor) of the capacitor are to be determined. When the bridge is balanced, the following equations must hold true for this circuit:

$$R_X = \frac{C_B}{C_A} R_C \tag{7.29}$$

$$C_X = \frac{R_B}{R_A} C_A \tag{7.30}$$

The Q of the capacitor C_X can be determined by the equation

$$Q = \frac{1}{\omega C_X R_X} \tag{7.31}$$

$$Q = \frac{1}{\omega C_B R_B} \tag{7.31a}$$

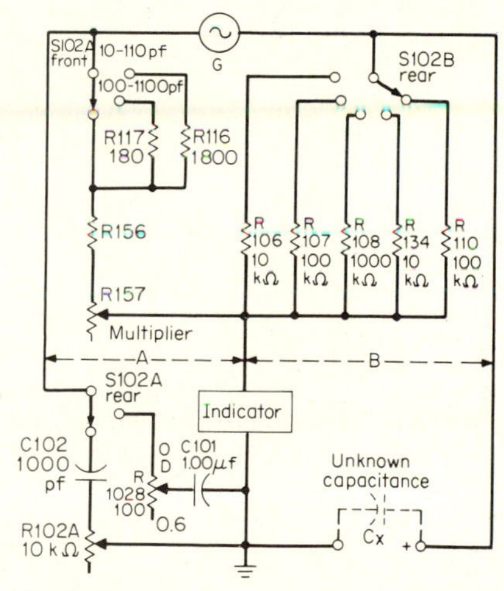

Fig. 7.42 Circuit for a manufactured Wien bridge.

Since a Schering bridge is easily calibrated to read the dissipation factor of the dielectric directly (by a dial connected to R_B), this circuit is sometimes used for comparing the properties of dielectrics. Two parallel plates are connected in the position of C_X, and various dielectrics are inserted between the plates. The bridge is balanced in each case. The dissipation factors are then tabulated.

Figure 7.42 shows a schematic diagram of a commercially manufactured capacitance bridge. In this case the Wien bridge circuit is used. Except for the fact that switching arrangements are provided for changing the range of capacitance values, the circuit is identical with the one shown in Fig. 7.41*a*.

Chapter 8

Semiconductor Devices and Transistors

8.1 INTRODUCTION

The variety of semiconductor devices available to the technician and engineer is prodigious. Included are rectifying, regulating, tunnel, light-emitting, light-sensing, switching, r-f, and microwave diodes. Diodes are used for converting alternating current to a unidirectional current (rectification), regulating voltage levels, detecting and mixing r-f signals, switching, clamping, in displays, and in the generation of microwave power. Some devices based partially on diode action are the silicon-controlled rectifier (SCR) and the unijunction transistor (UJT). The SCR and UJT find wide use in controlling efficiently power delivered to a load.

Transistors fall into two broad categories: the *bipolar junction transistor* (BJT) and the *field-effect transistor* (FET). Bipolar junction transistors, usually referred to simply as *transistors*, are available as npn and pnp types. Field-effect devices can be divided into two basic types: the *junction* FET (JFET) and the *metal-oxide semiconductor* FET (MOSFET). The MOSFET is also referred to as the *insulated-gate* FET (IGFET). Figure 8.1 provides a sampling of the available shapes and packages of discrete diodes and transistors.

In this chapter we shall examine the operation of semiconductor devices and transistors, and their characteristics, and define the significant parameters used for their characterization. A discussion of understanding diode and transistor data sheets concludes the chapter. To comprehend the operation of semiconductor devices, however, an elementary knowledge of semiconductor theory is vital.

8.2 ELEMENTARY SEMICONDUCTOR THEORY

Electrons appear to exhibit a dual nature, behaving as *particles* at one time and as *waves* at other times. The particle picture is useful in describing the operation of transistors and most diodes. For a few devices, such as the tunnel diode, the concept of electrons being wavelike in nature seems best suited for describing their operation.

ATOMIC STRUCTURE According to the particle picture, the structure of an atom is analogous to a miniature solar system. In the center of the atom is the *nucleus* which contains positive-charge particles called *protons* and, for many atoms, electrically neutral particles referred to as *neutrons*. Rotating around the nucleus are *electrons*, which have a negative charge. Because the number of electrons equals the number of protons, the atom is normally electrically neutral. The mass of the proton and neutron is approximately the same; the mass of the electron is about 1/2000th that of a proton.

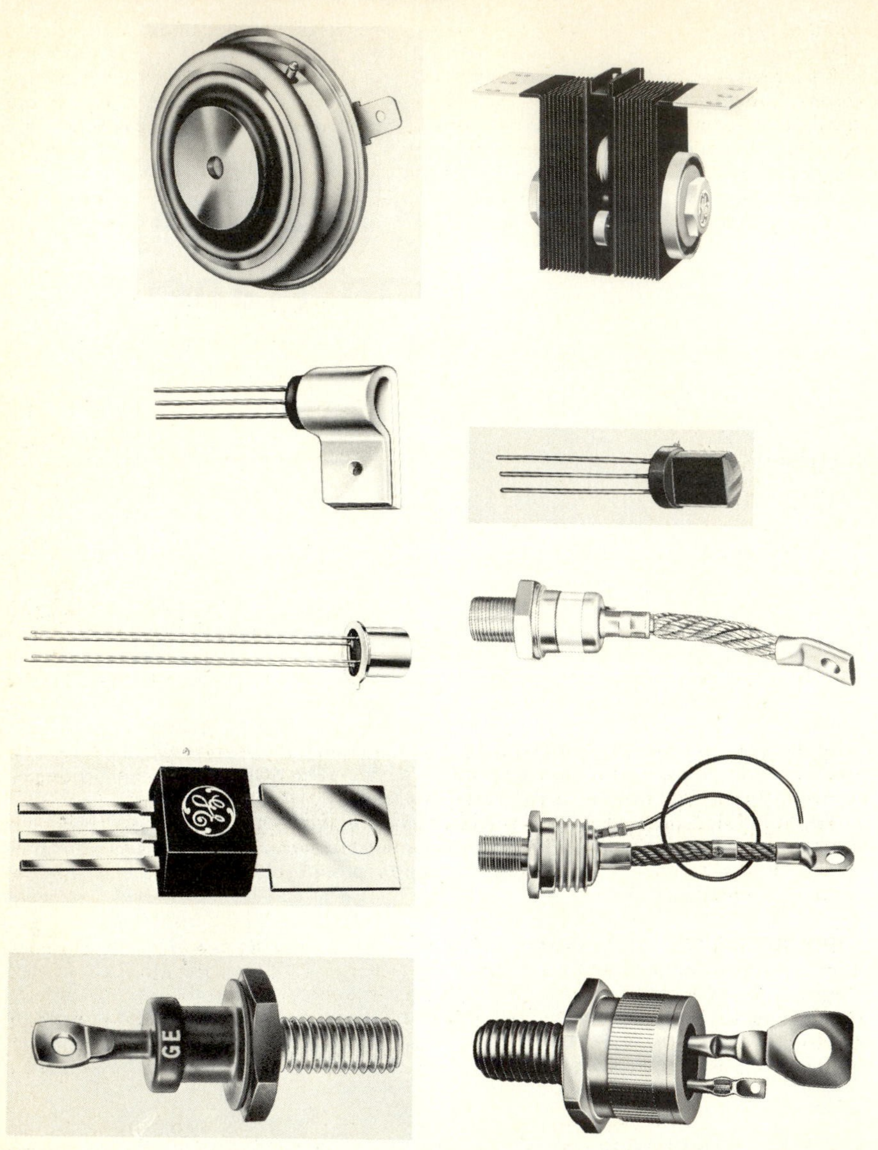

Fig. 8.1 Discrete diodes and transistors are available in many sizes and shapes. (*Courtesy General Electric Company, Semiconductor Department, Syracuse, N.Y.*)

The simplest atomic structure is hydrogen (H), depicted in the two-dimensional sketch of Fig. 8.2*a*. Its *atomic number* Z is 1 because a single electron orbits around a nucleus containing a single proton. The orbit, or *shell*, nearest to the nucleus can hold no more than two electrons. Helium (He), for example, has an atomic number $Z = 2$. This denotes that exactly two electrons are rotating around the nucleus (which contains two protons), and the shell is completely filled. For this reason helium is *chemically inert* and does not react with other elements. An element whose shell is unfilled, such as hydrogen, reacts with other elements. The number of electrons in an unfilled shell is called the *valence* electrons; for example, hydrogen has one valence electron.

For atoms of greater structural complexity, the electrons are contained in shells located farther away from the nucleus. The second shell can hold up to 8 electrons, the third up to 18, and so on. The atomic structure of silicon (Si), the basic material used for discrete semiconductor devices and integrated circuits, is illustrated in Fig. 8.2*b*. Its atomic number Z = 14; consequently, the third shell is unfilled and contains *four* valence electrons. Germanium (Ge), another semiconductor material, also has four valence electrons. Elements containing four valence electrons are said to be *tetravalent.*

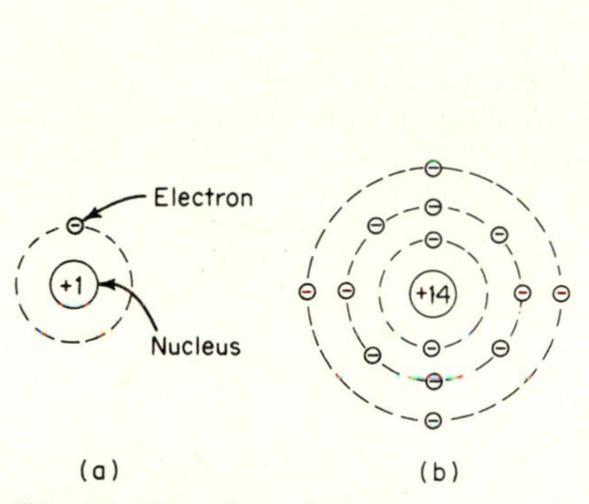

Fig. 8.2 Examples of atomic structures: (*a*) Hydrogen, H. (*b*) Silicon, Si.

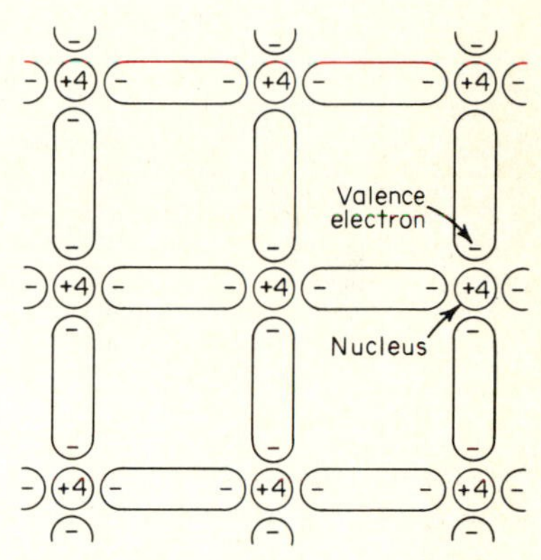

Fig. 8.3 Representation of covalent bonding.

Silicon for semiconductor devices is grown to have a *single-crystal* structure. In simple terms, a single crystal has an ordered array of atoms. In absolutely pure, or *intrinsic,* silicon, each atom shares an electron with its four neighboring atoms, as illustrated in Fig. 8.3. This type of electron sharing results in *covalent bonds.* Because only the valence electrons are involved in covalent bonds, the silicon atom is represented simply by four valence electrons around a nucleus of +4 positive charge for electrical neutrality.

ENERGY BANDS The resistance of a semiconductor is greater than that of a conductor, but considerably less than that of an insulator. With the aid of an *energy-band diagram,* shown in Fig. 8.4*a*, one can readily understand the differences in conductors, semiconductors, and insulators. According to modern atomic theory, the energies of electrons in a material, such as silicon, are grouped in *energy bands.* The *valence band* of energies corresponds to electrons that are attracted strongly by the nucleus. As a

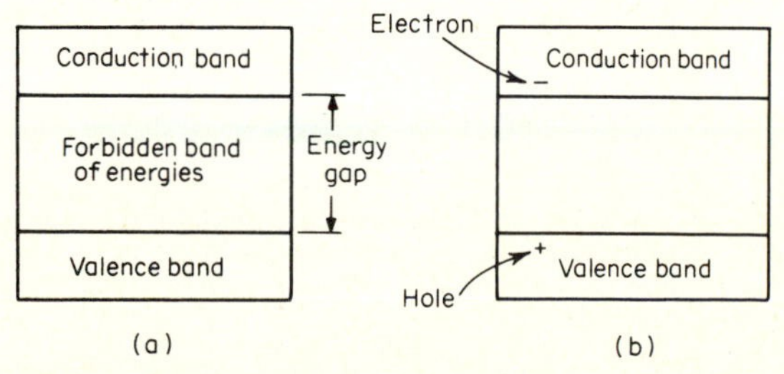

Fig. 8.4 Energy-band diagrams: (*a*) Basic energy diagram. (*b*) An electron reaching the conduction band of energies leaves a hole in the valence band of energies.

result, these electrons cannot serve as carriers of current. To become carriers of current, the electrons must occupy a higher band of energies, referred to as the *conduction band.* No electron, however, can ever occupy a region, referred to as the *forbidden band of energies,* which separates the valence and conduction bands. An electron in the valence band, therefore, must have a minimum energy before it occupies the conduction band. This minimum energy is the *energy gap* of the material.

For conductors such as copper the valence and conduction bands overlap. Consequently, there is a plentiful supply of electrons to serve as carriers of current. In an insulator, such as diamond, the energy gap is so great that there are virtually no electrons in the conduction band. The energy gap for semiconductors, however, is such that some electrons are found in the conduction band.

An electron that has reached the conduction band of energies in a semiconductor leaves a vacancy, or *hole,* in the valence band of energies, as illustrated in Fig. 8.4*b*. What is interesting is that the hole behaves like a positively charged particle whose charge is equal and opposite to the electron. It is influenced by an electric field, like an electron, but travels in an opposite direction and more slowly than an electron.

Because of the effects of energy, such as heat or light, equal numbers of electrons and holes are produced in a semiconductor. This action is referred to as the *intrinsic generation of electron-hole pairs.* As the temperature of a semiconductor rises, the electrons in the valence band gain energy and the generation of electron-hole pairs increases rapidly. In fact, this is what limits the operating temperature of silicon devices to 150°C. Because the energy gap of germanium is nearly one-half that of silicon, the maximum operating temperature of germanium is approximately 85°C.

DOPING Intrinsic silicon, which contains equal numbers of electrons and holes, is useless for semiconductor devices. One needs to alter intrinsic silicon so that it has either an excess of electrons or of holes. This is accomplished by introducing suitable impurities in the silicon. The process is called *doping,* and the impurity is referred to as a *dopant.*

n-TYPE SILICON Assume that an impurity, such as phosphorus (P) which has five valence electrons, is introduced in silicon. (Elements with five valence electrons are called *pentavalent* elements.) The impurity concentration required for semiconductor devices is extremely minute, in the order of 1 impurity atom for 10^8 silicon (*host*) atoms. The impurity atom replaces a silicon atom in the single-crystal lattice. Because for every 10^8 silicon atoms there is only one impurity atom, the probability is high that the impurity atom will be surrounded by silicon neighbors in the lattice. Silicon doped with a pentavalent impurity is called *n-type silicon.*

A two-dimensional view of n-type silicon is illustrated in Fig. 8.5*a*. Because phosphorus is a pentavalent dopant, it has five valence electrons. Four of the valence electrons form covalent bonds with four neighboring silicon atoms. The fifth valence electron is free and serves as a carrier of current. When a voltage is impressed across an

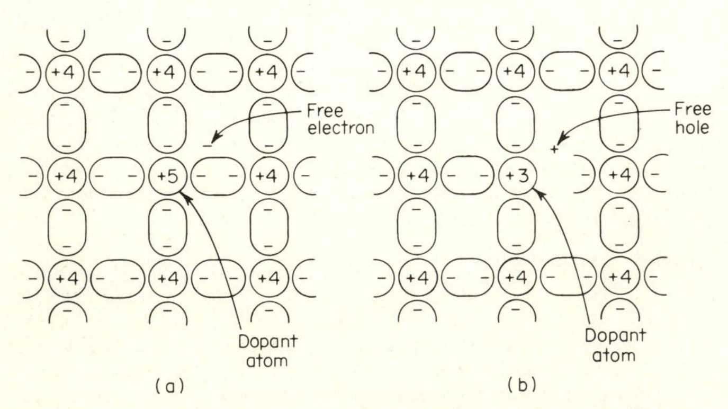

Fig. 8.5 Covalent bonding in a doped semiconductor: (*a*) n-type. (*b*) p-type.

n-type specimen, the free electrons are attracted toward the positive terminal, and current flows. If the doping is increased, more free electrons exist, and greater current flows.

Electrons in an n-type semiconductor are referred to as *majority carriers.* Regardless of the fact that the semiconductor is doped, the intrinsic generation of electron-hole pairs still continues, unabated. Hence, in addition to electrons, there are a few holes. Holes in an n-type material are referred to as *minority carriers.* At normal operating temperatures, the minority carriers are negligible. As the temperature rises, however, the minority carriers increase.

p-TYPE SILICON Assume that an element having three valence electrons (*trivalent* element), such as boron (B), is introduced in silicon. The result is pictured in Fig. 8.5*b*. The trivalent impurity now shares its three valence electrons with three neighboring silicon atoms. The deficiency of a fourth electron results in a hole. As mentioned earlier, the hole acts like a particle having a positive charge, equal and opposite to that of an electron, and serves as a carrier of current.

When a voltage is impressed across a p-type semiconductor, the holes are attracted to the negative terminal of the source. As for an n-type semiconductor, the greater the doping is, the more holes are present and greater current flows. Holes in a p-type semiconductor are now the majority carriers, and electrons the minority carriers.

Having this basic knowledge of semiconductor theory, we can proceed now to consider the operation of diodes and transistors.

8.3 JUNCTION DIODE

The simplest semiconductor device is the junction diode. Its basic physical construction and electrical symbol are shown in Fig. 8.6. The junction diode is formed by doping one-half of intrinsic silicon (or germanium) with a p-type dopant and the other half with an n-type dopant. The boundary at the p and n regions, where all the action occurs, is called the *pn junction*. Connections made to the p and n regions are referred to as the *anode* and *cathode*, respectively.

When the anode is made positive with respect to the cathode, as in Fig. 8.7*a*, the diode is said to be *forward-biased.* Electrons in the n region are attracted toward the anode end, and holes in the p region toward the cathode end of the diode. The electrons and holes, therefore, cross the junction, and current flows in the circuit.

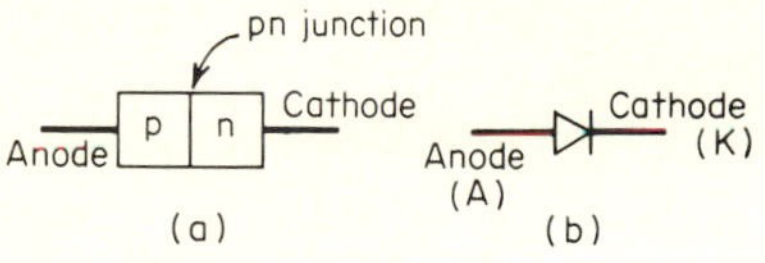

Fig. 8.6 Junction diode: (*a*) Basic physical construction. (*b*) Electrical symbol.

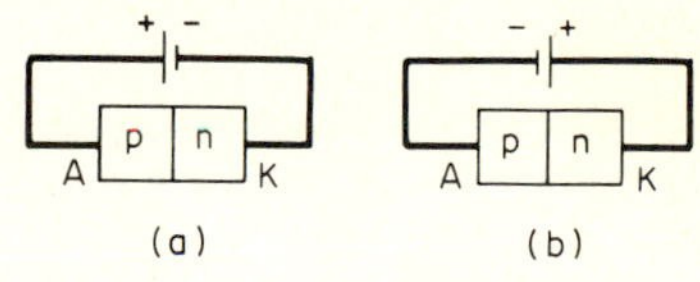

Fig. 8.7 Biasing a junction diode: (*a*) Forward-biased. (*b*) Reverse-biased.

In Fig. 8.7*b* the anode is made negative with respect to the cathode. Under this condition, the diode is *reverse,* or *back, biased.* Minority holes and electrons near the junction are initially attracted to the negative and positive terminals of the battery, respectively. As a result, the junction becomes depleted of carriers, and no further current flows. This region of depleted carriers is referred to as the *depletion region,* or *layer.* (The *voltage-variable capacitor,* or *varactor,* whose basis of operation is the depletion layer, is described in Chap. 2.)

DIODE CHARACTERISTICS In the previous section we saw that the junction diode behaves like a "valve." When forward-biased, the diode permits current to flow; in the reverse-biased state, it prevents the flow of current. If it were possible to make the *perfect,* or *ideal, diode,* its characteristics might appear as shown in Fig. 8.8*a*.

When the diode is forward-biased, current I_F (dc forward current) flows and voltage V_F (dc forward voltage) is zero, regardless of the forward current. (*Note:* It is common practice to use V instead of E in denoting voltage for diode and transistor characteristics.

This will be adhered to in this and other chapters for describing the characteristics of semiconductor devices.) When the diode is reverse-biased, I_R (dc reverse current) is zero, regardless of the value of V_R (dc reverse voltage).

A typical characteristic curve for a *physical diode* appears in Fig. 8.8*b*. In forward bias, for voltages less than the threshold voltage V_t, virtually no current flows. (At room temperature, $V_t = 0.6$ V for silicon and 0.25 V for germanium. The threshold voltage decreases at the rate of 2 mV for each degree Celsius rise in temperature.) For forward voltages greater than V_t, the diode conducts, and the voltage across the diode is somewhat greater than the threshold value.

In reverse bias, a minute reverse current, in the order of microamperes for silicon,

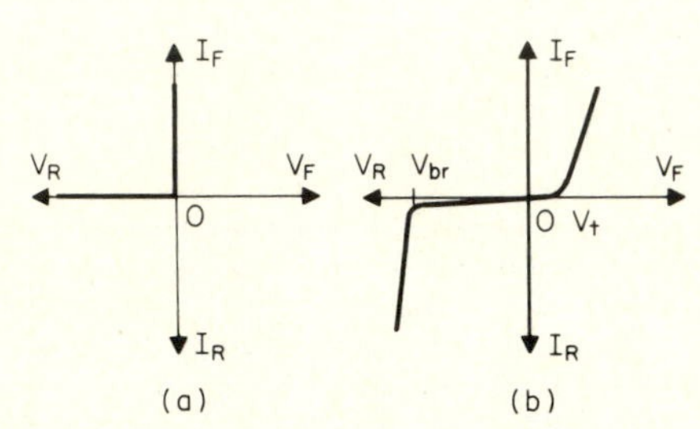

Fig. 8.8 Voltampere characteristics: (*a*) For an ideal diode. (*b*) For a physical diode.

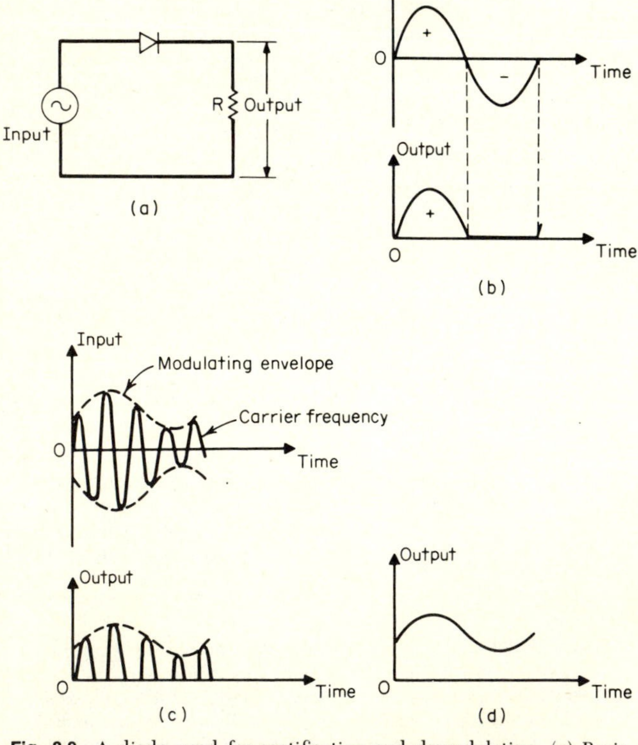

Fig. 8.9 A diode used for rectification and demodulation: (*a*) Basic circuit. (*b*) Input and output waveforms for rectification. (*c*) Input and output waveforms for demodulation. (*d*) Modulating signal obtained from demodulator. (See Example 8.1.)

flows. The reverse current increases as the reverse bias voltage is increased. At a voltage V_{br} (the *breakdown voltage*), the reverse current increases rapidly for a very slight increase in reverse voltage. This characteristic is used in voltage-reference and regulating diodes, to be considered in the next section.

RECTIFIER, R-F, AND SWITCHING DIODES A diode used for rectification, or in r-f and switching circuits, is basically a pn junction diode. What distinguishes it is its physical size, geometry, and encapsulation. Because it handles many milliamperes or amperes of current, a rectifier diode is generally larger in size than a r-f or switching diode. A larger physical size implies a large junction area and capacitance. As a result, rectifier diodes cannot be used effectively in high-frequency or switching applications.

R-f diodes are physically small and designed to minimize capacitance and inductance. Switching diodes also exhibit small physical size and are designed to switch rapidly from the reverse- to the forward-biased state. Recently switching diodes have been developed that are capable of switching a few amperes in the order of a tenth of a microsecond.

example 8.1 Show how a junction diode may be used for (*a*) rectification of alternating current, and (*b*) demodulation (detection) of an amplitude-modulated (AM) signal.

solution A basic circuit for both applications is illustrated in Fig. 8.9*a*.

(*a*) Referring to Fig. 8.9*b* for rectification, the input to the circuit is typically a 60-Hz sine wave. During the positive half-cycle, the diode is forward-biased (the anode is positive with respect to the cathode), and current flows. Neglecting the small forward voltage across the diode, the output is equal to the input. On the negative half-cycle, the diode is reverse-biased (the anode is negative with respect to the cathode), and no current flows. The output is *unidirectional*, that is, always positive. With suitable filtering (see Chap. 16), the unidirectional waveform is converted to a dc voltage.

(*b*) An example of an amplitude-modulated input signal is shown in Fig. 8.9*c*. As in rectification, on the positive half-cycle the diode conducts, and on the negative half-cycle the diode does not conduct. The output, therefore, is a group of unidirectional waves of different amplitudes. If a suitable capacitor is placed in parallel with resistor R in Fig. 8.9*a*, the modulating signal is obtained, as shown in Fig. 8.9*d*.

DIODE PARAMETERS Commonly used parameters for characterizing junction diodes are summarized in Table 8.1. In specifying their values, the manufacturer must

TABLE 8.1 Commonly Used Parameters for Junction Diodes

Parameter	Symbol	Meaning
Dc forward voltage	V_F	Voltage across a forward-biased diode (anode positive with respect to cathode)
Dc forward current	I_F	Direct current flowing in a forward-biased diode
Dc reverse voltage	V_R	Voltage across a reverse-biased diode (anode negative with respect to cathode)
Dc reverse current	I_R	Leakage current flowing in reverse-biased diode
Reverse breakdown voltage	V_{br}, B_V, PRV, PIV	Maximum reverse voltage across diode before it breaks down
Power dissipation	P	Maximum power that may be dissipated in a diode
Operating junction temperature	T_j	Temperature of the pn junction
Capacitance	C	Capacitance across diode in its forward- or reverse-biased state
Reverse recovery time	t_{rr}	Time required for reverse current or voltage to reach a specified value after switching diode from forward- to reverse-biased state
Forward recovery time	t_{fr}	Time required for forward voltage or current to reach a specified value after switching diode from its reverse- to forward-biased state
Noise figure	NF_O	Ratio of rms output noise power of receiver in which diode is used to that of an ideal receiver of same gain and bandwidth
Conversion loss	L_C	Power lost in mixer diode when converting an r-f signal to an i-f (intermediate frequency) signal
Video resistance	R_v	Low-level impedance of a detector diode

state the test conditions and ambient (operating) temperature at which measurements were made.

8.4 ZENER DIODES

The reverse characteristics of a physical-junction diode illustrated in Fig. 8.8*b* is redrawn in Fig. 8.10*a*. It was noted that as the reverse voltage increases, at a certain level, V_{br}, the diode breaks down. In the operating region of reverse current from I_{ZK} to I_{ZM}, the voltage across the diode is essentially constant.

Current I_{ZK}, the knee current, is the minimum current, and I_{ZM} is the maximum current limited by device dissipation. The voltage below the knee V_Z is the zener voltage of the device. Over the operating range of reverse current, the voltage across the diode is nearly constant and equal to the zener voltage.

Diodes made to exhibit specific zener voltages are referred to as *zener diodes;* their symbol is shown in Fig. 8.10*b*. Zener diodes are available having zener voltages of approximately 2 to 200 volts.

A major application for zener diodes is their use as a *voltage regulator*, shown in Fig. 8.11*a*. The zener diode is connected across the load, which is represented by

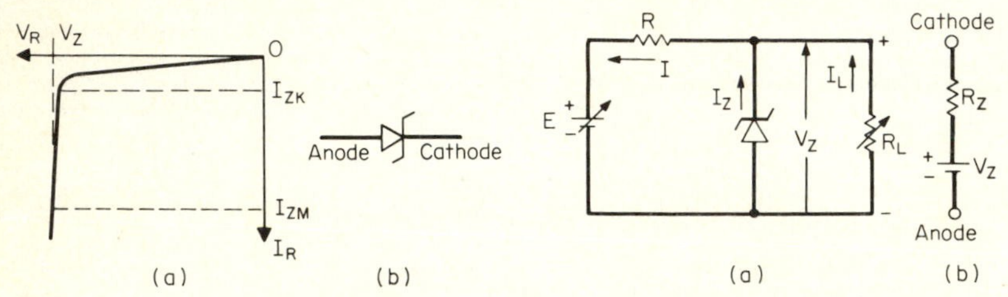

Fig. 8.10 The zener diode: (*a*) Voltampere characteristics. (*b*) Electrical symbol.

Fig. 8.11 A simple voltage regulator: (*a*) Circuit. (*b*) Model of a zener diode.

variable resistance R_L. Connected across R_L is the zener diode, which is reverse-biased (its cathode is positive with respect to the anode). The zener diode may be represented by a battery of V_Z volts in series with zener resistance R_Z, as shown in Fig. 8.11*b*. If the zener diode were ideal, its resistance would be zero, and the voltage across R_L would always equal the zener voltage. Because of the resistance, however, the voltage across the diode increases slightly with increasing zener current.

In Fig. 8.11*a*, current I flowing in resistance R always equals the sum of the zener and load currents, I_Z and I_L, respectively: $I = I_Z + I_L$. Assume that with E fixed, R_L is reduced. As a result, load current I_L increases, and the zener current decreases so that their sum is equal to I. If R_L is increased, the load current decreases and the zener current increases. During the changes in zener current, the zener voltage is hardly affected, and the voltage across the load resistance is nearly constant.

If the voltage source E should increase or decrease, the zener current will change accordingly. As before, the load voltage will hardly change its value.

example 8.2 For the voltage regulator of Fig. 8.11*a*, the zener diode regulates at 20 V over a current range of 5 to 30 mA. (*a*) If voltage source $E = 50$ V, determine the value of R if I_L varies from 0 to its maximum value. (*b*) What is the maximum load current?

solution (*a*) For zero load current, current I is equal to the maximum zener current I_{ZM}; hence, $I = I_{ZM} = 30$ mA. The value of R equals the voltage across it ($50 - 20 = 30$ V) divided by its current (30 mA); hence, $R = 30/(30 \times 10^{-3}) = 1\ \text{k}\Omega$.

(*b*) The maximum load current equals the value of I less the minimum zener current. Therefore, $I_{L,\max} = I - I_{ZK} = 30 - 5 = 25$ mA.

ZENER AND AVALANCHE BREAKDOWN A reverse-biased zener diode appears to break down by two different mechanisms. For zener voltages below approximately 6 V, the high electric field across the pn junction results in the rupture of covalent bonds and a rapid increase of electron-hole pairs. This process is referred to as zener breakdown.

In avalanche breakdown a few carriers, having acquired sufficient energy, collide with silicon atoms in the crystal lattice. As a result of the collision, new carriers are released which in turn collide with other atoms, releasing more carriers. Thus, a rapid increase of free electrons and holes occurs.

For either mechanism, the diode may be referred to as a zener or avalanche device. True zener diode action exhibits a negative temperature coefficient (breakdown voltage decreases with increasing temperature). True avalanche diode action displays a positive temperature coefficient (breakdown voltage increases with increasing temperature).

REGULATOR AND REFERENCE DIODES Zener diodes designed to maintain a relatively constant voltage over a wide range of load current, as in the simple regulator of Fig. 8.11*a*, are sometimes referred to as regulator diodes. A diode designed to exhibit a relatively fixed voltage and fixed current, regardless of temperature changes, is referred to as a reference diode.

DOUBLE-ANODE REGULATOR Physically, the double-anode regulator contains a single cathode which is common to two anodes (Fig. 8.12). This device is used for

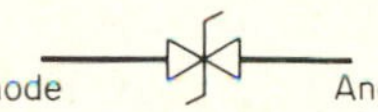

Fig. 8.12 Electrical symbol for double-anode regulating diode.

symmetrical clipping of an ac waveform and in protective circuits where negative and positive overloads may occur.

ZENER DIODE PARAMETERS Commonly used parameters for characterizing zener diodes are summarized in Table 8.2.

TABLE 8.2 Commonly Used Parameters for Zener Diodes

Parameter	Symbol	Meaning
Zener voltage	V_Z	Nominal voltage at which zener diode regulates
Knee current	I_{ZK}	Minimum current necessary for operation of zener diode
Maximum zener current	I_{ZM}	Maximum current that can flow in zener diode
Zener impedance	Z_Z	Indicates change in zener voltage for small changes in zener current with respect to a specified test current I_{ZT}.

8.5 BIPOLAR JUNCTION TRANSISTOR

The bipolar junction transistor (BJT), usually referred to simply as a transistor, is a device with three terminals. It comes in two types: npn and pnp. In the npn device, a "sandwich" of a thin p region is surrounded by two thicker n regions as shown in Fig. 8.13*a*. In the pnp device, illustrated in Fig. 8.13*b*, a thin n region is surrounded by two thicker p regions. Electrical symbols for both types of transistors are given in Fig. 8.14.

Connected to the inner (p or n) region is the *base* (*B*) lead. To the outer n (or p) regions are connected the *collector* (*C*) and *emitter* (*E*) leads. For either the npn or the pnp transistor, two junctions exist: the *base-emitter* and *collector-base junctions.*

In normal operation, as for an amplifier, the BJT is connected so the base-emitter

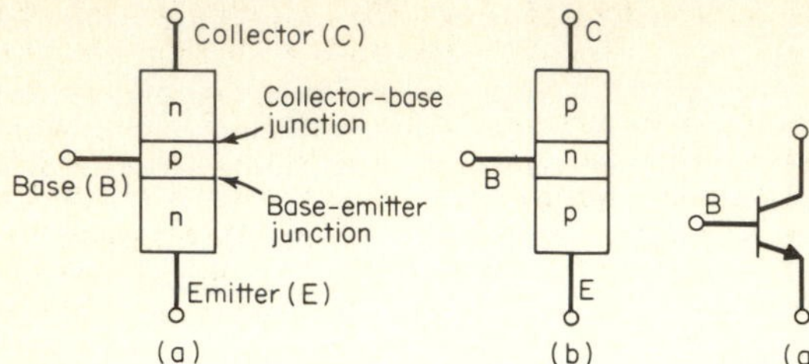

Fig. 8.13 Basic physical construction of a bipolar junction transistor (BJT): (*a*) npn type. (*b*) pnp type.

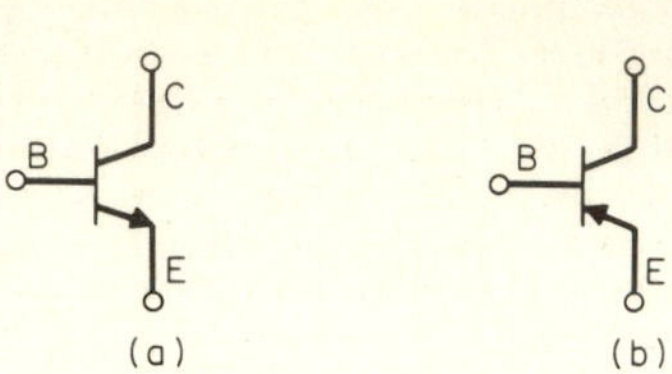

Fig. 8.14 Electrical symbols for a transistor: (*a*) npn type. (*b*) pnp type.

junction is *forward*-biased and the collector-base junction is *reverse*-biased. Referring to Fig. 8.15*a*, for an npn transistor the negative terminal of battery E_{BB} is connected to the emitter, and the positive terminal to the base. The base-emitter junction is therefore forward-biased like a pn junction diode. To the positive terminal of battery E_{CC} is connected the collector, and to the negative terminal the base. Similar to a reverse-biased *pn* junction diode, the collector-base junction is thereby reverse-biased.

To bias properly a pnp transistor, batteries E_{BB} and E_{CC} are reversed, as illustrated in Fig. 8.15*b*. In this case, the positive terminal of E_{BB} is connected to the emitter,

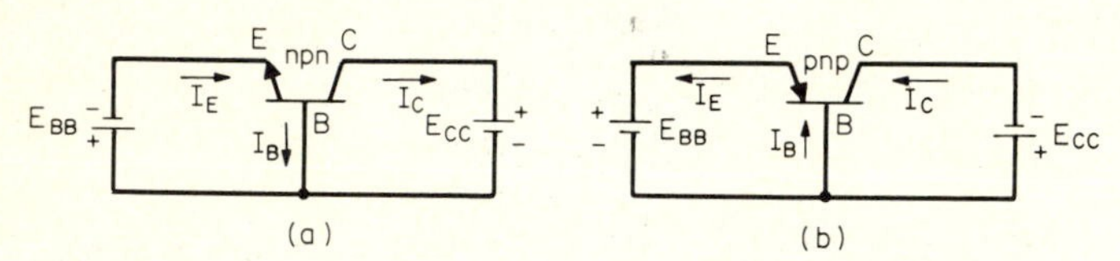

Fig. 8.15 For amplifier operation, the base-emitter junction is forward-biased and the collector-base junction is reverse-biased: (*a*) For an npn transistor. (*b*) For a pnp transistor.

and the negative terminal to the base. The collector is connected to the negative terminal of E_{CC}, and the positive terminal is returned to the base.

For either the npn or pnp transistor, the emitter current I_E always equals the sum of the base I_B and collector I_C currents:

$$I_E = I_B + I_C \tag{8.1}$$

OPERATION OF THE BJT To explain the operation of a bipolar junction transistor, the npn device will be considered. The operation of the pnp transistor is identical with the npn when the word "hole" is substituted for the word "electron" in the following description.

Consider the elementary transistor amplifier of Fig. 8.16. Load resistance R_C which, for example, can represent a loudspeaker, is connected to the collector in series with E_{CC}. The output voltage is designated by E_o. Signal source E_s, which may be the output of a phonograph cartridge, is connected in series with the base and E_{BB}. From a

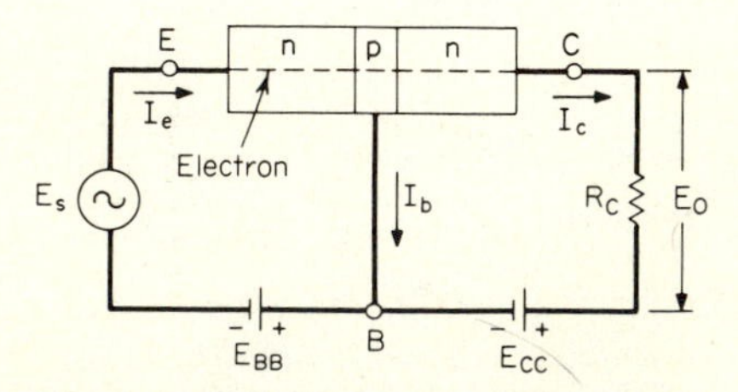

Fig. 8.16 Illustrating the operation of a common-base amplifier using an npn transistor.

signal standpoint, the batteries act like short circuits. The base, therefore, is common to the input E_s and output E_o signals. Such a configuration is called a *common-base* (CB) amplifier.

Because the base-emitter junction is forward-biased, electrons from signal E_s are readily injected into the emitter. The electrons then cross the base-emitter junction into the base region. The base, which is quite thin, has a much greater resistivity than either the emitter or the collector region. Hence, very few electrons are lost in the base, and most of them cross the base-collector junction into the collector region. The electrons then flow through resistance R_C and are attracted by the positive terminal of battery E_{CC}.

The voltage gain of an amplifier A_v is defined as the signal output voltage E_o divided by the signal input voltage E_s:

$$A_v = \frac{E_o}{E_s} \tag{8.2}$$

Output voltage E_o may be expressed as the product of output signal current I_c and load resistance R_C:

$$E_o = I_c R_C \tag{8.3}$$

Voltage E_s equals the product of input signal current I_e and the resistance of a forward-biased base-emitter junction. Letting the value of the diode resistance be designated by R_d, we have

$$E_s = I_e R_d \tag{8.4}$$

Substitution of Eqs. (8.3) and (8.4) in Eq. (8.2) yields

$$A_v = \left(\frac{I_c}{I_o}\right)\left(\frac{R_C}{R_d}\right) \tag{8.5}$$

The ratio I_c/I_e is generally designated by the Greek letter *alpha* (α) and is called the *short-circuit current gain* of a transistor in the common-base configuration. (Another symbol commonly used for α is h_{fb}). Substituting α for I_c/I_e, Eq. (8.5) appears as

$$A_v = \frac{\alpha R_C}{R_d} \tag{8.6}$$

Because I_c is always slightly less than I_e, α can never exceed unity; a typical value of α is 0.95. The value of R_C, however, is generally much greater than that of R_d. Therefore, even though alpha can never exceed unity, the voltage gain can be much greater than one.

example 8.3 In the amplifier of Fig. 8.16, assume that $\alpha = 0.95$, $R_C = 1000\ \Omega$, and $R_d = 20\ \Omega$. Find the voltage gain.

solution Substituting the given values in Eq. (8.6), we obtain

$$A_v = 0.95 \times \frac{1000}{20} = 47.5$$

The most useful connection for the BJT is the *common-emitter* (CE) configuration of Fig. 8.17. In this circuit, emitter terminal E is common to the input and output

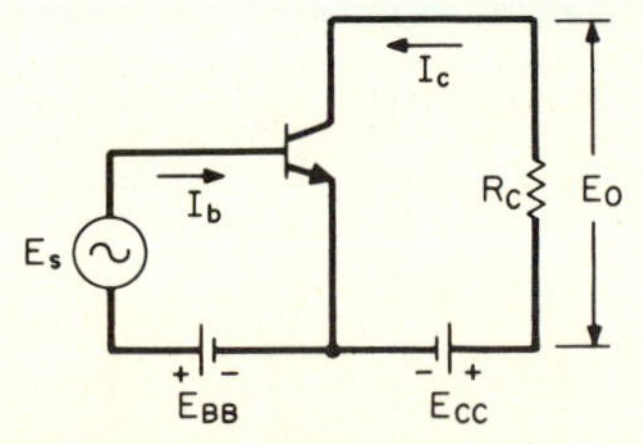

Fig. 8.17 An elementary common-emitter amplifier using an npn transistor.

signals. Because the base current I_b is minute, the current gain I_c/I_b is much greater than unity. The ratio, defined as the short-circuit current gain of a transistor in the common-emitter configuration, is designated by the Greek letter *beta* (β) or by the commonly used symbol, h_{fe}.

BJT PARAMETERS AND CURVES The most commonly used parameters for characterizing the BJT are summarized in Table 8.3. Bipolar devices are classified as

TABLE 8.3 Commonly Used Parameters for Bipolar Junction Transistors

Parameter	Symbol	Meaning
Collector-base voltage (emitter open)	V_{CBO}	Maximum voltage that can be impressed across collector and base of a transistor with emitter open
Collector-emitter voltage (base short-circuited to emitter)	V_{CES}	Maximum voltage that can be impressed across collector and emitter of a transistor with base short-circuited to emitter. Its value is in the order of one-half of V_{CBO}
Collector-emitter voltage (with specified resistor between base and emitter)	V_{CER}	For this condition, maximum collector-emitter voltage is greater than V_{CES} but less than V_{CBO}
Emitter-base voltage (collector open)	V_{EBO}	Maximum voltage that can be impressed across emitter and base of a transistor with collector open
Collector saturation voltage	$V_{CE,\text{sat}}$	Collector-emitter voltage of a transistor that is fully conducting, as in a transistor switch
Small-signal input resistance	h_{ib}, h_{ie}	Input resistance of a transistor with output short-circuited for the signal. This and other *h* parameters are called *hybrid parameters*. The second subscript refers to the transistor configuration: *b* for common base and *e* for common emitter.
Small-signal output admittance	h_{ob}, h_{oe}	Output admittance of a transistor with input open-circuited for the signal
Small-signal reverse-voltage transfer ratio	h_{rb}, h_{re}	Ratio of voltage developed across input of a transistor to voltage present at output, with input open-circuited for the signal
Small-signal forward-current gain	h_{fb}, h_{fe}	Ratio of output to input signal currents with output short-circuited
Dc forward-current gain	h_{FE}	Ratio of dc collector current to dc base current for transistor in common-emitter configuration
Collector dissipation	P_C	Power dissipated in collector of a transistor. It is equal to the product of the dc collector current and dc collector-emitter voltage.
Gain-bandwidth product	f_T	Frequency at which common-emitter forward current gain is unity
Cutoff frequency	f_{hfb}, f_{hfe}	Frequency at which h_{fb} (or h_{fe}) is 0.707 times its value at 1 kHz
Collector-cutoff current (emitter open)	I_{CBO}, I_{CO}	Reverse saturation (leakage) current flowing between collector and base with emitter open
Collector-cutoff current (base open)	I_{CEO}	Reverse saturation current flowing between collector and emitter with base open
Collector-base capacitance	C_{ob}, C_{cb}	Transistor capacitance across collector and base. This capacitance influences, to a great degree, the high-frequency performance of a transistor amplifier

small-signal, medium-power, power, r-f, and switching transistors. Several available transistors can operate at collector voltages and currents as high as 500 V and 100 A, respectively.

A useful set of curves is the common-emitter *collector characteristics* of Fig. 8.18. On the y axis is plotted the dc collector current i_C, and along the x axis the dc collector-emitter voltage, v_{CE}. A family of curves, for different values of dc base current I_B, is drawn in the figure. The reason for plotting base current is that the BJT is a *current-operated* device. As mentioned earlier, the base-emitter junction is forward-biased for normal transistor operation. Base current therefore flows and is a key quantity in establishing the operating point of a transistor.

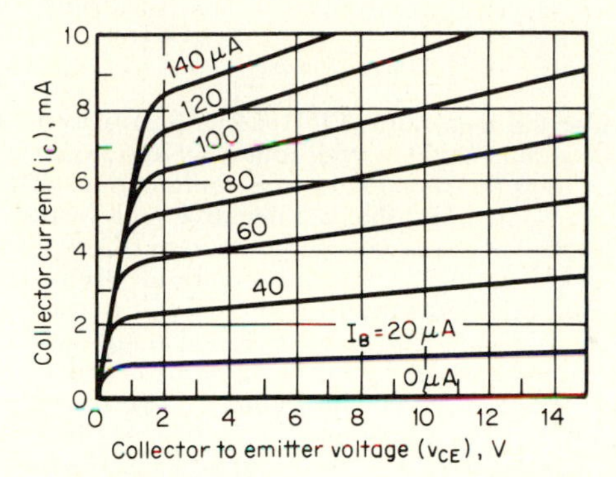

Fig. 8.18 Typical common-emitter collector characteristics.

8.6 FIELD-EFFECT TRANSISTOR

The field-effect transistor (FET) is basically a three-terminal semiconductor device having characteristics similar to that of a pentode vacuum tube. Unlike the bipolar transistor, the FET is a *voltage-operated* device. Instead of being biased by a current, the FET is biased by a voltage, and no input current flows. Its input resistance, therefore, is virtually infinite.

There are two kinds of field-effect devices: the *junction* FET (JFET) and the *metal-oxide semiconductor* FET (MOSFET). The MOSFET is also referred to as an *insulated-gate* FET (IGFET) in the literature.

JUNCTION FET A cross-sectional view of an elementary *n-channel* JFET is shown in Fig. 8.19*a*. It contains two p-type regions in an n-type silicon bar. The two p regions are connected and called the *gate*, *G*. Ohmic connections are made to each end of the silicon bar for the *source* (*S*) and *drain* (*D*) leads.

A cross-sectional view of a *p-channel* JFET is shown in Fig. 8.19*b*. In this structure, the two n-type regions for the gate are formed in a p-type silicon bar. The biasing voltages of a p-channel JFET are opposite to those of the n-channel type (similar to npn and pnp transistors). Electrical symbols for n- and p-channel JFET's are given in Fig. 8.20.

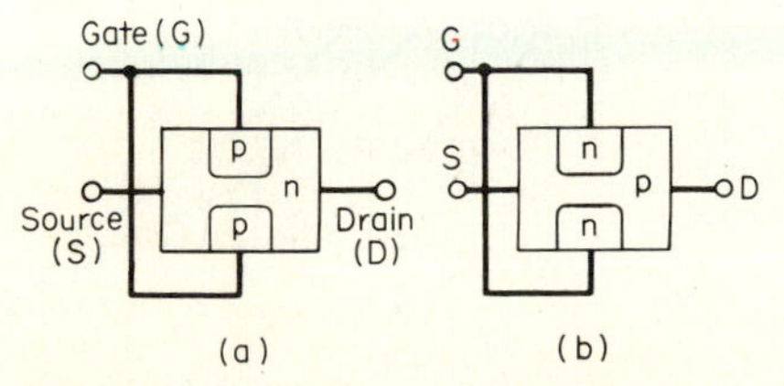

Fig. 8.19 Basic physical construction of an elementary JFET: (*a*) n-channel. (*b*) p-channel.

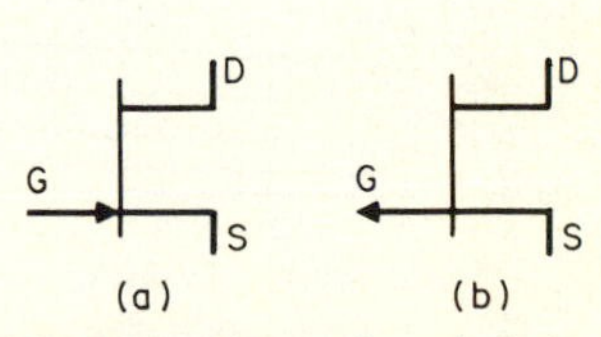

Fig. 8.20 Electrical symbols for a JFET: (*a*) n-channel. (*b*) p-channel.

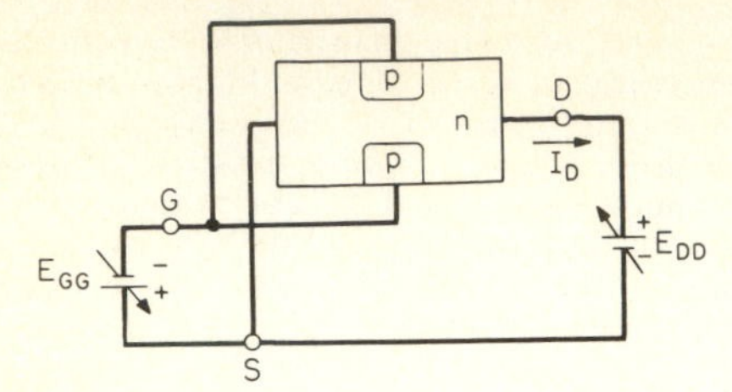

Fig. 8.21 Biasing an n-channel JFET in the common-source configuration.

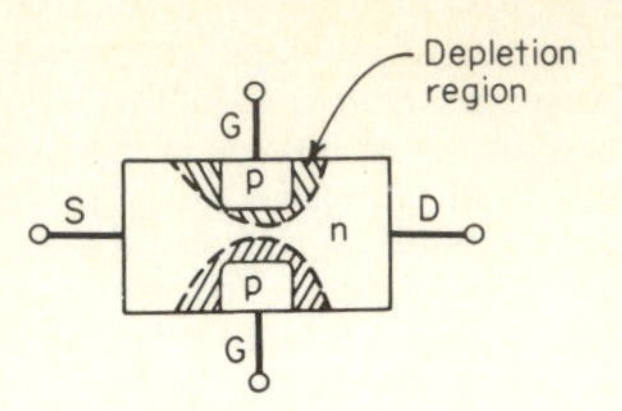

Fig. 8.22 Depletion region developed along the channel of a JFET.

Operation Consider the n-channel JFET of Fig. 8.21 biased for normal operation. Because source S is common to the output (drain D) and input (gate G) voltages, the connection is referred to as the *common-source* configuration. The drain is held positive with respect to the source, and the gate negative with respect to the source. Assume that the gate-source voltage is initially set to zero ($E_{GG} = 0$).

As the drain-source voltage E_{DD} is increased, more electrons are attracted from the source to the drain, and the drain current I_D increases. Because of the voltage drop along the n channel, a depletion region is developed along the pn junction formed by the p-type gate and n-type channel regions (Fig. 8.22). This is the same kind of depletion region that is formed in a pn junction diode. As you recall, a depletion region is devoid of current carriers. Because the channel becomes more depleted of carriers as E_{DD} is increased, the drain current levels off at a maximum value I_{DSS} (the drain current with $E_{GG} = 0$), as illustrated in Fig. 8.23. Voltage V_{PO} may be defined as the minimum value of drain-source voltage, V_{DS}, for which maximum drain current flows.

As the gate-source voltage is made increasingly negative, the depletion region is

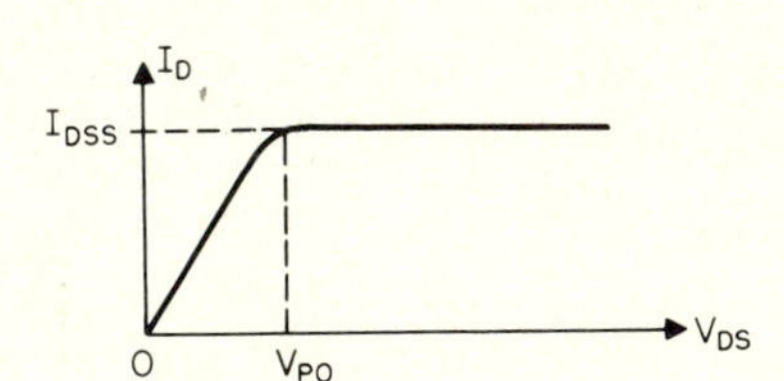

Fig. 8.23 Drain current I_D as a function of drain-source voltage V_{DS}.

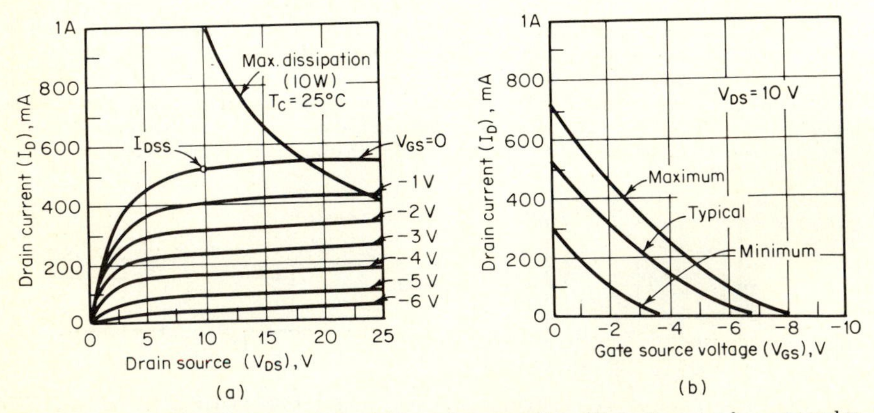

Fig. 8.24 Typical curves for a JFET: (*a*) Drain characteristics. (Operation must be restricted to the area below dissipation curve.) (*b*) Transfer characteristics. (*Courtesy Siliconix Incorporated.*)

formed more rapidly than with zero bias. As a result, V_{PO} occurs at a lower value of V_{DS}. A typical family of *drain characteristics* for different values of bias voltage V_{GS} is given in Fig. 8.24*a*. Plotted on the *y* and *x* axes are the drain current and voltage, respectively. Because the FET is a voltage-operated device, curves for different values of gate-source bias, V_{GS}, are plotted in the figure.

Another useful curve for field-effect transistors is the *transfer characteristics* of Fig. 8.24*b*. This is a plot of drain current I_D as a function of gate-source bias voltage V_{GS} for a given drain-source voltage, V_{DS}.

MOSFET In the MOSFET, the gate is a metallic electrode separated from the channel by a thin insulating material, such as silicon dioxide (SiO_2). There are two types of MOSFET's: *depletion* and *enhancement*. A cross-sectional view of an n-channel depletion-type MOSFET is shown in Fig. 8.25*a*. The silicon bar, referred to as the substrate, is p-type material. To one of the n+ regions (the + sign denotes a heavily doped region) is connected the source lead, and to the other the drain lead. Between the n+ regions is an n region which forms the channel.

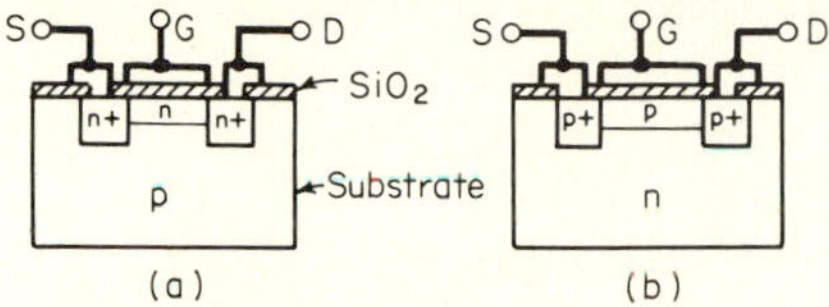

Fig. 8.25 Cross-sectional views of a depletion-type MOSFET: (*a*) n-channel. (*b*) p-channel.

Fig. 8.26 Electrical symbols for a depletion-type MOSFET: (*a*) n-channel. (*b*) p-channel.

A cross-sectional view of a p-channel MOSFET is shown in Fig. 8.25*b*. The substrate in this case is n-type material, and the drain and source connections are made to p+-type material. The channel is p-type. Electrical symbols for both types of depletion MOSFET's are shown in Fig. 8.26. The substrate (SUB) is generally connected to the source terminal.

Operation of depletion-type MOSFET The gate, insulator, and channel of a MOSFET act like a capacitor. When, for example, a positive charge is on the gate, an opposite (negative) and equal charge is induced in the channel. If the gate has a negative charge, an equal positive charge is induced in the channel.

Consider the n-channel depletion-type MOSFET in Fig. 8.27*a* biased for normal operation. (A p-channel depletion-type MOSFET is biased with opposite polarities.) The drain is positive with respect to the source. If the gate is biased negatively with respect to S, a positive charge is induced in the n channel, as illustrated in Fig. 8.27*b*. This results in a reduction of free electrons in the n channel, and drain current I_D is reduced. As the gate is made more negative with respect to the source, the drain current is reduced further. This is illustrated in the typical drain and transfer characteristics of Fig. 8.28.

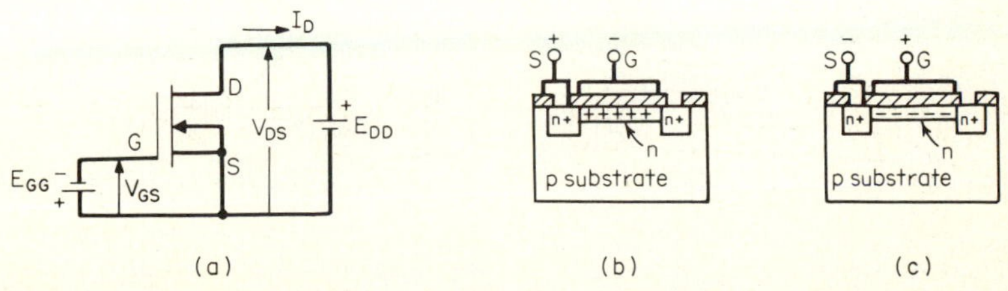

Fig. 8.27 Operation of an n-channel depletion-type MOSFET: (*a*) Circuit. (*b*) Gate biased negatively with respect to source. (*c*) Gate biased positively with respect to source.

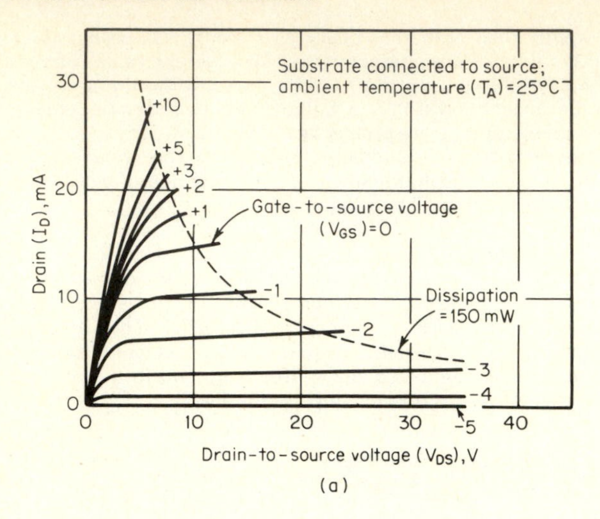

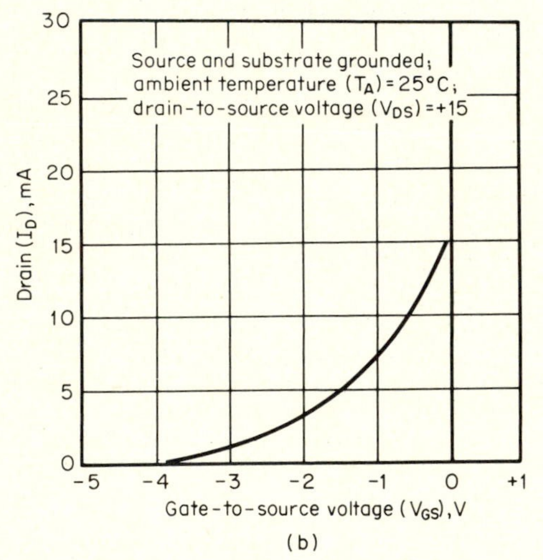

Fig. 8.28 Typical curves for an n-channel depletion-type MOSFET: (*a*) Drain characteristics. (*b*) Transfer characteristics. (*Courtesy RCA.*)

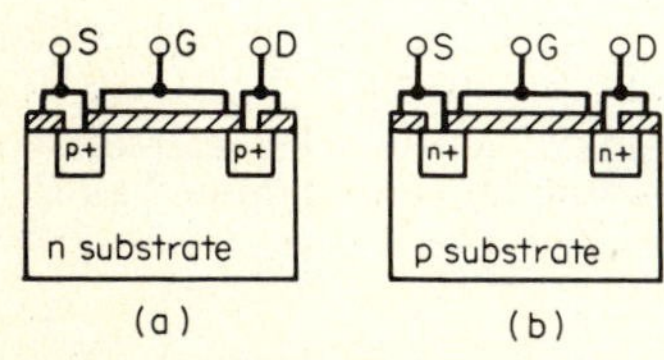

Fig. 8.29 Cross-sectional views of an enhancement-type MOSFET: (*a*) p-channel. (*b*) n-channel.

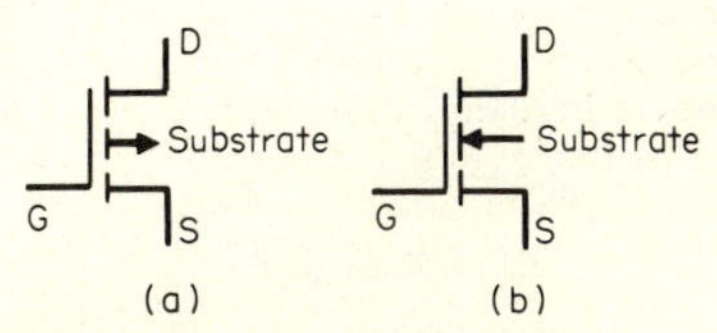

Fig. 8.30 Electrical symbols for an enhancement-type MOSFET: (*a*) p-channel. (*b*) n-channel.

If the gate is now made positive with respect to the source, as in Fig. 8.27*c*, electrons are induced in the n channel. Conduction is *enhanced*, and greater drain current flows (see Fig. 8.28). Because of the insulated gate electrode, the input resistance of a MOSFET, whether the gate is positive or negative with respect to the source, approaches infinity.

Operation of enhancement-type MOSFET The depletion-type MOSFET may be regarded as a *normally ON* device. The reason for this designation is that for $V_{GS} = 0$, appreciable drain current flows (see Fig. 8.28). For many applications a *normally OFF* MOSFET is desirable, so that, for $V_{GS} = 0$, no drain current flows. This kind of device is realized in the *enhancement-type* MOSFET.

Cross-sectional views of p- and n-channel enhancement-type MOSFET's are illustrated in Fig. 8.29; their electrical symbols are given in Fig. 8.30. In either type, no physical channel, as in the depletion MOSFET, exists. The device, therefore, is

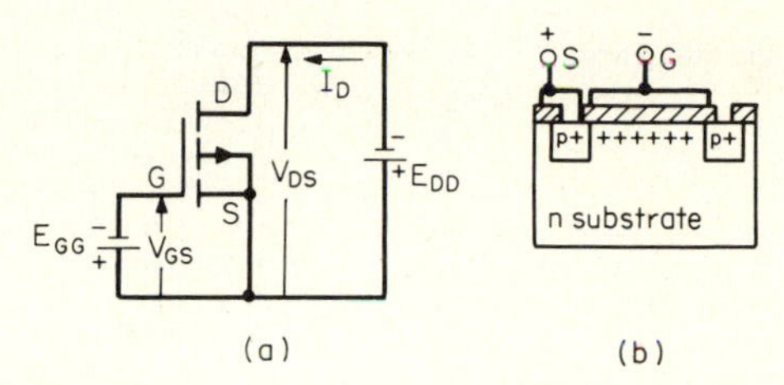

Fig. 8.31 Operation of a p-channel enhancement-type MOSFET: (*a*) Circuit. (*b*) Gate biased negatively with respect to the source.

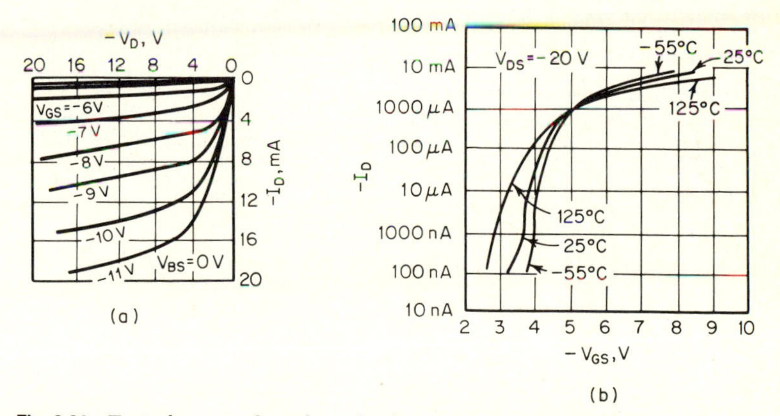

Fig. 8.32 Typical curves of a p-channel enhancement-type MOSFET: (*a*) Drain characteristics. (*b*) Transfer characteristics. (*Courtesy General Instrument Corporation.*)

normally OFF. Before a current can flow between the drain and source, a suitable bias is required across the gate and source terminals.

A p-channel enhancement MOSFET biased for normal operation is illustrated in Fig. 8.31. (For an n-channel device, the polarities of sources E_{DD} and E_{GG} are reversed.) The gate is held negative with respect to the source. Positive charges (holes) are induced in the region below the gate electrode, and drain current flows. As V_{GS} is increased, the drain current also increases. Typical drain and transfer characteristics for the enhancement MOSFET are illustrated in Fig. 8.32.

FET PARAMETERS Commonly used parameters for characterizing FET devices are summarized in Table 8.4.

TABLE 8.4 Commonly Used Parameters for the JFET and MOSFET

Parameter	Symbol	Meaning
Drain current for zero bias	I_{DSS}	Drain current flowing when gate is short-circuited to source ($V_{GS} = 0$)
Gate reverse current	I_{GSS}	Leakage current flowing between gate and source for a specified reverse bias across gate and source terminals
Drain cutoff current	$I_{D,\mathrm{off}}$	Drain current flowing when device is biased in its OFF state
Gate-source breakdown voltage	BV_{GSS}	Maximum reverse voltage that may be impressed across gate and source terminals without damaging the device
Gate-source pinchoff voltage	V_P	Gate-to-source voltage which reduces I_{DSS} to 1% or less of maximum value at a specified drain-to-source voltage
Small-signal forward transconductance	g_{fs}, g_m	Ratio of a small change in signal drain current to a small change in signal gate-to-source voltage in common-source configuration. Parameter g_{fs} is an indication of the gain for the device.
Dc drain-source ON resistance	r_{DS}	Ratio of dc drain-source voltage to the dc drain current, generally measured at $V_{GS} = 0$
Input capacitance	C_{iss}	Small-signal input capacitance for device in the common-source configuration with $V_{DS} = 0$
Reverse transfer capacitance	C_{rss}	Capacitance between drain and gate in common-source configuration with $V_{DS} = 0$

8.7 SILICON-CONTROLLED RECTIFIER

The silicon-controlled rectifier (SCR), also referred to as a ***thyristor***, is a four-layer pnpn "sandwich" device having three terminals, as illustrated in Fig. 8.33*a*. Connected to the outer p region is the *anode*, and to the outer n region the *cathode*. The *gate* terminal is connected to the inner p region. The electrical symbol for the SCR is given in Fig. 8.33*b*.

What makes the SCR unique and useful is its switching characteristics. With the anode held positive with respect to the cathode, a small pulse of current (*trigger*) is applied to the gate terminal. The SCR turns on (*conducts*). In its conducting state, it behaves like a rectifier diode. Furthermore, it remains conducting even after the trigger is reduced to zero.

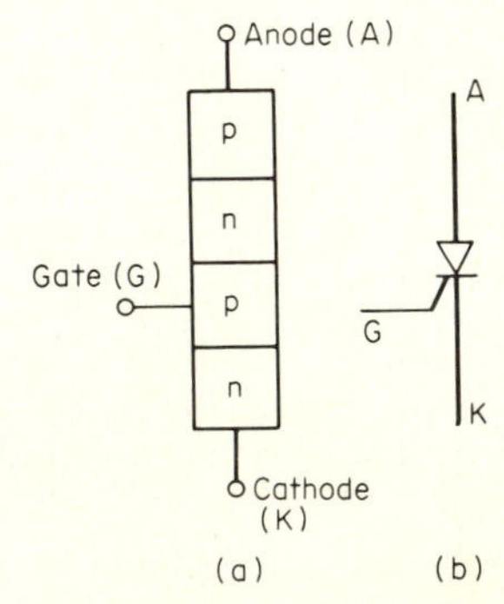

Fig. 8.33 Silicon-controlled rectifier (SCR): (*a*) pnpn "sandwich" construction. (*b*) Electrical symbol.

The SCR turns off (becomes *nonconducting*) only when the current flowing between the anode and the cathode is reduced to a level that is less than a minute critical value of current, called the *holding current*. Because of these characteristics, the SCR is extremely efficient in controlling, for example, the power to a motor or in the dimming of theater lights.

example 8.4 Show how a SCR may be used to rectify alternating current and control the power delivered to a dc motor.

solution A simplified circuit for accomplishing this is illustrated in Fig. 8.34*a*, and the waveforms are shown in Fig. 8.34*b* and *c*. The point at which the trigger is applied during the positive half-cycle of the ac source is controlled by suitable circuitry (to be described later) in the box labeled *phase control*. Once the SCR turns on, it conducts like a rectifier diode. It stops conducting when the anode current falls below the holding current, which is close to zero. During the negative half-cycle, the SCR is reverse-biased, and no current flows.

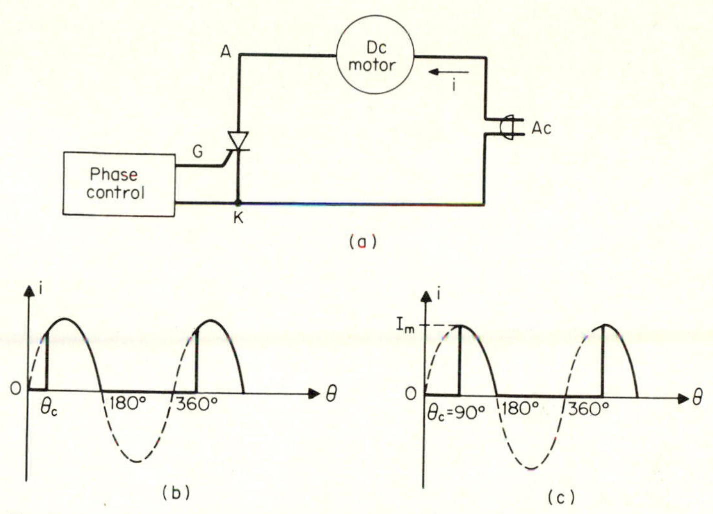

Fig. 8.34 A SCR controlling power delivered to a dc motor: (*a*) Circuit. (*b*) Typical current waveform. (*c*) Current waveform for a phase angle = 90°. (See Example 8.4.)

The angle at which conduction occurs, *phase angle* θ_c (Greek letter *theta*), may be anywhere between approximately 0 and 180° (Fig. 8.34*b*). In the example of Fig. 8.34*c*, where $\theta_c = 90°$, it is extremely simple to calculate the average (dc) current flowing in the motor. In a half-wave rectifier, current flows for 180° and the average current is $0.318I_m$ amperes, where I_m is the peak current. If current flows only during half of a positive cycle (90°), the average current is therefore equal to $0.318I_m/2 = 0.159I_m$ A.

OPERATION OF THE SCR For an understanding of SCR operation, it is convenient to visualize the pnpn sandwich structure of Fig. 8.33*a* as shown in Fig. 8.35*a*. The SCR is depicted here as a pnp transistor connected to an npn transistor and shown schematically in Fig. 8.35*b*.

Assume that anode *A* is positive with respect to cathode *K*, as indicated in Fig. 8.35*b*. With no positive trigger, both transistors are nonconducting, and anode current I_A is equal to a minute leakage current. Upon the application of a small positive trigger to the base of Q_2 (gate terminal *G*), the npn transistor begins to conduct. Collector current in the npn transistor is equal to the current flowing toward the base of pnp transistor Q_1. Hence, $I_{B1} = I_{C2}$, and Q_1 begins to conduct. Both transistors assume proper bias levels to ensure that the device conducts after the termination of the trigger. As mentioned earlier, the SCR turns off when the anode current is reduced to a value less than the holding current.

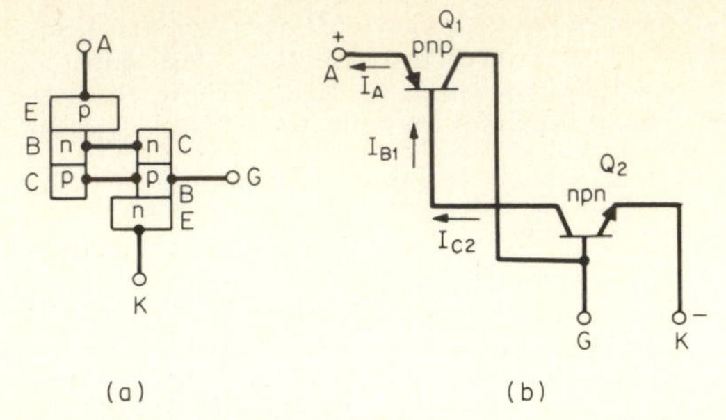

Fig. 8.35 Viewing the SCR as being comprised of an npn and a pnp transistor: (*a*) Connections of the transistors. (*b*) Schematic representation.

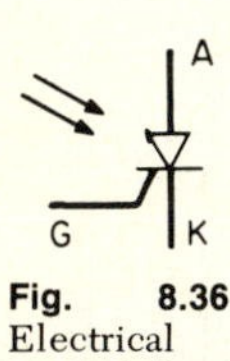

Fig. 8.36 Electrical symbol for a LASCR.

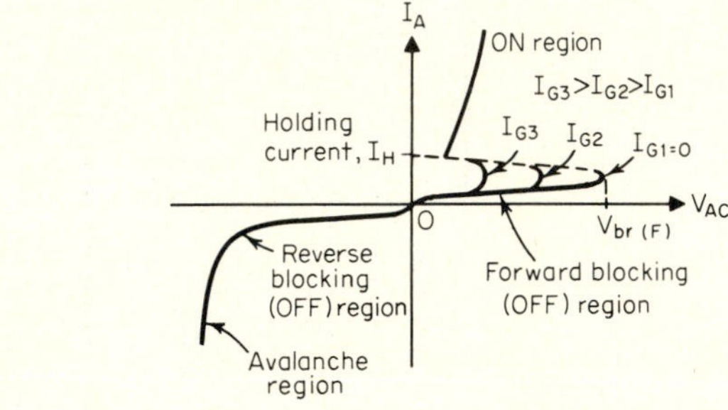

Fig. 8.37 Voltampere characteristics of a typical SCR.

LIGHT-ACTIVATED SCR A variation in the SCR is the light-activated SCR, referred to as the LASCR. Light passing through a translucent window in the package containing the device acts as a positive trigger. The LASCR, in other respects, behaves as a SCR. The electrical symbol for a LASCR is illustrated in Fig. 8.36.

CHARACTERISTIC CURVES Typical voltampere characteristics of a SCR are given in Fig. 8.37. Plotted along the vertical axis is anode current I_A and along the horizontal

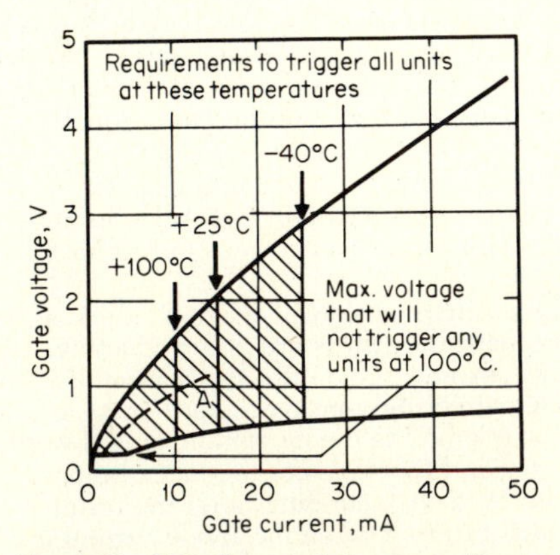

Fig. 8.38 Gate-characteristic curves for a typical SCR. (*Courtesy RCA.*)

axis the anode-cathode voltage V_{AC}. When V_{AC} is positive, the SCR is forward-biased. Depending on the value of gate current I_G, the device turns on, or *fires*, at a specific *forward breakdown voltage*, $V_{(br)F}$. Note that for $I_{G1} = 0$, the breakdown voltage is greater than for I_{G2} or I_{G3}, which are greater than zero. When the anode current is less than the holding current, the SCR reverts to its forward blocking (OFF) region. In the forward conducting and reverse regions of operation, the SCR resembles a junction diode.

Because the gate trigger current and voltage required vary with temperature and with units of the same type, manufacturers often supply *gate-characteristic curves* for their devices. An example of gate-characteristic curves is illustrated in Fig. 8.38. Gate voltage versus gate current is plotted for different temperatures. Note that smaller trigger voltages and currents are required at higher temperatures. For example, the maximum voltage that does not trigger the SCR at 100°C is approximately 0.2 V. Curve *A* represents the characteristics of a typical SCR.

SCR PARAMETERS Commonly used parameters for characterizing the SCR are summarized in Table 8.5.

TABLE 8.5 Commonly Used Parameters for the SCR and TRIAC

Parameter	Symbol	Meaning
Forward breakdown voltage	$V_{(br)F}$ V_{BO}	Forward voltage at which device fires
Reverse breakdown voltage	$V_{(br)R}$	Maximum reverse voltage that causes device to go into avalanche
On-state voltage	V_T, V_F	Voltage across device when it is conducting (ON state)
On-state current	I_T, I_F	Current flowing between anode and cathode in ON state
Holding current	I_H	Minimum current for device to be in ON state
Latching current	I_L	Minimum current to maintain device in ON state, after switching from OFF to ON state with trigger removed
Gate trigger current	I_{GT}	Minimum gate current for switching device from OFF to ON state
Gate trigger voltage	V_{GT}	Gate voltage needed to produce required gate current
Gate turn-on time	t_{on}	Time for device to turn on
Commutated turn-off time	t_{off}, t_q	Time for device to turn off

8.8 BIDIRECTIONAL TRIODE THYRISTOR

The electrical symbol for the bidirectional triode thyristor, or TRIAC, is shown in Fig. 8.39*a*, and its equivalent circuit in Fig. 8.39*b*. Referring to Fig. 8.39*b*, the TRIAC is viewed as two SCR's in parallel with the anode of one connected to the cathode of the other. The voltampere characteristics of a TRIAC (Fig. 8.40) exhibit the same forward blocking characteristics of the SCR when main terminal one is positive with respect to main terminal two, or vice versa. The parameters for characterizing the TRIAC are essentially the same as for the SCR, summarized in Table 8.5.

example 8.5 Show how a TRIAC may be used as a lamp dimmer.

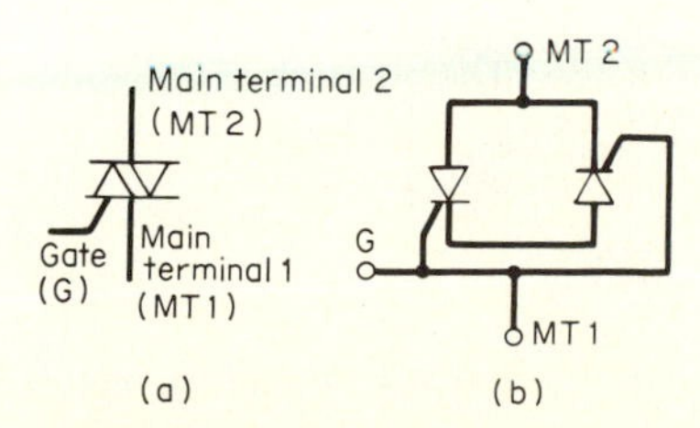

Fig. 8.39 The TRIAC: (*a*) Electrical symbol. (*b*) Equivalent circuit.

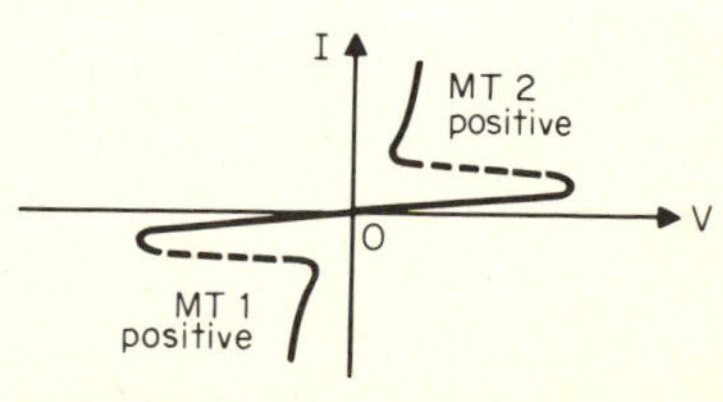

Fig. 8.40 Voltampere characteristics of a typical TRIAC.

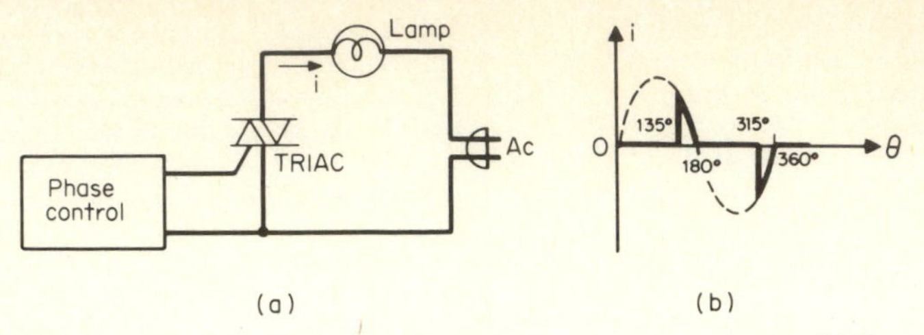

Fig. 8.41 The TRIAC used as a lamp dimmer: (*a*) Basic circuit. (*b*) Current waveform. (See Example 8.5.)

solution A basic circuit of a lamp dimmer is given in Fig. 8.41*a*. Depending on the setting of the phase control, the firing angle, as in the SCR, varies from approximately 0 to 180° in each half-cycle. For example, in Fig. 8.41*b* current flows only for 45° (180–135) during the positive half-cycle and another 45° (360–315) during the negative half-cycle.

8.9 UNIJUNCTION TRANSISTOR

The unijunction transistor (UJT) is a three-terminal device that differs from the BJT in two important respects:

1. It has only one junction.
2. It exhibits a region of *negative resistance.* A device is said to have negative resistance when, for increasing current, the voltage across the device decreases. Because of its negative resistance, the UJT finds application in oscillator, timing, and SCR trigger circuits.

The basic physical construction of a UJT is illustrated in Fig. 8.42*a*, and its electrical symbol is given in Fig. 8.42*b*. Two ohmic (nonrectifying) contacts, B_1 and B_2, are made to the ends of an n-type silicon bar. The resistance of the bar, referred to as the *interbase resistance* R_{BB}, is typically of the order of 4000 to 10 000 Ω. A rectifying pn junction is formed close to terminal B_2 to which is connected emitter lead E.

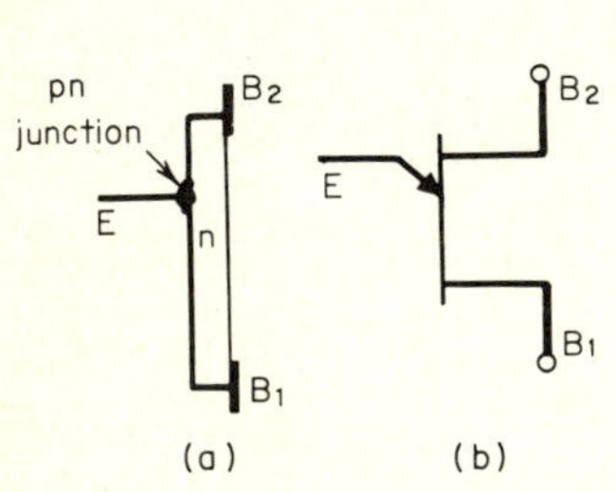

Fig. 8.42 The unijunction transistor (UJT): (*a*) Basic physical construction. (*b*) Electrical symbol.

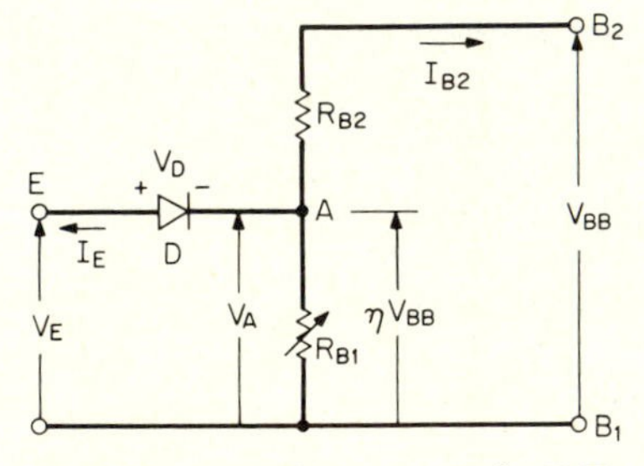

Fig. 8.43 Equivalent circuit of a UJT.

OPERATION OF THE UJT An equivalent circuit of a UJT is shown in Fig. 8.43. Between B_2 and B_1 are two resistances, R_{B1} and R_{B2}. Their sum is equal to the interbase resistance: $R_{B1} + R_{B2} = R_{BB}$. Because the value of R_{B1} varies as a function of emitter current I_E, it is represented as a variable resistance. The pn junction is denoted by diode D and the voltage across it by V_D. The voltage V_A at point A with respect to B_1, by voltage division, is

$$V_A = \frac{R_{B1}V_{BB}}{R_{B1} + R_{B2}} \tag{8.7}$$
$$= \eta V_{BB}$$

where η (Greek letter *eta*) is referred to as the *intrinsic standoff ratio* of the device

$$\eta = \frac{R_{B1}}{R_{B1} + R_{B2}} = \frac{R_{B1}}{R_{BB}} \tag{8.8}$$

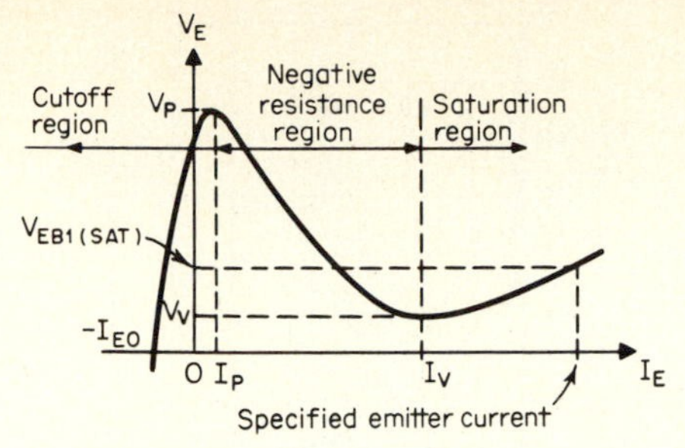

Fig. 8.44 Voltampere characteristics of a typical UJT.

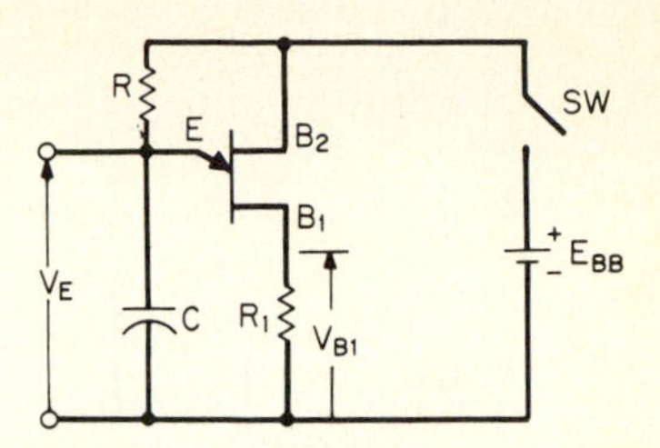

Fig. 8.45 A circuit using a UJT that generates sawtooth and trigger waveforms. (See Example 8.6.)

A plot of emitter voltage versus emitter current is illustrated in Fig. 8.44. For $V_E = 0$, diode D is reverse-biased, and a small reverse saturation current, $-I_{EO}$, flows. As V_E is increased, the diode becomes less reverse-biased and I_E less negative. At a sufficiently high emitter voltage the diode conducts and holes are injected into the bar region denoted by R_{B1}. Because of the excess holes, the value of R_{B1} decreases and I_E increases. For example, when $I_E = 0$, $R_{B1} = 4000\ \Omega$, and when $I_E = 50$ mA, $R_{B1} = 40\ \Omega$. Owing to the reduced resistance (increased conductance) of R_{B1}, V_E decreases as I_E increases. This behavior is referred to as negative resistance.

The voltage and current at the peak point of the characteristic curve are referred to as the *peak voltage* V_P and the *peak current*, I_P, respectively. Voltage V_P is the emitter voltage at which the UJT makes a transition from the cutoff region to the negative resistance region. Peak current I_P is the minimum current required to turn on the UJT.

At the valley point (V_V = *valley voltage*, I_V = *valley current*), the UJT makes a transition from the negative resistance region to the saturation region. In the saturation region R_{B1} acts as a positive resistance; that is, as I_E increases, V_E increases, too. Voltage $V_{BE1,\text{sat}}$ is the voltage across the emitter and base B_1 corresponding to a specified emitter current. An important application of the UJT and the significance of the peak and valley points are illustrated in Example 8.6.

example 8.6 Analyze the operation of the UJT circuit of Fig. 8.45, and determine the waveforms across capacitor C and resistor R_1.

solution Assume that the capacitor is initially uncharged. When the switch is closed, the capacitor begins to charge up, through R, toward E_{BB} volts. For V_E less than V_P, the UJT is in the cutoff region. As soon as the capacitor voltage reaches V_P, the UJT is turned on and R_{B1} is reduced to a small value. The capacitor discharges rapidly through R_{B1} and R_1 toward ground. But, when $V_E = V_V$, the UJT is turned off. The capacitor begins to charge up again and the cycle is repeated, as illustrated in Fig. 8.46*a*. The resulting waveform is referred to as a *sawtooth*, and its peak-to-peak amplitude is equal to the difference between the peak and valley voltages ($V_P = V_V$).

The waveform across R_1 is shown in Fig. 8.46*b*. During the time that C discharges, a current

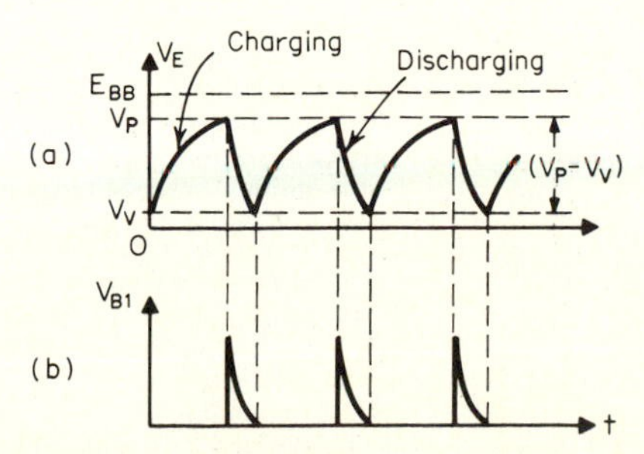

Fig. 8.46 Waveforms obtained from the UJT circuit of Fig. 8.45: (*a*) Across capacitor C. (*b*) Across resistor R_1.

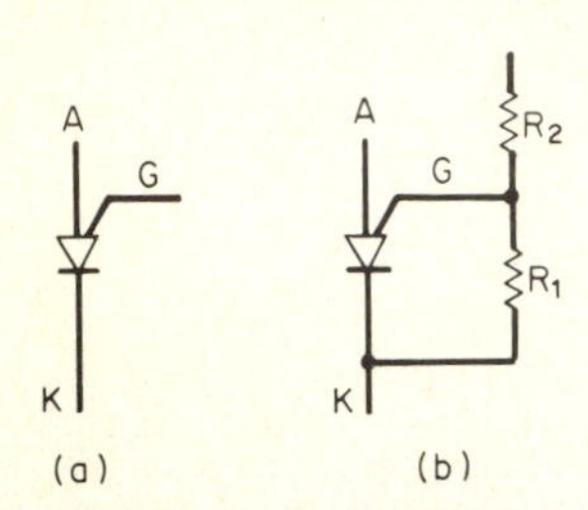

Fig. 8.47 Programmable unijunction transistor (PUT): (*a*) Electrical symbol. (*b*) Connection of resistors R_1 and R_2 to alter device parameters.

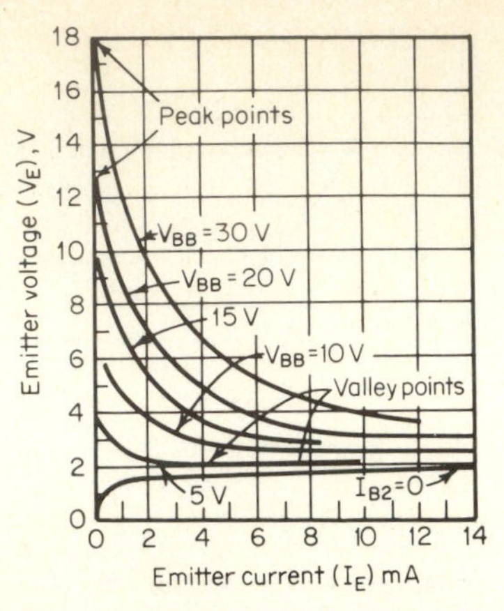

Fig. 8.48 Static emitter characteristics of a typical UJT. (*Courtesy General Electric Company.*)

flows in R_1, and a trigger voltage is obtained which is suitable for firing a SCR. The amplitude of the trigger is determined primarily by the value of R_1. By changing the values of C and R, the time between triggers is made variable. The UJT circuit of Fig. 8.45 is commonly used as the phase control in the circuits of Figs. 8.34*a* and 8.41 for the SCR and TRIAC.

PROGRAMMABLE UNIJUNCTION TRANSISTOR The electrical symbol for the programmable unijunction transistor (PUT) is given in Fig. 8.47*a*. A device that is similar to the SCR, the gate of the PUT is brought out near the anode instead of near the cathode. By selecting suitable values for resistors R_1 and R_2 shown connected to the PUT in Fig. 8.47*b*, parameters such as η, R_{BB}, I_P, and I_V can be tailored to meet the user's needs. The PUT is used often in long-duration timer circuits.

UJT CURVES AND PARAMETERS A useful family of curves for the UJT is the *static emitter characteristics* of Fig. 8.48. Emitter voltage is plotted along the y axis and emitter current along the x axis. A family of curves for different values of interbase voltage V_{BB} is drawn in the figure. The greater the value of V_{BB}, the greater are the values of V_P, I_P, V_V, and I_V. Table 8.6 summarizes the important parameters for characterizing the UJT.

TABLE 8.6 Commonly Used Parameters for the UJT

Parameter	Symbol	Meaning
Peak emitter voltage	V_P	Maximum emitter voltage before UJT enters negative-resistance region
Peak emitter current	I_P	Maximum emitter current before UJT enters negative-resistance region. It may also be thought of as minimum emitter current to turn on a UJT.
Valley emitter voltage	V_V	Emitter voltage at valley point
Valley emitter current	I_V	Emitter current at valley point
Interbase voltage	V_{BB}	Voltage between base one and base two
Emitter saturation voltage	$V_{BE1,\text{sat}}$	Voltage across emitter and base one at a specified emitter current and interbase voltage
Interbase resistance	R_{BB}	Dc resistance between base one and base two with emitter open-circuited
Intrinsic standoff ratio	η	May be defined as ratio of emitter–base-one resistance R_{B1} to interbase resistance R_{BB}: $\eta = R_{B1}/R_{BB}$.

8.10 LIGHT-EMITTING DIODE

In a forward-biased pn junction, electrons traveling from the n region are injected into the p region. Some of the injected electrons recombine with the holes in the p region. As a result of the recombination, there is a release of radiant energy. For good efficiency and greater released energy, materials such as gallium (Ga), arsenic (As), and phosphorus (P) are used instead of silicon or germanium. A semiconductor diode designed to emit light is called a *light-emitting diode* (LED); its electrical symbol is shown in Fig. 8.49.

The wavelength of emitted light depends on the energy gap of the material. For example, gallium arsenide phosphide (GaAsP) emits red light, and gallium phosphide (GaP) emits green light. Light-emitting diodes are available in numeral and alphabetic characters measuring in height from $\frac{1}{8}$ to approximately $\frac{3}{4}$ in. Their greatest application is as indicators in, for example, electronic calculators.

CHARACTERISTIC CURVES A number of typical characteristic curves for a LED are illustrated in Fig. 8.50. In operation, the LED is forward-biased, like the junction

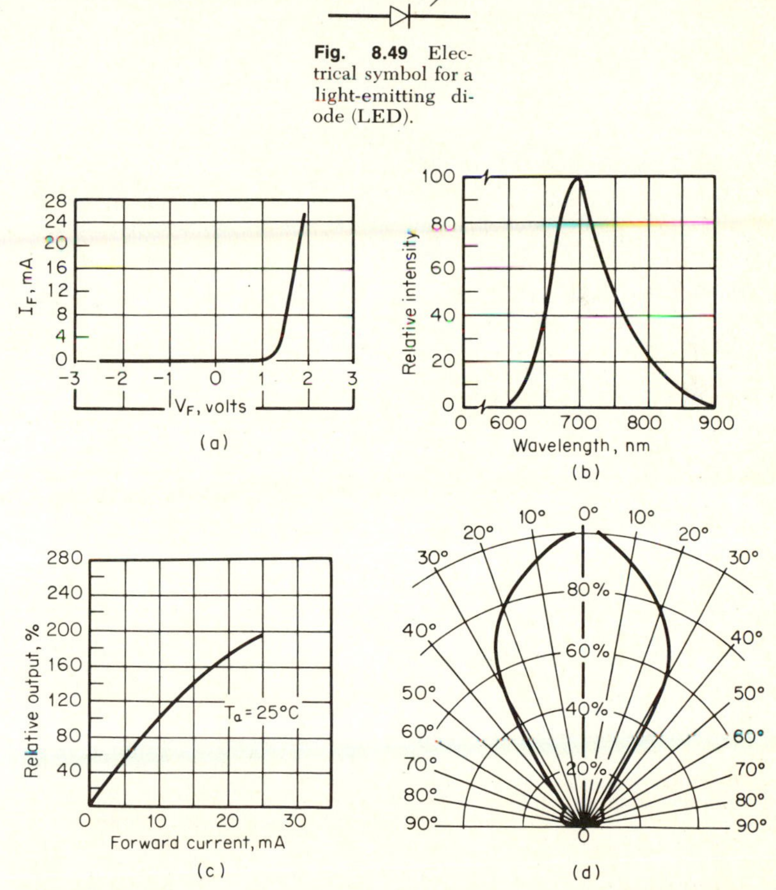

Fig. 8.49 Electrical symbol for a light-emitting diode (LED).

Fig. 8.50 Typical characteristic curves for a LED: (*a*) Voltampere characteristics. (*b*) Spectral distribution curve. (*c*) Visible output as a function of forward diode current. (*d*) Relative candlepower distribution curve. (*Courtesy General Electric Company.*)

diode. An example of the voltampere characteristics of a LED is shown in Fig. 8.50*a*. The threshold voltage is approximately 1.5 V.

Another curve of interest is the spectral distribution of light emanating from a LED. Such a curve is given in Fig. 8.50*b*. Relative intensity of the emitted light is plotted as a function of wavelength in nanometers (10^{-9} m). In this example, maximum intensity is obtained at a wavelength of 700 nm.

The dependency of the brightness of the emitted light and the forward current is illustrated by the curve of Fig. 8.50*c*. The curve of Fig. 8.50*d* shows the distribution of light for different viewing angles. *Candlepower* refers to the luminous intensity, that is, the brightness of the emitted light.

LED PARAMETERS Commonly used parameters for characterizing the LED are summarized in Table 8.7.

TABLE 8.7 Commonly Used Parameters for the LED

Parameter	Symbol	Meaning
Forward voltage	V_F	Dc forward voltage across an LED
Candlepower	CP	Measure of luminous intensity (brightness) of emitted light
Radian power output	P_o	Light power, or brightness, of an LED
Peak spectral emission	λ_{peak}	Wavelength of brightest emitted color. (Symbol λ is the Greek letter *lambda*.)
Spectral bandwidth	BW_λ	Indication of how concentrated is the emitted color

8.11 OTHER SEMICONDUCTOR DEVICES

In addition to the devices described, there is a host of other semiconductor devices that are of interest. In this section, many of these devices are described and some of their important characteristics examined.

TUNNEL DIODE The tunnel diode, also referred to as the *Esaki diode* after its inventor, is a two-terminal device that exhibits a region of negative resistance similar to that of a UJT. Materials used in its manufacture include silicon, germanium, gallium arsenide, and gallium antimony. Typical characteristics and the electrical symbol for the tunnel diode are illustrated in Fig. 8.51. The tunnel diode may be used as an amplifier, an oscillator, or a switch, and operates at very high frequencies (in the gigahertz region).

Referring to Fig. 8.51*a*, I_P and V_P are the peak current and voltage, respectively; I_V and V_V are the valley current and voltage, respectively. Two important parameters for the tunnel diode are the peak-to-valley current ratio, I_P/I_V, and the frequency where the

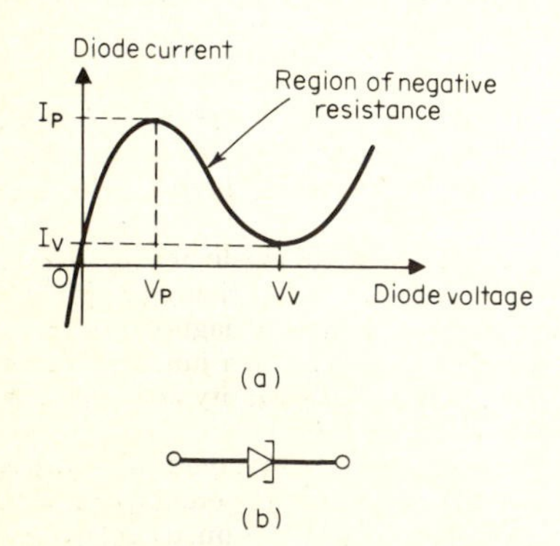

Fig. 8.51 Tunnel diode: (*a*) Typical voltampere characteristics. (*b*) Electrical symbol.

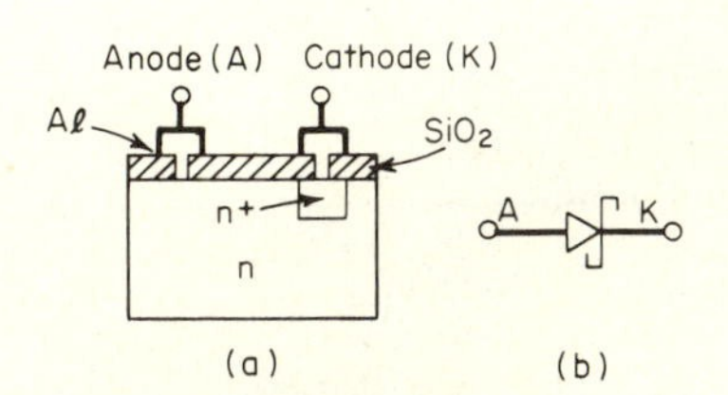

Fig. 8.52 Schottky diode: (*a*) Cross-sectional view of its basic physical construction. (*b*) Electrical symbol.

negative resistance is reduced to zero, f_{ro}. For switching, the device should exhibit a high I_P/I_V ratio; for high-frequency amplifier or oscillator operation, f_{ro} should be high.

SCHOTTKY BARRIER DIODE The Schottky barrier diode, also referred to as a *hot carrier diode*, achieves rectification by a metal semiconductor junction. A cross-sectional view of a basic Schottky diode is illustrated in Fig. 8.52*a*, and its electrical symbol in Fig. 8.52*b*. Contacts to the substrate are aluminum, a p-type impurity. The anode *A* is formed by the aluminum in contact with the n-type material. A pn junction is formed, which is referred to as a Schottky barrier. The cathode is a nonrectifying contact.

The Schottky diode is an example of a *majority carrier device* because only electrons are involved in its operation. (Both electrons and holes flow in a pn junction diode.) The Schottky diodes exhibit extremely fast switching capabilities and are also useful as detectors and mixers at microwave frequencies.

Typical forward and reverse characteristics of a Schottky diode are illustrated in Fig. 8.53. Currents I_F and I_R are the forward and reverse diode currents, respectively. Voltages V_F and V_R are the forward and reverse voltages across the diode, respectively. Note that at room temperature (25°C), the threshold voltage of a Schottky diode is approximately 0.1 V; for a silicon junction diode it is 0.6 V.

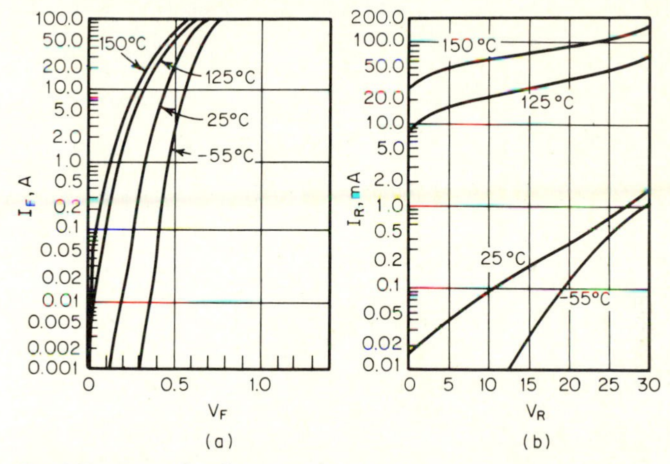

Fig. 8.53 Forward and reverse characteristics of a power Schottky diode: (*a*) Forward characteristics. (*b*) Reverse characteristics. (*Courtesy TRW Semiconductor Division.*)

PHOTODIODES There are two basic kinds of photodiodes: the *photovoltaic* and the *photoconductive* types. For the photovoltaic type, no external bias source is required; an external bias, however, is needed for the photoconductive types. The electrical symbol for the photodiode is given in Fig. 8.54.

From our earlier discussion of the operation of the pn junction diode, it was seen that its reverse saturation current (electron-hole pairs) increased with rising temperature (heat). Heat is a form of energy in the infrared region of the electromagnetic spectrum. Light, in the visible region of the spectrum, striking an exposed pn junction also results in the generation of electron-hole pairs. The greater the intensity and higher the frequency of the light, the greater is the diode current.

The diode is sensitive to certain colors, and its sensitivity is displayed by a spectral distribution curve similar to Fig. 8.50*b*. Sensitivity depends on the energy gap of the material used for the photodiode. In addition to germanium and silicon, cadmium selenide (CdSe), cadmium sulfide (CdS), and cadmium telluride (CdTe) are commonly used materials.

In connecting the conductive-type photodiode, the pn junction is reverse-biased, as

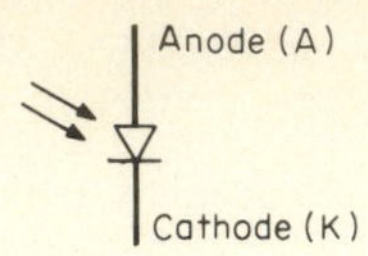

Fig. 8.54 Electrical symbol for a photodiode.

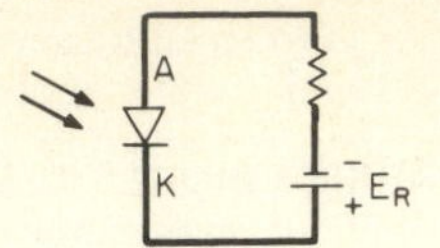

Fig. 8.55 Biasing a conductive-type photodiode.

illustrated by the simple circuit of Fig. 8.55. Variation in diode current with reverse voltage for different levels of illumination (expressed in footcandles, fc, a unit of illumination) is illustrated in Fig. 8.56. The *dark current* is a minute leakage current that flows for no incident radiation.

PHOTOTRANSISTOR The phototransistor operates like a photodiode, and, at the same time, the current generated by the light is amplified. The electrical symbol of a phototransistor is shown in Fig. 8.57*a*. One may view the device as an npn transistor with the incident radiation replacing the base of a BJT. Typical collector characteristics of a phototransistor are given in Fig. 8.57*b* for different levels of illumination. In this set of curves, illumination is expressed in milliwatts per square centimeter.

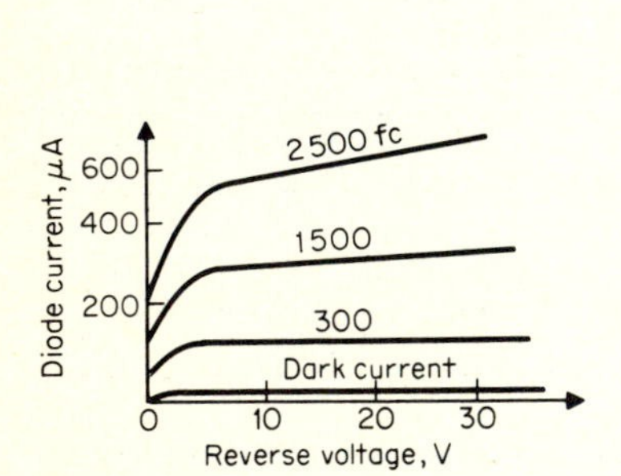

Fig. 8.56 Voltampere characteristics of a typical conductive-type photodiode.

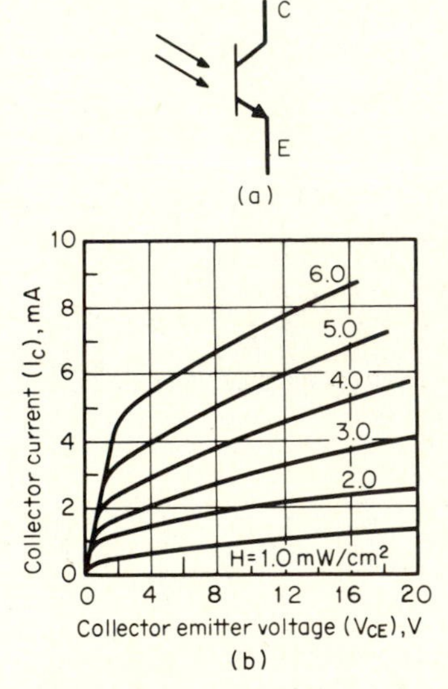

Fig. 8.57 Phototransistor: (*a*) Electrical symbol. (*b*) Typical collector characteristics. (*Courtesy Motorola Semiconductor Products, Inc.*)

OPTO-ISOLATOR The opto-isolator, also referred to as an *optoelectronic coupler*, generally consists of an infrared LED and a silicon phototransistor combined in a single package. A significant advantage of this component is its high isolation resistance (in the order of $10^{11}\ \Omega$) between its input and output. Applications for the opto-isolator include the interfacing of different types of logic circuits and their use in level- and position-sensing circuits.

CHARGE-COUPLED DEVICE A cross-sectional view of a charge-coupled device (CCD) is illustrated in Fig. 8.58*a*. Structurally, it consists of a p- (or an n-) type silicon

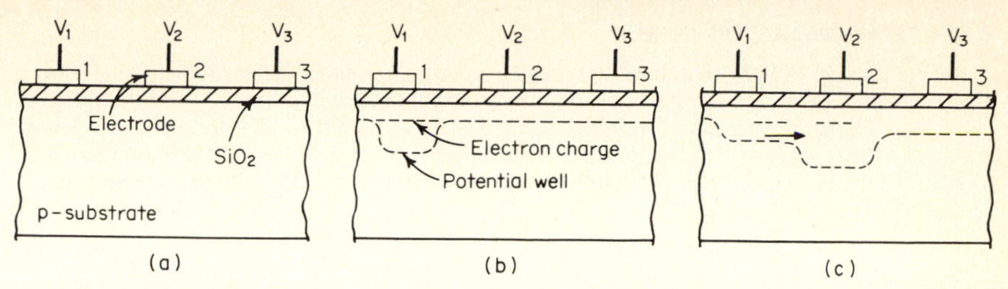

Fig. 8.58 Charge-coupled device (CCD): (*a*) Cross-sectional view of its basic physical construction. (*b*) Potential well and electron charge under electrode 1. (*c*) Electron charge being transferred to under electrode 2.

substrate over which is a layer of silicon dioxide (SiO_2). On top of the silicon dioxide is an array of metallized electrodes which are connected to signal voltages. In Fig. 8.58, V_1, V_2, and V_3 constitute a three-phase system of voltages.

Assume that V_1 is greater than either V_2 or V_3. A *potential level*, or *well*, is formed and an electron charge exists below electrode 1, as shown in Fig. 8.58*b*. If V_2 is made greater than V_1, the charge moves and resides below electrode 2, as illustrated in Fig. 8.58*c*. In this manner, the charge in the substrate is transferred under one electrode to the next, and so on.

The CCD finds application as a dynamic shift register in computers and in solid-state imaging, such as video cameras. The packing density (the number of devices occupying the substrate) is in the order of 100 times as great as for other semiconductor devices. The CCD consumes very little power and is capable of operating at high frequencies.

MICROWAVE POWER DIODES There are two basic families of diodes used in the generation of microwave frequencies: the *avalanche-type* and the *transferred-electron bulk* device. An example of a diode that operates in the avalanche mode is the IMPATT (IMPact Avalanche and Transit Time). The LSA (Limited Space-charge Accumulation) is representative of the transferred-electron bulk device.

A cross-sectional view of an IMPATT diode is given in Fig. 8.59. An n region is sandwiched between a thin n+ and p+ region. Generally the device is fabricated from gallium arsenide. In operation, the diode is reverse-biased, and, at a sufficiently high dc level, the n material is divided into two regions: the drift and avalanche regions. Because of the high electric field, electron-hole pairs are generated essentially in the avalanche region. The flow of carriers is proportional to the voltage in the drift region.

At the oscillating frequency, the phase shift between the two regions is 90°, and the device exhibits a negative resistance. When it is connected to a tuned circuit, the negative resistance may be thought of as canceling out the positive resistance of the tuned circuit, and the combination oscillates.

A cross-sectional view of an LSA diode is illustrated in Fig. 8.60. An n-type region is sandwiched between two alloyed metal contacts; the n-type material is gallium arsenide. Above a given threshold voltage, a charge of electrons (space charge) in the material is dissipated before it can build up appreciably. The current, therefore, tends to decrease as the impressed voltage is increased. This results in the device displaying negative resistance and serves as an oscillator, as for the IMPATT diode.

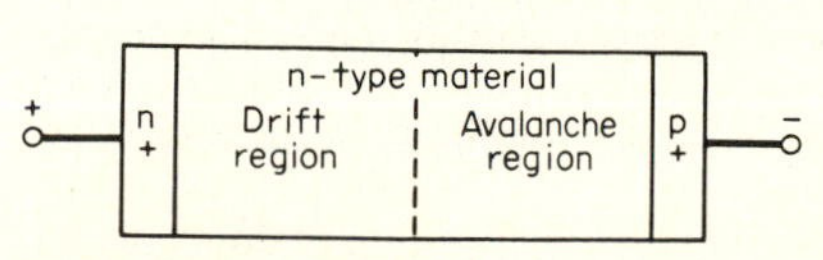

Fig. 8.59 Cross-sectional view of a basic IMPATT diode.

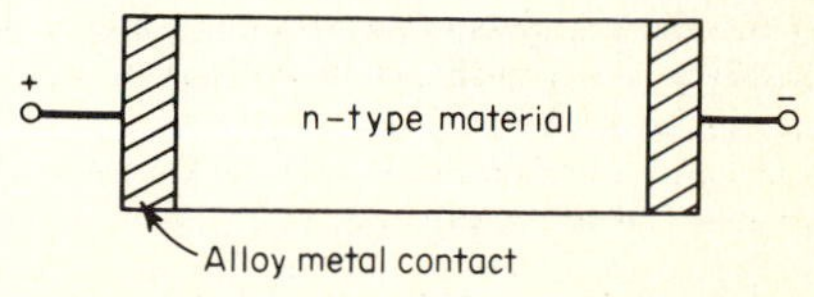

Fig. 8.60 Cross-sectional view of a basic LSA diode.

8.12 SEMICONDUCTOR CHIPS

With the growth of hybrid technology (see Chap. 9), *unencapsulated* diodes, transistors, and even monolithic integrated circuits are available for use in hybrid microelectronic circuits. The unencapsulated devices are referred to as *chips*. There are three basic types of chips: *flip chip, beam lead,* and *leadless inverted devices.* Miniature encapsulated devices, which may measure less than 0.1 × 0.1 × 0.05 in, are also used occasionally.

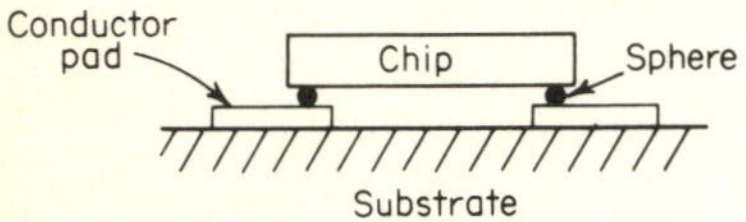

Fig. 8.61 Connecting flip chip to conductor pads on a substrate.

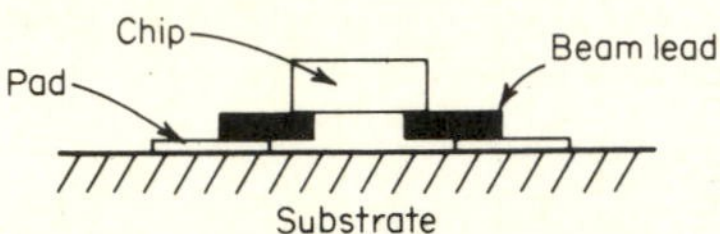

Fig. 8.62 Connecting beam-lead device to conductor pads on a substrate.

FLIP CHIP Contacts on the flip chip are on one side of the device and may take the form of solder-coated metal pillars or spheres about 0.05 in. in diameter. Figure 8.61 shows how a flip chip is connected to conductor pads on a substrate. By heating the pad in contact with the chip, solder flows between the pad and sphere and an electric connection is made.

BEAM-LEAD DEVICE In a beam-lead device, relatively long leads, or beams, are attached to the semiconductor chip (see Fig. 8.62). The beam-lead device is connected to the pads on a substrate in a manner similar to that for a flip chip.

LEADLESS INVERTED DEVICE The leadless inverted device (LID), also referred to as a *ceramic flip chip*, is a ceramic block in which is mounted the semiconductor chip (Fig. 8.63). Wires from the chip are bonded to the metallized lands. The LID is positioned over the pads on the substrate, and a connection is made with the application of heat.

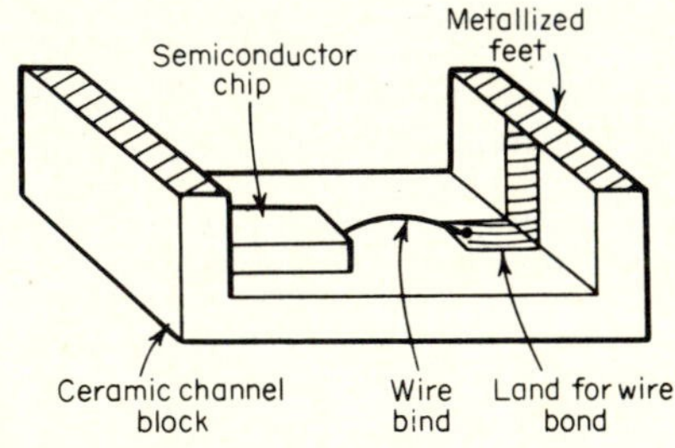

Fig. 8.63 Constructional features of an LID.

8.13 UNDERSTANDING DATA SHEETS

To develop an understanding of data sheets for semiconductor devices, three typical examples will be examined. The data sheets cover a medium-current silicon rectifier (Fig. 8.64), a germanium pnp transistor (Fig. 8.65), and an n-channel JFET (Fig. 8.66). The parameters specified on the sheets have been defined in appropriate tables contained in the chapter.

A study of the sheets reveals a commonality in the specifications of data and types of curves. Each data sheet has these common features:

1. A brief statement of the intended application and outstanding features of the device. For example, the rectifier is ". . . desirable in industrial applications that require high surge capabilities." The transistor is ". . . intended for industrial general-

Silicon Rectifier

IN3569-74

MEDIUM CURRENT

The 1N3569-74 series stud mounted device in the popular DO-4 package is designed especially for reliability and long life. This series is especially desirable in industrial applications that require high surge capabilities. In addition to this feature the 1N3569-74 units have:

- **Hermetically Sealed Housing**
- **Transient PRV Rating**
- **Low Forward Drop**
- **Operating Temperature up to 165°C**

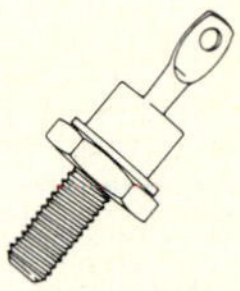

OUTLINE DRAWING

SYMBOL	INCHES		MILLIMETERS		NOTES
	MIN.	MAX.	MIN.	MAX.	
A		.405		10.29	
ϕD		.424		10.77	
E	.424	.437	10.77	11.10	
F	.075	.175	1.91	4.45	
J		.800		20.32	
M		.250		6.35	1
N	.422	.453	10.72	11.51	
ϕT	.060		1.52		
W					2

NOTES
1. Angular orientation of this terminal is undefined.
2. 10-32 UNF-2A. Maximum pitch diameter of plated threads shall be basic pitch diameter (.1697", 4.29 MM). Ref. (Screw thread standards for Federal Services 1957) Handbook H28, P1

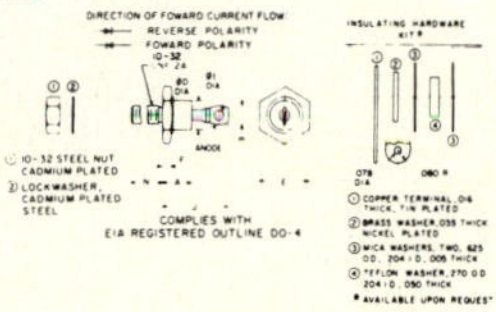

COMPLIES WITH EIA REGISTERED OUTLINE DO-4

RATINGS AND SPECIFICATIONS
60 cps Resistive or Inductive Load

	1N3569	1N3570	1N3571	1N3572	1N3573	1N3574
**Maximum Allowable Transient Peak Reverse Voltage (non-recurrent, 5 millisecond duration)	150	275	400	525	650	775 V
Maximum Allowable Repetitive Peak Reverse Voltage (PRV)	100	200	300	400	500	600 V
Maximum Allowable RMS Voltage	70	140	210	280	350	420 V
Maximum Allowable DC Blocking Voltage	100	200	300	400	500	600 V

*Maximum Allowable DC Output Current
- at 85°C Stud Temperature ← 3.50 → Amps
- at 150°C Stud Temperature ← 1.25 → Amps

**Maximum Allowable Peak One Cycle Surge Current (60 cps single phase, non-recurrent)—See Fig. 5 ← 35 → Amps

**Minimum I^2t Rating (non-recurrent)—See Fig. 6 ← 3 → Amp² sec

Maximum Full Load Voltage Drop (full cycle average at rated load) 150°C Ambient ← 0.5 → V

Maximum Reverse Current (full cycle average at rated voltage) 150°C Ambient ← 0.4 → ma

Maximum Operating Temperature ← −65°C to +165°C →

Maximum Storage Temperature ← −65°C to +175°C →

Thermal Characteristics
Typical Thermal Impedance: Junction to Stud ← 10°C/W →

Notes: *For current ratings on fins and at other stud temperatures, see Figures 1 and 2. Stud temperature refers to temperature on outside flat of hex.

**Rectifier must return to normal thermal equilibrium before reapplication of these overloads.

Fig. 8.64 Typical data sheet for a rectifying diode. (*Courtesy General Electric Company.*)

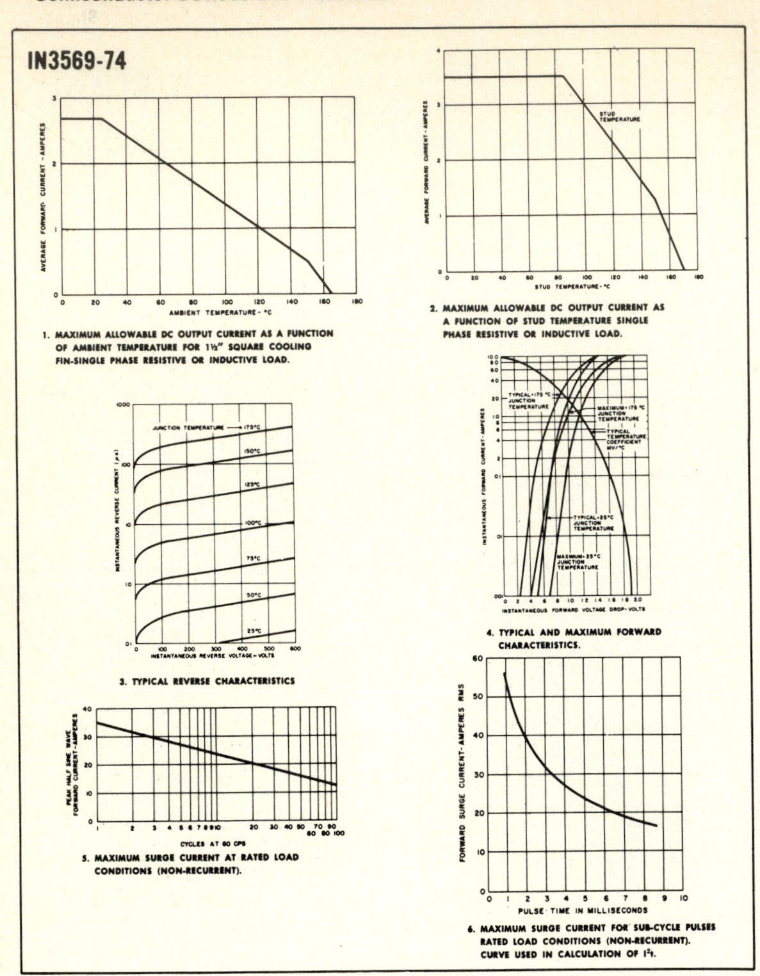

Fig. 8.64 *(Continued.)*

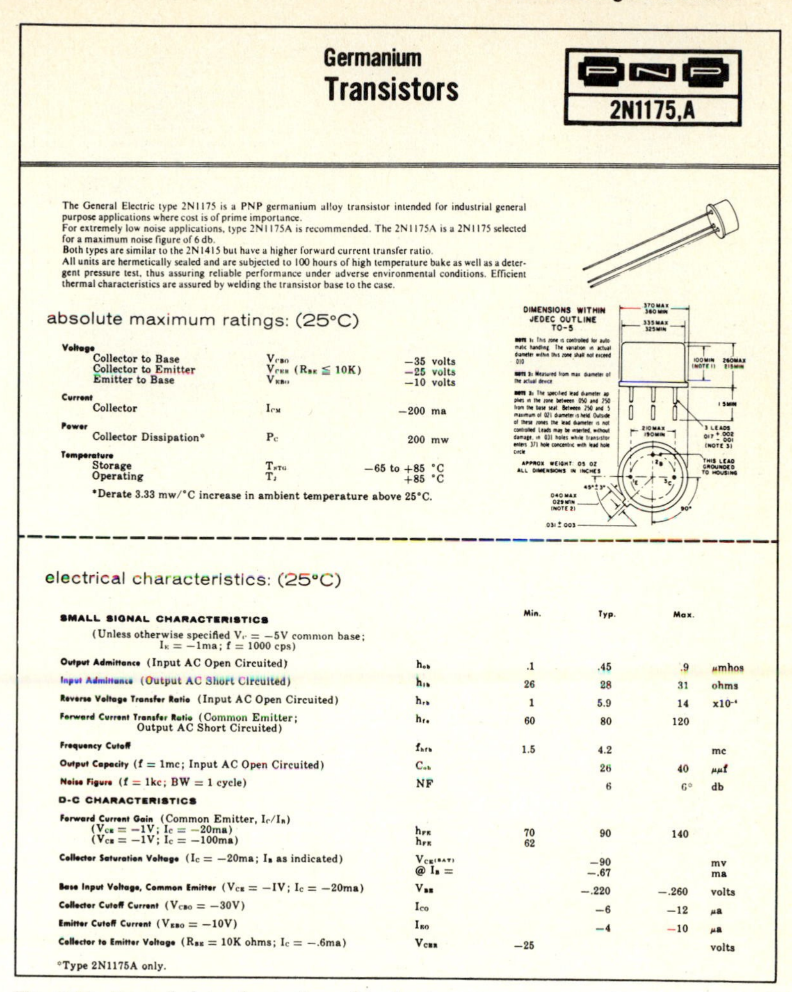

Germanium Transistors

PNP 2N1175,A

The General Electric type 2N1175 is a PNP germanium alloy transistor intended for industrial general purpose applications where cost is of prime importance.
For extremely low noise applications, type 2N1175A is recommended. The 2N1175A is a 2N1175 selected for a maximum noise figure of 6 db.
Both types are similar to the 2N1415 but have a higher forward current transfer ratio.
All units are hermetically sealed and are subjected to 100 hours of high temperature bake as well as a detergent pressure test, thus assuring reliable performance under adverse environmental conditions. Efficient thermal characteristics are assured by welding the transistor base to the case.

absolute maximum ratings: (25°C)

Voltage		
Collector to Base	V_{CBO}	−35 volts
Collector to Emitter	V_{CER} ($R_{BE} \leq 10K$)	−25 volts
Emitter to Base	V_{EBO}	−10 volts
Current		
Collector	I_{CM}	−200 ma
Power		
Collector Dissipation*	P_C	200 mw
Temperature		
Storage	T_{STG}	−65 to +85 °C
Operating	T_J	+85 °C

*Derate 3.33 mw/°C increase in ambient temperature above 25°C.

electrical characteristics: (25°C)

		Min.	Typ.	Max.	
SMALL SIGNAL CHARACTERISTICS (Unless otherwise specified $V_C = -5V$ common base; $I_E = -1ma$; f = 1000 cps)					
Output Admittance (Input AC Open Circuited)	h_{ob}	.1	.45	.9	μmhos
Input Admittance (Output AC Short Circuited)	h_{ib}	26	28	31	ohms
Reverse Voltage Transfer Ratio (Input AC Open Circuited)	h_{rb}	1	5.9	14	$\times 10^{-4}$
Forward Current Transfer Ratio (Common Emitter; Output AC Short Circuited)	h_{fe}	60	80	120	
Frequency Cutoff	f_{hfb}	1.5	4.2		mc
Output Capacity (f = 1mc; Input AC Open Circuited)	C_{ob}		26	40	μμf
Noise Figure (f = 1kc; BW = 1 cycle)	NF		6	6°	db
D-C CHARACTERISTICS					
Forward Current Gain (Common Emitter, I_C/I_B) ($V_{CE} = -1V$; $I_C = -20ma$)	h_{FE}	70	90	140	
($V_{CE} = -1V$; $I_C = -100ma$)	h_{FE}	62			
Collector Saturation Voltage ($I_C = -20ma$; I_B as indicated)	$V_{CE(SAT)}$		−90		mv
	@ I_B =		−.67		ma
Base Input Voltage, Common Emitter ($V_{CE} = -1V$; $I_C = -20ma$)	V_{BE}		−.220	−.260	volts
Collector Cutoff Current ($V_{CBO} = -30V$)	I_{CO}		−6	−12	μa
Emitter Cutoff Current ($V_{EBO} = -10V$)	I_{EO}		−4	−10	μa
Collector to Emitter Voltage (R_{BE} = 10K ohms; $I_C = -.6ma$)	V_{CER}	−25			volts

°Type 2N1175A only.

Fig. 8.65 Typical data sheets for a bipolar junction transistor. (*Courtesy General Electric Company.*)

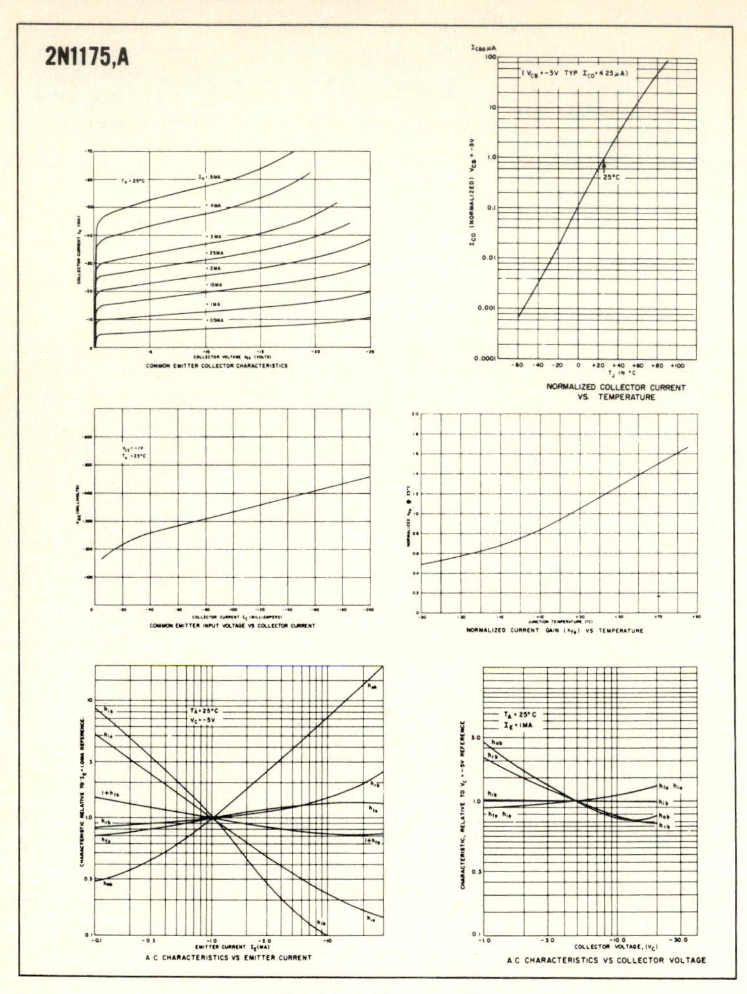

Fig. 8.65 *(Continued.)*

U244

N-CHANNEL SILICON JUNCTION FIELD-EFFECT TRANSISTOR

FOR POWER AMPLIFIERS AND SWITCHES

- High Power Capability — 10 W
- High g_{fs} — 80,000 μmhos Minimum
- High I_{DSS} — 300 mA Minimum
- No Thermal Runaway
- No Secondary Breakdown

All leads isolated from case.
TO-60

PRODUCT CONDITIONING

Units receive the following treatment before final electrical tests:

High Temp Storage: 24 Hours at 150°C
Thermal Shock: +100 to 0°C for 5 Cycles
25,000g Acceleration/Impact in the Y_1 Plane
Helium and/or Gross Leak Tests for Hermeticity

ABSOLUTE MAXIMUM RATINGS (25°C)

Gate-Drain Voltage and Gate-Source Voltage (Note 1) . . . . -25 V
Drain Current . . . . 900 mA
Total Device Dissipation at 25° C Case Temperature (Note 2) . . . . 10 W
Storage Temperature Range . . . . -65 to +150° C
Operating Junction Temperature Range . . . . -55 to +150° C

ELECTRICAL CHARACTERISTICS (25°C unless otherwise noted)

Characteristic		Test Conditions		Min	Max	Unit
I_{GSS}	Gate Reverse Current	V_{GS} = -15 V, V_{DS} = 0	25° C		-1	nA
			150° C		-1	μA
$I_{D(OFF)}$	Drain Cutoff Current	V_{DS} = 5 V, V_{GS} = -10 V	25° C		1	nA
			150° C		1	μA
BV_{GSS}	Gate-Source Breakdown Voltage	I_G = -1 μA, V_{DS} = 0		-25		V
V_P	Gate-Source Pinch-Off Voltage	V_{DS} = 10 V, I_D = 1 mA		-3.5	-8	V
I_{DSS}	Drain Current at Zero Gate Voltage (Note 3)	V_{DS} = 10 V, V_{GS} = 0		300	900	mA
g_{fs}	Common-Source Forward Transconductance (Note 3)	V_{DS} = 10 V, V_{GS} = 0, f = 100 kHz		80,000	200,000	μmho
r_{DS}	Static Drain-Source On Resistance	V_{GS} = 0, I_D = 10 mA			10	ohms
C_{iss}	Common-Source Input Capacitance	V_{DS} = 0, V_{GS} = -10 V, f = 1 MHz			35	pF
C_{rss}	Common-Source Reverse Transfer Capacitance	V_{DS} = 0, V_{GS} = -10 V, f = 1 MHz			15	pF

NOTES: NI
1. Due to symmetrical geometry, these units may be operated with source and drain leads interchanged.
2. Derate linearly to 150° C case temperature at rate of 80 mW/° C.
3. Pulse test: duration 2 ms.

These devices are manufactured to meet or exceed the requirements of MIL-S-19500.

Fig. 8.66 Typical data sheets for a field-effect transistor. (*Courtesy Siliconix Incorporated.*)

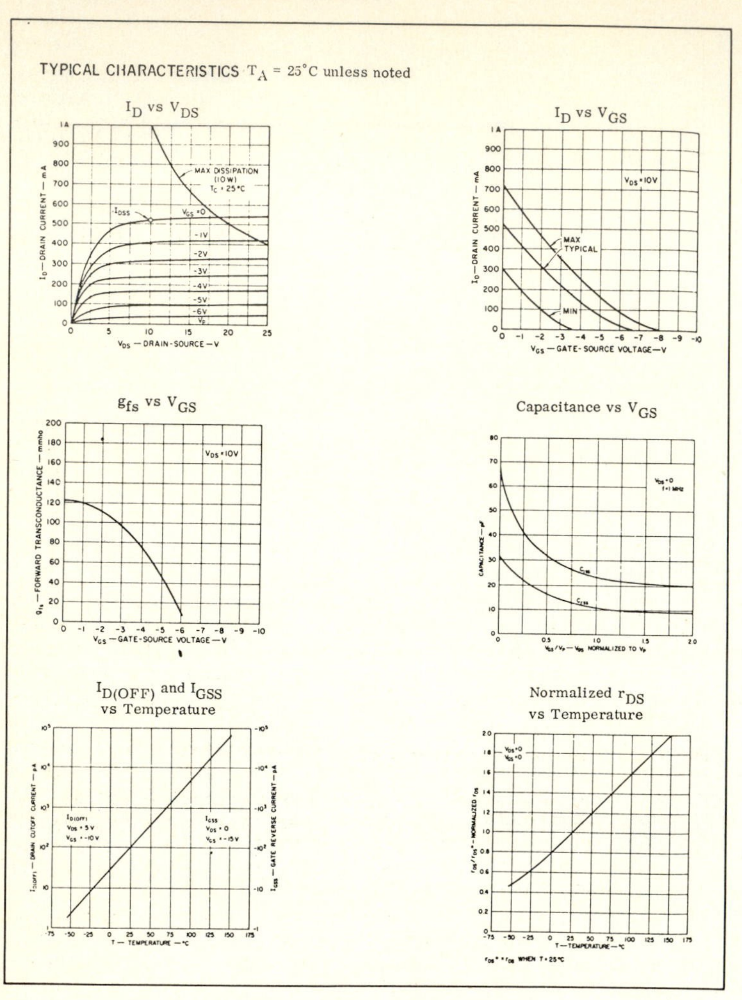

Fig. 8.66 *(Continued.)*

purpose applications where cost is of prime importance." The JFET is designed for power amplifier and switching applications.

2. A sketch with dimensions of the device: identification of connecting leads.

3. Maximum ratings, such as voltage, current, and temperature, that should never be exceeded.

4. Characteristics curves, such as the forward and reverse characteristics for a diode, collector characteristics for a BJT, and drain characteristics for a JFET.

5. Various curves of interest as a function of temperature, voltage, current, or frequency. For example, power derating and surge current curves are included for the diode. Interesting curves included for the BJT are the variation of the small-signal hybrid parameters with emitter current and collector voltage. The transfer characteristic (I_D versus V_{GS}) is of interest for the JFET.

6. The spread of parameters by denoting their minimum, typical, and maximum values.

In making a selection, the data sheets for the device should be carefully studied and understood. Test condition for each parameter should be stated on the sheets. If data that are important for the intended application are missing, the data should be obtained.

Chapter 9

Integrated Circuit Technology

9.1 INTRODUCTION

Since the appearance of the integrated circuit (IC) in the early 1960s, the electronics industry has experienced monumental changes in design philosophy and manufacturing methods. This has inspired a host of new products, especially in the consumer sector, such as hand-held electronic calculators and electronic watches. In the future, besides innovations in the industrial market, many other products incorporating IC's will be developed for consumer use.

The IC evolved from attempts by semiconductor manufacturers to improve transistor performance and reduce costs. These efforts culminated in the late 1950s in the development of the planar diffused transistor. The techniques used in processing this device are essentially the same as those employed in making the most widely produced integrated circuit, the *monolithic* IC. Monolithic is derived from a Greek word meaning "single stone." In the monolithic IC, all components of the integrated circuit are formed in a single chip of silicon.

A family tree of integrated circuits manufactured is shown in Fig. 9.1. In addition to the monolithic IC, there is the *hybrid* which is available in two forms: *thin* and *thick film*. In either one, resistors, capacitors, and conductors are deposited on a substrate, such as ceramic. In the thin-film process, resistors, capacitors, and conductors are deposited on the substrate by evaporation of a suitable material in a vacuum. In the thick-film process, the components may be thought of as being "painted on" the substrate. Thin- and thick-film, as well as monolithic technology, will be described in this chapter.

Because of the initial high "tooling up" costs, monolithic technology is suited for mass production of IC devices. For small to moderate requirements, hybrid processing is profitable. If, for technical reasons, it is not feasible to develop a monolithic IC for a given circuit, the hybrid is made, regardless of quantity.

9.2 DIMENSIONING INTEGRATED CIRCUITS

A number of different dimensional units are utilized in defining the geometry of integrated circuits and their components. A comparison of these units is given in Table 9.1. Generally, *microns* are used for IC device dimensions and *mils* for the overall chip.

9.3 MONOLITHIC TECHNOLOGY

In the manufacture of IC's, hundreds to thousands of circuits are *processed simultaneously*. For this reason, the cost per circuit is low, making it possible for many IC am-

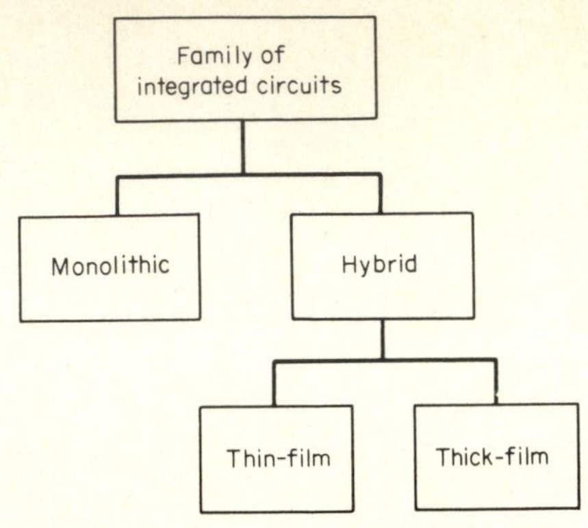

Fig. 9.1 A family tree of integrated circuits.

TABLE 9.1 Comparison of Dimensional Units for Integrated Circuits

	Mil	Centimeter	Micron	Angstrom
Mil	1	2.54×10^{-3}	25.4	2.54×10^{5}
Centimeter, cm	3.94×10^{2}	1	10^{4}	10^{8}
Micron, μ	3.94×10^{-2}	10^{-4}	1	10^{-4}
Angstrom, Å	3.94×10^{-6}	10^{-8}	10^{4}	1

For example, 1 mil $= 2.54 \times 10^{-3}$ cm $= 25.4\ \mu$; 1 $\mu = 10^{-4}$ cm; and 1 Å $= 10^{-4}\ \mu$.

plifiers to sell for less than a single transistor sold for a few years ago. The basic silicon material used in monolithic processing comes in a wafer, 1.5 to 3 in diameter and 6 mils thick, as shown in Fig. 9.2. The flat is used for reference, ensuring that the proper orientation of the wafer is maintained during the different processing steps.

Each of the squares in Fig. 9.2 is a chip that contains a complete integrated circuit. A typical size of a chip is $\frac{1}{16}$ by $\frac{1}{16}$ of an inch. Upon completion of processing, the wafer is *scribed* and then *fractured* along the vertical and horizontal lines. The chip is tested and packaged. Because of defects and loss of material along the periphery of the wafer, the number of good chips is less than the number theoretically possible from the wafer. The ratio of good chips to the maximum number of chips available from a wafer is referred to as the *yield* of the process.

BASIC PROCESSING STEPS To learn how a circuit is integrated, we shall first describe the processing steps required in integrating a single transistor (Fig. 9.3). Once

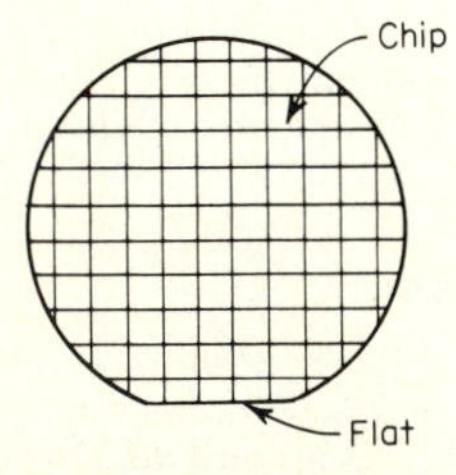

Fig. 9.2 A silicon wafer. Each chip contains a complete integrated circuit. The flat is used as a reference in the processing sequence of the wafer.

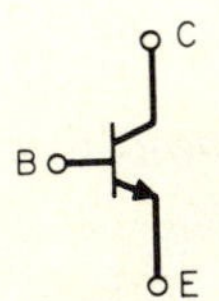

Fig. 9.3 A npn transistor to be integrated.

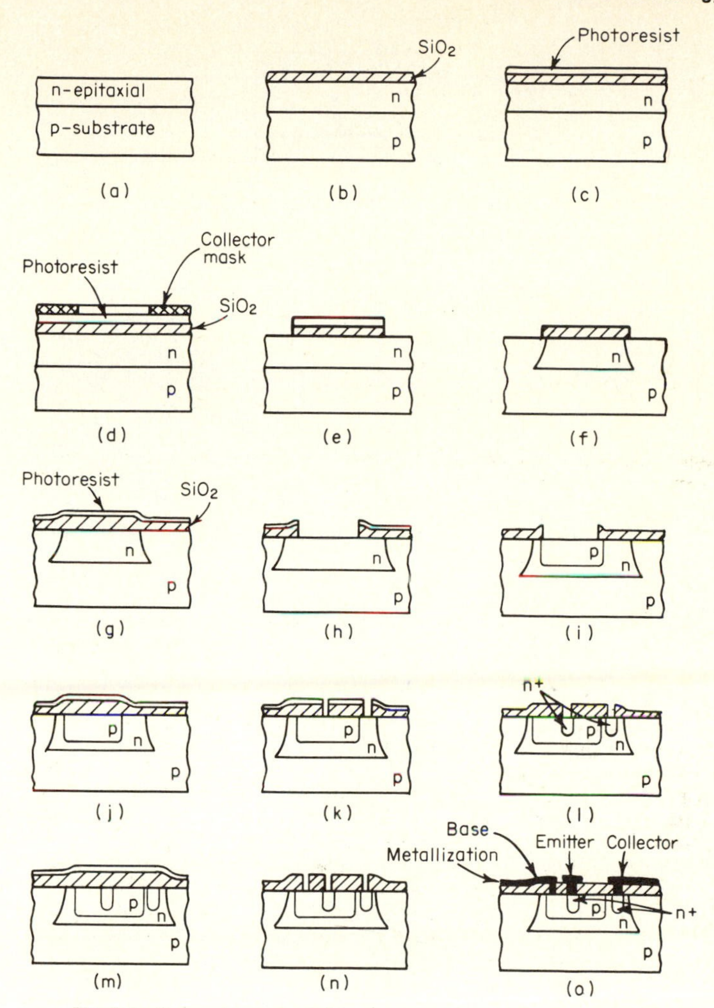

Fig. 9.4 Processing steps for integrating a transistor. (See text.)

the processing steps are understood, we shall see how an amplifier is integrated. Reference will be made to Fig. 9.4 in the following discussion.

P-type substrate The starting material is generally p-type silicon having a resistivity of about 10 Ω-cm. The material has a *single-crystal* structure. Single crystal denotes a material in which the atoms are arranged in an ordered pattern. Material in which the atoms are not in an ordered pattern is referred to as *polycrystalline*.

Epitaxial layer Upon the p-type silicon a well-defined n-type silicon region is grown in an epitaxial furnace. *Epitaxial growth* means that the n region formed has the same single-crystal structure as the p-type silicon (Fig. 9.4*a*).

Oxide layer The silicon wafer, seated in a *quartz boat*, is placed in an *oxidation furnace* of Fig. 9.5 and exposed to steam or dry oxygen at a temperature between 1000 and 1200°C. The surface of the silicon is oxidized, and a thin layer of silicon dioxide (SiO_2), which is glass, is formed (Fig. 9.4*b*).

An oxide layer has two important functions:

1. It serves as the surface for a light-sensitive film.
2. It protects the circuit against contamination, a serious problem in processing integrated circuits. This property is referred to as *passivation*.

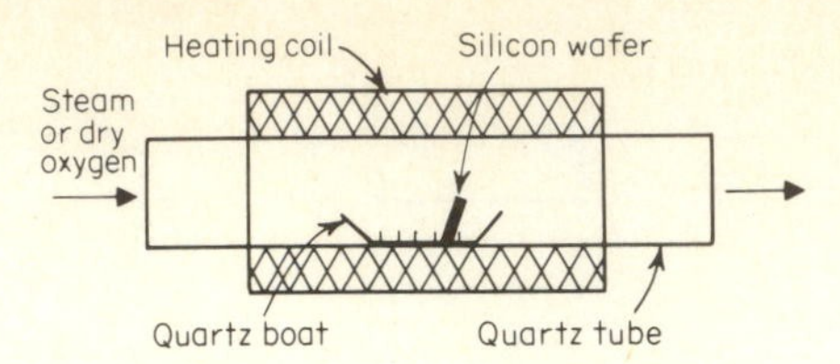

Fig. 9.5 Cross-sectional view of a basic oxidation furnace.

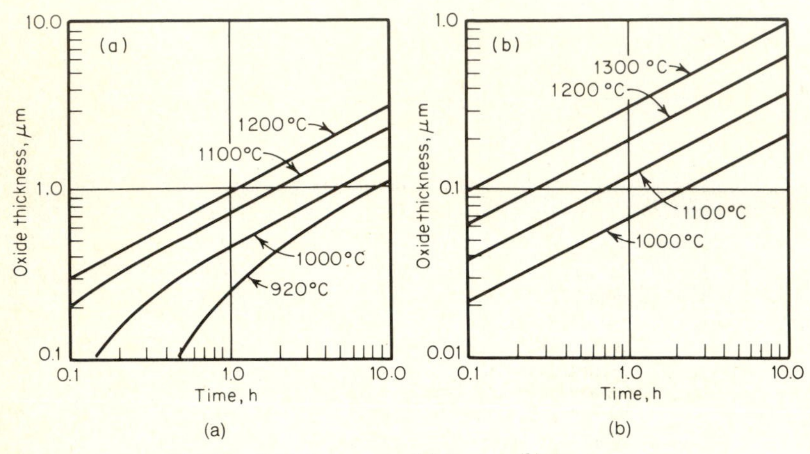

Fig. 9.6 Oxidation curves: (*a*) Steam. (*b*) Dry oxygen.

Curves of SiO_2 growth in steam and dry oxygen as a function of time for different temperatures are illustrated in Fig. 9.6. The application of the curves is illustrated in our first example.

example 9.1 Using the oxidation curves of Fig. 9.6, estimate the time needed to grow a 0.2-μ thickness of SiO_2 on silicon, using steam and dry oxygen. The temperature of the oxidation furnace is 1100°C.

solution It is noted that the vertical and horizontal scales of the oxidation curves are not linear. To cover a large range of values (10/0.1 = 100:1) in a compact manner, *logarithmic scales* are used. Logarithmic scales in electronics are nothing new. For example, the frequency scale for the frequency-response curve of an amplifier is also logarithmic. Referring to the 1100°C steam curve of Fig. 9.6*a*, for an oxide thickness of 0.2 μ, 0.1 h (6 min) is required. If dry oxygen is used, from the 1100°C curve of Fig. 9.6*b*, 3 h (180 min) is necessary. Because it takes less time to grow an oxide layer, steam is generally used.

Photoresist An organic substance that is sensitive to ultraviolet (UV) light, called *photoresist*, is applied over the oxide layer (Fig. 9.4*c*). A commonly used resist is *Kodak Photo Resist* (KPR). The wafer is held by vacuum on a *spinner*. Some drops of KPR are applied to the surface, and the wafer is spun at a few thousand rpm for 10 to 30 s. This ensures a uniform deposition of KPR over the wafer's surface.

Upon exposure to UV, the photoresist becomes *polymerized*. When this occurs, the molecular structure of the exposed resist is altered. The change results in the exposed material becoming *resistant* to chemicals used in subsequent processing such as hydrofluoric acid (HF) for etching and phosphorus (P) and boron (B) for doping. The unexposed areas, however, are affected by these materials.

Artwork Each chip on the wafer of Fig. 9.2, which may measure $^1/_{16}$ by $^1/_{16}$ in or less, contains perhaps a dozen or more transistors and an equal number of resistors. It becomes apparent that the regions on a chip must be precisely defined. For an npn transistor, three regions are needed: collector, base, and emitter. The three regions are cut as masks, illustrated in Fig. 9.7. A Mylar base, which is dimensionally stable with

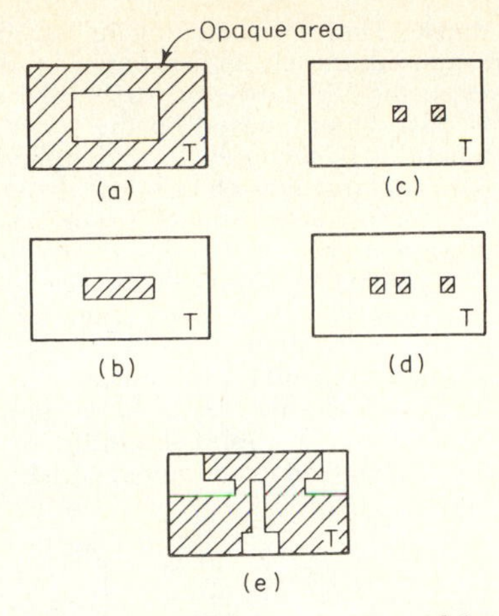

Fig. 9.7 Set of five masks required for integrating the transistor of Fig. 9.3. (See text.)

temperature and humidity, on which is a peelable red tinted layer, is used for the mask material. Commercially the material is referred to as *Rubylith.*

The pattern is formed by cutting through the tinted layer (and not the Mylar base) and peeling away the unwanted layer. To achieve the required precision, the pattern is cut on a mask of the size in the order of 20 by 30 in. This is accomplished on a *coordinatograph* (Fig. 9.8), a machine that provides the required cutting accuracy. The tee in each mask is used in mask alignment (to be described later).

The mask is ultimately reduced to the actual size of the chip (for example, $^1/_{16}$ by $^1/_{16}$-in chip size). In addition, the pattern must be repeated to cover the whole wafer. This is done in a *step-and-repeat camera,* which reduces the mask to its final size and repeats the pattern on a high-resolution emulsion glass plate. A plate is made for each of the masks.

Fig. 9.8 A coordinatograph is used for the precision cutting of masks. (*Courtesy Owen Jones Equipment Sales.*)

Exposure of collector mask The cross-hatching in Fig. 9.7 denotes the opaque areas of the mask. When exposed, the UV is prevented by the opaque areas on the mask from reaching the resist; the UV, however, strikes the resist through the clear mask areas. The collector mask superimposed on the wafer is shown in Fig. 9.4*d*.

Diffusion The key to monolithic IC technology is the diffusion of impurities, called *dopants*, into silicon. In diffusion, p-type such as boron (B), and n-type, such as phosphorus (P), dopants in high concentration are brought in contact with the silicon surface. The process occurs in a diffusion furnace, which is similar in basic construction to the oxidation furnace of Fig. 9.5. Diffusion occurs in the temperature range of 1100 to 1200°C. The dopant is typically in vapor or gaseous form. The penetration depth and concentration of impurity in the silicon depend on the type of dopant, time, and temperature of diffusion, and on how the dopant is introduced.

Silicon, like other materials, can absorb only a finite number of impurity atoms, called the *solid solubility* of the material. Solid solubility is the maximum number of atoms of one material that can be dissolved in another material.

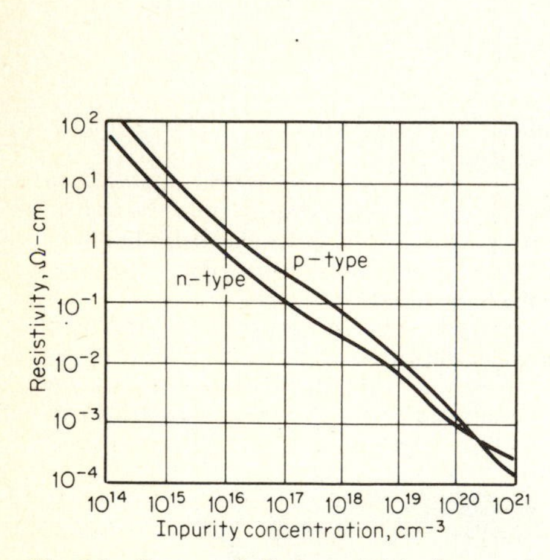

Fig. 9.9 Curves of silicon resistivity for p- and n-type impurity concentrations at room temperature (25°C).

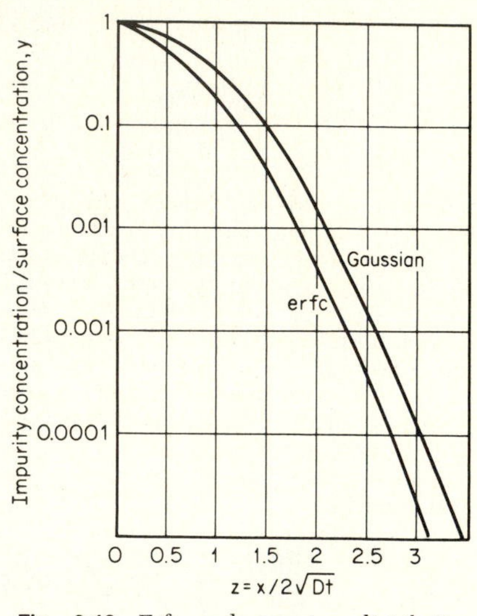

Fig. 9.10 Erfc and gaussian distribution curves.

Curves of silicon resistivity (ohm-centimeters) for p- and n-type *impurity concentrations* (impurity atoms per cubic centimeter of silicon) are given in Fig. 9.9. It is seen that, as the impurity concentration is increased, the resistivity of the silicon is decreased.

There are two methods for applying dopants to the wafer in the diffusion furnace. In the first method, the wafer *always "sees"* a *constant* concentration, N_o atoms/cm^3, of dopant. The resulting distribution of the impurity in silicon is described by what is called the complementary error function (erfc) curve of Fig. 9.10. In the second method, the wafer sees a *fixed* amount of dopant atoms. The resulting distribution is gaussian and is also shown in Fig. 9.10.

For determining the location of a pn junction in silicon, Fig. 9.10 is extremely useful. Both curves are plotted as N_x/N_o versus $z = x/2\sqrt{Dt}$. Term N_x is the actual concentration of impurity atoms per cubic centimeter at a distance x from the surface of the silicon; N_o has already been defined. Term D is the diffusion coefficient in square centimeters per second; it indicates how fast the dopant spreads out in the silicon. Plots of D for different dopants as a function of temperature are given in Fig. 9.11.

example 9.2 A uniformly doped p-type silicon wafer is used in the manufacture of an integrated circuit. The resistivity of the wafer measures 10 Ω-cm. It is placed in a diffusion furnace at 1200°C and is exposed to a *constant source* of phosphorus, $N_o = 10^{19}$ atoms/cm³. How long does it take for a pn junction to form 3 μ away from the surface?

solution Referring to the resistivity curves of Fig. 9.9, for p-type silicon of 10 Ω-cm resistivity, the impurity concentration is about 1.4×10^{15} atoms/cm³. This figure is called the *background concentration*, N_B, of impurity. In this example, then, $N_B = 1.4 \times 10^{15}$ atoms/cm³.

A constant source of dopant (phosphorus) implies a complementary error function (erfc) distribution. A pn junction is formed when the concentration of the n-type dopant (phosphorus) equals the p-type dopant (background concentration). Hence, $N_x = N_B = 1.4 \times 10^{15}$.

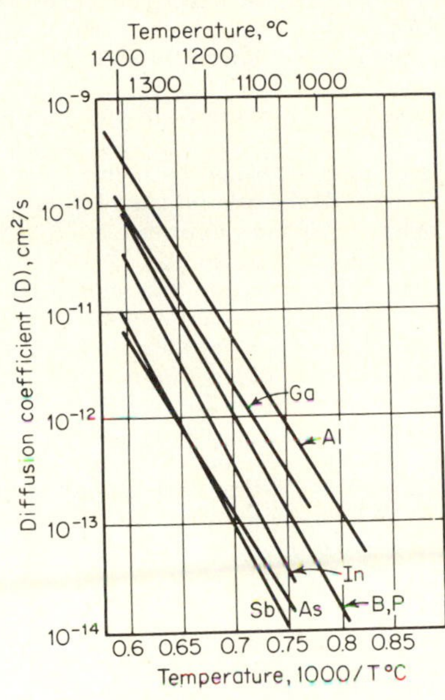

Fig. 9.11 Diffusion coefficients for various impurities in silicon.

The ratio $N_x/N_o = 1.4 \times 10^{15}/10^{19} = 1.4 \times 10^{-4} = 0.00014$. From the erfc curve of Fig. 9.10, it is estimated that $z = 2.6$.

At 1200°C, the diffusion coefficient for phosphorus, from Fig. 9.11, is $D_p \simeq 3 \times 10^{-12}$ cm²/s. The junction depth x is 3 $\mu = 3 \times 10^{-4}$ cm (see Table 9.1). Substitution of these values in z yields

$$2.6 = \frac{3 \times 10^{-4}}{2\sqrt{3 \times 10^{-12} t}}$$

Squaring both sides of the equation, we obtain

$$6.75 = \frac{9 \times 10^{-8}}{4 \times 3 \times 10^{-12} t}$$

Solving for t,

$$t = \frac{9 \times 10^{-8}}{4 \times 3 \times 10^{-12} \times 6.75}$$

$$\simeq 1100 \text{ s} \simeq 0.3 \text{ h}$$

Etching The unexposed SiO_2 is etched away by hydrofluoric acid (HF) (Fig. 9.4*e*).

p-type diffusion The wafer is placed in a diffusion furnace, and sufficient p-type impurity, such as boron, is introduced. The impurity penetrates the silicon which was etched away by HF. After a length of time, an n region (or *island, tub*) for the collector is formed (Fig. 9.4*f*). Because the impurity diffuses *horizontally* (laterally), as well as vertically, the profile of Fig. 9.4*f* is the result. Note that some SiO_2 overlaps the n region. This is desirable because it helps to minimize contamination. The photoresist is removed, or *stripped*, after each diffusion.

Oxidation and resist The wafer is placed again in the oxidation furnace, and a new layer of SiO_2 is grown. Photoresist is also applied (Fig. 9.4*g*). The wafer is now ready for exposure to the base mask.

Exposure to base mask The base mask of Fig. 9.7*b* is now exposed to UV. The unexposed regions are etched away, and the result is an opening of a "window" for base diffusion (Fig. 9.4*h*).

To ensure perfect alignment between the base mask and the wafer, a *mask aligner* is used. With this equipment, the operator looks through a microscope and adjusts the position of the mask so that the tee of the base mask aligns with the tee on the wafer from its exposure to the collector mask.

Base diffusion A p-type impurity is diffused through the window to form the base of the transistor (Fig. 9.4*i*).

Oxidation and resist A new layer of oxide is grown and photoresist applied (Fig. 9.4*j*).

Forming the emitter The emitter is formed by using a very high concentration of an n-type dopant, denoted by n+. (If p+ was indicated, this would mean a very high concentration of a p-type dopant.) Further, to ensure a *nonrectifying* or, *ohmic, contact* to the n-collector region, n+ doping is also used.

The connections between the diffused regions in a monolithic IC are generally made by a thin film of conducting material, such as aluminum. Aluminum is a p-type material. If a p material is in contact with an n material, a pn junction diode is formed. An n+ region in contact with aluminum, however, provides a nonrectifying contact.

Exposure to n+ mask The mask of Fig. 9.7*c* has cutouts for the emitter region and for a contact to the collector region. The mask is exposed to UV, and, as in previous processing steps, the unexposed regions are etched away (Fig. 9.4*k*).

n+ diffusion A high concentration of n-type dopant, such as phosphorus, is diffused through the windows to form the n+ regions (Fig. 9.4*l*).

Oxidation and resist A new layer of oxide is grown and photoresist applied (Fig. 9.4*m*).

Exposure to contact mask Windows must now be opened to permit electric connections to be made to the emitter, base, and collector regions of the transistor. The contact mask of Fig. 9.7*d* is exposed to UV. The unexposed regions are etched away, and the resist is stripped from the wafer (Fig. 9.4*n*).

Metallization To provide connections between the diffused regions, a metallic

Fig. 9.12 An example of a prober. (*Courtesy Electroglas, Inc.*)

material is deposited over the entire wafer. This process is called *metallization*. Typically, aluminum is used and is deposited as a thin film on the substrate, which is accomplished in a vacuum. (Vacuum deposition of materials is considered later in the chapter.)

Exposure to interconnection mask The deposited aluminum is covered with photoresist and exposed to the interconnection mask of Fig. 9.7*e*. The unexposed regions are etched away, and the resist is removed from the wafer. The result is a diffused (integrated) transistor (Fig. 9.4*o*).

Testing and packaging The integrated circuit is tested and packaged. In testing, a *prober* (Fig. 9.12) is used. The wafer is vacuum-held and, under a microscope, fine points are brought into contact with the diffused regions. Through external electric connections to the points, voltages are impressed across the regions and measurements of device performance are made.

In summary, five different masks and three diffusions were required in the fabrication of a monolithic transistor. As we shall see, the same number of masks and diffusions are all that is generally required in the monolithic integration of a complete circuit. This feature is what makes monolithic technology so attractive for large-volume production of integrated circuits.

9.4 ELECTRICAL ISOLATION

In the monolithic IC all circuit elements, such as transistors and resistors, share the same silicon chip. It is necessary, therefore, that the elements be *electrically isolated* from one another. At present, three methods are used for isolation:

1. Junction
2. Dielectric
3. Beam lead

JUNCTION ISOLATION We have seen how a single transistor is integrated. To illustrate junction isolation, consider the two transistors in a silicon chip shown in Fig. 9.13. Note that the p substrate is returned to a negative potential and the n regions to a positive potential. Because the n and p regions form pn junctions, they are *reverse-biased*, similar to a back-biased semiconductor diode. For this condition, no current flows between the n-collector regions through the p substrate, and electrical isolation is obtained.

DIELECTRIC ISOLATION An example of two transistors dielectrically isolated is illustrated in Fig. 9.14. Surrounding each n-collector region and, separating it from the substrate, is a layer of silicon dioxide. Because SiO_2 is an insulator, electrical isolation is realized. The substrate does not have to be single-crystal silicon; instead, polycrystalline silicon is used.

BEAM-LEAD ISOLATION In this method of isolation, the circuit elements are diffused as in the junction-isolated IC. The metallization deposited on the chip, however, is very thick (in the order of 1 mil). The superfluous silicon of the p substrate is re-

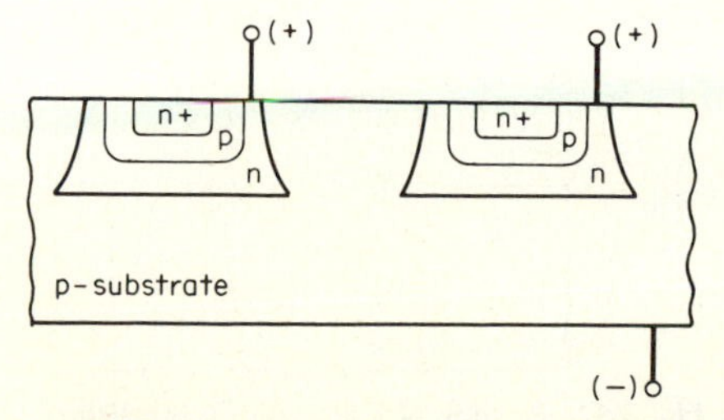

Fig. 9.13 Junction isolation for two diffused transistors.

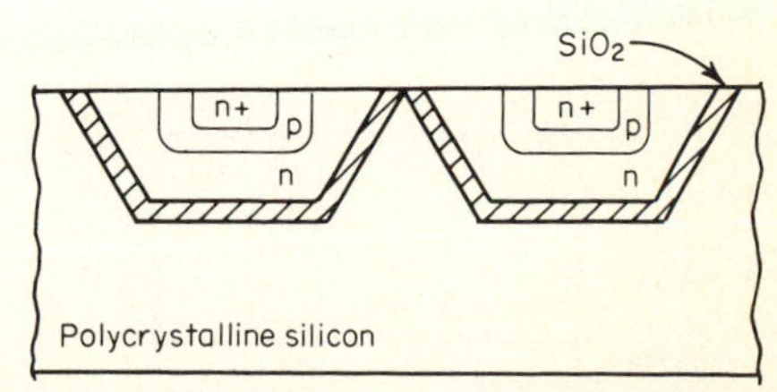

Fig. 9.14 Dielectric isolation for two diffused transistors.

moved. The resulting structure is a semirigid connected circuit with all elements physically separated from one another.

Owing to lower production costs, most monolithic circuits use junction isolation.

PARASITICS A problem with junction isolation is the existence of parasitics. A parasitic may be defined as an *undesired element*. Referring to Fig. 9.13, we saw that

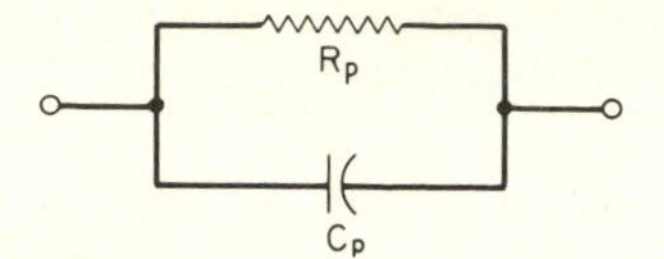

Fig. 9.15 Parasitic elements existing across a reverse-biased pn junction.

isolation was obtained by reverse-biasing the n-collector and p-substrate regions. In reverse-biasing a pn junction, a capacitance C_p, in parallel with a high-valued resistance R_p, exist as illustrated in Fig. 9.15. These two elements are examples of parasitics and, at very high frequencies, may create problems. Parasitics also exist for diffused resistors and capacitors in the junction-isolated IC.

9.5 PROPERTIES AND CHARACTERISTICS OF DIFFUSED ELEMENTS

BIPOLAR JUNCTION TRANSISTOR (BJT) A diffused BJT has typical values of short-circuit current gain h_{fe} as high as 300 and gain-bandwidth product f_T as high as 500 MHz. Because the gain of a junction transistor depends on the width of its base, the short-circuit current gain can be increased by reducing the base thickness. Transistors with very narrow base widths, exhibiting short-circuit current gains as high as 5000, are called *super-beta* and *punch-through* transistors. Owing to the narrow base width, the collector-emitter voltage for this type of transistor is limited to about 0.5 V. For a conventional BJT, the collector-emitter voltage is in the order of 20 to 30 V.

The gain-bandwidth product may be increased by decreasing the resistance of the collector region. A method for realizing this, without affecting appreciably the collector-emitter voltage rating of the transistor, is shown in Fig. 9.16. A heavily doped region, called an *n+ buried layer*, is diffused between the bottom of the n-collector region and the p substrate. Because the n+ layer is in parallel with the n-collector region, the effective collector resistance is reduced.

LATERAL pnp TRANSISTOR The diffusion of npn transistors is very compatible with the monolithic process. There are occasions, however, when a pnp transistor is needed in addition to an npn device. A structure that requires no additional diffusion steps is the *lateral pnp transistor* of Fig. 9.17.

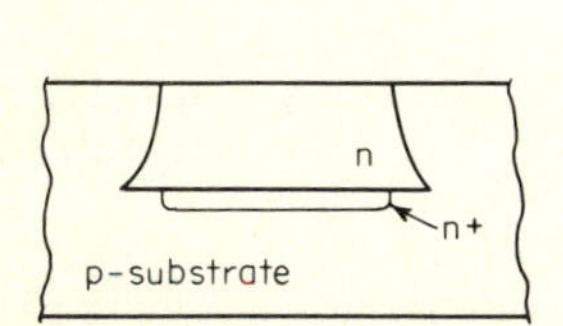

Fig. 9.16 An n^+ buried layer for reducing collector resistance.

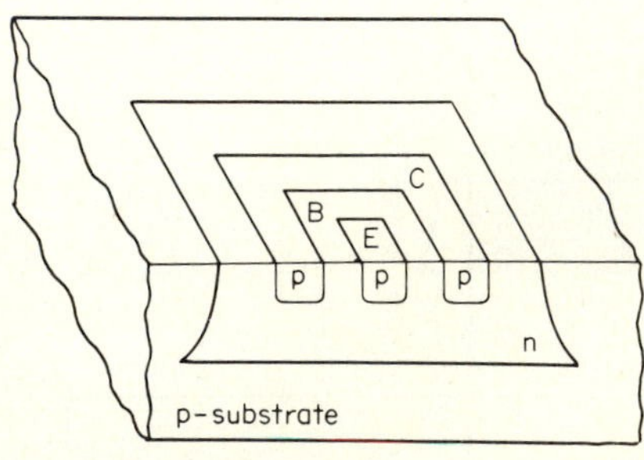

Fig. 9.17 A view of a lateral pnp transistor.

Two p-type regions are diffused in the n island (tub). One of the p regions serves as the emitter and the other as the collector. Because the effective base width is greater than that of an npn transistor, the short-circuit current gain and gain-bandwidth product of a lateral pnp transistor are appreciably less. Typical maximum values are $h_{fe} = 20$ and $f_T = 2$ MHz.

FIELD-EFFECT TRANSISTOR (FET) In the integration of field-effect transistors, the enhancement-mode MOSFET (IGFET) is invariably chosen because

1. It is self-isolating.
2. It can be formed by a single diffusion.
3. Because it is self-isolating, it has a much greater packing density than the BJT.

A cross-sectional view of a p-channel enhancement-mode MOSFET is shown in Fig. 9.18. Two p-type regions are diffused simultaneously in the n substrate. The gate is formed by depositing aluminum over the oxide layer between the p regions. In operation, the substrate is returned to the most positive potential in the circuit. The source, gate, and drain are so biased that they are negative with respect to the substrate. Thus, reverse-biased pn junctions are formed and electrical isolation is achieved.

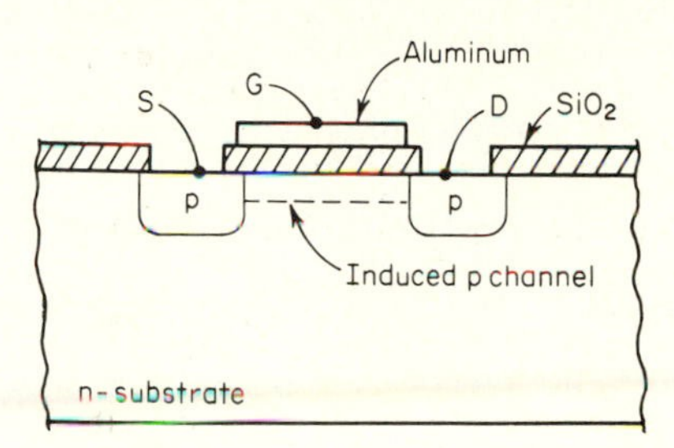

Fig. 9.18 A cross-sectional view of a p-channel enhancement MOSFET.

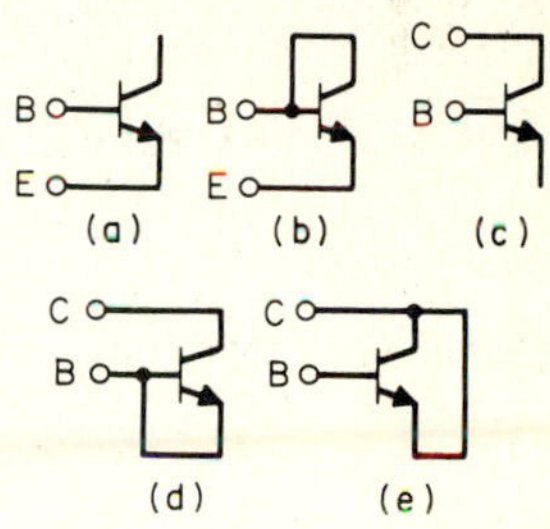

Fig. 9.19 A transistor may be connected in five ways to act like a diode (see text).

DIODE By utilizing only two terminals of a diffused npn transistor, five different diode configurations may be realized. Referring to Fig. 9.19, these are:

a. Base-emitter junction diode; collector is open. Series diode resistance and reverse breakdown voltage are low.

b. Base-emitter junction diode; collector short-circuited to base. Series diode resistance and reverse breakdown voltage are low.

c. Base-collector junction diode; emitter open. Series diode resistance and reverse breakdown voltage are high.

d. Base-collector junction diode; emitter short-circuited to base. Series diode resistance and reverse breakdown voltage are high.

e. Base-collector junction diode; emitter short-circuited to collector. Series diode resistance is high, and reverse breakdown voltage is low.

A series diode resistance which is low is less than 50 Ω; a high series diode resistance is greater than 50 Ω. Low reverse breakdown voltage is in the order of 7 V; high reverse breakdown voltage is about 40 V. Because of its good properties, the diode of Fig. 9.19*b* is commonly used.

RESISTOR Integrated resistors are generally formed by the diffusion of p-type material in an n tub, as illustrated in Fig. 9.20. This diffusion is done *simultaneously* with the diffusion of the p-base region of an npn transistor.

Resistance, R (ohms), of a material is expressed by

$$R = \rho\left(\frac{L}{A}\right) \tag{9.1}$$

where ρ = resistivity of material, Ω-cm; L = length of material, cm; and A = cross-

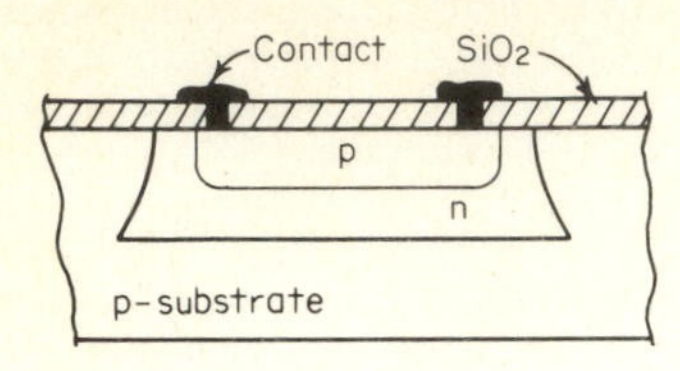

Fig. 9.20 A cross-sectional view of a p-diffused resistor.

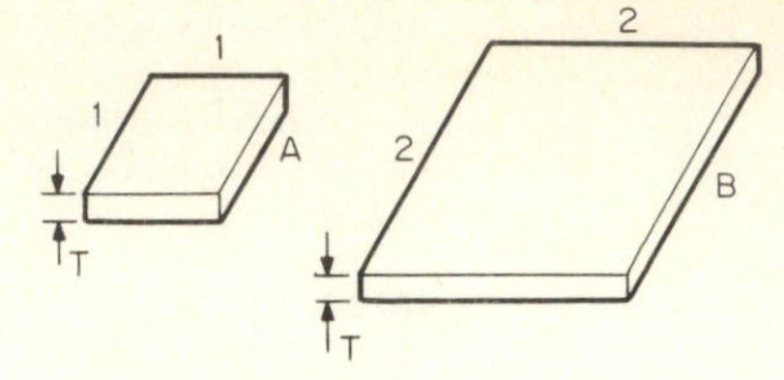

Fig. 9.21 The resistance measured across any two parallel edges is the same for either square.

sectional area of material, cm^2. The area may be expressed by the product of its width W and its thickness, T: $A = WT$. Substitution of this value for A in Eq. (9.1) yields

$$R = \frac{\rho}{T}\frac{L}{W} \tag{9.2}$$

If $L = W$, we have a *square* of material, and Eq. (9.2) reduces to

$$R = \frac{\rho}{T} \tag{9.3}$$

Assuming that thickness T is constant, which is reasonable, then the ratio ρ/T is also constant. Letting this quantity be denoted by R_S, the *sheet resistance*, we obtain

$$R = R_S \text{ ohms/square} \tag{9.4}$$

For any length and width of a diffused resistor,

$$R = R_S\left(\frac{L}{W}\right) \tag{9.5}$$

where L/W is the *aspect ratio*.

Consider the two square areas of Fig. 9.21. Both squares have the same thickness T. Although the surface area of square B is four times as large as that of square A, the resistance measures across their edges are the same and equal to the sheet resistance of the material R_S.

The last statement can be verified by noting that for square A, the cross-sectional area is $T(1) = T$. For square B, the area is $T(2) = 2T$. The length of A is 1 unit, and of B, 2 units. Hence, the length to cross-sectional areas are equal: $1/T = 2/2T = 1/T$. The resistance for each square is identical.

example 9.3 For a diffused resistor of 2000 Ω, $R_S = 200$ Ω/square. Determine the aspect ratio and, if $L = 20\ \mu$, the width of the resistor.

solution From Eq. (9.5), $R/R_S = L/W = 2000/200 = 10{:}1$. If $L = 20\ \mu$, $W = 20/10 = 2\ \mu$.

The practical range of p-diffused resistors is from 20 Ω to 30 kΩ. Their absolute tolerance, however, is poor, being in the order of ±25 percent. The tolerance of the *ratio* of two diffused resistors can be as low as ±2 percent. Resistors made with n+ diffusion are seldom used.

FOUR-POINT PROBE The four-point probe of Fig. 9.22 permits a simple measurement of the sheet resistance. A dc voltage source in series with a resistor, which simulates a *current source*, is connected to the outer points in series with a dc ammeter, I. The points are spaced about 0.1 cm apart and make physical contact with a silicon specimen whose sheet resistance is to be determined. The inner points are connected to a dc voltmeter which reads the voltage V developed across the points. Sheet resistance, in ohms per square, is given by

$$R_S = 4.5\frac{V}{I} \tag{9.6}$$

CAPACITOR Capacitors found in a monolithic IC are basically of two kinds: *junction* and *thin film*. A junction capacitor may be realized by using an npn transistor

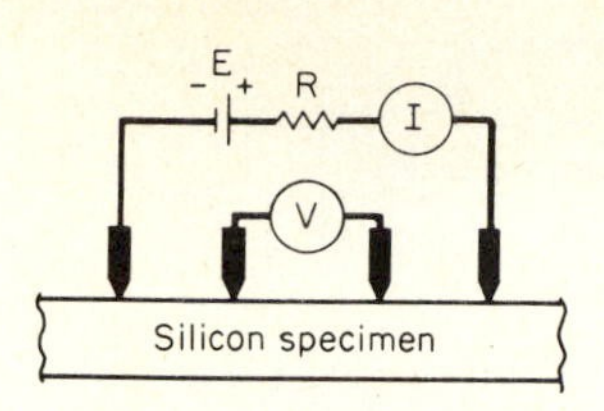

Fig. 9.22 A four-point probe used for measuring sheet resistance.

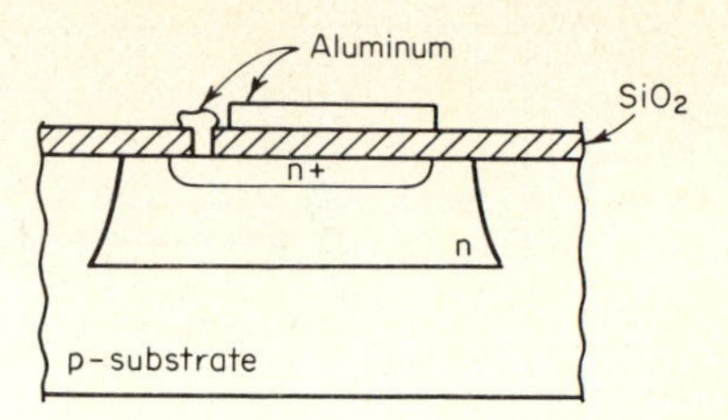

Fig. 9.23 A cross-sectional view of a thin-film monolithic capacitor.

connected as a diode (Fig. 9.19). Application of a reverse voltage across any two terminals results in a variation in capacitance: the greater the reverse voltage, the less is the capacitance.

A cross-section view of a thin-film monolithic capacitor is shown in Fig. 9.23. An n+ region, diffused in the n tub, acts as one plate of the capacitor. A growth of silicon dioxide serves as the dielectric, and the deposition of aluminum over the oxide acts as the other plate of the capacitor.

Capacitors occupy a relatively large area on a chip and are limited to small values of capacitance, typically 200 pF. For these reasons, then, they are avoided wherever possible.

INDUCTOR Currently there are no practical methods for diffusing inductors in a monolithic IC. Inductors, therefore, are avoided in the design of monolithic integrated circuits.

9.6 MAKING A MONOLITHIC IC

In this section we shall see how a circuit is transformed into a monolithic IC. Because it is most common, junction isolation will be assumed. In determining how many isolation tubs are required, the following guidelines are offered:

1. Transistor collectors at different potentials must be in separate isolation tubs.
2. Resistors connected to circuit points at different potentials can share the same isolation tub. The n tub should be connected to the most positive voltage available in the circuit.
3. Diffused capacitors generally require individual tubs.
4. Base-emitter diodes can generally share the same isolation tub.

Consider the two-stage direct-coupled amplifier of Fig. 9.24, which is to be integrated. Because the collector potentials of Q_1 and Q_2 are different, each transistor requires a separate isolation tub. Resistors R_1 and R_2 can share one tub. Therefore, three isolation regions are required.

A layout of the amplifier is given in Fig. 9.25*a*. Other layouts are also possible. A cross-sectional view of the integrated circuit is given in Fig. 9.25*b*. As in the integra-

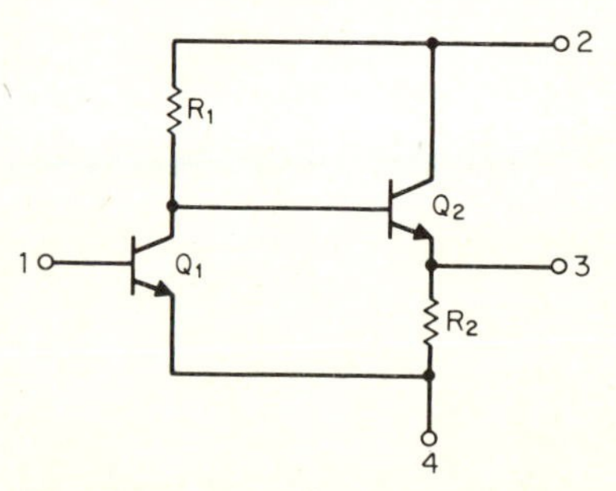

Fig. 9.24 A two-stage amplifier to be integrated.

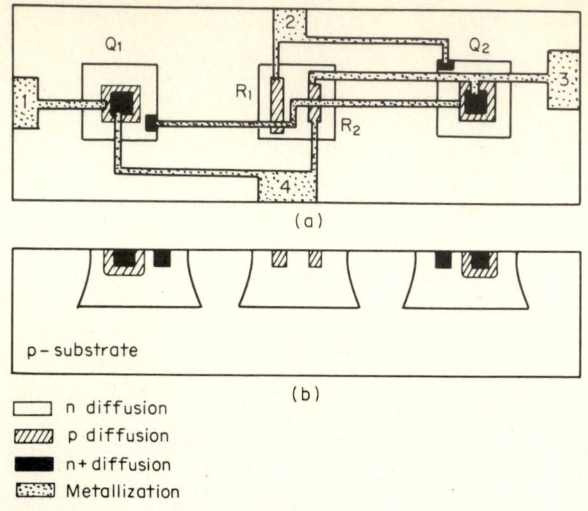

Fig. 9.25 Monolithic integration of the amplifier of Fig. 9.24: (*a*) Layout. (*b*) Cross-sectional view.

tion of a transistor, a p-type substrate with a grown n-epitaxial layer is the starting material. Also, five masks are required; referring to Fig. 9.26, these are:

Isolation mask Sufficient p-type impurity is diffused in the n-epitaxial region to ensure that the three n-isolation tubs are surrounded by p regions (Fig. 9.25*b*). The opaque areas of the isolation mask prevent UV from reaching the photoresist on the silicon wafer; consequently, windows are only opened for the diffusion of the p-type impurity.

p-diffusion mask This mask has windows for the transistor bases and resistors. The diffusion of the base and resistors occurs simultaneously.

n+ diffusion mask Windows are cut out for the diffusion of emitters and regions in n-type material to ensure nonrectifying contacts.

Contact and interconnection masks The contact mask has cutouts for connections, and the interconnection mask has the connection pattern for the diffused elements.

As for the integration of a transistor, described earlier in the chapter, oxidation, application of photoresist, etc., are used in similar sequence for integration of a circuit. An example of a commercial monolithic IC is illustrated in Fig. 9.27.

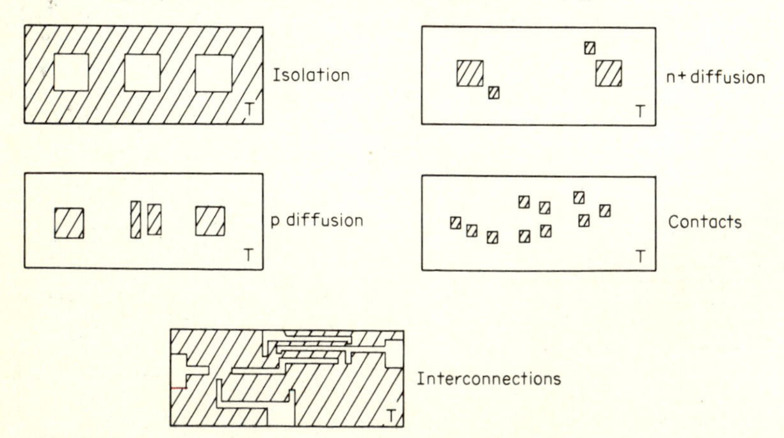

Fig. 9.26 Set of five masks required for the integration of the amplifier of Fig. 9.24 (see text).

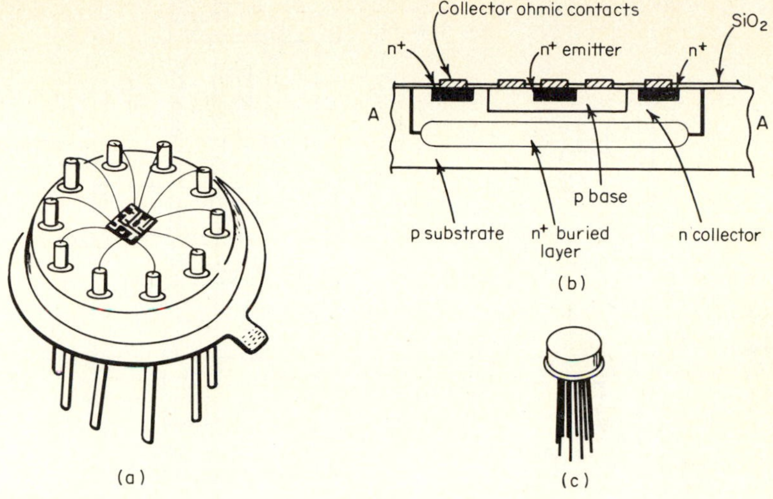

Fig. 9.27 An example of a commercial IC: (*a*) Constructional details. (*b*) Cross-sectional view of transistor. (*c*) Approximate size of package.

BURIED CROSSOVER In the layout of an integrated circuit it is found occasionally that two conducting paths, which cannot be eliminated, intersect each other. To correct this situation, a buried crossover, illustrated in Fig. 9.28, is required.

An n+ region is diffused in a separate n-isolation tub. Conductor 1 completes its path through the n+ buried region. Conductor 2, shown perpendicular to conductor 1, is insulated from the buried region by the SiO_2 layer.

example 9.4 The circuit of Fig. 9.29 is to be integrated. Draw a layout of the circuit, showing the required isolation tubs; the devices in each tub may be shown schematically. The numbering of terminals must be maintained in sequence on the periphery of the layout.

solution The layout, shown in Fig. 9.30, requires four isolation tubs. Because they share a common connection, the three diodes are placed in one tub. Owing to different collector potentials, each transistor requires its own isolation tub. The three resistors can share a single tub.

example 9.5 A magnified drawing of an integrated-circuit layout is given in Fig. 9.31. Draw a schematic diagram of the circuit, and identify all components and the values of resistors. The sheet resistance $R_S = 200\ \Omega$/square.

Assume that the transistor is saturated: that is, $E_{CE} = 0$ V. If the voltage across a forward-biased base-emitter junction or diode is 0.7 V, determine the base, emitter, and collector currents in the transistor.

solution The complete circuit is illustrated in Fig. 9.32. Resistance values were obtained by counting the number of squares in each p-diffused resistor in the isolation tub. Diode D is always reverse-biased. The base current, therefore, is $I_B = (6 - 0.7)/5.6 = 0.95$ mA. Col-

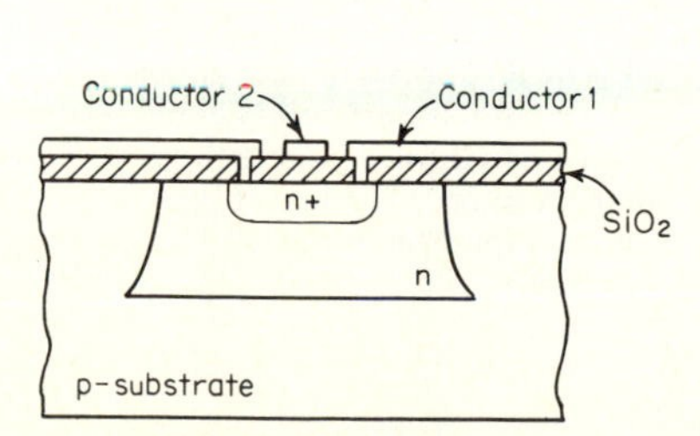

Fig. 9.28 A buried crossover.

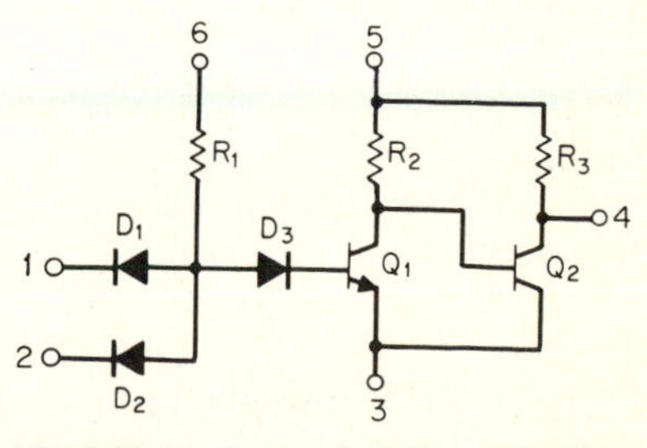

Fig. 9.29 A circuit to be integrated. (See Example 9.4.)

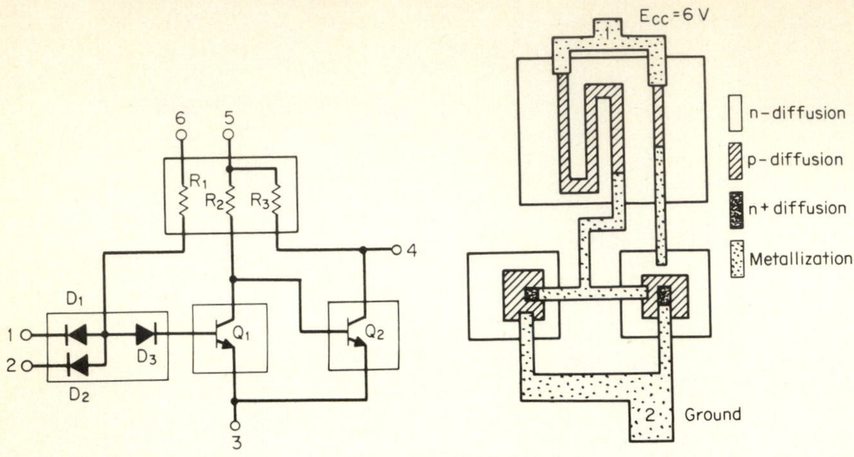

Fig. 9.30 Schematic layout of the circuit of Fig. 9.29.

Fig. 9.31 A given layout of an integrated circuit. (See Example 9.5.)

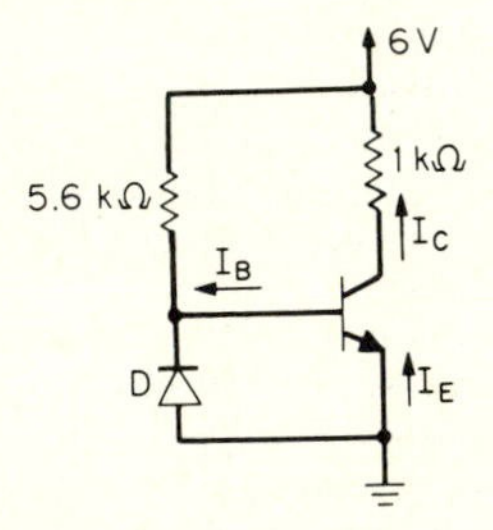

Fig. 9.32 Schematic diagram of circuit determined from the IC layout of Fig. 9.31.

lector current, $I_C = (6 - 0)/1 = 6$ mA. The emitter current I_E is equal to the sum of base and collector currents: $I_E = I_B + I_C = 0.95 + 6 = 6.95$ mA.

9.7 LARGE-SCALE INTEGRATION (LSI)

Large-scale integration (LSI) is an extension of monolithic technology where 100 or more interconnected circuits are in a single chip of silicon. (Integrated circuits containing fewer than 100 interconnected circuits are often referred to as *medium-scale integration*, MSI.) Because of such a large number, invariably some diffused components are defective. What is necessary, therefore, is the elimination of the defective units before the interconnections are made. One commonly used technique for accomplishing this is called *discretionary wiring*.

In discretionary wiring, a number of circuits, called *unit cells*, in excess of the actual number required, are integrated. Generally, LSI is best suited where identical circuits are used, such as logic gates and computer memories. Each unit cell is checked on an automatic tester. Coupled to the tester is a computer which remembers the good cells. The computer generates an interconnection pattern that includes only good cells. Because the location of defective cells will be different for other chips of the same type, the testing and computer routine is repeated for each chip.

9.8 THIN-FILM TECHNOLOGY

In thin- and thick-film technology, resistors, small-valued capacitors, and inductors are deposited and interconnected on a substrate, such as ceramic. Transistors, large-value capacitors, and even monolithic IC's may be connected to the substrate to form a *hybrid* integrated circuit. Thin films generally refer to a layer thickness of 5 angstroms (1 Å = 10^{-8} cm) to 10 μ (1 μ = 10^{-4} cm). The thickness of thick films is 10 μ or greater.

PRODUCING THIN FILMS In the thin-film process, the deposition is generally accomplished by *vacuum evaporation* or *cathode sputtering*. A basic vacuum-evaporation system is illustrated in Fig. 9.33. The bell jar is evacuated to a pressure in the order of 10^{-5} *torr* (1 torr = 1 mm of mercury). A tungsten element, in which is placed the material to be evaporated, and the substrate are mounted inside the bell jar. Approximately 20 to 25 A of current is passed through the element, and the material, such as aluminum, is melted and vaporized instantaneously and deposited on the substrate. To ensure good adhesion of the evaporated film, the substrate is sometimes heated in the bell jar.

Cathode sputtering is used when materials, such as tantalum, are not easily melted. An elementary sputtering system is illustrated in Fig. 9.34. The stand holding the substrate acts as the *anode*, and the material to be deposited as the *cathode* of a gas

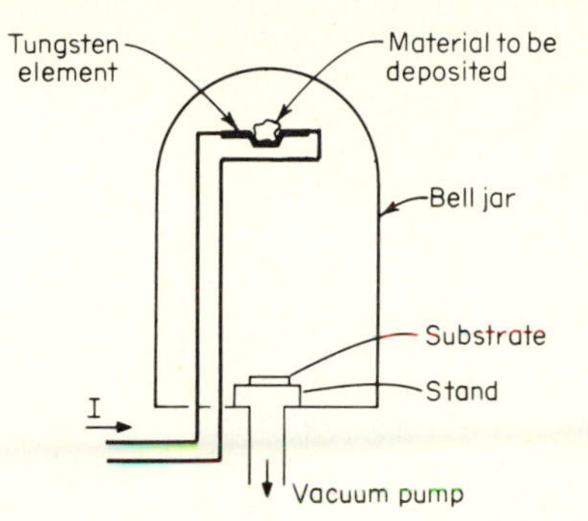

Fig. 9.33 A basic vacuum evaporation system for producing thin films.

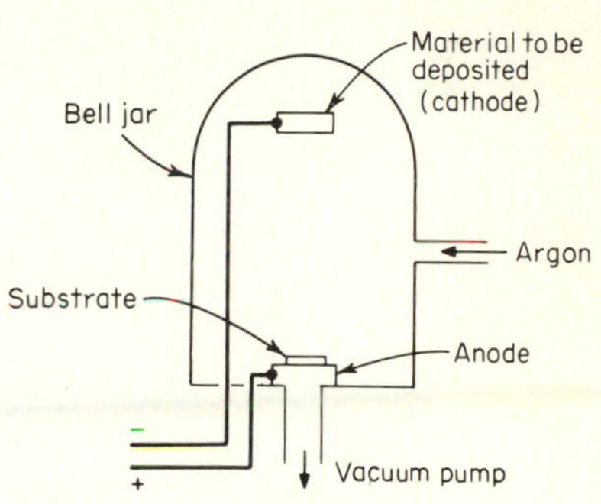

Fig. 9.34 An elementary cathode sputtering system.

discharge tube. The bell jar is initially evacuated, as in vacuum evaporation. Argon, or another inert gas, is introduced in the jar until the pressure increases to approximately 0.05 torr.

Upon the application of 4 to 6 kV across the anode and cathode terminals, a glow discharge is initiated. The argon atoms become positively ionized and are attracted toward the cathode (which is negative). When the ions strike the cathode, atoms of the material to be deposited are liberated and attracted by the substrate.

The thickness of a deposited film can be measured with a quartz crystal oscillator. A quartz crystal, which is connected to an oscillator circuit, is placed near the substrate in the bell jar. The material deposited on the substrate is also deposited on the crystal. Because of the deposition, the mass of the crystal is increased and its frequency of oscillation reduced. The actual frequency is calibrated to read directly film thickness.

THIN-FILM COMPONENTS The electrical characteristics of thin-film components depend primarily on the film material and the geometry of the deposited component.

Resistor Commonly used materials for resistors include Nichrome (R_S = 10 to 400 Ω/square) and tantalum nitride (R_S = 50 to 500 Ω/square). The higher resistance values are obtained with thin deposits of material, and the low values with thick deposits.

Nichrome, which is deposited by vacuum evaporation, exhibits a good temperature coefficient and good adhesion properties. Sputtering is used for the deposition of tantalum nitride. This material has a good temperature coefficient, is highly stable, and adheres well to the substrate.

By means of the folded geometry of Fig. 9.35 (which is also used for diffused resistors), resistance values in megohms may be realized that occupy less than 0.5 by

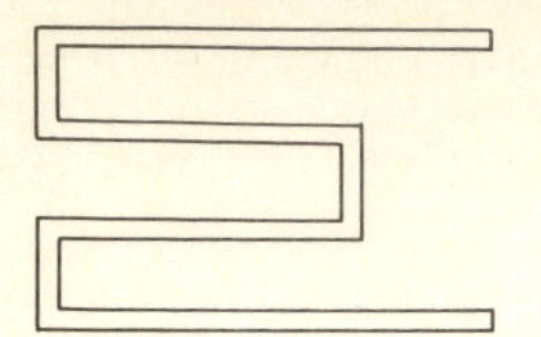

Fig. 9.35 A folded geometry resistor allows a large-valued resistor to occupy a relatively small substrate area.

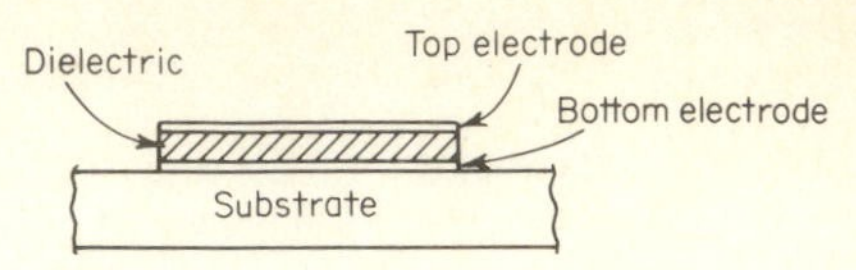

Fig. 9.36 A "sandwich" thin-film capacitor.

0.5 in on a substrate. The tolerance of thin-film resistors is in the order of 5 to 10 percent. With suitable trimming, such as laser trimming, tolerances better than one percent may be achieved. Thin-film resistors deposited on a glazed substrate can dissipate approximately one watt per square inch of substrate area.

Capacitor A thin-film capacitor is a "sandwich" of a bottom electrode, dielectric, and top electrode (Fig. 9.36). Generally, the top and bottom electrodes are evaporated aluminum. The dielectric used determines the value of capacitance. For example, if the deposited dielectric is a 1000-A layer of silicon dioxide, approximately 350 pF/mm^2 of surface area is realized.

The tolerance of thin-film capacitors before trimming is about 10 to 20 percent; after trimming, it is less than 5 percent. The quality factor Q seldom exceeds 200.

Inductor Thin-film inductors are generally deposited on the substrate in the form of a spiral. Because they occupy a large surface area, they are practical only for circuits operating in the gigahertz frequency range.

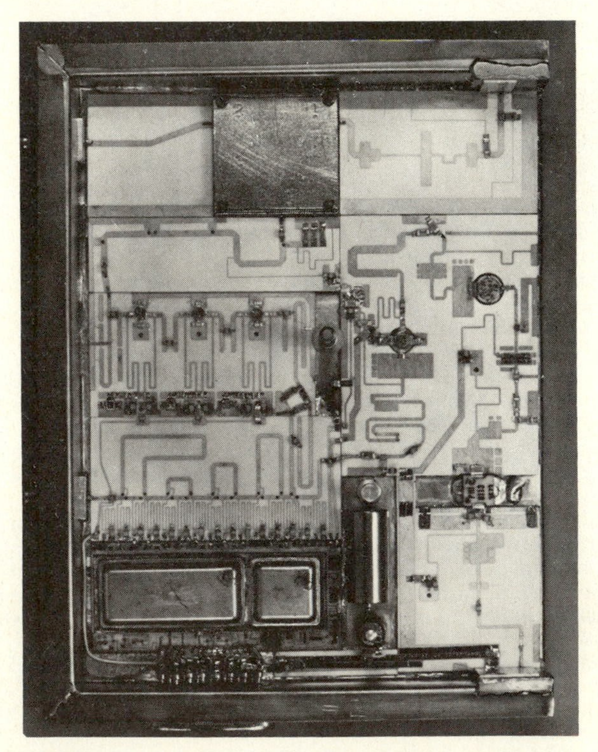

Fig. 9.37 An example of thin-film circuits used in a 1-GHz transceiver. (*Courtesy RCA*.)

THIN-FILM PROCESSING Many of the steps used in thin-film processing are similar to those used in monolithic technology. A layout of the circuit to be integrated is made. Four masks are generally required: resistor, bottom capacitor electrode, top capacitor electrode, and interconnection pattern masks.

The resistor material is deposited over the entire substrate. This is followed by a coating of photoresist which is exposed through the resistor mask to UV. The unexposed resist is dissolved, and the result is a resistor pattern on the substrate. Metal, such as aluminum, is now deposited over the entire substrate, and resist is applied. The resist is exposed to a mask containing the pattern for the bottom capacitor electrode. The same procedure is repeated for the top capacitor electrode, dielectric, and interconnection pattern.

Upon completion of the preceding steps, transistors, large-value capacitors and inductors, and occasionally monolithic IC chips are attached to the circuit. Transistors and capacitors for this purpose come in the form of a *flip chip*. A flip chip is a component having an upside-down mounting, making it relatively simple for attaching to a substrate. Finally, the completed hybrid circuit is tested and encapsulated. An example of a commercial thin-film circuit is shown in Fig. 9.37.

9.9 THICK-FILM TECHNOLOGY

Thick-film technology is a method for depositing resistors, capacitors, and interconnections on a substrate which is based on *screen-process printing*. One may think of the process as "painting on" components over a substrate. The advantages of thick- over thin-film technology are:

1. Lower cost of capital equipment.
2. Easier to train manufacturing personnel.
3. High production rates are realized.
4. Thick-film sheet resistance of 10 MΩ/square are possible.
5. Greater dissipation of thick-film resistors.

On the debit side, thick-film resistors are noisier, have poorer tolerance, and are wider than thin-film resistors.

SCREEN-PROCESS PRINTING The starting point for the process is a fine-mesh screen. Generally, a stainless steel screen mounted on an aluminum frame is used (Fig. 9.38). The *mesh count*, which equals the number of wires per inch of screen, comes in various sizes (80 to 400). Typically, a 200-mesh screen is used for resistors and a 300-mesh screen for conductors.

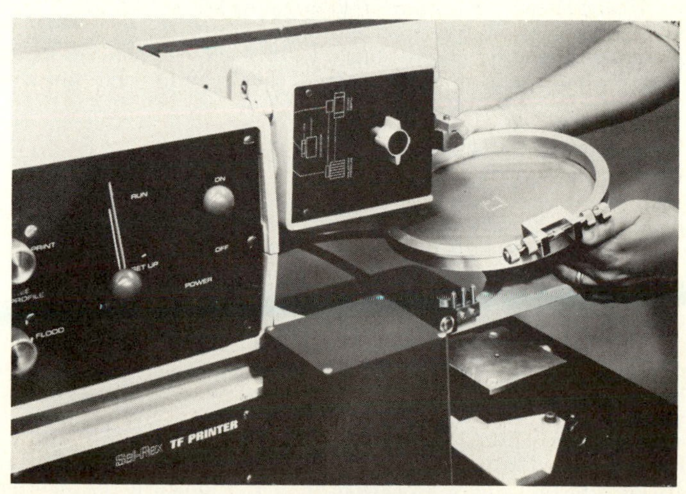

Fig. 9.38 An example of a round stainless steel screen being inserted in a screen printer. *(Courtesy The Sel-Rex Company.)*

Separate masks are needed for conductors, resistors (one for each different sheet resistance used), and dielectric for thick-film capacitors. The masks are reduced, step-and-repeated, and a positive is made of the pattern. The dark areas on a positive correspond to what is printed on the screen.

A number of methods are available for transferring the mask pattern to the screen (referred to as a *screen stencil*). The *direct-emulsion* method is most commonly used today. It has good life expectancy and is inexpensive and easy to prepare.

A direct-emulsion stencil is made by first applying a light-sensitive emulsion to the screen. After drying, a positive of the mask pattern is placed in contact with the emulsion and exposed to UV. The exposed regions are polymerized, and the unexposed regions are washed away. Areas on the screen which were polymerized are therefore sealed, and the remaining areas are opened.

To print the pattern, the substrate is positioned and secured on a printing platform (Fig. 9.39). The screen is then located parallel to and about 20 mils above the substrate. Paste (ink) is applied and a squeegee is moved at a constant rate across the screen. The paste is forced through the openings in the screen onto the substrate.

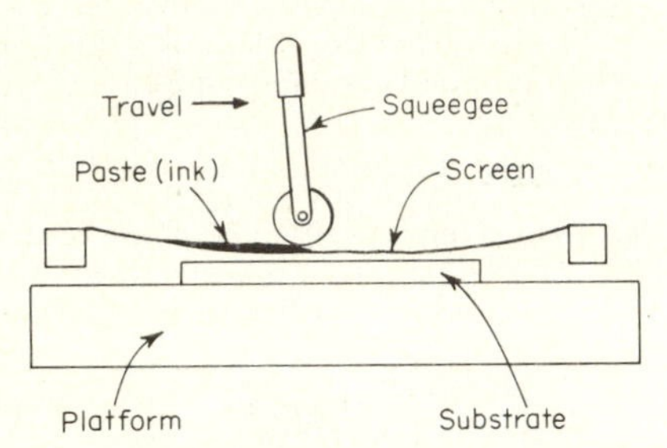

Fig. 9.39 Printing a screen pattern on a substance.

In practice, a *screen printer* is used to ensure uniform patterns on the substrate. An example of a commercial screen printer is shown in Fig. 9.40.

After printing, the substrate remains at room temperature for a few minutes to allow the paste to level and even out impressions left by the screen. It is then dried for approximately 10 min at 100 to 150°C to drive volatile solvents out of the paste. Finally, the substrate is placed in a furnace between 700 and 1000°C for 30 min for *firing*. In addition to burning off the organic binder (filler) in the paste, firing forces the particles in the deposited conductors and resistors into a solid mass.

Upon the completion of firing, trimming may be used to tighten resistance values. Transistors and other chips are then connected to the circuit, as in the thin-film process, and the completed circuit is tested and encapsulated.

THICK-FILM PASTES (INKS) Pastes, or inks, for thick-film circuits are a blend of three materials:

1. Base material for conductor, resistor, or dielectric
2. Organic binder to carry the ink through the mesh openings
3. Volatile solvent to control the flow (viscosity) of the paste

Because they have low resistivity and do not oxidize, noble metals are used in pastes for conductors. Examples of conductive pastes are silver, palladium-silver, gold, and palladium-gold.

Basically, a mixture of finely divided glass and metal forms a resistor after firing. Four families of resistor pastes and their firing temperatures commonly used are:

1. Platinum: 980°C
2. Ruthenium oxide: 760°C
3. Palladium-silver: 760°C
4. Thallium oxide: 580°C

Sheet resistances of available pastes range from 1 Ω/square to 10 MΩ/square. By proper blending of material, nearly any sheet resistance can be obtained. Manufacturers of resistive pastes provide *blending curves* for this purpose.

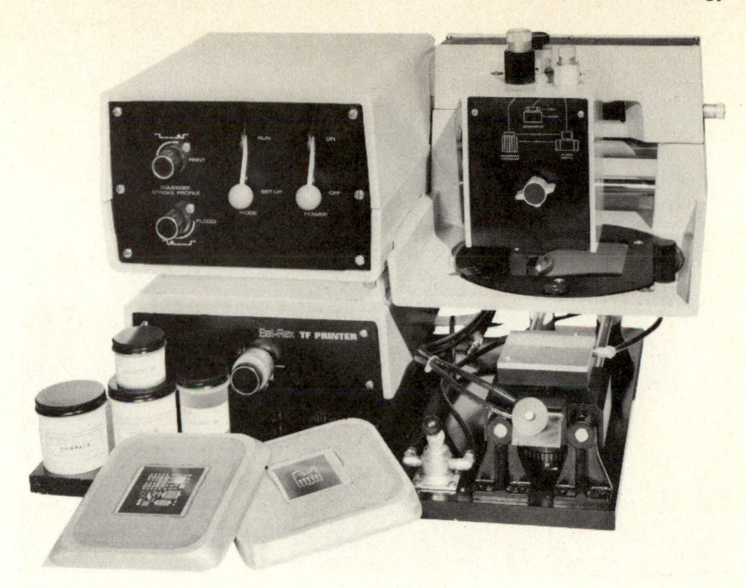

Fig. 9.40 An example of a commercial screen printer. Also shown are various pastes and rectangular screens. (*Courtesy The Sel-Rex Company.*)

The geometry of a thick-film capacitor is similar to that of the thin-film capacitor of Fig. 9.36. The lower electrode is screened on the substrate. A dielectric paste, having a high glass content, is then screened on top of the lower electrode. Finally, the screening of the upper electrode over the dielectric completes the capacitor.

THICK-FILM COMPONENTS Resistance values range from a few ohms to tens of megohms. Power dissipation may be as high as 25 W/in^2 of surface. Tolerance before trimming is in the order of ± 20 percent; after trimming, better than 1 percent.

The useful range for thick-film capacitors is 10 to 300 pF; tolerance is ± 20 percent. As in thin-film circuits, thick-film inductors are not too practical.

Chapter 10

Tuned Circuits

10.1 INTRODUCTION

In previous chapters, the subject of *reactance* was discussed. It was noted that inductors (*coils*) have a quality that is called *inductive reactance*, the magnitude of which varies with the applied frequency. It was also noted that capacitors have a quality called *capacitive reactance*, the magnitude of which also varies as a function of frequency.

Of particular importance is the fact that inductive reactance, which is generally considered as being *positive* reactance, is vectorially opposite to capacitive reactance, generally considered as being *negative* reactance. That is, when they are combined in a circuit, inductive reactance tends to *cancel* capacitive reactance. Also, inductive and capacitive reactance behave oppositely with respect to changes in applied frequency. For a fixed value of inductance, inductive reactance *increases* as the applied frequency is increased, but the capacitive reactance of a fixed value of capacitance *decreases* as the applied frequency is increased.

The reference to inductive reactance as being positive and capacitive reactance as being negative should not be misinterpreted. In an inductive circuit, the voltage leads the current and, in a capacitive circuit the voltage lags behind the current. Therefore, the effect of an inductor is opposite to the effect of a capacitor. As a convenient way of distinguishing between the effects of the two reactances, we call one positive and the other negative.

Another way of distinguishing between the effects of inductive reactance and capacitive reactance is to represent them graphically as shown in Fig. 10.1. This illustration shows that the effects of X_L and X_C are in opposite directions. If the values of X_L and X_C are equal, their effects cancel, leaving only resistance in the circuit. If the three quantities shown in Fig. 10.1 (X_L, X_C, and R) are added vectorially, the resultant will be the circuit *impedance*. By definition, impedance is the combined opposition to the flow of alternating current that is offered by the inductance, capacitance, and resistance in that circuit.

The qualities of inductive and capacitive reactance give rise to an important area of applications in what are referred to as tuned circuits. The name *tuned circuits* generally refers to a combination of inductive and capacitive reactances arranged so that the impedance of the combination changes abruptly at some specific frequency, or narrow range of frequencies. Tuned circuits are made with inductors and capacitors. As with any circuit, they also have resistance. Resistance is frequently useful in establishing the desired performance characteristics of a tuned circuit, and it is present in all circuits. The frequency at which the tuned circuit exhibits its sharp change in impedance is referred to as the *resonant frequency*. There is also a term *antiresonant frequency* which, in some applications, should be used instead of the term resonant frequency

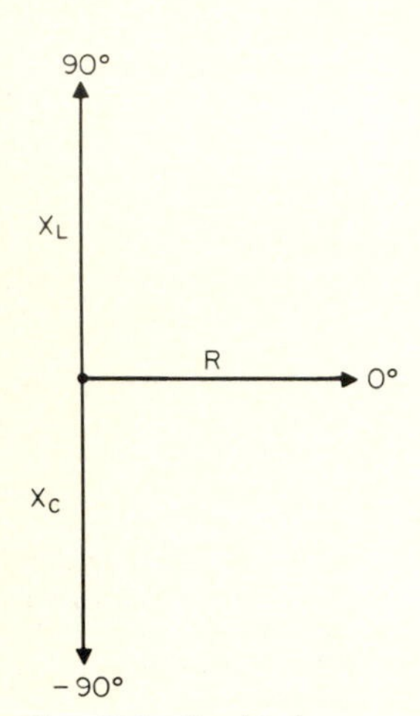

Fig. 10.1 Graphical representation of X_L, X_C, and R.

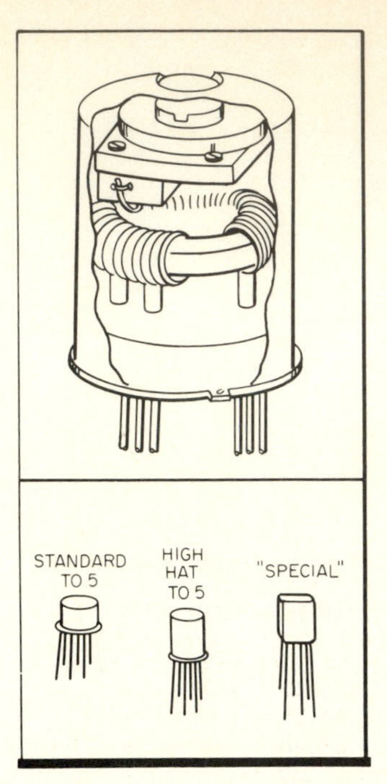

Fig. 10.2 A miniaturized tuned circuit, used for printed-circuit boards. In the upper view the microminiature, variable ceramic capacitor is at the top. Beneath this is the inductor, wound on a powdered iron toroidal core. (*Courtesy JFD Electronics Corp.*)

when referring to certain types of tuned circuits. There is a fine distinction between the terms which will be covered later in this chapter.

ANALOGIES There are many analogies between electronic tuned circuits and other physical phenomena. The relationship between the length and period of swing for a pendulum is one example. A tuning fork is another good example of a tuned mechanical device. A tuning fork will vibrate at only one frequency, regardless of how hard it is struck, or how it is set into vibration. In a like manner, an electrical tuned circuit will operate at only one frequency, or at only one narrow band of frequencies.

SOME USES OF TUNED CIRCUITS Tuned circuits are an extremely important factor in the overall art and science of electronics, as they make it possible to perform many functions that would otherwise be either impossible or extremely difficult. These functions include *selective amplification of certain frequencies* (or bands of frequencies) without amplifying other unwanted frequencies, and *rejection of unwanted frequencies* or bands of frequencies. Tuned circuits are utilized in radio broadcast receivers as the means of selecting the one desired station to be received from among the large number of stations that are simultaneously delivering signals to the receiver. Tuned circuits are employed in radio transmitters to establish the transmitter carrier frequency. In television receivers they are used to select the desired channels. In electronic circuits there are many other applications.

The importance of tuned circuits in the overall field of electronics can hardly be overestimated. It can safely be said that, were it not for tuned circuits, the art and science of electronics as we know it today would not exist. Figure 10.2 shows an example of a tuned circuit so small that it can be packaged into a transistor case. Much larger tuned circuits are used in transmitters where much larger powers are involved.

10.2 TYPES OF TUNED CIRCUITS

There are many different methods of classifying tuned circuits. For example, they may be classified as *tunable* or *fixed*—depending on whether the frequency of the tuned circuit can be varied, or whether it is always set at the same frequency. They may be called *narrow-band* or *wideband*—depending on the shape of their characteristic curves. They are referred to as being either *series-tuned* or *parallel-tuned* circuits—according to whether their components are series- or parallel-connected. Further, a given tuned circuit can be classified according to more than one of the several different methods of distinction. That is, a tuned circuit may be a tunable, narrow-band, parallel-tuned circuit.

A few examples of commonly used tuned circuits are illustrated in Fig. 10.3. The most often used circuit consists of a variable capacitor either in series or in parallel with an inductor (Fig. 10.3*a* and *b*). The minimum and maximum values of capacitance for the variable capacitor are chosen to ensure complete coverage of a band of frequencies. Where a number of different frequency bands must be tuned, as in a shortwave receiver, a switchable inductor may be used (Fig. 10.3*c*). If, for example, the switch is in position one, the capacitor and resulting inductance covers the lower-band (broadcast) frequencies. In position two or three, less inductance is present across the capacitor; for these positions the circuit is tuned to the shortwave frequencies.

The distinction between *fixed* (sometimes called *fixed-tuned*) and *tunable* (sometimes called *variable*) circuits refers to whether the circuit is tuned permanently to a single (*fixed*) frequency, or whether the frequency at which the circuit exhibits its unique impedance-response characteristic can be altered, or varied, according to the needs and desires of the user.

The many different methods of classifying tuned circuits, and the fact that a given tuned circuit can fit simultaneously into more than one classification, may seem confusing at first. However, there is a reason for each of the different methods of classification, and each has its own importance. This will become further apparent after the circuits have been discussed in greater detail.

FIXED-TUNED CIRCUITS Fixed-tuned circuits are set permanently to a single frequency. They find many applications in electronic equipment where it is desired to respond to a known, predetermined frequency, either by enhancing the equipment response to that particular frequency as compared to the response to other frequencies, or by producing an especially *low* response to (rejecting) the frequency to which the circuit is resonant.

Some nominally fixed-tuned circuits may, however, be slightly *adjustable*. The term *adjustable* generally refers to the fact that one or more of the reactive elements of the tuned circuit can be varied (adjusted) over a narrow range. This adjustment is most often provided for the purpose of compensating for minor variations or imperfections

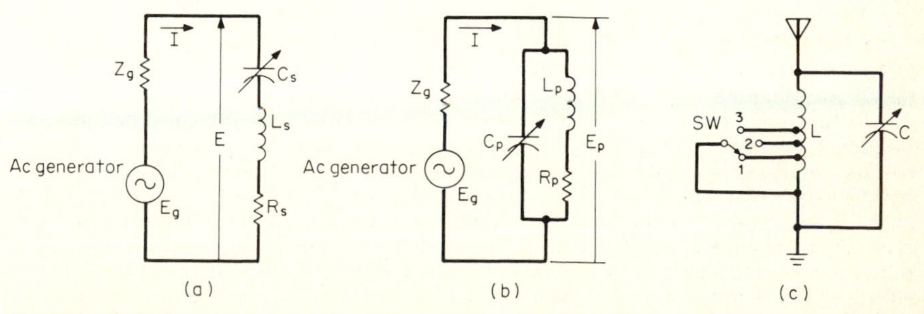

Fig. 10.3 Examples of tuned circuits: (*a*) A series-tuned circuit. (*b*) A parallel-tuned circuit. (*c*) Tuned circuit used in radio receiver circuit. Switch permits the receiver to be tuned to more than one band of frequency.

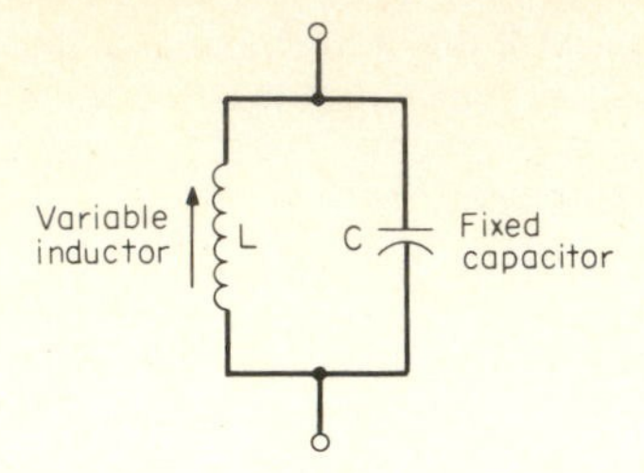

Fig. 10.4 A tuned circuit in which the resonant frequency is adjusted with a variable inductor.

of the circuit elements (or in some cases, minor variations in frequency of the input signal that is desired to be tuned). The frequency of the tuned circuit in Fig. 10.4 is adjusted with a slug-tuned inductor.

TUNABLE CIRCUITS Tuned circuits which are slightly adjustable circuits should not be confused with tunable circuits, as the latter have a very different kind of application. In general, tunable, or variable, circuits are used in applications in which it is desirable to be able to change the frequency for which the circuit exhibits its characteristic *selective* response. Such changing may be done frequently and with relative ease. Perhaps the most common example of the use of tunable circuits is in radio broadcast receivers, in which tunable input circuits can be varied by the user to select the particular one station to be received out of the many that are broadcasting simultaneously on different frequencies.

CIRCUIT BANDWIDTH Another method of classifying tuned circuits is as "narrow-band" or "wideband"—whether the circuit responds only to frequencies within a very narrow range or *band* about the resonant frequency, or whether the response is to frequencies in a relatively wide band about the resonant frequency. (The terms *narrow band* and *wide band* are, of course, very general, and more sophisticated methods for describing the bandwidth of tuned circuits are needed in most practical situations.)

The fact that a tuned circuit is narrow-band or wideband may sometimes be a matter of necessity brought on by the practical limitations of the components with which the equipment designer must work. More often, however, it is a matter of deliberate design to fit the requirements of the application. In a radio broadcast receiver, for example, the tuned circuits that determine which of the many available stations is actually received are narrow-band circuits designed so that the one desired station can be received clearly, without interference such as might be produced by other stations broadcasting on frequencies near that of the desired station.

In many applications, on the other hand, including television broadcast receivers, radar receivers, and other equipment that must handle a wide band of signals, the tuned circuits are deliberately made so as to respond effectively to signals within a relatively wide band on both sides of the center frequency to which the tuned circuit is resonant.

SINGLE-TUNED AND DOUBLE-TUNED CIRCUITS The distinction between single-tuned and double-tuned (or more generally, multiple-tuned) circuits refers to certain details of tuned-circuit configuration and/or adjustment. Generally speaking, the distinction concerns whether the tuned circuit is "resonant" to only one frequency (single-tuned), or whether it is resonant *simultaneously* to two or more frequencies (double- or multiple-tuned).

Multiple tuning of circuits may be done for a variety of reasons, but most often it is a means of obtaining wideband response without sacrificing other important circuit performance characteristics. One of these other performance characteristics is *selectivity*, which is a measure of the ability of a tuned circuit to respond to a desired frequency or band of frequencies, while *rejecting* frequencies outside this band, even though close to it.

SERIES-TUNED AND PARALLEL-TUNED CIRCUITS Perhaps the most important single method of classifying tuned circuits is based on the circuit configuration. As we shall see in the next section, a tuned circuit with the components connected in series may have a completely different function from one with the components connected in parallel.

10.3 COMPARISON OF SERIES-TUNED AND PARALLEL-TUNED CIRCUITS

The terms "series circuit" and "parallel circuit" refer to the arrangement of the capacitive and inductive elements that make up the tuned circuit. The distinction can readily be understood with the aid of Fig. 10.3. Figure 10.3*a* shows a simple series-tuned circuit, and Fig. 10.3*b* shows a parallel-tuned circuit.

In Fig. 10.3*a* the capacitor C_s and the inductor L_s that make up the tuned circuit are in *series* across the source of input voltage (shown schematically as a generator in Fig. 10.3*a*). Also in series with the other circuit elements are the resistances of the tuned circuit, shown schematically as a single lumped resistance R_s and the internal impedance of the generator, or *source generator*, Z_g.

In Fig. 10.3*b* the capacitor C_p and the inductor L_p are in parallel across the "source" generator with its internal impedance Z_g. In this parallel-tuned circuit the symbol R_p is used to denote the *total* effective resistance of the circuit. This lumping of resistance into a single equivalent-circuit element is common practice, and reflects the practical fact that, in the great majority of instances, the resistance of the inductor is so much greater than that of the capacitor that the latter is, in fact, negligible.

Figure 10.3, then, illustrates the basic configurations of both series- and parallel-tuned circuits. The origin of the terminologies for describing these two types of tuned circuits is apparent from the illustration. Note that the distinguishing feature of the series circuit is that all the generator current *must* flow through *R*, *L*, and *C*. In the parallel-tuned circuit, the generator current divides so that each branch receives only part.

10.4 SERIES-TUNED CIRCUITS

APPLICATIONS Series-tuned circuits have applications wherever it is desired to pass a signal of one frequency with minimum attenuation while rejecting signals of all other frequencies.

As one example, consider a radio transmitter that is to be coupled to its antenna in such a way that the desired transmitter output frequency is transmitted without attenuation, but the harmonics—that is, multiples—of the output frequency that may also be present do not reach the antenna. This can be accomplished by using a series-tuned circuit, resonant at the desired output frequency, between the transmitter output and the antenna, as shown in Fig. 10.5. The series-tuned circuit, being resonant at the desired carrier frequency, will have minimum effect on the transmitter output at this frequency. At higher frequencies, however, such as at the second harmonic (second multiple) of the desired output frequency, the series circuit will present an impedance approximately equal to the reactance of the inductor at the frequency in question, thus effectively preventing harmonics from reaching the antenna.

Another example of the use of series-tuned circuits may be found in what are frequently referred to as "trap" circuits. These are circuit applications in which a series-

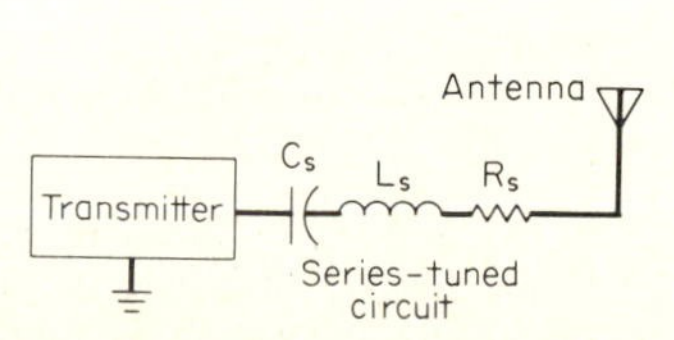

Fig. 10.5 Use of a series-tuned circuit in transmitter output.

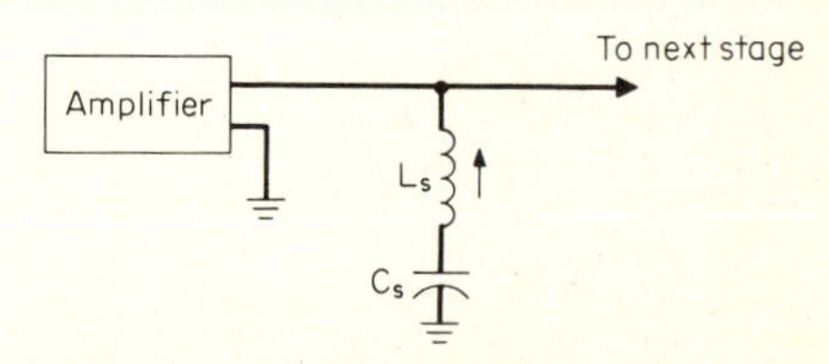

Fig. 10.6 An example of a series-tuned circuit used as a trap across the output of an amplifier stage.

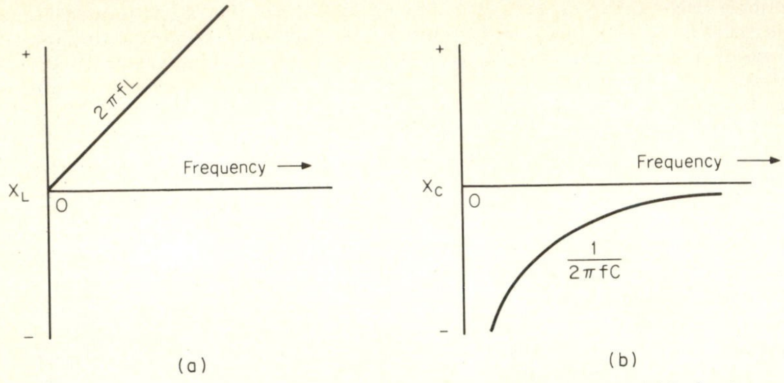

Fig. 10.7 Comparison of inductive and capacitive reactance curves: (*a*) Inductive reactance versus frequency. (*b*) Capacitive reactance versus frequency.

tuned circuit is used in parallel, or shunt, across a source of voltage, for the purpose of short-circuiting, or "trapping," signals of one frequency, while having negligible effect on signals of other frequencies. A specific example of the use of a series-tuned circuit as a trap circuit is shown in Fig. 10.6. This type of trap can be found in many television receivers.

The characteristic curves for inductors and capacitors are shown in Fig. 10.7.

FUNDAMENTALS The *reactance* of an *inductor* varies with frequency according to the equation

$$X_L = 2\pi fL \tag{10.1}$$

where X_L = inductive reactance, Ω
f = applied frequency, Hz
L = value of inductance, H
π = 3.14 (approximately)

Equation (10.1) states that for a given value of inductance L the inductive reactance X_L varies directly with frequency. This is shown graphically in Fig. 10.7*a*. At zero frequency (dc) the reactance is zero and the inductor acts as a short circuit. As the frequency increases, the inductive reactance increases in a direct manner.

The *reactance* of a *capacitor* varies with frequency according to the equation

$$X_C = -\frac{1}{2\,\pi fC} \tag{10.2}$$

where X_C = capacitive reactance, Ω
C = value of capacitance, F

Equation (10.2) states that the capacitive reactance varies *inversely* with frequency; that is, as the frequency increases the capacitive reactance decreases—opposite to that for an inductor. Referring to Fig. 10.7*b*, at frequencies close to zero it can be seen that the capacitive reactance has a very large value, approaching infinity. The capacitor at zero frequency (dc), therefore, acts as an open circuit. At higher frequencies X_C approaches zero, and the capacitor is said to act as a short circuit.

The right side of Eq. (10.2) is preceded by a minus sign to denote the fact that capacitive reactance is opposite in effect to inductive reactance, and is considered to be *negative* reactance, whereas inductive reactance is considered to be *positive*. Equations (10.1) and (10.2) and also Fig. 10.7*a* and *b* both show that inductive reactance and capacitive reactance are oppositely affected by frequency. That is, *inductive reactance increases* as frequency increases, whereas the *value of capacitive reactance decreases* as frequency increases.

For a given combination of values of inductance L and capacitance C in a series circuit there will be some one frequency for which the inductive reactance X_L has the same magnitude as the capacitive reactance X_C. If we remember that inductive reactance and capacitive reactance are opposites, that is, the former is positive while the latter is negative, it follows that they tend to cancel each other when they are equal in value.

The frequency for which the value of inductive reactance X_L equals the value of capacitive reactance X_C is the *resonant* frequency of a tuned circuit consisting of an inductance of L henrys and a capacitance of C farads. In other words, the resonant frequency of an LC series circuit is the frequency for which

$$X_L = X_C \tag{10.3}$$

The negative sign is omitted because, by definition, the resonant frequency is that frequency for which the inductive reactance is *numerically* equal to the capacitive reactance.

Substituting $2\pi fL$ for X_L, and $\frac{1}{2\pi fC}$ for X_C, we get

$$2\pi f_r L = \frac{1}{2\pi f_r C} \tag{10.4}$$

where f_r = resonant frequency, Hz.

Solving Eq. (10.4) for f_r, we obtain

$$f_r = \frac{1}{2\pi\sqrt{LC}} \tag{10.5}$$

Equation (10.5) states that the resonant frequency f_r is inversely proportional to the square root of the LC product. Thus, for a high resonant frequency the LC product is small, and for a low resonant frequency it is large.

The behavior of the series-resonant circuit is summarized in Fig. 10.8. Figure 10.8*a*

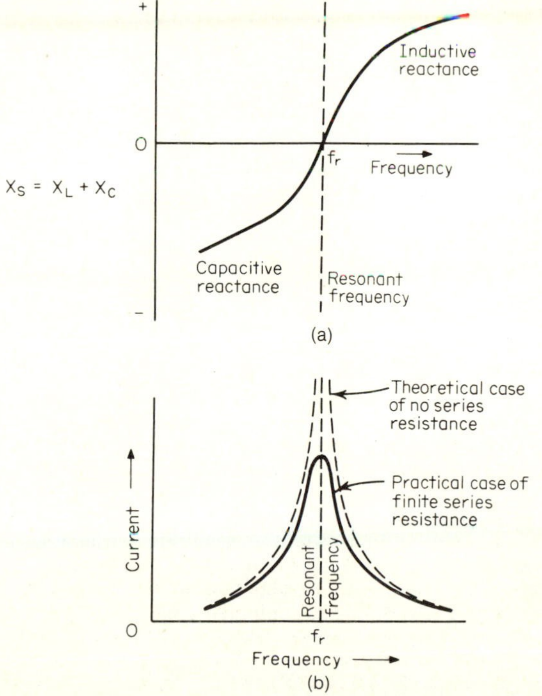

Fig. 10.8 Characteristic curves for a series-resonant circuit: (*a*) Reactance of a series-tuned circuit versus frequency. (*b*) Current through a series-tuned circuit at frequencies in the vicinity of resonance.

shows that at the resonant frequency the net reactance is zero, that is, $X_L = X_C$. Below the resonant frequency, capacitive reactance predominates; at frequencies above resonance, inductive reactance is predominant.

From Fig. 10.8*b* it is seen that the current in a series-tuned circuit is maximum at the resonant frequency. For frequencies less than or greater than the resonant frequency, the current falls and has a lower value. The magnitude of the current at resonance depends on the series resistance present. If there is zero resistance, the current approaches infinity; as the series resistance increases, the current at resonance decreases.

example 10-1 A series-tuned circuit, like the one shown in Fig. 10.3*a*, has an inductance of 3 mH, a capacitance of 30 μF, and a resistance of 300 Ω. What is the resonant frequency of this tuned circuit?

solution In a *series*-tuned circuit the amount of resistance in the circuit can be disregarded. (This is not always true in a parallel-tuned circuit.) To solve the problem, the inductance and capacitance values are expressed in henrys and farads.

$$3\text{ mH} = 3 \times 10^{-3}\text{ H}$$
$$30\ \mu\text{F} = 3 \times 10^{-5}\text{ F}$$

Substituting these values into Eq. (10.5), we get

$$f_r = \frac{1}{2\pi\sqrt{LC}} \tag{10.5}$$

$$= \frac{1}{(2 \times 3.14)\ \sqrt{3 \times 10^{-3}\ 3 \times 10^{-5}}}$$

$$= \frac{1}{6.28\ \sqrt{9 \times 10^{-8}}}$$

$$= \frac{1}{6.28 \times 3 \times 10^{-4}} = 530\text{ Hz}$$

example 10.2 It is desired to design a series-tuned circuit to have a resonant frequency of 5000 Hz. If the capacitor to be used has a value of 0.1 μF, what value of inductance is required?

solution Substitution of the given values in Eq. (10.4) yields

$$2\pi f_r L = \frac{1}{2\pi f_r C} \tag{10.4}$$

or

$$2\pi(5000)L = \frac{1}{2\pi(5000) \times 0.1 \times 10^{-6}}$$

Solving for L,

$$L = \frac{1}{4\pi^2 \times 25 \times 10^6 \times 0.1 \times 10^{-6}}$$

$$= \frac{1}{10\pi^2} = 0.010\ 14\text{ H} = 10.14\text{ mH}$$

The simple tuned circuit in Example 10.1 will be resonant at a frequency of 530 Hz. At this frequency, the inductive and capacitive reactances *cancel*. This means that the only opposition to current flow will be what is offered by Z_g and R_s (see Fig. 10.3*a*). If these values are small, then there will be very little opposition to current flow, and the circuit current will be very high.

Equation (10.5) applies to series-tuned circuits with or without a series resistance R_s. However, the equation will only apply to parallel circuits if the resistances in the L_p and C_p branches are small enough to be neglected. If, for example, the resistance value of R_p in Fig. 10.3*b* is more than about 10 percent of the value of X_L, then Eq. (10.5) *cannot be used* to determine the value of resonant frequency for the circuit. The same is true if a resistance is placed in series with C_p in the circuit.

SERIES-TUNED VOLTAGES AT RESONANCE Another property of series-tuned circuits that deserves attention concerns the voltages that exist across the circuit elements at frequencies at or near the resonant frequency of the circuit.

If we remember that, at the resonant frequency, the capacitive reactances X_C of the inductor are equal and opposite to each other; it follows then that the net reactance of the series circuit will be zero at the resonant frequency. The current flowing through the circuit at the resonant frequency is thus given by the simple Ohm's law relationship,

$$I = \frac{E}{R_s} \tag{10.6}$$

where E = voltage applied across the entire series-tuned circuit, that is, the voltage developed by the voltage source less the voltage drop across the internal impedance of the source (Z_g in Fig. 10.3*a*)

I = current through the series-tuned circuit at resonance

R_s = total series resistance of the series-tuned circuit

If the total series resistance R_s is reasonably low, as it will be in a well-designed series-tuned circuit for most applications, the current I through the circuit will be quite large at resonance.

The voltage across the capacitor at resonance is given by the Ohm's law relationship,

$$E_C = I \times X_C \tag{10.7}$$

where E_C = voltage across the capacitor at resonance.

Similarly, the voltage across the inductor at resonance will be

$$E_L = I \times X_L \tag{10.8}$$

where E_L = voltage across the inductor at resonance.

The last two equations state that for a series-tuned circuit at resonance, the voltage across the capacitor (or inductor) is equal to the product of current I at resonance [defined by Eq. (10.6)] and the capacitive reactance (or inductive reactance). The practical significance of these effects will be considered shortly. Equations (10.7) and (10.8) both assume that whatever resistance is present in the series-tuned circuit is considered as a separate element. However, this assumption does not in any way affect the validity of the above equations.

Bearing in mind that, in a series-tuned circuit with a typically low value of series resistance R_s, the current through the circuit at resonance is quite large [see Eq. (10.6)], it follows that the voltages across the capacitive and inductive elements of the circuit are also quite large at resonance. These voltages, which will be equal in magnitude, will be greater than the voltage applied across the series-tuned circuit by a factor of X_C/R_s, or X_L/R_s. That is, if the voltage applied across the total circuit at resonance is E, the voltage across the capacitor and inductor may be also expressed as

$$E_C = E\left(\frac{X_C}{R_s}\right) \tag{10.9}$$

and

$$E_L = E\left(\frac{X_L}{R_s}\right) \tag{10.10}$$

The high voltages across the reactive components affect the voltage ratings. They must be considered if the circuit is to operate without arcing or short circuiting. The ratio of capacitive or inductive reactance at resonance to the equivalent series resistance may have values in the range of from 10 to perhaps 100; so it follows that the capacitor and inductor must be constructed to withstand voltages from 10 to 100 times as great as the voltage applied across the series-tuned circuit. (This refers, of course, to the r-f or ac voltage applied to the circuit, and not to any dc voltage that may be present across the series-tuned circuit.)

In low-voltage applications, such as a series-tuned circuit in a radio receiver, this is not a matter of great consequence. In other applications, such as in transmitters where the applied voltage itself may be reasonably large, the requirement that the individual elements of the series-tuned circuit withstand voltages that are several times the applied voltage can represent an important design requirement. This is illustrated in the next example.

example 10.3 In a series-tuned circuit R_s = 10 Ω, and the magnitude of the capacitive reactance is 1000 Ω at the resonant frequency. If the voltage source $E_g = E$ = 10 V, $Z_g \approx 0$, what are the voltages across the inductor, capacitor, and resistor at the resonant frequency?

solution From Eq. (10.6),

$$I = \frac{E}{R_s} = \frac{10}{10} = 1 \text{ A}$$

At resonance,

$$X_C = X_L = 1000\ \Omega$$

Using Eqs. (10.7) and (10.8), we find

$$\begin{aligned} E_C &= I \times X_C \\ &= 1 \times 1000 \\ &= 1000 \text{ V} \end{aligned} \qquad \text{and} \qquad \begin{aligned} E_L &= I \times X_L \\ &= 1 \times 1000 \\ &= 1000 \text{ V} \end{aligned}$$

Also,

$$E = I \times R_s = 1 \times 10 = 10 \text{ V}$$

The significance of this example is that, with practical values of circuit elements, the individual voltages across the capacitor and inductor at resonance are *100 times* the voltage applied across the series-tuned circuit.

10.5 PARALLEL-TUNED CIRCUITS

In the preceding section, the characteristics and behavior of series-tuned circuits were described and discussed. We now consider the second major category of tuned circuits, the parallel-tuned circuit, which is also sometimes referred to as an *antiresonant circuit.*

In Fig. 10.3*b*, the parallel-tuned circuit consists essentially of a capacitor C_p and an inductor L_p connected in parallel (or *shunt*) across a source of voltage, a *generator.* The circuit in Fig. 10.3*b* shows resistance in the inductive branch, but not in the capacitive branch. This is a practical consideration because the resistance of the wires in the inductor is usually large compared to the ESR (Equivalent Series Resistance) of the capacitor. However, this is not to say that resistance will never be found in the capacitive branch. The equation for parallel resonance will be developed later in this chapter for all possible cases.

APPLICATIONS Applications of parallel-tuned circuits are based on their impedance versus frequency characteristics, as illustrated in Fig. 10.9. At the resonant frequency the impedance is maximum, and at frequencies below and above resonance the impedance falls off. In many applications parallel-tuned circuits are used to obtain selective filtering of specific frequencies.

Figure 10.10 illustrates the use of a parallel-tuned circuit to obtain selective amplification of a desired frequency in a radio or television receiver. The antenna is connected to the parallel-tuned circuit, to which the input terminal of the first amplifier stage is also connected.

When signals arrive from the antenna at the resonant frequency of the parallel-tuned circuit, this presents a high impedance to ground. Thus, the signal current supplied

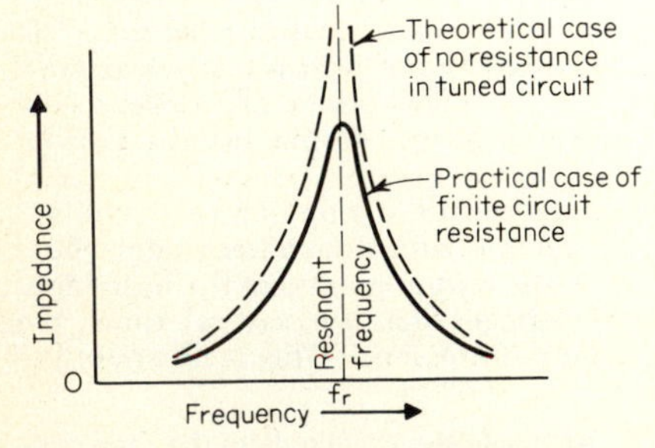

Fig. 10.9 Impedance of a parallel-tuned circuit at frequencies in the vicinity of resonance.

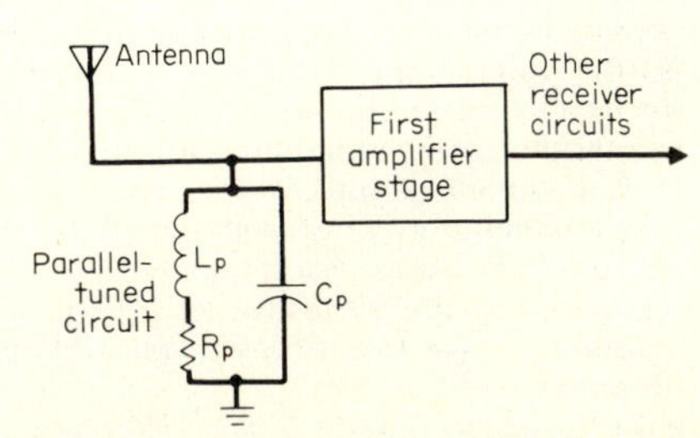

Fig. 10.10 Use of a parallel-tuned circuit for selective amplification at the receiver input.

by the antenna develops a relatively high voltage across the tuned circuit, this voltage being amplified by the first amplifier stage.

For signals of other frequencies, that is, frequencies relatively far from the resonant frequency, the impedance is much lower. The voltage developed in response to the current from the antenna, therefore, will also be lower, and the signals will not be amplified to any appreciable degree. (In fact, the circuit of Fig. 10.10 may be viewed as one that has the effect of *shunting* all signals (other than those at the resonant frequency) to ground. Since they are shunted to ground, they are not presented to the amplifier.

A similar application of parallel-tuned circuits is illustrated in Fig. 10.11. A parallel-tuned circuit is used at the output of an amplifier, instead of at the input as in Fig. 10.10.

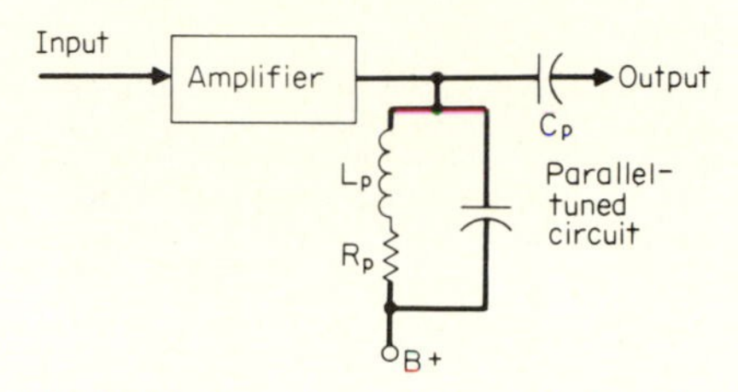

Fig. 10.11 Use of a parallel-tuned circuit for selective amplification at the amplifier output.

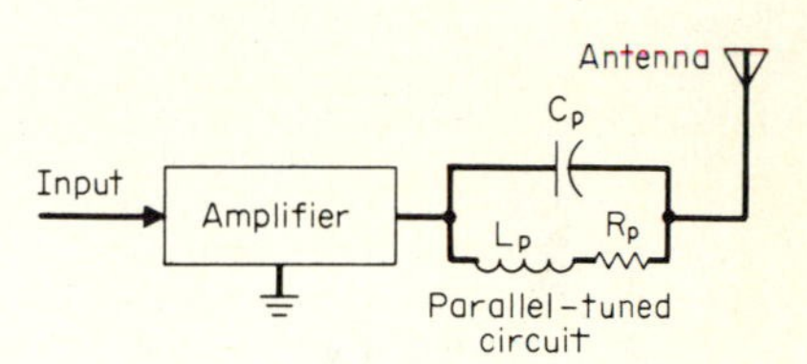

Fig. 10.12 Use of a parallel-tuned circuit in series with amplifier output.

The amplifier in Fig. 10.11 may be a circuit in a transmitter, or in a receiver. In either case, the purpose of the circuit is to cause the amplifier to respond primarily to the resonant frequency of the tuned circuit, and to minimize the amplification of signals at other frequencies.

The voltage gain of an amplifier is directly proportional to the impedance of the load. In the circuit of Fig. 10.11, the parallel-tuned circuit is the amplifier load. At the resonant frequency this impedance will be quite high, and therefore the gain will be high. At all other frequencies the impedance of the parallel-tuned circuit will be low, and as a result the gain will be low. To summarize, the parallel-tuned circuit permits the amplifier to produce a large voltage gain at a specific frequency.

Another type of application of parallel-tuned circuits is illustrated in Fig. 10.12. In this circuit the parallel-tuned circuit is in series with an amplifier load (which in this case is an antenna). Its purpose is to prevent signals of an undesired frequency from reaching the load or the next stage. This is accomplished because the parallel-tuned circuit presents a high impedance, the signals of the (undesired) frequency to which it is tuned, while presenting a relatively low impedance to signals of the desired frequency. The circuit of Fig. 10.12 is used frequently in radio transmitters to prevent the second harmonic (second multiple) of the desired output frequency from reaching the antenna, from which it could be radiated and cause interference.

FUNDAMENTALS The equation for parallel resonance (which is sometimes called *antiresonance*) will be obtained for the circuit of Fig. 10.13. A few mathematical derivations are given in this section, but the overall results are summarized for those who want a less mathematical approach to the subject of parallel-tuned circuits. Note that the resistances in both the inductive and capacitive branches will be taken into consideration. Then the equation will be simplified for use in certain specialized cases such as the one shown in Fig. 10.3*b*.

When a circuit is in resonance, the current and the voltage in the circuit must be in phase. This is true for both series and parallel resonant circuits. Another way of stating the condition for resonance is that the circuit must have *unity power factor*.

The only way that unity power factor can occur in the circuit of Fig. 10.13 is if the phasor sum of the currents in the two branches produces a total current that is in phase with the applied voltage V. The mathematics required to establish the values of R_C, C, R_L, and L that are needed to produce resonance would be very tedious. Fortunately, there is a shortcut method.

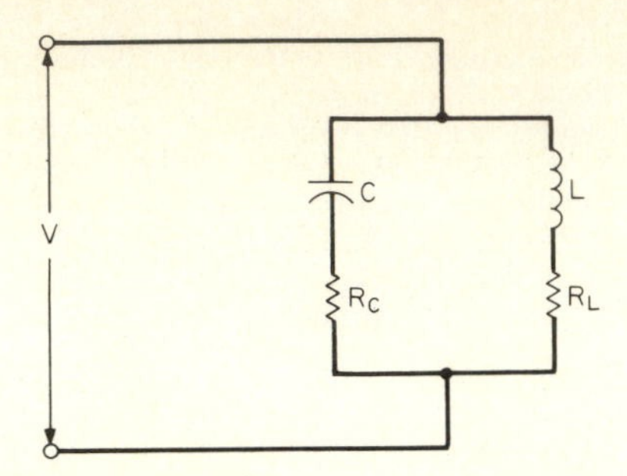

Fig. 10.13 A general circuit for determining parallel resonance.

The two series branches of Fig. 10.13 can be converted to the four parallel branches shown in Fig. 10.14. The equations required for finding the conductances G_C and G_L and the susceptances B_C and B_L are shown in the illustration. It is important to understand that the conversion from the circuit of Fig. 10.13 to the one in Fig. 10.14 is a standard procedure used for simplifying and solving parallel circuits. In other words, this is not a special case, but rather it is a standard procedure.

In the equivalent circuit of Fig. 10.14 there is no resistance in the capacitive and inductive branches. Resonance will occur when $B_C = B_L$, and this is the basis for finding the equation for parallel resonance. (Compare this with the solution of the series-resonant equation which is based on $X_C = X_L$.)

$$B_L = B_C$$

$$\frac{X_L}{R_L^2 + X_L^2} = \frac{X_C}{R_C^2 + X_C^2}$$

$$\frac{2\pi f L}{R_L^2 + (2\pi f L)^2} = \frac{1/2\pi f C}{R_C^2 + (1/2\pi f C)^2}$$

When this equation is solved for f_r, which is the resonant frequency of the circuit in Fig. 10.13, the result is

$$f_r = \frac{1}{2\pi\sqrt{LC}}\sqrt{\frac{L - CR_L^2}{L - CR_C^2}} \tag{10.11}$$

where f_r = resonant frequency, Hz
L = circuit inductance, H
C = circuit capacitance, F
R_C = resistance in capacitive branch, Ω
R_L = resistance in inductive branch, Ω

This is the equation for the resonant frequency of the circuit in Fig. 10.13, even though it is derived from the equivalent circuit of Fig. 10.14.

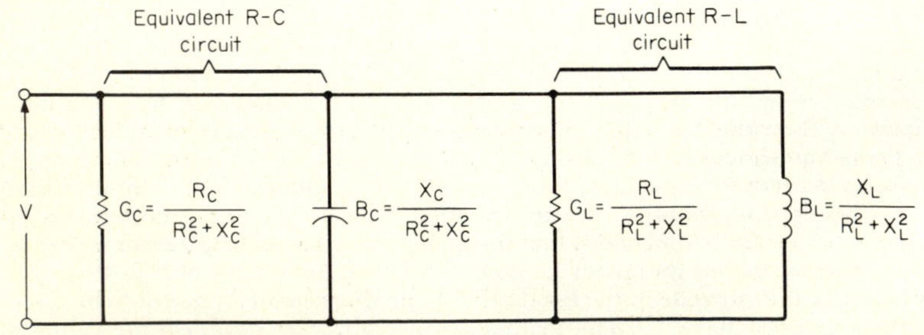

Fig. 10.14 This circuit has the same impedance and will produce the same phase angle as the circuit in Fig. 10.13.

Several very important facts about parallel resonance can be determined from Eq. (10.11).

1. When the resistance in the two branches is negligible, the value zero can be substituted for both R_C and R_L in the equation. The result is

$$f_r = \frac{1}{2\pi\sqrt{LC}} \tag{10.12}$$

where f_r = resonant frequency, Hz
L = inductance, H
C = capacitance, F

In other words, the same equation can be used for both series and parallel resonance if there is negligible resistance in the parallel branches. In actual practice, if X_L is at least ten times the value of R_L, and X_C is at least ten times the value of R_C, the resistances are considered to be negligible.

2. Equation (10.11) shows that a parallel-resonant circuit can actually be tuned by varying the resistance in one of the branches—that is, by varying either R_C or R_L. This follows from the fact that changing the value of either R_C or R_L in the equation will change the value of f_r. (Of course, f_r can also be varied by changing either L or C.)

If we solve Eq. (10.11) for the value of R_L, the result is

$$R_L = \sqrt{\frac{X_L}{X_C} R_C^2 - X_L^2 + \frac{L}{C}} \tag{10.13}$$

where the letters have the same meanings as in Eq. (10.11).

If we solve Eq. (10.11) for the value of R_C, the result is

$$R_C = \sqrt{\frac{R_L^2}{(2\pi f)^2 LC} - \frac{1}{\omega^2 C^2} + \frac{L}{C}} \tag{10.14}$$

where the letters have the same meanings as in Eq. (10.11).

Equations (10.13) and (10.14) give the value of R_L or the value of R_C needed to bring the circuit of Fig. 10.13 into resonance.

3. It was mentioned earlier that the circuit of Fig. 10.3*b* is often given as a practical circuit because the dc resistance of the inductive branch cannot always be ignored. The equation for resonance in this circuit can be obtained by setting $R_C = 0$ in Eq. (10.11). The result is

$$f_r = \frac{1}{2\pi\sqrt{LC}} \sqrt{\frac{L - CR_L^2}{L}} \tag{10.15}$$

4. In a series LC circuit there is always a resonant frequency because there is always some frequency at which $X_L = X_C$. However, in the parallel-resonant circuit, it is possible to have a combination of components that are not resonant at any frequency. To understand how this is possible, look at the numerator and denominator under the radical in Eq. (10.11):

Numerator: $L - CR_L^2$
Denominator: $L - CR_C^2$

A very important point to consider is that the value under the radical will be negative if

CR_L^2 is greater than L
or
CR_C^2 is greater than L

If the value under the radical is negative, no value of f_r exists because the result is an *imaginary* value. To summarize, then, it is possible to have a parallel circuit that has no resonant frequency!

5. Another important feature of parallel-resonant circuits is that they may be designed to be resonant at all frequencies. This can be demonstrated with a short mathematical derivation.

In the derivation of Eq. (10.11) this equation was given:

$$\frac{2\pi fL}{R_L^2 + (2\pi fL)^2} = \frac{1/2\pi fC}{R_C^2 + (1/2\pi fC)^2}$$

To simplify equations of this type, the symbol ω (omega) is substituted for $2\pi f$

$$\frac{\omega L}{R_L^2 + \omega^2 L^2} = \frac{1/\omega C}{R_C^2 + 1/\omega^2 C^2}$$

Setting the product of the means equal to the product of the extremes (sometimes called "cross-multiplying") gives

$$\omega L R_C^2 + \frac{\omega L}{\omega^2 C^2} = \frac{R_L^2}{\omega C} + \frac{\omega L^2}{C}$$

The lowest common denominator is ωC^2. Multiplying every term by this value gives

$$\omega^2 L C^2 R_C^2 + L = R_L^2 C + \omega^2 L^2 C$$

Now, if either the numerator or the denominator under the radical in Eq. (10.11) is equal to zero, then the resonant frequency cannot be given a numerical value.

For the numerator:
$$L - C R_L^2 = 0$$
$$R_L = \sqrt{\frac{L}{C}}$$

For the denominator:
$$L - C R_C^2 = 0$$
$$R_C = \sqrt{\frac{L}{C}}$$

Substituting the radical equivalent values for R_L and R_C in the above equations gives

$$\frac{\omega L^2}{C} + \frac{L}{\omega C^2} = \frac{\omega L^2}{C} + \frac{L}{\omega C^2}$$

Note that this equation is valid for ***every value*** of ω (and therefore every value of frequency f)—*except the value of* $\omega = 0$. The importance of this derivation is that the circuit is resonant at *all frequencies* whenever $R_L = R_C = \sqrt{L/C}$.

In a series-tuned circuit, if the applied frequency is above resonance, there is a greater voltage drop across the inductor than across the capacitor. This makes the circuit behave like an inductor. If the applied frequency is below the resonant frequency, on the other hand, the circuit behaves like a capacitor because the larger voltage drop is across the capacitor.

TABLE 10.1 Comparison of Series- and Parallel-Tuned Circuits

	Series-tuned circuit	Parallel-tuned circuit
Impedance	Minimum at the resonant frequency	Maximum at the resonant frequency
Current	Maximum at resonant frequency	Minimum at resonant frequency
Voltage	Minimum voltage across the circuit at resonant frequency	Maximum voltage across the circuit at resonant frequency
Resonance	There is one resonant frequency for each *LC* combination. Resistance does not affect resonance for all practical purposes.	a. Resistance affects resonant frequency. b. Circuit may resonate at all frequencies. c. Circuit may not have a resonant frequency.
When applied frequency is below resonance	Circuit is capacitive.	Circuit is inductive.
When applied frequency is above resonance	Circuit is inductive.	Circuit is capacitive.
Power factor at resonance	1.0	1.0
Special feature	Voltage across a component may be many times applied voltage.	Current through a component may be many times main current.

When the frequency of the applied signal across a parallel-tuned circuit is above the resonant frequency, more current flows through the capacitor because its reactance is lower than that of the inductor. This means that the circuit will be capacitive. Conversely, if the applied frequency is below resonance, the greater current flows through the inductor, and the circuit is inductive.

The characteristics of series- and parallel-tuned circuits are summarized in Table 10.1.

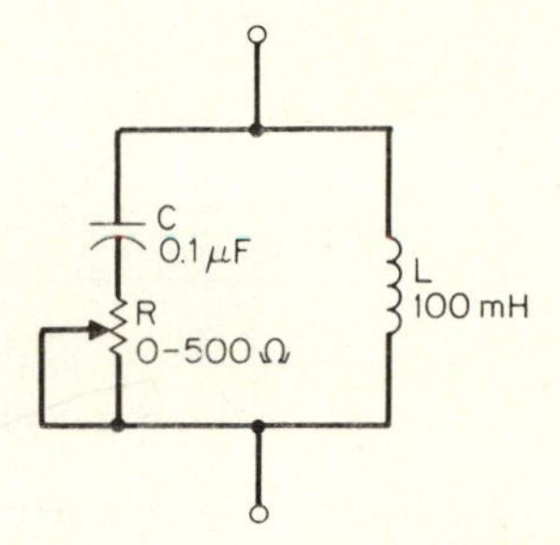

Fig. 10.15 Circuit for Example 10.5.

example 10.4 In the parallel-tuned circuit of Fig. 10.13 assume that $L_p = 100$ mH, $C_p = 0.1\ \mu$F, and $R_C = R_L = 0$. What is the resonant frequency of the circuit?

solution From Eq. (10.12),

$$f_r = \frac{1}{2\pi\sqrt{LC}}$$

$$= \frac{1}{6.28\sqrt{0.1 \times 0.1 \times 10^{-6}}} = 1590 \text{ Hz} \qquad (10.12)$$

example 10.5 In the circuit of Fig. 10.15, a variable resistor is placed in the capacitive branch. This resistor is used for varying the resonant frequency. What is the range of frequencies through which R_C can tune the circuit?

solution When $R_C = 0$,

$$f_r = \frac{1}{2\pi\sqrt{LC}} = 1590 \text{ Hz} \qquad \text{(See Example 10.4.)}$$

When $R_C = 0.5$ kΩ,

$$f_r = \frac{1}{2\pi\sqrt{LC}}\,\frac{L}{L - CR_C^2}$$

$$= 1590\sqrt{\frac{100 \times 10^{-3}}{100 \times 10^{-3} - (0.1 \times 10^{-6})\,(0.5 \times 10^3)^2}} = 1828 \text{ Hz}$$

Thus a change of 500 Ω results in a frequency change of 238 Hz.

10.6 SELECTIVITY, BANDWIDTH, AND Q FACTOR

The ability of a tuned circuit to respond differently to signals of different frequencies is referred to as its *selectivity*. For quantitative purposes, the selectivity of a tuned circuit—that is, its ability to discriminate between signals of different frequencies—is measured in terms of a parameter called *bandwidth*.

BANDWIDTH The bandwidth of a tuned circuit is a range of frequencies between two points on its selectivity curve—one on each side of the resonant frequency—at which the circuit response is a stated fraction of the response at the resonant frequency. The curve of Fig. 10.16 is an example of a selectivity curve. Although the fractional response point on the selectivity curve at which the tuned circuit bandwidth is determined can be stated arbitrarily, it is almost universal practice to measure bandwidth at points on the selectivity curve at which the circuit power response is one-half the

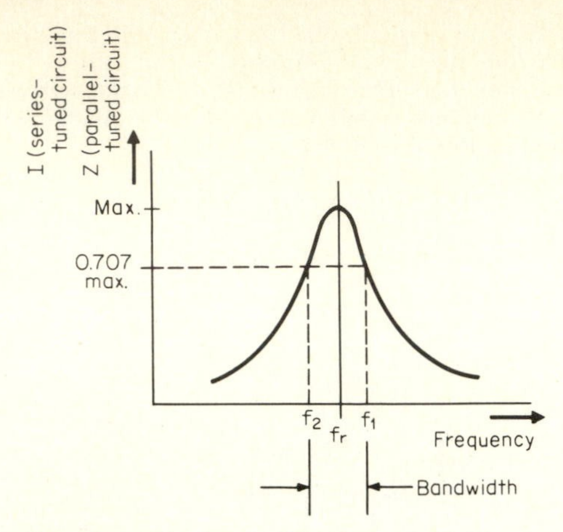

Fig. 10.16 The bandwidth is taken between the points where the current, voltage, or impedance is 70.7 percent of maximum, depending on whether it is a series- or parallel-tuned circuit. It may also be taken between points where the power is 50 percent of maximum.

power that is produced at resonance. The bandwidth so measured is called the *half-power* bandwidth, or the $-3dB$ bandwidth. The latter term is derived from the fact that one-half power corresponds to a power level three decibels below maximum power.

Although bandwidth is defined in terms of power, it is generally more convenient to measure it in terms of other circuit performance characteristics. The following relationships will be found especially useful in practice:

1. In a series-tuned circuit, the half-power bandwidth is the width (in frequency) of the portion of the selectivity curve between points at which the current through the circuit is equal to $1/\sqrt{2} = 0.707$ times the current at resonance. (This relationship should be self-evident when it is remembered that power P is proportional to the square of current, $P = I^2R$.)

2. In a parallel-tuned circuit, the half-power bandwidth is the width of the portion of the selectivity curve between points at which the effective parallel impedance of the circuit is equal to $1/\sqrt{2} = 0.707$ times the impedance at resonance.

It is evident that the bandwidth relationships for series-tuned and parallel-tuned circuits correspond, with the difference that bandwidth of a series-tuned circuit is expressed in terms of current flow while bandwidth of a parallel-tuned circuit is expressed in terms of effective parallel impedance. Figure 10-16 summarizes the definitions of bandwidth for series- and parallel-tuned circuits.

Q FACTOR. The bandwidth of a tuned circuit is determined by a circuit parameter called Q or Q factor. Circuit Q is a very important parameter used for describing the performance characteristics of a tuned circuit.

The Q of a tuned circuit is equal to the reactance of either the capacitor or the inductor at the resonant frequency of the circuit; the reactance of the capacitor and inductor are, as explained previously, equal at the resonant frequency, divided by the effective series resistance of the circuit. This relationship applies to both the series- and parallel-tuned circuits of Fig. 10.3. For the series-tuned circuit,

$$Q = \frac{2\pi f_r L_s}{R_s} \tag{10.16}$$

Similarly, for the parallel-tuned circuit with negligible resistance,

$$Q = \frac{2\pi f_r L_p}{R_p} \tag{10.17}$$

Expressions similar to Eqs. (10.16) and (10.17) can be written in terms of the capacitive reactances at resonance in place of the inductive reactances. However, expressing Q in terms of the inductive reactance is usually more convenient.

Another way of expressing Q is in terms of the half-power bandwidth of the tuned circuit as related to its resonant frequency:

$$Q = \frac{f_r}{f_1 - f_2} = \frac{f_r}{\Delta f} \tag{10.18}$$

where f_1 = frequency above the resonant frequency at which the half-power relationship exists (see Fig. 10.16)

f_2 = frequency below the resonant frequency at which the half-power relationship exists

f_r = resonant frequency of a tuned circuit (either series or parallel)

$\Delta f = f_1 - f_2$ = bandwidth of the circuit

Equation (10.18) is especially convenient for determining the Q of a tuned circuit experimentally. It is necessary only to measure the frequencies at which the circuit exhibits a half-power response from which information, together with knowledge of the resonant frequency, the Q is readily calculated. (The required frequency measurement must, however, be made with some care for circuits with high Q's because errors in measuring either f_1 or f_2 can lead to serious errors in determining bandwidth Δf, and therefore in calculating the Q of the circuit.)

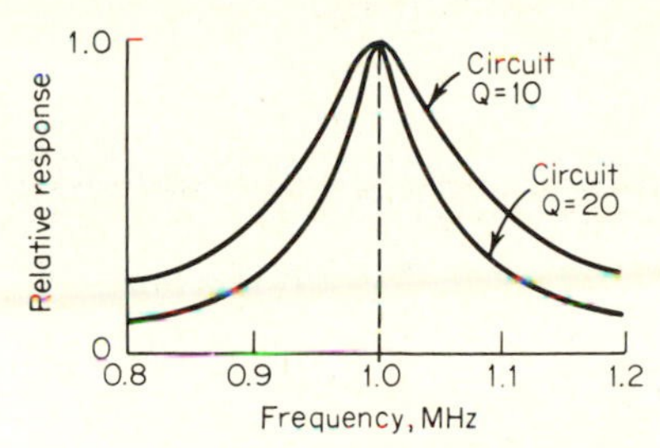

Fig. 10.17 Effect of circuit Q on bandwidth.

Measurements based on Eq. (10.18) are frequently superior to a method of determining Q according to Eq. (10.16) or (10.17). This is true because of the practical difficulty in determining the true value of R_s or R_p, inasmuch as their values are influenced by high-frequency losses in a manner that is not always easy to determine.

For use at many frequencies, however, there are instruments (usually called *Q meters*) that can measure the Q of an inductor, a capacitor, or a tuned circuit directly. This avoids the difficulty of determining the effective resistance by other means, and also the problems that sometimes arise in attempting to determine Q by frequency measurements.

It is evident from Eq. (10.16) or (10.17) that the resistance in a tuned circuit in relation to the reactance at resonance is the parameter that determines the circuit Q, and therefore the bandwidth. The effect of circuit Q on bandwidth is illustrated in Fig. 10.17. The higher the Q, the smaller the bandwidth.

The current in an inductor or a capacitor in a parallel-tuned circuit at resonance is Q times the external circuit current. Likewise, for a series-tuned circuit, the voltage across the inductor or capacitor at resonance is Q times the applied circuit voltage.

example 10.6 In a parallel-tuned circuit, $L = 100\ \mu\text{H}$, $R_L = 10\ \Omega$, $f_r = 1$ mHz. Determine the values of (*a*) Q, and (*b*) bandwidth.

solution (*a*) From Eq. (10.17),

$$Q = \frac{2\pi f_r L_p}{R_p}$$

$$= \frac{6.28 \times 10^6 \times 100 \times 10^{-6}}{10} = 62.8 \tag{10.17}$$

(There are no units for measuring Q.)

(*b*) Solving Eq. (10.18) for the bandwidth Δf,

$$\Delta f = \frac{f_r}{Q} \tag{10.19}$$

Therefore

$$\Delta f = \frac{10^6}{62.8} \approx 16 \text{ kHz}$$

example 10.7 Design a parallel-tuned circuit to cover the a-m broadcast band from 500 to 1600 kHz. At 500 kHz the minimum bandwidth should be 25 kHz. The minimum capacitance of the variable tuning capacitor is 30 pF, and stray wiring capacitance in parallel with the capacitor is on the order of 20 pF. Determine the value of L_p, Q, and the capacitance range of the variable capacitor.

solution The value of L_p is determined at the upper frequency of 1600 kHz and at minimum capacitance of the capacitor, including stray capacitance (30 + 20 = 50 pF). From Eq. (10.12),

$$L = \frac{1}{4\pi^2 f_r^2 C}$$

Substituting known values,

$$\frac{1}{4\pi^2(1600 \times 10^3)^2 \times 50 \times 10^{-12}} \approx 200 \ \mu\text{H}$$

From Eq. (10.18),

$$Q = \frac{f_r}{\Delta f} = \frac{500}{25} = 20$$

By using the value of $L = 200\ \mu$H, the maximum value of C is determined at the lower frequency of 500 kHz. Solving Eq. (10.12) for C gives

$$C = \frac{1}{4\pi^2 f_r^2 L}$$

and substituting known values,

$$= \frac{1}{4\pi^2(500 \times 10^3)^2 \times 200 \times 10^{-6}} \approx 500 \text{ pF}$$

But 20 pF of stray capacitance is already present; therefore, the maximum value of capacitance is 500 − 20 = 480 pF, and the capacitor range is 30 to 480 pF.

APPLICATION REQUIREMENTS In some applications for tuned circuits the highest possible degree of selectivity—that is, the lowest bandwidth and the highest Q—is desired. This permits the circuit to do the best job of rejecting signals of unwanted frequencies, even though they may be very close to the desired frequency to which the circuit is tuned. In such cases, the tuned circuit is designed and constructed to preserve the highest possible Q by keeping the effective circuit resistance as low as possible.

There are also cases, namely in so-called wideband applications (of which television receiver circuits are an example), in which care must be taken lest the bandwidth of the tuned circuit be too narrow. This would result in some signals of desired frequencies being rejected, along with signals of undesired frequencies that may be farther from the resonant, or center, frequency of the circuit.

One method of keeping a tuned circuit from having a bandwidth that is too narrow is to deliberately introduce resistance into the circuit, thereby lowering the Q and producing a selectivity curve similar to the wider of the two curves in Fig. 10.17.

10.7 DOUBLE-TUNED (MULTIPLE-TUNED) CIRCUITS (ALSO CALLED STAGGER-TUNED)

As indicated in the preceding section, there are applications of tuned circuits where the circuit must be capable of responding to a wide band of frequencies about the center frequency (the term *center frequency* corresponds to *resonant frequency*, the latter term being used primarily in connection with circuits for narrow-band applications). In such applications, one solution is to resort to the use of double-tuned circuits, or more generally, to multiple-tuned circuits.

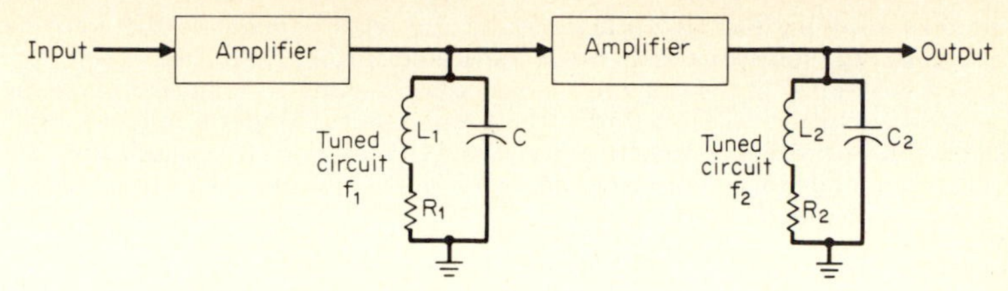

Fig. 10.18 Double-tuned (stagger-tuned) circuits.

The terms double-tuned and multiple-tuned circuits are used to refer to combinations of two (double-tuned) or more (multiple-tuned) circuits in which the individual circuits, instead of all being tuned to the same frequency, are tuned to slightly different frequencies for the purpose of producing the desired wideband response. Figure 10.18 illustrates such a circuit configuration. It is noted that the tuned circuits are isolated from each other by amplifier stages (other means of isolation are possible). Although such isolation is not absolutely necessary, if the circuits are allowed to interact with each other directly (that is, if they are *coupled*), the response characteristics are somewhat different. Coupled tuned circuits are discussed in the next section.

Figure 10.19 illustrates the response of the double-tuned circuit of Fig. 10.18. Both response curves (selectivity curves) of the individual tuned circuits and their combined response are shown in Fig. 10.19. This illustration clearly exhibits the desired wideband characteristic.

It is important to note that obtaining wideband response by using multiple-tuned circuits produces an overall selectivity curve which, while being broad at the top (to obtain uniform response), also has steep sides. The response is said to have good *skirt selectivity*, giving the circuit the ability to reject unwanted frequencies, even though they may be very close to the desired frequency band. The name skirt selectivity comes from the fact that the sides of a response curve are called "skirts."

The combination of wide bandwidth and good skirt selectivity is the primary reason why multiple-tuned circuits are often preferred over single-tuned circuits with low Q when a wideband response is required. The advantage of multiple-tuned circuits, in terms of skirt selectivity, is apparent from a comparison of the overall response curve of Fig. 10.19 with the wider of the two curves of Fig. 10.17, which is the selectivity curve of a single-tuned circuit where wide bandwidth is obtained by reducing circuit Q.

There are applications for which the required bandwidth is so wide that two circuits used in a double-tuned configuration are not adequate. This is because as the spacing between the separate frequencies to which the two circuits are tuned is increased to enlarge the overall bandwidth, a gap, or *valley*, begins to develop in the region between the frequencies to which the two circuits are individually tuned. That this can occur is

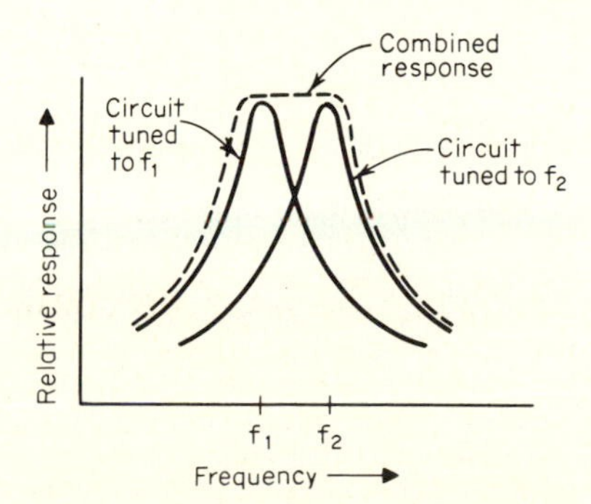

Fig. 10.19 Response of double-tuned (stagger-tuned) circuits.

evident from studying Fig. 10.19 and visualizing what happens to the combined response curve if the individual curves are separated further from each other.

To avoid this valley in the response curve, a three-circuit configuration can be used (triple-tuned circuit), in which a third circuit, tuned to the center frequency, is utilized to "fill in" the valley that otherwise exists when the original two circuits are separated further in frequency to increase overall bandwidth. Figure 10.20 illustrates the response of a triple-tuned circuit.

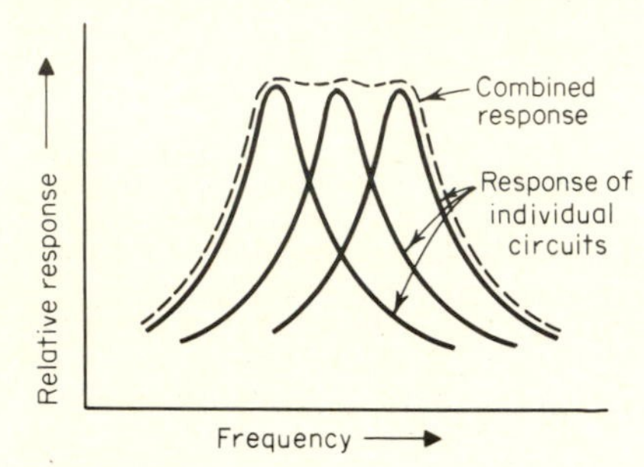

Fig. 10.20 Response of a triple-tuned (stagger-tuned) circuit.

Theoretically, of course, the concept of obtaining increased bandwidth through the use of multiple-tuned circuits can be extended to almost any number of circuits. In fact, arrangements of more than three such circuits are common. However, five circuits are about the usual practical limit. This is because, when bandwidths greater than those obtained with five circuits (quintuple-tuned circuits) are required, other methods exist which are superior from a practical standpoint.

10.8 COUPLED TUNED CIRCUITS

There are numerous applications of tuned circuits in which, instead of the individual circuits being isolated from each other by amplifiers or otherwise, they are deliberately coupled together in such a way that they interact directly on each other. Although the complete mathematical theory of coupled circuits is complex, there are certain phenomena that will be described as being of importance from a practical standpoint.

Figure 10.21 shows the general example of two parallel-tuned circuits coupled together. The symbol M denotes the *mutual inductance* that exists by virtue of the fact that the two circuits are inductively coupled. Although this mutual inductance does not correspond directly to any physical element in the circuit, as do the other circuit parameters L_1, L_2, C_1, and C_2, it is nonetheless important to the performance of the coupled circuit, and cannot be overlooked. Mathematically, mutual inductance is given by the expression

$$M = k\sqrt{L_1L_2} \qquad (10.20)$$

where M = mutual inductance, expressed in the same inductance units (e.g., henrys, millihenrys, etc.) as L_1 and L_2
L_1 = inductance of the *primary* of the coupled circuit
L_2 = inductance of the *secondary* of the coupled circuit
k = coefficient of coupling. The coefficient of coupling is a measure of how many of the primary flux lines actually link with the secondary

It is important to note that the coefficient of coupling, k, has a value between 0 and 1. Thus, the mutual inductance can never be greater than the square root of the product of the two individual inductances, and is ordinarily less, inasmuch as k almost always has a value substantially less than 1.0. (In fact, a coefficient of coupling closely approaching the maximum value of 1.0 is usually found only in low-frequency transformers, in which magnetic cores are used. Coefficients of coupling approaching 1.0 are not found in high-frequency coupled circuits.)

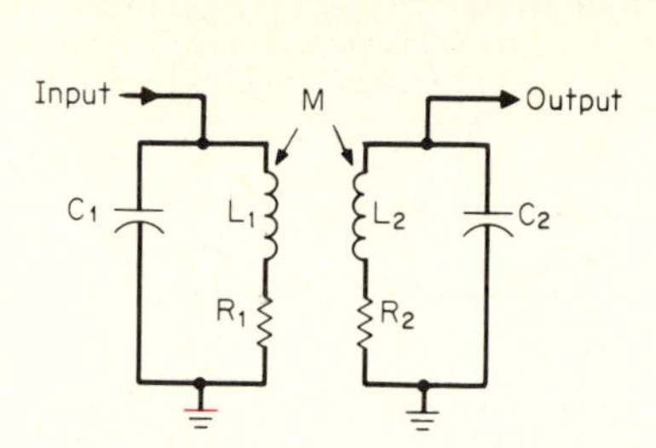

Fig. 10.21 Coupled tuned circuits.

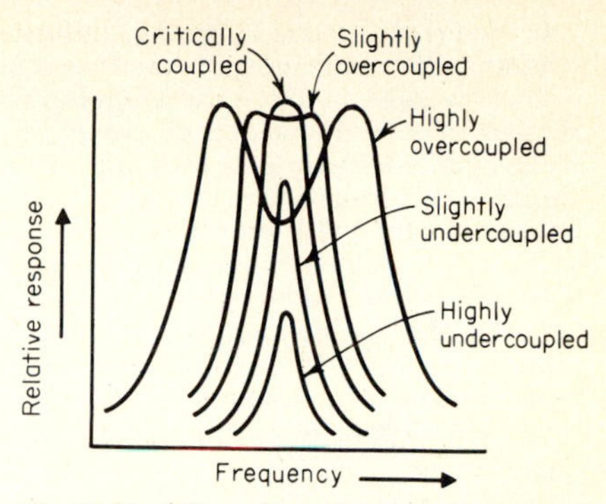

Fig. 10.22 Effect of coupling on selectivity of coupled tuned circuits.

CRITICAL COEFFICIENT OF COUPLING A related concept of importance is the critical coefficient of coupling k_c.

$$k_c = \frac{1}{\sqrt{Q_1 Q_2}} \tag{10.21}$$

where k_c = critical coefficient of coupling
Q_1 = Q factor of the primary circuit
Q_2 = Q factor of the secondary circuit

The critical coefficient of coupling is that value of k which will produce a maximum current in the secondary winding of the transformer in Fig. 10.21. With the usual values of Q, the critical coefficient of coupling will be quite small.

example 10.8 Assume that the two identical circuits of Example 10.6, each with a Q of 62.8, are coupled together. What is the value of the critical coefficient of coupling k_c?

solution From Eq. (10.21),

$$k_c = \frac{1}{\sqrt{Q_1 Q_2}} = \frac{1}{\sqrt{62.8^2}} = 0.016$$

Practically, critical coupling is the condition for maximum energy transfer between the primary and secondary coupled circuits. This is of importance in many (but not all) applications. (For example, in transmitters or other applications where significant amounts of power are being handled, it is desired to transfer the maximum power from one stage to the next. Furthermore, critical coupling is the largest value of the coupling coefficient for which two circuits tuned to the same frequency will exhibit a single-peaked response. The significance of this point will become more apparent from the discussion that follows.

In some applications of coupled tuned circuits, the application is essentially one involving a transformer consisting of a primary winding L_1 and a secondary winding L_2. The primary objective is to transfer signals from one stage to the next. The transformer is chosen as the most convenient means of accomplishing this signal transfer, and the fact that the two transformer windings are tuned is hardly more than incidental.

In such a case, the coefficient of coupling is usually kept somewhere below the critical value, and the two tuned circuits behave more or less independently of each other, almost as though they are isolated by an intervening amplifier stage. Tuned interstage transformers, such as used in standard broadcast receivers, are an example of coupled tuned circuits with low values of coefficient of coupling.

As the coupling coefficient increases, the effective overall bandwidth also increases. When the coefficient of coupling is at the critical value, the overall bandwidth is the widest possible without departing from the single-peaked response that is typical of a single-tuned circuit. If the coefficient of coupling is increased to a value greater than the critical value, the overall bandwidth continues to increase, but it now has a two-peaked, or *double-hump,* shape. The effect of coefficient of coupling on the selectivity response (bandwidth) of coupled tuned circuits is illustrated in Fig. 10.22.

It is seen from the foregoing discussion and from Fig. 10.22 that coupled tuned circuits, with coefficients of coupling at or above the critical value, represent another means (in addition to the use of multiple-tuned circuits as discussed in the preceding section) of obtaining wideband circuit response while preserving good skirt selectivity. Even with coefficients of coupling so high as to produce a distinct double-humped response, the lack of flatness may not pose a serious overall performance problem. The increased bandwidth thus obtained may be well worth whatever slight performance sacrifice is involved by virtue of the lack of an absolutely flat response.

In other cases, where wide bandwidth is needed but the overall response must be kept flat, the double-humped response of the *overcoupled* (that is, coefficient of coupling in excess of the critical value) circuit can be combined with an isolated single-tuned circuit to produce an overall flat response. The result resembles the triple-tuned circuit response of Fig. 10.20.

10.9 TEMPERATURE COMPENSATION OF TUNED CIRCUITS

Inductors exhibit a *positive temperature coefficient.* This means that as temperature increases, the value of inductance also increases. Temperature coefficient, abbreviated TC, is generally expressed as a change in the value of a component in parts per million per degree Celsius change in temperature (ppm/°C). Capacitors can have a positive, negative, or zero temperature coefficient. The value of a component with a negative temperature coefficient *decreases* with increasing temperature and increases with decreasing temperature. A zero temperature coefficient means that there is no significant change in component value with reasonable changes in temperature.

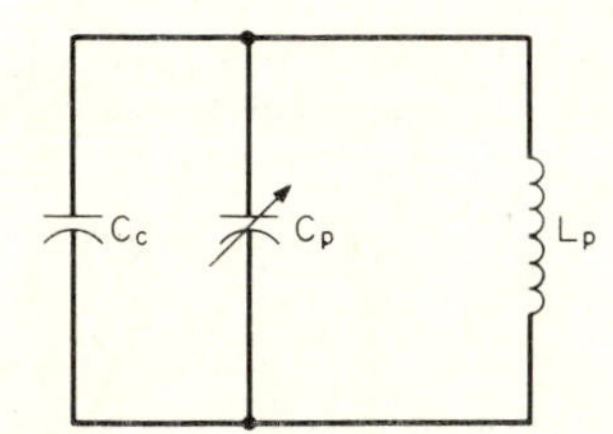

Fig. 10.23 Temperature-compensating capacitor C_c across a parallel-tuned circuit.

With the proper application of temperature-coefficient capacitors in a tuned circuit, the resonant frequency of the circuit does not change with temperature. This is called *temperature compensation.* If the tuned circuit is not compensated, the values of L and C change with temperature, and this results in a change in resonant frequency. This undesired circuit behavior is referred to as *temperature drift.* Temperature drift is very detrimental to the performance of precisely tuned circuits, such as those used in oscillators.

If the temperature coefficient is positive, the letter P precedes the value of the coefficient. If the temperature coefficient is negative, the letter N precedes the coefficient. For zero temperature coefficient, designation NPO is used.

Consider the parallel-tuned circuit of Fig. 10.23 where it is assumed that there is negligible resistance. Capacitor C_c is a temperature-compensating capacitor. Let the total parallel capacitance be $C_T = C_p + C_c$. Including the effect of temperature coefficient, the value of the parallel capacitor combination is expressed by

$$C_T(1 + \mathrm{TC}_T)$$

and the inductor by

$$L_p(1 + TC_L)$$

where TC_T = temperature coefficient for the parallel capacitors
TC_L = temperature coefficient for the inductor

If these values are substituted in Eq. (10.12),

$$f_r = \frac{1}{2\pi \sqrt{LC}}$$

we obtain

$$f_r = \frac{1}{2\pi \sqrt{L_p(1 + TC_L)C_T(1 + TC_T)}} \tag{10.22}$$

Expanding the term under the radical sign of Eq. (10.22), we obtain

$$L_pC_T(1 + TC_L + TC_T + TC_LTC_T) \tag{10.23}$$

Because TC_L and TC_T are extremely small (ppm), their product TC_LTC_T is much less than either TC_L or TC_T and, therefore, may be neglected in Eq. (10.23). This results in

$$L_pC_T(1 + TC_L + TC_T) \tag{10.24}$$

If $TC_L = -TC_T$, Eq. (10.24) is equal to L_pC_T, and the resonant frequency is *independent* of temperature.

SELECTING A TEMPERATURE-COMPENSATING CAPACITOR Capacitors may be combined in series or parallel, as shown in Fig. 10.24 to obtain a desired temperature

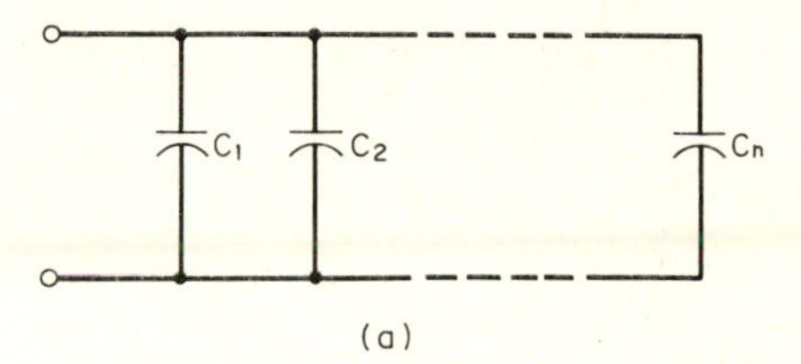

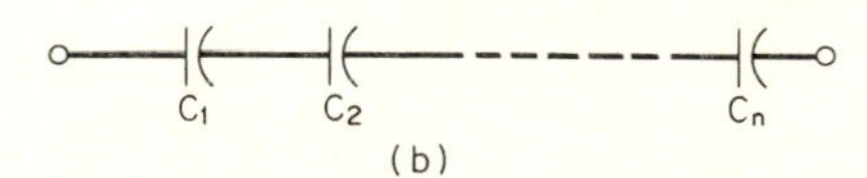

Fig. 10.24 Capacitors in parallel and in series: (*a*) Capacitors in parallel. (*b*) Capacitors in series.

compensation. For a number of capacitors in parallel (Fig. 10.24*a*), the equivalent temperature coefficient TC_P for the resulting combination is

$$TC_P = \frac{C_1(TC_1) + C_2(TC_2) + \cdots + C_n(TC_n)}{C_1 + C_2 + \cdots + C_n} \tag{10.25}$$

For the capacitors in series of Fig. 10.24*b*, the equivalent temperature coefficient TC_S is

$$TC_S = C_T\left(\frac{TC_1}{C_1} + \frac{TC_2}{C_2} + \cdots + \frac{TC_n}{C_n}\right) \tag{10.26}$$

where TC_1 = temperature coefficient of capacitor C_1
TC_2 = temperature coefficient of capacitor C_2, etc.

$$C_T = 1/\left(\frac{1}{C_1} + \frac{1}{C_2} + \cdots + \frac{1}{C_n}\right)$$

example 10.9 Two capacitors connected in parallel have the following values: $C_1 = 50$ pF, $TC_1 = P200$, $C_2 = 30$ pF, and $TC_2 = N500$. Find the temperature coefficient for the combination and the total capacitance.

solution Remember that P refers to a positive (+) temperature coefficient, and N refers to a negative (−) temperature coefficient. For two capacitors in parallel, Eq. (10.25) reduces to

$$\text{TC}_P = \frac{C_1(\text{TC}_1) + C_2(\text{TC}_2)}{C_1 + C_2} \tag{10.27}$$

Substitution of the given problem values in Eq. (10.27) yields

$$\text{TC}_P = \frac{50 \times 200 + 30(-500)}{50 + 30}$$

$$= \frac{-5000}{80} = -62.5 \text{ ppm} = N62.5$$

The total capacitance $C_T = C_1 + C_2 = 50 + 30 = 80$ pF.

example 10.10 For the parallel-tuned circuit of Fig. 10.23, assume that inductor L_p has a positive temperature coefficient of +30 ppm, $C_p = 200$ pF, $\text{TC}_p = +50$ ppm, and $C_c = 30$ pF. Determine the temperature coefficient TC_c for the temperature-compensating capacitor.

solution From Eq. (10.24) it was found that $\text{TC}_T = -\text{TC}_L$ for temperature compensation of a tuned circuit. Substitution of this and the problem values in Eq. (10.27) yields

$$-30 = \frac{20(\text{TC}_c) + 200(50)}{220}$$

Solving for TC_c, we obtain

$$-30 \times 220 = 20(\text{TC}_c) + 10\ 000$$

or

$$\text{TC}_c = \frac{16\ 600}{20} = -830 \text{ ppm} = N830$$

A 20 pF, N830 capacitor is therefore required for temperature compensation.

PRACTICAL CONSIDERATIONS Ceramic capacitors are available in a wide range of temperature coefficient values, and they are, therefore, generally used for the temperature compensation of tuned circuits. Other types of capacitors have tolerances on their values, and their temperature coefficient may not be constant with changes in temperature value. It is usually required to make further calculations or take temperature runs to determine the tracking of a compensated circuit with temperature.

Chapter 11

Filters

11.1 INTRODUCTION

The term *filter* is used to describe a wide variety of circuits that are frequency-selective. Certain frequencies will pass through a given filter while others will be attenuated. Electric filters are analogous to lint filters on automatic dryers that allow air to pass through but trap the lint. Another analogy can be made with the fuel-line filter of an automobile. It traps dirt, but allows gasoline to go to the carburetor. By way of comparison, electric filters will pass some frequencies and trap (or attenuate) others.

Knowing which frequencies are to pass and which are to be rejected, the circuit designer must decide which filter *configuration* is to be employed, and he must calculate the values of the components used. (The configuration is the method of electrically interconnecting the parts.) In this chapter we will discuss the advantages and disadvantages of the different configurations, and the methods used for determining their component values.

11.2 CLASSIFICATION OF FILTERS

There are many different ways of classifying or identifying filters. For example, they are sometimes identified by the layout of the components. Thus, a *pi filter* looks like the Greek letter π, and a *ladder circuit* looks like a ladder. Filters may be identified by the method used to design them. A *constant-k* filter is designed on the basis of a constant k which is related to the impedances of the circuit. An *m-derived filter* is based on a constant m which is derived from the constant k. Another way of identifying filters is by the shape of their characteristic curves. Thus, a *sharp cutoff filter* makes a very sharp distinction between which frequencies will pass through it and which will not. In some cases, the name of the person who originated the design or method of calculation is used for identifying the filter circuit. *Chebyshev filters* and *Butterworth filters* are examples of these.

One widely used scheme for classifying filters depends on the range of frequencies that the filter passes and rejects. According to this view, there are only four basic filter types: *low-pass, high-pass, bandpass,* and *band rejection.* All filter configurations can be placed in one of these categories.

The characteristic curves of Fig. 11.1 show the differences among the four basic categories. The low-pass filter of Fig. 11.1*a* permits all frequencies below f_c to pass, but rejects all frequencies above f_c. The high-pass filter (Fig. 11.1*b*) rejects all frequencies below f_c but passes all others. The bandpass filter (Fig. 11.1*c*) allows only a limited range of frequencies to pass. The band reject filter (Fig. 11.1*d*) permits all frequencies, except those in a limited range, to pass.

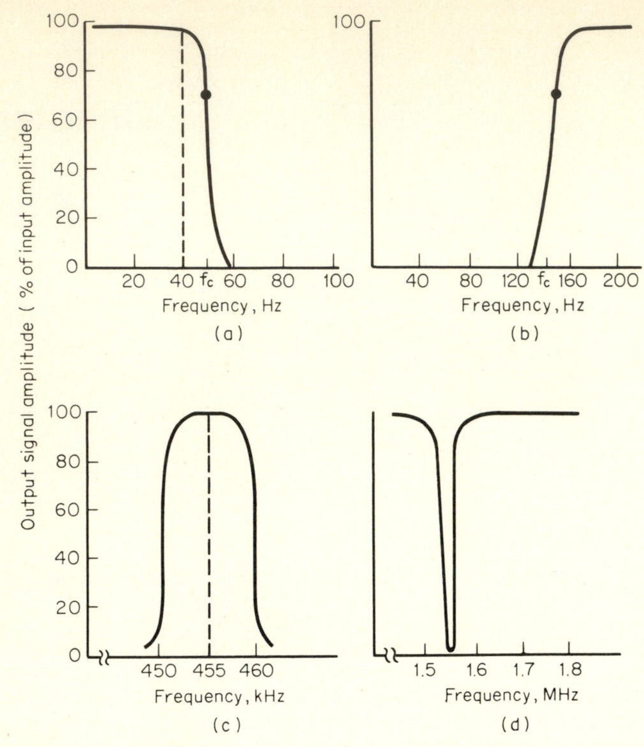

Fig. 11.1 Characteristic curves for four basic types of filters: (*a*) Low-pass filter. (*b*) High-pass filter. (*c*) Bandpass filter. (*d*) Band-reject filter; also called bandstop filter.

Filters composed of *R*, *L*, and *C* elements may be referred to as *passive filters*. Another type that utilizes a high-gain operational amplifier is the *active filter*. Besides an amplifier, an active filter also requires an external power source. The most commonly used active filter contains *R* and *C* elements.

The principal advantage of the active filter over the passive type is that inductors are not required in the former. This is especially important at the lower frequencies where inductors become excessively large and lossy.

APPLICATIONS The applications for filters are so numerous in electronics that only a few examples will be cited here. In terms of frequency range, filters are used at frequencies from less than a hertz in seismology to many gigahertz in microwaves.

In the rectification of alternating current to a unidirectional current it is required to filter the ripple in order to obtain an output approximating pure direct current. The simplest filter for this purpose is a capacitor across the supply output terminals, as shown in Fig. 11.2. In this application, a large-value capacitor is placed in parallel with the load resistance (Fig. 11.2*a*). The capacitor appears as an "open" circuit to direct current, while to alternating current it looks almost like a "short" circuit. The capacitor filter is most effective for small load currents (large values of load resistance). An example of such an application is the high-voltage power supply for a cathode-ray tube in which the anode current is typically on the order of a few hundred microamperes. As load current increases, the ripple increases and the regulation becomes poor.

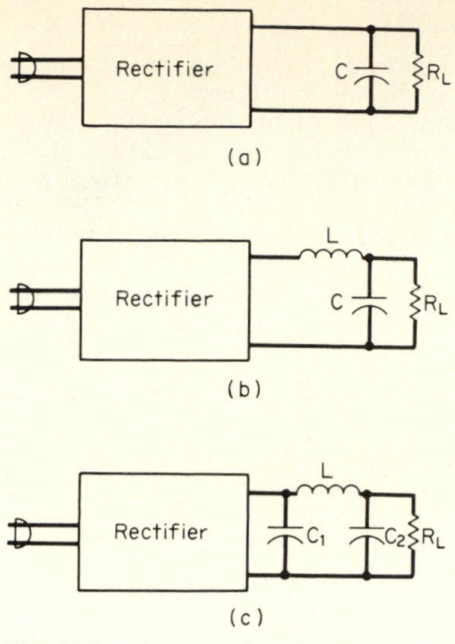

Fig. 11.2 Filters used with power supplies: (*a*) Pure *C*. (*b*) L section. (*c*) Pi.

Where appreciable load current is drawn from the supply, an L-section or pi filter (Fig. 11.2*b* and *c*) is required. In the L-section filter a large value of inductance, typically 10–30 H, referred to as a *choke*, is placed in series with the load resistance.

Placing a capacitor before the L-section configuration yields the pi filter of Fig. 11.2*c*. Greater ripple attenuation and dc output voltage are realized with this circuit than with the L-section filter.

Figure 11.3 illustrates some examples of filters found in communications circuits.

The *RC*-coupled network of Fig. 11.3*a*, employed in cascading ac amplifier stages, behaves as a filter. Low frequencies and direct current are attenuated, while the higher frequencies pass with minimum attenuation. Another example of a filter is the capacitor bypassing the emitter resistance in Fig. 11.3*b*. This circuit also tends to attenuate the lower frequencies. In fact, the low-frequency response of an ac amplifier is largely determined by the coupling and bypass networks which behave as filters.

At higher frequencies, the shunt capacitance of the active device, such as a transistor, and shunt stray and wiring capacitances, limit the high-frequency response of the amplifier. The shunting capacitance is lumped as capacitor C_s in Fig. 11.3*c*. Capacitor C_s and its associated resistors behave as a filter that attenuates the higher frequencies. This accounts for the falling off of high frequencies in an amplifier.

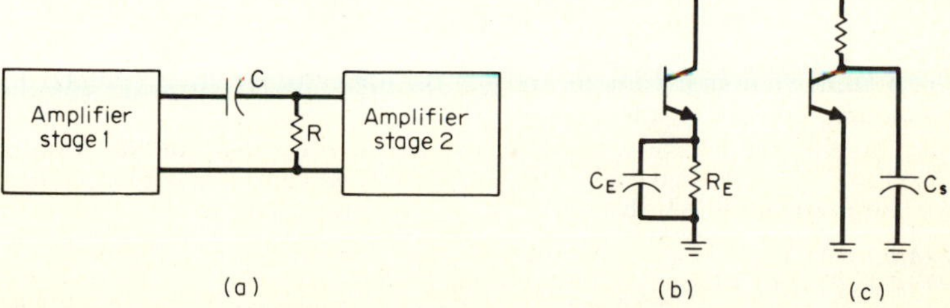

Fig. 11.3 Examples of filters found in communications circuits: (*a*) *RC* coupling. (*b*) Bypassing emitter resistor. (*c*) Shunt capacitance across amplifier output.

11.3 FILTER TERMINOLOGY

Before making a detailed study of filter theory, we will first review some of the more important terms that are applicable to this area.

The *configuration* of a circuit is simply the layout of its parts in relation to one another. An example is the pi filter shown in Fig. 11.2*c*. The parts are arranged on the schematic diagram in the form of the Greek letter π, and we can say that the filter has that configuration.

All filter circuits can be arranged into one of two possible configurations: the *lattice* and the *ladder*. These are illustrated in Fig. 11.4. The filter *elements* (that is, the inductors, capacitors, and resistors) are shown as separate impedances labeled Z. The most important difference between the ladder and the lattice is symmetry. The lattice,

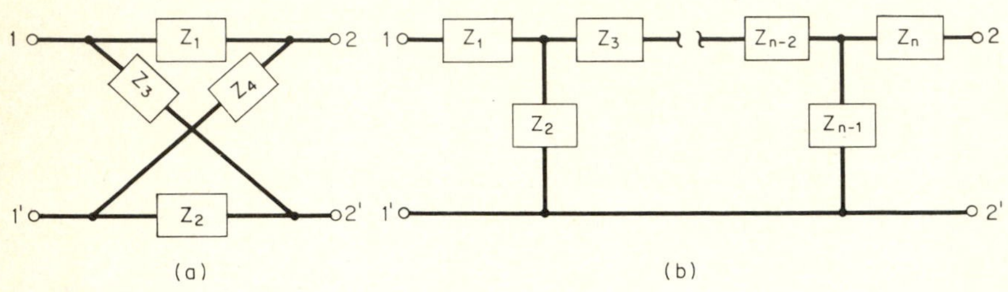

Fig. 11.4 Typical filter configurations: (*a*) The bridge, or lattice, circuit. (*b*) The ladder circuit.

or bridge, is *symmetrical*. This means that each terminal has a mirror-image counterpart on the opposite side of the circuit. An example is the equivalence of Z_1 and Z_2 or Z_3 and Z_4 in Fig. 11.4*a*. The ladder circuit does not have this property of symmetry. Because the lattice is symmetrical, it may drive (or be driven by) balanced loads, such as center-tapped transformers. The ladder can only drive unbalanced loads, such as a grounded resistor, tube grid, or transistor base.

The term *balance* always means with respect to ground. Figure 11.4*b* shows a ladder-circuit configuration. As an additional way of distinguishing between the circuits, any of the terminals of the lattice may be grounded without affecting the circuit performance while only terminals 1′ and 2′ of the ladder can be grounded.

Although there may be variations of these configurations, the ladder and the lattice are the basis of all filter structures. The type of element may change, i.e., capacitive, inductive, or resistive, but the configuration will not.

Insertion loss is the loss in signal strength that the frequencies in the passband experience in passing through the filter. (The *passband* of a filter is the range of frequencies that are supposed to be able to pass through the circuit.) If the filter were absent, and the source and load were connected directly to each other, the output signal would increase by the amount of the insertion loss.

The bandwidth of a filter is the frequency difference between two points on the filter-response curve that have a specified insertion loss. The points on the curve generally used to define bandwidth are the half-power, or −3-dB, points. These are the points where the filter response is down 3 dB from the maximum point on the curve. On the curve of Fig. 11.5, the bandwidth is between f_1 and f_2, where f_1 is the lower frequency at a point 3 dB down from maximum, and f_2 is the upper frequency at a point 3 dB down from maximum. The formula for bandwidth is

$$\text{Bandwidth, Hz} = f_2 - f_1 \tag{11.1}$$

where f_1 and f_2 are at the −3-dB points.

example 11.1 Two points on a response curve that are at the −3-dB points are at 10 and 1000 Hz for f_1 and f_2 respectively. What is the bandwidth?

solution

$$\begin{aligned}\text{Bandwidth, Hz} &= f_2 - f_1 \\ &= 1000 - 10 \text{ Hz} = 990 \text{ Hz}\end{aligned}$$

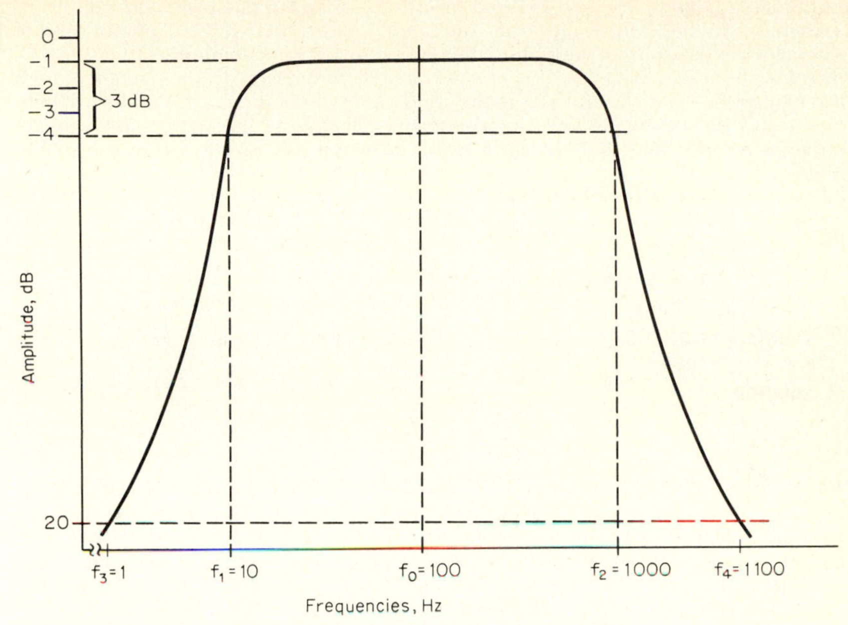

Fig. 11.5 A typical bandpass filter response.

Instead of expressing the bandwidth as the difference between two frequencies, it is sometimes expressed as a percent of frequency (*percent bandwidth*). For example, if a 100-kHz bandpass filter has a bandwidth of 10 percent, its bandwidth is 10 percent of 100 kHz, or 10 kHz. This can be expressed as ±5 percent, or ±5 kHz with respect to 100 kHz.

The *center frequency* F_0 may be defined as either the *arithmetic* or the *geometric center*. For narrow bandwidths these are practically equal. The formulas are:

$$F_0 = \frac{f_1 + f_2}{2} = \begin{pmatrix}\text{Arithmetic center}\\ \text{frequency, Hz}\end{pmatrix} \tag{11.2}$$

$$F_0 = \sqrt{f_1 \times f_2} = \begin{pmatrix}\text{Geometric center}\\ \text{frequency, Hz}\end{pmatrix} \tag{11.3}$$

Using our previous example,

$$F_0 = \frac{1000 + 10}{2} = 505 \text{ Hz} \qquad \text{(arithmetic center)}$$

$$F_0 = \sqrt{1000 \times 10} = 100 \text{ Hz} \qquad \text{(geometric center)}$$

If a filter-response curve is plotted on graph paper having the abscissa (horizontal axis) marked in evenly spaced frequencies, then the arithmetic center frequency is exactly halfway between the −3-dB points. As an example, for the response curve of Fig. 11.1*c* the arithmetic center is 455 kHz.

If a filter-response curve is plotted on graph paper having a logarithmic abscissa, then, the geometric center frequency will be exactly halfway between the −3-dB points. For the response curve of Fig. 11.5 the abscissa is logarithmically marked. Note that the same distances are used for 10 to 100 Hz as for 100 to 1000 Hz. The geometric center on this curve is 100 Hz, which is seen to be halfway between the −3-dB points.

The arithmetic center is used only for very narrow bandwidths; then the arithmetic and geometric centers are equal. In any case, the geometric center is always correct, and when we speak of the center frequency, we shall mean the geometric center.

Selectivity is a measure of the effectiveness of a filter in performing its function. If the filter were perfect, it would have no insertion loss in the passband and infinite attenuation at all other frequencies (the *stop band*). In practice, it takes a finite band of frequencies to change from very little attenuation to a greater amount.

One way to measure selectivity is by the *shape factor*. Shape factor is often defined as the ratio of the −3-dB bandwidth to the bandwidth where the insertion loss is 20 dB. (The −6- and −60-dB bandwidths are also used.) The closer this ratio is to a value of 1.0, the greater the selectivity of the filter. The shape factor for the response of Fig. 11.5 is

$$\text{Shape factor} = S = \frac{f_4 - f_3}{f_2 - f_1} \tag{11.4}$$

where f_4, f_3 = points on the response curve that are down 20 dB from maximum
f_2, f_1 = points on the response curve that are down 3 dB from maximum

example 11.2 Using the values of f_1, f_2, f_3, and f_4 in Fig. 11.5, what is the shape factor of the selectivity curve?

solution

$$S = \frac{f_4 - f_3}{f_2 - f_1} = \frac{1100 - 1}{1000 - 10} = 1.11 \qquad \textit{Answer}$$

When lumped components are used to make a filter, there are a number of undesirable distributed elements—called *parasitic elements*—that cause attenuation of the *desired* frequencies. Parasitic elements may also change the shape of the response curve from the desired characteristic. A few examples of parasitic elements are: losses in iron cores of chokes, distributed wiring capacitance, capacitance between windings of a choke, resistance of connecting wires, and leakage inductance. Parasitic elements cannot be eliminated completely. In some cases, they may have to be considered during the design procedure.

This is not a complete list of all terms used to describe filters, but it includes the most important ones for evaluating filters.

11.4 FILTER COMPONENTS

Passive filters may be classed as *resistive-reactive* type or *totally reactive*. The resistive-reactive type is called *RC* or *RL*, depending whether an inductor and resistor (*RL*) or capacitor and resistor (*RC*) are used. The totally reactive filter uses capacitors and inductors, with resistance present only as an undesired distributed component. These filters are also called *LC* types because of the components utilized in their construction.

The *RL* and *RC* types are sometimes employed to generate high-pass or low-pass responses, but special modifications must be made to these in order to generate the bandpass or band-rejection response. Even then they are not as efficient as the *LC* type, owing to their inherent resistance dissipation.

11.5 METHODS OF DESIGNING LC FILTERS

Two different approaches are used for designing *LC* filters. One method is to treat the filter as a transmission line. This approach is called the *image-parameter design* method. For this method, the lumped components (which in *LC* filters are inductors and capacitors) are described in terms of the distributed components in a transmission line. The other approach to filter design is called the *network method of design*. (It is also called the *insertion loss method*, or *polynomial method*, or the *exact method*.) It deals directly with the filter as a circuit having parameters instead of dealing with the transmissionline.

IMAGE-PARAMETER DESIGN The image-parameter method of design was developed during the conduction of early experiments on telephone transmission lines. It was found that the attenuation of telephone lines increased with frequency, owing to the large amount of distributed capacitance, especially in the early types of lines. Inserting inductors reduced the attenuation and made it fairly constant over a range of frequencies. It was also noted that the attenuation increased very quickly at a particular frequency called the *cutoff frequency*.

As the requirements for bandwidth in the transmission line increased, various schemes were used to simultaneously increase bandwidth and keep the attenuation low. As a result of experiments, image-parameter filters were developed. When the equivalent circuits of loaded transmission lines were constructed with lumped components, it was found that they had different characteristics from those of the loaded line. The bandwidth and attenuation were not the same as in the transmission line because the filter is only an approximation of a transmission-line circuit. This is one of the difficulties in designing such filters.

The image-parameter filter has evolved into two basic types: *the constant-k type*, and *the m-derived type*. The constant-k type is the result of the first attempt to build an artificial transmission line. When component values of the constant-k type are modified in a certain way so as to vary the impedance and the shape factor, the result is an m-derived filter.

NETWORK METHOD OF DESIGN Modern mathematics and the advent of the computer have done much to change filter design. Tables and nomographs obtained through the use of computers make it possible to obtain a desired filter characteristic with relative ease. Using only simple mathematics, it is possible to design a filter for almost any application and shape factor.

The image-parameter method of designing a filter is based on transmission lines. It attempts to simulate a transmission line with inductors and capacitors, that approximates the distributed constants in a coaxial cable, and obtain a certain impedance and frequency response simultaneously. For a long time it has been known that filters could be designed from response-curve formulas. The difficulty was that the equations were so long they were nearly impossible to solve. A computer was required for their solution, but at that time computers were not available for such applications.

When more and more computers became available, the required equations were solved, and tables of elements were generated that could be used in many different filter designs. It was desirable to make the tables useful for *any* frequency or impedance level; so they were developed in such a way that this information could be added later. Such tables are said to be *normalized*. The values are referenced to 1 Ω and 2π Hz (1 rad/s), meaning that the impedance and the cutoff frequency are 1 Ω and 2π Hz, respectively, for the filter elements listed. This is equivalent to the mechanical operation of *scaling* an object. To change the size of a scaled object, you simply divide or multiply by the number of times bigger or smaller that the object is to be. For example, if a scale model of a boat is $1/25$ of the actual size, then each and every dimension on the model must be multiplied by 25 to obtain its actual size. In a similar manner, the tables used for the network method of design can be considered to be scaled down. In order to get the desired values, it is necessary to multiply by the desired impedance in ohms and the desired frequency in hertz. By making the tables this way, they apply to *any* impedance and frequency.

We will give examples of each method of design, starting with the image-parameter approach. The constant-k and m-derived filters are designed by this method.

11.6 THE SIGNIFICANCE OF FILTER IMPEDANCE

Constant-k filters can be made in any of the three configurations shown in Fig. 11.6. Each of these is a form of ladder network, and is identified according to the alphabet letter that it forms. Thus we have an L, a T, and a Π section. A number of these sections can be added to obtain a sharper cutoff (or other desired characteristic).

An important property of constant-k filters—and for that matter, *all* types of filters—is their impedance. This is because of the *maximum power-transfer theorem*, which states that the maximum possible amount of power that can be delivered to a load occurs when the load impedance "matches" the internal impedance of the generator. For ac generators, the match must be achieved by using a *conjugate impedance*. Thus, if the internal impedance of the generator is $7 + j11$ Ω, then maximum power will be delivered to the load impedance when its value is $7 - j11$ Ω.

Figure 11.7 shows the effect of the filter on the impedance match. Suppose we have

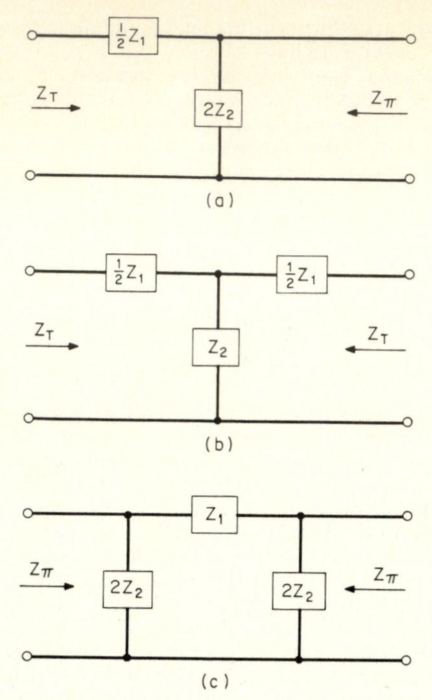

Fig. 11.6 Configurations of the basic filter sections: (*a*) An L section. (*b*) A T section. (*c*) A pi section.

a circuit like that of Fig. 11.7*a*, in which the load impedance Z_L is matched to the internal impedance Z_i of the generator *G*. In accordance with the maximum power-transfer theorem, the greatest amount of power (for this generator) is being delivered to the load. Now if we insert a filter circuit between the generator and load, as shown in Fig. 11.7*b*, then the presence of the filter can result in an improper match of impedances. (It is important to note that inserting a transmission line between a generator and a load can also produce a mismatch, and therefore transmission-line impedance and filter impedance are both important, and have a similar effect on the circuit into which they are inserted.)

Suppose that the internal impedance and the load impedance of the circuit in Fig.

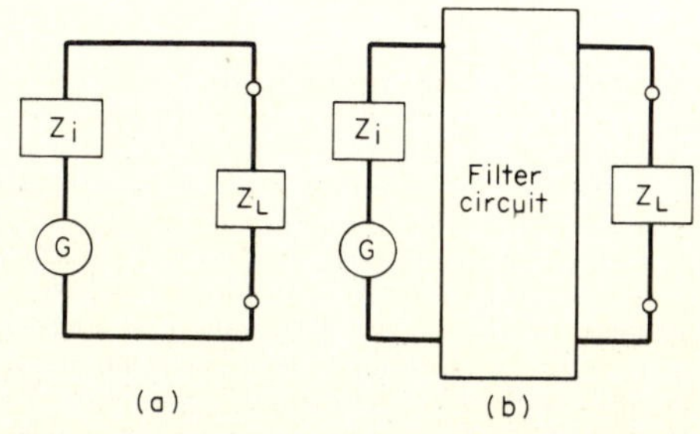

Fig. 11.7 Illustrating the importance of filter circuit impedance: (*a*) A circuit designed for maximum power transfer. (*b*) Unless properly designed, the filter circuit could unbalance the impedance match.

11.7*a* are conjugates, and they both have the same impedance — say 500 Ω. If the input impedance to the filter circuit is also 500 Ω — looking from either end — then, no impedance mismatch will occur between the generator and the load.

Now suppose the internal impedance of the generator is different from the load impedance. If the filter is designed in such a way that the generator operates into the correct impedance, *and* the load operates into the correct impedance, then the filter is actually serving two functions: elimination of unwanted frequencies and matching of impedances. Whenever a filter is designed in such a way that the generator "sees" the conjugate of its own internal impedance when looking into the filter output terminals, then the filter is said to be operating on an *image impedance basis*.

The *characteristic impedance* of a filter, and also of a transmission line, is the value of load impedance that must be connected across the output terminals in order to match the load to the filter (or line). This is also known as the *surge impedance*. If a filter, or transmission line, is terminated in a resistance that is equal to its impedance, then *all* the energy moving through the filter (or through the line) will be dissipated by that resistance.

The characteristic impedance Z_o of a filter or transmission line is given by the equation

$$Z_o = \sqrt{Z_{sc}Z_{oc}} \tag{11.5}$$

where Z_{sc} = input impedance with the output terminals short-circuited together
Z_{oc} = input impedance with the output terminals open

11.7 DESIGN OF CONSTANT-k FILTERS

The application of the constant-*k*, as well as other *LC* filters, is generally practical over a range of frequency up to 100 mHz. Their applications in communication circuits include the limiting of the signal spectrum to a receiver (preselector), separating or combining the spectrum so that parts of it may be individually processed (diplexers and diversity receiver combiners), separating r-f signals, and reducing interference.

Although constant-*k* filters are easy to design and make, they have a number of shortcomings. Their attenuation is low near its cutoff frequency, and their image impedance is not constant with frequency. Also, the phase shift in the passband is not proportional to frequency.

It is apparent when we examine the circuits of Fig. 11.6 that the impedance of a filter circuit will normally change with frequency. As an example, assume that Z_1 is an inductor and that both impedances marked $2Z_2$ are capacitors in the circuit of Fig. 11.6*c*. If the input frequency increases, then the inductive reactance increases, and at the same time the capacitive reactance decreases. The result is a change in filter impedance.
reactance decreases. The result is a change in filter impedance.

If the inductance and capacitance are chosen so that the increase in inductive reactance is exactly offset by the decrease in capacitive reactance, then their product will be a constant, which is called the constant *k*. Expressed mathematically,

$$k = \sqrt{Z_1Z_2} \tag{11.6}$$

where k = an impedance independent of frequency
Z_1, Z_2 = series and shunt impedances of Fig. 11.6

The attenuation of all frequencies will be a constant up to the cutoff frequency. If the *Q* of the reactive components in the filter is high, then the cutoff will be quite sharp. As the *Q* is made lower and lower, the reduction in attenuation beyond cutoff becomes more and more gradual.

LOW-PASS FILTER DESIGN Figure 11.8 shows three possible configurations for a constant-*k* low-pass filter. To design such a filter, it is necessary to know the cutoff frequency f_c — that is, the frequency the filter is to pass. This is designated f_c in Fig. 11.9. It is also necessary to know the impedance of the load to be connected across the output terminals of the filter. This load, which is designated as Z_o, must be the conjugate of the characteristic impedance of the filter in order for maximum power transfer to occur. Looking in from the output terminals, this is also the image impedance.

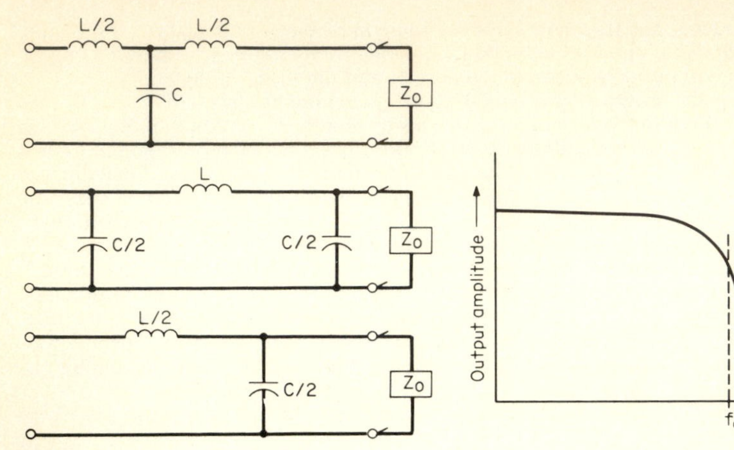

Fig. 11.8 Three forms of constant-k low-pass filters.

Fig. 11.9 Characteristic curve of a constant-k low-pass filter.

By knowing the value of f_c and Z_o, it is possible to design the filter from the following equations which are given here without derivation:

$$L = \frac{Z_o}{3.14 f_c} \tag{11.7}$$

$$C = \frac{1}{3.14 f_c Z_o} \tag{11.8}$$

where L = inductance, H
C = capacitance, F
Z_o = load impedance, Ω

Once we have calculated the required values of L and C, we can verify that this is a constant-k filter by substituting the values into the equation

$$Z_o = k = \sqrt{\frac{L}{C}} \tag{11.9}$$

where Z_o, L, C = values in Eqs. (11.8) and (11.7)
k = constant k, and is equal to the image impedance for all frequencies that the filter is designed to pass

The following example illustrates the design of a low-pass constant-k filter. The cutoff frequency is given as 1000 Hz. All frequencies above 1000 Hz are attenuated. A filter of this characteristic might be used in narrow-band transmission of information.

example 11.3 Design a constant-k low-pass filter that will attenuate frequencies above 1000 Hz (f_c = 1000 Hz). The load across the filter output terminals, Z_o, is a pure resistance of 600 Ω. In order to obtain sharp cutoff characteristics, use a three-section ladder configuration.

solution The greater the number of filter sections used, the sharper the cutoff—hence the three-section ladder. Not all applications could justify the cost of components, and in many cases it is not necessary that the cutoff be sharp.

Figure 11.10 shows the circuit. The three-section filter can be made by combining two L sections with a T section, as shown in Fig. 11.10*a*. The first step is to calculate the required value of L from Eq. (11.7).

$$L = \frac{Z_o}{3.14 f_c}$$

$$= \frac{600}{3.14 \times 1000} = 0.191 \text{ H} \tag{11.7}$$

As shown in Fig. 11.10*a*, each inductor in this illustration actually has a value of $L/2$, or 0.095 H.

The next step is to calculate the required value of C from Eq. (11.8).

$$C = \frac{1}{3.14 f_c Z_o}$$

$$= \frac{1}{3.14 \times 1000 \text{ Hz} \times 600 \ \Omega} = 0.53 \ \mu\text{F} \qquad (11.8)$$

Each capacitor in Fig. 11.10*a* actually has a value of $C/2$, or 0.265 μF.

Figure 11.10*b* shows the circuit with the calculated component values. When the sections are combined to produce the required ladder filter, the two halves of L are combined to produce $L(L/2 + L/2 = L)$ as shown in Fig. 11.10*c*.

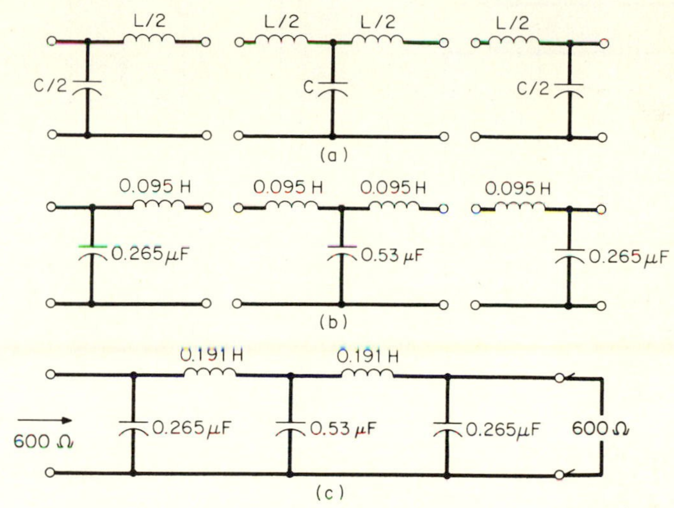

Fig. 11.10 An example of a constant-*k* filter design: (*a*) Prototype. (*b*) Element values. (*c*) Composite filter design.

If this is truly a constant-*k* filter, then the characteristic impedance will be 600 Ω, as calculated from Eq. (11.9).

$$Z_o = k = \sqrt{\frac{L}{C}} = \sqrt{\frac{0.191}{0.53 \times 10^{-6}}}$$

$$\simeq \sqrt{360\,000} \simeq 600 \ \Omega$$

It will not be possible to obtain capacitors and inductors with the precise computed values for the filter. Any compromise in component values will also result in a compromise in shape factor and other characteristics.

The impedance of the filter must be identical with the load impedance because the circuit is the same when viewed from either end. In practice, an improved impedance match to the source and load can be obtained by adding an *m*-derived half-section at each end of the constant-*k* filter of Fig. 11.10*c*. These added half-sections would be designed to match the image impedance of the filter to the resistive terminations of the source and the load.

HIGH-PASS CONSTANT-k FILTER DESIGN The arrangement of components for a high-pass constant-*k* filter design and the resulting characteristic curve are shown in Fig. 11.11. Figure 11.11*a* shows the three possible configurations for a constant-*k* high-pass filter. As with the low-pass filters, these configurations are named according to the letter of the alphabet that they are similar to: T, π, and L. The equations for calculating the value of L and C are given here without derivation.

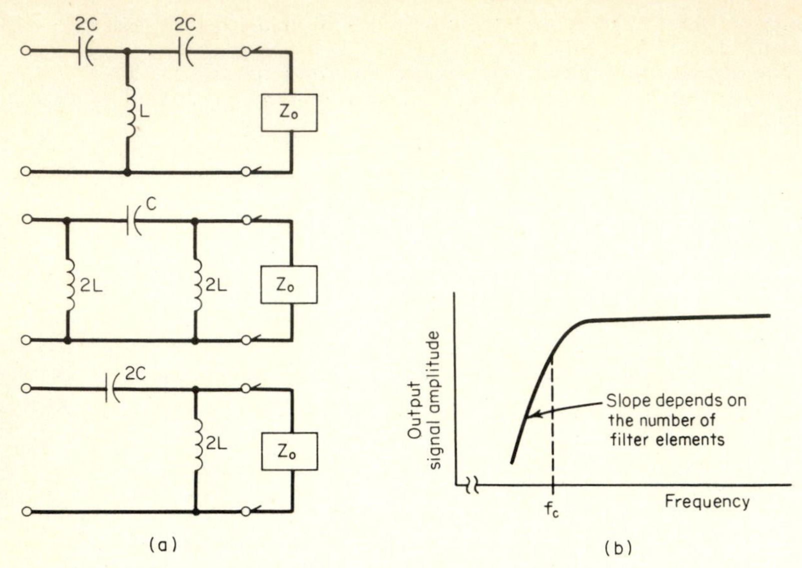

Fig. 11.11 Circuits and characteristic curves of constant-*k* high-pass filter: (*a*) Three forms of constant-*k* high-pass filter circuits. (*b*) Characteristic curve.

$$L = \frac{Z_o}{12.6 f_c} \tag{11.10}$$

$$C = \frac{1}{12.6 f_c Z_o} \tag{11.11}$$

where L = inductance, H
C = capacitance, F
Z_o = load impedance, Ω

Components L and C are identified in the circuit of Fig. 11.11*a*. The characteristic impedance is Z_o, which is also the value of k, and also the value of the load impedance. The cutoff frequency is f_c and its meaning is indicated in Fig. 11.11*b*.

BANDPASS FILTER DESIGN The constant-*k* bandpass filter circuit and its characteristic curve are shown in Fig. 11.12.

The series circuits in Fig. 11.12*a* (comprised of L_1 and C_1) are resonant to the frequency being passed. Actually, when L_1 and C_1 are tuned to the center frequency of the passband, there is very little opposition at that frequency. The parallel circuit

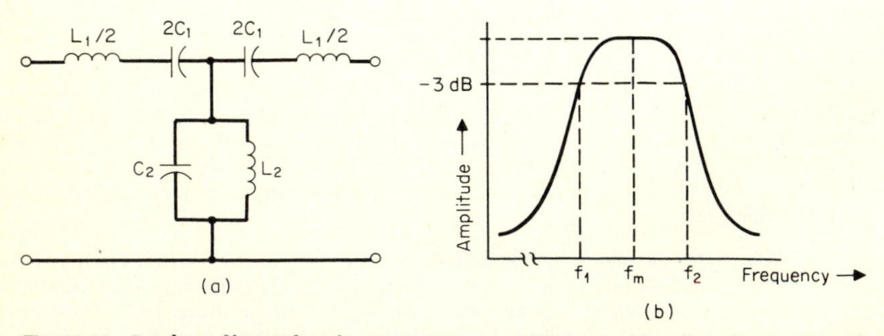

Fig. 11.12 Bandpass filter and its characteristic curve: (*a*) Constant-*k* bandpass filter. (*b*) Bandpass filter characteristic curve.

(comprised of L_2 and C_2) is virtually an open circuit to the center frequency—which is also the parallel resonant frequency.

The characteristic curve of this filter circuit, shown in Fig. 11.12*b*, also defines the frequencies used in the calculations of inductance and capacitance. The equations for L_1, L_2, C_1, and C_2 are given here without derivation.

$$L_1 = \frac{Z_o}{3.14(f_2 - f_1)} \tag{11.12}$$

$$L_2 = \frac{(f_2 - f_1)Z_o}{12.6 f_1 f_2} \tag{11.13}$$

$$C_1 = \frac{f_2 - f_1}{12.6 f_1 f_2 Z_o} \tag{11.14}$$

$$C_2 = \frac{1}{3.14(f_2 - f_1)Z_o} \tag{11.15}$$

where L_1, L_2 = are identified in Fig. 11.12*a* and are measured in henrys
C_1, C_2 = are identified in Fig. 11.12*a* and are measured in farads
f_1, f_2 = limits of bandwidth at the −3-dB points
Z_o = characteristic impedance, Ω

The constant k is equal to the image impedance, and is defined by the equation

$$k = \sqrt{\frac{L_1}{C_2}} \tag{11.16a}$$

$$= \sqrt{\frac{L_2}{C_1}} = Z_o \tag{11.16b}$$

The geometric center frequency f_m is given by the equation

$$f_m = \sqrt{f_1 f_2}$$

$$f_m = \frac{1}{6.28\sqrt{L_1 C_1}}$$

$$= \frac{1}{6.28\sqrt{L_2 C_2}} \tag{11.17}$$

In the next example, the design of a bandpass filter is illustrated. Such a filter may find application in a spectrum analyzer, where a band of frequencies is to be viewed at the exclusion of all other frequencies.

example 11.4 (*a*) Design a bandpass filter for passing frequencies between 90 and 110 kHz. The load is 1000 Ω. (*b*) Calculate its center frequency.

solution (*a*) The bandpass filter circuit is shown in Fig. 11.12*a*. The load impedance is 1000 Ω, and this must equal the characteristic impedance Z_o. The lower frequency limit f_1 is 90 kHz, and the upper frequency limit is 110 kHz. With these values, the required inductances, L_1 and L_2, and the capacitance values, C_1 and C_2, can be calculated.

$$L_1 = \frac{Z_o}{3.14(f_2 - f_1)}$$

$$= \frac{1000\ \Omega}{3.14(110\text{ kHz} - 90\text{ kHz})} = 15.9\text{ mH} \tag{11.12}$$

$$L_2 = \frac{(f_2 - f_1)Z_o}{12.6 f_1 f_2}$$

$$= \frac{(110\text{ kHz} - 90\text{ kHz})(1000\ \Omega)}{12.6(90\text{ kHz})(110\text{ kHz})} = 160\ \mu\text{H} \tag{11.13}$$

$$C_1 = \frac{f_2 - f_1}{12.6 f_1 f_2 Z_o}$$

$$= \frac{110\text{ kHz} - 90\text{ kHz}}{12.6(90\text{ kHz})(110\text{ kHz})(1\text{ k}\Omega)} = 160\text{ pF} \tag{11.14}$$

$$C_2 = \frac{1}{3.14(f_2 - f_1)Z_o}$$

$$= \frac{1}{3.14(20 \text{ kHz})(1 \text{ k}\Omega)} = 0.0159 \ \mu\text{F} \qquad (11.15)$$

If these values are correct, then the value of the constant k will be equal to the characteristic impedance of the filter. This can be checked with Eq. (11.16a):

$$k = \sqrt{\frac{L_1}{C_2}} = \sqrt{\frac{15.9 \times 10^{-3}}{0.0159 \times 10^{-6}}}$$

$$k = Z_o = 1000 \ \Omega$$

The value of k can also be computed from Eq. (11.16b):

$$k = \sqrt{\frac{L_2}{C_1}} = \sqrt{\frac{160 \times 10^{-6}}{160 \times 10^{-12}}}$$

$$k = Z_o = 1000 \ \Omega$$

The value of k has been calculated in two different ways, using L_1 and C_2, and also L_2 and C_1. In both cases it is equal to the characteristic impedance required to match the filter to the load. This, in turn, is the image impedance as seen by the load.

(b) The center frequency of the band of frequencies passed by the filter is obtained from Eq. (11.3).

$$F_0 = \sqrt{f_1 \times f_2} = \sqrt{90 \times 110 \text{ kHz}} = 99.5 \text{ kHz}$$

This is the geometric center frequency which is sometimes designated f_m, the subscript m standing for midfrequency. The arithmetic frequency, when calculated from Eq. (11.2) is 100 kHz, which is close to the geometric mean.

BAND-REJECTION FILTER DESIGN Band-rejection filters are sometimes called *band-stop filters*, or *notch filters*.

Figure 11.13 shows a constant-k band-rejection filter circuit and its characteristic curve. The parallel-tuned circuits, L_1 and C_1, allow all frequencies to pass through the

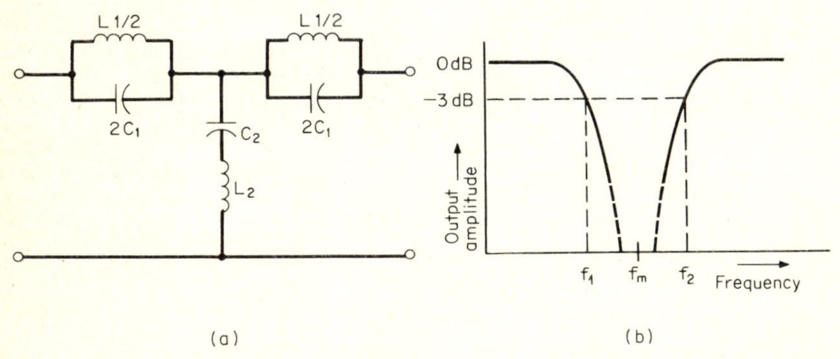

Fig. 11.13 Band-stop filter and its characteristic curve: (a) Band-rejection filter. (b) Filter response.

filter except the frequencies near resonance. The resonant frequency of these tuned circuits is the center frequency of the band being rejected. The series-resonant circuit, L_2 and C_2, short-circuits the unwanted frequency to ground. The width of the band being rejected is a direct function of the circuit Q. The slope of the curve—that is, the sharpness of tuning—is increased by adding filter sections.

The equations for L_1, L_2, C_1, and C_2 are given here without derivation.

$$L_1 = \frac{(f_2 - f_1)Z_o}{3.14 f_1 f_2} \qquad (11.18)$$

$$L_2 = \frac{Z_o}{12.6(f_2 - f_1)} \qquad (11.19)$$

$$C_1 = \frac{1}{12.6(f_2 - f_1)Z_o} \qquad (11.20)$$

$$C_2 = \frac{f_2 - f_1}{3.14 f_1 f_2 Z_o} \qquad (11.21)$$

where L_1, L_2 = are identified in Fig. 11.13*a* and are measured in henrys
C_1, C_2 = are identified in Fig. 11.13*a* and are measured in farads
f_1, f_2 = limits of the bandwidth being rejected as measured at the 3-dB points
Z_o = characteristic impedance, Ω

The constant k is equal to the image impedance and is defined by the equation

$$k = \sqrt{\frac{L_1}{C_2}} \qquad (11.22a)$$

$$k = \sqrt{\frac{L_2}{C_1}} = Z_o \qquad (11.22b)$$

The geometric center frequency f_m is given by the equation

$$f_m = \sqrt{f_1 f_2} = \frac{1}{6.28\sqrt{L_1 C_1}} = \frac{1}{6.28\sqrt{L_2 C_2}} \qquad (11.23)$$

11.8 SUMMARY OF CONSTANT-k FILTER CHARACTERISTICS

In each of the relationships given for the design of constant-k filters, it was necessary to know the value of the load impedance to be placed across the filter output terminals. The values of inductances and capacitances are always dependent upon (among other things) this impedance value. Use of the constant k was not necessary, and in our examples the only use of constant k was to show that it is equal to the image impedance. However, the derivations of the equations for L and C, which are rather tedious and were not given, *do* depend on use of the constant k.

The shapes of the characteristic curves for the constant-k filters cannot be controlled by the design engineer. He can, of course, add more sections to increase the selectivity, but the additional cost may not be justified. The sides of the selectivity curves of bandpass and band-stop filters are referred to as the *skirts*, and the shapes of these curves—as measured by the shape factor—are often referred to as the *skirt selectivity*. This is simply an indication of how sharply the circuit tunes, as is the shape factor. Constant-k filters are characterized by poor skirt selectivity when compared to other types of filters.

As we shall see in the next section, the first step in designing an m-derived filter is to design a constant-k filter. Then, the constant-k filter is modified by adding a reactance to either the series or shunt leg. Since the constant-k filter is designed first, it is often referred to as the *prototype*.

11.9 m-DERIVED FILTERS

Besides having improved skirt selectivity, the frequency of infinite attenuation can be readily selected in the m-derived filter. It still shares with the constant-k filter the disadvantage of not having a constant image impedance throughout its passband. The combination of constant-k and m sections, however, provides a flexibility in filter design not realized with the constant-k filter alone.

The constant-k filter has a rather gradual *roll-off*: that is, the amplitude of the signal does not change rapidly at the skirts. There are some applications where a sharper response is needed. This can be accomplished by adding reactance to either the series or parallel (shunt) branch. Figure 11.14 shows three examples with a reactance added to the shunt path. Note that the same configurations (T, π, and L) are employed as were used for constant-k filters. In fact, these are identical with the constant-k low-pass filters except that an inductor has been added to resonate with the shunt capacitance. The series elements L_1 are unchanged, and, therefore, these configurations are referred to as series-type m-derived filters. The m-derived filter, then, is

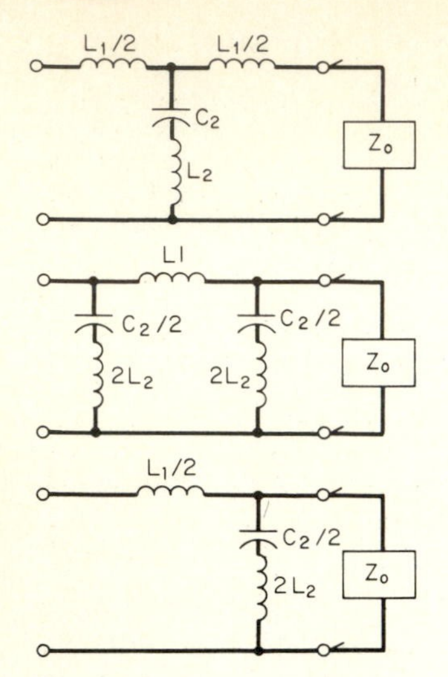

Fig. 11.14 Low-pass series-type m-derived filter configurations.

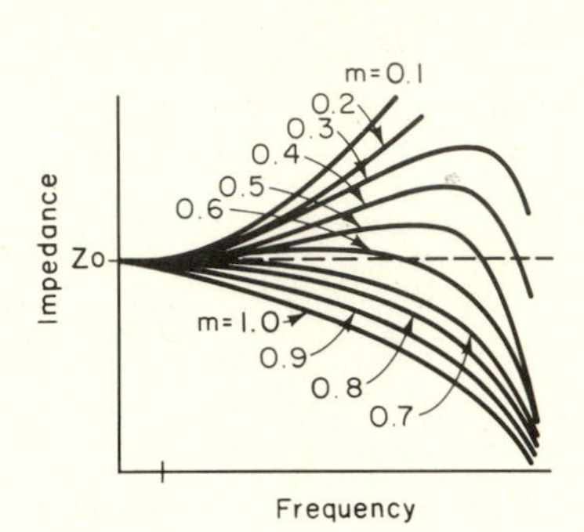

Fig. 11.15 The effect of m on filter impedance.

named for the branch that is *unchanged* from the constant-k type. If the shunt paths C_2 are unchanged, and reactive components are added to the series elements, the configuration is then called a shunt-type m-derived filter.

Returning again to the circuits of Fig. 11.14, we see that the shunt path is now a resonant circuit. There will be a resonant frequency at which the circuit presents a short across the line. That frequency, called the *infinite attenuation frequency* f_∞, and the cutoff frequency f_c do *not* have the same value. Cutoff occurs at the point where the output signal amplitude is down 3 dB from the maximum value, just as in the constant-k filter.

The value of m in the m-derived filter is always less than 1, and may be obtained mathematically from one of the two equations given below.

For low-pass, m-derived filters:

$$m = \sqrt{1 - \frac{f_x}{(f_c)^2}} \tag{11.24a}$$

For high-pass, m-derived filters:

$$m = \sqrt{1 - \frac{f_\infty}{f_c}} \tag{11.24b}$$

where f_c = cutoff frequency Hz
f_∞ = infinite attenuation frequency, Hz
m = a constant used in determining the values of elements

A special case occurs for $m = 1.0$. If the value of 1.0 is substituted into the equations for m-derived filters, the filter is converted into a constant-k filter. This means that the constant-k filter may be considered to be a special type of m-derived filter for which $m = 1.0$.

Choice of the value of m is not entirely dependent on the relationship between f_c and f_∞, as might be supposed from Eqs. (11.24a) and (11.24b). Different values of m cause the filter to have different impedance characteristics. The relationship between the value of m and the value of filter impedance is illustrated in the graph of Fig. 11.15. In fact, this is one of the important advantages of m-derived filters: the designer can choose the particular impedance relationship that he wants. In the constant-k filter the designer must accept the particular impedance relationship for that type of filter.

In Fig. 11.15, the constant-k filter impedance is represented by the curve marked $m = 1.0$.

In the graph of Fig. 11.15 the curve marked $m = 0.6$ provides the impedance versus frequency relationship which is nearest to being ideal. This curve is considered to be the best choice for many applications, and, therefore, $m = 0.6$ is a value that is frequently chosen in practical filter design. Of course, the greater the range of frequencies over which the impedance is constant, the greater the range of frequencies over which maximum power transfer can occur.

LOW-PASS m-DERIVED FILTER DESIGN (SERIES TYPE) The equations for L_1, L_2, and C_2 are given here without derivation. In each case, it may be necessary to alter the calculated value in accordance with the specifications marked on Fig. 11.14. For example, suppose you are solving for L_1 in the T configuration, and you determine that $L_1 = 10$ mH from the equations below. According to the illustration, this value is divided by 2, so that the actual value of each of the two inductors will be $L_1/2 = 10/2 = 5$ mH.

$$L_1 = m \frac{Z_o}{3.14 f_c} \tag{11.25}$$

$$L_2 = \frac{1 - m^2}{4m} \times \frac{Z_0}{3.14 f_c} \tag{11.26}$$

$$C_2 = m \frac{1}{3.14 Z_o f_c} \tag{11.27}$$

where L_1, L_2 = series and shunt inductances, respectively, H
C_2 = shunt capacitance, F
m = a constant obtained from the use of Eq. (11.24a)
f_c = cutoff frequency, Hz
Z_o = characteristic impedance of the filter circuit, Ω

If the value of m is chosen as unity ($m = 1.0$), then the three equations become

$$L_1 = \frac{Z_o}{3.14 f_c}$$

$$L_2 = 0$$

$$C_2 = \frac{1}{3.14 Z_o f_c}$$

The results mean that no inductance is used for the shunt leg ($L_2 = 0$), and the values of L_1 and C_2 are identical with those of the constant-k filter as given by Eqs. (11.7) and (11.8). This proves that a constant-k filter is a special case of an m-derived filter for which $m = 1.0$.

LOW-PASS m-DERIVED FILTER DESIGN (SHUNT TYPE) Instead of the reactive component being placed in the shunt leg (as was done for the series-type m-derived filter), it may be placed in the series leg, as shown in Fig. 11.16. In this case a parallel circuit is formed by L_1 and C_1 which offers high impedance at the frequency to which they are tuned. If the inductor and capacitor were perfect elements—that is, if they contained no resistance—then the opposition to signal flow at the resonant frequency of L_1 and C_1 would be infinite. For this reason, the resonant frequency is called the infinite attenuation frequency, as it was in the series type of Fig. 11.14.

The equations for L_1, C_1, and C_2 are given here without derivation. As before, it may be necessary to modify the values of the elements as for the series type. Thus, L_1 becomes $L_1/2$ in the T type, C_2 becomes $C_2/2$ in the π type, etc.

$$L_1 = m \frac{Z_o}{3.14 f_c} \tag{11.28}$$

$$C_1 = \left(\frac{1 - m^2}{4m}\right)\left(\frac{1}{3.14 Z_o f_c}\right) \tag{11.29}$$

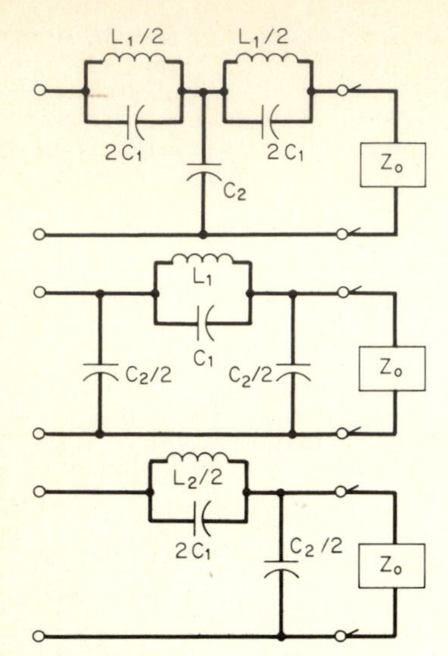

Fig. 11.16 Low-pass shunt-type m-derived filter configurations.

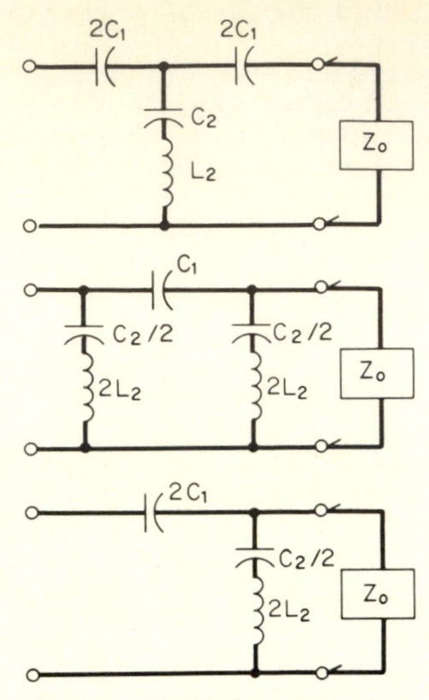

Fig. 11.17 High-pass series-type m-derived filter configurations.

$$C_2 = m\frac{1}{3.14Z_o f_c} \tag{11.30}$$

where L_1 = inductance in the series leg, H
C_1 = capacitor that resonates with L_1, F
C_2 = capacitance in the shunt leg
f_c = cutoff frequency, Hz
Z_o = characteristic impedance of the filter, Ω

HIGH-PASS m-DERIVED FILTER DESIGN (SERIES TYPE) High-pass m-derived filters can assume the same three basic configurations as are used for low-pass filters (T, π, and L). Also, each of these configurations may be designed as a series type or shunt type. Like the low-pass filter, a high-pass m-derived filter is a series type when the shunt leg is modified, and it is a shunt type when the series leg is modified. Figure 11.17 shows the three configurations for the series-type high-pass m-derived filters. The equations for C_1, C_2, and L_2 are given without derivation.

$$C_1 = \frac{1}{12.6 f_c Z_o m} \tag{11.31}$$

$$C_2 = \left(\frac{4m}{1-m^2}\right)\left(\frac{1}{12.6 f_c Z_o}\right) \tag{11.32}$$

$$L_2 = \frac{Z_o}{12.6 f_c m} \tag{11.33}$$

where C_1, C_2 = series and shunt capacitance values, F
L_2 = inductance in the shunt leg that resonates with C_2, H
Z_o = characteristic impedance of the filter circuit, Ω
f_c = center frequency, Hz
m = a constant that determines the rate at which the filter characteristic rolls off.

In the circuits of Fig. 11.17 the series-resonant circuit comprised of C_2 and L_2 provides a short circuit to the frequency of infinite attenuation, thus preventing any signal at this frequency from passing through the filter.

HIGH-PASS m-DERIVED FILTER DESIGN (SHUNT TYPE) Figure 11.18 shows the three configurations for an *m*-derived shunt-type high-pass filter. In this case a parallel-tuned circuit comprised of L_1 and C_1 is tuned to the frequency of infinite impedance. The parallel-tuned circuit prevents any signal at that frequency from passing. The equations for the values of elements C_1, L_1, and L_2 are given here without derivation.

$$C_1 = \frac{1}{12.6 f_c Z_o} \tag{11.34}$$

$$L_1 = \left(\frac{4m}{1 - m^2}\right)\left(\frac{Z_o}{12.6 f_c}\right) \tag{11.35}$$

$$L_2 = \frac{Z_o}{12.6 f_c m} \tag{11.36}$$

where L_1, L_2 = series and shunt inductances, respectively, H
C_1 = capacitance that resonates with L_1, F
f_c = cutoff frequency, Hz
Z_0 = characteristic impedance of the filter circuit, Ω
m = constant arrived at in the same manner as in other *m*-derived filters

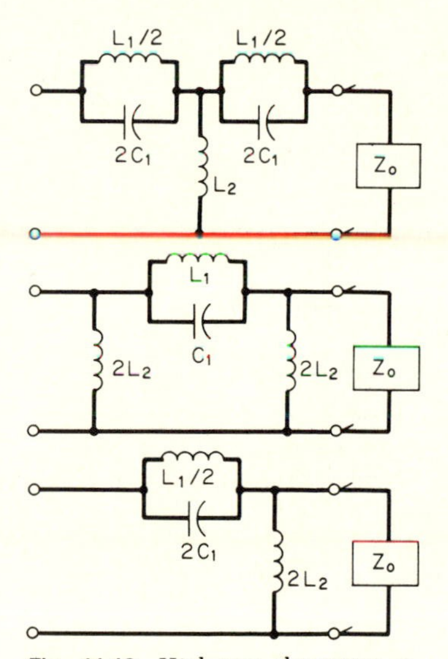

Fig. 11.18 High-pass shunt-type *m*-derived filter.

11.10 SUMMARY AND DESIGN EXAMPLE OF m-DERIVED FILTERS

A number of *m*-derived filter sections can be combined to form a ladder-type filter. In some cases an *m*-derived section may be combined with a constant-*k* section to obtain a particular overall characteristic. At the input and output ends of the filter, L sections may be used to obtain an impedance match.

Before an example of *m*-derived filter design is given, a few of the more important facts about *m*-derived filters will be summarized. The first step in designing an *m*-derived filter is to design the constant-*k* prototype. A value of *m* is decided upon next. If the circuit is to match a load impedance over a range of frequencies, then a value of $m = 0.6$ is generally used. The smaller the value of *m*, the sharper the cutoff characteristic of the filter and the greater the change in impedance with frequency. A constant-*k* filter section added to the filter will ensure that there is attenuation beyond the cutoff point of the filter.

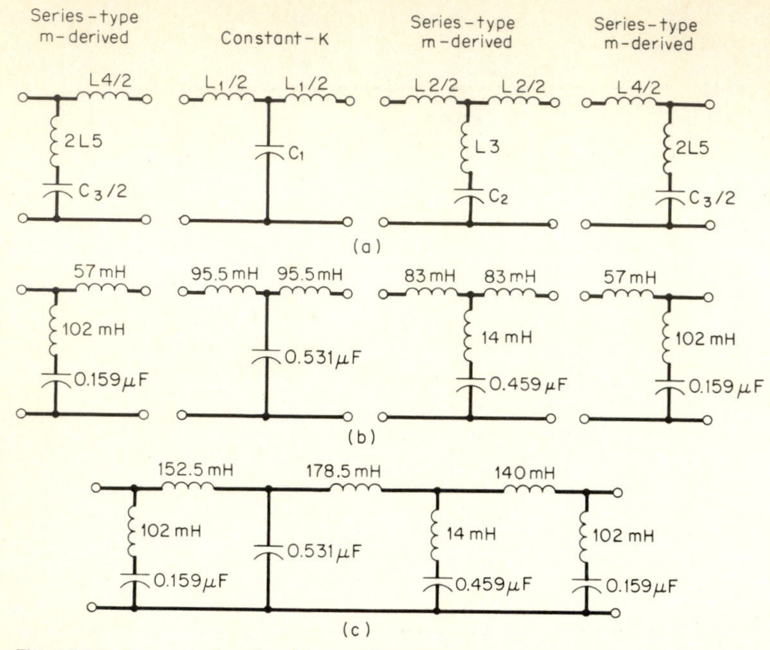

Fig. 11.19 An example of *m*-derived filter design: (*a*) Filter prototype. (*b*) Filter sections with calculated values. (*c*) Composite filter.

example 11.5 Design an *m*-derived ladder-type low-pass filter. The cutoff frequency f_c is to be 1000 Hz and the load impedance is to be 600 Ω.

solution To get a sharp cutoff, we shall use two full T sections and two L sections. See Fig. 11.19. Figure 11.19*a* shows the circuit. The input and output L sections are series-type *m*-derived filters. The first T section (reading left-to-right) is a simple constant-*k* filter, and the other T section is an *m*-derived series section.

The first step is to find the value of L_1 and C_1 for the constant-*k* filter.

$$L_1 = \frac{Z_o}{3.14 f_c} = \frac{600}{3.14 \times 1000} \tag{11.7}$$

$$= 0.191 \text{ H (or 191 mH)}$$

$$C_1 = \frac{1}{3.14 f_c Z_o}$$

$$= \frac{1}{3.14 \times 1000 \times 600} = 0.531\ \mu\text{F} \tag{11.8}$$

The values of L_1 and C_1 for the constant-*k* filter are shown in Fig. 11.19*b*. As required by the specification of Fig. 11.19*a*, the value of L_1 has been divided by two. The next step is to determine the values of L_2, L_3, and C_2 of the *m*-derived T section. The subscripts are changed from the original equations so that the various sections will not be confused.

$$L_2 = m \frac{Z_o}{3.14 f_c} \tag{11-25}$$

in which the term $(Z_o/3.14f_c)$ is equal to the constant-*k* prototype value of L_1.

A value of $m = 0.866$ will be used for the *m*-derived T section in order to get adequate attenuation. A value of $m = 0.6$ will be used for the end sections in order to get an impedance match over a range of frequencies.

$$L_2 = \frac{0.866 \times 600}{3.14 \times 1000} = 0.166 \text{ H (or 166 mH)}$$

$$L_2 = \frac{0.3 \times 0.191}{2} = 29 \text{ mH}$$

$$L_3 = \frac{1 - m^2}{4m} \left(\frac{Z_o}{3.14 f_c}\right) \qquad (11.26)$$

$$= \frac{1 - 0.75}{3.46} (0.19) = 0.014 \text{ H (or 14 mH)}$$

$$C_2 = \frac{m}{3.14 Z_o f_c} \qquad (11.27)$$

$$= \frac{0.866}{1.884 \times 10^6} = 0.459\ \mu\text{F}$$

The values of L_2 L_3, and C_2 are shown in Fig. 11.19*b*. The values of L_4, L_5, and C_3 will now be calculated.

$$L_4 = \frac{m Z_o}{3.14 f_c} \qquad (11.28)$$

$$= \frac{0.6 \times 600}{3.14 \times 1000} = 0.114 \text{ H (or 114 mH)}$$

$$L_5 = \left(\frac{1 - m^2}{4m}\right) \left(\frac{Z_o}{3.14 f_c}\right) \qquad (11.26)$$

$$= \frac{(1 - 0.36) \times 0.190}{2.4} = 0.051 \text{ H (or 51 mH)}$$

$$C_3 = \frac{m}{3.14 Z_o f_c} \qquad (11.27)$$

$$= 0.6 \times 0.531 \times 10^{-6} = 0.319\ \mu\text{F}$$

The end sections are shown with the calculated values in Fig. 11.19*b*. The value of C_3 is divided by two, and the value of L_5 is multiplied by 2, as required by the specification shown in Fig. 11.19*a*.

In combining the four sections of Fig. 11.19*b*, the series inductances are added to make one single inductance value, thus reducing the number of components required. For example, L_4 is added to $L_1/2$ in combining the first and second sections. The final composite filter design is shown in Fig. 11.19*c*.

Although these filters are easy to design, they suffer from certain shortcomings. The most important is that the terminations must be such that at the center frequency of the filter the source and load must be pure resistances equal to the image impedance. As the frequency increases, the impedance must decrease to zero at the cutoff frequency. This cannot be done with lumped circuits, and so the actual band shape with resistive terminations is never exactly as shown on the graphs.

11.11 SOME ADDITIONAL TYPES OF FILTERS

Constant-*k* and *m*-derived filters are easy to design (in comparison to other types), and they have reasonably good roll-off characteristics. Many other methods of designing *LC* filters are also possible, each with its own advantages and disadvantages.

Filter manufacturers may make use of a computer to design a variety of filters having different attenuation and phase characteristics. The designers choose the filter with the characteristic most nearly equal to their required characteristics. This is comparable to the situation with tubes or transistors in which case the designer purchases the tube or transistor most nearly suited to his needs.

Since the computer has been made available to filter designers, there is a trend toward obtaining "tailor-made" filters. The designer specifies the characteristics he wants, and the computer performs the computation. What used to amount to hours and hours of tedious equation solving can now be accomplished in minutes. Also, more complex designs can now be obtained—designs that were not economically feasible when calculated without computers.

The designer of today, then, must know the characteristics, advantages, and disadvantages of the various types of filters. This enables him to specify filters so as to get the best possible design for his needs.

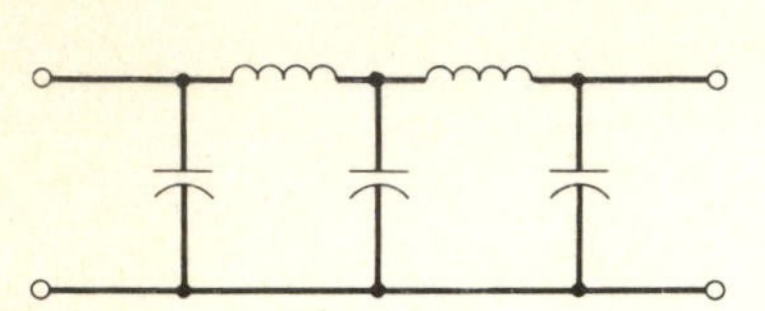

Fig. 11.20 The configuration of a typical Butterworth or Chebyshev filter.

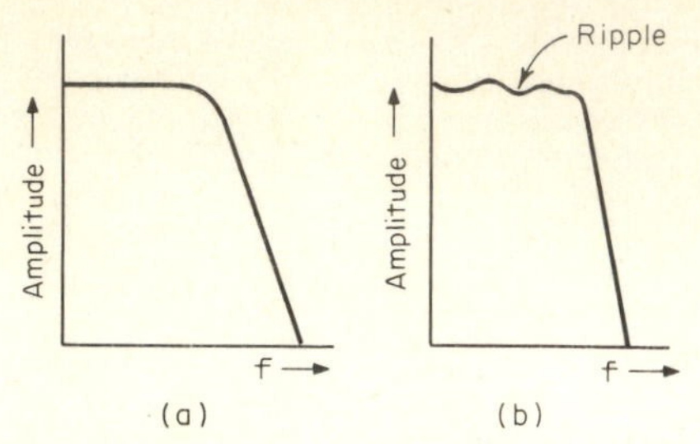

Fig. 11.21 Typical attenuation response curves for Butterworth and Chebyshev filters: (*a*) Butterworth filter. (*b*) Chebyshev filter.

BUTTERWORTH FILTER This type of filter is also called a *maximally flat filter.* It is a form of *LC* filter that has a relatively flat response—that is, it offers a relatively constant impedance to signals with frequencies in its passband range.

The configuration of a typical Butterworth filter is shown in Fig. 11.20. You will note that it is like a constant-*k* filter. However, the values of the components are different. This is to be expected because they are arrived at by a different method.

The Butterworth filter has a good phase response. This is a measure of the amount of phase shift that occurs when a signal passes through the filter. It also has good amplitude response, which is another way of saying that its response is flat. However, it has a rather gradual roll-off, which makes it unsuitable in applications where it is necessary to sharply cut off adjacent frequencies.

CHEBYSHEV FILTER The Chebyshev filter is also a form of *LC* filter in the same configuration as the Butterworth filter. The design of the Chebyshev filter is based on the assumption that all the frequencies to be passed are equally important. In this respect, it is different from the filters previously discussed which are designed on the assumption that zero frequency (dc) is the most important factor in design.

The cutoff characteristic of a Chebyshev filter is much sharper than that of the Butterworth or other types discussed (constant-*k* and *m*-derived). An important disadvantage of the Chebyshev is that there is a ripple in its passband response. This is another way of saying that its response curve is not flat in the passband region. Figure 11.21 compares the attenuation response curves of the Butterworth filter (which has no ripple) with that of the Chebyshev filter (which has ripple).

CRYSTAL FILTER (SOMETIMES CALLED "QUARTZ FILTER") Figure 11.22 shows two equivalent circuits for crystals in their holders. The equivalent circuit of a piezo-

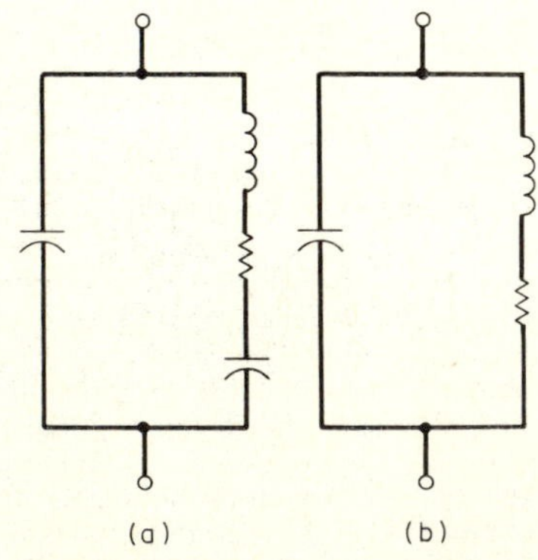

Fig. 11.22 Equivalent circuits for a crystal: (*a*) This is the equivalent circuit that is usually shown. (*b*) Mathematically, this circuit can be obtained from the one shown in (*a*).

electric crystal that is usually shown is in Fig. 11.22*a*. The extremely high Q of a crystal causes it to behave as a bandpass filter with a very narrow bandwidth.

Mathematically, it is possible to convert the equivalent circuit of Fig. 11.22*a* into another equivalent circuit like the one shown in Fig. 11.22*b*. This one is easier to visualize, and calculations for this type of circuit were discussed in Chap. 10 of this book.

Figure 11.23 shows a crystal in filter circuit configurations. If a coil L_1 with a resistance R_1 is placed in series with the crystal, as shown in Fig. 11.23*a*, the circuit behaves as a simple parallel-resonant tuned circuit. It will pass all frequencies except the one to which it is tuned. Because of the high circuit Q, a resistor or coil may be placed in parallel with the circuit. This will lower the Q and broaden the frequency response. Also, a variable capacitor may be placed across the crystal circuit to tune it through a narrow range of frequencies. Figure 11.23*b* shows the practical circuit for a crystal filter, and Fig. 11.23*c* shows a typical response curve.

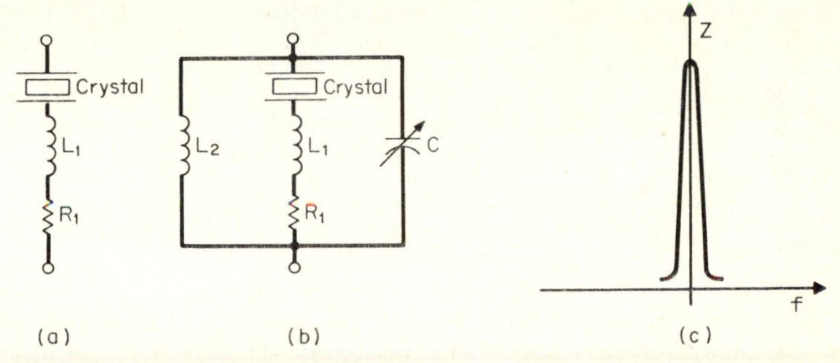

Fig. 11.23 Crystal circuits used for filters: (*a*) A coil L_1 and its dc resistance R_1 in series with the crystal makes it behave as a simple parallel-resonant circuit with a high Q. (*b*) In this circuit the variable capacitor C tunes the circuit through a narrow range, and coil L_2 lowers the circuit Q for a broader response.

Crystal filters are used in applications where a parallel-resonant *LC* circuit response curve is applicable. For example, crystal filters are used in i-f amplifier coupling circuits to provide a narrow passband with sharp skirt selectivity. A number of crystals may be used in a stagger tuning arrangement to obtain a wider response than is normally obtainable with one crystal circuit. The very narrow bandwidth of a crystal filter is useful in CW receivers because their sharp selectivity makes possible separate signals of adjacent carrier frequencies. *Q multipliers* in the i-f stage may employ crystals for this purpose.

Disadvantages of crystal filters include their cost, which may be high compared to that of an *LC* filter. (However, if the *LC* filter was designed to produce exactly the same resonant curve as the crystal filter, it would be much bulkier and far more expensive.) Crystal filters cannot be tuned over a wide range of frequencies. They are bulkier than the ceramic filters that will be described next.

CERAMIC FILTER Ceramic filters have characteristics similar to those of quartz filters, but are much smaller in size. Instead of using a piece of quartz for the vibrating element, the ceramic filter uses the piezoelectric properties of lead zirconate–lead titanate resonators.

Figure 11.24 shows the construction of a simple three-terminal ceramic filter. The dot electrode terminal (in the center of the ceramic disk) is the one to which the input signal is delivered. The input signal causes the ceramic disk to vibrate at its mechanical resonant frequency. If the input signal frequency is other than the mechanical resonant frequency of the ceramic disk, no vibration will take place. When the ceramic disk is vibrating, voltages are generated around its edges which are picked off by the ring electrode terminal. The common terminal, which is simply a silver plate across the bottom of the ceramic disk, is normally operated at ground potential. As an indication of the very-small physical size of the ceramic filter, a ceramic disk 0.2 in. in diameter by 0.1 to 0.4 in thick will resonate at a frequency of 455 kHz.

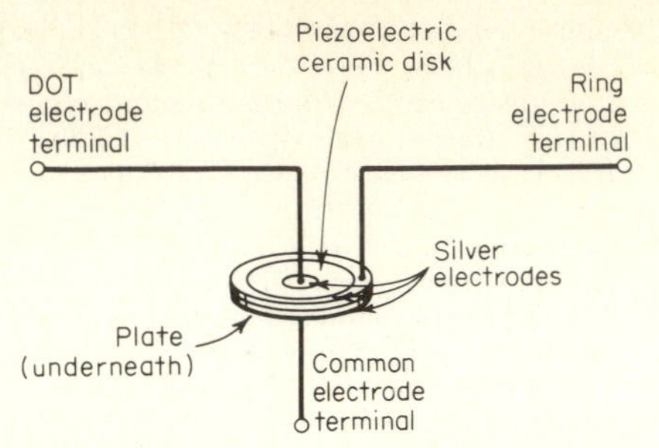

Fig. 11.24 Construction details of a three-terminal ceramic filter.

A two-terminal ceramic filter is made by placing a silver plate on both sides of the piezoelectric ceramic disk. There is no common electrode terminal. In this type of filter, the signal is fed to one of the plates and taken from the other.

To get a sharper bandpass characteristic, a number of the two-terminal disks can be stacked to produce what is known as a *ceramic ladder filter*. Figure 11.25 presents this type of filter. Figure 11.25*a* shows its physical appearance, and Fig. 11.25*b* indicates how this type of filter is constructed by stacking ceramic wafers. Figure 11.25*c* gives the response curve of a ladder filter compared with that of an ideal filter (shown in dotted lines). Note that the shape of the ceramic ladder filter response curve is very near to the shape of the ideal curve.

Figure 11.26 demonstrates an i-f amplifier stage Q_1 coupled to another i-f amplifier stage Q_2 with ceramic filter coupling. Note that this is a three-terminal type filter, but it may also be a stacked filter. In other words, it is possible to stack three-terminal ceramic disks just as it is possible to stack the two-terminal type in order to get the desired bandpass characteristics.

The input signal to the filter in Fig. 11.26 is taken from collector-load resistor R_3, passed through the filter. The output signal is delivered to the base of Q_2 at the junction of its base-bias voltage divider comprised of R_5 and R_6. Although the circuit looks like a resistance-capacitance amplifier, remember that the filter is equivalent to an *LC* filter circuit.

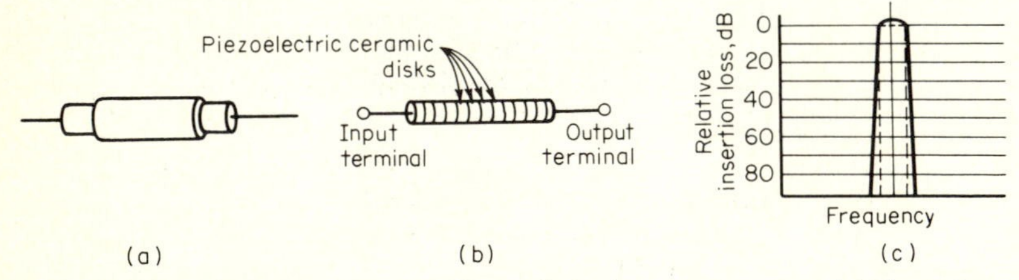

Fig. 11.25 The ladder-type ceramic filter and its characteristic curve: (*a*) Physical appearance of a ceramic ladder filter. (*b*) Construction of a ceramic ladder filter. (*c*) Response of a ceramic ladder filter.

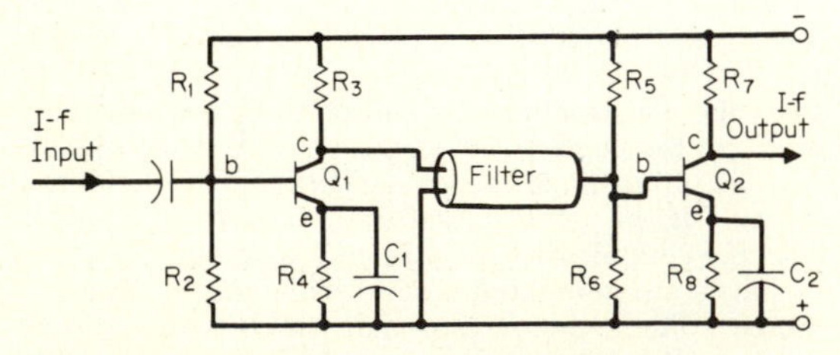

Fig. 11.26 An i-f stage that uses a ceramic filter for coupling.

MECHANICAL FILTERS (ALSO CALLED "MAGNETOSTRICTIVE FILTERS") Both the quartz crystal and ceramic filters employ the piezoelectric effect to produce a mechanical vibration in the filter material. This mechanical vibration, which determines the frequency of the filter, is directly related to its physical size and can be held to a very close frequency tolerance.

Instead of the combination of piezoelectric principle and mechanical resonance, it is also possible to use the magnetostrictive principle. The two magnetostrictive effects that are related to the use of mechanical filters are the *Joule effect* and the *Vallari effect*. The Joule effect relates to the change in physical length of a ferromagnetic material in the presence of an applied magnetic field. The Vallari effect relates to the change in magnetization of a material when an external stress is applied to that material. Both these effects are usually combined and referred to as the magnetostrictive effect.

Figure 11.27 presents a mechanical filter and its equivalent *LC* circuit. The construction details of a mechanical filter are shown in Fig. 11.27*a*. The electric input signal produces a varying magnetic field around a magnetostrictive driving rod. This varying field causes the driving rod to change in length at a rate that is directly equal to the rate at which the field varies. The varying length of the rod produces a mechanical vibration in the resonant mechanical section. Note that the resonant section is comprised of metal disks which are manufactured to be precisely resonant at a certain frequency. These disks are coupled together by rigid coupling rods.

Any frequency other than the resonant frequency cannot pass through the mechanical filter because the resonant frequency of the disks is quite precise—that is, they are manufactured to very close physical dimensions. Thus, the mechanical filter cannot be set into vibration by frequencies other than the mechanical frequencies. The mechanical oscillations are converted back to electric oscillations by a transducer output coil.

The bias magnets at each end produce a restraining force on the motion of the disks in order to make the filter vibrate only once for each input cycle. Remember that one complete cycle of ac signal consists of two peak currents. Each of these peaks will magnetize the rod and cause it to change its length. Thus, two vibrations would occur for each cycle if it were not for the biasing magnet. The magnetic field from the bias magnet holds the resonant disks so that they can move easily in one direction, but cannot move easily in the opposite direction. With this arrangement the output frequency is the same as the input frequency.

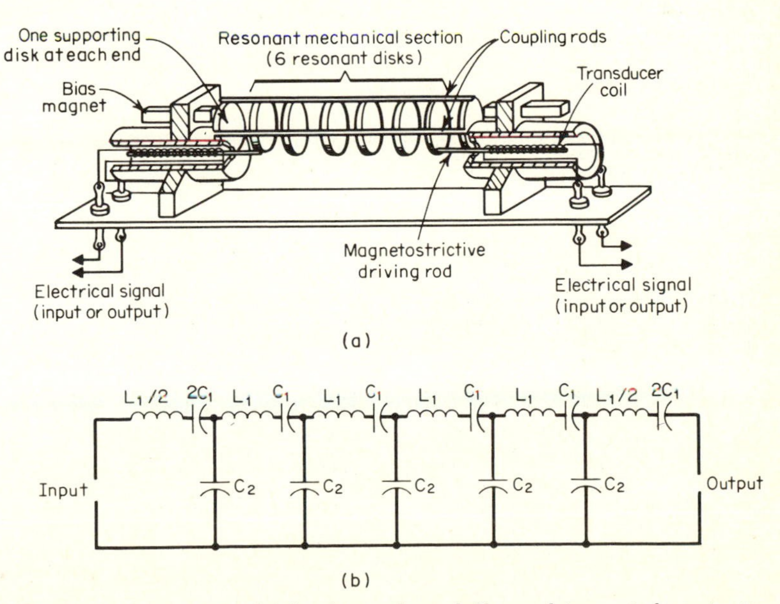

Fig. 11.27 Construction details of a mechanical filter and its equivalent circuit: (*a*) Elements of a mechanical filter. (*b*) Electrical analogy of a mechanical filter.

The electric equivalent circuit in Fig. 11.27*b* shows that the mechanical filter is comparable to a ladder-type *LC* filter.

A variation in design of the mechanical filter, which is used at much lower frequencies, employs a tuning fork instead of vibrating mechanical metal disks. The input signal is electromagnetically coupled to one prong of the tuning fork which sets it into vibration. The output signal is taken from the other prong. Mechanical filters of this type are used at audio frequencies.

11.12 ACTIVE FILTERS

As mentioned in the beginning of the chapter, an active filter contains, in addition to op amps, resistors and capacitors. Inductors, a component in *LC* filters, are thereby eliminated. Because of their relative large size and cost, the elimination of inductors in filters is indeed a considerable achievement. This is especially true for filters having cutoff frequencies of 100 Hz or less.

There are many methods one may use for the realization of active filters. In this section we shall concentrate on an active filter that provides the Butterworth response of Fig. 11.21. Two basic low-pass circuits for realizing Butterworth active filters are illustrated in Fig. 11.28.

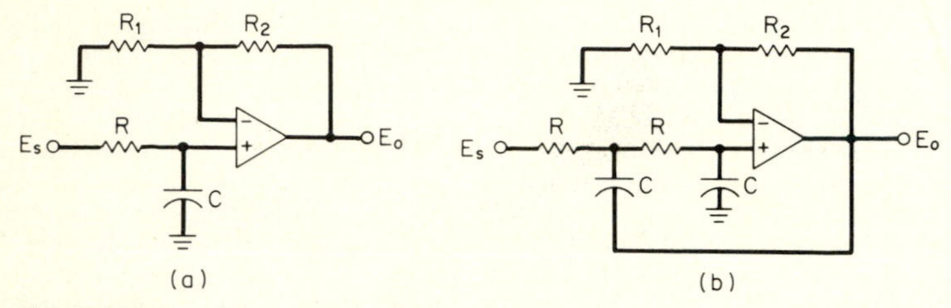

Fig. 11.28 Basic low-pass Butterworth filters: (*a*) First-order section. (*b*) Second-order section.

A first-order low-pass section is shown in Fig. 11.28*a* and a second-order low-pass section in Fig. 11.28*b*. The voltage gain, A_v, of each section, from Chap. 13, is determined by resistors R_1 and R_2 connected to the inverting (−) terminal of the op amp:

$$A_v = 1 + \frac{R_2}{R_1} \tag{11.37}$$

The half-power cutoff frequency f_c for each section is determined from the following expression,

$$f_c = \frac{0.159}{RC} \tag{11.38}$$

where f_c is in hertz, R in ohms, and C in farads. The basic sections of Fig. 11.28, as

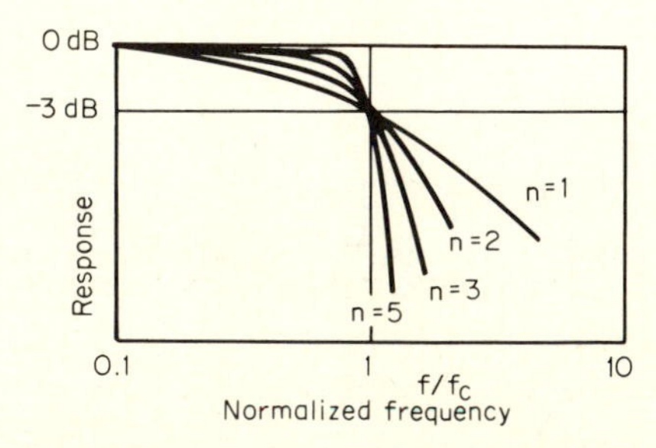

Fig. 11.29 Frequency-response curves of a Butterworth low-pass filter for different orders, *n*.

will be explained shortly, may also be used to realize high-pass, bandpass, and band-reject filters.

Figure 11.29 provides plots of frequency response of a low-pass Butterworth filter for different orders, n. A particular value of n is obtained by cascading the basic sections of Fig. 11.28. As n increases, the shape factor improves. The frequency scale is normalized by plotting the ratio of f/f_c. At $f = f_c$ ($f/f_c = 1$), the response of the filter is down by -3 dB.

For a first-order ($n = 1$) Butterworth low-pass filter, the first-order section of Fig. 11.28*a* is used. For a second-order ($n = 2$) filter, the second-order section of Fig. 11.28*b* is employed. If a third-order ($n = 3$) filter is necessary, the first-order and second-order sections of Fig. 11.28 are cascaded. A fourth-order ($n = 4$) filter is realized by cascading two second-order sections, and so on.

The gain of the first-order section is arbitrary. The gain of a second-order section, however, is defined. It is related by coefficient a, listed in Table 11.1, for different values of n. The relationship between A_v and a is

$$A_v = 3 - a \tag{11.39}$$

Example 11.6 illustrates the use of the preceding equations and Table 11.1 in a filter design problem.

TABLE 11.1 Coefficients for a Second-Order Butterworth Section

n	a
2	1.414
3	1
4	0.765, 1.848
5	0.618, 1.618
6	0.518, 1.414, 1.932

patch p. 11-27

example 11.6 Design a fifth-order ($n = 5$) Butterworth low-pass filter having a cutoff frequency of 1 kHz. For each section, $R = 1$ kΩ and $R_1 = 5$ kΩ. The first-order section has a gain of 10.

solution The filter consists of a cascaded connection of a first-order and two second-order sections, as illustrated in Fig. 11.30. By Eq. (11.28), for $R = 1$ kΩ,

$$C = \frac{0.159}{f_c R}$$

$$= \frac{0.159}{10^3 \times 10^3} = 0.159 \times 10^{-6}\text{ F} = 0.159\ \mu\text{F}$$

For a gain of 10 for the first-order section, by Eq. (11.37),

$$10 = 1 + \frac{R_2}{5\text{ k}}$$

Solving, $R_2 = 45$ kΩ.

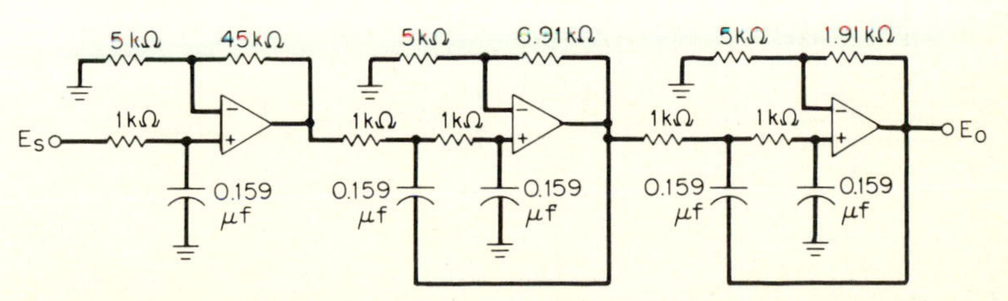

Fig. 11.30 An example of a fifth-order ($n = 5$) 1-kHz low-pass Butterworth active filter. (See Example 11.6.)

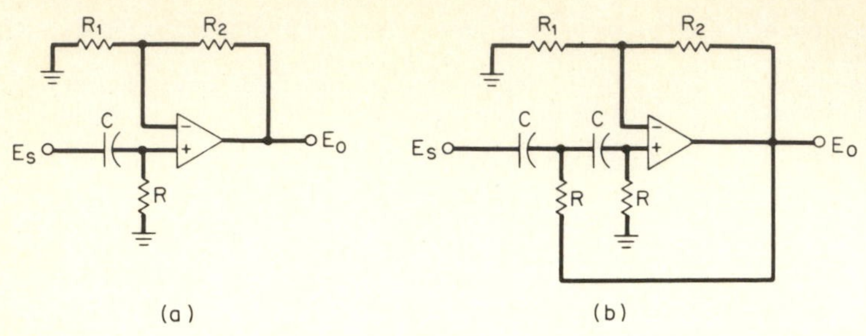

Fig. 11.31 Basic high-pass Butterworth filters: (*a*) First-order section. (*b*) Second-order section.

Referring to Table 11.1, for the first second-order section in Fig. 11.30, $a = 0.618$. By Eq. (11.39),

$$A_{vl} = 3 - 0.618 = 2.382$$

By Eq. (11.37) then,

$$2.382 = 1 + \frac{R_2}{5\text{ k}}$$

and $R_2 = 6.91$ kΩ.

For the remaining second-order section, from Table 11.1, $a = 1.618$. Hence,

$$A_{v2} = 3 - 1.618 = 1.382$$

$$1.382 = 1 + \frac{R_2}{5\text{ k}}$$

Solving, $R_2 = 1.91$ kΩ.

If, for example, a sixth-order ($n = 6$) filter were required, three second-order sections of Fig. 11.28*b* would be cascaded. In calculating the values of R_2, from Table 11.1, $a =$ 0.518 would be used for the first section, 1.414 for the second, and 1.932 for the last section.

High-pass filter If the *R*'s and *C*'s in Fig. 11.28 are interchanged, first-order and second-order high-pass sections are obtained. These are illustrated in Fig. 11.31. Table 11.1 and Eqs. (11.37) to (11.39) also apply to the high-pass sections. This is illustrated in the following example.

example 11.7 Design a fifth-order ($n = 5$) Butterworth high-pass filter having a cutoff frequency of 1 kHz. For each section, $R = 1$ kΩ and $R_1 = 5$ kΩ. The first stage has a voltage gain of 10.

solution The filter, shown in Fig. 11.32, uses the basic high-pass sections of Fig. 11.31. The values of *C* and R_2 were obtained in the manner illustrated in Example 11.6 for the fifth-order low-pass filter.

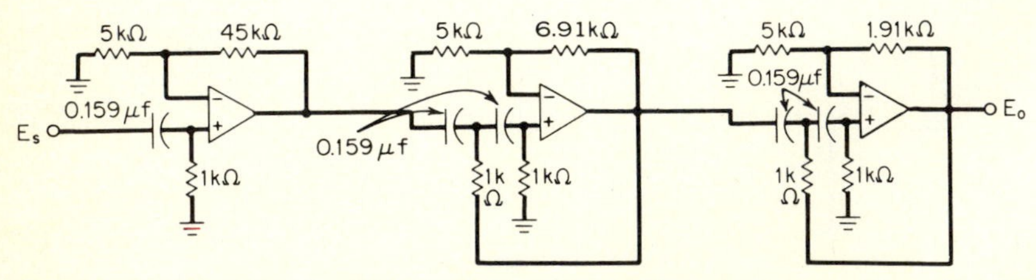

Fig. 11.32 An example of a fifth-order l-kHz high-pass Butterworth active filter. (See Example 11.7.)

Bandpass filter An active bandpass filter may be obtained by cascading high- and low-pass active filters, as illustrated in the block diagram of Fig. 11.33. The cutoff frequency of the low-pass filter must be greater than the cutoff frequency of the high-pass filter. If the cutoff frequencies for the high-pass and low-pass filters are f_H and f_L, respectively, the bandwidth of the bandpass filter is $f_L - f_H$.

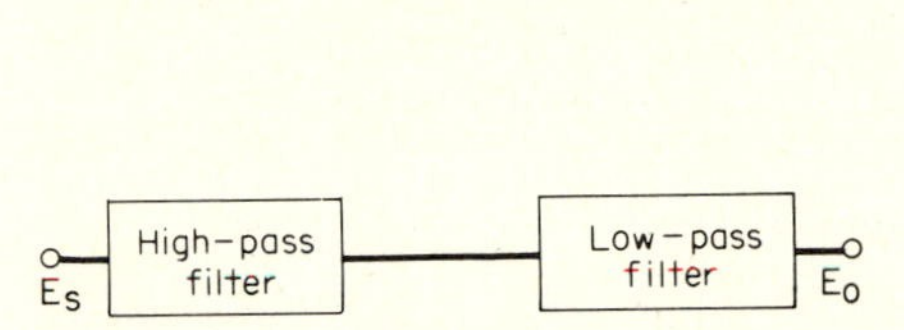

Fig. 11.33 Block diagram of a bandpass active filter.

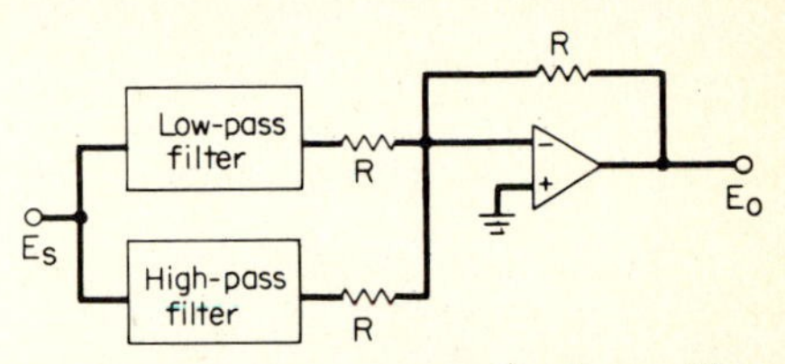

Fig. 11.34 Realizing an active band-reject filter by connecting a low-pass and a high-pass filter to a summer.

Band-reject filter An active band-reject filter may be realized by connecting together the inputs of a low-pass and a high-pass filter and their respective outputs to a summing amplifier. This is illustrated in Fig. 11.34. For this filter, f_L must be less than f_H. The stop band is then equal to $f_H - f_L$.

Chapter 12

Transistor Amplifiers and Oscillators

12.1 INTRODUCTION

An amplifier increases the magnitude of, or *amplifies*, an electric signal. The signal may be derived, for example, from a TV antenna, a phono cartridge, or a strain gage. Of all components, the amplifier is by far the most widely used building block in electronics systems.

An oscillator generates a waveform without the presence of an external signal. If the waveform is a sine wave, the oscillator is referred to as a *sinusoidal* oscillator. For a waveform other than a sine wave, such as a square or sawtooth wave, the oscillator is called a *nonsinusoidal*, or *relaxation*, oscillator. Sinusoidal oscillators are used, for example, to test the performance of circuits, to generate radio frequencies, and in receivers.

In this chapter we shall consider how amplifiers are classified, their characteristics, biasing, how to calculate their gain and frequency response, and the effects of utilizing more than one amplifier stage. The various parameters used for characterizing transistors are defined in Chap. 8. Feedback, where a portion or all of the output signal is returned to the input of an amplifier, and the operation of sinusoidal oscillators, conclude the chapter.

12.2 CLASSIFICATION OF AMPLIFIERS

An amplifier may be represented by the simple block diagram of Fig. 12.1. The input signal voltage is denoted by E_s and the input signal current by I_s. Across the output terminals is load resistor R_L. Resistance R_L may represent, for example, the resistance of a loudspeaker coil, a motor winding, or the input of a connected amplifier stage. Voltage E_o is the output (load) voltage, and current I_o the output (load) current. Symbols P_i and P_o are the input and output signal powers, respectively.

If the amplifier is optimized to amplify voltage signals, it is called a *voltage* amplifier. For current signals, it is referred to as a *current* amplifier. If it is to develop output power of one watt or more, it is generally classified as a *power* amplifier.

Another method of classification is based on the amplitude of signals. For small signals, in the order typically of microvolts and millivolts, the amplifier may be classified as *small-signal*, and for larger signals as a *large-signal* amplifier. Simple algebra is all that is required for calculating the performance of small-signal amplifiers. Graphical methods, however, are generally used in the analysis of large-signal amplifiers.

Amplifiers are also classified in terms of load current flow. Figure 12.2*a* shows one cycle (360°) of a sine wave input signal to an amplifier. If the load current also flows

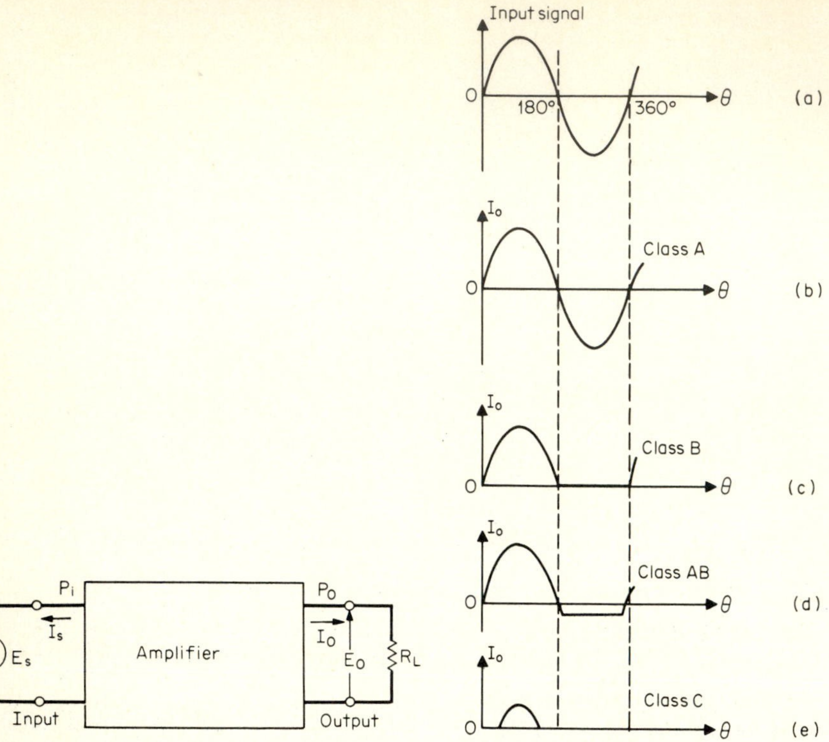

Fig. 12.1 Block diagram of an amplifier.

Fig. 12.2 Classes of amplifier operation: (*a*) Input signal. (*b*) Load current flows 360° for class A operation. (*c*) Load current flows 180° for class B operation. (*d*) In class AB operation, load current flow is greater than 180° but less than 360°. (*e*) Class C operation, where load current flows for less than 180°.

for 360°, as indicated in Fig. 12.2*b*, the amplifier is operating as *class A*. *Class B* operation, illustrated in Fig. 12.2*c*, exists when the load current flows for only 180°. If the load current flows for more than 180 but less than 360°, *class AB* operation is realized (Fig. 12.2*d*). *Class C* operation is obtained if the load current flows for less than 180° (Fig. 12.2*e*).

Class A operation is used for small-signal and single-transistor (referred to as *single-ended*) power amplifiers. If class B, AB, or C operation is used in a single-ended amplifier, the output waveform is badly distorted and the amplifier is worthless. For this reason, for class B or AB operation the amplifier contains two transistors, referred to as a *push-pull* amplifier. Class C operation is often used in the amplification of r-f signals in a circuit containing a tuned load.

Amplifiers are also classified according to their intended operation. Examples include *audio, r-f, video, microwave,* and *pulse* amplifiers.

12.3 CHARACTERISTICS OF AMPLIFIERS

The performance of amplifiers is characterized by the following terms:

Voltage gain
Current gain
Power gain
Input resistance

Output resistance
Bandwidth
Distortion
Slewing rate

VOLTAGE GAIN The voltage gain of an amplifier A_v is defined as the ratio of the output signal voltage E_o to the input signal voltage E_s:

$$A_v = \frac{E_o}{E_s} \tag{12.1}$$

CURRENT GAIN The current gain of an amplifier A_i is defined as the ratio of the output signal current I_o to the input signal current I_s:

$$A_i = \frac{I_o}{I_s} \tag{12.2}$$

POWER GAIN The power gain of an amplifier A_p is defined as the ratio of the output signal power P_o to the input signal power P_i:

$$A_p = \frac{P_o}{P_i} \tag{12.3a}$$

An alternative definition is that the power gain is equal to the product of the voltage and current gains:

$$A_p = A_v A_i \tag{12.3b}$$

INPUT RESISTANCE The input resistance (or impedance) of an amplifier R_i (or Z_i) is the resistance (or impedance) across the input terminals of an amplifier.

OUTPUT RESISTANCE The output resistance (or impedance) of an amplifier R_o (or Z_o) is the resistance (or impedance) across the output terminals of an amplifier with the signal source set to zero (see Chap. 6).

BANDWIDTH The bandwidth of an amplifier BW is a band of frequencies that is amplified with modest attenuation in gain. A typical bandwidth curve is illustrated in Fig. 12.3. Gain, expressed in decibels (dB), is generally plotted along the y axis, and frequency along the x axis. To cover conveniently a large range of frequencies, a logarithmic scale is used for the frequency axis.

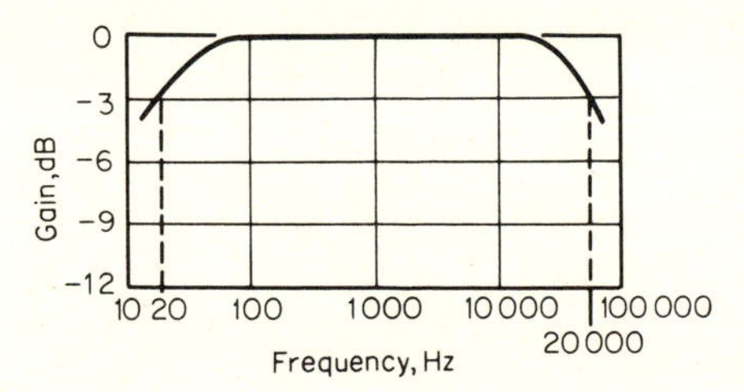

Fig. 12.3 An example of bandwidth characteristics of an amplifier.

The nominal gain over a large range of frequencies is expressed as 0 dB. At frequencies $f_L = 20$ Hz and $f_H = 20$ kHz, the gain is down by −3 dB. Therefore, frequency f_L is referred to as the *lower* −3-dB frequency, and f_H as the *upper* −3-dB frequency. Other terms used include the lower and upper *break*, *cutoff*, *corner*, and *half-power* frequencies. The bandwidth is equal to the difference between the upper and lower −3-dB frequencies:

$$BW = f_H - f_L \tag{12.4}$$

From Fig. 12.3, BW = 20 000 − 20 ≃ 20 000 Hz = 20 kHz.

DISTORTION Because the characteristics of the BJT and FET are not strictly linear (see Chap. 8), the output of an amplifier contains frequencies that were not present in the input signal. Assume, for example, that a 1-kHz sine wave is impressed across the input terminals of an amplifier. The output will contain, in addition to a 1-kHz frequency, referred to as the *fundamental frequency*, integrally related frequencies of 2, 3, 4, . . . , kHz. These frequencies are referred to as *harmonics:* 2 kHz is the *second harmonic*, 3 kHz is the *third harmonic*, and so on.

Harmonic distortion Harmonic distortion is defined as the ratio of the amplitude of a harmonic component to the amplitude of the fundamental frequency. For example, if the 1-kHz output is 10 V and the 2-kHz output is 0.5 V, the *second-harmonic distortion* D_2 is $0.5/10 = 0.05 = 5$ percent.

Intermodulation distortion Assume that a 1000-Hz and a 400-Hz signal are fed into an amplifier. The output, in addition to the 1000- and 400-Hz signals, will contain frequencies equal to their *difference* ($1000 - 400 = 600$ Hz) and to their *sum* ($1000 + 400 = 1400$ Hz). The sum and difference frequencies is the result of intermodulation distortion in an amplifier.

SLEWING RATE The slewing rate of an amplifier indicates how well it responds to a rapidly changing waveform. To measure the response, a rectangular pulse is applied to the input of the amplifier (Fig. 12.4). Because of internal capacitances of

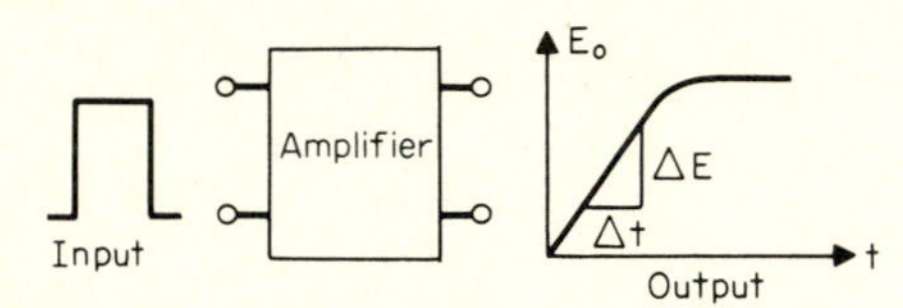

Fig. 12.4 Determining the slewing rate of an amplifier. The value of the slewing rate equals $\Delta E/\Delta t$.

the amplifier, it takes a finite time for the output to reach its maximum value. How fast it reaches it is the slewing rate of the amplifier.

As indicated in Fig. 12.4, an increment of time Δt (Δ is the Greek letter *delta*) is required for the output voltage to increase by the increment ΔE along its linear ascent. The ratio $\Delta E/\Delta t$ is equal to the slewing rate. Its unit is generally expressed in volts per microsecond.

12.4 BIASING AND STABILIZATION

Because of its wide use, in this section we consider how the BJT and FET amplifiers are biased and stabilized for class A operation.

BIASING THE COMMON-EMITTER AMPLIFIER A typical common-emitter (CE) junction transistor amplifier is illustrated in Fig. 12.5. A single bias source, E_{CC}, sup-

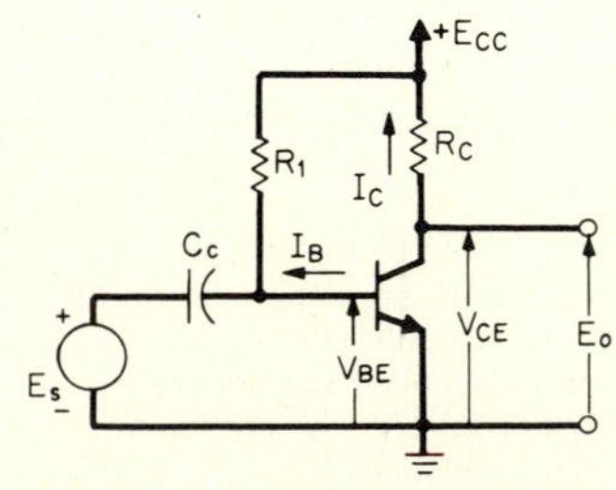

Fig. 12.5 Basic class A common-emitter amplifier biased from a single dc source, E_{CC}.

plies the necessary collector and base currents for class A operation. Capacitor C_c, called a *coupling* capacitor, blocks any direct current from the base that may be in signal source E_s. Output signal voltage E_o is taken across the collector and ground.

A simple method of finding the *quiescent operating point* Q is by graphical analysis. First, summing the dc voltages in the collector circuit of Fig. 12.5, we have

$$E_{CC} = R_C I_C + V_{CE} \tag{12.5a}$$

Solving Eq. (12.5*a*) for I_C,

$$I_C = -\frac{V_{CE}}{R_C} + \frac{E_{CC}}{R_C} \tag{12.5b}$$

Expression (12.5*b*) is an equation of a straight line of the form $y = mx + b$, where m is the slope and b is the y intercept. Comparing the equation of a straight line with Eq. (12.5*b*), we see that $y = I_C$, $m = -1/R_C$, $x = V_{CE}$, and $b = E_{CC}/R_C$. Because only two points determine a straight line, it is convenient to superimpose Eq. (12.5*b*) on the collector characteristics of a transistor. The superimposed line is referred to as a *load line*.

To obtain the two points, Eq. (12.5*b*) is first solved for $I_C = 0$; then

$$V_{CE} = E_{CC} \tag{12.6a}$$

This is one point needed to plot the load line. The second point is obtained by setting $V_{CE} = 0$. For this condition, Eq. (12.5*b*) reduces to

$$I_C = \frac{E_{CC}}{R_C} \tag{12.6b}$$

The two points and the load line drawn between them on the collector characteristics are illustrated in Fig. 12.6. The slope of the load line is $-1/R_C$. Location of the Q

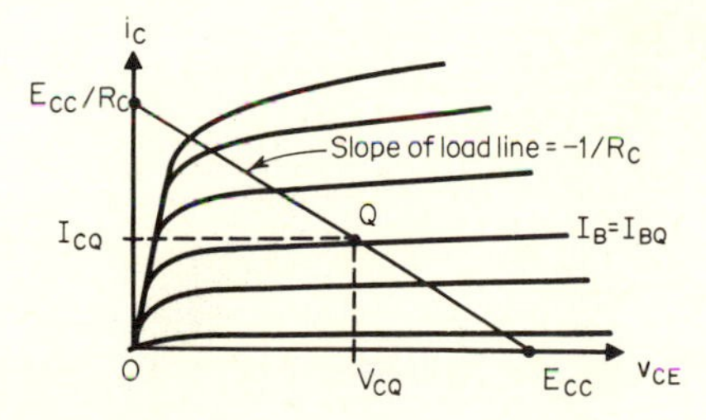

Fig. 12.6 Superimposing load line on collector characteristics for location of the Q point.

point is at the intersection of the load line and base current curve I_{BQ}. For class-A operation the Q point is located to be approximately at the center of the load line. The quiescent collector voltage and current are denoted by V_{CQ} and I_{CQ}, respectively.

Returning to Fig. 12.5, the equation for the dc quantities in the input circuit is

$$E_{CC} = I_B R_1 + V_{BE} \tag{12.7a}$$

where, at room temperature, $V_{BE} = 0.7$ V for a silicon transistor and 0.3 V for a germanium transistor (see Chap. 8).

Because the bipolar junction transistor is a *current-operated* device, it is biased by a base current I_B. Solving Eq. (12.7*a*) for I_B, we obtain

$$I_B = \left(\frac{E_{CC} - V_{BE}}{R_1}\right) \tag{12.7b}$$

The application of this and the other expressions derived in the section is illustrated in Example 12.1.

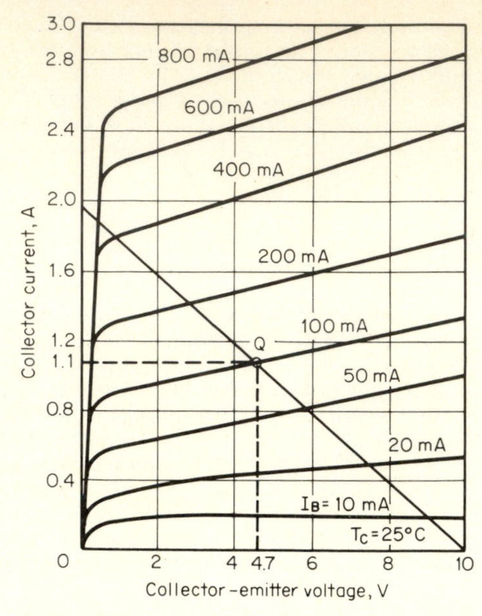

Fig. 12.7 Locating the Q point in Example 12.1.

example 12.1 In the circuit of Fig. 12.5, $E_{CC} = 10$ V, $R_C = 5$ Ω, $I_B = 100$ mA, and $V_{BE} =$ 0.7 V. Assume that the collector characteristics of Fig. 12.7 are appropriate for the transistor. (*a*) Write the equation for the load line. (*b*) Superimpose the load line on the collector characteristics. (*c*) Locate the Q point. (*d*) Determine the value of R_1.

solution (*a*) Substitution of $R_C = 5$ Ω and $E_{CC} = 10$ V in Eq. (12.5*b*) yields $I_C = -V_{CE}/5 + 10/5 = -V_{CE}/5 + 2$, which is the equation of the load line.

(*b*) For $I_C = 0$, by Eq. (12.6*a*), $V_{CE} = 10$ V. For the second point, letting $V_{CE} = 0$, by (12.6*b*), $I_C = {}^{10}\!/_5 = 2$ A. The load line drawn between the two points is shown in Fig. 12.7.

(*c*) From Fig. 12.7, the Q point is located at the intersection of the load line and the $I_B =$ 100 mA curve: $V_{CQ} \simeq 4.7$ V and $I_{CQ} \simeq 1.1$ A.

(*d*) Solving Eq. (12.7*b*) for R_1,

$$R_1 = \frac{E_{CC} - V_{BE}}{I_B}$$

Substitution of the given values in the equation yields

$$R_1 = \frac{10 - 0.7}{0.1} = 93\ \Omega$$

If collector characteristics are not available, the Q point can be estimated. The procedure is illustrated in the next example.

example 12.2 Using the values given in Example 12.1 and the fact that the dc current gain, $h_{FE} = 10$, find the Q point of the amplifier of Fig. 12.5.

solution Collector current I_C equals the product of base current I_B and the dc current gain h_{FE}: $I_C = I_B h_{FE}$. From Example 12.1, $I_B = 100$ mA $= 0.1$ A; hence, $I_C = I_{CQ} = 0.1 \times 10 = 1$ A. Solving Eq. (12.5*a*) for V_{CE}, we obtain

$$V_{CE} = E_{CC} - R_C I_C$$

Therefore, $V_{CE} = V_{CQ} = 10 - 5 \times 1 = 5$ V. The Q point is located at $V_{CQ} = 5$ V and $I_C = 1$ A. These values are reasonably close to those obtained graphically in Example 12.1.

BIASING THE COMMON-BASE AMPLIFIER Generally, two dc sources are used for biasing the common-base (CB) amplifier. Required are E_{CC} for the output circuit and E_{BB} for the input circuit, as shown in Fig. 12.8*a*. Because the current gain of a transistor in the common-base configuration is close to unity, the common-base collector

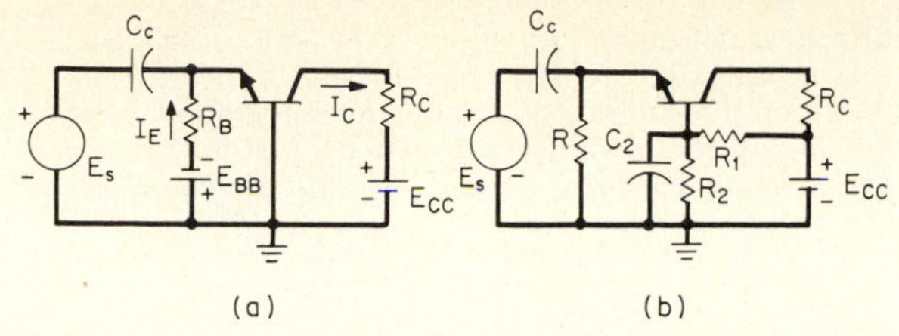

Fig. 12.8 Biasing a common-base amplifier: (*a*) Using two dc sources, E_{CC} and E_{BB}. (*b*) Using a single source, E_{CC}.

characteristics are not very informative. The Q point is therefore calculated in a manner similar to that illustrated in Example 12.2. The dc current gain h_{FB} of a CB amplifier is taken to be equal to one: $h_{FB} = 1$.

A method for biasing a common-base amplifier from a single dc source E_{CC} is shown in Fig. 12.8*b*. *Bypass capacitor* C_2 is chosen so that its reactance to the lowest signal frequency is at least one-tenth the value of the resistance of R_2. This ensures that the *base is common to the input and output signals.* For direct current, however, the capacitor acts like an "open circuit," and the transistor is biased as a common-emitter amplifier. Resistor R provides a dc return path between the emitter and ground.

BIASING THE EMITTER FOLLOWER A basic emitter follower, also referred to as a *common-collector* (CC) amplifier, is illustrated in Fig. 12.9. The collector is connected directly to dc bias source E_{CC}. Output signal voltage E_o is taken across the emitter resistor R_E. Because the emitter and collector currents are approximately the same in a transistor, the common-emitter collector characteristics can be used in the analysis of the emitter follower.

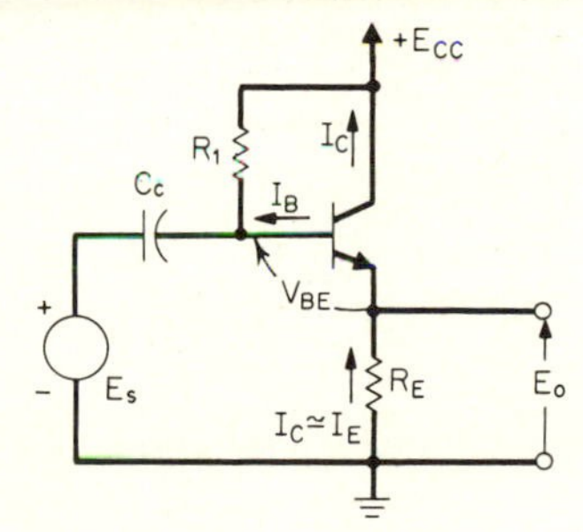

Fig. 12.9 Basic emitter follower circuit.

Writing an equation for the dc voltages around the input circuit, we have

$$E_{CC} = R_1 I_B + V_{BE} + R_E I_C$$

assuming that I_C is approximately equal to the emitter current I_E. Solving for I_C,

$$I_C = \frac{E_{CC} - R_1 I_B - V_{BE}}{R_E} \tag{12.8}$$

The dc voltage across the emitter resistor $E_{o(\mathrm{dc})}$ is

$$E_{o(\mathrm{dc})} = R_E I_C \tag{12.9}$$

STABILIZING THE COMMON-EMITTER AMPLIFIER Because of variations in reverse saturation current, current gain, and the base-emitter voltage with temperature and aging, it is necessary to *stabilize* the Q point of most common-emitter amplifiers. (The common-base and emitter follower amplifiers may be considered to be inherently stable.) A number of stabilization techniques exist and are examined in this section.

Current-feedback stabilization An example of current-feedback stabilization is illustrated in Fig. 12.10*a*. Resistors R_1 and R_2 establish the *Q* point. Emitter resistor R_E, connected between the emitter of the transistor and ground, provides stabilization. In analyzing the circuit, it is convenient to replace resistors R_1 and R_2 connected to E_{CC} with a Thévenin equivalent circuit (see Chap. 6). Thévenin resistance R_B is

$$R_B = R_1 \| R_2 = \frac{R_1 R_2}{R_1 + R_2} \tag{12.10a}$$

The Thévenin voltage E_B is

$$E_B = \frac{E_{CC} R_2}{R_1 + R_2} \tag{12.10b}$$

The Thévenin circuit is given in Fig. 12.10*b*.

Assume that the dc current gain increases beyond its initial value. As a result, the collector current tends to rise, and the voltage across the emitter resistor E_E increases.

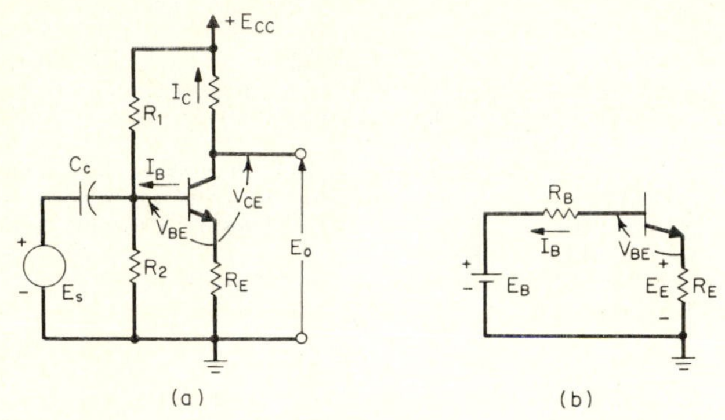

Fig. 12.10 Common-emitter amplifier with current-feedback stabilization: (*a*) Circuit. (*b*) Biasing network replaced by its Thévenin equivalent circuit.

Because voltage E_E opposes E_B, the base-emitter voltage tends to decrease. Less base current flows, and the collector current is reduced to approximately its initial value.

If, for some reason, the collector current tends to decrease, voltage E_E also decreases. The base-emitter voltage increases, and more base (and collector) current flows.

From the previous description, current-feedback stabilization provides "automatic regulation" of the location of the *Q* point. A useful criterion for the effectiveness of current-feedback stabilization is the *current stability factor* S_i. It relates changes in collector current to changes in the reverse saturation current. A conservative approximate expression for S_i is

$$S_i \simeq 1 + \frac{R_B}{R_E} \tag{12.11}$$

where, referring to Fig. 12.10*b*, R_B is the resistance of the parallel combination of R_1 and R_2 and R_E is the emitter resistance.

It is noted that by Eq. (12.11), if R_B is zero or R_E is infinite, $S_i = 1$, which represents perfect stabilization. For either of these values, however, the signal gain would be zero. Practical values for S_i are in the range of 4 to 10.

example 12.3 If, in Fig. 12.10*a*, $R_1 = 60$, $R_2 = 12$, and $R_E = 2$ kΩ, calculate the value of S_i.

solution By Eq. (12.10*a*), $R_B = R_1 \| R_2 = (60 \times 12)/(60 + 12) = 10$ kΩ. By Eq. (12.11), $S_i = 1 + {}^{10}\!/_2 = 1 + 5 = 6$.

Voltage-feedback stabilization Another stabilization technique, voltage-feedback stabilization, is illustrated in Fig. 12.11. The feedback resistor R_F connected between the collector and base of the transistor is the key element.

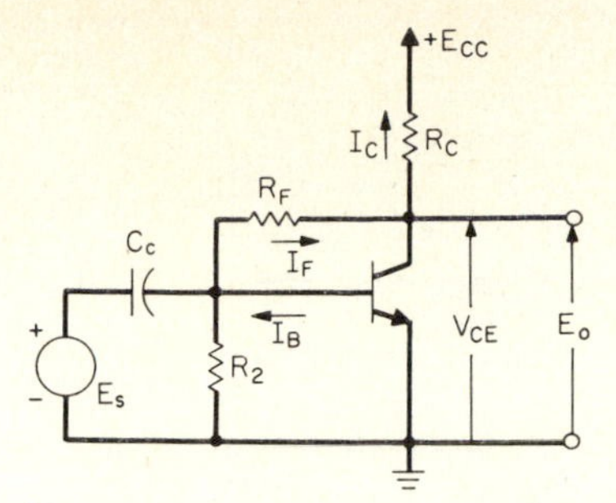

Fig. 12.11 Common-emitter amplifier with voltage-feedback stabilization.

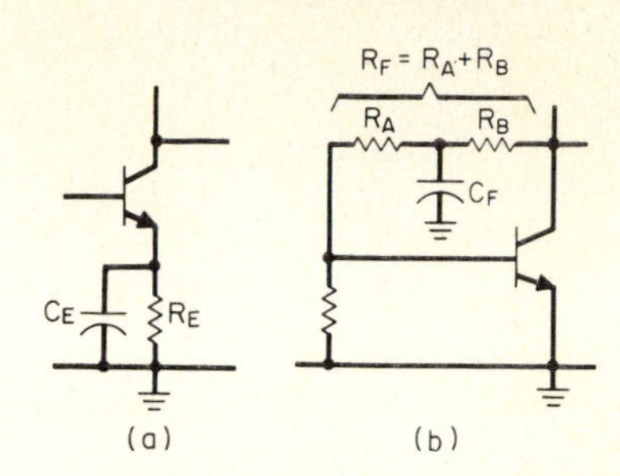

Fig. 12.12 Offsetting the reduction in signal gain of an amplifier with current- or voltage-feedback stabilization: (*a*) Bypassing emitter resistance. (*b*) Decoupling feedback network.

Assume that, because of high-temperature operation, the base-emitter voltage has been reduced. Base current and, as a result, collector current increase. Consequently, the collector-emitter voltage V_{CE} is reduced. Because the feedback current I_F in the feedback resistor is approximately equal to the collector-emitter voltage divided by the feedback resistance ($I_F \simeq V_{CE}/R_F$), the feedback current falls. This, in turn, causes the base and collector currents to decrease. The Q point, therefore, tends to return to its initial value. Changes in reverse saturation current or dc current gain, whether increasing or decreasing, are stabilized in a similar manner.

The current stability factor for voltage-feedback stabilization is given by

$$S_i \simeq 1 + \frac{R_F}{R_C} \tag{12.12}$$

For good stability R_F should be low and R_C high in value. A low value of feedback resistance, however, reduces the signal gain of the amplifier.

Bypass and decoupling capacitors To realize a low value of stability factor without appreciable deterioration in signal gain, *bypass* and *decoupling* capacitors are employed. In Fig. 12.12*a*, bypass capacitor C_E is in parallel with emitter resistor R_E. If the reactance of C_E is one-tenth or less of the value of R_E at the lowest signal frequency, the signal gain of the amplifier is hardly affected.

In Fig. 12.12*b*, resistor R_F is replaced by two resistors, R_A and R_B. Their sum is made equal to R_F ($R_F = R_A + R_B$). Often, R_A and R_B are chosen to be equal to one-half the value of R_F. Decoupling capacitor C_F is connected between the junction of R_A, R_B and ground. For minimum effect on signal gain, C_F is chosen so its reactance is one-tenth the value of R_A or R_B, whichever is smaller, at the lowest signal frequency.

Diode stabilization A method employing a junction diode for stabilizing a common-emitter amplifier is illustrated in Fig. 12.13. The cathode of junction diode D is connected to the emitter of the transistor. Source E_D forward-biases the diode. It is reasonable to expect that variation in the diode junction voltage will track and oppose variations in the base-emitter voltage of the transistor. Consequently, changes in base-emitter voltage are canceled. In a silicon transistor, the most commonly used BJT, changes in the Q point are very dependent on the base-emitter voltage.

BIASING JFET AND DEPLETION-TYPE MOSFET AMPLIFIERS Because the FET is a voltage-operated device, *self-biasing* may be used. This is the prime method for biasing the JFET or depletion-type MOSFET. An example of an n-channel FET, self-biased by source resistor R_S connected between the source terminal of the FET and ground, is shown in Fig. 12.14. Drain current I_D results in a voltage drop across R_S which is equal to $I_D R_S$. The polarity of the voltage drop is such that the source is positive with respect to the ground.

Resistor R_G (typical value is 100 kΩ) is connected between the gate terminal and ground. It provides a dc return path for the voltage across the source resistor. As a

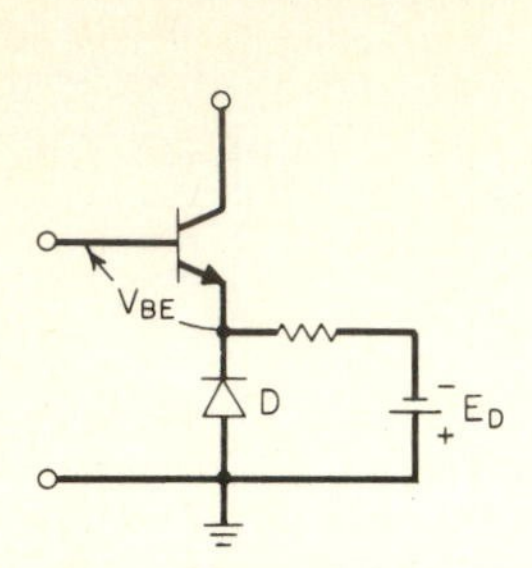

Fig. 12.13 Example of diode stabilization for a common-emitter amplifier.

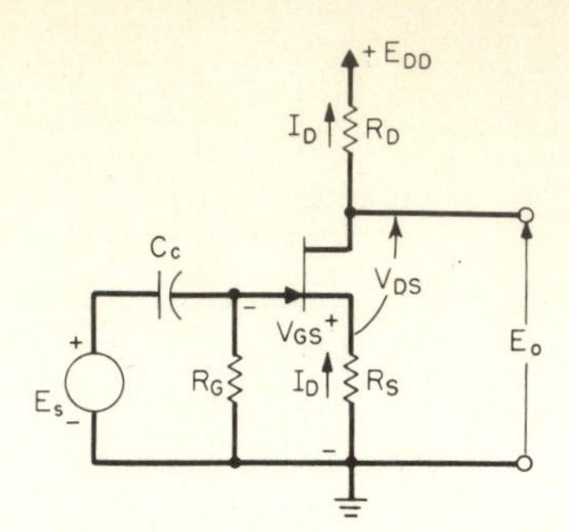

Fig. 12.14 An n-channel common-source FET amplifier which is self-biased by source resistor R_S.

result, the gate is biased *negatively* with respect to the source by $-I_DR_S$ volts. For a p-channel device, the polarity of E_{DD} is reversed and the gate-source voltage is $+I_DR_S$ volts.

In addition to providing a bias voltage, the source resistor also stabilizes the operation of the amplifier. Its behavior is analogous to the emitter resistor used in current feedback stabilization of a BJT amplifier. To minimize the decrease in signal gain, R_S is bypassed with a capacitor, as in the common-emitter amplifier.

Writing an equation for the dc voltages in the output side of Fig. 12.14, we obtain

$$E_{DD} = I_D(R_D + R_S) + V_{DS}$$

Solving for I_D,

$$I_D = -\frac{V_{DS}}{R_D + R_S} + \frac{E_{DD}}{R_D + R_S} \tag{12.13}$$

Equation (12.13), similar to Eq. (12.5*b*), is the equation of a load line. Its use is illustrated in the next example.

example 12.4 Assume that in Fig. 12.14 the sum of R_D and R_S is 40 Ω and $E_{DD} = 20$ V. If the Q point is located at $V_{DQ} = 10$ V, determine (*a*) the quiescent current and the gate-source voltage, and (*b*) the values of R_D and R_S. Use the drain characteristics of Fig. 12.15.

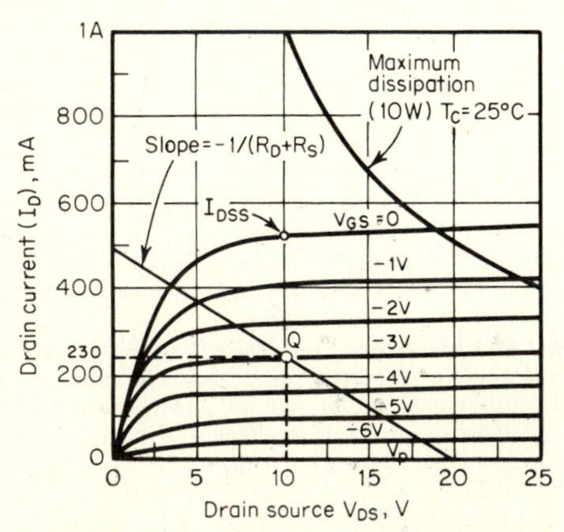

Fig. 12.15 Determining the Q point of a FET amplifier. (See Example 12.4.)

solution (*a*) Substitution of the given values in Eq. (12.13) yields $I_D = -V_{DS}/40 + 20/40 = -V_{DS}/40 + 0.5$. For $I_D = 0$, $V_{DS} = 20$ V, which is one point on the load line. For $V_{DS} = 0$, $I_D = 0.5$ A $= 500$ mA, the second point on the load line. The load line is drawn between the two points in Fig. 12.15.

For $V_{DQ} = 10$ V, the load line intersects the −3-volt bias curve. Therefore, the quiescent current I_{DQ} is approximately 230 mA, and the gate-source bias voltage is −3 V.

(*b*) Because the gate-source voltage is −3 V, $-I_D R_S = -3$. The quiescent drain current was found in (*a*) to be 230 mA = 0.23 A. Solving for R_S, we have $R_S = 3/0.23 \simeq 13\ \Omega$. The value of R_D is $40 - 13 = 27\ \Omega$.

BIASING THE ENHANCEMENT-TYPE MOSFET An example of biasing an n-channel enhancement-type MOSFET is illustrated in Fig. 12.16. Feedback resistor R_F is

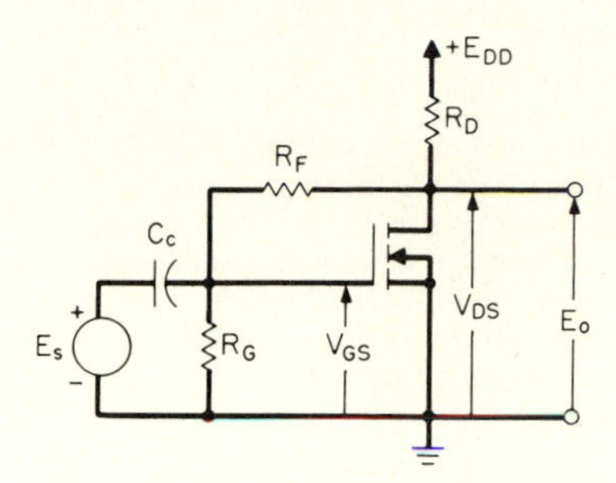

Fig. 12.16 Biasing an n-channel enhancement-type MOSFET.

connected between the drain and gate terminals of the transistor. By voltage division, the gate-source voltage V_{GS} equals the product of the drain-source voltage V_{DS} and the ratio $R_G/(R_G + R_F)$:

$$V_{GS} = \frac{V_{DS} R_G}{R_G + R_F} \tag{12.14}$$

For a p-channel enhancement-type MOSFET, the polarity of E_{DD} is reversed.

In addition to providing a bias voltage, the feedback resistor also stabilizes the Q point of the transistor. To minimize the attenuation in signal gain, R_F may be decoupled as in the common-emitter amplifier using voltage-feedback stabilization.

12.5 SMALL-SIGNAL AMPLIFIERS

The performance of small-signal amplifiers is determined with the aid of a few simple equations. These equations are based on *models*, or *equivalent circuits*, of the amplifier in question. In this section the performance of small-signal BJT and FET amplifiers is considered. The basic parameters for these devices are defined in Chap. 8.

In the analysis of amplifiers it is desirous to consider specific frequency ranges of signals. This allows the development of simple models and provides insight into amplifier performance. In this section it is assumed that the range of frequencies is less than 5000 Hz, referred to the *midfrequency range*. At midfrequencies, the inherent transistor and stray wiring capacitances may be neglected.

ANALYZING THE BJT AMPLIFIER The use of hybrid (h) parameters in the analysis of the small-signal BJT amplifier leads to a simple and generalized model. Let the transistor be represented by a circle with three leads numbered 1, 2, and 3, as shown in Fig. 12.17*a*. For example, lead 1 may be the base, lead 2 the collector, and lead 3 the emitter. A *hybrid model* of the BJT, suitable for midfrequencies, is shown in Fig. 12.17*b*.

Parameter h_i is the short-circuit input resistance; h_r is the reverse transfer ratio; h_f is the short-circuit forward current gain; and h_o is the output admittance. The unit for h_i is ohms and for h_o it is mhos. Parameters h_f and h_r have no units; that is, they are

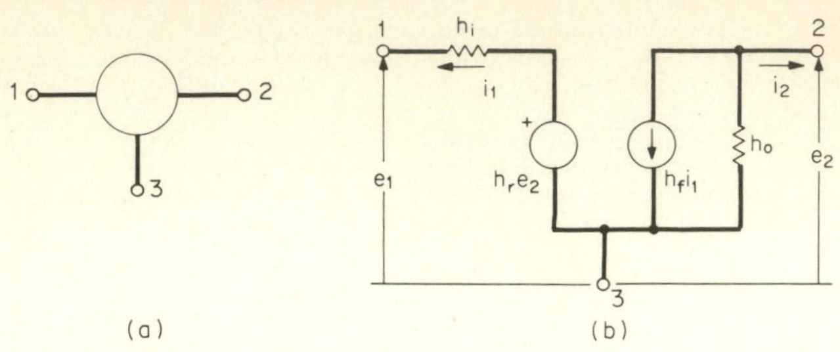

Fig. 12.17 Developing a small-signal model of the BJT: (*a*) General representation of a transistor. (*b*) Hybrid model.

numerics. Sources $h_r e_2$ and $h_f i_1$ are examples of *dependent*, or *controlled*, voltage and current sources, respectively (see Chap. 6).

The three useful configurations for the BJT are the common-emitter (CE), common-base (CB), and common-collector (CC), or emitter follower amplifiers. To distinguish the configuration, a second letter is added to each subscript. For example, if the transistor is used in the common-emitter configuration, h_i becomes h_{ie}, h_r becomes h_{re}, and so on. The notation is summarized in Table 12.1.

TABLE 12.1 Hybrid Parameter Notation

	Configuration		
h parameter	CE	CB	CC
h_i	h_{ie}	h_{ib}	h_{ic}
h_r	h_{re}	h_{rb}	h_{rc}
h_f	h_{fe}	h_{fb}	h_{fc}
h_o	h_{oe}	h_{ob}	h_{oc}

Because the CE configuration is widely used, the manufacturer normally provides only common-emitter hybrid parameters on his data sheets. If the common-base or common-collector parameters are needed, they can be obtained by the approximate equations summarized in Table 12.2. In the table the common-base and common-collector parameters are expressed in terms of the common-emitter parameters.

TABLE 12.2 Approximate Hybrid Parameter Conversions

$h_{ib} \simeq h_{ie}/(1 + h_{fe})$	$h_{ic} \simeq h_{ie}$
$h_{rb} \simeq h_{ie}h_{oe}/(1 + h_{fe}) - h_{re}$	$h_{rc} \simeq 1$
$h_{fb} \simeq -h_{fe}/(1 + h_{fe})$	$h_{fc} \simeq -(1 + h_{fe})$
$h_{ob} \simeq h_{oe}/(1 + h_{fe})$	$h_{oc} \simeq h_{oe}$

example 12.5 The common-emitter *h* parameters of a particular transistor are: $h_{ie} = 1000$ Ω, and $h_{re} = 10^{-4}$, $h_{fe} = 49$, $h_{oe} = 10^{-5}$ mho. Calculate the *h* parameters for the transistor in the (*a*) common-base and (*b*) common-collector configurations.

solution (*a*) By the equations of Table 12.2, $h_{ib} = 1000/(1 + 49) = 20$ Ω; $h_{rb} = 10^3 \times 10^{-5}/(1 + 49) - 10^{-4} = 2 \times 10^{-4} - 10^{-4} = 10^{-4}$; $h_{fb} = -49/(1 + 49) = -49/50 = -0.98$; and $h_{ob} = 10^{-5}/(1 + 49) = 0.2 \times 10^{-6}$ mho.

(*b*) From Table 12.2, $h_{ic} = 1000$ Ω; $h_{rc} = 1$; $h_{fc} = -(1 + 49) = -50$; and $h_{oc} = 10^{-5}$ mho.

Simplified hybrid model For most small-signal amplifiers, the simplified model of Fig. 12.18 may be used. Comparing Fig. 12.18 with Fig. 12.17*b*, $h_r e_2$ and h_o have been eliminated in the simplified hybrid model. The model for each of the three configurations is illustrated in Fig. 12.19. In each case the numerals have been replaced by letters *E*, *B*, and *C*. It is interesting to note that the same basic model is used for all three configurations.

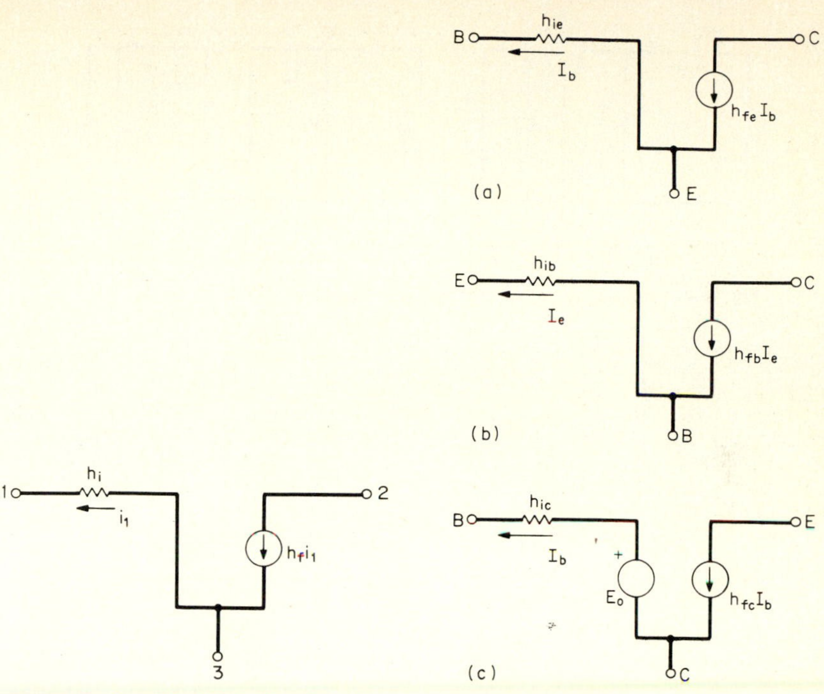

Fig. 12.18 Simplified hybrid model of a BJT.

Fig. 12.19 Hybrid models for the three transistor configurations: (*a*) Common emitter. (*b*) Common base. (*c*) Common collector (emitter follower).

Application of the simplified hybrid model We consider first the common-emitter amplifier of Fig. 12.20*a*. Resistance R_s is the resistance of the source (source resistance). It is reasonable to assume that the reactance of capacitors C_c and C_E are zero: that is, they act like short circuits. Because our concern here is with the signal behavior of an amplifier, the dc sources (E_{CC} in this case) are set to zero. The resultant model of the amplifier is given in Fig. 12.20*b*. Resistance R_B is equal to the parallel combination of bias resistors R_1 and R_2.

Figure 12.21*a* shows a common-base amplifier, and its model is given in Fig. 12.21*b*. The emitter follower and its model are illustrated in Fig. 12.22. In deriving the models for these two configuration, the dc sources were set to zero and the coupling capacitor was assumed to act as a short circuit.

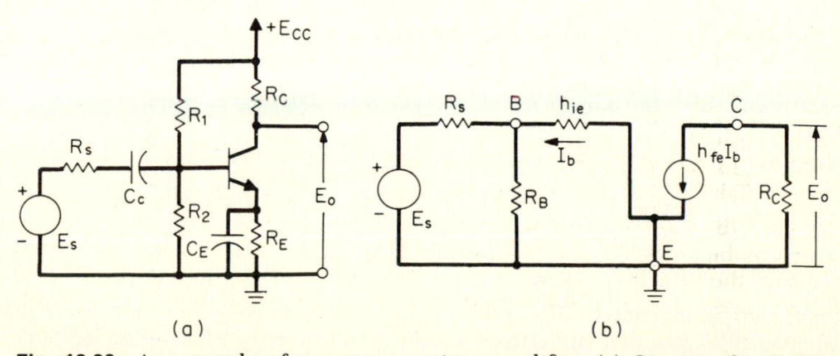

Fig. 12.20 An example of a common-emitter amplifier: (*a*) Circuit. (*b*) Model.

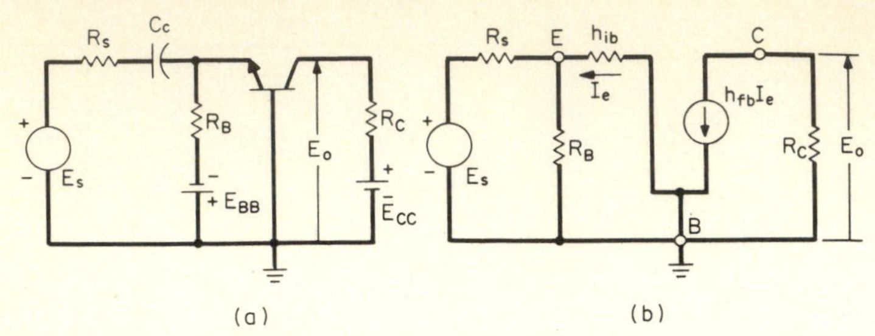

Fig. 12.21 An example of a common-base amplifier: (*a*) Circuit. (*b*) Model.

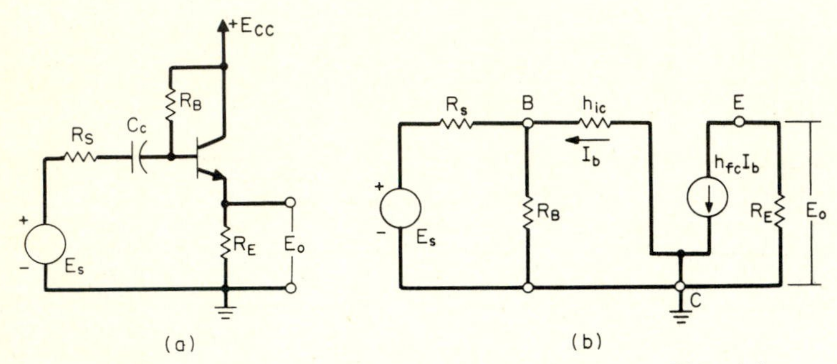

Fig. 12.22 An example of an emitter follower: (*a*) Circuit. (*b*) Model.

Equations for current gain A_i, input resistance R_i, voltage gain A_v, and output resistance R_o, for each of the three configurations are summarized in Table 12.3. The equations, expressed in terms of the common-emitter hybrid parameters, are based on:

1. The simplified hybrid model.
2. It is assumed that R_B is much greater than the input resistance and that R_C is equal to or less than 5000 Ω.

The power gain of an amplifier may be taken as the product of the voltage and current gains.

TABLE 12.3 Expressions for Evaluating the Performance of a BJT Amplifier

	CE	CB	CC
A_i	h_{fe}	$-h_{fe}/(1+h_{fe})$	$-(1+h_{fe})$
R_i	h_{ie}	$h_{ie}/(1+h_{fe})$	$h_{ie}+(1+h_{fe})R_E$
A_v	$-h_{fe}R_C/(R_s+R_i)$	$R_C/(R_s+R_i)$	1
R_o	∞	∞	$(R_s+h_{ie})/(1+h_{fe})$

example 12.6 The transistor used in the amplifiers of Figs. 12.20 to 12.22 have an h_{fe} of 49 and an h_{ie} of 900 Ω. Assume that $R_C = 5000$, $R_E = 1000$, and $R_s = 100$ Ω. With the aid of Table 12.3, calculate A_i, R_i, A_v, and R_o for each amplifier.

solution For the CE amplifier of Fig. 12.20, $A_i = 49$; $R_i = 900$ Ω; $A_v = -49 \times 5000/(100 + 900) = -245$; and $R_o = \infty$.

For the CB amplifier of Fig. 12.21, $A_i = -49/50 = -0.98$; $R_i = 900/(1+49) = 900/50 = 18$ Ω; $A_v = 5000/(100+18) = 5000/118 = 42.5$; and $R_o = \infty$.

For the CC (emitter follower) amplifier of Fig. 12.22, $A_i = -(1+49) = -50$; $R_i = 900 + (1 + 49) \times 1000 = 50\,900$ Ω, $A_v = 1$; and $R_o = (100+900)/(1+49) = 1000/50 = 20$ Ω.

Practical considerations The negative sign before an expression for voltage or current gain indicates that the output signal is 180° *out of phase* with respect to the input signal. Referring to Fig. 12.23, during the positive half-cycle of the input signal,

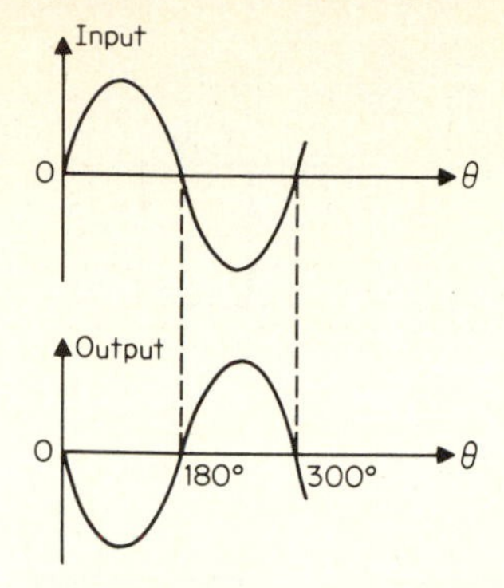

Fig. 12.23 Input and output signal waveforms 180° out of phase.

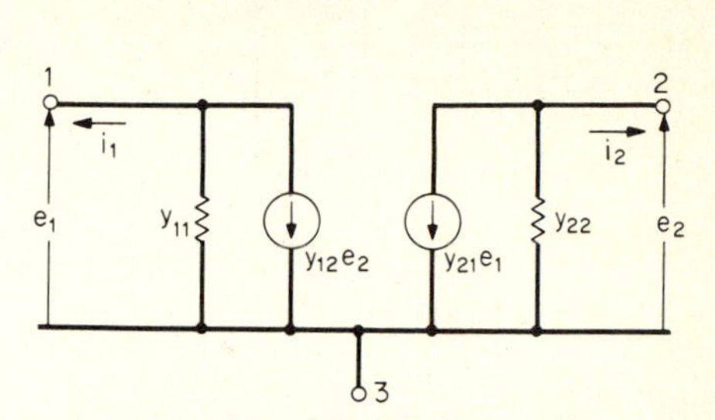

Fig. 12.24 Small-signal admittance (y) model of a FET operating at midfrequencies.

the output is a negative half-cycle. As the input signal goes through its negative half-cycle, the output is positive-going.

Although the voltage gain of an emitter follower never exceeds unity, its usefulness lies in its relatively *high input resistance* and *low output resistance.* (In Example 12.6 we found for the emitter follower that $R_i = 50\,900$ and $R_o = 20\ \Omega$.) For example, the emitter follower is used if a voltage source having a large source resistance is required to deliver power to a low-resistance load. If an emitter follower is interposed between the source and load, virtually full power is delivered to the load.

The equations summarized in Table 12.3 assume that R_B is much greater than the input resistance of the amplifier. If this is not true, because of *current division,* a portion of the input signal current is drained through R_B, and less base current flows. The method for considering current division is covered in Chap. 6.

If, in Fig. 12.20a, emitter resistor R_E is not suitably bypassed, the equations for the CE amplifier of Table 12.3 must be modified. For an unbypassed emitter resistor, the input resistance becomes

$$R_i = h_{ie} + (1 + h_{fe})R_E \tag{12.15}$$

Substitution of Eq. (12.15) for R_i in the equations for the CE amplifier of Table 12.3 yields the correct results for an unbypassed emitter resistor.

ANALYZING THE FET AMPLIFIER In a manner analogous to the development of the small-signal model for the BJT, the basic model for a FET, or MOSFET, operating at midfrequencies is illustrated in Fig. 12.24. For the field-effect device, *admittance* (y) *parameters* are used. The equivalent circuit is referred to as a y, or an admittance, model. Table 12.4 lists the y parameters for midfrequency operation for the three configurations of the FET or MOSFET: common source (CS), common gate (CG), and common drain (CD), or source follower, amplifiers. Parameter g_{fs} is the common-source forward transconductance and g_d the drain-source admittance. Their unit is the *mho.*

The three amplifier configurations and their models for the field-effect transistor are

TABLE 12.4 Small-Signal *y* Parameters for the Field-Effect Transistor

Parameter	CS	CG	CD
y_{11}	0	$g_{fs} + g_d$	0
y_{12}	0	$-g_d$	0
y_{21}	g_{fs}	$-(g_{fs} + g_d)$	$-g_{fs}$
y_{22}	g_d	g_d	$g_{fs} + g_d$

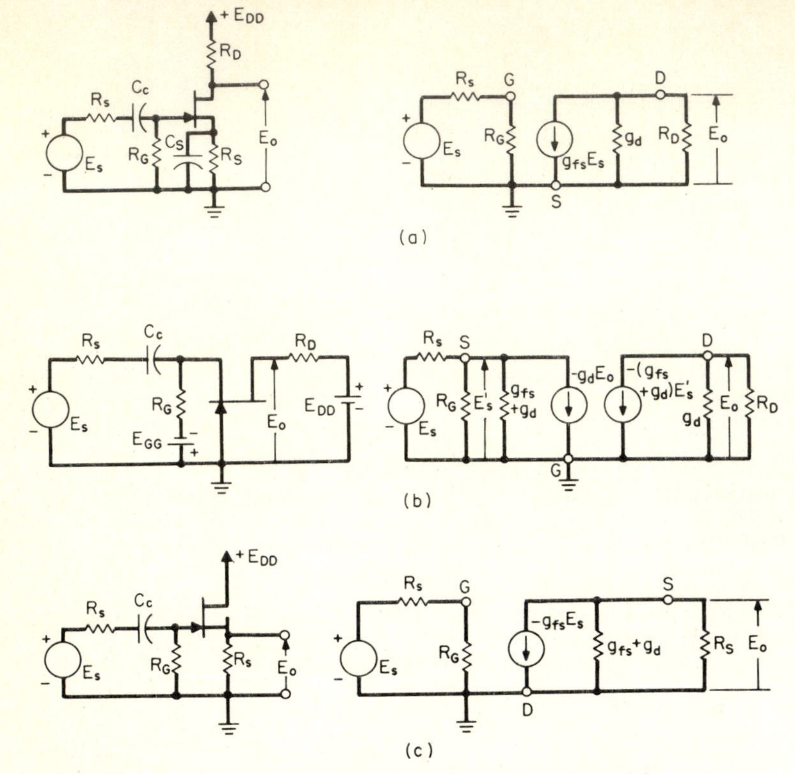

Fig. 12.25 Three basic FET amplifiers and their small-signal models: (*a*) Common source. (*b*) Common gate. (*c*) Common drain (source follower).

given in Fig. 12.25. Table 12.5 summarizes the equations for voltage gain, etc., for each circuit. The equations are based on the following assumptions:

1. $R_G >> R_s$.
2. $g_{fs} >> g_d$.
3. Capacitors C_c and C_S act like short circuits.
4. Input resistance does not include R_G.
5. Midfrequency operation.

example 12.7 A FET has a $g_{fs} = 0.1$ mho and $g_d = 10^{-6}$ mho. If $R_D = 10$ kΩ and $R_S = 1$ KΩ, calculate the values of A_v and R_o the common-source, common-gate, and common-drain amplifiers of Fig. 12.25.

solution Substituting the given values in the appropriate expressions of Table 12.5, we obtain

TABLE 12.5 Expressions for Evaluating the Performance of a FET (or MOSFET) Amplifier

	CS	CG	CD
A_v	$-g_{fs}/(1/R_L + g_d)$	$g_{fs}/(1/R_L + g_d)$	$g_{fs}/(1/R_L + g_{fs})$
A_i	∞	-1	∞
R_i	∞	$\dfrac{1}{g_{fs} - g_d g_{fs}(1/R_L + g_d)}$	∞
R_o	$1/g_d$	$1/g_d$	$1/g_{fs}$

CS: $A_v = -g_{fs}/(1/R_L + g_d) = -0.1/(1/10^4 + 10^{-6}) = -0.1/(10^{-4} + 10^{-6}) \simeq -0.1/10^{-4} = -1000$. $R_o = 1/g_d = 1/10^{-6} = 1\ \text{M}\Omega$.

CG: $A_v = g_{fs}/(1/R_L + g_d) = 0.1/(1/10^4 + 10^{-6}) \simeq 1000$. $R_o = 1/g_d = 1/10^{-6} = 1\ \text{M}\Omega$.

CD: $A_v = g_{fs}/(1/R_L + g_{fs}) = 0.1/(1/10^4 + 0.1) \simeq 1$. $R_o = 1/g_{fs} = 1/0.1 = 10\ \Omega$.

Like the emitter follower, the source follower has a high (infinite) input resistance and a low (10 Ω in this example) output resistance. Its voltage gain can never exceed unity.

If the emitter resistance of the common-source amplifier (Fig. 12.25*a*) is not suitably bypassed, the equation for A_v in Table 12.5 cannot be used. Instead, the voltage gain is expressed by

$$A_v = \frac{-g_{fs}}{1/R_L + g_d + g_{fs}R_S/R_L}$$

12.6 FREQUENCY RESPONSE OF AN AMPLIFIER

In the previous section, effects of the inherent transistor capacitances and the coupling capacitor on amplifier performance were neglected. Transistor capacitance attenuates high-frequency signals and the coupling capacitor attenuates low-frequency signals. These effects, for the common-emitter and common-source amplifiers, are examined in this section.

HIGH-FREQUENCY MODEL OF COMMON-EMITTER AMPLIFIER The widely used *hybrid-pi model* of a transistor in the CE configuration at high-frequency operation is provided in Fig. 12.26*a*. Resistance $r_{bb'}$, the *base-spreading resistance*, is the ohmic

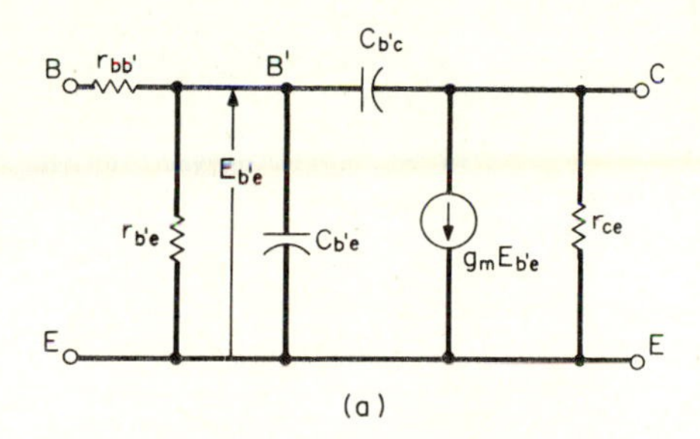

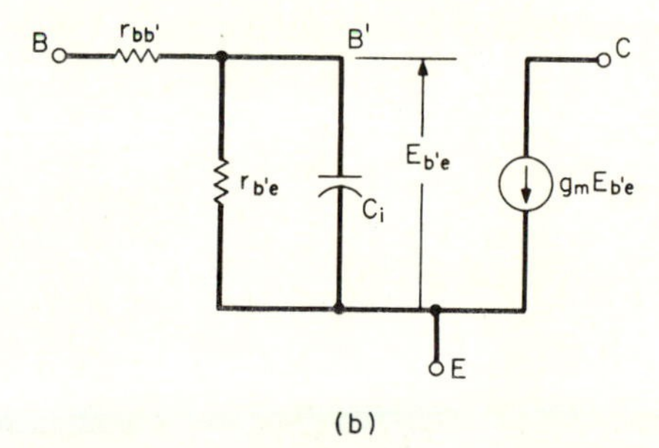

Fig. 12.26 Hybrid-pi model of a BJT in the common-emitter configuration for high-frequency operation: (*a*) Actual. (*b*) Simplified.

resistance of the base region. Its value is in the order of 100 Ω. *Base-emitter resistance* $r_{b'e}$ varies with the collector current: the higher the current, the lower its value, and vice versa. The sum of $r_{bb'}$ and $r_{b'e}$ is equal to h_{ie}:

$$h_{ie} = r_{bb'} + r_{b'e} \tag{12.16}$$

The dependent source is $g_m E_{b'e}$, where g_m is the *forward transconductance* of the transistor, and $E_{b'e}$ is the voltage across points B' and E. The transconductance is related to h_{fe} and $r_{b'e}$ by

$$g_m = \frac{h_{fe}}{r_{b'e}} \tag{12.17a}$$

In general, $r_{bb'}$ is much less than $r_{b'e}$; hence, $r_{b'e} \simeq h_{ie}$, and

$$g_m \simeq \frac{h_{fe}}{h_{ie}} \tag{12.17b}$$

The transconductance also depends on the dc collector current I_C. At room temperature (25°C), the relationship is

$$g_m = \frac{I_C}{26} \text{ millimhos} \tag{12.18}$$

where I_C is expressed in milliamperes.

The *collector-emitter resistance* r_{ce} is equal to one divided by h_{oe}:

$$r_{ce} = \frac{1}{h_{oe}} \tag{12.19}$$

It is interesting to note that the noncapacitive elements in the hybrid-pi model are related to the hybrid parameters.

Capacitances $C_{b'e}$ and $C_{b'c}$ are referred to the *diffusion* and *depletion*, or *barrier*, *capacitances*, respectively. Values of $C_{b'e}$ range from 100 to 1000 pF (picofarad $= 10^{-12}$ F) and $C_{b'c}$ is typically a few picofarads.

Simplified hybrid-pi model Owing to a phenomenon, called the *Miller effect*, capacitance $C_{b'c}$ suitably modified may be placed in parallel with $C_{b'e}$. Capacitance $C_{b'c}$ is multiplied by one plus the product of g_m and R_C. The result C_M, referred to as the *Miller capacitance*, is

$$C_M = (1 + g_m R_C) C_{b'c}$$

Addition of C_M to $C_{b'e}$ yields the effective input capacitance C_i:

$$C_i = C_{b'e} + (1 + g_m R_C) C_{b'c} \tag{12.20}$$

The *simplified hybrid-pi model* containing C_i is shown in Fig. 12.26*b*. Because r_{ce} is usually much greater than R_C, it is omitted from the figure. The high-frequency model of the common-emitter amplifier of Fig. 12.20*a* appears as in Fig. 12.27.

Because at increasing frequencies the reactance of C_i decreases, the voltage across it, $E_{b'e}$, also drops. As a result, the collector signal current I_c, which equals the product of g_m and $E_{b'e}$, and the output voltage E_o are reduced. The voltage gain, therefore, falls with increasing frequency.

The upper −3-dB frequency f_H at which the voltage gain falls to approximately 70 percent of its value at midfrequencies (usually taken at 1 kHz) is determined as follows. Signal source E_s is set to zero. (As explained in Chap. 6, in setting a voltage source to zero, it is short-circuited; in setting a current source to zero, it is open-circuited.) A

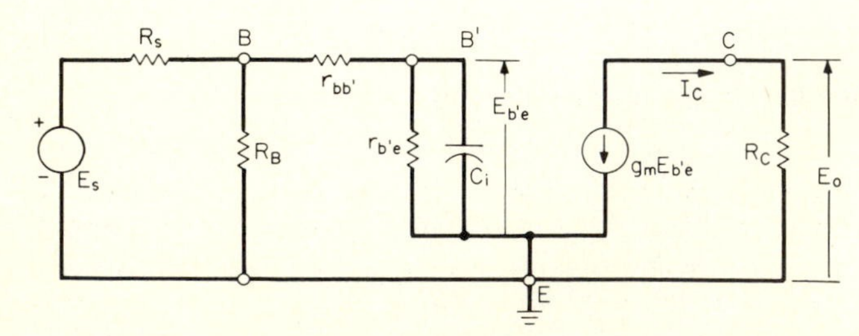

Fig. 12.27 High-frequency model of a common-emitter amplifier.

reasonable assumption is that R_B is much greater than source resistance R_s. Therefore, neglecting R_B, the equivalent resistance R_{eq} across C_i is $(R_s + r_{bb'})$ in parallel with $r_{b'e}$ (Fig. 12.28). Hence,

$$R_{eq} = \frac{(R_s + r_{bb'})r_{b'e}}{R_s + r_{bb'} + r_{b'e}} \qquad (12.21)$$

At the upper −3-dB frequency f_H, R_{eq} is equal to the reactance of $C_i = 1/2\pi f_H C_i$. Therefore,

$$R_{eq} = \frac{1}{2\pi f_H C_i}$$

Solving for f_H,

$$f_H = \frac{1}{2\pi R_{eq} C_i} = \frac{0.159}{R_{eq} C_i} \qquad (12.22)$$

since $1/2\pi = 0.159$.

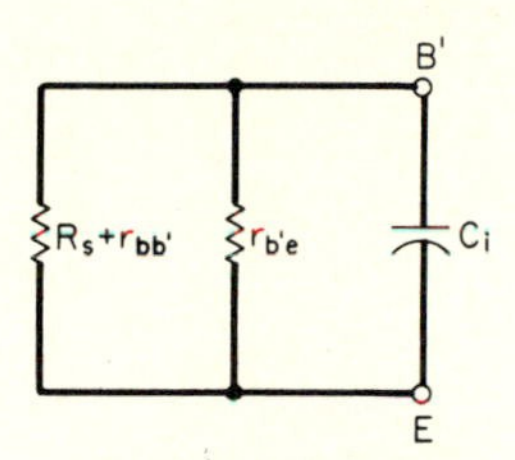

Fig. 12.28 Model of input circuit for determining the upper −3-dB frequency, f_H.

example 12.8 In Fig. 12.27, $g_m = 0.1$ mho, $C_{b'e} = 300$ pF, $C_{b'c} = 4$ pF, $R_C = 1000\ \Omega$, $R_s = 900\ \Omega$, $r_{bb'} = 100\ \Omega$, and $r_{b'e} = 1000\ \Omega$. Neglecting R_B, calculate (*a*) C_i, (*b*) R_{eq}, and (*c*) f_H.

solution (*a*) Substituting the given values, by Eq. (12.20), $C_i = 300 + (1 + 0.1 \times 1000) \times 4 = 300 + 101 \times 4 = 704$ pF.

(*b*) By Eq. (12.21), $R_{eq} = (900 + 100) \times 1000/(900 + 100 + 1000) = 500\ \Omega$.

(*c*) By Eq. (12.22), $f_H = 0.159/(500 \times 704 \times 10^{-12}) = 452{,}000$ Hz $= 452$ kHz $= 0.452$ MHz.

Figures of merit Three commonly used figures of merit for comparing the high-frequency performance of bipolar transistors are:

1. *Short-circuit beta cutoff frequency* f_β is the frequency at which h_{fe} is equal to approximately 70 percent of its value at midfrequencies. It is expressed by

$$f_\beta = \frac{0.159}{r_{b'e}(C_{b'e} + C_{b'c})} \qquad (12.23)$$

The lower the values of $r_{b'e}$, $C_{b'e}$, and $C_{b'c}$, the higher is f_β.

2. *Short-circuit gain-bandwidth product* f_T is the frequency at which h_{fe} is equal to one. It is given by

$$f_T = \frac{0.159 g_m}{C_{b'e} + C_{b'c}} \qquad (12.24)$$

The relationship of h_{fe}, f_T, and f_β is given by

$$f_T = h_{fe} f_\beta \qquad (12.25)$$

3. *Short-circuit alpha cutoff frequency* f_α is the frequency at which h_{fb} is equal to approximately 70 percent of its value at midfrequencies. It is expressed by

$$f_\alpha = \frac{0.159 g_m}{C_{b'e}} \qquad (12.26)$$

Because $C_{b'c}$ is generally much less than $C_{b'e}$, comparing Eqs. (12.24) and (12.26), $f_\alpha \simeq f_T$.

HIGH-FREQUENCY MODEL OF COMMON-SOURCE AMPLIFIER A high-frequency model of a FET in the common-source configuration is given in Fig. 12.29*a*. Capacitances C_{gs} and C_{gd} are depletion capacitances between gate and source, and gate and drain, respectively. Capacitance C_{ds} is the drain-source capacitance of the channel.

By using the Miller effect, as was done for the common-emitter amplifier, the high-

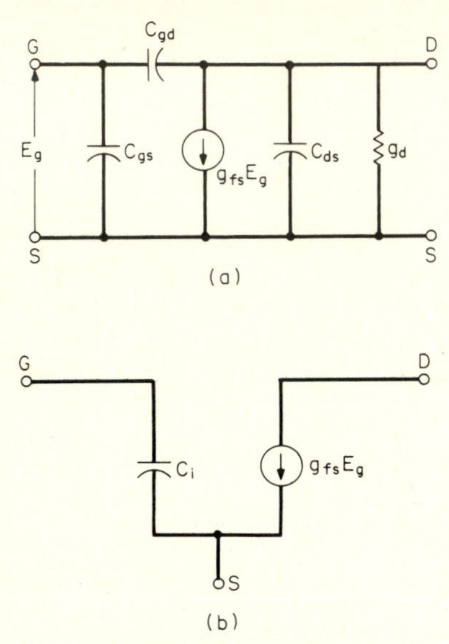

Fig. 12.29 High-frequency model of a FET (or MOSFET) in the common-source configuration: (*a*) Actual. (*b*) Simplified.

frequency model is simplified, as shown in Fig. 12.29*b*. Input capacitance C_i is expressed by

$$C_i = C_{gs} + (1 + g_mR_D) \tag{12.27}$$

The value of $1/g_d$ is generally much greater than practical values of load resistance R_D; therefore it is omitted from the simplified model. Because C_i is the influencing capacitance in determining the high-frequency response, C_{ds} is also omitted from the figure.

The high-frequency model of the common-source amplifier of Fig. 12.25*a* is illustrated in Fig. 12.30. Similar to the analysis of the CE amplifier, the upper −3-dB frequency f_H is determined by equating source resistance R_s to the reactance of C_i:

$$R_s = \frac{0.159}{f_HC_i}$$

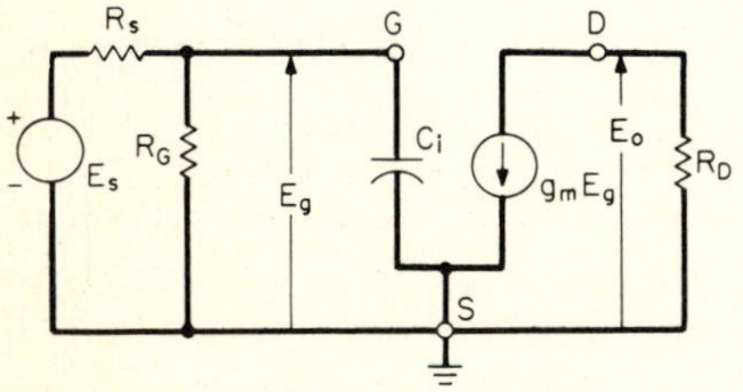

Fig. 12.30 High-frequency model of a common-source amplifier.

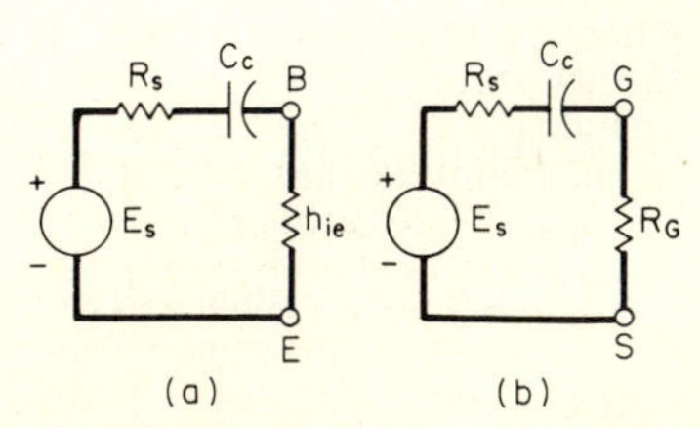

Fig. 12.31 Input circuits at low frequency for (*a*) a CE amplifier and (*b*) a CS amplifier.

Solving for f_H,

$$f_H = \frac{0.159}{R_s C_i} \tag{12.28}$$

LOW-FREQUENCY RESPONSE Assume that in a common-emitter amplifier the emitter resistor is suitably bypassed. (For a common-source amplifier, it is assumed that the source resistor is suitably bypassed.) The low-frequency response is then determined by coupling capacitor C_c. This is illustrated in Fig. 12.31 where the input circuits for the CE and CS amplifiers at low-frequency operation are shown.

The reactance of a capacitor, as mentioned earlier, decreases with increasing frequencies and increases with decreasing frequencies. Thus, at high frequencies, the reactance is so much less than the series resistance in the circuit that it acts like a short circuit. As the frequency is reduced, however, the reactance is increased. As a result, the voltage across the base and emitter (or gate and source) terminals is reduced and the gain of the amplifier falls.

The lower −3-dB frequency f_L is determined by setting the total series resistance in the input circuit equal to the reactance of the coupling capacitor at f_L. For the CE amplifier,

$$R_s + h_{ie} = \frac{0.159}{f_L C_c}$$

Solving for f_L,

$$f_L = \frac{0.159}{(R_s + h_{ie})C_c} \tag{12.29a}$$

For the CS amplifier,

$$R_s + R_G = \frac{0.159}{f_L C_c}$$

and

$$f_L = \frac{0.159}{(R_s + R_G)C_c} \tag{12.29b}$$

A similar approach is used in calculating the low-frequency response of the other amplifier configurations.

12.7 CASCADING AMPLIFIER STAGES

To realize greater signal gain than obtainable from a single amplifier stage, a number of stages are *cascaded*, as illustrated in the block diagram of Fig. 12.32. In a cascaded amplifier, the output of one amplifier stage is *coupled* to the input of the following stage. If the stages are *noninteracting*, that is, if there is no appreciable change in the gain or other parameters of an individual stage when cascaded, the overall gain A_t is equal to the product of the individual gains,

$$A_t = A_1 \times A_2 \times \cdots \times A_n \tag{12.30}$$

where A_1 is the gain of the first stage, A_2 the gain of the second stage, and A_n the gain of the nth stage.

Owing to its nearly infinite input resistance, the FET amplifier may be considered as being noninteracting. The BJT, because its input resistance is relatively low, interacts and Eq. (12.30) cannot be used. Instead, working from the output toward the input stage, the gain of each stage must be considered separately, taking into account loading effects of the cascaded stages.

COUPLING CASCADED STAGES Two basic methods are used for coupling cascaded stages:

1. Resistance-capacitance (*RC*) coupling.
2. Direct coupling (*DC*).

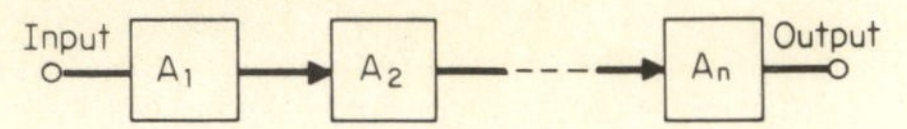

Fig. 12.32 Block diagram of a cascaded amplifier.

Examples of resistance-capacitance and direct-coupled BJT stages are illustrated in Fig. 12.33. (If a FET is substituted for a BJT, the circuits are identical.)

In *RC* coupling, the dc collector voltage of the preceding stage is blocked by the coupling capacitor from the base of the following stage. In direct coupling, the collector of one transistor is coupled directly to the base of the following transistor. Because the dc operating point of a transistor tends to drift, special circuit precautions are taken in *DC* amplifier design. The greatest use of direct coupling is in integrated-circuit amplifiers (see Chap. 13). Unless direct current or slowly varying signals are to be amplified, *RC* coupling is invariably used in discrete amplifiers.

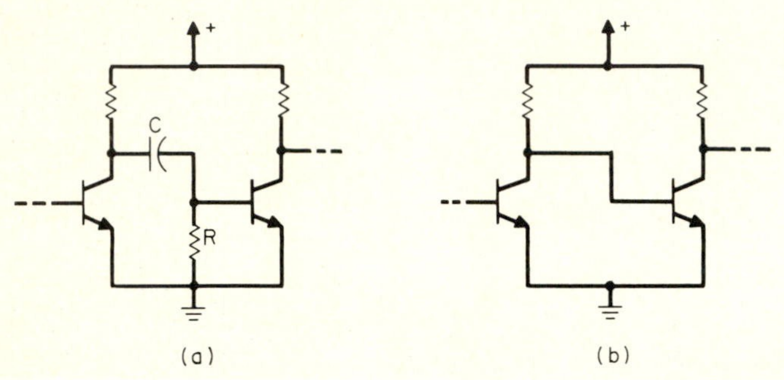

Fig. 12.33 Coupling cascaded stages: (*a*) Resistance-capacitance (*RC*) coupling. (*b*) Direct coupling (*DC*).

FREQUENCY RESPONSE OF A CASCADED AMPLIFIER The overall frequency response of a cascaded amplifier is less than that for an individual stage. For *noninteracting identical stages*, the upper −3-dB frequency of an n-stage cascaded amplifier f'_H is

$$f'_H = f_H \sqrt{2^{1/n} - 1} \tag{12.31a}$$

The lower −3-dB frequency f'_L is

$$f'_L = f_L / \sqrt{2^{1/n} - 1} \tag{12.31b}$$

where f_H and f_L are the upper and lower −3-dB frequencies of an individual stage.

example 12.9 An individual stage of a two-stage *RC*-coupled amplifier has $f_H = 20$ kHz and $f_L = 200$ Hz. Its bandwidth, therefore, is 20 000 − 200 = 19 800 Hz = 19.8 kHz. Assuming that the stages are identical and noninteracting, calculate the bandwidth of the cascaded amplifier.

solution By Eq. (12.31*a*), $f'_H = 20\sqrt{2^{1/2} - 1}$. The quantity $2^{1/2} = \sqrt{2} = 1.414$; hence, $\sqrt{1.414 - 1} = \sqrt{0.414} = 0.64$. Therefore, $f'_H = 20 \times 0.64 = 12.8$ kHz. By Eq. (12.31*b*), $f'_L = 200/0.64 = 313$ Hz.

The bandwidth of the cascaded amplifier is $f'_H - f'_L = 12\,800 - 313 \simeq 12.5$ kHz. With respect to an individual stage, the bandwidth of the cascaded amplifier was reduced in excess of 60 percent.

12.8 POWER AMPLIFIERS

A useful criterion for comparing power amplifiers is the conversion efficiency η (Greek letter *eta*). The conversion efficiency is defined as the signal power delivered to the load P_{ac} to the dc power P_{dc} supplied to the collector or drain circuit of a transistor. Expressed as a percentage,

$$\eta = \frac{P_{ac}}{P_{dc}} \times 100\% \tag{12.32}$$

For example, the maximum conversion efficiency realized with class A operation is 50 percent and with class B operation 78.4 percent. Class A and B power amplifiers are examined in the following sections.

CLASS A POWER AMPLIFIER A typical class A power amplifier is illustrated in Fig. 12.34. Load R_L, which may be the coil of a loudspeaker, is coupled to the collector circuit of the transistor by *coupling transformer T*. The turns ratio of the transformer n is the ratio of N_1 to N_2 turns:

$$n = \frac{N_1}{N_2} \tag{12.33}$$

The value of R_L "seen" by the transistor is modified. The modified, or *reflected*, value R_L' is given by the product of n squared and R_L:

$$R_L' = n^2 R_L \tag{12.34}$$

The output power is generally determined graphically. Two load lines are drawn on the collector (or drain) family of curves. The dc, or *static*, load line represents the dc resistance of transformer winding N_1. Because its resistance is extremely low, it is taken to be zero. A resistance value of zero ohms appears as a *vertical line* perpendicular to the voltage axis at E_{CC} (or E_{DD}), as shown in Fig. 12.35.

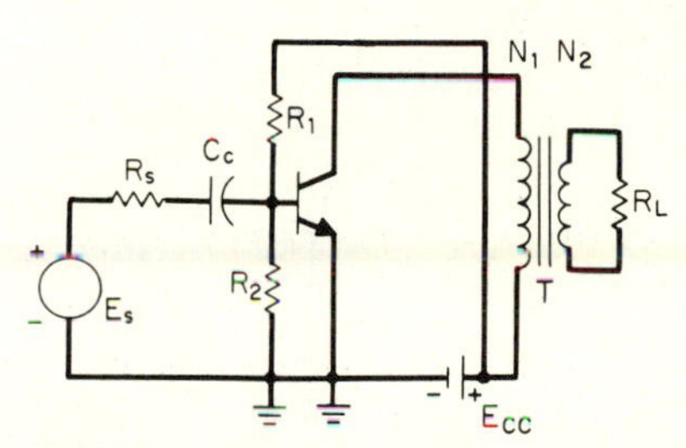

Fig. 12.34 Typical class A power amplifier.

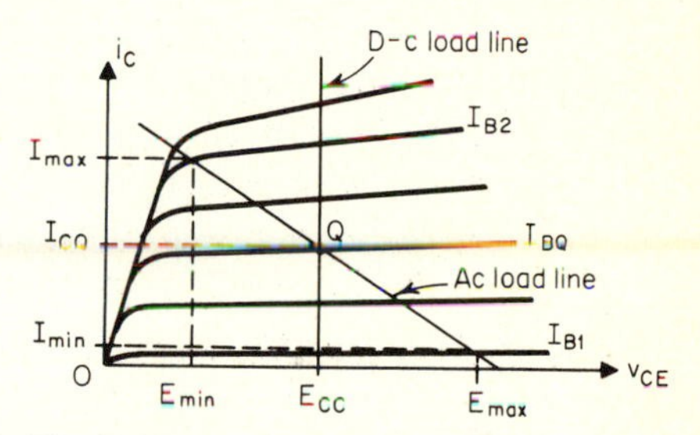

Fig. 12.35 Superimposing the dc and ac load lines on the collector characteristics.

The Q point is determined by the intersection of the dc load line and base current I_{BQ}. The value of the base current depends on the values of R_1 and R_2 of the biasing network (Fig. 12.34). Intersecting the Q point and having a slope of $-1/R_L'$ is the ac, or *dynamic*, load line.

Assume that the input sine wave signal has a positive peak value of I_{B2} and a negative peak value of I_{B1}. The corresponding maximum and minimum values of collector current and voltage are found by the intersection of the ac load line and the I_{B1} and I_{B2} curves. As indicated in Fig. 12.35, these values are denoted by $I_{\max}$, $E_{\max}$, $I_{\min}$, and $E_{\min}$, respectively. The output power P_{ac} in watts is

$$P_{ac} = \frac{(E_{\max} - E_{\min})(I_{\max} - I_{\min})}{8} \tag{12.35}$$

COLLECTOR DISSIPATION The power dissipated as heat in the collector is referred to as collector dissipation P_C. (For a FET, the dissipated power in the drain is called the *drain dissipation* P_D.) The value of P_C is

$$P_C = E_{CC} I_{CQ} - P_{ac} \tag{12.36a}$$

A class A amplifier operating without a signal is said to be *idling*. When idling, $P_{ac} = 0$, and the amplifier is *dissipating maximum power*, $P_{C,\max}$:

$$P_{C,\max} = E_{CC} I_{CQ} \tag{12.36b}$$

It is important that the maximum dissipated power never exceed the value given by the manufacturer. Further, from Fig. 12.35, it is seen that the maximum collector signal voltage E_{max} exceeds E_{CC}. The value of E_{max} must always be less than the collector (or drain) breakdown voltage of the transistor specified by the manufacturer.

SECOND-HARMONIC DISTORTION Second-harmonic distortion of an amplifier provides an indication of how faithful the amplified output signal is with respect to the input signal. Using the values found graphically in determining the output power, second-harmonic distortion D_2 is

$$D_2 = \frac{|I_{max} + I_{min} - 2I_{CQ}|}{2|I_{max} - I_{min}|} \times 100\% \tag{12.37}$$

The vertical lines indicate the *magnitude of the quantity*.

example 12.10 The dc and ac load lines are superimposed on the collector characteristics of Fig. 12.36. The Q point for class A operation is located at approximately $E_{CC} = 4$ V and $I_{CQ} = 1$ A. For an input signal swing between $I_B = 200$ mA and 10 mA, the following values are estimated from the intersection of the ac load line and the base current curves: $E_{max} \simeq 7.3$ V, $E_{min} \simeq 2.7$ V, $I_{max} \simeq 1.4$ A, and $I_{min} \simeq 0.2$ A. Find (*a*) P_{ac}; (*b*) η; (*c*) P_C and $P_{C,max}$; (*d*) D_2.

solution (*a*) By Eq. (12.35), $P_{ac} = (7.3 - 2.7)(1.4 - 0.2)/8 = 4.6 \times 1.2/8 \simeq 0.69$ W.

(*b*) $P_{dc} = E_{CC}I_{CQ} = 4 \times 1 = 4$ W. By Eq. (12.32), $\eta = (0.69/4) \times 100\% = 17.2\%$. This is considerably less than the maximum conversion efficiency of 50 percent for class A operation.

(*c*) By Eqs. (12.36*a*) and (12.36*b*), $P_C = 4 - 0.69 = 3.31$ W; $P_{C,max} = 4$ W.

(*d*) By Eq. (12.37), $D_2 = \dfrac{|1.4 + 0.2 - 2 \times 1|}{2|1.4 - 0.2|} \times 100\% = 0.4/2.4 \times 100\% = 16.6\%$.

PUSH-PULL POWER AMPLIFIERS Examples of push-pull amplifier circuits are illustrated in Fig. 12.37. An advantage of push-pull over single-ended operation is that,

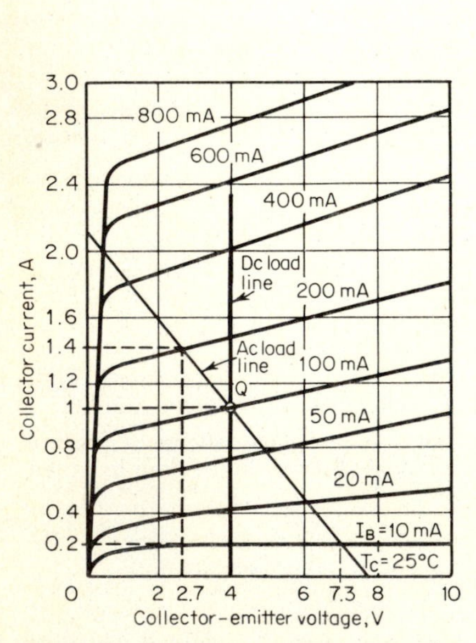

Fig. 12.36 Determining output power of a class A single-ended amplifier. (See Example 12.10.)

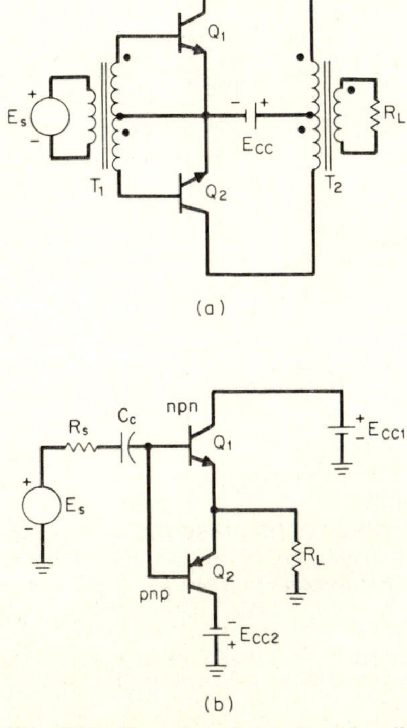

Fig. 12.37 Examples of class B push-pull power amplifiers: (*a*) Conventional type. (*b*) Complementary symmetry type.

assuming identical transistors, all even harmonics of distortion are canceled. If the transistors are operated as class B, conversion efficiencies as high as 78.4 percent are realized. Further, when the amplifier is idling, the collector dissipation is zero. The discussion to follow is limited to class B operation.

The circuit of Fig. 12.37*a* requires two center-tapped transformers, T_1 and T_2. Transformer T_1 is referred to as a *phase-inverting input transformer*. The voltage polarity at the dotted terminals is always the same, either plus or minus. If the base of transistor Q_1 is positive, the base of transistor Q_2 is negative, and vice versa. Output transformer T_2 couples load resistance R_L across the collectors of Q_1 and Q_2.

Assume that the signal is a sine wave. On the positive half-cycle, the base of Q_1 is positive and the base of Q_2 negative. Transistor Q_1 conducts and Q_2 is cut off. During the negative half-cycle, the polarities are reversed. Now, Q_1 is cut off and Q_2 conducts current. During a cycle, therefore, each transistor in a class B push-pull amplifier conducts current half the time.

The circuit of Fig. 12.37*b*, referred to as a *complementary symmetry amplifier*, does not require input and output transformers. (It may be regarded as a *transformerless* power amplifier.) Transistor Q_1 is npn and Q_2 is pnp. For good operation, Q_1 and Q_2 should be closely matched in their electrical characteristics. During the positive half-cycle, Q_1 conducts and Q_2 is cut off. On the negative half-cycle the situation is reversed: Q_2 conducts and Q_1 is cut off.

CROSSOVER DISTORTION The input resistance of an amplifier increases with decreasing collector current and decreases with increasing collector current. Consequently, in class B operation, the sine wave is flattened, as illustrated in Fig. 12.38. The flattening is referred to as *crossover distortion*.

To reduce crossover distortion, power amplifiers are generally operated as class AB. Under this mode of operation, a small collector current flows for zero signal current. The conversion efficiency of a class AB push-pull amplifier is in the order of 60 to 70 percent.

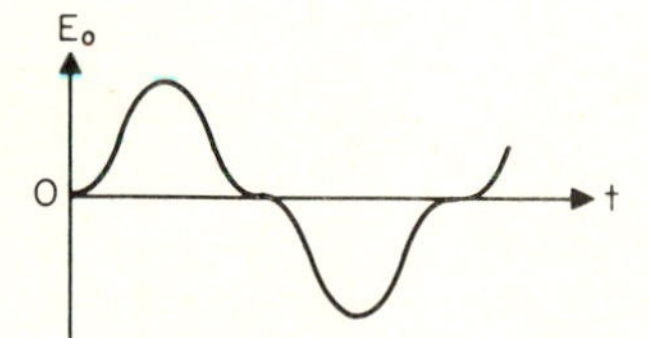

Fig. 12.38 Illustration of crossover distortion.

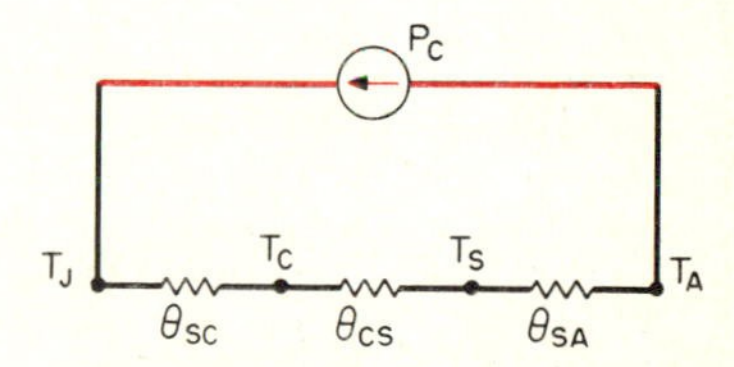

Fig. 12.39 Thermal equivalent circuit of a transistor mounted on a heat sink.

HEAT SINKS Because of possible large collector dissipation, power transistors (as well as power diodes and other semiconductor devices) are often mounted on heat sinks. A heat sink is a metal plate or structure, typically aluminum, which aids in dissipating the heat away from the device.

One can draw a *thermal equivalent circuit* of a transistor mounted on a heat sink, as shown in Fig. 12.39. This circuit is similar to a series circuit of resistors where temperature T is analogous to voltage; thermal resistance Θ (Greek letter *theta*) is analogous to electrical resistance; and dissipated power P_C is analogous to current. By Ohm's law, $I = E/R$; for the thermal equivalent circuit, one can write $P_C = T/\Theta$.

The temperature of the collector-base junction is designated by T_J; T_C is the temperature of the transistor case; T_S is the temperature of the heat sink; and T_A is the ambient (environment) temperature. The unit of temperature is degrees Celsius.

Thermal resistance Θ_{JC} is the resistance to the flow of heat between the junction and case of the device; Θ_{CS} is the thermal resistance between the case and heat sink; and Θ_{SA} is the thermal resistance between the heat sink and the ambient. The unit for Θ is degrees Celsius per watt and P_C is in watts.

Applying Ohm's law to the thermal equivalent circuit, the allowable dissipated power is equal to the difference in the junction and ambient temperature divided by the sum of the thermal resistances:

$$P_C = \frac{T_J - T_A}{\Theta_{JC} + \Theta_{CS} + \Theta_{SA}} \tag{12.38}$$

The application of Eq. (12.38) is illustrated in the following example.

example 12.11 A silicon transistor is mounted on a heat sink. If $T_J = 125°C$ and $T_A = 25°C$, determine the thermal resistance of the heat sink needed in order for the transistor to dissipate 50 W. Assume that $\Theta_{JC} = 0.5°C/W$ and $\Theta_{CS} = 0.3°C/W$.

solution Substitution of the given values in Eq. (12.38) yields $50 = (125 - 25)/(0.5 + 0.3 + \Theta_{SA}) = 100/(0.8 + \Theta_{SA})$. Solving, $0.8 + \Theta_{SA} = 100/50 = 2$. Hence, $\Theta_{SA} = 2 - 0.8 = 1.2°C/W$.

12.9 FEEDBACK AMPLIFIERS

Examples of feedback have already been considered in this chapter. It was seen that current- and voltage-feedback stabilization helped maintain the Q point of an amplifier relatively fixed, regardless of changes in temperature or other parameters. In general, an amplifier with feedback (*feedback amplifier*) provides gain stability, increases frequency response, and decreases distortion.

The operation of a feedback amplifier is influenced by its output signal, in addition to its input signal. This is realized by returning, or feeding back, a portion or all of the output signal to the input of the amplifier. In *voltage feedback*, the feedback signal is proportional to the load voltage; in *current feedback*, the returned signal is proportional to the load current.

If the feedback signal subtracts from the input signal, *negative feedback* is said to exist. For *positive feedback*, the feedback signal adds to the input signal. Negative feedback is used in amplifiers; positive feedback is the basis for the operation of sinusoidal oscillators.

PROPERTIES OF NEGATIVE FEEDBACK To highlight some important properties of negative feedback, reference is made to the block diagram of a feedback amplifier in Fig. 12.40. Output voltage E_o, which is across load resistor R_L, is also impressed across the input terminals of the feedback network. The output voltage of the feedback network E_f is connected in series with signal source E_s. The gain of the amplifier block is A_v, and the gain of the feedback network is B.

Because the feedback voltage is proportional to the output voltage, voltage feedback is obtained. Furthermore, the feedback voltage is in series with the input signal. For these reasons, the configuration of Fig. 12.40 is referred to as a *voltage-feedback series-input* feedback amplifier.

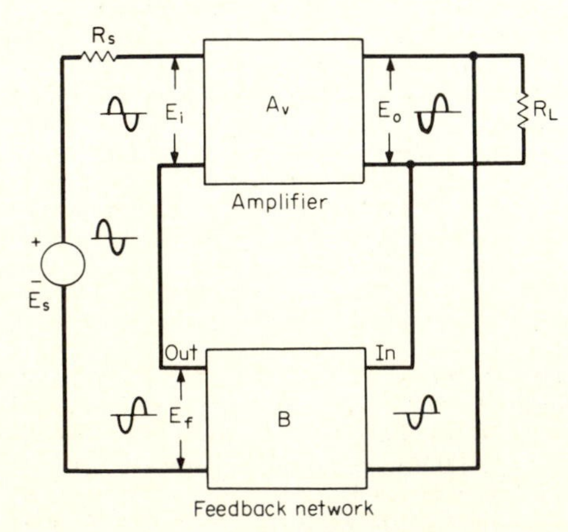

Fig. 12.40 Block diagram of a feedback amplifier.

For the verification that negative feedback indeed exists, voltage waveforms at the input and output terminals of the amplifier and feedback network are indicated in Fig. 12.40. Assume that the input sine wave to the amplifier E_i is positive-going. Also, the numerical value of A_v is negative, indicating that the output is 180° out of phase with respect to the input. As a result, output E_o is negative-going.

The input to the feedback network, which is equal to E_o, is also negative-going. Because the feedback network is generally comprised of resistors, there is no phase inversion, and its output E_f is negative-going. Negative feedback exists since E_f opposes E_s.

We now derive an expression for voltage gain with feedback A_{vf}. Referring back to Fig. 12.40, input voltage E_i is equal to the algebraic sum of the signal voltage E_s and feedback voltage E_f:

$$E_i = E_s + E_f \tag{12.39a}$$

The feedback voltage is equal to the product of output voltage E_o and the gain of the feedback network B:

$$E_f = BE_o \tag{12.39b}$$

Substitution of Eq. (12.39*b*) in Eq. (12.39*a*) yields

$$E_i = E_s + BE_o \tag{12.39c}$$

The gain of the amplifier block A_v is equal to the ratio of its output to input voltages

$$A_v = \frac{E_o}{E_i} \tag{12.39d}$$

Substituting Eq. (12.39*c*) for E_i in Eq. (12.39*d*) results in

$$A_v = \frac{E_o}{E_s + BE_o} \tag{12.39e}$$

Solving Eq. (12.39*e*) for A_{vf}, which is equal to output voltage E_o divided by the signal voltage E_s, we obtain

$$A_{vf} = \frac{A_v}{1 - BA_v} \tag{12.40}$$

It is common practice to refer to A_{vf} as the *closed-loop gain*, A_v as the *open-loop gain*, and the product BA_v as the *loop gain*.

Equation (12.40) illustrates some very important facts about feedback. If $B = 0$ (no feedback present), the closed-loop and open-loop gains are equal ($A_{vf} = A_v$). For negative feedback, the product BA_v is negative. Consequently, the denominator is greater than one and the voltage gain with negative feedback is less than without feedback. This is a very important property of negative feedback.

If BA_v is much greater than one, the one can be neglected in the denominator of Eq. (12.40). Consequently,

$$A_{vf} \simeq \frac{A_v}{-BA_v}$$

The A_v terms cancel and

$$A_{vf} \simeq -\frac{1}{B} \tag{12.41}$$

Equation (12.41) demonstrates that if $BA_v \gg 1$, the voltage gain depends essentially on the gain of the feedback network. This means that, as long as B is constant, even if the gain of the amplifier changes owing to variation in temperature or in device parameters, the gain of the feedback amplifier is hardly affected. The gain of the feedback network, because it is usually comprised of resistors, is easy to stabilize. Another important property of negative feedback, therefore, is *gain stability*.

example 12.12 Negative feedback is to be introduced around an amplifier with an open-loop voltage gain of $A_v = -200$. It is required that the voltage gain with feedback also equal

−200 ($A_{vf} = -200$). Because of negative feedback, the open-loop gain must be increased to realize $A_{vf} = -200$. If $B = 4 \times 10^{-3}$, determine the new value of A_v.

solution Solving Eq. (12.40) for A_v yields

$$A_v = \frac{A_{vf}}{1 + BA_{vf}} \tag{12.42}$$

Substituting the given values in Eq. (12.42), we have $A_v = -200/(1 - 4 \times 10^{-3} \times 200) = -200/0.2 = -1000$.

Bandwidth Another advantage of negative feedback is that it extends the bandwidth of an amplifier. Letting the upper −3-dB frequency with feedback be f_{HF} and the lower −3-dB frequency be f_{LF}, we have

$$f_{HF} = f_H(1 - BA_v) \tag{12.43a}$$

$$f_{LF} = \frac{f_L}{1 - BA_v} \tag{12.43b}$$

For example, if $f_H = 10$ kHz and $f_L = 100$ Hz, then for $1 - BA_v = 10$, $f_{HF} = 10 \times 10$ kHz = 100 kHz and $f_{LF} = {}^{100}/_{10} = 10$ Hz. The bandwidth has been extended by a factor slightly greater than ten.

Distortion Negative feedback also reduces distortion in an amplifier. Letting D_F equal the distortion with feedback, and D the distortion without feedback, then

$$D_F = \frac{D}{1 - BA_v} \tag{12.44}$$

For example, if the second-harmonic distortion without feedback D_2 is 10 percent, then the second-harmonic distortion with feedback $D_{F2} = {}^{10}/_{10}$ is 1 percent. The distortion has been reduced by a factor of ten.

Instability At very low and high frequencies, the loop gain BA_v may become equal to, or close to, one. If a one is substituted for BA_v in Eq. (12.40), we obtain

$$A_{vf} = \frac{A_v}{1 - 1} = \frac{A_v}{0} = \infty$$

The gain of the feedback amplifier, therefore, approaches infinity. What this means in practical terms is that the amplifier is *unstable* and sine-wave oscillations are generated. This is undesirable for amplifiers although, as we shall see, a necessary condition for a sine-wave oscillator.

Whenever negative feedback is introduced in an amplifier, an opportunity exists for the feedback amplifier to become unstable. There are criteria, such as the *Nyquist criterion*, for determining the stability of a feedback amplifier. *Compensation networks* can be used to convert an unstable feedback amplifier to a stable one.

EFFECTS ON INPUT AND OUTPUT RESISTANCE In addition to providing voltage and current gain stability, increased bandwidth, and reduced distortion, negative feedback also modifies the input and output resistances of an amplifier. Depending on how feedback is introduced, it can either decrease or increase the input and output resistances of a feedback amplifier. There are four basic types of feedback, as illustrated in the examples of Fig. 12.41.

a. Voltage feedback, series input. Input resistance, R_{iF}, is *greater* than input resistance without feedback, R_i: $R_{iF} > R_i$. Output resistance, R_{oF}, is *less* than output resistance without feedback, R_o: $R_{oF} < R_o$.

b. Voltage feedback, shunt input: $R_{iF} < R_i$; $R_{oF} < R_o$.

c. Current feedback, series input: $R_{iF} > R_i$; $R_{oF} > R_o$.

d. Current feedback, shunt input: $R_{iF} < R_i$; $R_{oF} > R_o$.

12.10 SINUSOIDAL OSCILLATORS

In the discussion of instability in feedback amplifiers, it was pointed out that if the loop gain $BA_v = 1$, the amplifier oscillates. The condition that the loop gain be unity for

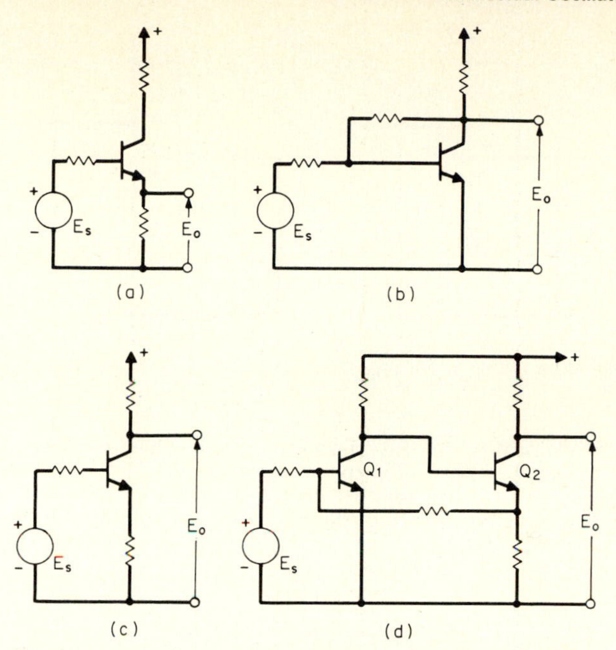

Fig. 12.41 Four basic types of feedback: (*a*) Voltage feedback, series input. (*b*) Voltage feedback, shunt input. (*c*) Current feedback, series input. (*d*) Current feedback, shunt input.

sine-wave oscillations is called the *Barkhausen criterion.* In this section a number of commonly used sine-wave oscillators are examined.

PHASE-SHIFT OSCILLATOR An example of a phase-shift oscillator using a FET is illustrated in Fig. 12.42. Biased for class A operation, the output of the FET is connected to three cascaded *RC* sections. In turn, the output of the three *RC* sections, which serves as the feedback network, is connected to the input (gate) of the amplifier.

The amplifier and the *RC* network each contributes 180° phase shift at the frequency of oscillation. Total phase shift, therefore, is 180 + 180 = 360°. This ensures that positive feedback exists and, if A_v and B are chosen properly, the Barkhausen criterion is satisfied. The magnitude of the amplifier gain must be at least 29 and the frequency of oscillation f in hertz is

$$f = \frac{0.159}{(\sqrt{6}RC)} \tag{12.45}$$

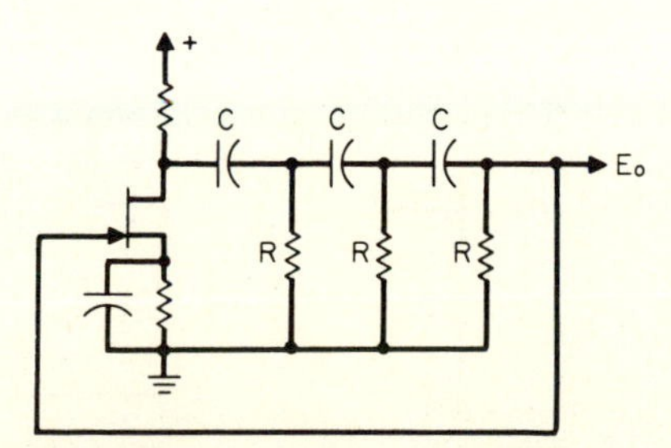

Fig. 12.42 An example of a phase-shift oscillator.

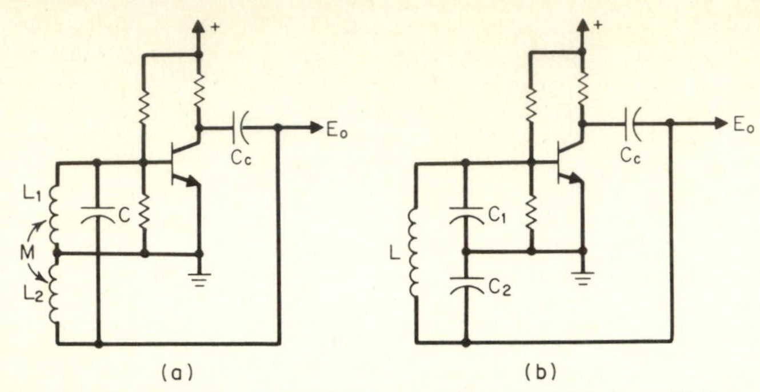

Fig. 12.43 Examples of tuned-circuit oscillators: (*a*) Hartley. (*b*) Colpitts.

TUNED-CIRCUIT OSCILLATORS Two examples of tuned-circuit, or *resonant circuit*, oscillators using a BJT are illustrated in Fig. 12.43. In the *Hartley* oscillator of Fig. 12.43*a*, appropriate phase shift to satisfy the Barkhausen criterion is obtained by the mutual coupling M between inductors L_1 and L_2. The output of the amplifier is coupled to the input side by coupling capacitor C_c. Assuming small mutual coupling, an approximate expression for frequency of oscillation is

$$f = \frac{0.159}{\sqrt{C(L_1 + L_2)}} \tag{12.46}$$

In the *Colpitts* oscillator of Fig. 12-43*b*, suitable phase shift is obtained by capacitors C_1, C_2, and inductor L. As in the Hartley oscillator, C_c couples the output voltage of the BJT to the input side. Because capacitor C_1 is across the base-emitter junction of the transistor, the input capacitance C_i of the transistor should be included. Letting $C_1' = C_1 + C_i$, the frequency of oscillation is given by

$$f = \frac{0.159}{\sqrt{LC_1'C_2/(C_1' + C_2)}} \tag{12.47}$$

FREQUENCY STABILITY The frequency of oscillation tends to vary, or *drift*, with variations in temperature or aging. *Frequency stability* refers to the ability of an oscillator to maintain its frequency constant over long periods of time. To ensure frequency stability, various methods are employed. One method uses a capacitor with a negative-temperature coefficient (see Chaps. 2 and 10) to cancel the positive temperature coefficient of the inductor. Another approach is to use a very high-Q tuned circuit. An example of a high-Q circuit is the quartz crystal, which is the basis of the crystal oscillator.

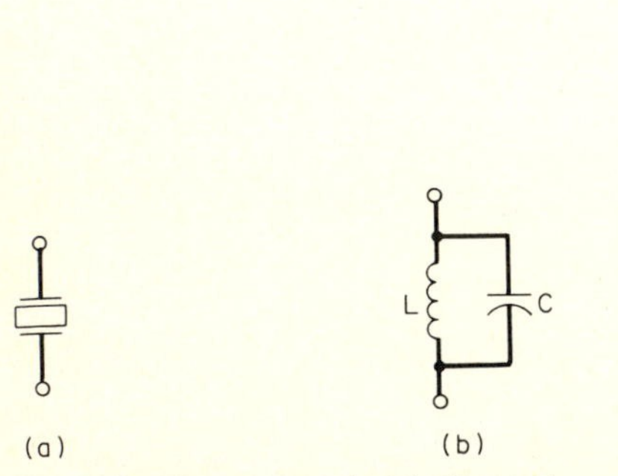

Fig. 12.44 Quartz crystal: (*a*) Electrical symbol. (*b*) Approximate model.

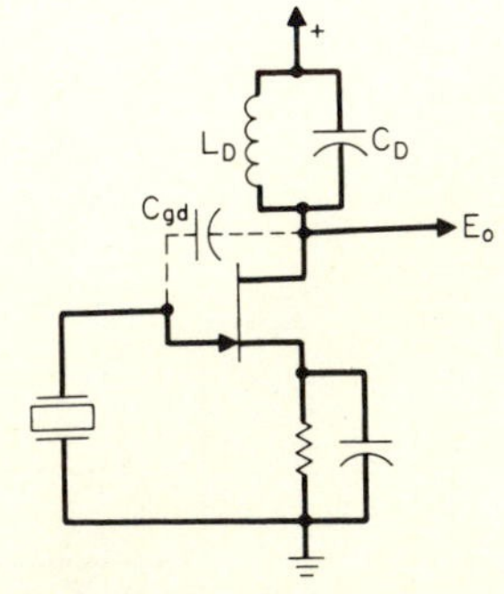

Fig. 12.45 An example of a crystal oscillator.

CRYSTAL OSCILLATOR Quartz crystal has the property of becoming mechanically deformed when a voltage is impressed across its faces. Or, if mechanically strained, it produces a voltage. This phenomenon is referred to as the *piezoelectric effect*. The electrical symbol and approximate model of a quartz crystal are shown in Fig. 12.44. An examination of the model shows that the quartz crystal acts like a parallel LC tuned circuit. Its Q may be as high as 10 000, thereby ensuring excellent frequency stability.

An example of a FET crystal oscillator is illustrated in Fig. 12.45. The crystal is connected to the input circuit of the amplifier, and an L_DC_D circuit tuned to the frequency of the crystal is in the output circuit. Suitable coupling to satisfy the Barkhausen criterion is obtained through the internal drain-gate capacitance C_{gd}, and stray capacitance. The frequency of oscillation is determined by the crystal. If, for example, a 2.5-MHz oscillator is needed, a 2.5-MHz crystal is inserted in the circuit.

Chapter 13
Operational Amplifiers

13.1 INTRODUCTION

The phenomenal development of monolithic *IC* technology has made the operational amplifier (*op amp*) perhaps the most versatile component in electronics. With proper selection of feedback elements, the op amp may be used as a precision gain-voltage amplifier, buffer, summer, current source, convertor, oscillator, and in many more applications. Its field is generally limited by the user's ingenuity and imagination.

Actually, the operational amplifier is not new. It was, and still is, used in analog computers. (An analog computer is employed to simulate physical systems.) Because of IC technology, it is possible to obtain an op amp, containing a dozen transistors, in a TO-5 package, which is used for packaging a single discrete transistor. Also, the cost of some IC operational amplifiers are less than what one paid for a single transistor just a few years ago.

13.2 THE BASIC OP AMP

In this section we treat the op amp as a *black box* with ideal characteristics. By considering the op amp as an ideal component, we can easily derive results of its application in different circuits. Fortunately, the derivations also hold, with an error of a few percent, for the majority of commercially available op amps.

The basic op amp is represented by the symbol of Fig. 13.1. The input denoted by the negative sign is referred to as the *inverting input;* the plus sign denotes the *noninverting input.* For simplicity, ground, power-supply, and frequency-compensation (to be considered later) connections are omitted from the figure.

If a signal is impressed across the inverting terminal with respect to ground, the output signal is 180° out of phase with respect to the input signal, as in the common-emitter amplifier. For a signal applied across the noninverting terminal with respect to ground, the output is in phase with the input. If signals are impressed across both the inverting and noninverting terminals with respect to ground, the output is proportional to the difference of the input signals.

IDEAL CHARACTERISTICS The attributes of the ideal op amp, which many physical op amps approach, are:

1. Voltage gain is infinite.
2. Input impedance is infinite.
3. Output impedance is zero.
4. The frequency response is flat (constant) for all frequencies.
5. The output signal is zero if the input signal is zero.

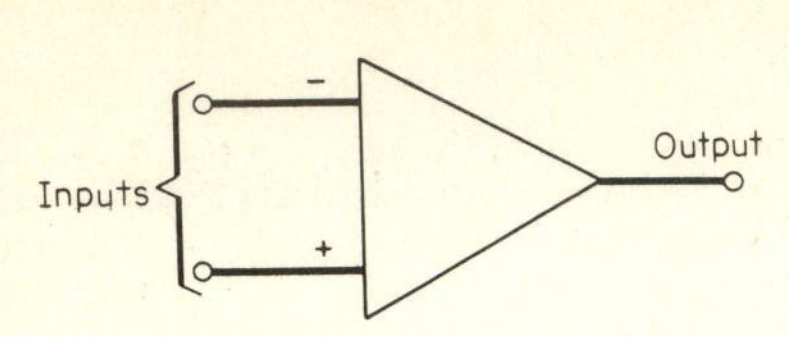

Fig. 13.1 Symbol for the basic op amp.

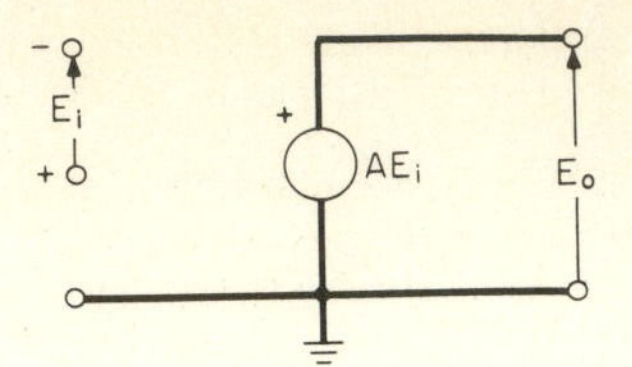

Fig. 13.2 Model of an ideal op amp.

Based on these attributes, a model of the ideal op amp is drawn in Fig. 13.2. The output voltage E_o equals the product of the gain A and the net input signal across the inverting and noninverting terminals E_i.

FEEDBACK Physical op amps can exhibit voltage gains as high as 10^9. If the input voltage is finite, but close to zero, the output voltage of such a high-gain amplifier would soar to infinity. Actually, the output would be limited by the power-supply voltages. Under this condition, the amplifier is said to be *saturated*, and useless. To make the op amp useful and versatile, *feedback* is introduced around the amplifier.

Consider the operational circuit, called an *inverting amplifier*, of Fig. 13.3. The noninverting terminal is returned to ground and a feedback resistor R_F is connected between the inverting and output terminals. A resistor R_1 is in series with the signal source E_s and the inverting terminal.

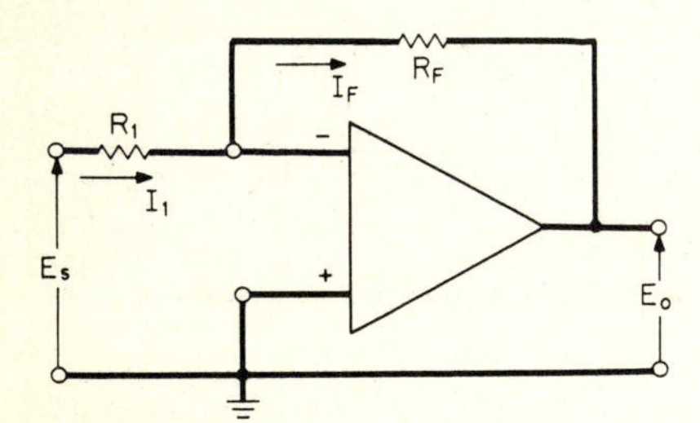

Fig. 13.3 An inverting amplifier.

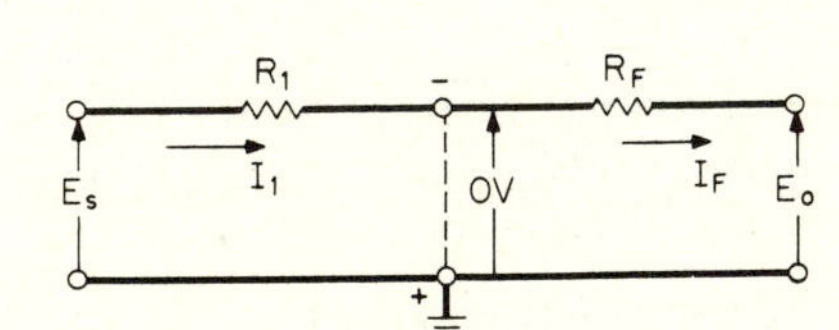

Fig. 13.4 The concept of the virtual ground.

As mentioned earlier, the ideal op amp exhibits infinite input impedance and voltage gain. The latter property implies that the voltage across the inverting and noninverting terminals is zero. Because the input impedance is infinite, no current flows into the inverting terminal. This condition is referred to as a *virtual ground* (Fig. 13.4). From the figure it is seen that I_1, the current flowing in R_1, is equal to I_F, the current flowing in R_F. Current $I_1 = (E_s - 0)/R_1$ and $I_F = (0 - E_o)/R_F$. Hence,

$$\frac{E_s}{R_1} = \frac{-E_o}{R_F}$$

Solving for the voltage gain with feedback, $A_f = E_o/E_s$, we have

$$A_f = -\frac{R_F}{R_1} \tag{13.1}$$

What is striking and significant about Eq. (13.1) is that the voltage gain is only a function of resistors R_F, R_1, and is independent of the amplifier. For example, if $R_F = 100$ and $R_1 = 1\ \mathrm{k}\Omega$, $A_f = {-100}/{1} = -100$. If R_F and R_1 are precision resistors, the gain of -100 will hold, regardless of variations in amplifier characteristics, power-supply voltage, etc. This behavior is typical of high-gain commercial op amps with suitable feedback.

13.3 APPLICATIONS

The inverting amplifier has already been analyzed. In this section, examples of other commonly used applications of the op amp are considered.

NONINVERTING AMPLIFIER The noninverting amplifier is shown in Fig. 13.5. In this circuit, the signal is fed directly to the noninverting terminal of the op amp. Resistors R_F and R_2 comprise the feedback network. Because of the virtual ground, the voltage at the inverting and noninverting terminals is equal to E_s. Owing to the

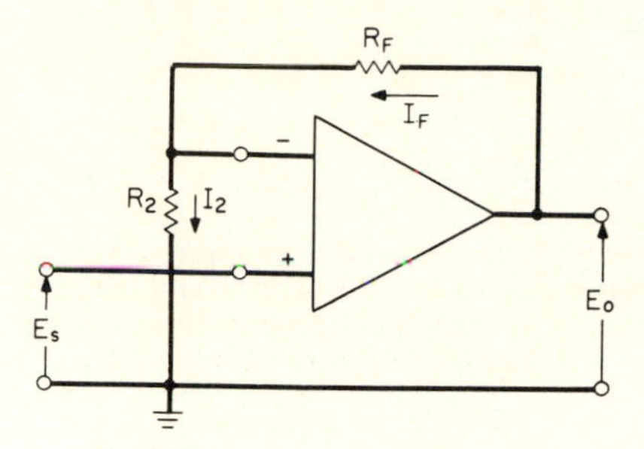

Fig. 13.5 A noninverting amplifier.

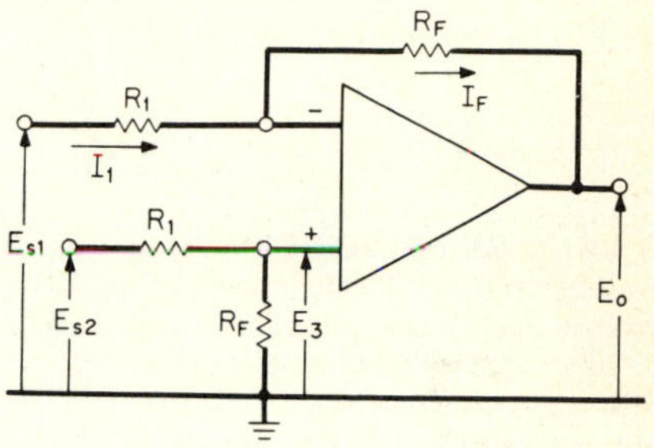

Fig. 13.6 A difference amplifier.

input impedance being infinite, $I_F = I_2$. Current $I_F = (E_o - E_s)/R_F$ and current $I_2 = E_s/R_2$. Hence,

$$\frac{E_o - E_s}{R_F} = \frac{E_s}{R_2}$$

Solving for $A_f = E_o/E_s$,

$$A_f = \frac{(R_2 + R_F)}{R_2} = 1 + \frac{R_F}{R_2} \tag{13.2}$$

For the noninverting amplifier, the output and input voltages are in phase; for the inverting amplifier, they are 180° out of phase.

DIFFERENCE AMPLIFIER The difference amplifier of Fig. 13.6 provides an output that is proportional to the difference of the input signals, $E_{s1} - E_{s2}$. Using the concept of the virtual ground, the voltage at the inverting and noninverting terminals is equal to voltage E_3. Current $I_1 = (E_{s1} - E_3)/R_1$ and $I_F = (E_3 - E_o)/R_F$. Because $I_1 = I_F$,

$$\frac{E_{s1} - E_3}{R_1} = \frac{E_3 - E_o}{R_F}$$

By voltage division, $E_3 = E_{s2}R_F/(R_1 + R_F)$. Substitution of this expression in the preceding equation yields

$$\frac{E_{s1} - E_{s2}R_F/(R_1 + R_F)}{R_1} = \frac{E_{s2}R_F/(R_1 + R_F) - E_o}{R_F}$$

Solution for the last equation for the output voltage E_o gives

$$E_o = -\frac{R_F}{R_1}(E_{s1} - E_{s2}) \tag{13.3}$$

Thus the output voltage is proportional to the difference in the input voltages.

example 13.1 An input signal feeding an inverting amplifier has 5 mV of 60-Hz hum. How can a difference amplifier be used to cancel the hum?

solution The circuit using a difference amplifier is shown in Fig. 13.7. The variable-voltage 60-Hz source is adjusted until no 60-Hz hum appears in the output E_o. This condition is monitored on a scope connected across the output terminal and ground of the op amp.

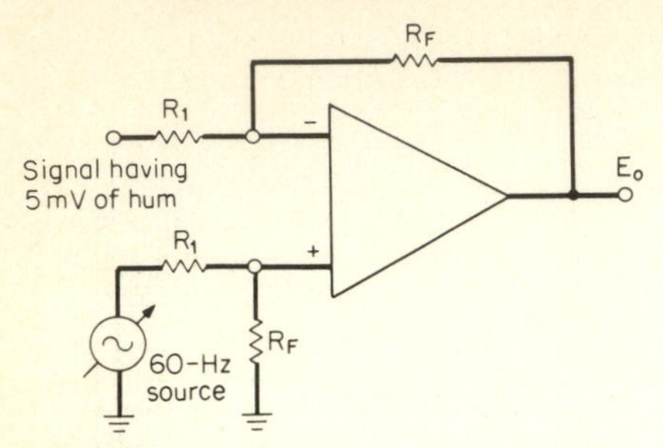

Fig. 13.7 Cancellation of 60-Hz hum present in a signal.

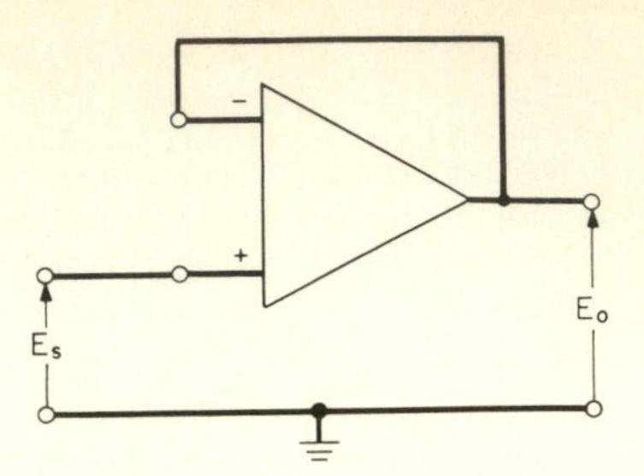

Fig. 13.8 A voltage follower.

VOLTAGE FOLLOWER An op amp circuit which has unity voltage gain, called a voltage follower, is shown in Fig. 13.8. Voltage followers, like emitter and source followers, are used for matching a high-impedance source to a low-impedance load.

The output terminal of the op amp is connected directly to the inverting terminal; the signal is impressed between the noninverting terminal and ground. Because of the virtual ground, the voltage at the inverting terminal is equal to the voltage at the noninverting terminal E_s. Since the output is connected to the inverting terminal, $E_o = E_s$, and

$$A_f = \frac{E_o}{E_s} = 1 \tag{13.4}$$

SUMMING AMPLIFIER The summing amplifier of Fig. 13.9 provides an output voltage proportional to the sum of the input voltages, $E_1, E_2, \ldots, E_n$. If resistors $R_1 = R_2 = \cdots = R_F$, then the output is equal to the negative sum of the inputs. If the resistors are unequal, then each input is summed with a *scale factor*.

Owing to the virtual ground, $I_1 = E_1/R_1$, $I_2 = E_2/R_2, \ldots$, $I_n = E_n/R_n$ and $I_F = -E_o/R_F$. Applying Kirchhoff's current law (KCL), the algebraic sum of the currents at the inverting terminal equals zero. Hence,

$$\frac{E_1}{R_1} + \frac{E_2}{R_2} + \cdots + \frac{E_n}{R_n} = \frac{-E_o}{R_F}$$

Solving for E_o, we obtain

$$E_o = -\left(\frac{R_F}{R_1}E_1 + \frac{R_F}{R_2}E_2 + \cdots + \frac{R_F}{R_n}E_n\right) \tag{13.5}$$

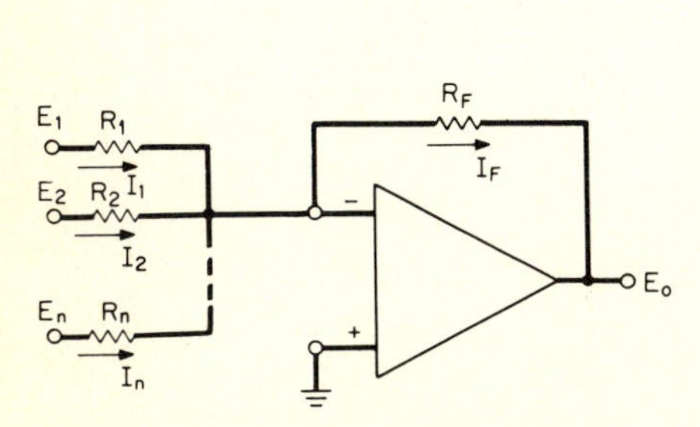

Fig. 13.9 A summing amplifier.

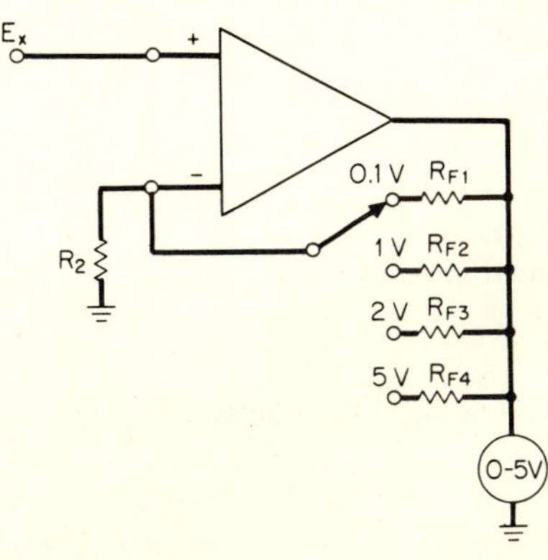

Fig. 13.10 An example of a multirange dc voltmeter. (See Example 13.3.)

example 13.2 Three input voltages are to be summed in the summing amplifier of Fig. 13.9. The feedback resistor $R_F = 100$ kΩ, $E_1 = 0.1$ V, $E_2 = 0.2$ V, and $E_3 = 0.05$ V. Determine E_o if (*a*) $R_1 = R_2 = R_3 = 100$ kΩ and (*b*) $R_1 = 20$, $R_2 = 50$, $R_3 = 10$ kΩ.

solution (*a*) For equal resistors, Eq. (13.5) reduces to $E_o = -(E_1 + E_2 + E_3) = -(0.1 + 0.2 + 0.05) = -0.35$ V.

(*b*) For unequal input resistors, each input voltage, from Eq. (13.5), is multiplied by a scale factor: R_F/R_1, R_F/R_2, and R_F/R_r. Hence,

$$\begin{aligned} E_o &= -(^{100}/_{20} \times 0.1 + {}^{100}/_{50} \times 0.2 + {}^{100}/_{10} \times 0.05) \\ &= -(5 \times 0.1 + 2 \times 0.2 + 10 \times 0.05) \\ &= -(0.5 + 0.4 + 0.5) = -1.4 \text{ V} \end{aligned}$$

example 13.3 A precision multirange dc voltmeter is to be constructed. The voltage ranges desired are zero to 0.1, 1, 2, and 5 V full scale. Using an op amp and a 0- to 5-V dc meter, design the circuit.

solution A design of such a meter is shown in Fig. 13.10. The unknown voltage to be measured, E_x, is impressed across the noninverting terminal and ground. To realize the four voltage ranges, a four-position switch and four resistors, $R_{F1}, \ldots, R_{F4}$, are connected in the feedback path. Resistor R_2, connected to the inverting terminal and the switch, completes the feedback circuit. A reasonable value for R_2 is 5 kΩ.

When $E_x = 0.1$ V, the 0- to 5-V meter must read 0.1 V full scale. The gain of the noninverting amplifier is, therefore, equal to $5/0.1 = 50$. From Eq. (13.2), $A_f = 1 + R_F/R_2$; hence, $50 = 1 + R_{F1}/5$. Solving, $R_{F1} = (50 - 1) \times 5 = 245$ kΩ.

Following the same procedure, for the 0- to 1-V scale, $A_f = 5/1 = 5$. Therefore, $R_{F2} = (5 - 1) \times 5 = 20$ kΩ. For the 0- to 2-V scale, $A_f = 5/2 = 2.5$; $R_{F3} = (2.5 - 1) \times 5 = 7.5$ kΩ. Finally, for the 0- to 5-V scale, $A_f = 5/5 = 1$; $R_{F4} = (1 - 1) \times 5 = 0$.

LOGARITHMIC AMPLIFIER This circuit provides an output voltage that is proportional to the logarithm of the input voltage. A simple version of a logarithmic amplifier

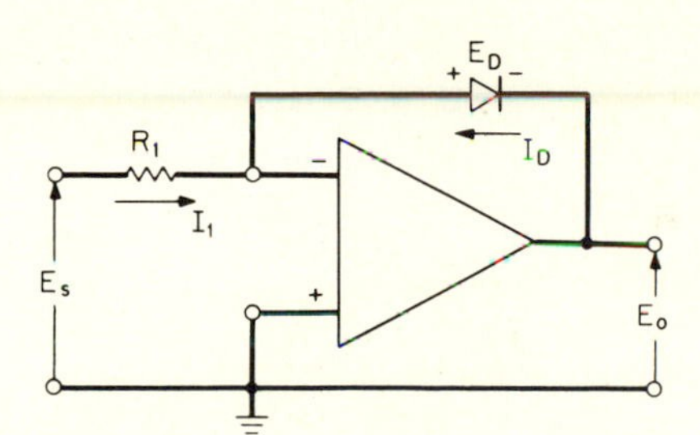

Fig. 13.11 An elementary logarithmic amplifier.

is obtained by replacing the feedback resistor R_F in the inverting amplifier of Fig. 13.3 by a diode (Fig. 13.11). The diode current I_D may be approximated by

$$I_D = I_o \epsilon^{+E_D/k} \tag{13.6}$$

where I_o is the diode reverse saturation current
E_D is the voltage across the diode
k is a term that includes physical constants and temperature

Because of the virtual ground, $I_1 = E_s/R_1 = I_D$. Also, $E_D = 0 - E_o = -E_o$ V. Making these substitutions in Eq. (13.6), we obtain

$$\frac{E_s}{R_1} = I_o \epsilon^{-E_o/k} \tag{13.7}$$

Taking the natural logarithm (ln) of Eq. (13.7) yields

$$\ln \frac{E_s}{R_1} = \ln I_o - \frac{E_o}{k}$$

Therefore,

$$E_o = -k\left(\ln \frac{E_s}{R_1} - \ln I_o\right) \tag{13.8}$$

Equation (13.8) demonstrates that the output voltage is indeed proportional to the natural log of the input voltage.

CURRENT-TO-VOLTAGE CONVERTER Devices, such as photomultipliers and photocells, act like an ideal current source. The output current of these devices is essentially constant, independent of the load. The current-to-voltage converter provides an output voltage which is proportional to the current.

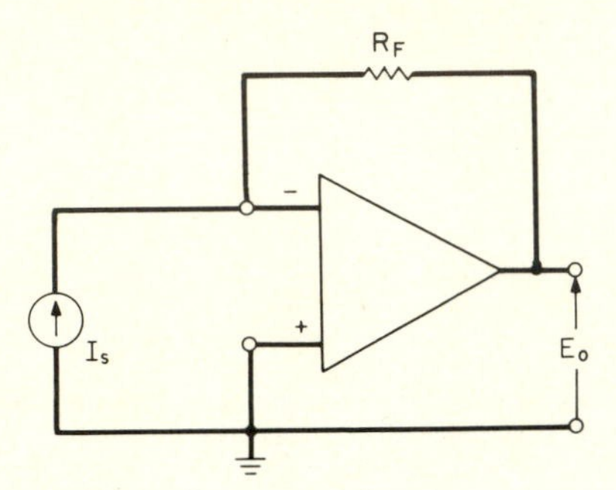

Fig. 13.12 A current-to-voltage converter.

An example of such a converter is illustrated in Fig. 13.12. The device is represented by the ideal current source I_s. Because of the virtual ground,

$$I_s = \frac{-E_o}{R_F}$$

or

$$E_o = -I_s R_F \tag{13.9}$$

The output voltage is, therefore, proportional to the input current.

13.4 PRACTICAL CONSIDERATIONS

To this point we represented the op amp as having only three terminals: the inverting and noninverting input terminals and an output terminal. In addition to these, there are terminals for ground, power supply, and compensating network connections (Fig. 13.13).

Generally negative ($-E$) and positive ($+E$) power supplies are required. There are at least two terminals to which are connected a compensating network. Compensation is required to ensure stable amplifier operation. The compensating network usually consists of a small capacitor in series with a resistor.

FREQUENCY RESPONSE AND COMPENSATION A frequency response plot of an *uncompensated* operational amplifier may take the form shown in Fig. 13.14 (solid lines). Along the vertical axis is plotted the magnitude of the voltage gain without feedback, or the *open-loop gain*, in decibels (dB). A voltage gain of 100 dB corre-

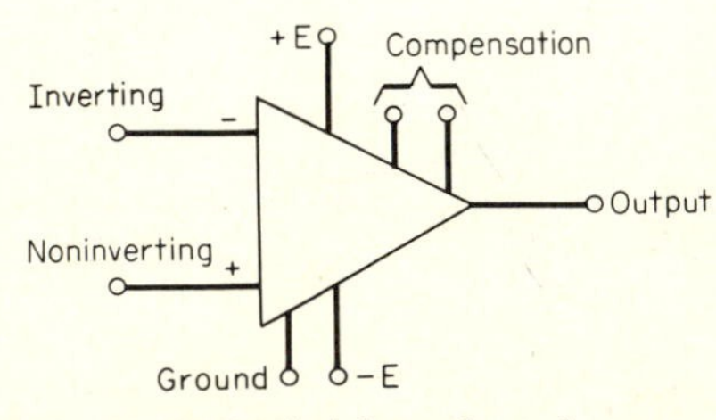

Fig. 13.13 Symbol for a physical op amp.

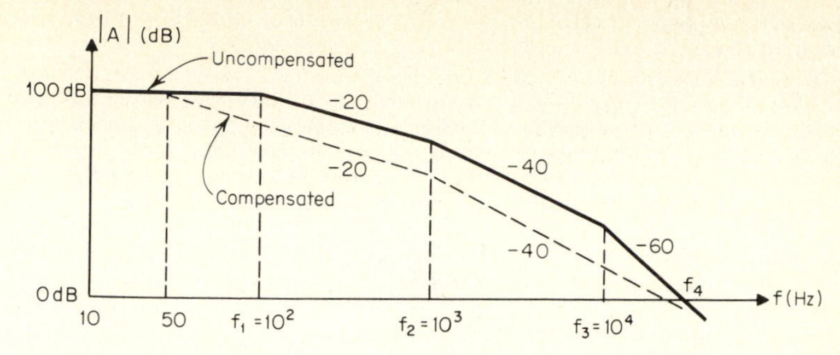

Fig. 13.14 Bode response plots of an uncompensated and a compensated op amp.

sponds to an amplifier gain of 100 000. Frequency is plotted to a logarithmic scale along the horizontal axis. This permits a large range of frequencies to be plotted in a compact manner.

The actual curve is approximated by straight-line segments. Such a representation is referred to as a *Bode*, or an *asymptotic*, plot. The Bode plot is easy to draw and provides useful information.

In Fig. 13.14 frequency $f_1 = 100$ Hz is called a *break frequency*, where the gain is actually 3 dB less than at lower frequencies. For frequencies between f_1 and f_2, the gain decreases, or *rolls off*, −20 dB for a decade of frequency. This means that for a 10-to-1 increase in frequency, the voltage gain falls by 20 dB. An *equivalent expression*, −6 dB/octave, means that as the frequency is *doubled*, the gain decreases by 6 dB. At frequency f_2 the gain begins to roll off at −40 dB/decade (−12 dB/octave), and at f_3 it rolls off at −60 dB/decade (−18 dB/octave). Frequency f_4, which intersects the frequency axis at unity voltage gain (0 dB), is the *crossover frequency*.

An amplifier that exhibits an open-loop response of −60 dB/decade (or −18 dB/octave) at the crossover frequency is *unstable*. It can break into oscillation and generate sine waves. By the proper selection of a compensating network, the amplifier is made stable. For this condition, indicated by the dashed lines in Fig. 13.14, the slope is −40 dB/decade (−12 dB/octave) at the crossover frequency.

The specific compensation network to use is recommended by the manufacturer of the op amp. It may, for example, be a 2200-pF capacitor connected in series with a 10-kΩ resistor.

From Fig. 13.14 the dominant frequency of the compensated amplifier is only 50 Hz. It might appear, therefore, that the op amp is really not too useful. Because of the use of feedback, which increases the bandwidth of an amplifier, this is not true.

Consider an op amp with an open-loop gain of 100 000 (100 dB). If, for an inverting amplifier, $A_f = 100$ (which equals $20 \log 100 = 40$ dB), the difference between the open-loop gain and the gain with feedback, referred to as the *closed-loop gain*, is $100 - 40 = 60$ dB at low frequencies. This is illustrated in Fig. 13.15. Note that the

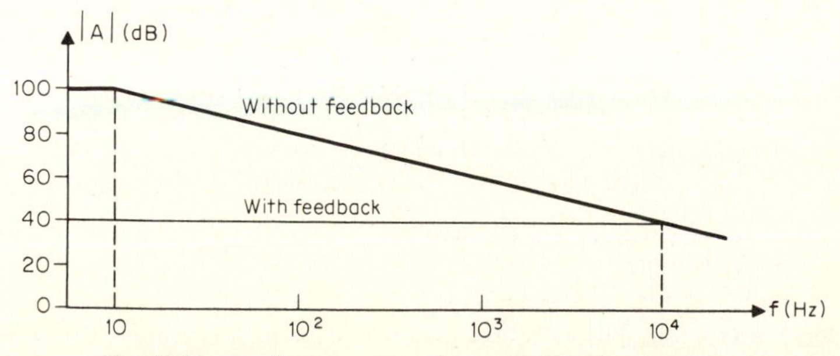

Fig. 13.15 Feedback increases the bandwidth of an amplifier.

bandwidth is much greater than 10 Hz; in this example it is 10 kHz. For frequencies greater than 10 kHz, the open- and closed-loop gains are equal.

OFFSET For a zero input, an ideal op amp has exactly zero output. In physical op amps, because of variations in transistor characteristics, there is a nonzero output for a zero signal. To compensate for this, a dc voltage is applied to the amplifier such that, for a zero input signal, the output is also zero. Before we see how this is accomplished, some terms used in the literature and on data sheets are defined.

Input offset current The input offset current I_{io} is the difference in dc bias currents of the inverting and noninverting transistors in an op amp.

Input offset voltage The input offset voltage E_{io} is the voltage impressed across the input of an op amp to ensure that the output is zero when the input signal is also zero.

Input offset current drift The input offset current drift is the ratio of input offset current change ΔI_{io} to a change in temperature ΔT. It is expressed by $\Delta I_{io}/\Delta T$.

Input offset voltage drift The input offset voltage drift is the ratio of input offset voltage change ΔE_{io} to a change in temperature ΔT. It is expressed by $\Delta E_{io}/\Delta T$.

OFFSET CORRECTION A circuit for offset correction in an inverting amplifier is illustrated in Fig. 13.16. Connected to the noninverting input is a voltage divider

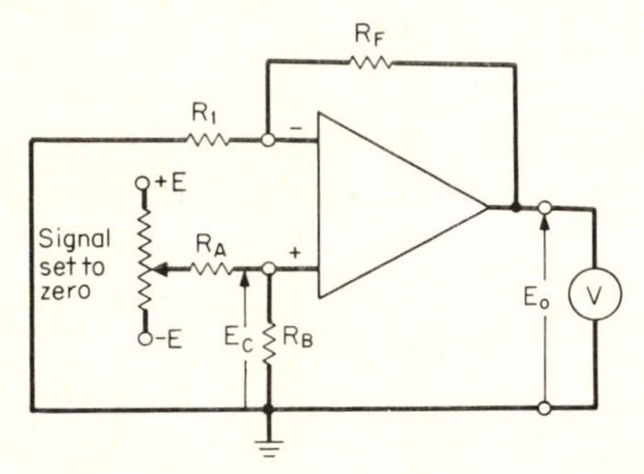

Fig. 13.16 Offset correction of an inverting amplifier.

consisting of resistors R_A and R_B. One end of resistor R_A is connected to the wiper arm of the potentiometer (pot). A positive voltage ($+E$) is connected to one end of the pot, and the other end is connected to a negative ($-E$) voltage. This arrangement makes available both positive and negative voltages for offset correction. The value of the pot is in the order of 20 kΩ.

Offset correction is done as follows: With the signal set to zero, a sensitive dc voltmeter is connected across the output to monitor output voltage E_o. The pot is then adjusted until $E_o = 0$. It must be emphasized that, because offset voltage and current drift with temperature, the correction obtained is only valid for a *given operating temperature*. Variations in offset current and voltage with temperature are illustrated in Fig. 13.17.

example 13.4 Referring to Fig. 13.16, assume that $\pm E = \pm 20$ V, $R_A = 250$ kΩ, and $R_B = 100$ Ω. Determine the maximum value of the offset correction voltage E_C.

solution By voltage division, $E_C = \pm 20 \times 100/(250\,000 + 100) \simeq \pm 2000/250\,000 = \pm 8$ mV.

A circuit for correcting offset in a noninverting amplifier is given in Fig. 13.18. As for the inverting amplifier, with a dc voltmeter connected across the output, the pot is adjusted until $E_o = 0$.

SLEWING RATE An important characteristic of an op amp is how well it follows a rapidly changing waveform. Assume that the input to a noninverting amplifier, with a

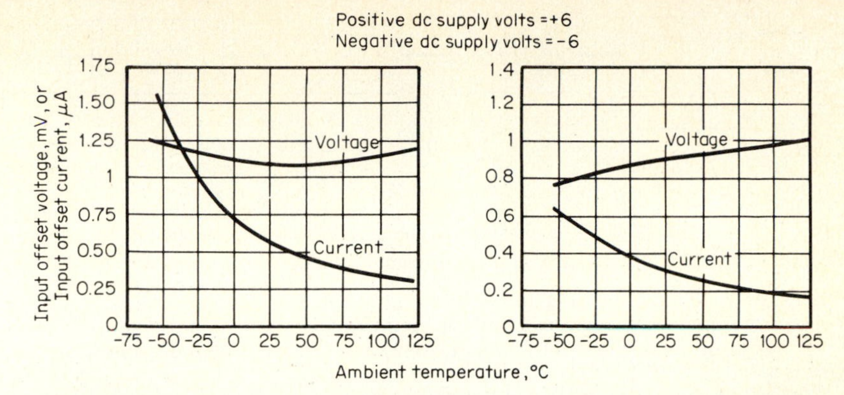

Fig. 13.17 Examples of how input offset current and voltage vary with temperature. (*Courtesy RCA.*)

gain of 10, is the step input signal of Fig. 13.19*a*. The output is shown in Fig. 13.19*b*. Note that the output voltage of 10 V is not reached instantaneously. The rate of change of the rising voltage ΔE_o with respect to time Δt is the *slewing rate:*

$$\text{Slewing rate} = \frac{\Delta E_o}{\Delta t} \tag{13.10}$$

The unit of the slewing rate is usually expressed as volts per microsecond. A high slewing rate indicates an op amp with a good response to rapidly changing waveforms.

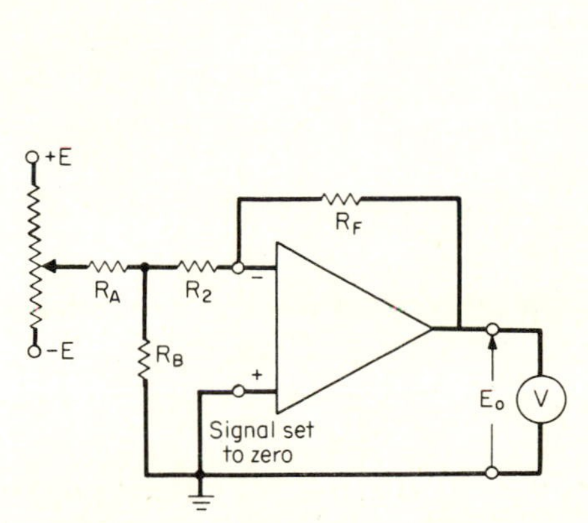

Fig. 13.18 Offset correction of a noninverting amplifier.

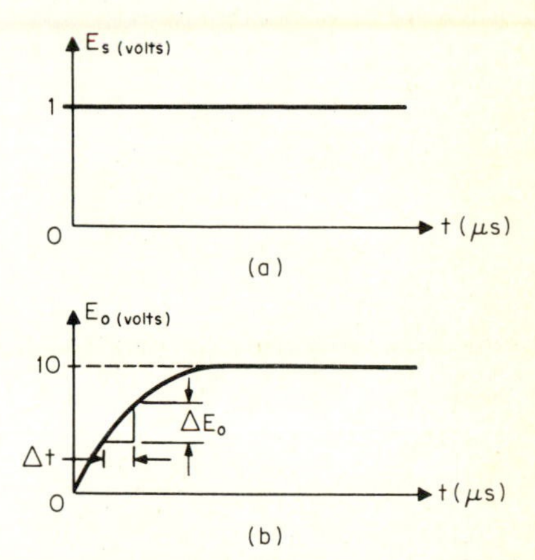

Fig. 13.19 Slewing rate of an op amp: (*a*) Step input to a noninverting amplifier. (*b*) Output of amplifier.

13.5 OP AMP CIRCUITRY

An operational amplifier contains a number of stages, each of which performs a specific function. This is illustrated in the block diagram (Fig. 13.20) of a typical op amp. Each of the circuits will be first briefly described and then examined in detail.

The differential amplifier in the first stage provides the inverting and noninverting inputs, gain, and is very stable in operation. The constant-current sink may be thought of as simulating a high emitter resistance with good efficiency. For additional gain, a second amplifier stage is required. Often, the second stage is another differential amplifier.

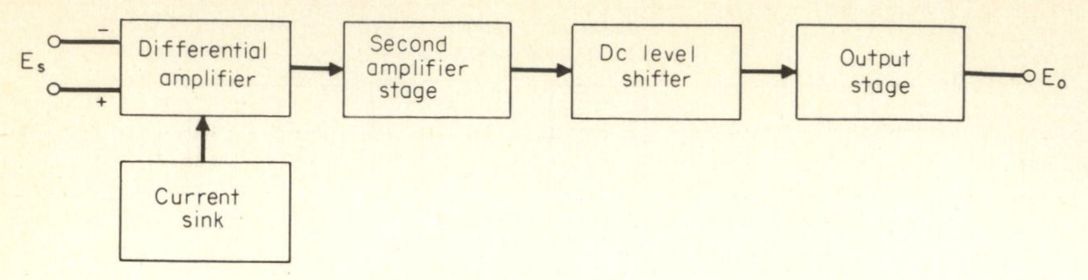

Fig. 13.20 General block diagram of an operational amplifier.

Because the circuits in an op amp are direct-coupled, as the signal progresses from the input to the output there is a buildup in dc voltage. The dc level shifter cancels out the built-up dc voltage. This ensures that, neglecting offset, the output is zero for zero input signal.

The output stage may be an emitter follower. For high-level signals, however, a push-pull type of output stage is used.

DIFFERENTIAL AMPLIFIER A basic differential amplifier is illustrated in Fig. 13.21*a*. A striking feature of the circuit is its *symmetry*. Because IC processing yields devices that are closely matched, the differential amplifier may be viewed as a Wheatstone bridge shown in Fig. 13.21*b*. Two power supplies are generally required for its operation: $+E_{CC}$ and $-E_{EE}$ volts. Input signals E_1 and E_2, or either one alone, are applied to the transistors, Q_1 and Q_2, respectively.

Assume that the input signals are equal: $E_1 = E_2$. For this condition, the output voltages are also equal ($E_{o1} = E_{o2}$) and, viewed as a bridge, the bridge is said to be balanced. If input E_1 is greater than E_2, more collector current flows in Q_1 and less collector current in Q_2. Hence, $E_{o1} < E_{o2}$. On the other hand, if $E_2 > E_1$, more collector current now flows in Q_2 than in Q_1, and $E_{o1} > E_{o2}$.

Practical signals may be thought of as containing two components: the *common-mode* (E_c) and *difference-mode* (E_d) signals. The common-mode signal is *common to both inputs*. Examples of this kind of signal include noise, hum from a power supply, and effects of temperature change. The difference-mode signal is the signal that is to be amplified. For an ideal differential amplifier, therefore, the gain to the common-mode signal would be zero, and the gain to the differential-mode signal would be very high.

Neglecting h_{re} and h_{oe} of the transistors, the common-mode signal gain A_c is given by

$$A_c \simeq \frac{-R_C}{2R_E} \tag{13.11}$$

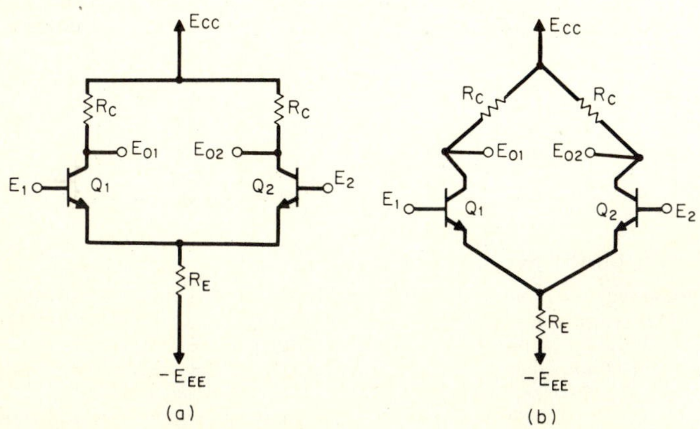

Fig. 13.21 Basic differential amplifier: (*a*) Circuit. (*b*) Amplifier viewed as a bridge.

and the difference-mode signal gain A_d is

$$A_d \simeq \frac{-h_{fe}R_C}{2h_{ie}} \tag{13.12}$$

From Eq. (13.11), for a given value of R_C, a small common-mode gain requires that R_E should be large. For very practical reasons, R_E cannot be made too large. For one, the practical limit of diffused resistors is 30 kΩ. Furthermore, for a large value of R_E, excessive power is wasted as heat. For these reasons a current sink, which simulates electronically a large value of resistance, is used in place of R_E. (The current sink is treated later in this chapter.)

Referring to Eq. (13.12), for a high value of difference-mode gain, h_{fe} and R_C should be large and h_{ie} low. High-gain transistors can be fabricated by making the bases exceedingly thin. One such device, called the *super beta transistor*, has an h_{fe} of a few thousand. For practical reasons, the maximum value of R_C is 30 kΩ. The value of h_{ie} varies inversely with collector bias current. Because the transistors in a differential amplifier are biased at a low quiescent current, a typical value of h_{ie} is a few kilohms.

A figure of merit for differential (and operational) amplifiers is the *common-mode rejection ratio*, CMRR. It is defined as the ratio of the difference-mode gain to the common-mode gain:

$$\text{CMRR} = \frac{A_d}{A_c} \tag{13.13}$$

The CMRR is generally expressed in decibels. The greater the value of CMRR, the better is the differential (or operational) amplifier.

example 13.5 Referring to the differential amplifier of Fig. 13.21*a*, assume that $R_C = R_E = 10$ kΩ, $h_{fe} = 100$, and $h_{ie} = 4$ kΩ; h_{re} and h_{oe} are negligible. Determine (*a*) A_c, (*b*) A_d, and (*c*) CMRR.

solution (*a*) From Eq. (13.11), $A_c \simeq -R_C/2R_E = -10/(2 \times 10) = -0.5$.
(*b*) From Eq. (13.12), $A_d \simeq -h_{fe}R_C/2h_{ie} = -100 \times 10/(2 \times 4) = -125$.
(*c*) $\text{CMRR} = A_d/A_c = -125/(-0.5) = 250$. In decibels, $\text{CMRR} = 20 \log (250) \simeq 48$ dB.

CURRENT SINK In the previous section it was seen that for a low value of common-mode gain, resistor R_E should be large. An efficient method for simulating a high-value resistance is by use of a *current sink*. An example of such a circuit connected to the emitters of a differential amplifier is shown in Fig. 13.22*a*.

Transistor Q_3 acts like a constant-current source. Resistors R_1 and R_2 bias the transistor and R_3 stabilizes its quiescent operating point. The circuit of Fig. 13.22*a* simulates a resistance of a few hundred kilohms. A general symbol for a current sink, often found in the literature, is illustrated in Fig. 13.22*b*.

An example of a commercially integrated differential amplifier, the Fairchild μA730, is shown in Fig. 13.23. Transistors Q_1 and Q_2 comprise the differential amplifier; Q_5 is the current sink. For low output impedance, Q_3 and Q_4 are connected as emitter followers.

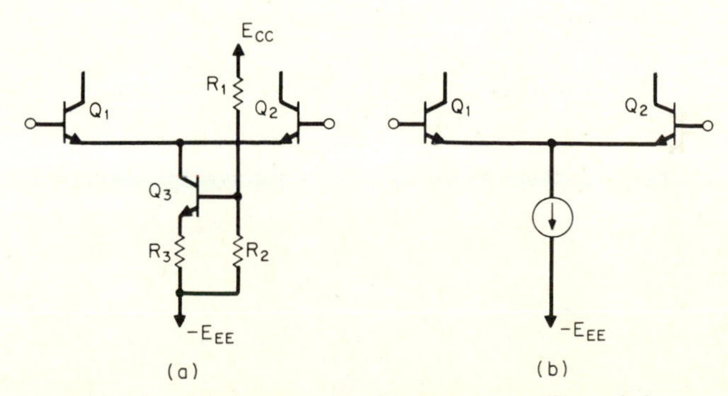

Fig. 13.22 Constant-current sink: (*a*) Circuit. (*b*) Symbol.

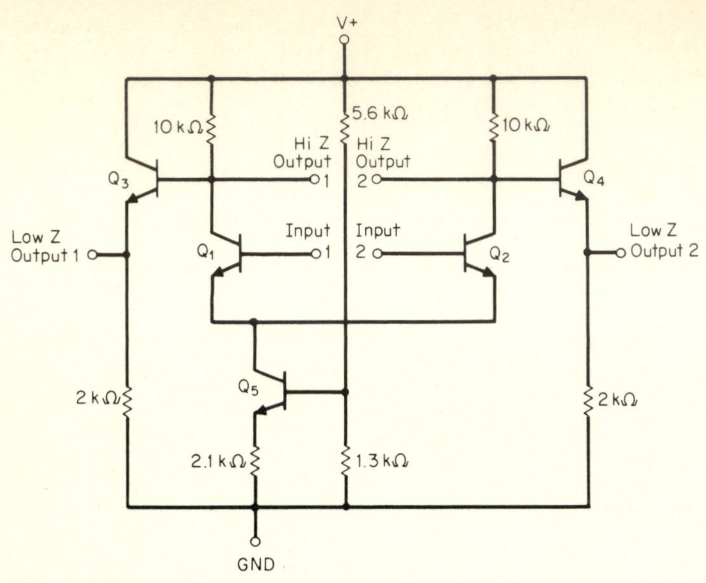

Fig. 13.23 Example of a commercial IC (μA730) differential amplifier. *(Courtesy Fairchild Semiconductor.)*

CASCADING STAGES To achieve high gain, an additional amplifier stage is cascaded to the differential amplifier. An example is the cascaded pair of differential amplifiers shown in Fig. 13.24. Transistors Q_1 and Q_2 and the constant current sink comprise the first differential amplifier stage. The second differential-amplifier stage contains transistors Q_3 and Q_4. Because only a single-ended output E_o from Q_4 is required, no collector resistor is necessary for Q_3. Also, an emitter resistor R_E, instead of a current sink, is adequate for the second stage.

Another circuit for increasing gain is illustrated in Fig. 13.25. Transistors Q_1 and Q_2 constitute the differential amplifier; Q_3 and Q_4 provide increased gain and convert the double-ended output of the differential amplifier to a single-ended output. The bases of Q_3 and Q_4 are fed from a common voltage point P through equal resistors R_{C1}. Transistor Q_3 inverts the output of Q_1 and compares it with the output of Q_2. The full differential output of the differential amplifier, therefore, is impressed across Q_4. Transistor Q_4 amplifies the differential signal and provides a single-ended output.

Another interesting feature of the circuit is that it is relatively insensitive to changes in the positive supply voltage E_{CC}. If, for example, E_{CC} should rise, the collector cur-

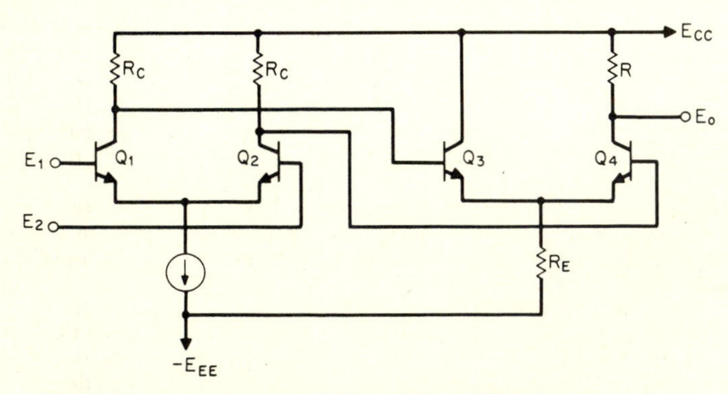

Fig. 13.24 A cascaded pair of differential amplifiers for increased gain.

rents in Q_3 and Q_4 also rise. Consequently, the collector voltages of Q_3 and Q_4 tend to decrease. As their voltage falls, point P becomes less positive, thereby reducing the base current to Q_4. If the base current to a transistor is reduced, its collector voltage increases. Thus we have a regulating (feedback) mechanism in operation that ensures the collector voltage of Q_4 remaining essentially constant.

DC LEVEL SHIFTER In monolithic integrated circuits, coupling capacitors are not practical. Instead, direct coupling is used and, as the signal travels from the input to the output stage, there is a buildup in dc voltage. If this built-up voltage is not canceled, a nonzero output is obtained for zero signal.

A circuit for canceling the built-up dc voltage in an op amp is called a *dc level shifter*. There are many circuits for accomplishing this; a basic dc level shifter is shown in Fig. 13.26.

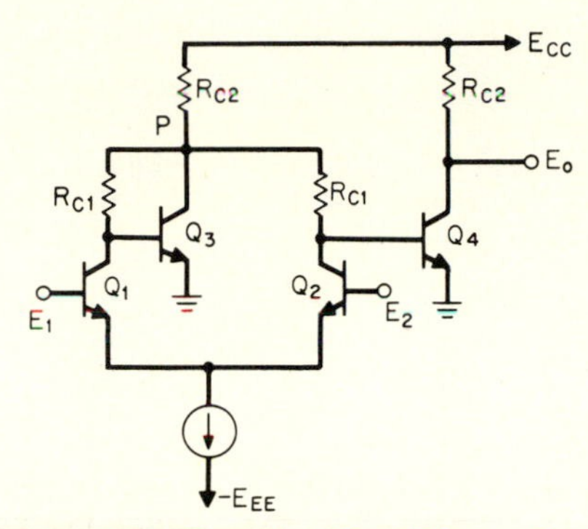

Fig. 13.25 Transistors Q_3 and Q_4 provide additional gain and convert the double-ended output of the differential amplifier (Q_1, Q_2) into a single-ended output voltage, E_o.

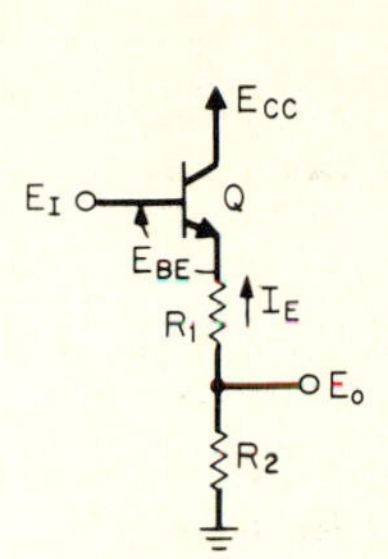

Fig. 13.26 A basic dc level shifter.

In this circuit, transistor Q is connected as an emitter follower. The emitter resistance consists of two resistors: R_1 in series with R_2. The dc base-emitter voltage E_{BE} is nominally 0.7 V at room temperature. Let E_I be the dc voltage at the base of Q and E_O the output dc voltage. The shifted voltage is equal to $E_I - E_O$ and is given by

$$E_I - E_O = E_{BE} + I_E R_1 \tag{13.14}$$

For example, if $I_E = 2$ mA and $R_1 = 600\ \Omega$, then the dc voltage shift is, by Eq. (13.14), $0.7 + 2 \times 0.6 = 1.9$ V.

OUTPUT STAGE The output stage of an op amp must exhibit low output impedance and be capable of providing the required load power, and its output signal should be limited only by the supply voltages for a maximum input signal. For low output power, an emitter follower is often used as the output stage in an op amp. If the power requirements are large, however, excessive power is dissipated in the emitter resistor. For this case, the circuits of Fig. 13.27 are generally employed as the output stage of an op amp.

In the circuit of Fig. 13.27*a*, transistor Q_1 operates as a *phase splitter*. If $R_C = R_E$, the signals are equal and 180 out of phase at the collector and in phase at the emitter with respect to input signal E_i. Transistors Q_2 and Q_3 operate as a push-pull amplifier, often referred to as a *totem-pole amplifier*.

Assume, for example, that the base voltage to Q_2 is decreasing and the base voltage to Q_3 is increasing. Transistor Q_3, therefore, conducts more current than Q_2. The difference in transistor currents is the load current, which flows in load resistor R_L. In many integrated circuits, emitter resistor R_E of Q_1 is replaced by a diode.

The circuit of Fig. 13.27*b*, called a *complementary pair* output stage, eliminates the

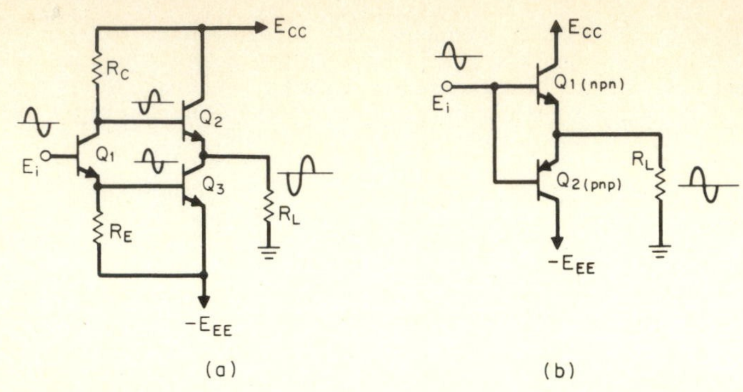

Fig. 13.27 Examples of output stages used in op amps: (*a*) Totem pole. (*b*) Complementary pair.

need for a phase splitter. Transistors Q_1 and Q_2 are of opposite types: Q_1 is npn and Q_2 is pnp. For efficient operation, Q_1 and Q_2 should be matched.

On the positive half-cycle of input signal E_i, Q_1 is conducting (because it is npn) and Q_2 is nonconducting (because it is pnp). Current, therefore, flows in the positive direction through Q_1 to load resistor R_L. On the negative half-cycle, Q_1 is nonconducting and Q_2 is conducting. Current flows in the negative direction through the pnp transistor Q_2 to R_L, thereby completing the output waveform.

13.6 EXAMPLES OF PRACTICAL OP AMPS

Two examples of commercially available op amps will now be analyzed. The first is the Motorola MC-1530 op amp of Fig. 13.28. Transistors Q_2 and Q_4 comprise the first differential amplifier stage, which is cascaded to a second differential amplifier containing Q_6 and Q_7. A current sink, Q_3, is used for the first stage and an emitter resistor, R_6, for the second stage.

Dc level shifting is achieved primarily by transistors Q_8 and Q_{10}. A phase splitter, Q_{11}, drives the totem-pole amplifier containing transistors Q_{12} and Q_{13}. Note that a diode is used instead of an emitter resistor in the phase splitter.

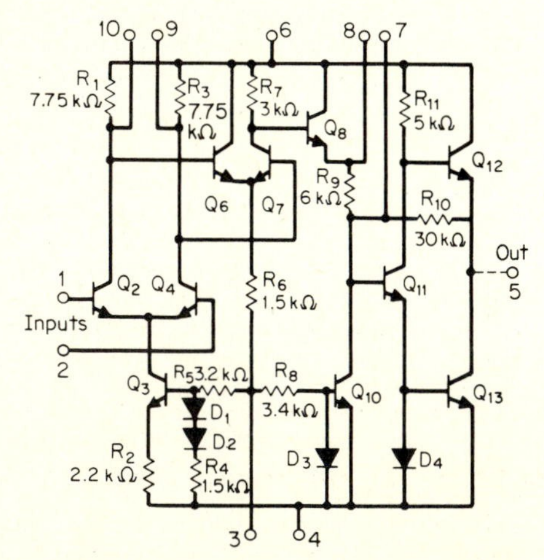

Fig. 13.28 Circuit of the Motorola MC-1530 op amp. (*Courtesy Motorola Semiconductor Products, Inc.*)

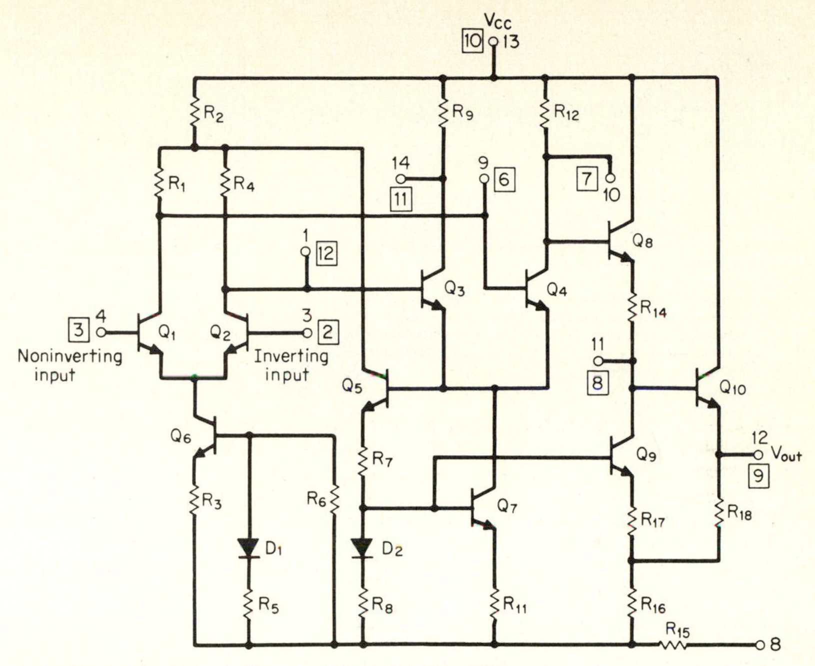

Fig. 13.29 Circuit diagram of the RCA CA 3008 op amp. (*Courtesy RCA.*)

As a second example, consider the RCA CA 3008 op amp of Fig. 13.29. Transistors Q_1 and Q_2 constitute the first differential amplifier stage which is coupled to a second differential amplifier (Q_3, Q_4). Current sinks are used for both stages (Q_6, Q_7), and diodes D_1 and D_2 provide thermal stabilization for their operating points.

Transistor Q_5 compensates the effect of common-mode signals. Assume, for example, that the dc power-supply voltage decreases. As a result, the voltage at the emitters of Q_3 and Q_4 also decrease. Less collector current flows in Q_5, as well as in Q_7 and Q_9. Because of the decrease in collector current of Q_7, which acts as a sink, less current flows in Q_3 and Q_4. This results in an increase in their collector voltage, thereby canceling the decrease in the supply voltage.

Transistors Q_8 and Q_9 provide dc level shifting. The output stage is an emitter follower. Note that emitter resistor R_{18} of the follower is returned to the junction of resistors R_{16} and R_{17}, instead of to the negative supply voltage. This connection provides *positive feedback* and increases the gain of the emitter follower from less than 1 to about 1.5.

The current flowing in R_{18} is added to the current in R_{16}. The voltage drop across R_{16} is thereby increased. Because the power-supply voltages are fixed, the voltage across R_{14} is reduced. This results in the collector voltage of Q_9, which is connected to the base of the emitter follower, to increase. The greater input voltage to the emitter follower results in a higher output voltage of the op amp. Resistor R_{15} limits the signal load current in case output terminal 12 is short-circuited to terminal 8.

13.7 UNDERSTANDING THE DATA SHEET FOR AN OP AMP

In selecting an op amp for a particular application, the user must consult the data (specs) sheet provided by the manufacturer of the device. Representative data sheets for an op amp, the Fairchild μA709C, are reproduced in Fig. 13.30. Invariably, the data included in the sheets are:

1. General description A cogent statement that stresses special features and application areas of the op amp under consideration. For example, the μA709C features

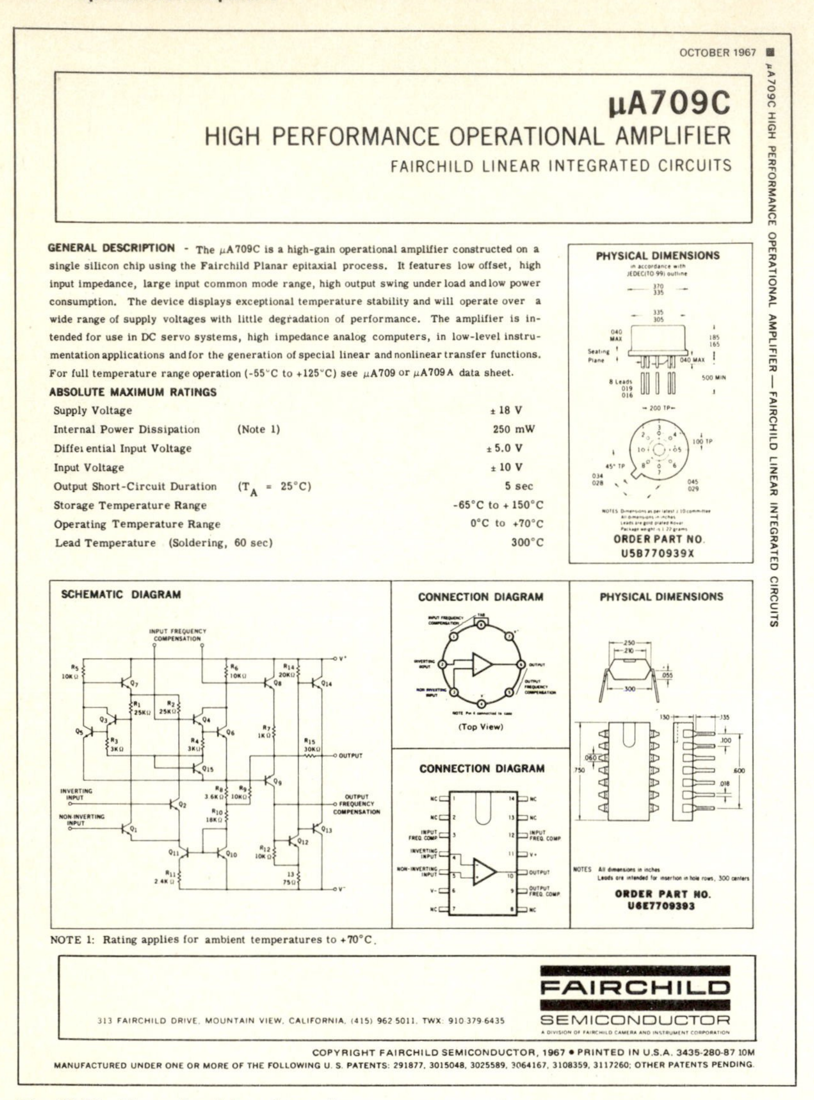

OCTOBER 1967

μA709C

HIGH PERFORMANCE OPERATIONAL AMPLIFIER

FAIRCHILD LINEAR INTEGRATED CIRCUITS

GENERAL DESCRIPTION - The μA709C is a high-gain operational amplifier constructed on a single silicon chip using the Fairchild Planar epitaxial process. It features low offset, high input impedance, large input common mode range, high output swing under load and low power consumption. The device displays exceptional temperature stability and will operate over a wide range of supply voltages with little degradation of performance. The amplifier is intended for use in DC servo systems, high impedance analog computers, in low-level instrumentation applications and for the generation of special linear and nonlinear transfer functions. For full temperature range operation (-55°C to +125°C) see μA709 or μA709A data sheet.

ABSOLUTE MAXIMUM RATINGS

Supply Voltage		± 18 V
Internal Power Dissipation	(Note 1)	250 mW
Differential Input Voltage		± 5.0 V
Input Voltage		± 10 V
Output Short-Circuit Duration	(T_A = 25°C)	5 sec
Storage Temperature Range		-65°C to + 150°C
Operating Temperature Range		0°C to +70°C
Lead Temperature (Soldering, 60 sec)		300°C

PHYSICAL DIMENSIONS

ORDER PART NO. U5B770939X

SCHEMATIC DIAGRAM

CONNECTION DIAGRAM

(Top View)

CONNECTION DIAGRAM

PHYSICAL DIMENSIONS

NOTES All dimensions in inches. Leads are intended for insertion in hole rows, 300 centers

ORDER PART NO. U6E7709393

NOTE 1: Rating applies for ambient temperatures to +70°C.

FAIRCHILD SEMICONDUCTOR

313 FAIRCHILD DRIVE, MOUNTAIN VIEW, CALIFORNIA, (415) 962 5011, TWX: 910 379 6435

μA709C HIGH PERFORMANCE OPERATIONAL AMPLIFIER — FAIRCHILD LINEAR INTEGRATED CIRCUITS

Fig. 13.30 Example of data sheets for an op amp provided by a manufacturer. (*Courtesy Fairchild Semiconductor.*)

low offsets, high input impedance, etc. The amplifier is intended for such applications as dc servos, analog computers, low-level instrumentation, and the generation of special functions.

2. Absolute maximum ratings Included under this heading are such items as supply voltage, internal power dissipation, and operating temperature range. For reliable operation and long life, the absolute maximum ratings must never be exceeded.

3. Physical dimensions Provides detailed dimensions of the package for mounting. Additional examples of IC packages are illustrated in Fig. 13.31.

4. Schematic diagram The schematic diagram is a circuit description of the op amp. The μA709C has some interesting features. The first differential stage (Q_1, Q_2)

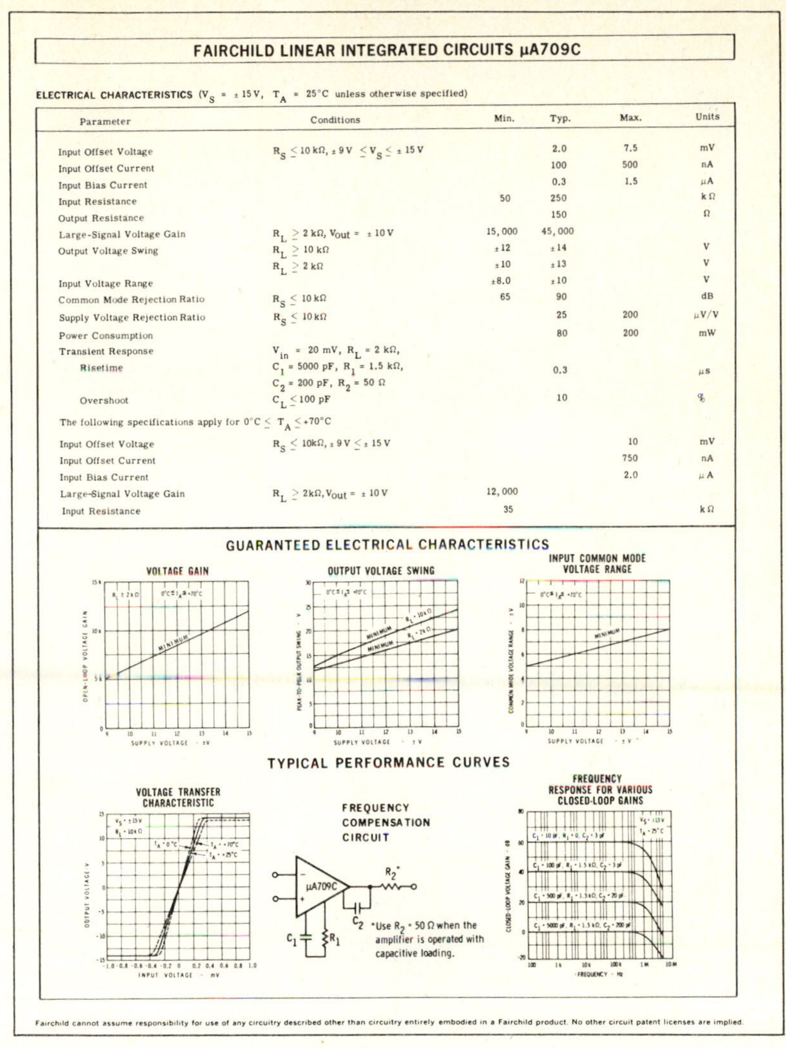

FAIRCHILD LINEAR INTEGRATED CIRCUITS μA709C

ELECTRICAL CHARACTERISTICS (V_S = ±15 V, T_A = 25°C unless otherwise specified)

Parameter	Conditions	Min.	Typ.	Max.	Units
Input Offset Voltage	$R_S \leq 10\,k\Omega$, $\pm 9\,V \leq V_S \leq \pm 15\,V$		2.0	7.5	mV
Input Offset Current			100	500	nA
Input Bias Current			0.3	1.5	μA
Input Resistance		50	250		kΩ
Output Resistance			150		Ω
Large-Signal Voltage Gain	$R_L \geq 2\,k\Omega$, $V_{out} = \pm 10\,V$	15,000	45,000		
Output Voltage Swing	$R_L \geq 10\,k\Omega$	±12	±14		V
	$R_L \geq 2\,k\Omega$	±10	±13		V
Input Voltage Range		±8.0	±10		V
Common Mode Rejection Ratio	$R_S \leq 10\,k\Omega$	65	90		dB
Supply Voltage Rejection Ratio	$R_S \leq 10\,k\Omega$		25	200	μV/V
Power Consumption			80	200	mW
Transient Response	V_{in} = 20 mV, R_L = 2 kΩ,				
Risetime	C_1 = 5000 pF, R_1 = 1.5 kΩ, C_2 = 200 pF, R_2 = 50 Ω		0.3		μs
Overshoot	$C_L \leq 100$ pF		10		%
The following specifications apply for $0°C \leq T_A \leq +70°C$					
Input Offset Voltage	$R_S \leq 10k\Omega$, $\pm 9\,V \leq \pm 15\,V$			10	mV
Input Offset Current				750	nA
Input Bias Current				2.0	μA
Large-Signal Voltage Gain	$R_L \geq 2k\Omega$, $V_{out} = \pm 10\,V$	12,000			
Input Resistance		35			kΩ

GUARANTEED ELECTRICAL CHARACTERISTICS

TYPICAL PERFORMANCE CURVES

Fairchild cannot assume responsibility for use of any circuitry described other than circuitry entirely embodied in a Fairchild product. No other circuit patent licenses are implied.

Fig. 13.30 Continued.

is cascaded to a second differential stage which uses *Darlington pairs* (Q_3, Q_5 and Q_4, Q_6) for the transistors. A Darlington pair provides a very high input impedance. (In some op amps, field-effect transistors are used in the first differential amplifier stage to achieve a high input impedance.) The output stage (Q_{13}, Q_{14}) is a complementary pair circuit.

5. Connection diagram Identifies the terminals on the package.

6. Electrical characteristics Provides ranges of parameter values such as input offset voltage, gain, and power consumption. Test conditions for the parameters must always be explicit. The *supply-voltage rejection ratio* is the ratio of a change in offset voltage for a corresponding change in a supply voltage.

7. Guaranteed electrical characteristics Curves show how voltage gain, output voltage swing, and input common-mode voltage range vary with the supply voltage.

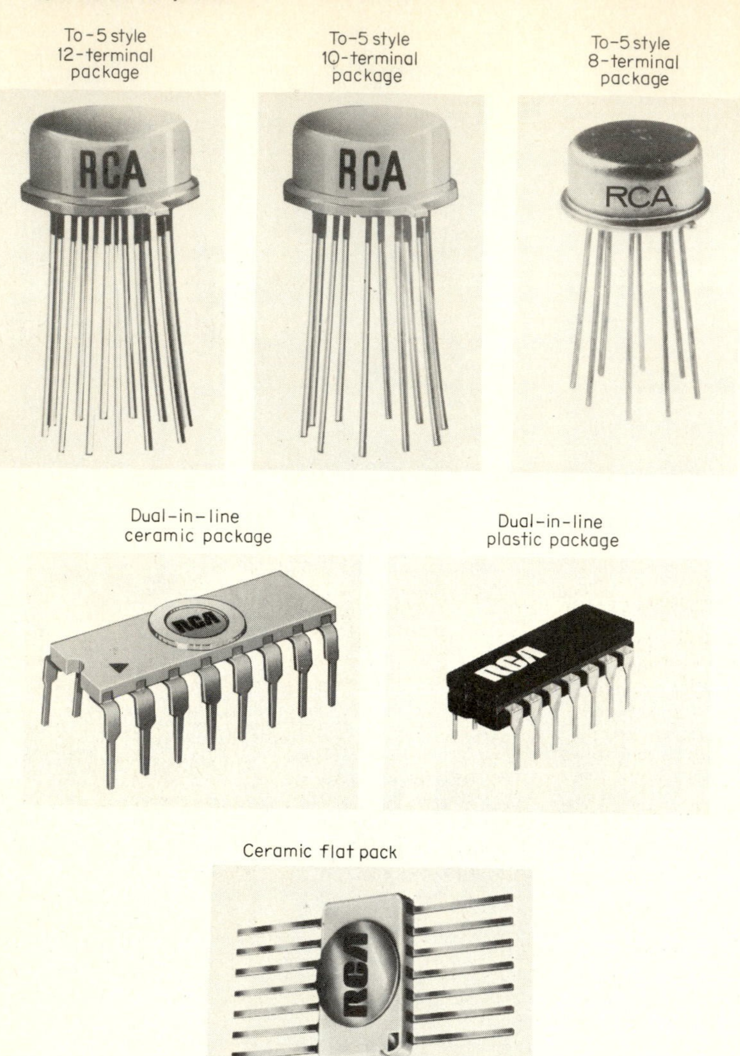

Fig. 13.31 Examples of commonly used IC packages. (*Courtesy RCA.*)

The *common-mode voltage range* is the maximum input voltage that may be applied to either the inverting or noninverting terminal of an op amp for safe operation.

8. Typical performance curves The *voltage-transfer characteristic* relates the output voltage to the input voltage over the operating temperature range of the op amp. The frequency compensation circuit and response curves are extremely important. As mentioned earlier, frequency compensation is required to ensure stable operation of the amplifier. It is noted that the closed-loop gain is dependent on the values of the capacitors and resistors used in the compensation network.

Chapter 14

Digital Circuit Fundamentals

14.1 INTRODUCTION

The development of the junction diode and transistor has been the chief catalyst in the rapid growth of computers and digital systems in general. Semiconductor devices can be made to act like nearly perfect switches, making them ideal for digital circuits. By the use of a few elementary circuits over and over again, all the necessary functions in the operation of a digital system are realized.

Because a switch is either ON or OFF, it is said to exhibit two stable states. (A switch, for this reason, is sometimes referred to as a *bistable device*.) For semiconductor switches to perform arithmetic and logic operations, it therefore becomes necessary to represent numbers by two digits, or *bits*, 0 and 1. In this manner, a 0 may be represented by a closed switch and a 1 by an open switch. Such a number system using only 0s and 1s, called the *binary number system*, is employed in digital systems.

In this chapter we consider first how numbers are expressed in the binary number system. It is also necessary to know how to convert a decimal number into a binary number and back again to decimal. Basic logic functions, such as AND, OR, NOT, NAND, and NOR, are examined next. These functions are described completely by the *truth table*, which is discussed. Operation of multivibrators is also considered. The chapter concludes with a study of the response of switching devices to rapidly changing waveforms, referred to as *dynamic response*.

14.2 BINARY NUMBERS

Because we are born with ten fingers, it is by sheer accident that we count by 10s. The *base*, or *radix*, of our number system is ten, referred to as the *decimal number system*. A zero and nine different symbols are needed to represent decimal numbers: 0, 1, 2, 3, 4, 5, 6, 7, 8, 9. In some past societies people used both their fingers and toes to count. It is not surprising to learn that the base of their number system was twenty. If we were born with only two hands and no fingers, it is a safe bet that we would count by 2s.

Number systems are *positional*. The value of a number depends not only on its symbols but also on their position in the number. For example, 367 and 673 are two numbers having the same symbols but different positions. One may regard a decimal number, such as 367, as a shorthand for

$$3 \times 10^2 + 6 \times 10^1 + 7 \times 10^0 = 300 + 60 + 7 = 367$$

where $10^0 = 1$, $10^1 = 10$, and $10^2 = 100$. Thus, 367 is equal to 7 units (10^0), 6 tens (10^1), and 3 hundreds (10^2). All decimal numbers are expressed in this manner, regardless of their magnitude.

The binary number system is also positional. For example, 1101 is equal to

$$1 \times 2^3 + 1 \times 2^2 + 0 \times 2^1 + 1 \times 2^0 = 8 + 4 + 0 + 1 = 13$$

where $2^0 = 1$, $2^1 = 2$, $2^2 = 4$, and $2^3 = 8$. Thus, binary number 1101 is equal to 1 unit (2^0), 0 twos (2^1), 1 four (2^2), and 1 eight (2^3).

In comparing numbers having different base systems, it is customary to place a subscript, equal to the number base, after the number. Hence, one would write

$$13_{10} = 1101_2$$

DECIMAL-TO-BINARY CONVERSION To convert a whole decimal number to a binary number, the following procedure is used:

1. Divide the number by 2; the remainder is either a 0 or a 1.
2. Place the remainder to the right of the partial quotient obtained in step 1.
3. Divide the partial quotient of step 1 by 2, placing the remainder to the right of the new partial quotient.
4. Repeat the preceding steps until a quotient of zero is obtained.
5. The binary number is equal to the remainders arranged so that the first remainder is the least significant bit (LSB) and the last remainder is the most significant bit (MSB) of the binary number.

example 14.1 Convert the following decimal numbers to binary numbers: (*a*) 13; (*b*) 103.

solution (*a*)

2)13	
2)6	1
2)3	0
2)1	1
0	1

(Remainders)

Rearranging remainders, $13_{10} = 1101_2$. (This equality was verified in the preceding section.)

(*b*)

2)103	
2)51	1
2)25	1
2)12	1
2)6	0
2)3	0
2)1	1
0	1

Hence, $103_{10} = 1100111_2$. To verify the answer, the binary number is expanded into its constituent parts: $1100111_2 = 1 \times 2^6 + 1 \times 2^5 + 0 \times 2^4 + 0 \times 2^3 + 1 \times 2^2 + 1 \times 2^1 + 1 \times 2^0 = 1 \times 64 + 1 \times 32 + 0 \times 16 + 0 \times 8 + 1 \times 4 + 1 \times 2 + 1 \times 1 = 64 + 32 + 4 + 2 + 1 = 103_{10}$.

DECIMAL-TO-BINARY FRACTION CONVERSION To convert a decimal fraction to a binary fraction, the following procedure is used:

1. Multiply the decimal fraction by 2.
2. If, as a result of step 1, a number equal to or greater than 1 is obtained, the 1 is placed to the right of the partial product. For example, multiplying 0.6 by 2 yields 1.2, which is a number greater than 1. The 1 is placed to the right of the partial product 0.2. If the result obtained is less than one, a zero is placed to the right of the partial product.
3. The partial product obtained in step 2 is multiplied by 2. The process is repeated until the partial product is 0 or the resulting binary fraction is to the required places of the binary point (which corresponds to the decimal point of a decimal fraction).
4. The 1s and 0s in the order obtained are equal to the binary fraction.

example 14.2 Convert the following decimal fractions to binary fractions: (*a*) 0.375; (*b*) 0.68.

solution (*a*)

$2 \times 0.375 = 0.750$	0
$2 \times 0.750 = 1.500$	1
$2 \times 0.500 = 1.000$	1
$2 \times 0.000 = 0.000$	

Hence, $0.375_{10} = 0.011_2$. To verify the answer, 0.011 is expanded into its constituent parts: $0.011 = 0 \times 2^{-1} + 1 \times 2^{-2} + 1 \times 2^{-3} = 0/2 + 1/4 + 1/8 = 0.25 + 0.125 = 0.375$.

(*b*)

$$2 \times 0.68 = 1.36 \qquad 1$$
$$2 \times 0.36 = 0.72 \qquad 0$$
$$2 \times 0.72 = 1.44 \qquad 1$$
$$2 \times 0.44 = 0.88 \qquad 0$$
$$2 \times 0.88 = 1.76 \qquad 1$$

and so on. In this example the binary fraction is unending. To five *binary places*, $0.68_{10} \simeq 0.10101_2$.

BINARY-TO-DECIMAL CONVERSION To convert a binary whole number to a decimal number, the following procedure is used:

1. Multiply the most significant bit (MSB) by 2.
2. If the next to the MSB is a one, add a 1 to the partial product obtained in step 1; if a zero, add a 0.
3. Multiply the result obtained in step 2 by 2. Continue the process until the least significant bit (LSB) is included in the conversion.

example 14.3 Convert the following binary numbers to decimal numbers: (*a*) 11010; (*b*) 110101.

solution (*a*)

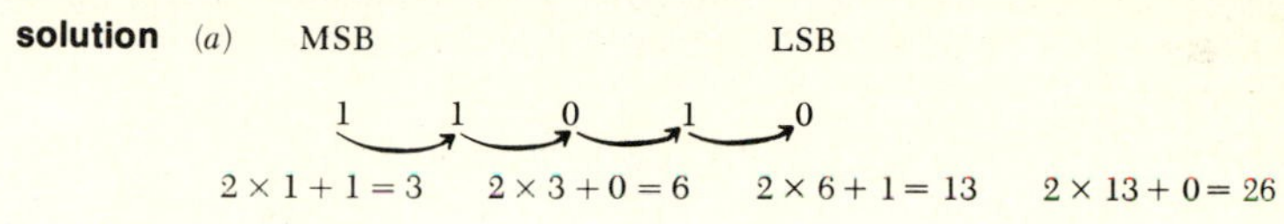

Therefore, $11010_2 = 26_{10}$.

(*b*) 1 1 0 1 0 1

$2 \times 1 + 1 = 3 \qquad 2 \times 3 + 0 = 6 \qquad 2 \times 6 + 1 = 13 \qquad 2 \times 13 + 0 = 26 \qquad 2 \times 26 + 1 = 53$

Hence, $110101_2 = 53_{10}$.

BINARY-TO-DECIMAL FRACTION CONVERSION To convert a binary fraction to a decimal fraction, the following procedure is used:

1. Divide the LSB by 2.
2. If the next to the LSB is a 1, add it to the result of step 1.
3. Divide the result of step 2 by 2. If the next bit is a 1, add it to the result.
4. Continue the process until the binary point is reached.

example 14.4 Convert the following binary fractions to decimal fractions: (*a*) 0.0111; (*b*) 0.01001.

solution (*a*) Dividing the LSB of 0.0111 by 2 yields $1/2 = 0.5$. Because the next bit is a 1, it is added to 0.5. Hence, the first ratio, R_1, is

$$R_1 = 0.5 + 1 = 1.5$$

Dividing 1.5 by 2 and adding 1 yields

$$R_2 = \frac{1.5}{2} + 1 = 1.75$$

Dividing 1.75 by 2 and, because the next bit is a zero, we have

$$R_3 = \frac{1.75}{2} + 0 = 0.875$$

Finally, the binary point is reached and $0.875/2 = 0.4375$. Hence, $0.0111_2 = 0.4375_{10}$.

(*b*)

$$R_1 = 0.5 + 0 = 0.5$$

$$R_2 = \frac{0.5}{2} + 0 = 0.25$$

$$R_3 = \frac{0.25}{2} + 1 = 1.125$$

$$R_4 = \frac{1.125}{2} + 0 = 0.5625$$

$$R_5 = \frac{0.5625}{2} = 0.28125$$

Hence, $0.01001_2 = 0.28125_{10}$.

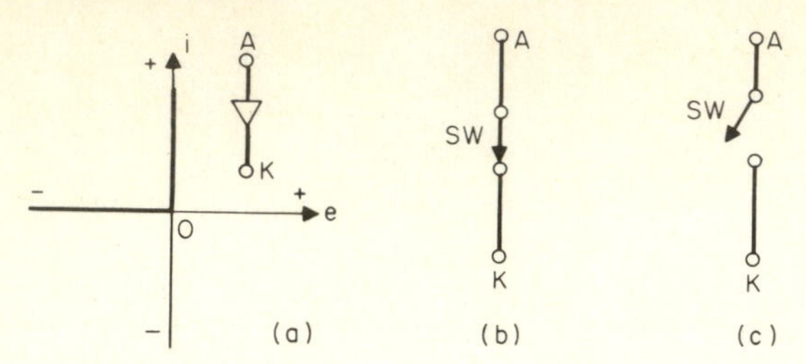

Fig. 14.1 Ideal diode: (*a*) Current-voltage characteristics. (*b*) Closed-switch model for a forward-biased ideal diode. (*c*) Open-switch model for a reverse-biased ideal diode.

14.3 DIODE AS A SWITCH

As explained in Chap. 8, a junction diode when forward-biased conducts current and when reverse-biased it does not conduct current. *It behaves like a switch.*

Consider the characteristics of an ideal diode, shown in Fig. 14.1*a*. When forward-biased (A positive with respect to K), the voltage across the ideal diode is zero, regardless of the current flowing. The forward-biased ideal diode, therefore, may be represented by the closed-switch model of Fig. 14.1*b*. On the other hand, when reverse-biased, the current is zero regardless of the magnitude of the reverse voltage. A reverse-biased ideal diode, therefore, can be represented by the open-switch model of Fig. 14.1*c*.

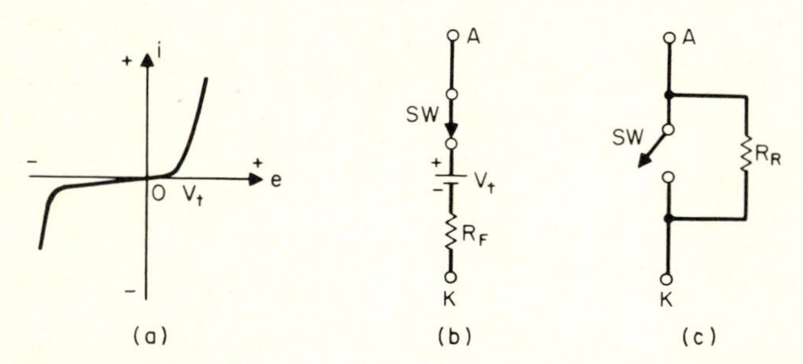

Fig. 14.2 Physical junction diode: (*a*) Current-voltage characteristics. (*b*) Model for a forward-biased physical diode. (*c*) Model for a reverse-biased physical diode.

The characteristics of a physical junction diode are illustrated in Fig. 14.2*a*. When forward-biased, before appreciable current flows, the forward voltage must exceed the threshold voltage V_t (0.6 V for silicon at room temperature). In conduction, the forward resistance of the diode, which can be as low as a few ohms, is designated by R_F. A forward-biased physical diode, therefore, acts like a closed switch in series with a battery of V_t volts and a resistance of R_F ohms, as illustrated in Fig. 14.2*b*.

When reverse-biased (the reverse voltage must not exceed the breakdown voltage of the diode), the reverse resistance is extremely high, in the order of megohms. Designating the reverse resistance by R_R, the reverse-biased diode is represented by a switch in parallel with R_R shown in Fig. 14.2*c*. In practice, the models for the ideal diode are used often to represent a physical diode.

14.4 TRANSISTOR AS A SWITCH

Figure 14.3*a* is an elementary circuit of an npn junction transistor operating as a switch. (Field-effect transistor switches are considered in Chap. 15.) The configuration is that

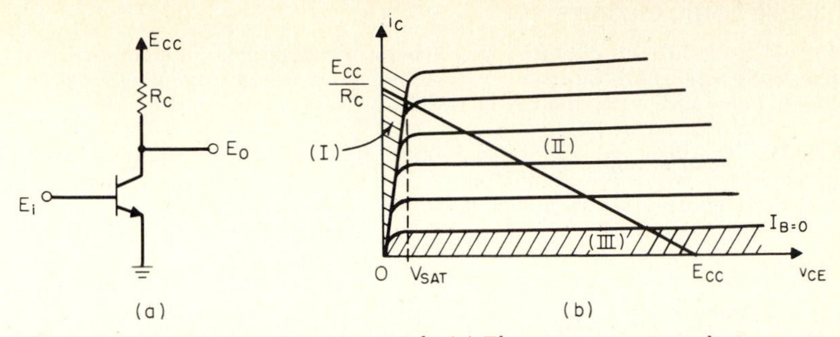

Fig. 14.3 Transistor operating as a switch: (*a*) Elementary circuit employing a npn transistor in the common-emitter configuration. (If a pnp transistor is used, the polarity of E_{CC} is reversed.) (*b*) Load line superimposed on the collector characteristics. Regions I, II, and III are the saturation, active, and cutoff regions, respectively.

of a common-emitter amplifier. Input E_i is applied to the base of the transistor. The collector is connected to resistance R_C in series with the collector supply voltage E_{CC}. Output voltage E_o is taken across the collector and ground.

The load line superimposed on the collector characteristics is shown in Fig. 14.3*b*. One may define three regions of operation, indicated by Roman numerals I, II, and III.

I. Saturation region In the saturation region, both the collector-base and base-emitter junctions are *forward-biased.* The voltage across the collector and emitter is referred to as the *saturation voltage,* V_{sat}. Typically, V_{sat} for most switching transistors is 0.1 V or less.

When a transistor is in the saturation region, it is considered to be ON. It may be represented by the model of a closed switch in series with a battery of V_{sat} volts, as in Fig. 14.4*a*. A transistor that when ON is in saturation is referred to as a *saturated*

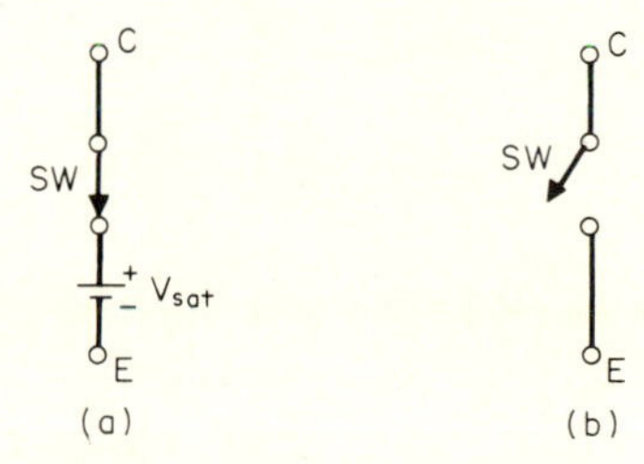

Fig. 14.4 Models for a common-emitter transistor switch: (*a*) For the ON state. (*b*) For the OFF state.

switch. If a transistor is ON and not in the saturated region, it is referred to as a *nonsaturated switch.* Because of greater efficiency, transistors are generally operated as saturated switches.

II. Active region In the active region, the collector-base junction is *reverse-biased* and the base-emitter junction is *forward-biased.* The *Q* point of an amplifier, for example, is located in the active region of the collector characteristics.

III. Cutoff region In the cutoff region, both the collector-base and base-emitter junctions are *reverse-biased.* When in the cutoff region, the transistor is considered to be OFF. It may be represented by the model of an open switch of Fig. 14.4*b*.

A transistor operated as a saturated switch is either in the saturation (ON) or cutoff (OFF) regions. In switching a transistor from ON to OFF (or from OFF to ON), it makes a transition through the active region. A *fast* switching transistor makes the transition rapidly.

14.5 BASIC LOGIC CIRCUITS

There are three basic logic circuits which, when used in various combinations, perform the arithmetic and control functions in a digital computer or system. These circuits are the OR gate, AND gate, and the INVERTER (NOT).

OR GATE The logic symbol for a two-input OR gate is given in Fig. 14.5. Output X is obtained when A *or* B, *or* both A and B, are present at the input. The preceding statement can be expressed by a *logic equation* as

$$X = A + B \tag{14.1}$$

where the plus sign (+) is interpreted as OR. Because the binary number system is used in logic systems, inputs A and B, and output X can assume only the values of 0 and 1.

An elementary example of an OR gate is the two parallel manual switches, A and B of Fig. 14.6. The input to both switches is voltage source E; the output X is a glowing lamp. The lamp glows if switch A *or* B *or* both A and B are closed.

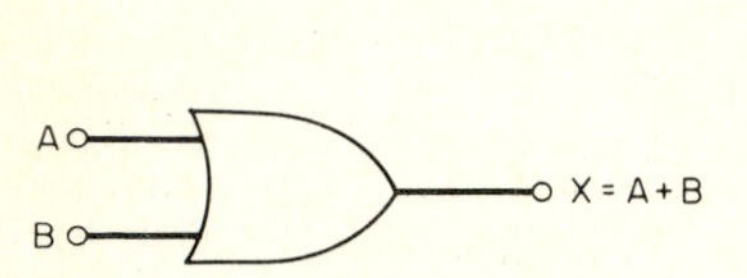

Fig. 14.5 Logic symbol for a two-input OR gate ($X = A + B$). OR gates having more than two inputs are also possible.

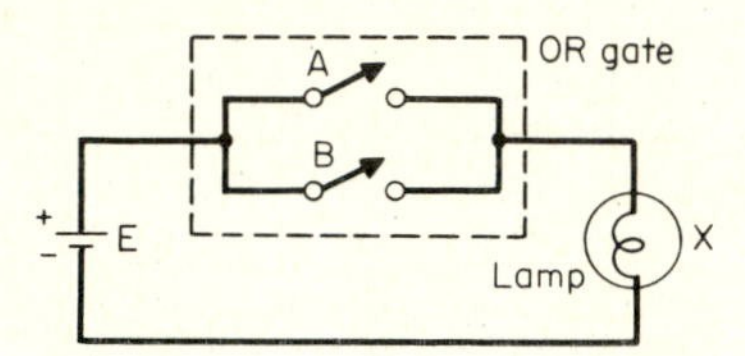

Fig. 14.6 An example of an OR gate consisting of two parallel manual switches. The lamp glows ($X = 1$) if switch A, or B, or A and B are closed.

The operation of a logic circuit can be described by a *truth table*. In a truth table, all possible combinations of inputs and their corresponding outputs are displayed in tabular form. Inputs A and B can assume only four possible combinations of values: 00, 01, 10, and 11. These values and their corresponding output X are listed in the truth table for the two-input OR gate in Fig. 14.7. It is read as an output X exists when $A = 0$ or $B = 1$; $A = 1$ or $B = 0$; or $A = B = 1$. No output exists when $A = B = 0$.

The number of different possible combinations assumed by n inputs is expressed by

$$\text{Combinations} = 2^n \tag{14.2}$$

As we saw in the preceding paragraph, for two inputs A and B four combinations were possible ($2^n = 2^2 = 4$).

example 14.5 Draw the truth table for a three-input (A, B, C) OR gate.

A	B	X
0	0	0
0	1	1
1	0	1
1	1	1

Fig. 14.7 Truth table for a two-input OR gate.

A	B	C	X
0	0	0	0
0	0	1	1
0	1	0	1
0	1	1	1
1	0	0	1
1	0	1	1
1	1	0	1
1	1	1	1

Fig. 14.8 Truth table for a three-input OR gate. (See Example 14.5.)

solution Because $n = 3$, by Eq. (14.2), $2^3 = 8$ combinations are possible. These are: 000, 001, 010, 011, 100, 101, 110, and 111. (A simple way of listing the combinations is to write the binary numbers corresponding to decimal numbers 0 to $2^n - 1$.) The combinations and the corresponding values of X for the three-input OR gate are listed in the truth table of Fig. 14.8.

Positive and negative logic In *positive logic,* the voltage that represents a 1 is greater than the voltage that represents a 0. The reverse is true in *negative logic:* A 1 is represented by a voltage less than the voltage denoting a 0. Letting E_1 be the voltage for a 1 and E_0 be the voltage for a 0, examples of positive and negative logic are illustrated in Fig. 14.9. The 1 and 0 are commonly referred to as *logic 1* and *logic 0.*

OR gate circuit An example of a two-input OR gate (for positive logic) using diodes is illustrated in Fig. 14.10. (In negative logic the diodes are reversed.) The circuit is an example of *diode logic.*

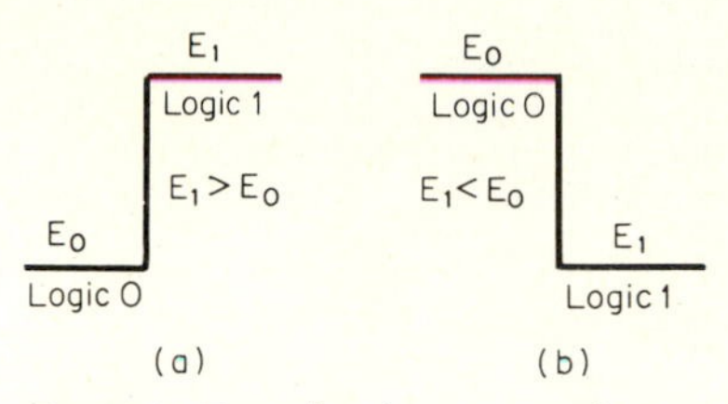

Fig. 14.9 Example of positive and negative logic levels: (*a*) Positive logic: $E_1 > E_0$. (*b*) Negative logic: $E_1 < E_0$.

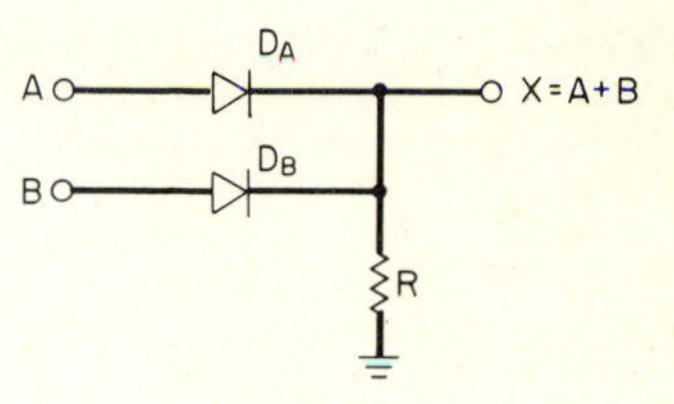

Fig. 14.10 An example of a two-input OR gate using diodes for positive logic. (In negative logic, the diode connections are reversed.)

Let $E_1 = 5$ V and $E_0 = 0$ V. Assuming ideal diodes, if $E_A = E_B = 0$ V, both diodes are nonconducting, and the output = 0 V ($X = 0$). If $E_A = 5$ V and $E_B = 0$ V, diode D_A conducts and diode D_B is reverse-biased; the output, therefore, equals 5 V ($X = 1$). If $E_B =$ 5 V and $E_A = 0$ V, diode D_B now conducts and D_A is reverse-biased; output equals 5 V. For $E_A = E_B = 5$ V, both diodes conduct and the output is at 5 V ($X = 1$).

Because of noise in digital systems, it is possible that voltages E_A and E_B are unequal. Suppose that $E_A = 6$ V and $E_B = 5$ V for a logic 1. If $E_A = 6$ V and $E_B = 0$ V, the output equals 6 V. If $E_A = 5$ V and $E_B = 0$ V, the output will be at 5 V. If both inputs are applied simultaneously, $E_A = 6$ V and $E_B = 5$ V, the output is equal to the *larger of the inputs.*

Assume that indeed $E_A = 6$ V and $E_B = 5$ V. The output is 6 V. Because the anode of diode D_B is at 5 V and the cathode at 6 V, diode D_B is reverse-biased and does not conduct.

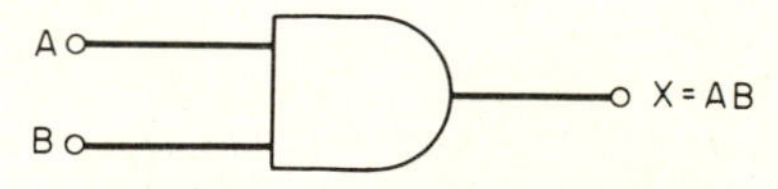

Fig. 14.11 Logic symbol for a two-input AND gate ($X = AB$). AND gates having more than two inputs are also possible.

AND GATE The logic symbol for a two-input AND gate is shown in Fig. 14.11. Output $X = 1$ when inputs A *and* B are present; otherwise, $X = 0$. Expressed by a logic equation,

$$X = A \text{ x } B \tag{14.3a}$$

where the x is interpreted as AND. As in algebra, the x may be eliminated and Eq. (14.3*a*) expressed by

$$X = AB \tag{14.3b}$$

A	B	X
0	0	0
0	1	0
1	0	0
1	1	1

Fig. 14.12 Truth table for a two-input AND gate.

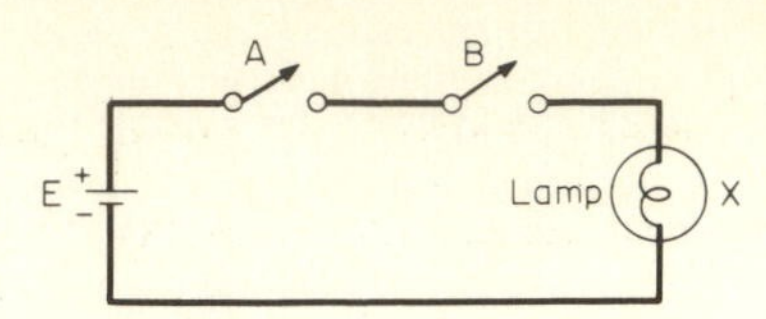

Fig. 14.13 An example of an AND operation. The lamp glows ($X = 1$) only if switches A and B are closed.

The truth table for a two-input AND gate is given in Fig. 14.12. It is seen that an output exists ($X = 1$) when *both* inputs equal 1; otherwise, $X = 0$. For a three-input AND gate, $X = 1$ only when all three inputs equal logic 1, and so on.

An elementary example of an AND operation is illustrated in Fig. 14.13. Switches A and B are connected in series with source E and the lamp. The lamp glows ($X = 1$) when switches A *and* B are closed. If either or both switches are open, the lamp does not glow ($X = 0$).

AND gate circuit An example of a two-input AND gate for positive logic using diodes is shown in Fig. 14.14. Resistance R is returned to voltage E_D which is equal to or greater than E_1. (For negative logic, the diodes and polarity of E_D are reversed.)

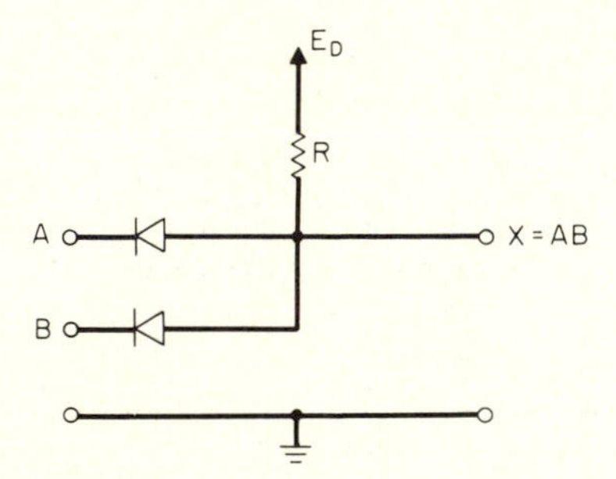

Fig. 14.14 An example of a two-input AND gate for positive logic. (In negative logic, the diode connections and supply voltage E_D are reversed.)

Assume ideal diodes, $E_1 = 5$ V, $E_0 = 0$ V, and $E_D = 5$ V. If either E_A or $E_B = 0$ V, diode D_A or D_B conducts; the output is zero volts ($X = 0$). If $E_A = E_B = 5$ V, both diodes are reverse-biased and the output equals $E_D = 5$ V ($X = 1$).

When E_A and E_B for logic 1 are unequal, the output is equal to the *lesser* voltage. Assume that $E_A = 5$ V and $E_B = 3$ V. Diode D_B becomes forward-biased because its anode was initially at $E_D = 5$ V and its cathode at $E_B = 3$ V. The output is equal to 3 V ($X = 1$). Since the anode of diode D_A is now at 3 V and its cathode at $E_A = 5$ V, diode D_A is reverse-biased.

INVERTER The logic symbol for the INVERTER, also referred to as a NOT, is given by either of the symbols of Fig. 14.15. Output X is equal to the *not* of A. Expressed by a logic equation,

$$X = \overline{A} \tag{14.4}$$

where the bar over A indicates a NOT operation. If $A = 1$, output $X = 0$; if $A = 0, X = 1$. The truth table for the inverter is given in Fig. 14.16.

INVERTER circuit An example of an inverter using an npn junction transistor is illustrated in Fig. 14.17. Resistance R_1 is in series with the transistor base and input A.

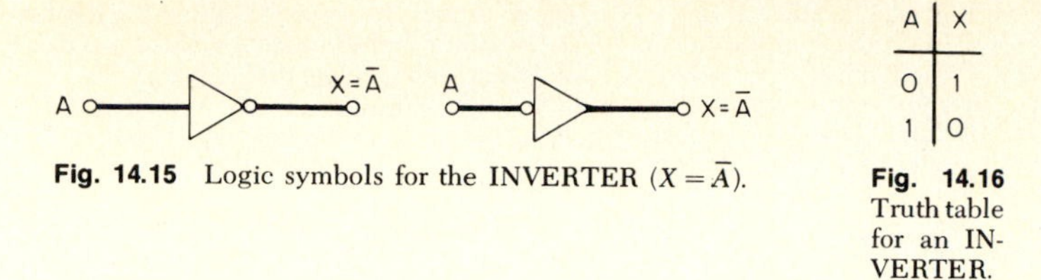

Fig. 14.15 Logic symbols for the INVERTER ($X = \bar{A}$).

Fig. 14.16 Truth table for an INVERTER.

Resistance R_2 is connected between the base and the negative-voltage supply, $-E_{BB}$ volts. This ensures that for $A = E_0$ (logic 0), the base-emitter junction is reverse-biased and the output $= E_{CC}$ volts (logic 1). When $A = E_1$ (logic 1), transistor Q conducts and the output is equal to V_{sat} volts (logic 0).

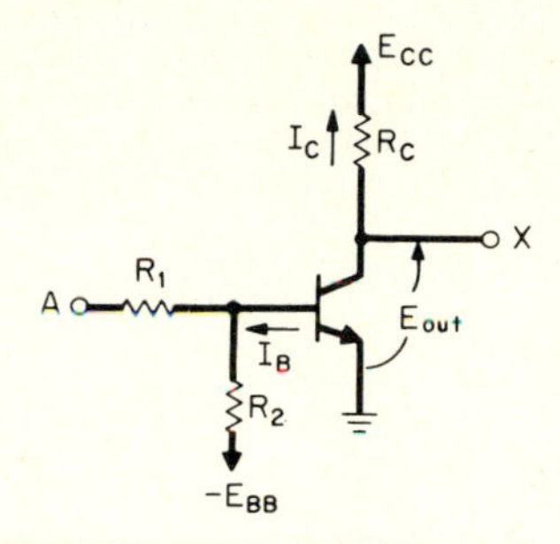

Fig. 14.17 An example of an INVERTER circuit using a npn transistor. For a pnp transistor the polarities of E_{CC} and E_{BB} are reversed.

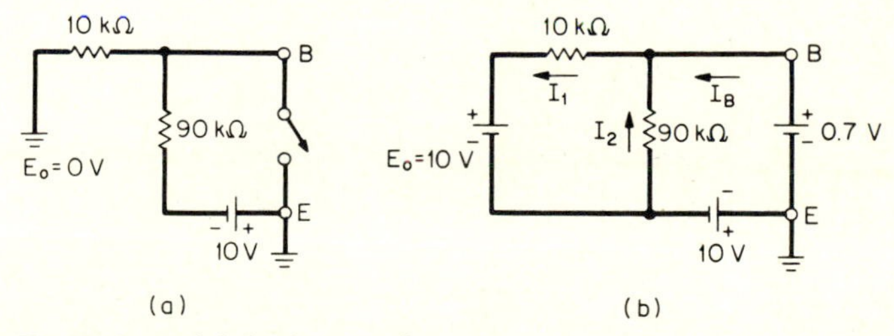

Fig. 14.18 Models for the input (base) circuit of a transistor switch: (*a*) For $E_0 = 0$ V. (*b*) for $E_1 = 10$ V. (See Example 14.6.)

example 14.6 For the INVERTER of Fig. 14.17, $E_{CC} = 10$ V, $-E_{BB} = -10$ V, $R_C = 1$ kΩ, $R_1 = 10$ kΩ, and $R_2 = 90$ kΩ. Logic signals $E_0 = 0$ V and $E_1 = 10$ V. (*a*) Verify that the circuit behaves as an INVERTER. (*b*) Determine the minimum dc current gain for the transistor, $h_{FE,\,\min}$ to ensure that the transistor operates as a saturated switch. Assume that $V_{sat} = 0.1$ V and that the base-emitter voltage is 0.7 V when the transistor is ON.

solution (*a*) For $E_0 = 0$ V, the base-emitter junction is reverse-biased and is represented by the model of an open switch, shown in Fig. 14.18*a*. The voltage across the base-emitter junction, by voltage division, is $-10 \times 10/(10 + 90) = -1$ V. A reverse bias of -1 V is more than necessary for a transistor to be OFF (nonconducting).

For $E_1 = 10$ V, the model for the base circuit is illustrated in Fig. 14.18*b*. The base-emitter junction is represented by a 0.7-V battery. To calculate the base current I_B, superposition is used. Setting E_{BB} to zero, the current owing to E_1 is $I_1 = (10 - 0.7)/10 = 0.93$ mA. Setting E_1 to zero, the current due to E_{BB} is $I_2 = (10 + 0.7)/90 = 0.12$ mA. The base current is equal to the difference of I_1 and I_2: $I_B = I_1 - I_2$. Hence, $I_B = 0.93 - 0.12 = 0.81$ mA.

From Fig. 14.17, collector current I_C is equal to the difference of E_{CC} and V_{sat} divided by R_C: $I_C = (E_{CC} - V_{sat})/R_C$. Substitution of the given values in the equation yields $I_C = (10 - 0.1)/1 = 9.9$ mA. When input $A = E_1$ (10 V), the output equals V_{sat} (logic 0) if $I_B = 0.81$ mA.

(*b*) The minimum dc current gain is expressed by $h_{FE,\min} = I_C/I_B$. From part (*a*), $h_{FE,\min} = 9.9/0.81 = 12.2$. Transistor Q, therefore, must have a dc current gain of at least 12.2 to ensure that it operates as a saturated switch.

14.6 HALF-ADDER

To demonstrate the usefulness of the AND, OR, and INVERTER circuits, these building blocks are used to construct a basic component of a binary adder, the half-adder (HA). Besides addition, the binary adder can be used for subtraction, multiplication, and division. In subtraction, the subtrahend is inverted and added to the minuend. Multiplication may be performed by repeated addition, and division by repeated subtraction. The binary adder is indeed a universal circuit.

The addition table for binary numbers is extremely simple, as illustrated in Fig. 14.19. We see that $0 + 0 = 0$; $0 + 1 = 1$; $1 + 0 = 1$; and $1 + 1 = 10$ (decimal 2). The heart of the binary adder is the half-adder, shown symbolically in Fig. 14.20. It has two inputs A and B, and two outputs C (*carry*) and S (*sum*).

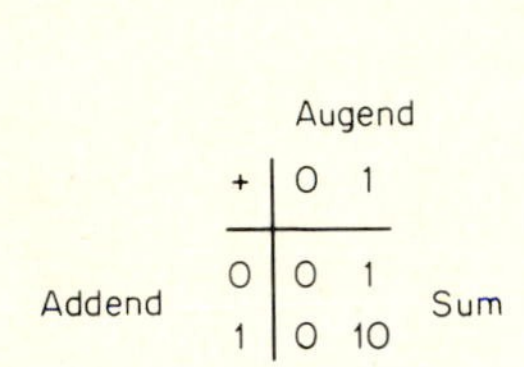

Augend

+	0	1
0	0	1
1	0	10

Addend — Sum

Fig. 14.19 Addition table for binary numbers.

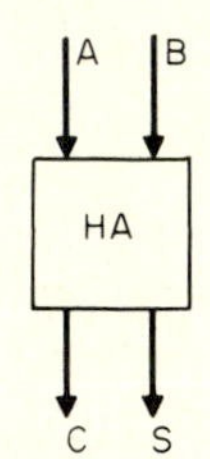

Fig. 14.20 Symbol for a half-adder. Bits to be added are A and B. Output S is the sum bit and C is the carry bit.

A	B	C	S
0	0	0	0
0	1	0	1
1	0	0	1
1	1	1	0

Fig. 14.21 Truth table for a half-adder.

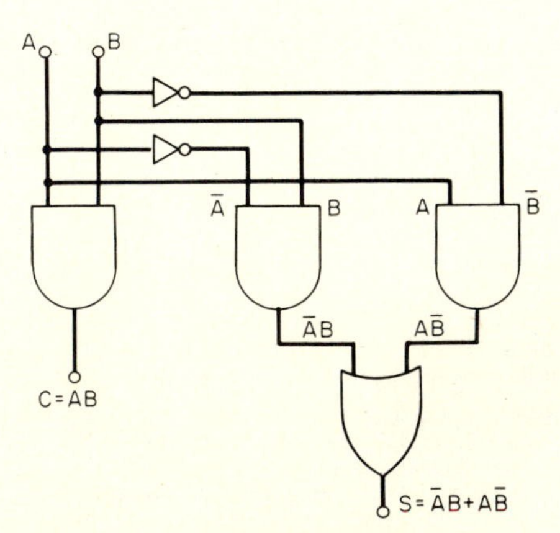

Fig. 14.22 A half-adder realized with AND and OR gates and INVERTERS.

The truth table for the half-adder is given in Fig. 14.21. A sum bit results when $A=0$ and $B=1$, or $A=1$ and $B=0$. When $A=1$ and $B=1$, a 0 results in the sum column and a 1 in the carry column.

The preceding results may be summarized by two logic equations. Sum $S=1$ when $A=0$ *and* $B=1$ *or* $A=1$ *and* $B=0$. Carry $C=1$ when A *and* $B=1$. Hence,

$$S=\overline{A}B+A\overline{B} \tag{14.5a}$$

$$C=AB \tag{14.5b}$$

Using AND and OR gates, and INVERTERS, the half-adder may be realized as shown in Fig. 14.22.

14.7 INHIBIT AND EXCLUSIVE OR GATES

The logic symbol for a three-input INHIBIT gate is illustrated in Fig. 14.23*a*. It is basically an AND gate with one input (C) negated by an INVERTER, as shown in Fig. 14.23*b*. This is indicated on the symbol by a circle drawn at input C.

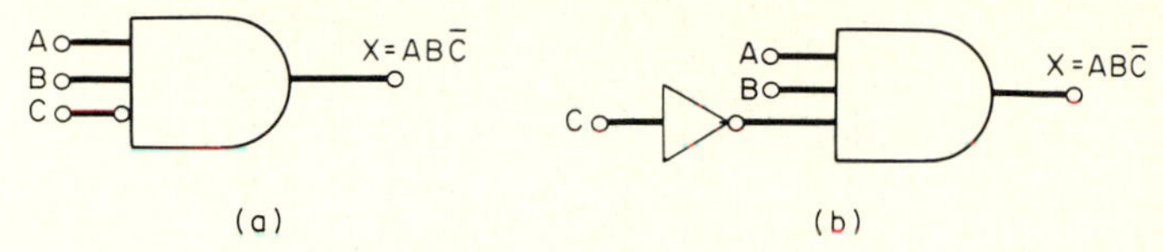

Fig. 14.23 An INHIBIT gate: (*a*) Logic symbol ($X=AB\overline{C}$). (*b*) It consists of an AND gate with one input negated by an INVERTER. INHIBIT gates having more than three inputs, as well as only two inputs, are possible.

The operation of an INHIBIT gate is such that $X=1$ when $A=B=1$ and $C=0$; if C also equals 1, $X=0$. Input C therefore acts to *inhibit* the gate. If $C=0$, it *enables* the gate, permitting it to perform an AND operation. The logic equation for the INHIBIT gate is

$$X=AB\overline{C} \tag{14.6}$$

The truth table is given in Fig. 14.24. Because three inputs are present, $n=3$. By Eq. (14.2), $2^3=$ eight possible combinations of A, B, and C exist. It is seen that $X=1$ only if $A=1$, $B=1$, and $C=0$.

The logic symbol for an EXCLUSIVE OR gate is shown in Fig. 14.25. For this gate, $X=1$ if $A=1$, $B=0$, or $A=0$, $B=1$. If $A=B=0$ or $A=B=1$, the output is zero. Expressed by a logic equation,

$$X=\overline{A}B+A\overline{B} \tag{14.7}$$

A	B	C	X
0	0	0	0
0	0	1	0
0	1	0	0
0	1	1	0
1	0	0	0
1	0	1	0
1	1	0	1
1	1	1	0

Fig. 14.24 Truth table for a three-input INHIBIT gate.

Fig. 14.25 Logic symbol for an EXCLUSIVE OR gate ($X=\overline{A}B+A\overline{B}$).

Equation (14.7) is identical with Eq. (14.5*a*) for the sum S of a half-adder. The circuit for S in Fig. 14.22 can therefore be used for realizing an EXCLUSIVE OR gate.

The truth table is illustrated in Fig. 14.26. Comparing this table with the truth table for the OR gate of Fig. 14.7, it is seen that in the OR gate, $X = 1$ even if $A = B = 1$. The OR gate, therefore, is sometimes referred to as an INCLUSIVE OR, because it provides an output when all inputs are at a logic 1.

A	B	X
0	0	0
0	1	1
1	0	1
1	1	0

Fig. 14.26 Truth table for an EXCLUSIVE OR gate.

14.8 NAND AND NOR GATES

Two universal gates, the NAND and NOR, are covered in this section. With either of these gates, the OR, AND, and NOT functions can be realized.

NAND GATE Logic symbols for the NAND gate are illustrated in Fig. 14.27. The NAND gate may be regarded as an AND gate followed by an INVERTER, as shown in

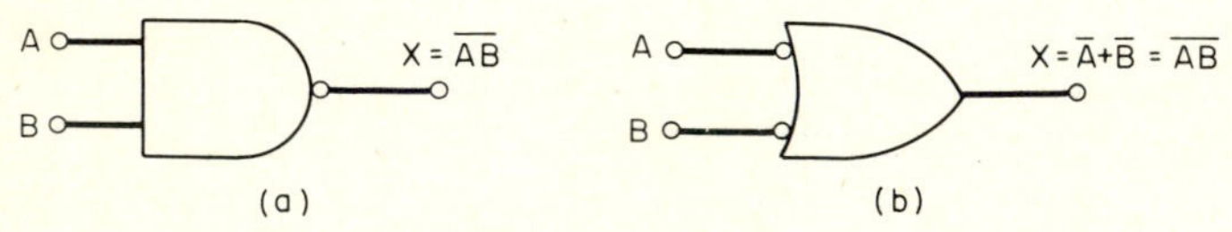

Fig. 14.27 Logic symbols for a two-input NAND gate ($X = \overline{AB} = \overline{A} + \overline{B}$). NAND gates with more than two inputs are also possible.

Fig. 14.28. Output $X = 1$ when $A = B = 0$; otherwise $X = 0$. Expressed by a logic equation,

$$X = \overline{AB} \tag{14.8a}$$

which is read as "X is equal to the *not* of A *and* B."

By a form of a very useful theorem, De Morgan's theorem, the *not* of A *and* B is equal to the *not* of A *or* the *not* of B: $\overline{AB} = \overline{A} + \overline{B}$. The logic equation for NAND operation may therefore also be expressed by

$$X = \overline{A} + \overline{B} \tag{14.8b}$$

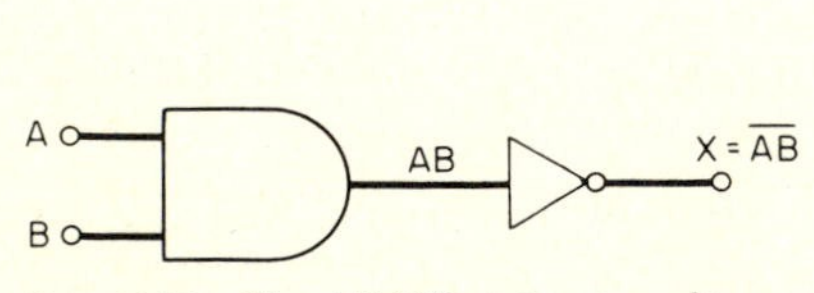

Fig. 14.28 The NAND gate viewed as an AND gate followed by an INVERTER.

A	B	X
0	0	1
0	1	1
1	0	1
1	1	0

Fig. 14.29 Truth table for a two-input NAND gate.

Based on Eq. (14.8*b*), an alternative symbol for the NAND gate is given in Fig. 14.27*b*, which shows negated inputs to an OR gate.

The truth table for a two-input NAND gate for positive logic is provided in Fig. 14.29. It is seen that $X = 0$ when $A = B = 1$; for all other inputs, $X = 1$.

If both inputs A and B are connected together, the NAND gate functions as an INVERTER. An INVERTER following a NAND yields an AND gate, as illustrated in Fig. 14.30. A *not of a not* ($\overline{\overline{AB}}$), indicated by two horizontal bars above AB, yields the function itself, AB. For negative logic, where a logic 1 is represented by a voltage that is less positive than for a logic 0, the positive NAND gate behaves as an OR gate.

example 14.7 Verify that for negative logic the positive NAND gate behaves as an OR gate.

solution In the truth table of Fig. 14.29, a 0 becomes a 1 and a 1 becomes a 0 for inputs A and B in negative logic. The new truth table, therefore, appears as in Fig. 14.31. This is identical with the truth table for the OR gate of Fig. 14.7.

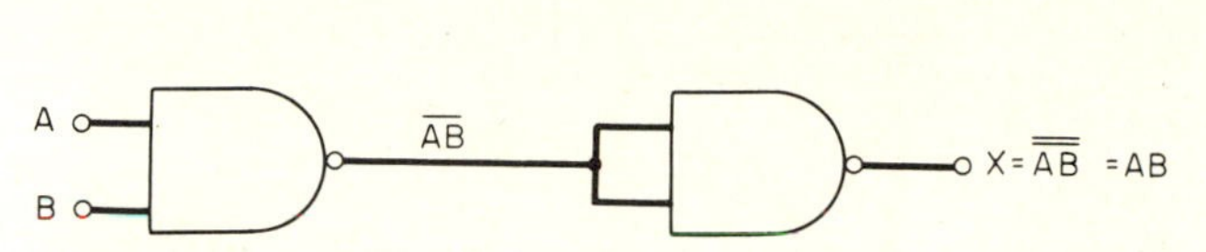

Fig. 14.30 A two-input AND gate realized by a NAND gate followed by a second NAND gate acting as an INVERTER.

A	B	X
1	1	1
1	0	1
0	1	1
0	0	0

Fig. 14.31 Truth table for a two-input positive NAND gate functioning as an OR gate for negative logic. (See Example 14.7.)

NOR GATE Logic symbols for the NOR gate are given in Fig. 14.32. The NOR gate may be viewed as an OR gate followed by an INVERTER, illustrated in Fig. 14.33. Output $X = 1$ only when $A = B = 0$; otherwise $X = 0$. Expressed by a logic equation,

$$X = \overline{A + B} \tag{14.9a}$$

which is read as "X is equal to the *not* of A *or* B."

By another form of De Morgan's theorem, $\overline{A + B}$ is equal to the *not* of A *and* the *not* of B. Therefore X also equals $\overline{A} \times \overline{B}$:

$$X = \overline{A} \times \overline{B} \tag{14.9b}$$

Based on Eq. (14.9*b*), an alternative symbol for the NOR gate is given in Fig. 14.32*b*, which shows negated inputs to an AND gate.

The truth table for a two-input NOR gate for positive logic is illustrated in Fig. 14.34. It is seen that output $X = 1$ only for $A = B = 0$; for all other pairs of inputs, $X = 0$.

If both inputs are tied together, the NOR gate acts as an INVERTER. An INVERTER

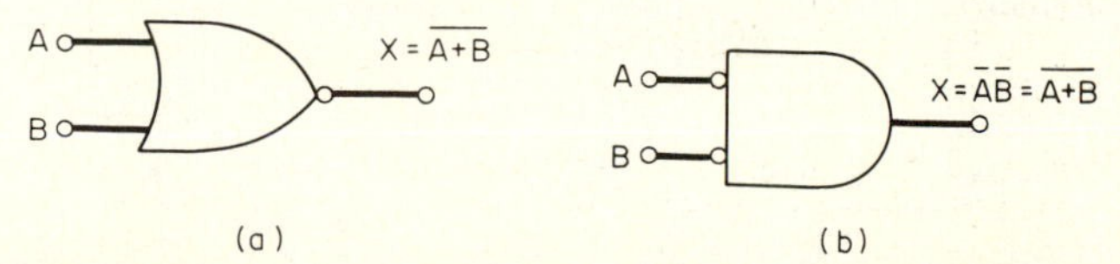

Fig. 14.32 Logic symbols for a two-input NOR gate ($X = \overline{A + B} = \overline{A} \times \overline{B}$). NOR gates with more than two inputs are also possible.

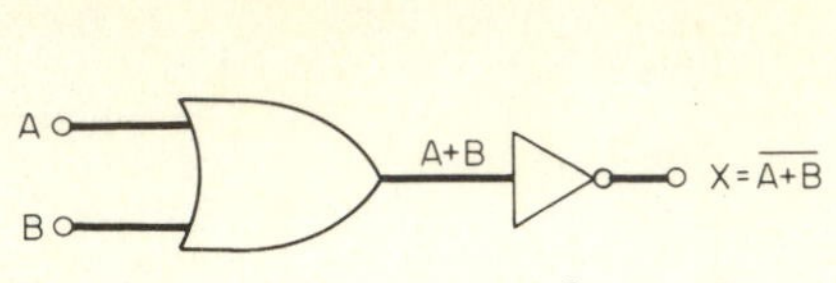

Fig. 14.33 A NOR gate viewed as an OR gate followed by an INVERTER.

A	B	X
0	0	1
0	1	0
0	1	0
1	1	0

Fig. 14.34 Truth table for a two-input NOR gate.

following a NOR yields an OR gate, as shown in Fig. 14.35. For negative logic, the positive NOR gate acts as an AND gate.

example 14.8 Verify that for negative logic the positive NOR gate acts as an AND gate.

solution In the truth table of Fig. 14.34, a 0 becomes a 1 and a 1 becomes a 0 for inputs *A* and *B* in negative logic. The resulting truth table appears in Fig. 14.36. This is identical with the truth table for the AND gate of Fig. 14.12.

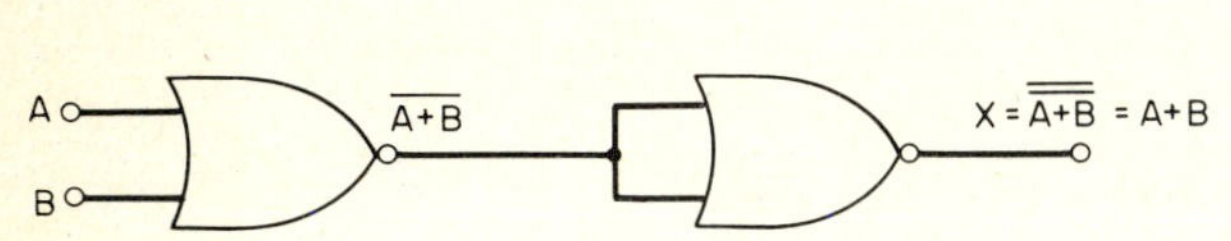

Fig. 14.35 A two-input OR gate realized by a NOR gate followed by a second NOR gate acting as an INVERTER.

A	B	X
1	1	1
1	0	0
0	1	0
0	0	0

Fig. 14.36 Truth table for a two-input positive NOR gate functioning as an AND gate for negative logic. (See Example 14.8.)

14.9 MULTIVIBRATORS

There are three kinds of multivibrators: the *bistable*, *monostable*, and *astable*. Their fundamental behavior and properties are explained initially with the aid of the block diagrams shown in Fig. 14.37.

a. The block diagram of a bistable multivibrator (MV), also referred to as a *flipflop*, *binary*, and an *Eccles-Jordan circuit*, is illustrated in Fig. 14.37*a*. The bistable MV has two inputs, labeled 1 and 2, and two outputs, X and $\overline{X}$. Output $\overline{X}$ (*not* X) is said to be the *complement* of X. If X is a logic 1, $\overline{X}$ is a logic 0; if $X = 0$, then $\overline{X} = 1$.

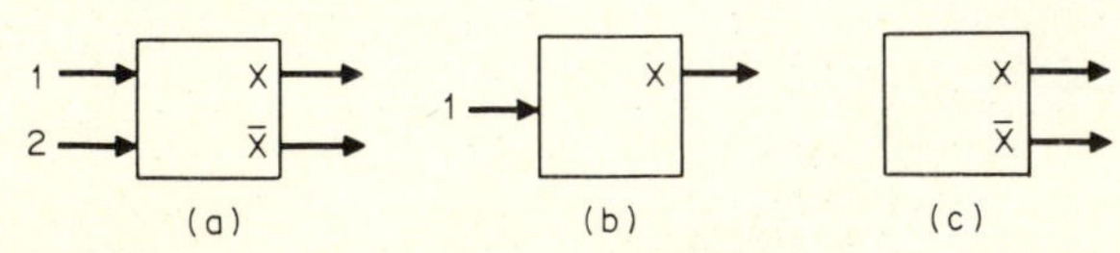

Fig. 14.37 Block diagrams of multivibrators: (*a*) Bistable. (*b*) Monostable. (*c*) Astable.

Assume that $X = 1$. The bistable MV remains in this state until a trigger is applied to, say, input 1; then $X = 0$. Output $X = 0$ until a trigger is applied to terminal 2; then $X = 1$ once again. Thus, the bistable MV exhibits two stable dc states: $X = 1$ and $X = 0$. It remains in one of these two states until a trigger is applied to either terminal 1 or 2 to change its state.

The bistable MV exhibits *memory*. Its output depends not only on the present input, but also on which input terminal the previous trigger was applied. Such a circuit is referred to as *sequential*. The OR, AND, etc., gates depend only on their present inputs and are said to be *combinational*.

The bistable MV is used in *shift registers* and *counters*. These are two important components employed in digital computers and systems. Shift registers are used to store information, such as data or an instruction. The counter, as its name implies, counts pulses.

b. The block diagram of a monostable MV, commonly referred to as a *one-shot*, is illustrated in Fig. 14.37*b*. It has one input and one output terminal. With no trigger applied, the output voltage is approximately 0 V (logic 0). It remains in this stable dc state until a trigger is applied to the input terminal. As a result of the trigger, a well-defined rectangular pulse appears at X. At the termination of the rectangular pulse, the output returns to its stable state, logic 0.

In addition to generating rectangular pulses, the one-shot is also used as a delay. If, for example, the output pulse of a one-shot is applied to terminal C of the INHIBIT gate of Fig. 14.23*a*, the output of the gate is delayed for a time equal to the length of the rectangular pulse.

c. The block diagram of an astable, or *free-running*, MV, is given in Fig. 14.37*c*. There are two output terminals, X and $\overline{X}$, and no input terminals. At the outputs are rectangular-type waveforms. The astable MV is an example of a *nonsinusoidal*, or *relaxation*, oscillator. It is used as a *clock* in digital computers to ensure that operations, such as addition, are synchronized with other operations in the system.

BISTABLE MV The basic circuit of a *collector-coupled* bistable MV employing npn transistors is illustrated in Fig. 14.38. Note the symmetrical nature of the circuit. The

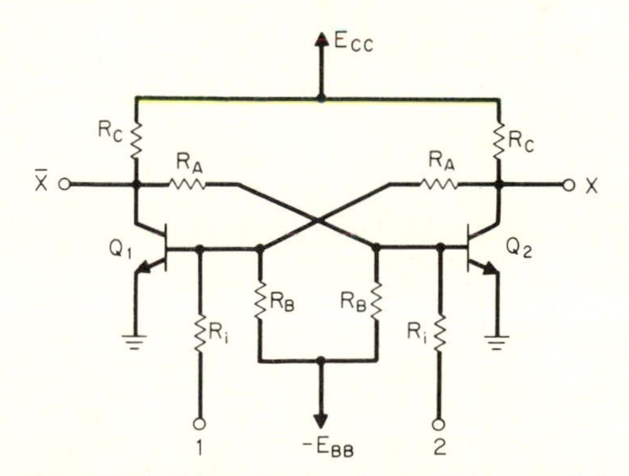

Fig. 14.38 Basic circuit of a collector-coupled bistable MV employing npn transistors. (For pnp transistors, the polarities of E_{CC} and E_{BB} for this and other multivibrator circuits are reversed.)

collector of one transistor is coupled to the base of the other transistor by resistor R_A. Resistor R_B is connected to the base and supply source, $-E_{BB}$ volts. Output X is taken at the collector of transistor Q_2 and $\overline{X}$ at the collector of transistor Q_1.

Assume that $X = 1$ and, therefore, $\overline{X} = 0$. In this state, Q_1 is conducting (ON) and Q_2 is nonconducting (OFF). The base-emitter voltage of Q_1 is approximately 0.7 V and the base-emitter voltage of Q_2 is negative, ensuring that it is cut off.

To change the state of the flipflop, a negative trigger is applied to terminal 1. (For best triggering of multivibrators, the trigger is selected to *turn off* the ON transistor.)

Transistor Q_1 is turned off and its collector voltage rises toward E_{CC} volts. The increasing collector voltage is coupled by resistor R_A to the base of Q_2. Its base-emitter voltage, therefore, becomes less negative and is finally equal to 0.7 V. Transistor Q_2 is now ON ($X = 0$), and Q_1 is cut off ($\overline{X} = 1$).

To obtain $X = 1$ and $\overline{X} = 0$, a negative trigger is applied to terminal 2. The chain of events described in the preceding paragraph occurs, forcing Q_1 to go ON and Q_2 to go OFF. Now, $X = 1$ and $\overline{X} = 0$.

MONOSTABLE MV The basic circuit of a collector-coupled monostable MV (one-shot) using npn transistors is shown in Fig. 14.39. As in the bistable MV, the collector of Q_2 is coupled to the base of Q_1 by resistor R_A. The collector of Q_1, however, is coupled to the base of Q_2 by coupling capacitor C.

In its dc stable state, Q_2 is ON ($X = 0$). To ensure that indeed Q_2 is ON, the maximum value of R, R_{max}, is equal to the net voltage across it (E_{CC} − base-emitter voltage E_{BE2}) divided by base current I_{BS} required for Q_2 to be in saturation:

$$R_{max} = \frac{E_{CC} - E_{BE2}}{I_{BS}} \tag{14.10}$$

To aid us in understanding the operation of the one-shot, waveforms at the bases and collectors of Q_1 and Q_2 are shown in Fig. 14.40. For time less than t_1, the one-shot is in

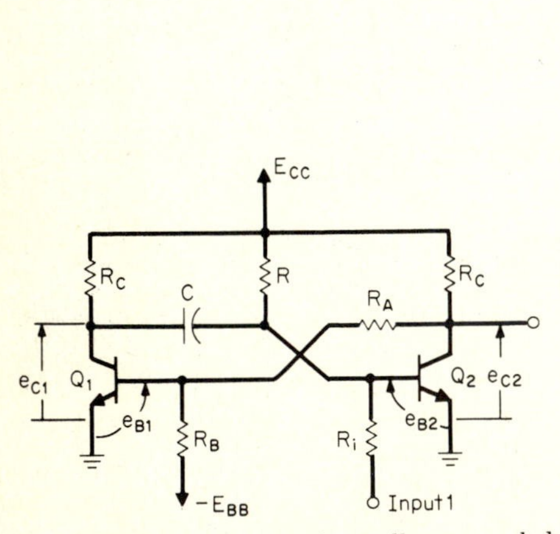

Fig. 14.39 Basic circuit of a collector-coupled monostable MV employing npn transistors.

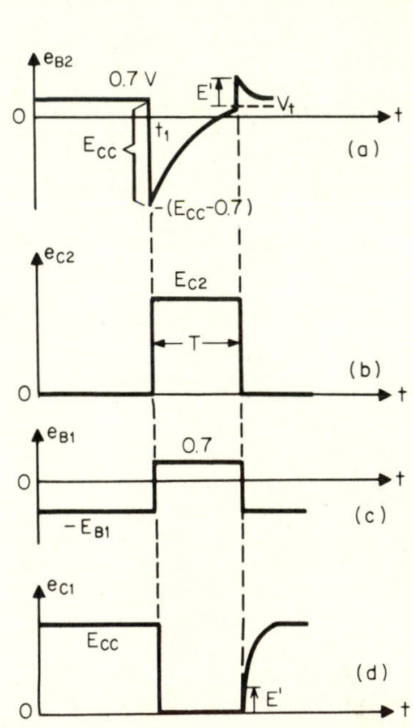

Fig. 14.40 Waveforms at the bases and collectors of transistors in a collector-coupled monostable MV: (*a*) Base of Q_2. (*b*) Collector of Q_2. (*c*) Base of Q_1. (*d*) Collector of Q_1.

its stable state where Q_1 is OFF and Q_2 is ON. The base-emitter voltage of Q_2, e_{B2}, is at approximately 0.7 V. Assuming that $V_{sat} = 0$, the collector-emitter voltage of Q_2 is, therefore, $e_{C2} = 0$. Because Q_1 is OFF, its base-emitter voltage is negative ($-E_{B1}$ volts) and its collector-emitter voltage, e_{C2}, is equal to the supply voltage, E_{CC} volts.

At $t = t_1$ assume that a negative trigger is applied to the input terminal. Transistor Q_1 turns ON and Q_2 turns OFF. For $V_{sat} = 0$, the voltage at the collector of Q_1 drops by E_{CC} volts. Because the voltage across a capacitor cannot change instantaneously, the drop in E_{CC} volts is transmitted to the base of Q_2. Hence, at $t=t_1$, $e_{B2}=-(E_{CC}-0.7)$ volts, and Q_2 is cut off.

Voltage e_{B2} begins to rise exponentially toward E_{CC} volts. When it is equal to the threshold voltage V_t, transistor Q_2 turns ON and Q_1 turns OFF. At the base of Q_2 an *overshoot* of E' volts occurs. Because of the coupling capacitor e_{C1} rises instantaneously by E' volts. During the time Q_2 is OFF, its collector-emitter voltage, e_{C2}, is equal to E_{C2} volts which is slightly less than the supply voltage, E_{CC} volts.

The length, or width, of the rectangular pulse T generated at the collector of Q_2 is given by

$$T \simeq 0.69RC \tag{14.11}$$

where T is in seconds, R in ohms, and C in farads. Equation (14.11) is an excellent approximation for T if E_{CC} is much greater than 0.7 V. The length of the negative trigger that initiates the one-shot is generally a small fraction of T.

ASTABLE MV The basic circuit of a collector-coupled astable MV using npn transistors is shown in Fig. 14.41. Note that the collector of each transistor is coupled by capacitor C to the base of the other transistor. The astable MV, therefore, has no dc stable states.

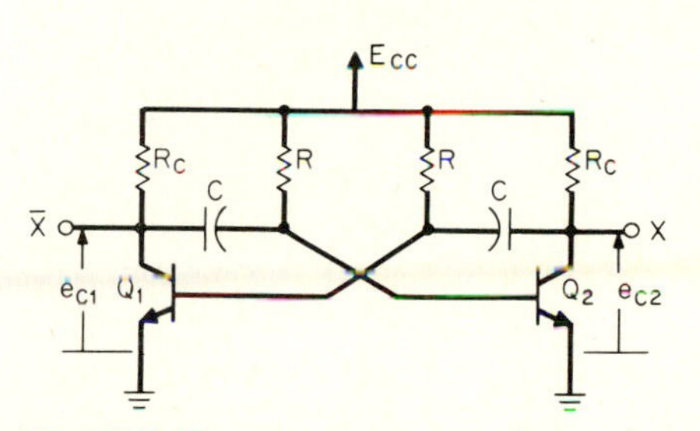

Fig. 14.41 Basic circuit of a collector-coupled astable MV employing npn transistors.

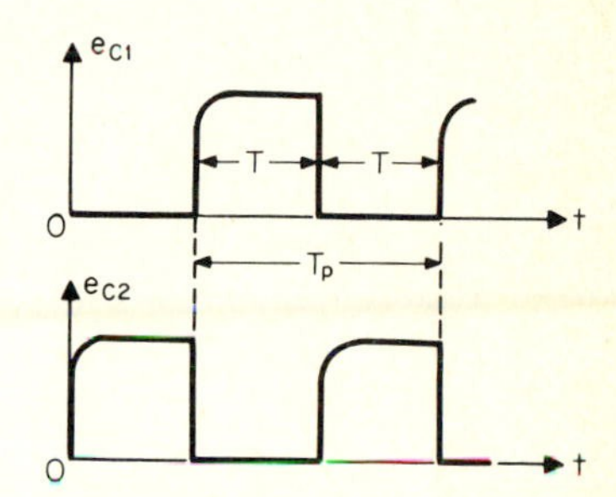

Fig. 14.42 Waveforms at the collectors of Q_1 and Q_2 of a collector-coupled astable MV.

Referring to the waveforms at the collectors of Q_1 and Q_2, e_{C1} and e_{C2}, respectively (Fig. 14.42), it is seen that each transistor is always generating a waveform. The shape of the waveforms is similar to that generated at the collector of Q_1 in the one-shot. The length T indicated in Fig. 14.42 is equal to $0.69RC$, as for the one-shot. The total period T_p is therefore equal to $2T$:

$$T_p = 1.38RC \tag{14.12a}$$

The frequency f is one divided by the total period. Hence the frequency of the generated square wave, in hertz, is

$$f = \frac{1}{1.38RC} \tag{14.12b}$$

14.10 DYNAMIC RESPONSE OF DIODE AND TRANSISTOR SWITCHES

Because of various factors, to be examined in this section, the response of a diode or transistor operating as a switch is delayed and the waveshape distorted. As a result, the rate at which a diode or transistor can be switched is limited.

RESPONSE OF A DIODE SWITCH Assume that the waveform of Fig. 14.43a is applied to a diode, such as a diode in a positive OR gate. During the interval of t_1 seconds, the diode is forward-biased and a current I_{D1} flows, as indicated in Fig. 14.43b.

At $t = t_1$, the input voltage is $-E$ volts, and the diode becomes reverse-biased. What one expects is that the current in a reverse-biased diode should be zero. Instead, an appreciable current, $-I_{D2}$, flows for a time t_s seconds, which eventually goes to zero.

To explain this anomalous behavior, it is necessary to examine the action of a pn junction (see Chap. 8). When a diode is forward-biased, there is a huge buildup of electrons in the p region and holes in the n region in the vicinity of the pn junction. When the diode is reverse-biased, it takes time for the electrons and holes to depart from the junction region. Consequently the diode conducts in the reverse direction for a time t_s, referred to as the *storage time*.

After t_s has elapsed, the reverse current gradually decreases and ultimately reaches zero. The reason for the gradual decay in reverse current is that it takes time for the diode depletion capacitance to charge to the reverse voltage, $-E$ volts. The time from t_1, when the diode is first reverse-biased, to the time when the reverse current equals a specified value, $-I_R$, is the *reverse-recovery time* t_{rr} of the diode. For good switching diodes, t_{rr} is in the order of nanoseconds (10^{-9} s).

RESPONSE OF A TRANSISTOR SWITCH Assume that the rectangular pulse (length $= t_p$) of Fig. 14.44*a* is impressed across the input terminals of an INVERTER.

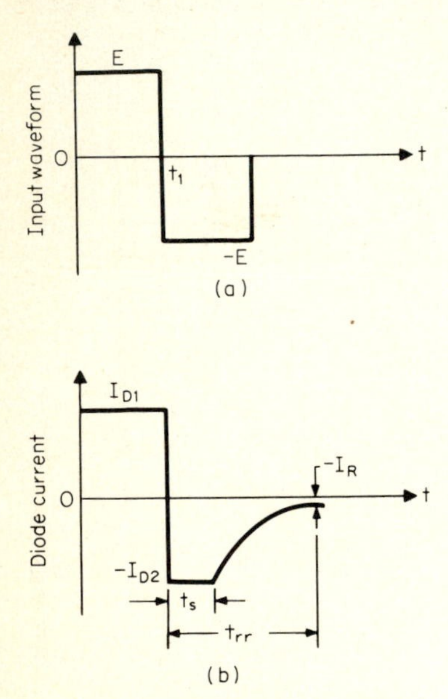

Fig. 14.43 Response of a diode switch: (*a*) Voltage waveform applied to diode. (*b*) Diode current flow in response to impressed voltage waveform.

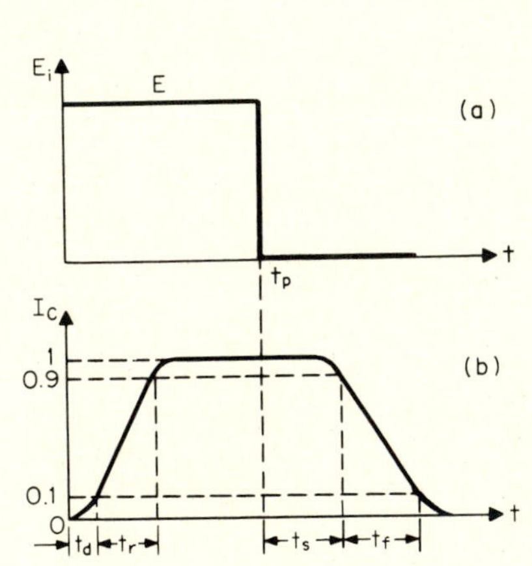

Fig. 14.44 Response of a transistor switch: (*a*) Pulse applied to the input of an INVERTER. (*b*) Collector current waveform in response to the input pulse.

The collector current appears as shown in Fig. 14.44*b*. The output waveshape is stretched and distorted. For convenience, the maximum value of collector current is normalized to one. Four specific times are indicated in the figure. These are:

1. *Delay time* t_d is defined as the time it takes the collector current to reach 0.1 (10 percent) of its maximum value. It corresponds to the time required to charge the junction capacitance plus the time it takes for the carriers (electrons or holes) to reach the collector region.

2. *Rise time* t_r is defined as the time it takes the collector current to rise from 0.1 (10 percent) to 0.9 (90 percent) of its maximum value. The rise time is the time needed to charge additional transistor capacitances.

3. *Storage time* t_s is defined as the time, measured from t_p to where the collector

current falls to 90 percent of its maximum value. In saturation, both the base-emitter and collector-base junctions are forward-biased. Consequently the collector, in addition to the emitter, injects electrons (or holes) into the base region. Similar to turning off a forward-biased diode, it takes time for the carriers to leave the base region. This phenomenon results in a storage time delay.

4. *Fall time* t_f is defined as the time required for the collector current to fall from its 90 to 10 percent maximum values. During this interval, the transistor capacitances are being discharged.

The sum of the delay and rise times, $t_d + t_r$, is equal to the *turn-on time* t_{on} of the transistor:

$$t_{on} = t_d + t_r \tag{14.13a}$$

Similarly, the sum of the storage and fall times, $t_s + t_f$, is equal to the *turn-off time*, t_{off}, of the transistor:

$$t_{off} = t_s + t_f \tag{14.13b}$$

To decrease the turn-on time, the base is driven hard. In this condition, more base current is supplied to the base than required for the transistor to saturate. Because of the excess base current the storage time, and hence the turn-off time, increases. For high-speed switching, the transistor may be operated as a nonsaturated switch. An example of a nonsaturating circuit is emitter-coupled logic (ECL), considered in Chap. 15.

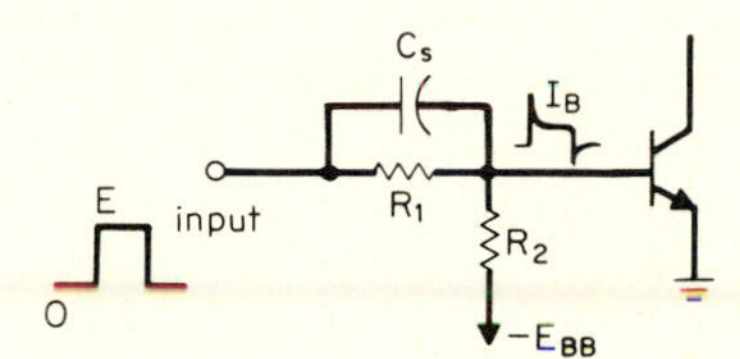

Fig. 14.45 The use of a speed-up (commutating) capacitor C_s across the series input resistor R_1 to improve the switching speed of a transistor.

A method used often in discrete circuits to decrease the turn-on and turn-off times is to place a *speed-up*, or *commutating*, *capacitor* across the resistor in series with the transistor base (Fig. 14.45). The value of the speed-up capacitor C_s is in the order of 50 pF. Initially, when the input rises from 0 to E volts, the capacitor acts as a short circuit. Maximum base current therefore flows and the turn-on time is reduced. At the termination of the input pulse, the capacitor also acts as a short circuit and aids in removing the excess carriers from the base region. As a result, the storage and turn-off times are also reduced.

Chapter 15

Digital Integrated Circuits

15.1 INTRODUCTION

In the development of the digital integrated circuit (DIC), families of logic evolved which exhibit well-defined properties. Three important families that enjoy wide use are:

1. *Transistor-transistor logic* (TTL, T^2L) is the most widely used logic. It has good speed, reasonably low power dissipation per gate, and is relatively low in cost.
2. *Emitter-coupled logic* (ECL), or *current-mode logic* (CML), is the fastest logic available today. Operating as a nonsaturated switch, it can switch frequencies as high as 500 MHz. Its power dissipation per gate, as well as its cost, is relatively high.
3. *Complementary metal-oxide semiconductor logic* (CMOS, COS/MOS) exhibits the lowest dissipation per gate. It is slower than T^2L or ECL, and its cost is moderate. Whereas T^2L and ECL employ the bipolar junction transistor in their circuits, CMOS uses the enhancement-type MOSFET. A comparison of the three major families of logic is provided in Table 15.1.

Owing to rapid advances in integrated circuit processing, logic circuits are available with an excess of 1000 logic gates on a single chip of silicon. One can define four levels of complexity in terms of the number of gates on a chip:

1. *Small-scale integration* (SSI): A DIC containing less than 12 gates on a chip
2. *Medium-scale integration* (MSI): A DIC containing more than 12, but less than 100 gates on a chip
3. *Large-scale integration* (LSI): A DIC containing more than 100, but less than 1000 gates on a chip
4. *Grand-scale integration* (GSI): A DIC containing more than 1000 gates on a chip

This chapter defines the terms used to characterize a DIC; discusses the operation of T^2L, ECL, and CMOS logic; and considers their application as flipflops in shift registers and counters. The chapter concludes with a discussion of semiconductor memories, digital-to-analog and analog-to-digital converters, and the microprocessor.

15.2 DIC TERMS AND PARAMETERS

In addition to the specification of required power-supply voltages and operating temperature range, a number of terms and parameters are used in characterizing the performance of a digital integrated circuit. Important specifications include:

FAN IN The maximum number of inputs to a gate.

TABLE 15.1 Comparison of IC Logic Families

Parameter	T^2L				ECL				CMOS
	Standard	Low-power	High-speed	Schottky	1-nS	2-nS	4-nS	8-nS	
Logic function	NAND	NAND	NAND	NAND	OR/NOR	OR/NOR	OR/NOR	OR/NOR	NOR/NAND
Supply voltage, V	5	5	5	5	−5.2	−5.2	−5.2	−5.2	3 to 18
Power dissipated/gate, mW	12	1	22	18	60	25	22	31	10^{-5}*
Propagation delay/gate, nS	10	33	6	3	1	2	4	8	70
Speed, MHz	35	3	50	125	500	200	165	30	5
Noise margin	Very good	Very good	Very good	Good	Fair	Fair	Fair	Fair	Excellent
Fan out (typical)	10	10	10	10	10	25	25	25	>50

* This is the *static* value. In switching, the power dissipation per gate at 1 MHz is approximately 1 mW and increases with frequency.

FAN OUT The maximum number of gates that may be connected to the output terminal of a single gate.

PROPAGATION DELAY The propagation delay t_{dp} is the difference in time between the application of an input signal and its presence at the output of a DIC. Its unit is generally nanoseconds.

SPEED This parameter indicates how fast a flipflop can change states. Its unit is megahertz.

GATE DISSIPATION The average dc power dissipated in a gate. Its unit is milliwatts or microwatts.

SPEED-POWER PRODUCT The product of propagation delay (in nanoseconds) and gate dissipation (in milliwatts). Because the product of power and time yields energy, the unit for the speed-power product is picojoules, pJ (1 pJ $= 10^{-12}$ J).

NOISE MARGIN Noise margin NM is the maximum extraneous voltage that causes a gate to change its state. Its unit is volts or millivolts.

15.3 T²L

A basic T²L circuit having a fan-in of two is illustrated in Fig. 15.1. Transistor Q_1 is unique; it has multiple emitters. Each base-emitter junction behaves like a diode.

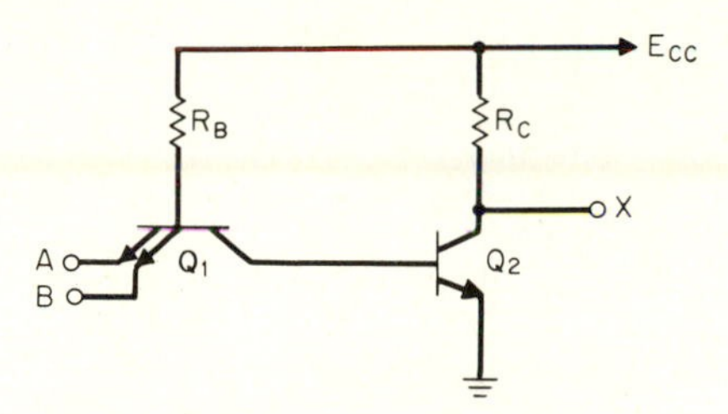

Fig. 15.1 Basic T²L gate shown with a fan-in of two. The circuit performs positive NAND logic.

The two base-emitter junctions, therefore, act like diodes of an AND gate to which are connected inputs A and B. The base-collector junction of Q_1 serves as a diode in series with the base of Q_2, which is connected as an INVERTER.

The logic performed by T²L is positive NAND logic. Assume that either input A or B is at logic 0. One of the base-emitter junctions becomes forward-biased. The base of Q_1 is therefore near ground, and the collector-base junction is reverse-biased. No base current flows in Q_2, and the transistor is OFF: $X = 1$.

If both inputs are at logic 0, the same condition exists as though either one is at logic 0 and output $X = 1$. When both A and B are at logic 1, the base-emitter diodes are reverse-biased. The collector-base junction of Q_1 is now forward-biased, and base current flows in Q_2. Output X, therefore, is at logic 0.

There exists a number of members in the T²L family. They include the standard, low-power, high-speed, and Schottky-clamped series. These circuits are examined in the following discussion.

STANDARD T²L The circuit for a two-input standard T²L NAND gate is shown in Fig. 15.2. Transistor Q_1 serves the same function as in the basic circuit of Fig. 15.1. Transistor Q_2 is connected as a *phase splitter*. When a logic 0 is at its base, its collector is at a logic 1 and its emitter at a logic 0. If the input is equal to a logic 1, the reverse is true. Now, the collector is at a logic 0 and the emitter at a logic 1.

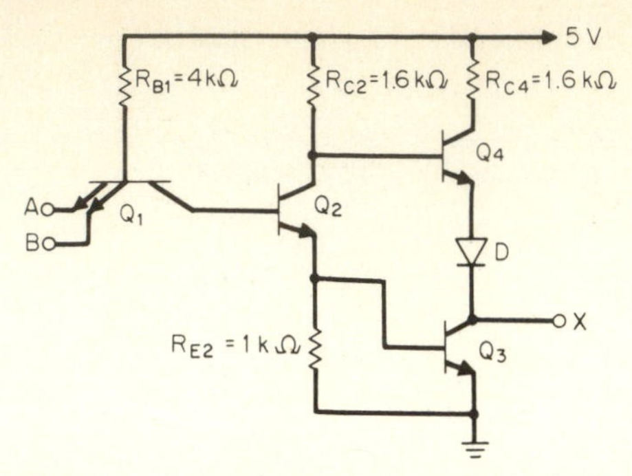

Fig. 15.2 Circuit for a two-input standard T²L NAND gate.

Transistors Q_3 and Q_4 are connected as a *totem-pole amplifier*. This configuration minimizes the effects of any capacitance present, such as stray, across the output. One serious effect of capacitance across the output of a gate is to reduce its switching speed. The function of diode D is to limit the collector current to a reasonable value when Q_3 is in the ON state.

Assume that inputs A and B are at logic 0. The base-emitter junctions of Q_1 are forward-biased, and a logic 0 appears at the base of Q_2. Owing to the operation of the phase splitter, at the collector of Q_2 there appears a logic 1 and at its emitter a logic 0. Transistor Q_4 is therefore ON and Q_3 is OFF. Output X is at a potential equal to E_{CC} less the voltage drops across R_{C4}, the collector-emitter of Q_4, and the diode; hence, $X = 1$.

If either input is at logic 0, the same condition described in the preceding paragraph prevails, and $X = 1$. Assume that both inputs are at logic 1. A logic 1 therefore appears at the base of Q_2. As a result, a logic 0 is at the collector and a logic 1 at the emitter of Q_2. Transistor Q_4 is now OFF, and Q_3 is ON; $X = 0$.

LOW-POWER T²L A version of T²L which dissipates less than one-tenth the power of standard T²L is illustrated in Fig. 15.3. Referred to as low-power T²L, the NAND gate is very similar to the standard circuit. To ensure low-power dissipation, the resistance values are increased appreciably. For example, $R_{B1} = 40$ kΩ, and $R_{C2} = 20$ kΩ; in the standard T²L, their corresponding values are 4 and 1.6 kΩ, respectively.

Transistors Q_3 and Q_4 constitute a Darlington pair. In addition to providing greater current gain than a single transistor, the Darlington pair also increases the switching speed of the gate.

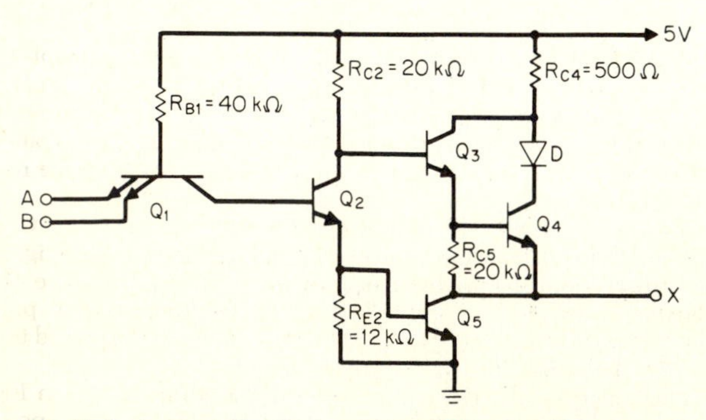

Fig. 15.3 An example of a low-power T²L gate. Transistors Q_3 and Q_4 are connected as a Darlington pair which provides more current gain than a single transistor.

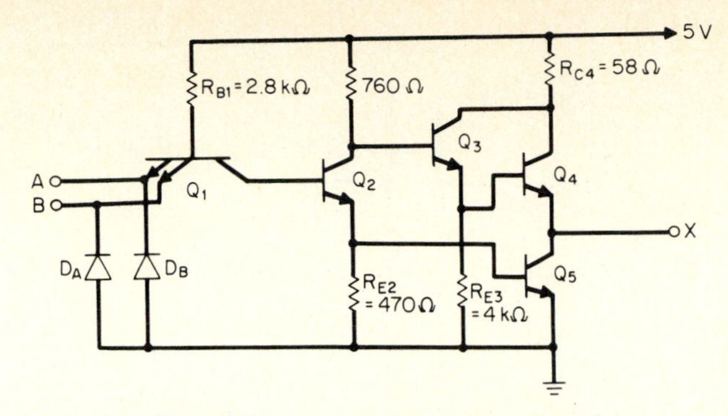

Fig. 15.4 High-speed T²L gate. It features low values of resistances. Diodes D_A and D_B protect the base-emitter junctions of transistor Q_1 against voltage breakdown.

HIGH-SPEED T²L Similar in configuration to low-power T²L, a high-speed T²L NAND gate is shown in Fig. 15.4. To achieve high-speed operation, the resistance values in the circuit are kept low. Because at high switching speeds stray inductance, in addition to stray capacitance, enters the picture, oscillations may be superimposed on the pulses. This phenomenon is referred to as *ringing*. The purpose of diodes D_A and D_B, referred to as *clamping diodes*, connected between each input and ground is to limit negative signal swings because of ringing. In this manner the base-emitter junctions are protected from avalanche or zener breakdown.

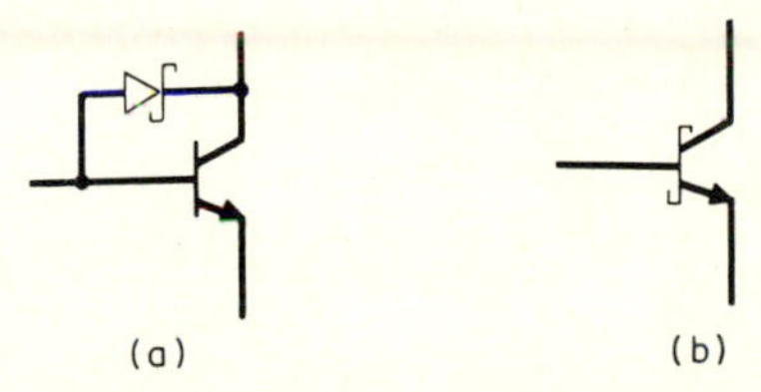

Fig. 15.5 Schottky-clamped transistor: (*a*) Schottky diode connected across the base and collector terminals of a junction transistor. (*b*) Symbol for a Schottky-clamped transistor, referred to as a Schottky transistor.

SCHOTTKY-CLAMPED T²L The transistors in the previous examples of transistor-transistor logic all operate in saturation when in the ON state. As explained in Chap. 14, a saturated switch gives rise to storage time which reduces its speed of operation. To increase switching speed, the transistor is operated as a nonsaturated switch. In this mode of operation, the transistor in the ON state is in the active region close to, but never in, saturation.

To achieve efficient nonsaturated switching, a Schottky diode (see Chap. 8) is connected across the collector and base of a transistor, as illustrated in Fig. 15.5*a*. When the transistor is turned on, some of the base current is diverted by the diode from the base. Consequently, less base current flows, and the transistor is prevented from saturating. The symbol for a Schottky-clamped transistor, also referred to as a *Schottky transistor*, is given in Fig. 15.5*b*.

An example of a Schottky-clamped T²L NAND gate is illustrated in Fig. 15.6. The circuit is similar to the high-speed T²L gate of Fig. 15.4. A low-power Schottky-clamped T²L gate is also available. As in the low-power T²L gate, high-resistance values are used in this version.

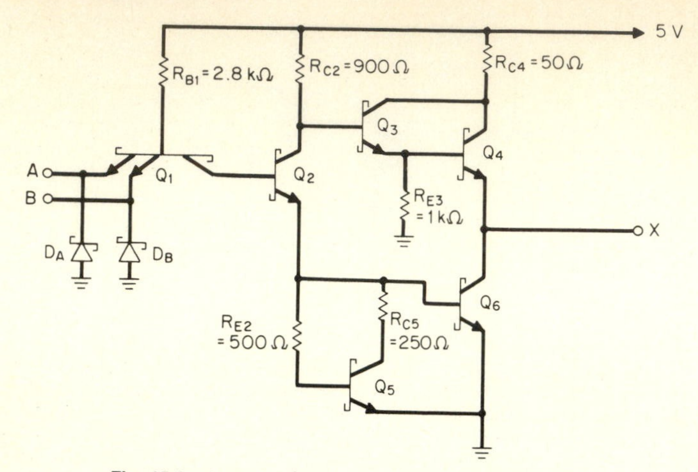

Fig. 15.6 An example of a Schottky-clamped T²L gate.

15.4 ECL

With its transistors operating as nonsaturated switches, emitter-coupled logic (ECL), also referred to as current-mode logic (CML), is the fastest family of logic circuits. It is available in four basic types having propagation delays of 8, 4, 2, and 1 ns (see Table 15.1). Emitter-coupled logic provides two logic functions: the NOR and OR. Its logic symbol is illustrated in Fig. 15.7.

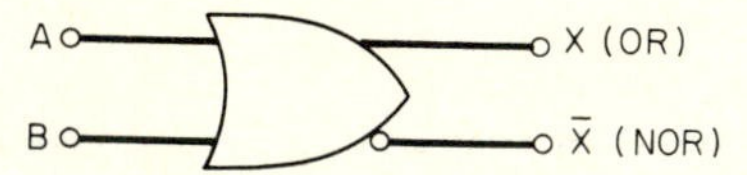

Fig. 15.7 Logic symbol for an ECL gate.

A recent member of the family is the 2-nS ECL gate. This version optimizes switching speed and gate dissipation (see Table 15.1). An example of a 2-nS gate having a fan-in of two is provided in Fig. 15.8. It is powered by a −5.2-V dc power supply. A logic 0 is approximately −1.7 V, and a logic 1 is equal to −0.9 V. Because a logic 1 is less negative than a logic 0, their values correspond to positive logic.

A basic circuit of ECL is the differential amplifier comprised of transistors Q_2 and Q_3

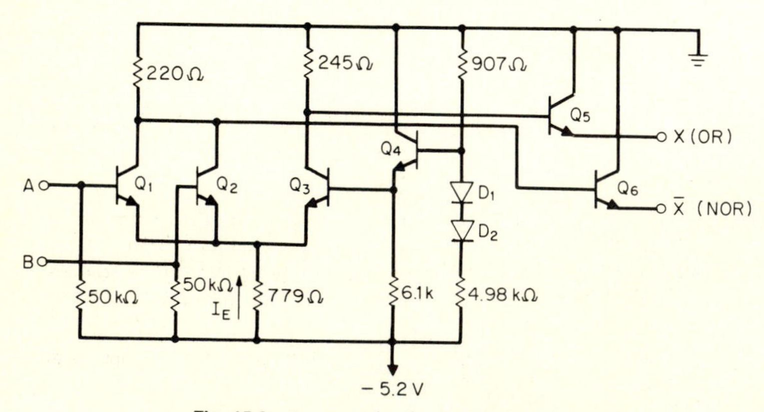

Fig. 15.8 An example of a 2-nS ECL gate.

(see Chap. 13). Emitter current I_E is essentially constant. The base of Q_3 is held at a constant dc reference voltage of 1.29 V. The reference voltage is stabilized with respect to temperature by transistor Q_4 and diodes D_1 and D_2. Switching occurs when the input to either transistor, Q_1 or Q_2, is approximately 0.1 V greater or less than the reference voltage. Emitter followers Q_5 and Q_6 provide dc level shifting to ensure that a logic 0 corresponds to −1.7 V and a logic 1 to −0.9 V.

Assume that inputs A and B are at logic 0. For this condition, Q_1 and Q_2 are OFF and Q_3 is ON (logic 0). The output of Q_5, therefore, is $X = 0$ (OR) and at Q_6, $\bar{X} = 1$ (NOR).

Suppose that A is at logic 0 and B at logic 1. Now Q_3 is OFF, Q_2 is ON; and $X = 1$, $\bar{X} = 0$. If both inputs are at logic 1, as in the preceding case, $X = 1$ and $\bar{X} = 0$. We see therefore that ECL provides an OR and a NOR output.

15.5 CMOS

Complementary metal-oxide semiconductor (CMOS) logic dissipates minute power and can operate over a power-supply range of 3 to 18 V. It is, however, slower than T^{2}L or ECL (see Table 15.1). In this logic family, no resistors are used. Instead, p-channel and n-channel enhancement-type MOSFETs are employed in complementary pairs. As a result, a much greater packing density is realized than with the bipolar junction transistor in monolithic integrated circuits.

INVERTER A CMOS inverter is illustrated in Fig. 15.9. Source S_2 of the p-channel device is returned to the supply voltage (E_{DD} volts), and source S_1 of the n-channel device is returned to ground. Output X is obtained at the junction of the drain terminals, D_1 and D_2. Both gates, which are connected together, constitute input terminal A. The substrate (Sub) is connected to the source of each transistor.

A logic 1 is equal to E_{DD} volts and a logic 0 to 0 V. Assume that A is at logic 1. Because G_1 is positive with respect to S_1, the n-channel MOSFET (Q_1) is ON. The gate-source voltage of the p-channel device (Q_2), however, is at 0 volts ($E_{DD} - E_{DD} = 0$); hence, Q_2 is OFF. Output $X = 0$.

If A is at logic 0, the n-channel transistor is OFF. The p-channel device, whose gate is now negative with respect to its source ($0 - E_{DD} = -E_{DD}$), is ON. The output in this condition may be thought of as being shifted to E_{DD} volts and $X = 1$.

In either case, one transistor is always ON and the other is always OFF. Since the transistors are connected in series, the drain current flowing is equal to their leakage current. Because of its extremely small value, the dissipation is in the order of 0.01 μW. (In switching, however, the gate dissipation is approximately 1 mW at 1 MHz and increases with frequency.)

NOR GATE An example of a two-input NOR gate employing CMOS logic is shown in Fig. 15.10. Two p-channel MOSFETs are connected in series and two n-channel

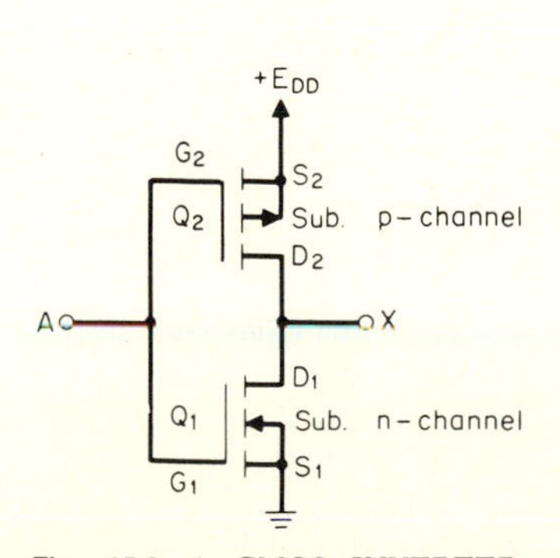

Fig. 15.9 A CMOS INVERTER. Transistor Q_1 is an n-channel enhancement-type MOSFET and Q_2 a p-channel device.

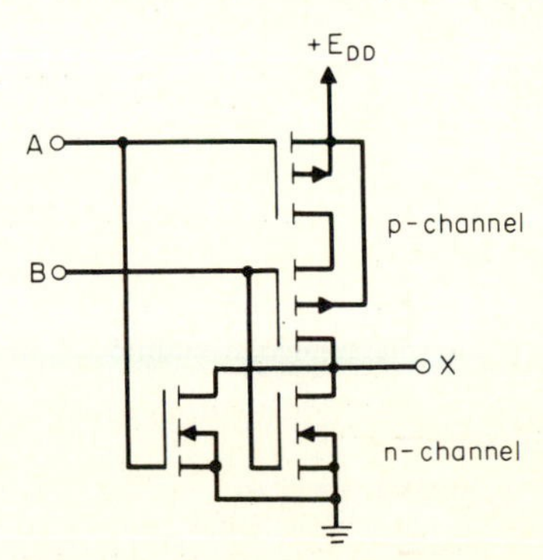

Fig. 15.10 A CMOS two-input NOR gate.

MOSFETs are connected in parallel. The gate of each n-channel transistor is connected to the gate of a p-channel transistor. The substrates of the p-channel devices are connected to E_{DD} and the substrates of the n-channel devices are returned to ground.

Assume that A and B are at logic 0. In this case both p-channel devices are ON, and the n-channel devices are OFF. Output $X = 1$ (the NOT of a zero is one). If A or B, or both, are at logic 1, one or both p-channel transistors are OFF. Also, one or both n-channel transistors are ON. Output $X = 0$ (the NOT of a one is zero).

example 15.1 Show how the two-input NOR gate of Fig. 15.10 may be expanded into a three-input NOR gate.

solution This is accomplished by adding a p-channel MOSFET in series and an n-channel MOSFET in parallel (Fig. 15.11). Their gates are connected together for the third input C. In such fashion, one may develop a four-, or greater, input NOR gate.

NAND GATE A two-input CMOS NAND gate is illustrated in Fig. 15.12. In contrast with the NOR gate of Fig. 15.10, the p-channel devices are in parallel and the n-channel devices are in series in the NAND gate. For each additional input, a p-channel MOSFET is connected in parallel and an n-channel MOSFET in series. Their gates are then joined together for the new input.

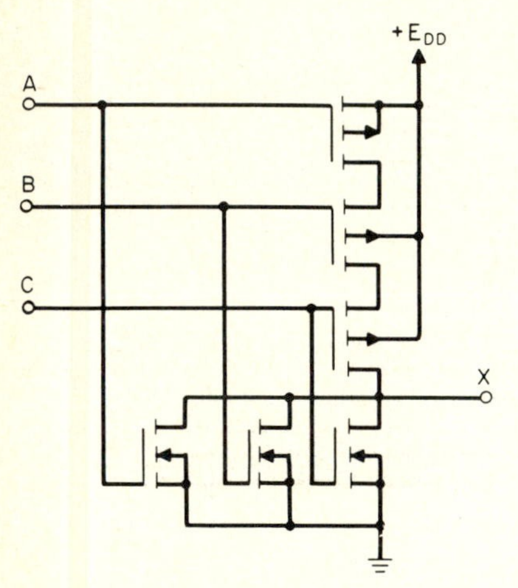

Fig. 15.11 A CMOS three-input NOR gate. (See Example 15.1.)

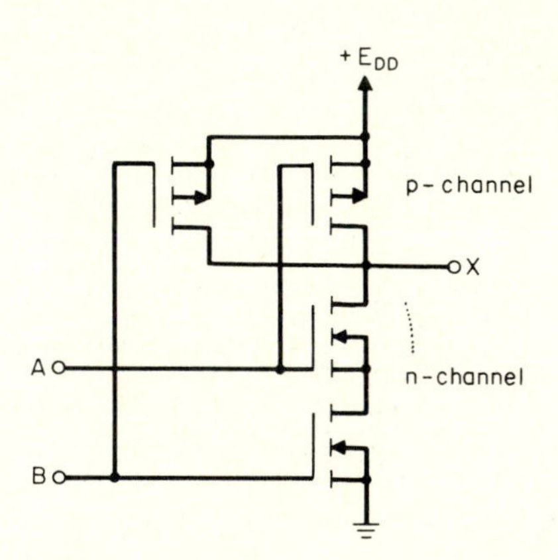

Fig. 15.12 A CMOS two-input NAND gate.

Assume that both A and B are at logic 1. The p-channel devices are OFF, and the *n*-channel devices are ON. Output $X = 0$ (the NOT of a one is zero). If either or both A and B are at logic 0, one or both p-channel transistors are ON and one or both n-channel transistors are OFF. Output $X = 1$ (the NOT of a zero is one).

15.6 FLIPFLOPS

The bistable MV, described in Chap. 14, is the basic circuit in a number of flipflops used in shift registers and counters. In this section, their behavior is illustrated by truth tables. Because the gates of logic families are generally either NAND or NOR (see Table 15.1), the implementation of flipflops is based on these configurations.

R-S FLIPFLOP (LATCH) An *R-S* (reset-set) flipflop, also referred to as a *latch,* implemented with NAND gates is illustrated in Fig. 15.13*a*. The output of each gate is connected to the input of the other gate. In the truth table of Fig. 15.13*b*, subscript $n + 1$ indicates the state of the flipflop after being triggered by a suitable pulse at either the R or S input terminals. Subscript n denotes the state of the flipflop before being triggered. The logic symbol for an *R-S* flipflop is provided in Fig. 15.14.

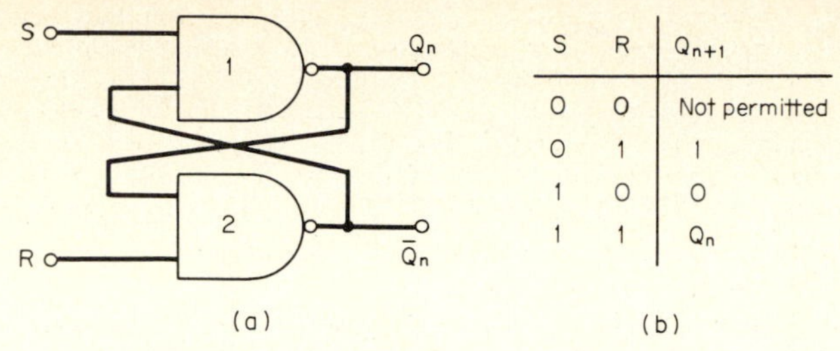

S	R	Q_{n+1}
0	0	Not permitted
0	1	1
1	0	0
1	1	Q_n

Fig. 15.13 An *R-S* flipflop (latch) using NAND gates: (*a*) Circuit. (*b*) Truth table.

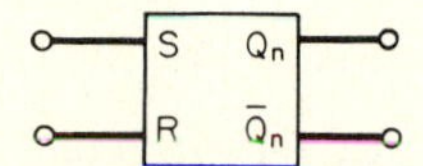

Fig. 15.14 Logic symbol for an *R-S* flipflop.

For no input trigger ($S = R = 0$), both outputs tend to go to a logic 1 (the NOT of a zero is one). Because this is an unstable state, this condition is not permitted.

Assume that $Q_n = 0$ and, therefore, $\overline{Q}_n = 1$. Because output $\overline{Q}_n$ is an input to NAND gate 1, one of its inputs, therefore, is a logic 1. If $S = 0$, the output of gate 1 is a one (the NOT of a zero is a one); hence, $Q_{n+1} = 1$. Similarly, if $Q_n = 1$ and $R = 0$, Q_{n+1} becomes a 0. For $S = R = 1$, the state of the flipflop is unchanged.

example 15.2 Show how an *R-S* flipflop may be implemented using NOR gates. Draw a truth table for the flipflop.

solution The circuit is shown in Fig. 15.15*a* and the truth table in Fig. 15.15*b*. For $S = R = 0$, there is no change in state ($Q_{n+1} = Q_n$). Assume that $Q_n = 1$; therefore $\overline{Q}_n = 0$. Output $\overline{Q}_n$ is one of the inputs to NOR gate 1. If $R = 1$, the NOT of $1 + 0 = 0$; hence $Q_{n+1} = 0$ and $\overline{Q}_{n+1} = 1$.

Now assume that $Q_n = 0$ ($\overline{Q}_n = 1$). Output Q_n is one of the inputs to NOR gate 2. If $S = 1$, the NOT of $0 + 1 = 0$; hence $Q_{n+1} = 1$ and $\overline{Q}_{n+1} = 0$. The condition $S = R = 1$ is not permitted.

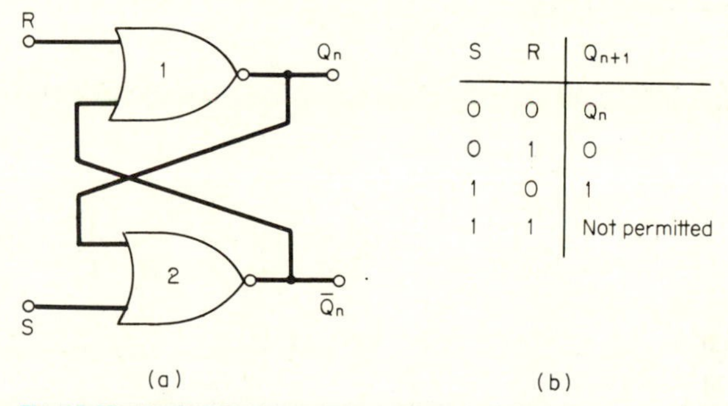

S	R	Q_{n+1}
0	0	Q_n
0	1	0
1	0	1
1	1	Not permitted

Fig. 15.15 An *R-S* flipflop using NOR gates: (*a*) Circuit. (*b*) Truth table. (See Example 15.2.)

CLOCKED R-S FLIPFLOP In many applications, such as shift registers, it is necessary that the operation of flipflops be synchronized by a master clock in the system. Such a flipflop is referred to as a clocked or *synchronous* flipflop. An example of a clocked *R-S* flipflop employing NAND gates and its truth table are illustrated in Fig. 15.16. The logic symbol for a clocked *R-S* flipflop is shown in Fig. 15.17.

NAND gates 1 and 2 are referred to as *steering gates*. A clock pulse *C* is one of the inputs to each steering gate. NAND gates 3 and 4 constitute the *R-S* flipflop of Fig.

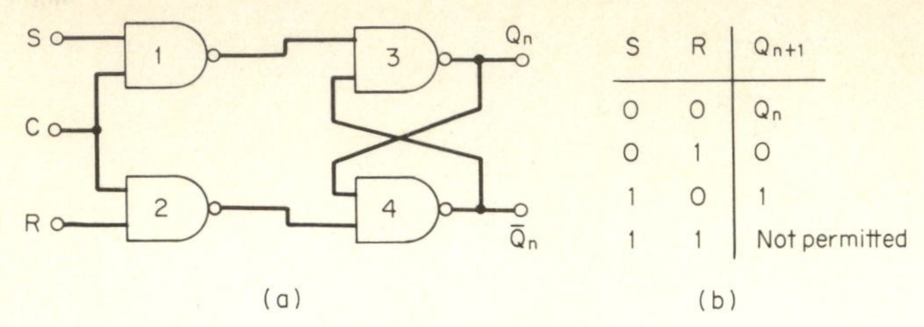

S	R	Q_{n+1}
0	0	Q_n
0	1	0
1	0	1
1	1	Not permitted

Fig. 15.16 A clocked *R-S* flipflop using NAND gates: (*a*) Circuit. (*b*) Truth table.

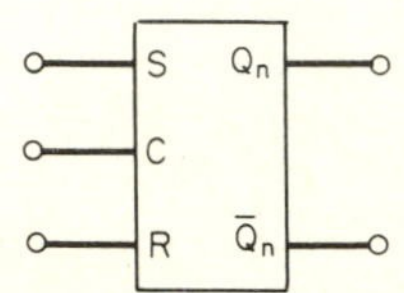

Fig. 15.17 Logic symbol for a clocked *R-S* flipflop.

15.13*a*. If $S = R = 0$, the flipflop remains in its original state. The condition $S = R = 1$ is not permitted.

Assume that $Q_n = 0$ ($\bar{Q}_n = 1$). A trigger at terminal *S* and a clock pulse to gate 1 result in a zero at its output. At the input of gate 3, therefore, there is a 0 from gate 1, and a 1 owing to the output of gate 4. The output of gate 3 is now a logic 1 ($Q_{n+1} = 1$). The flipflop has changed its state.

To return the flipflop to its original state ($Q_n = 0$), a trigger is applied to terminal *R* and a clock pulse to the other terminal of gate 2. The inputs to gate 4 are now a 0 from gate 2 and a 1 owing to the output of gate 3. Hence, $Q_{n+1} = 0$ and $\bar{Q}_{n+1} = 1$.

example 15.3 Show how a clocked *R-S* flipflop may be implemented using NOR and AND gates. Draw a truth table for the flipflop.

solution The circuit is given in Fig. 15.18; the truth table is identical with that of Fig. 15.16*b*.

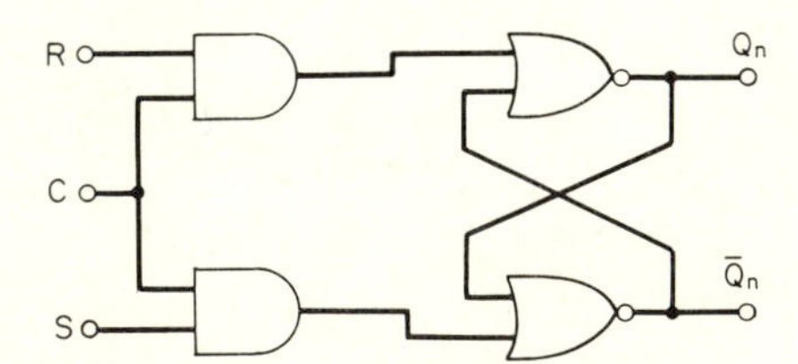

Fig. 15.18 A clocked *R-S* flipflop employing NOR and AND gates. (See Example 15.3.)

J-K FLIPFLOP The circuit of a clocked *J-K* flipflop and its truth table are provided in Fig. 15.19. The input terminals, in addition to the *C* terminal, are denoted by *J* and *K*. The basic circuit is the clocked *R-S* flipflop considered in the preceding section. The outputs of AND gates 1 and 2 are connected to the *S* and *R* terminals, respectively.

The first three entries in the truth table of Fig. 15.19*b* are identical with the clocked *R-S* flipflop. Although $S = R = 1$ is not permitted in the *R-S* flipflop, $J = K = 1$ is allowed in the *J-K* flipflop. An examination of the truth table reveals that for this input condition the output is complemented ($\bar{Q}_n$). That is, if $Q_n = 1$, $Q_{n+1} = 0$ and if $Q_n = 0$, $Q_{n+1} = 1$. The logic symbol for the *J-K* flipflop is given in Fig. 15.20.

example 15.4 Show how a *J-K* flipflop may be implemented using NAND gates.

solution This is accomplished by replacing the two-input NAND gates (1 and 2) of Fig. 15.16*a* by three-input NAND gates, as illustrated in Fig. 15.21.

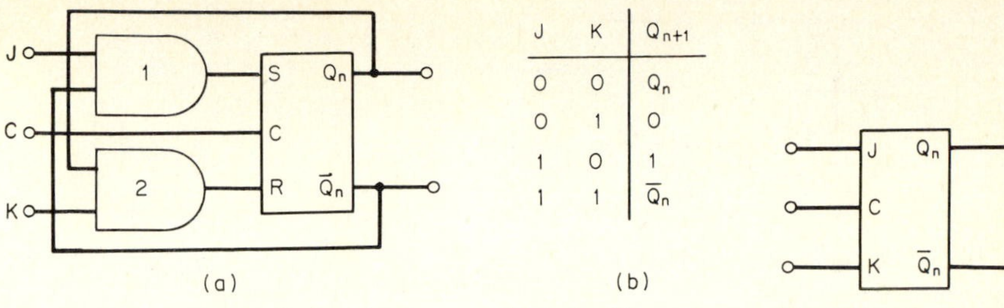

J	K	Q_{n+1}
0	0	Q_n
0	1	0
1	0	1
1	1	$\bar{Q}_n$

Fig. 15.19 The *J-K* flipflop: (*a*) Circuit. (*b*) Truth table.

Fig. 15.20 Logic symbol for a *J-K* flipflop.

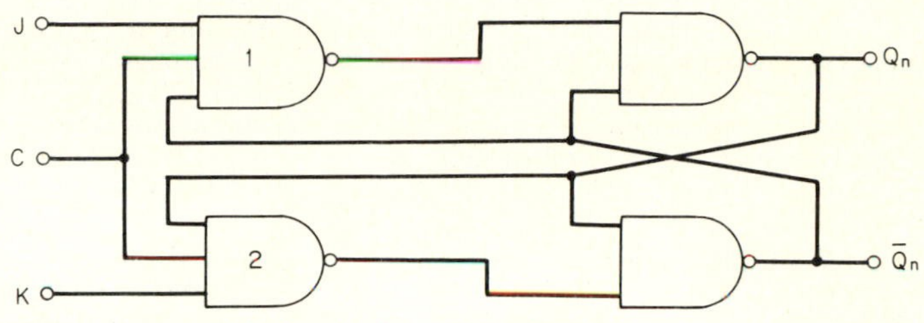

Fig. 15.21 A *J-K* flipflop implemented with NAND gates. (See Example 15.4.)

T FLIPFLOP This flipflop changes its state with each application of a trigger to a single input terminal *T*. Also referred to as a *toggle*, the *T* flipflop is used often in counters. An example of a clocked *T* flipflop implemented by using a *J-K* flipflop, and its truth table, are provided in Fig. 15.22. It is seen that the *T* input terminal is formed by joining together the *J* and *K* terminals. The logic symbol for the *T* flipflop is shown in Fig. 15.23.

example 15.5 Show how a *T* flipflop may be realized using AND gates and an *R-S* flipflop.

solution The circuit is given in Fig. 15.24.

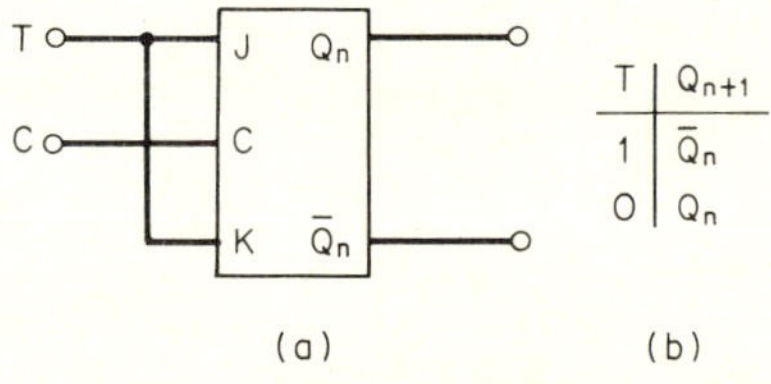

T	Q_{n+1}
1	$\bar{Q}_n$
0	Q_n

Fig. 15.22 A *T* (toggle) flipflop implemented with a *J-K* flipflop: (*a*) Circuit. (*b*) Truth table.

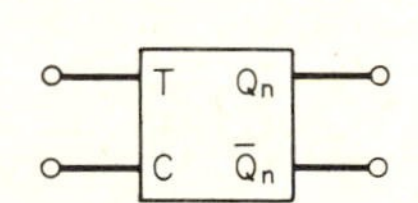

Fig. 15.23 Logic symbol for a *T* flipflop.

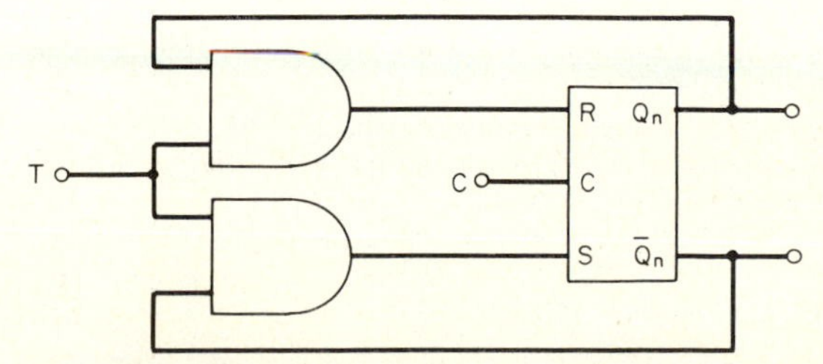

Fig. 15.24 A *T* flipflop realized by using AND gates and an *R-S* flipflop. (See Example 15.5.)

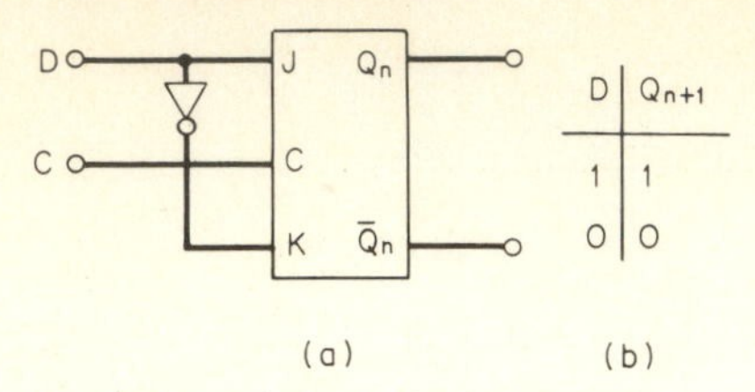

Fig. 15.25 A delay (*D*) flipflop implemented with a *J-K* flipflop and an INVERTER: (*a*) Circuit. (*b*) Truth table.

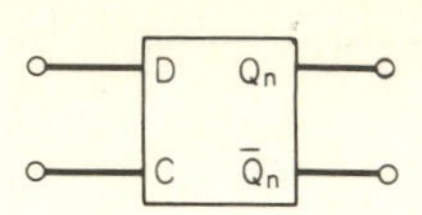

Fig. 15.26 Logic symbol for a *D* flipflop.

D FLIPFLOP An example of a clocked *D*, or *delay*, flipflop using a *J-K* flipflop and an INVERTER is given in Fig. 15.25*a*. Its truth table is provided in Fig. 15.25*b*. An *R-S* flipflop may be used instead of the *J-K* type. In this case the input to the INVERTER is connected to the *S* terminal and the output to the *R* terminal. The *D* terminal is also connected to *S*. The logic symbol for the *D* flipflop is shown in Fig. 15.26.

The *D* flipflop is used as a one-bit delay. When $D = 1$, $J = 1$ and, because of the INVERTER, $K = 0$. Referring to the truth table of Fig. 15.19*b* for the *J-K* flipflop, $Q_{n+1} = 1$. Also, if $D = 0$, $J = 0$, and $K = 1$; hence, from the truth table, $Q_{n+1} = 0$. Output Q_{n+1} therefore equals *D* delayed by one clock pulse, or bit.

CLEAR AND PRESET It is often necessary to *clear* a flipflop to its zero ($Q = 0$) or *preset* it to its one ($Q = 1$) state. An example of how this is realized is illustrated in Fig. 15.27. All four NAND gates have three inputs each. One input to gate 3 is the *preset*, Pr, input; and Cl to gate 4 is the *clear* input. The logic symbol for a clear-preset *J-K* flipflop is illustrated in Fig. 15.28.

In either clearing or presetting a flipflop, the clock pulse is at logic zero ($C = 0$). Assume that the flipflop is to be preset to one ($Q = 1$). Inputs $\text{Pr} = 0$, $\text{Cl} = 1$ and, as mentioned previously, $C = 0$. Because $\text{Pr} = 0$, even if the other two inputs to gate 3 are

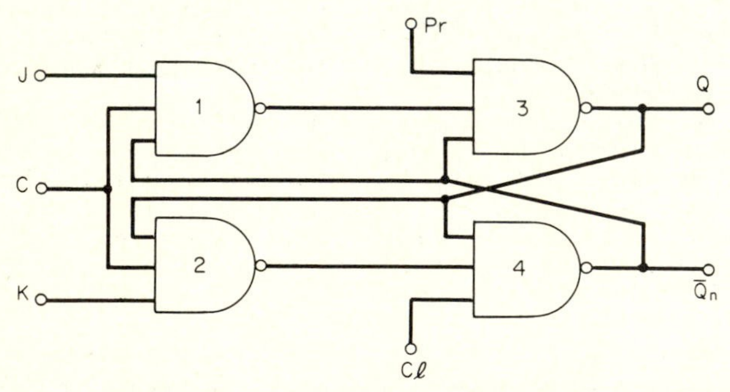

Fig. 15.27 Adding clear (Cl) and preset (Pr) inputs to a *J-K* flipflop.

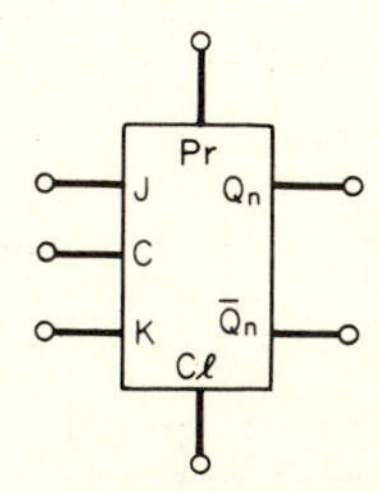

Fig. 15.28 Logic symbol for a clear-preset *J-K* flipflop.

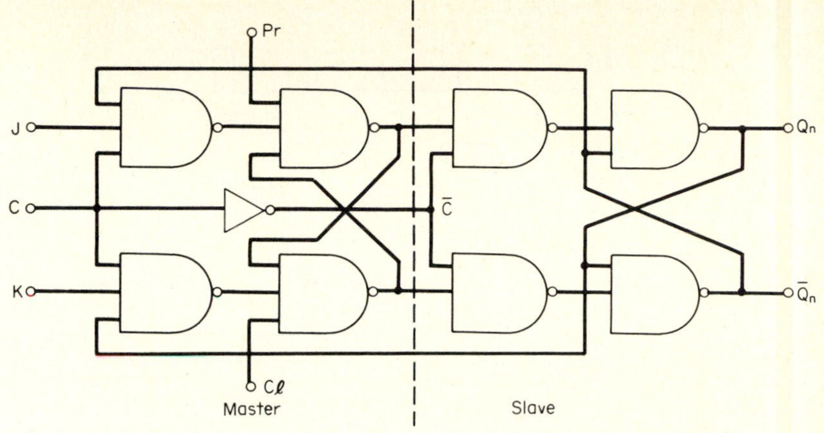

Fig. 15.29 A master-slave flipflop. The master is a *J-K*, and the slave an *R-S* flipflop.

at logic 1, the output $Q = 1$. Why? Because the NOT of 0 *and* 1 *and* 1 equals the NOT of zero which is a logic 1.

If the flipflop is cleared, $Q = 0$. For this condition, Cl = 0, Pr = 1, and $C = 0$. Because Cl = 0, the output of gate 4 is a logic 1; hence $\bar{Q} = 1$ and $Q = 0$. Once a flipflop has been cleared or preset, both the Cl and Pr inputs are maintained at a logic 1 before the next input bit arrives at the *J* or *K* terminal.

MASTER-SLAVE FLIPFLOP If the length of a trigger pulse is greater than the propagation delay of a gate, it may not be possible to change its state reliably. This problem is referred to as a *race condition*. Because propagation delays are generally very small (see Table 15.1), there is a reasonable chance of a race condition existing in digital integrated circuits. One technique for avoiding this is to use a *master-slave* flipflop.

An example of a master-slave flipflop is shown in Fig. 15.29. The first flipflop is called the *master* and the second the *slave*. The master is a *J-K* and the slave an *R-S* flipflop. An INVERTER is connected between the clock terminals of the *J-K* and *R-S* flipflops. In operation, Pr = Cl = 1.

When a clock pulse is present, $C = 1$ and, because of the INVERTER, $\bar{C} = 0$. The master is operative (enabled), and the slave is inoperative (inhibited). Consequently, during the clock pulse, the *J-K* flipflop can change state, but not the *R-S* flipflop.

In the interval when the clock pulse returns to zero, $C = 0$ and $\bar{C} = 1$. The *R-S* flipflop is now enabled, and the *J-K* is inhibited. During this interval, when the clock pulse is zero, the content of the *J-K* flipflop is transferred to the slave. Output *Q* corresponds to the output of the *J-K* flipflop.

15.7 COUNTERS

Flipflops connected to count pulses are referred to as counters. Counters have many applications in digital systems. They are used, for example, to count clock pulses (CP) in controlling the sequence of operations in a digital computer.

A basic counter, referred to as a *ripple counter*, is illustrated in Fig. 15.30. In this

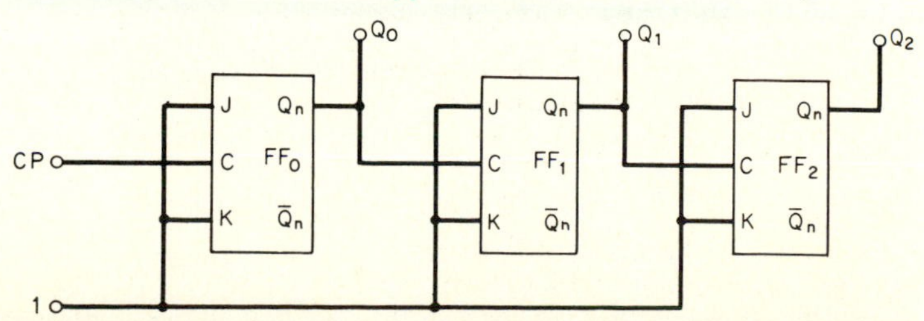

Fig. 15.30 An example of a three-stage ripple counter employing *T* flipflops.

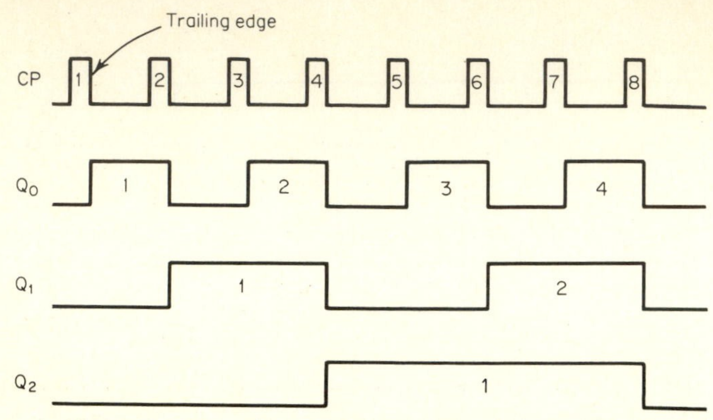

Fig. 15.31 Timing diagram for the ripple counter of Fig. 15.30.

example, a three-stage ripple counter implemented with T flipflops is shown. The J and K terminals of each flipflop are connected and returned to a logic 1. This connection converts the J-K to a T flipflop (see Fig. 15.22a). Because the J-K terminals are returned to a logic 1, output Q of each flipflop is at logic zero. The outputs of FF_0, FF_1, and FF_2 are designated as Q_0, Q_1, and Q_2, respectively. Output Q_0 is the LSB and Q_2 the MSB.

Assume that each flipflop changes its state on the falling (trailing), or *negative-going edge*, of a pulse. A timing diagram for the ripple counter of Fig. 15.30 is given in Fig. 15.31. On the falling edge of CP_1, Q_0 goes to a logic 1. Because each flipflop changes its state on the falling edge, FF_1 and FF_2 are unaffected; that is, $Q_1 = Q_2 = 0$. Thus, after one clock pulse, we have a binary count of 1:

Q_2	Q_1	Q_0
0	0	1

At the end of CP_2, Q_0 in going to zero triggers Q_1; Q_2 remains unchanged. Hence, we now have a binary count of 2:

Q_2	Q_1	Q_0
0	1	0

Proceeding in this manner, at the end of CP_7, we have a binary count of 7:

Q_2	Q_1	Q_0
1	1	1

At the end of CP_8, the counter resets itself: that is, $Q_0 = Q_1 = Q_2 = 0$. For a four-stage counter, the counter resets at the end of the 16th pulse, and so on.

It is interesting to note from the timing diagram that the output of FF_0 changes its state once for every two clock pulses; the output of FF_1 changes its state once for every four clock pulses, and so on. A ripple counter, therefore, may be used as a *divide-by* 2, 4, 8, etc., circuit.

OTHER TYPES OF COUNTERS The ripple counter of Fig. 15.30 is an example of an *asynchronous counter*. In an asynchronous counter, the individual flipflops are not clocked simultaneously. In a *synchronous counter*, all flipflops are clocked and change their states simultaneously. Definitions of other counters include:

Up counter An up, or *forward*, counter adds each input pulse to the count. The ripple counter of Fig. 15.30 is an example of an up counter.

Down counter A down, or *reverse*, counter subtracts 1 from a preset number during each input pulse.

Up-down counter An up-down, or *bidirectional*, counter can operate as either an up or a down counter.

Modulo counter A counter that counts other than binary multiples is referred to as a

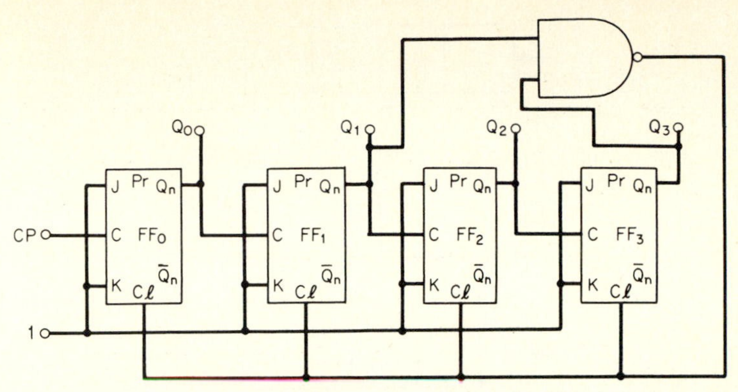

Fig. 15.32 An example of a decade counter using *J-K* flipflops. (See Example 15.6.)

modulo counter. For example, a counter that counts 1, 2, 3, 4 and resets itself to 0 on the 5th pulse is a modulo-5 counter. A *decade counter* that counts 1, . . . , 9 and resets itself to 0 on the 10th pulse is a modulo-10 counter.

example 15.6 Using *J-K* flipflops, design a decade counter.

solution A decade (modulo-10) counter employing *J-K* flipflops is illustrated in Fig. 15.32. Four flipflops are required. If three flipflops are used, a maximum count of only 7 (111) is possible, which is inadequate for a decade counter.

On the 10th input pulse, the decade counter must reset to zero. Decimal 10 equals binary 1010. Therefore, on the 10th pulse, $Q_3 = 1$ (MSB), $Q_2 = 0$, $Q_1 = 1$, and $Q_0 = 0$ (LSB). Outputs Q_3 and Q_1 are fed to the two-input NAND gate. When, on the count of 10, $Q_1 = Q_3 = 1$, the output of the NAND gate equals zero (the NOT of 1 *and* 1 = 0). The zero is returned to the clear (Cl) inputs of each flipflop, thereby resetting the counter to zero.

15.8 SHIFT REGISTERS

A shift register may be regarded as a component for storing a binary number or an instruction. For example, if a binary number or instruction has *n* bits, it is referred to as an *n*-bit word. An *n*-bit shift register consists of *n*-cascaded flipflops, indicated symbolically in Fig. 15.33. Flipflop FF_0 contains the least significant bit (LSB) and FF_{n-1} the most significant bit (MSB) of the word stored in the register.

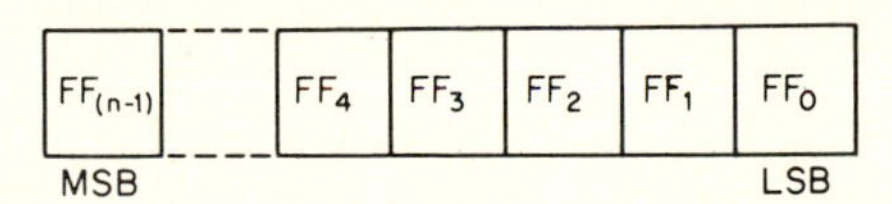

Fig. 15.33 A symbolic representation of an *n*-bit shift register.

There are a number of varieties of shift registers, as illustrated symbolically in Fig. 15.34. For simplicity, four-bit registers are shown.

Serial register (Fig. 15.34*a*) In this register the word is stored and read out serially, that is, in sequence. Because the word is shifted to the *right*, it is referred to as a *right-shift* register. A register that shifts to the *left* is called a *left-shift* register.

Serial-in, parallel-out register (Fig. 15.34*b*) In this register, the word is stored serially and read out in parallel. The serial-in, parallel-out register is also referred to as a *serial-to-parallel converter*.

Parallel-in, serial-out register (Fig. 15.34*c*) In this register the word is stored in parallel (each bit being stored simultaneously) and read out serially. The parallel-in, serial-out register is also referred to as a *parallel-to-serial converter*.

Parallel-in, parallel-out register (Fig. 15.34*d*) In this register the word is stored and read out in parallel.

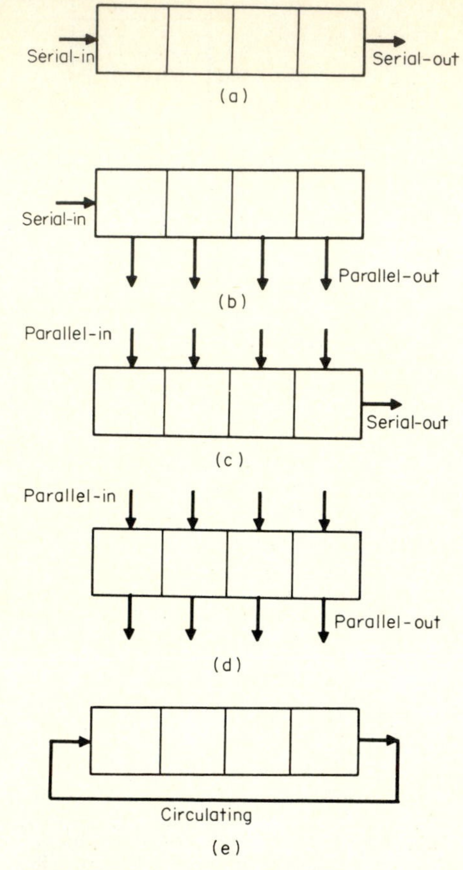

Fig. 15.34 Different types of shift registers. For simplicity, four-bit registers are illustrated: (*a*) Serial. (*b*) Serial-in, parallel-out (serial-to-parallel converter). (*c*) Parallel-in, serial-out (parallel-to-serial converter). (*d*) Parallel-in, parallel-out. (*e*) Circulating (dynamic shift register; shift-right read-only memory).

Circulating register (Fig. 15.34*e*) In this register the word circulates continuously. The circulating register is also referred to as a *dynamic shift register* and a *shift-right read-only memory*.

An example of a four-bit serial shift register employing *J-K* flipflops is illustrated in Fig. 15.35. During a clock pulse, each flipflop assumes the state of its preceding flip-

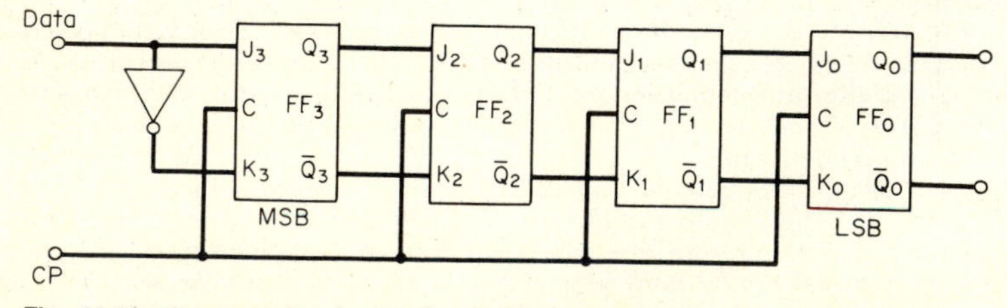

Fig. 15.35 An example of a four-bit serial shift register employing *J-K* flipflops.

flop. Assume that the word to be stored in the register is 1011. The sequence of operations is illustrated in Table 15.2.

Before CP_1, the state of each flipflop is at a logic zero ($Q_3 = Q_2 = Q_1 = Q_0 = 0$). At the termination of CP_1, the LSB, 1, is stored in FF_3; hence, $Q_3 = 1$ and $Q_2 = Q_1 = Q_0 = 0$. At the end of CP_2, the LSB has been shifted to FF_2 and the next bit, 1, is shifted to FF_3; hence, $Q_3 = Q_2 = 1$ and $Q_1 = Q_0 = 0$.

TABLE 15.2 Storing the Word 1011 in a Serial Shift Register

		Flipflop status			
CP	Bit	Q_3	Q_2	Q_1	Q_0
		0	0	0	0
1	1	1	0	0	0
2	1	1	1	0	0
3	0	0	1	1	0
4	1	1	0	1	1

Upon the termination of CP_3, the contents of FF_3 and FF_2 are shifted to FF_1 and FF_0, respectively; now, $Q_3 = 0$, $Q_2 = Q_1 = 1$, and $Q_0 = 0$. Finally, at the end of CP_4, the word 1011 is stored in the register with $Q_3 = 1$, $Q_2 = 0$, and $Q_1 = Q_0 = 1$.

To read out the word, four clock pulses are applied to the register. At the end of the fourth clock pulse, $Q_3 = Q_2 = Q_1 = Q_0 = 0$. Because the stored word is destroyed in reading out the shift register, the process is called *destructive readout* (DRO).

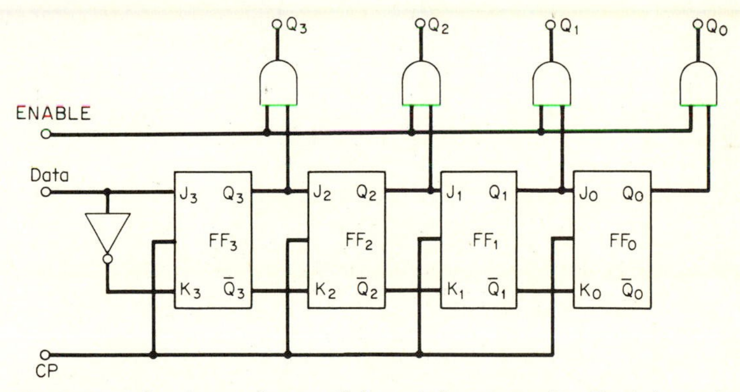

Fig. 15.36 A four-bit serial-in, parallel-out shift register. (See Example 15.7.)

example 15.7 Show how the serial shift register of Fig. 15.35 may be converted to a serial-in, parallel-out shift register (serial-to-parallel converter).

solution The circuit of a serial-in, parallel-out shift register is given in Fig. 15.36. Four AND gates are connected to the circuit. One input to each AND gate is connected to the output of a flipflop. The second input is connected to an *enable* signal. When the enable signal equals a logic 1, the content of each flipflop in the register is read simultaneously. Because the word in the shift register is not destroyed in reading out, the process is called *nondestructive readout* (NDR).

15.9 SEMICONDUCTOR MEMORIES

One important building block found in digital computers and systems is the memory. A memory is required for the storage of instructions (computer program), data, and results obtained in the processing of data. Whereas in the past ferrite cores were widely used, semiconductor memories have replaced cores in most digital systems. In a semiconductor memory, flipflops are used to store a 0 or a 1 bit. The transistors in the flipflop may be bipolar (BJT) or field-effect (MOSFET).

Memories are organized to store W words, each word being B bits in length. The storage capacity of a memory, therefore, is $W \times B$ bits. A typical capacity of a semiconductor memory chip is 4096 bits (1024 words × 4 bits/word). A group of bits is often referred to as a *byte*. Generally a byte is taken to be equal to 8 bits. Most semiconductor memories exhibit nondestructive readout.

TYPES OF MEMORIES Semiconductor memories may be categorized as:

1. Read-only memory (ROM).
2. Random-access memory (RAM).

In the ROM, the data are written in once and thereafter cannot be changed. This type of memory, for example, may be used to store a computer program. Information stored in the RAM can be readily changed in the processing of data.

A variation of the ROM is the *programmable* ROM (PROM). In the PROM, each memory element is in series with an aluminum or Nichrome strip that acts as a fuse link. The user can open the links by passing a specified current through the element. In this manner elements may be removed to represent a 0 or a 1 bit.

Memories are also categorized as being either *static* or *dynamic*. A static memory requires no clock pulses; a dynamic memory needs clock pulses. Generally, dynamic memories are less costly and consume less power than static memories.

ACCESS AND CYCLE TIMES Two important terms used in characterizing memories are *access* and *cycle time*. Access time is the time needed to read out a word from the memory. Cycle time is equal to one over the rate at which a word may be selected for reading or writing. Typical access and cycle times are 400 and 600 ns, respectively.

ROM An example of a ROM using p-channel depletion-type MOSFETs is illustrated in Fig. 15.37. For simplicity, only three-bit (B_0, B_1, B_2) and three-word (W_0, W_1,

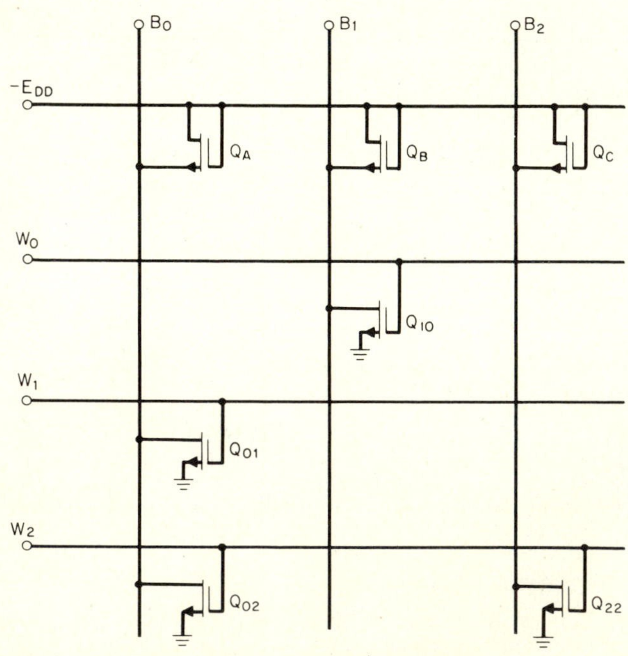

Fig. 15.37 An example of a ROM employing p-channel depletion-type MOSFETs. For simplicity, only three bit and word lines are shown.

W_2) lines are shown. Typically, a semiconductor ROM has 1024 word lines, each word being four bits in length.

Transistors Q_A, Q_B, and Q_C have their drains connected to the gate terminals. These transistors serve as the load resistors for the memory transistors. For example, Q_A is the load transistor for Q_{01} and Q_{02}, Q_B is the load resistor for Q_{10}, and Q_C is the load resistor for Q_{22}. Each drain of memory transistors Q_{01}, Q_{12}, etc., is connected to a bit line; their gates are connected to a word line. The presence of a transistor, such as Q_{10}, indicates a stored bit of data. An absence of a transistor indicates no stored bit of data.

For example, if word W_2 is to be read, a negative voltage is applied to word line W_2. Because the transistors are p-channel devices, Q_{02} and Q_{22} are turned on. Bit lines B_0 and B_2, therefore, read approximately 0 volts. Where a transistor is absent, a negative voltage is read. If a conducting transistor represents a 1 and an absent transistor a 0, the binary number read is 101.

RAM An example of a RAM cell using bipolar junction transistors is illustrated in Fig. 15.38. (Random-access memories using MOSFETs are also available.) Transistors Q_1 and Q_2 have multiple emitters, similar to that used in T^2L. The circuit is a bistable MV in which the collector of one transistor is directly coupled to the base of the other transistor.

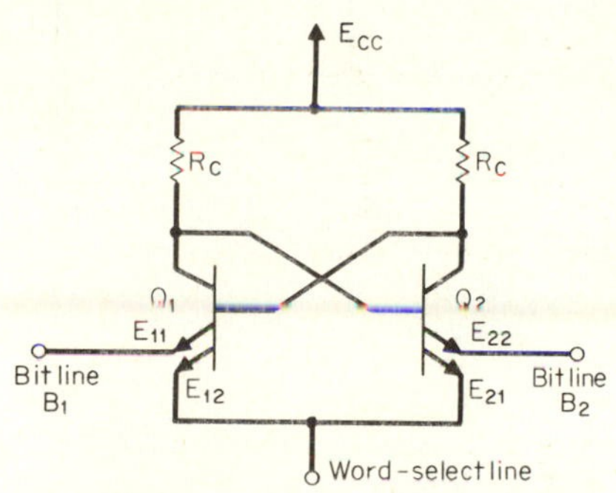

Fig. 15.38 A typical memory cell using bipolar junction transistors found in random-access memories. In a four-bit, 1024-word RAM, for example, 4096 of these cells are used in the memory.

Bit lines B_1 and B_2 are connected to an emitter of each transistor. The *word-select line* is connected to the remaining emitters. Assume that when Q_1 is conducting (Q_2 is therefore OFF), a binary 1 is stored in the cell. If Q_2 is ON (Q_1 is now OFF), a binary 0 is stored.

In its quiescent state, the word-select line is returned to ground or a low voltage. If a binary 1 is stored in the memory cell, transistor Q_1 is ON and Q_2 is OFF. Current, therefore, flows through emitter E_{12}.

To read the contents of the cell an appropriate positive voltage is applied to the word-select line. Emitter E_{12} becomes reverse-biased with respect to the base of Q_1. Emitter E_{11}, however, is forward-biased owing to a small potential present at bit line B_1. As a result, emitter current flows in bit line B_1, indicating that a binary 1 was stored in the cell. Because Q_2 is OFF, no current flows in bit line B_2 when it is sensed.

To write or store data in the cell, the word-select line is raised to a suitable positive potential. If a small voltage is applied to bit line B_2, transistor Q_2 is turned ON and Q_1 is turned OFF. A binary 0 is thereby stored in the cell. If a large positive voltage is applied to bit line B_2, the reverse is true: Q_1 is turned ON and Q_2 is turned OFF. This corresponds to a binary 1.

15.10 DIGITAL-TO-ANALOG AND ANALOG-TO-DIGITAL CONVERTERS

In data processing systems, there is the need to convert a digital signal to its analog equivalent or an analog signal to its digital equivalent. To cite an example, consider the digital voltmeter (DVM). To read a voltage, which is an analog quantity, the voltage is converted to a digital quantity by an analog-to-digital (A/D) converter. If the digital voltage is to be converted to an analog voltage, a digital-to-analog (D/A) converter is required. In this section, both types of converters are examined.

D/A CONVERTER An example of a D/A converter that converts a three-bit binary quantity to an equivalent analog quantity is illustrated in Fig. 15.39. The basic circuit of the D/A converter is a summing amplifier using an op amp (see Chap. 13). Output voltage E_o is equal to

$$E_o = \frac{-R'}{R} E_1 - \frac{R'}{2R} E_2 - \frac{R'}{4R} E_3 \tag{15.1}$$

For simplicity, mechanical single-pole, double-throw (SPDT) switches are shown in Fig. 15.39. When SW_1 is in position A, it reads the most significant bit (MSB). Switch SW_3 in position A reads the least significant bit (LSB). Any switch in position B is returned to ground.

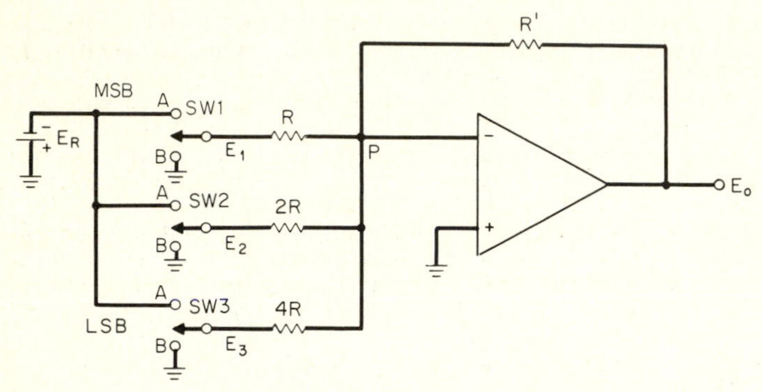

Fig. 15.39 An example of a weighted-resistor D/A converter. For simplicity, mechanical switches are shown.

Assume that the binary quantity N to be converted is $N = 001$. Let the feedback resistor R' be equal to $R/2.5$ and $E_R = -10$ V. For $N = 001$, switches 1 and 2 are returned to ground and switch 3 to $E_R = -10$ V. By Eq. (15.1),

$$E_{o(1)} = \frac{-R/2.5}{4R}(-10) = 1 \text{ V}$$

Suppose that $N = 011$ (decimal 3). Then, switches 2 and 3 are connected to $E_R = -10$ V, and SW_1 is returned to ground. By Eq. (15.1),

$$E_{o(3)} = \frac{-R/2.5}{2R}(-10) - \frac{-R/2.5}{4R}(-10) = 3 \text{ V}$$

Consider $N = 111$ (decimal 7). All three switches are now connected to $E_R = -10$ V, and

$$E_{o(7)} = \frac{-R/2.5}{R}(-10) - \frac{R/2.5}{2R}(-10) - \frac{R/2.5}{4R}(-10) = 7 \text{ V}$$

The preceding results indicate that the output voltage always corresponded to the binary input, and, indeed, a digital quantity was converted to an analog equivalent. For a four-bit quantity, another resistor equal to $8R$ in series with a switch is connected to point P. In general, for an n-bit quantity, n resistors are required. The resistance value for the least significant bit is equal to $2^{n-1} R$.

The D/A converter of Fig. 15.39 is referred to as a *weighted-resistor converter*, because its output voltage depends on the values of the input summing resistors. These resistances have different values (R, $2R$, and so on) and the resistor for the LSB can become excessively large ($2^{n-1} R$).

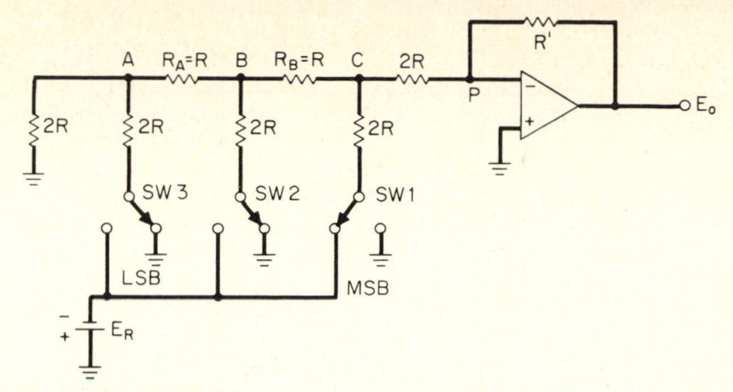

Fig. 15.40 A ladder-type *D/A* converter for a three-bit word. The values of resistors in this kind of converter are either *R* or 2*R*.

Owing to these facts, it is difficult for all the resistors to track with temperature. If good tracking is not realized, errors result when the converter is operated at different temperatures. To overcome these difficulties, the *ladder-type D/A* converter of Fig. 15.40 has been developed. Although it uses twice as many resistors as the weighted-resistor type *D/A*, the values of the resistors are only either *R* or 2*R*.

Assume that $N = 100$ (decimal 4). Switch SW_1 is connected to reference voltage E_R, and switches 2 and 3 are returned to ground. The resistance at node *A* with respect to ground is $2R \| 2R = R$. Addition of this resistance to $R_A = R$ yields 2*R*. The resistance at node *B* with respect to ground is therefore $2R \| 2R = R$. Addition of this resistance to $R_B = R$ yields 2*R*. Hence, looking to the left from node *C*, there is an equivalent resistance of 2*R* to ground, as shown in Fig. 15.41*a*. Because the input voltage to the op amp is approximately zero, point *C* also "sees" an equivalent resistance of 2*R* to its right.

Current *I* flowing in the 2*R* resistor connected to E_R is, therefore,

$$I = \frac{10}{2R + 2R \| 2R} = \frac{10}{3R}$$

The voltage at node *C* is current *I* multiplied by $2R \| 2R$:

$$\frac{10}{3R} \times 2R \| 2R = \frac{10}{3R} \times R = \frac{10}{3} \text{ V}$$

Using these results and letting $R' = 2.4R$, the model of Fig. 15.41*b* is drawn. By Eq. (15.1),

$$E_o = \frac{10}{3} \times \frac{2.4R}{2R} = 4 \text{ V}$$

which is the decimal equivalent of binary 100.

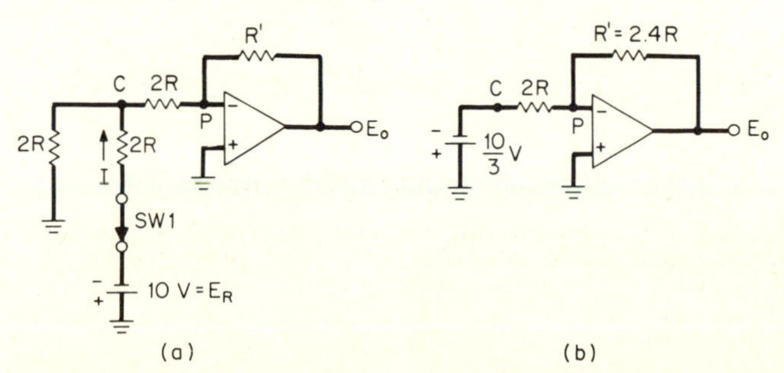

Fig. 15.41 A ladder-type *D/A* converter processing the binary number $N = 100$: (*a*) Resistance to the left or right of node *C* is equal to 2*R*. (*b*) Model for determining output voltage, E_o.

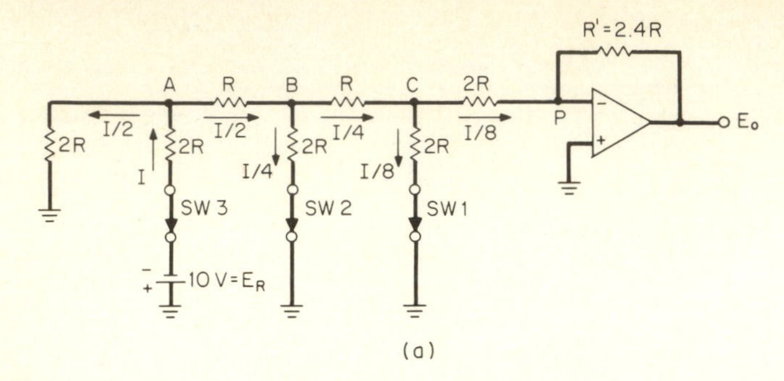

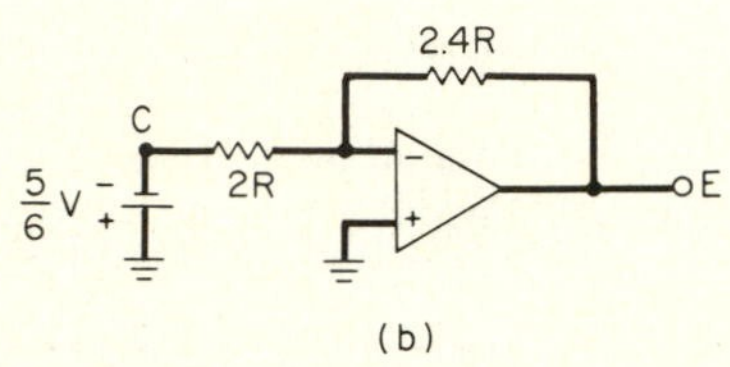

Fig. 15.42 Ladder-type *D/A* converter processing $N = 001$: (*a*) Circuit. (*b*) Model for determining E_o. (See Example 15.8.)

example 15.8 In the ladder-type *D/A* converter of Fig. 15.40, determine E_o when $N = 001$. Assume that $R' = 2.4R$.

solution For $N = 001$, switches 1 and 2 are returned to ground and SW_3 is connected to E_R. Referring to Fig. 15.42*a*, node *A* "sees" a resistance to its left of $2R$ and an equivalent resistance of $2R$ to its right. Because $2R \| 2R = R$, the current in SW_3 is $10/(2R + R) = 10/3R$ amperes.

At node *A*, half of current *I* ($I/2$) flows to the left, and the other half flows to the right. Reaching point *B*, the current is divided again by 2, and $I/4$ flows toward node *C*. At *C* it is split again, and $I/8$ flows through $2R$ in series with SW_1.

The voltage at node *C* with respect to ground is equal to the product of $I/8$ and $2R$:

$$\frac{I}{8} \times 2R = \frac{1}{8} \times \frac{(-10)}{3R} \times 2R = \frac{-5}{6} \text{ V}$$

This is illustrated in the model of Fig. 15.42*b*. By Eq. (15.1)

$$E_o = \left(\frac{-2.4R}{2R}\right)\left(\frac{-5}{6}\right) = 1 \text{ V}$$

Electronic switches are used in place of mechanical switches. To gain some idea how this is accomplished, consider the transistor switch of Fig. 15.43. As in the mechanical switch, point *A* goes to an input resistor and point *B* to ground.

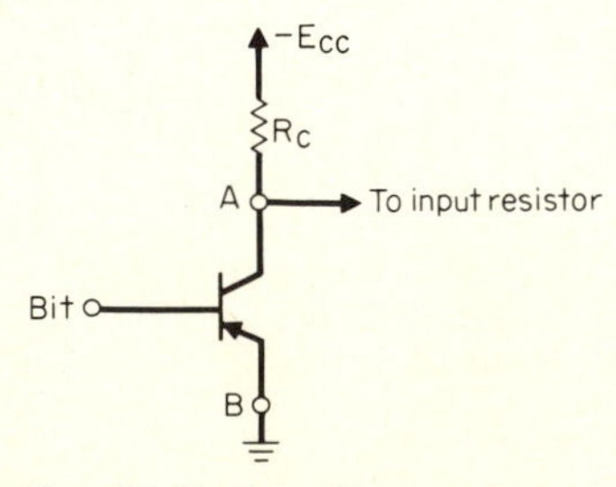

Fig. 15.43 An elementary pnp transistor switch used in place of mechanical switches in *D/A* converters.

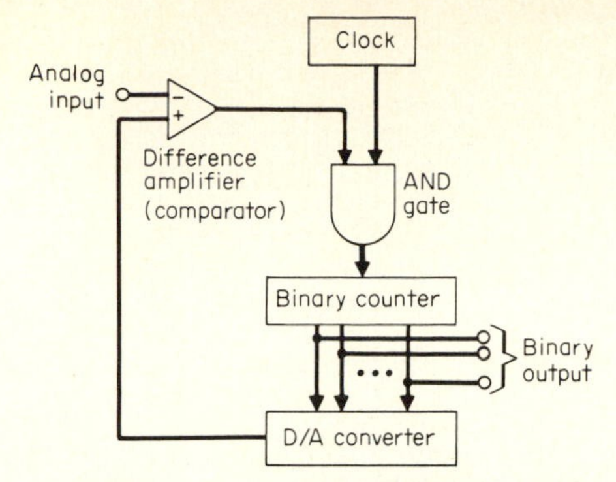

Fig. 15.44 An example of an A/D converter. The digital (binary) output is obtained from the binary counter.

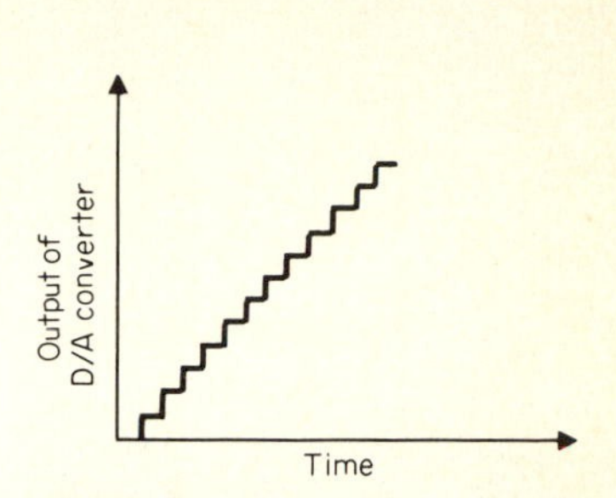

Fig. 15.45 A ramp (staircase) waveform at the output of the D/A converter of Fig. 15.44.

When a bit is present at the Bit terminal, assume that a positive voltage is impressed across the base of the transistor. Because the transistor is pnp, it is cut off, and the input resistor is connected to reference voltage $-E_{CC}$ through resistance R_C. If a bit is not present, a negative is impressed across the base. The transistor is turned ON, and the input resistor is returned to ground.

Because of collector resistor R_C, some error is introduced in the converter. Other configurations, containing a pair of bipolar or field-effect transistors, are used for minimizing the error.

A/D CONVERTER One example of a commonly used A/D converter is provided in Fig. 15.44. One input to the AND gate is derived from the clock, and the other input is obtained from the output of a difference amplifier. The difference amplifier compares the output of the D/A converter and the analog input signal. For this type of operation the difference amplifier is referred to as a *comparator.*

If the output of the D/A converter is less than the analog signal, the AND gate remains operative, and the binary counter continues counting clock pulses. When their difference is zero, the AND gate becomes inoperative, and the binary counter stops counting. The output of the counter is a binary number equal to the analog input signal.

The output of the D/A converter continuously increases by one bit as the binary counter counts. The resultant waveform, referred to as a *ramp*, or *staircase*, waveform is illustrated in Fig. 15.45. For this reason, the circuit of Fig. 15.44 is referred to as a ramp A/D converter. Although other methods are used in analog-to-digital conversion, the ramp type is fairly common.

15.11 THE MICROPROCESSOR

A microprocessor may be regarded as a general-purpose digital computer that is available on a few silicon chips. Currently, over a dozen IC manufacturers are producing more than two dozen different types of microprocessors. Although most microprocessors use MOSFETs, a few employ the BJT. Applications of the microprocessor include traffic control systems, numerically controlled machine tools, and the automatic testing of engines.

For illustrative purposes, the microprocessor chips manufactured by National Semiconductor Corporation are considered. National produces two basic microprocessor chips. These are the Register, Arithmetic, and Logic Unit (RALU); and the Control Read-Only Memory (CROM).

The RAUL performs the basic control and arithmetic functions in processing data. The CROM contains the instructions (program) that direct the execution of data by the RALU. In addition, National provides software, that is, programs for a given application of the microprocessor. By using a number of RALU and CROM chips, it is possible to realize a 16-bit word computer, referred to as a *microcomputer.*

Chapter 16

Power Supplies

16.1 INTRODUCTION

The function of an electronic power supply is to convert the available prime power source into a form required by a particular system. Typical prime power sources are rated at: 115 V, 60 Hz, single phase; 208 V, 60 Hz, three phase; 115 V, 400 Hz, single phase; and 12 or 28 V, dc.

The voltage requirements will vary depending on the type of system and also the portion of the system. Solid-state systems require considerably lower dc voltages than vacuum-tube systems. Some systems require regulated dc voltages, while others do not. In a television receiver, the r-f and video portions may require low to moderate values of dc voltages, while the picture tube may require high voltages in the order of 30 000 V. In computers, low dc voltages may be required for solid-state logic circuits, while intermediate values of dc voltages may be required for solid-state analog circuits.

Regardless of the type of electronic power supply involved, the basis of the supply is its rectifier system. We will begin our study of power supplies with a review of the basic types of rectifier systems.

16.2 RECTIFIER SYSTEMS

A block diagram of an electronic power supply is shown in Fig. 16.1. The ac power source is generally the power delivered by the power company, but in some applications other sources may be used. One example is in automotive electric systems that employ alternators. An alternator is basically an ac power source, and the ac power is converted to dc power by diodes which are mounted on the metal frame of the alternator. In this particular example the diodes are semiconductor types, but in other applications vacuum-tube diodes may be used.

Returning to the block diagram, the ac power is delivered to the primary of a power transformer. This transformer may be a step-up or a step-down type, depending on how much voltage is required by the load. Some electronic power-supply circuits operate without a transformer, and are called *transformerless power supplies.* Such supplies are less expensive to build, and this is considered to be an important advantage. However, the amount of voltage delivered to the load is a function of the voltage at the power source, and therefore, the use of such transformerless supplies is limited.

The output of the power-supply transformer is delivered to a section marked "diode rectifier circuit." The purpose of the rectifiers is to convert the ac voltage input to a pulsating dc voltage. Diodes function as rectifiers because they will allow current flow in only one direction—that is, from cathode to plate or anode. When an ac voltage is placed across the diode, it conducts during the half-cycles when the plate (or anode) is

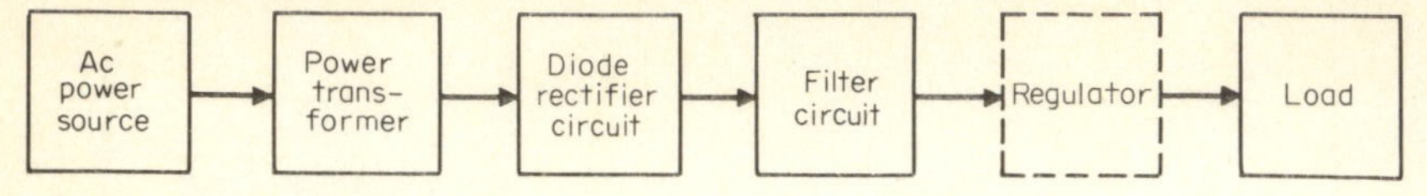

Fig. 16.1 Block diagram of an electronic power supply.

positive with respect to the cathode, but it will not conduct on the alternate half-cycles when the plate is negative with respect to the cathode.

The output of the diode rectifier circuit is a pulsating voltage. It cannot be used for operating transistor and vacuum-tube circuits because these circuits normally require an unvarying dc voltage. Therefore, a filter circuit is used to smooth out the voltage ripples. Usually filter circuits are *passive*, and are comprised of resistors and capacitors, or inductors and capacitors. (A passive filter is one that simply operates as a smoothing circuit.) There is always some loss of power in such a circuit, but this is not considered to be a serious disadvantage since the input stages of the power supply can be designed to deliver a sufficient amount of power for the filter loss, as well as for the load.

In the block diagram there is a dotted section marked *regulator* between the filter circuit and the load. Not all power supplies have regulators. Their purpose is to maintain a constant value of voltage across the load regardless of changes in the load resistance or ac input voltage. The regulator is usually an electronic circuit that accomplishes its function by sensing any change in the voltage across the load, and then feeding this change into the regulator circuit for use as a control to offset the change.

The last stage in the block diagram of Fig. 16.1 is the *load*. The load is the component or circuit that receives the dc power. In a radio or television set the load will be the vacuum tubes or transistors operating in the electronic circuitry. In other applications the load may be a dc motor, a battery that is being charged, or any other circuit or component requiring a dc source of power.

From the discussion of the block diagram, it can be seen that the term "electronic power supply" is somewhat misleading. This circuit does not *supply* the power, but rather, converts it from one form (ac) to another form (dc). However, the name is widely used in electronics. Transformers and filter circuits are covered in other chapters in this book. Our concern here will be primarily with various rectifier configurations, how the rectifier circuits operate, and the advantages and disadvantages of each circuit.

HALF-WAVE RECTIFIERS The simplest rectifier circuit is the one-diode configuration shown in Fig. 16.2. This is a half-wave rectifier, so named because the output voltage waveform for the circuit is only one-half of the ac input voltage waveform. It is an example of a transformerless power supply, and it is very popular in low-cost home-entertainment electronic systems such as table-model radios and record players.

The input voltage to the circuit is an ac waveform. On one-half cycle, which we have designated the *positive half-cycle*, the anode of the diode is made positive and electron current can flow through the diode. The path of electron current flow is shown by the arrows in Fig. 16.2*a*. It will be noted that the current is flowing upward through the load resistor R_L. This means that the voltage drop across R_L will be such that the negative polarity of the voltage is at the bottom and the positive polarity is at the top. The output waveform during the positive half-cycle is an exact reproduction of the positive half-cycle of input waveform.

During the next half-cycle, the input voltage is negative. (See Fig. 16.2*b*.) This causes a negative voltage to be applied to the anode of D_1. No electron current can flow through the diode during this half-cycle, and the output voltage will be zero, as shown by the waveform on the diagram. Figure 16.2*c* shows the waveform of the voltage across the load resistance.

The waveform of Fig. 16.2*c* is referred to as a *pulsating dc voltage*. It is a dc voltage because its polarity is always positive (or zero). It is pulsating because the amplitude changes from moment to moment. Such a waveform is not useful for operating tube or transistor circuits unless it is first filtered. The purpose of filtering is to take out the pulsations. Because the voltage rises from zero to its maximum value and back to zero, and then stays at zero for a certain period of time, makes this waveform difficult to filter.

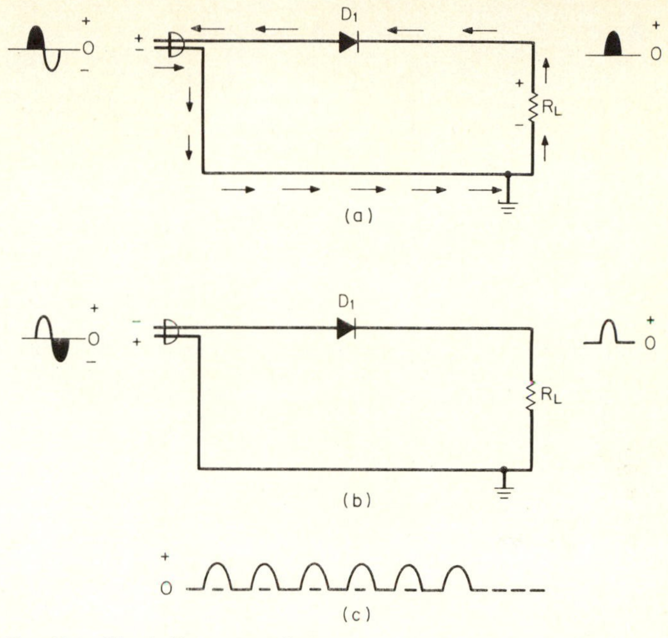

Fig. 16.2 The half-wave rectifier circuit with current flow and voltage waveforms: (*a*) Current flows during the positive half-cycle. (*b*) No current flows during the negative half-cycle. (*c*) Output voltage waveform of a half-wave rectifier circuit.

If this waveform is going to be converted to a steady dc voltage, then all the periods during which the voltage is zero must be filled in.

FILTERING The procedure for filtering the voltage usually involves charging a capacitor to the *peak* ac input voltage during a time when the diode is conducting and then having the capacitor discharge through the circuit in order to maintain the voltage across the load. To understand this, we will look at a simple filter circuit consisting only of a capacitor, as shown in Fig. 16.3. The only difference between this half-wave rectifier circuit and the one shown in the previous illustration is that a filter capacitor *C* has been added. This filter capacitor is an electrolytic type with a high capacitance value which enables it to store a considerable amount of energy during each half-cycle. The positive sign beside the capacitor symbol indicates that the capacitor is of the electrolytic type.

During the positive half-cycle of input voltage, the diode conducts as before. In this case, however, only a small part of the conducting current initially flows through the load, producing the output voltage waveform. The rest of the current flows into one plate of the capacitor and out of the other, thus charging the capacitor. This is shown in Fig. 16.3*a*. During the negative half-cycle, as shown in Fig. 16.3*b*, a negative input voltage is delivered to the plate of the diode, and it can no longer conduct. At this time the voltage across the capacitor causes current to flow through the load resistance. The capacitor discharge current is indicated by the arrow in Fig. 16.3*b*. This current prevents the voltage from dropping to zero, as it normally would without the presence of the capacitor.

The filter capacitor charges to the peak value of the ac input voltage. When the value of input voltage begins to decrease below the voltage across the capacitor, then the capacitor begins to discharge through the resistor. Figure 16.3*c* shows the filtered output waveform from the power supply in solid lines. The unfiltered output waveform (without the filter capacitor) is shown with dotted lines. Although the voltage drops

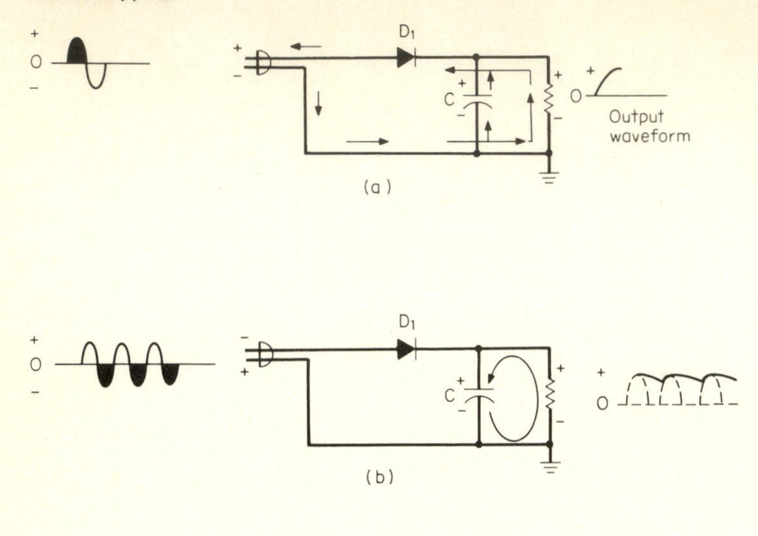

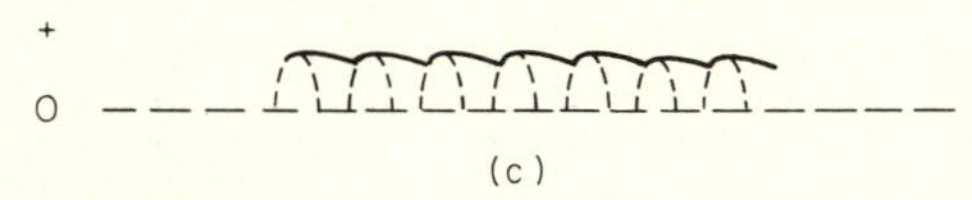

Fig. 16.3 The half-wave rectifier with a simple filter: (*a*) During the first positive half-cycle the capacitor charges. (*b*) During the next half-cycle the capacitor discharges. (*c*) The filtered output waveform.

somewhat during the negative half-cycles of input voltage, it does not drop all of the way to zero.

The operation of the simple capacitor filter shown in Fig. 16.3 is based on the fact that a capacitor can store energy. Basic theory tells us that a capacitor stores energy in the form of an *electrostatic field,* and an inductor (another device that can store energy) stores it in the form of an *electromagnetic field.* Passive filter circuits always rely on the ability of either a capacitor or an inductor (or both) to store energy and return the energy to the circuit at the proper time. Figure 16.4 shows a half-wave rectifier with a more elaborate filter. This circuit is popular with low-cost electronic circuits. The switch (SW) energizes or deenergizes the circuit. The pi filter, comprised of R_1, C_1, and C_2, smooths the dc pulsations from the half-wave rectifier. This circuit is used mainly for low-current applications, unless the capacitors are very large in value. In the latter case, the resistor can be kept to a reasonably low value, to minimize the voltage drop across it.

FULL-WAVE RECTIFIERS One of the most important disadvantages of the half-wave rectifier circuit is the difficulty in filtering its output waveform. The relatively

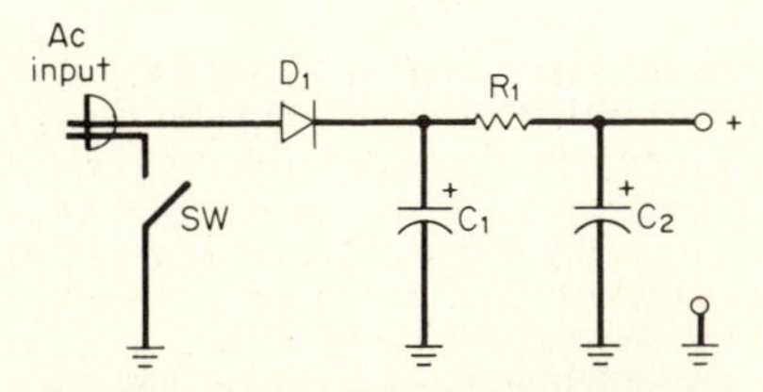

Fig. 16.4 Half-wave rectifier with a filter.

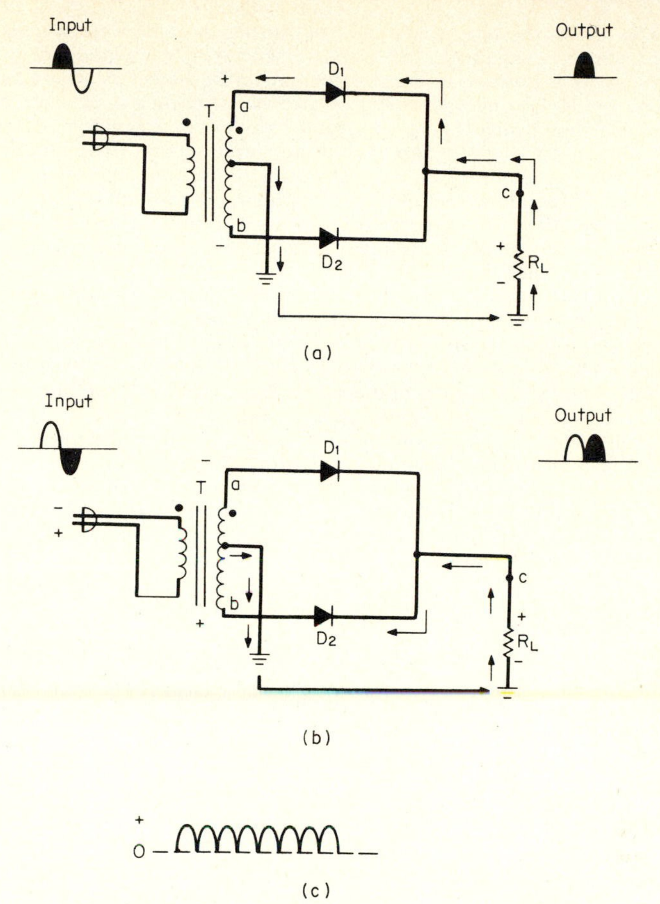

Fig. 16.5 The full-wave rectifier: (*a*) Full-wave rectifier showing current flow during one half-cycle. (*b*) Full-wave rectifier showing current flow during the next half-cycle. (*c*) Output waveform of the full-wave rectifier.

long period between positive voltage peaks makes it necessary for the filter circuit to supply a considerable amount of current flow through the load for a long period of time —that is, the discharge time is long compared to the charge time of the filter components. In spite of this disadvantage, the half-wave rectifier is a popular circuit because it is inexpensive and relatively trouble-free.

Figure 16.5 shows a full-wave rectifier circuit. This type of circuit *requires* a transformer T_1 with a center-tapped secondary winding. In addition to the need for the transformer, the circuit also requires the use of two diodes—shown as D_1 and D_2 in the circuit.

With an ac voltage across the primary of the transformer, the polarity of the voltage across the secondary will periodically reverse. The dot notation on the transformer is a standard way of indicating points of identical phase. When the voltage on the primary winding goes positive at the point where the dot is located, then at that instant the voltage at the secondary also goes positive at the point where the dot is located.

On one half-cycle, point *a* will be positive; and on the next half-cycle, point *a* will be negative. The voltage at point *b* will always have the opposite polarity of the voltage at point *a*. The center tap which is connected to the common, or *ground point*, will always be maintained at zero volts.

Figure 16.5*a* shows the operating condition during the half-cycle when point *a* is positive and point *b* is negative. The negative voltage on the plate of D_2 prevents it from conducting. The positive voltage on the plate of D_1, however, allows that tube to conduct. The conduction path is shown by arrows. If point *a* is positive, and if the center tap is zero, then it stands to reason that the center tap is negative with respect to point *a*. Starting at the negative terminal of the *voltage source* (which in this case is one-half the secondary winding) the electron current flows out of the center tap into the common circuit. It then flows out of the common circuit, through R_L, and to the junction of the two cathodes. The electron current cannot flow through D_2 because it is cut off with a negative voltage on its anode. Therefore, all the current flows through diode D_1. This completes the circuit.

During the first half-cycle, the current is seen to be flowing upward through the load resistor, making the voltage drop across it positive at point *c*. The output waveform shown on the illustration is seen to be a reproduction of the input waveform during this half-cycle. The amplitude of the output may be either smaller or larger than the input, depending on the turns ratio of the power transformer. If the turns ratio is 1:1, it means that the number of turns on the primary and on *one-half* of the secondary are equal. In such a case, and assuming there is no filter circuit, the output amplitude will be approximately equal to the amplitude to the input signal. It is necessary to say "approximately" because there is always some voltage drop across the diode itself. The actual value of this drop can be obtained by multiplying the current through the diode by the forward resistance of the diode. Although this voltage drop may be small enough to be neglected in most cases, it does, nevertheless, reduce the output amplitude to something less than the input voltage (measured across one-half of the transformer secondary winding). There is also a voltage drop across the diode in the half-wave rectifier circuit previously discussed.

In Fig. 16.5*b* the circuit is shown with the voltage polarity across the secondary reversed. In this case, the anode of D_1 is negative (with respect to ground) and the anode of D_2 is positive. Therefore, only D_2 can conduct during this half-cycle. Again, electron current leaves the center tap, flows into the common connection, and then through the load resistor R_L. The current flow is shown by arrows. It is important to note that the electron current is again flowing in the same direction as before, through R_L, making point *c* again positive with respect to ground. This means that a positive-going voltage will appear across the load for *both* half-cycles of input voltage.

Figure 16.5*c* shows the output waveform of a full-wave rectifier. Since there are no long periods during which zero voltage is generated across the load, this waveform is easier to filter than the half-wave rectifier output. That is one of the important advantages of the full-wave rectifier circuit.

TUBE RECTIFIERS Some full-wave rectifier power supplies are designed with diode tubes having filament emitters. Filament emitters are more efficient, in general,

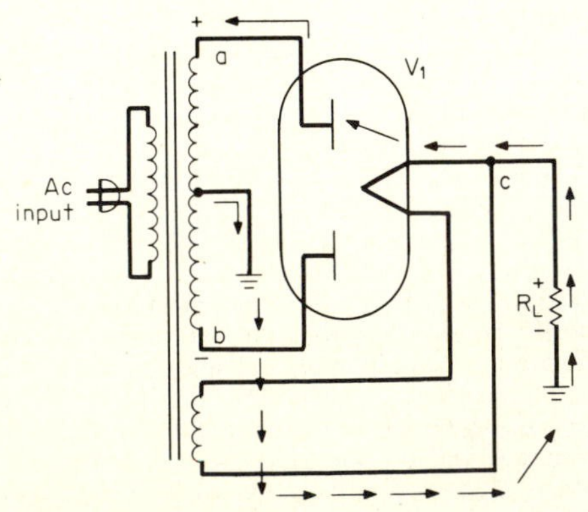

Fig. 16.6 A full-wave rectifier circuit using tubes with filament emitters.

than cathode-type emitters. Cathodes are used whenever it is necessary to isolate the cathode emitter from the heater circuit. Figure 16.6 shows a full-wave rectifier with the filament emitters. The filament voltage is obtained from a separate winding on the transformer. This is a common practice with full-wave rectifier circuits. In other rectifier circuits, there may be a separate filament transformer, or the filament may be in series with a number of other filaments which produce (when in series) a total voltage drop equal to the line voltage. When the latter method is used, only tubes with cathode emitters can be used, so that the ac filament circuit is isolated from the dc supply voltage. (The only purpose of the filament winding in the circuit of Fig. 16.6 is to provide current for heating the filament. It does not enter into the operation of a rectifier power supply in any other way.)

The circuit shows the current flow for one-half cycle, during which point *a* is positive and point *b* is negative. As before, only the plate with the positive voltage can attract electrons. Therefore, only that half of the tube conducts during this half-cycle. Electron current leaves the center tap, flows upward through the load resistor into the filament, and then from the filament to the plate of the upper diode. During the next half-cycle the upper diode will have a negative voltage on its plate, and the lower diode will have the positive voltage. Therefore, the current flow will be through the lower diode.

An important consideration here is the fact that the filament winding of the transformer will be above ground by the amount of voltage drop across R_L because of the common connection at point *c*. In some power supplies this may be 300 or 400 V (or more) above ground. In choosing a transformer for this type of circuit, it is important to determine that the transformer secondary can withstand the resulting voltage stress. If the filament winding is above ground by 400 V, and point *b* is negative by 400 V during one-half cycle, then the voltage difference between the filament winding and that half of the winding is 800 V. Although this is not an excessively high voltage, it does, nevertheless, put a voltage stress between windings within the transformer itself and the transformer core, which is grounded.

Although the output waveform of the full-wave rectifier is easier to filter, this circuit is not without disadvantages. The fact that the secondary winding of the transformer *must be center-tapped* means that only one-half of the total secondary voltage will appear across the load. As an example, in the circuit of Fig. 16.6 assume that the voltage between point *a* and point *b* is 1000 V. This would be the total secondary winding voltage. Under this condition the voltage across R_L would be only 500 V. (Again, we are neglecting the drop across the diode itself.)

Another disadvantage that must be taken into consideration is that two diodes are needed for the operation of the full-wave rectifier. This means an added expense and somewhat reduced circuit reliability.

BRIDGE RECTIFIERS A full-wave rectifier circuit that makes use of the *full* transformer secondary-winding voltage, and does not require the center tap, is shown in Fig. 16.7. It is called a bridge rectifier because of its similarity to the configuration of the Wheatstone bridge. Although a transformer is shown in the circuit, it may also be connected as a transformerless power supply.

On the first half-cycle, shown in Fig. 16.7*a*, point *a* becomes positive with respect to point *b*. If we consider the secondary winding of the transformer as the voltage source, we can trace the electron current flow (starting at point *b*) for this half-cycle. The arrows show the current path flowing from *b* through D_1, up through R_L, and through D_2 back to point *a* on the secondary winding. Diodes D_3 and D_4 are cut off during this half-cycle, because D_3 has a positive voltage on its cathode, and D_4 has a negative voltage on its anode. In order for a diode to conduct, it is necessary for the anode to be positive with respect to the cathode. If the cathode of a diode is made highly positive with respect to the anode, then this is the same as saying that the anode is negative with respect to the cathode. That is why the positive voltage on the cathode of D_3 prevents it from conducting.

The next half-cycle of operation is shown in Fig. 16.7*b*. The polarity of voltage across the secondary winding has now been reversed, making point *a* negative with respect to point *b*. Starting at the negative terminal of the source (which is now point *a* on the secondary winding), current flows through D_3, through R_L, through D_4, and back to point *b* on the secondary winding. Diode D_1 cannot conduct because of the positive

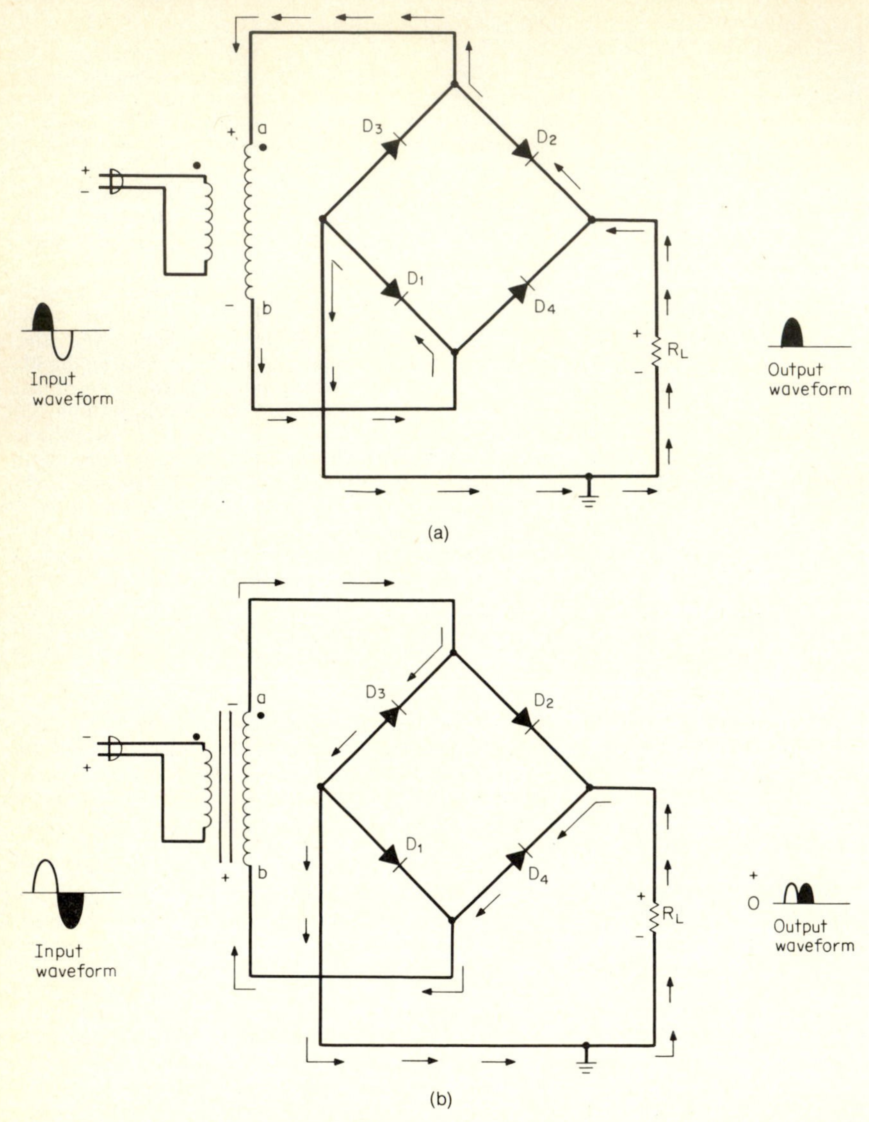

Fig. 16.7 A bridge rectifier circuit: (*a*) Current flow in the bridge rectifier circuit for the first half-cycle of operation. (*b*) Current flow during the second half-cycle.

voltage on its cathode, and D_2 cannot conduct because of the anode during this half-cycle.

The diodes in the bridge rectifier circuit may be thought of as switches which are operated by voltage polarities. During both half-cycles of operation, the current is seen to flow in the same direction through R_L. This means that the circuit produces a full-wave pattern like the one shown in Fig. 16.7*b*.

When semiconductor diodes are used for bridge rectifiers, they are often mounted in the same package for convenience.

The bridge rectifier circuit is found in many applications in the communications field and elsewhere. It is also used in some instrument circuits. Most meter move-

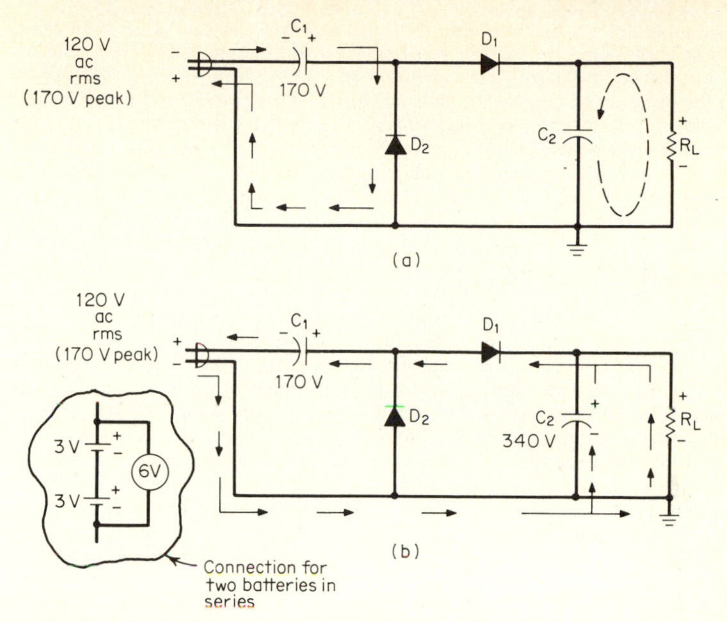

Fig. 16.8 The half-wave voltage doubler circuit: (*a*) This half-cycle charges C_1 to 170 V. (*b*) During this half-cycle, the 170 V across C_1 is in series with the 170-V input, making a total of 340 V; capacitor C_2 charges to 340 V.

ments on voltmeters respond only to a dc voltage. In order to measure an ac voltage, it is necessary to convert the alternating current into a direct current. A half-wave rectifier circuit *could* be used, but the bridge rectifier is generally employed because it is a full-wave rectifier. The input voltage—that is, the voltage being measured—is applied directly across the bridge at points *a* and *b*, and the meter movement is connected into the circuit at the position of R_L in the circuit of Fig. 16.7.

HALF-WAVE VOLTAGE DOUBLERS The half-wave voltage doubler circuit of Fig. 16.8 will provide an output voltage across R_L that is approximately twice the peak input ac voltage to the circuit. Thus we have a transformerless circuit which steps up the line voltage and converts it to a dc potential.

During one-half cycle of input power, the voltage polarity will be as shown in Fig. 16.8*a*. The positive voltage on the anode of D_2 causes it to conduct, and the conduction path is shown by the arrows. This charges C_1 to the peak voltage (170 V in this example). At the same time, diode D_1 cannot conduct because its anode is not positive with respect to its cathode.

The next half-cycle of operation is illustrated in Fig. 16.8*b*. This shows that the voltage across capacitor C_1 and the input voltage are in series. In other words, they are connected plus to minus as are the two batteries in series shown at the inset. The input voltage to the circuit during this half-cycle is equal to the voltage across C_1 (170 V) plus the peak voltage across the line, making a total of 340 V. The negative side of the line is connected to the anode of D_2 which prevents that diode from conducting during this half-cycle. The positive voltage on capacitor C_1 is applied to the anode of D_1, and therefore D_1 *can* conduct during this half-cycle. The conduction path is shown by the arrows.

It will be noted that the current in Fig. 16.8*b* divides: part of it flowing into capacitor C_2, and the remainder flowing through the load. Capacitor C_2 will charge to the full 340 V during this half-cycle (minus, of course, the drop across the diode, which can usually be neglected in this type of circuit).

On the next half-cycle, the condition of Fig. 16.8*a* again prevails. The input voltage recharges capacitor C_1 by its conduction through D_2. At the same time D_1 is cut off during this half-cycle. It will be remembered that capacitor C_2 was charged to the 340-V input during the half-cycle shown in Fig. 16.8*b*. Now capacitor C_2 will discharge through the load, as shown by the dotted arrow. This prevents the voltage across the load from dropping to zero, as it would without the filter capacitor.

The output voltage waveform is a filtered half-wave voltage like the one shown in Fig. 16.3*c*. If the value of load resistance in this circuit is small, then capacitor C_2 will discharge rapidly through it, causing the voltage across R_L to go to zero during some portion of the half-cycle when D_1 is not conducting. Therefore, this type of power supply cannot be used for circuits that require a high current flow—that is, circuits that load the supply. Both C_1 and C_2 are normally quite large in capacitance value, and are usually electrolytic capacitors. However, even with a large capacitance value for C_2, the power-supply circuit cannot sustain a heavy load current.

If R_L were a variable resistor, it would be found that the smaller the value of resistance, the lower the output voltage for the circuit. What is more important is that when the load current through R_L varies from moment to moment, the output voltage of the power supply also varies. An ideal power supply would be one that maintains a constant output voltage, regardless of the amount of load current being drawn. Such a power supply is not realizable in practice, but can be approached closely under certain conditions. The *regulation* of the power supply is a measure of how well it maintains its output voltage under variable load conditions. The disadvantage of the half-wave voltage doubler shown in Fig. 16.8 is that it has poor regulation. Therefore a half-wave voltage doubler is not generally used where the load varies widely.

FULL-WAVE VOLTAGE DOUBLERS The distinguishing feature of the half-wave voltage doubler just described is that it delivers power to the load only during one-half cycle of input alternating current. This, in part, accounts for the relatively poor regulation of the circuit. A full-wave doubler provides better regulation and is easier to filter.

A full-wave voltage doubler is shown in Fig. 16.9. On the first half-cycle, the input voltage will have a polarity as marked at the plug in Fig. 16.9*a*. A positive voltage, under this condition, is delivered to the cathode of D_1, and that diode is cut off. At the same time, the anode of D_2 receives a positive voltage, and that diode *will* conduct. The conduction path is shown by the arrows. Capacitor C_1 charges to the peak line voltage (170 V) during this period of time.

On the second half-cycle, the polarity of the voltage at the plug reverses—placing a negative voltage on the anode of D_2, and also a negative voltage on the cathode of D_1. (See Fig. 16.9*b*.) Under this condition, only D_1 will conduct. The conduction path is again shown by arrows. Capacitor C_2 charges to the peak line voltage of 170 V. The voltage across C_2 is added in series to the voltage across C_1 (which is obtained during the previous half-cycle). The two voltages, in series, combine to produce a 340-V output voltage. The load resistance is connected across the two series capacitors.

It might be argued that this is not really a full-wave rectifier system because the total voltage is not developed at the output capacitor circuit on *both* cycles of input. Instead, the voltage is delivered to one-half of the output circuit each time the input voltage reverses polarity. However, at least *some* voltage is delivered to the output capacitor circuit each half-cycle, and this makes it more of a full-wave rectifier than the one shown in Fig. 16.8.

Although the regulation of the full-wave voltage doubler circuit is somewhat improved over that of the half-wave doubler, the circuit is still unable to sustain a heavy load current without a serious drop in the output voltage.

VOLTAGE TRIPLERS AND OTHER VOLTAGE MULTIPLIERS In the voltage doubler circuits just discussed, the principle of operation is based on charging a capacitor in one-half cycle, and then adding the voltage across that capacitor to the circuit voltage (or to the voltage across another capacitor) during the next half-cycle. It should be obvious that any number of capacitors can be charged, and their voltages added. However, as the amount of voltage multiplication goes up, the amount of regulation of the power-supply circuit becomes poorer and poorer.

Figure 16.10 shows another example of a voltage multiplier. (Voltage multipliers are diode rectifier circuits that increase the input voltage. The *half-wave doubler*, and

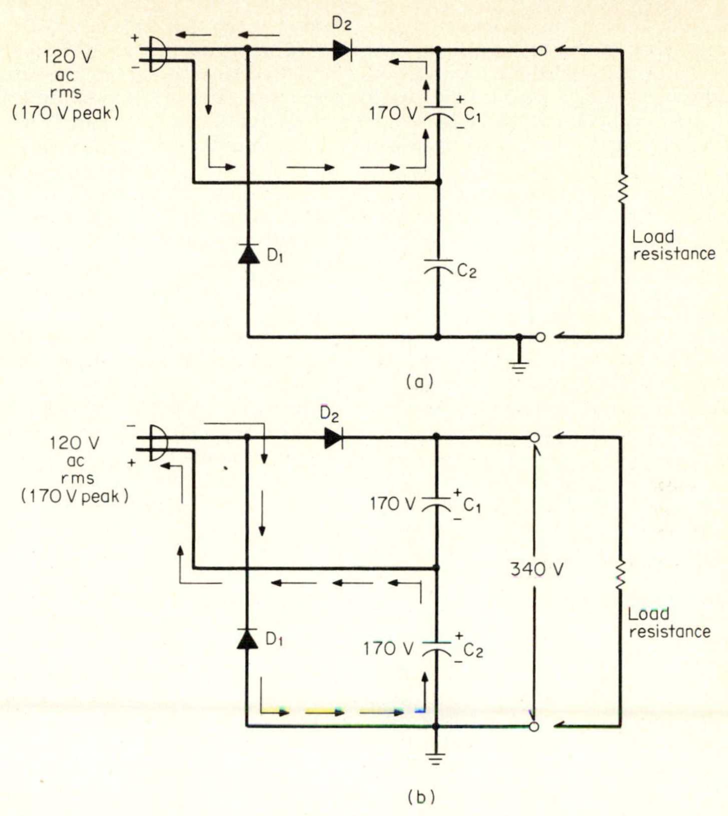

Fig. 16.9 The full-wave voltage doubler: (*a*) First half-cycle of operation. (*b*) Operation during the second half-cycle.

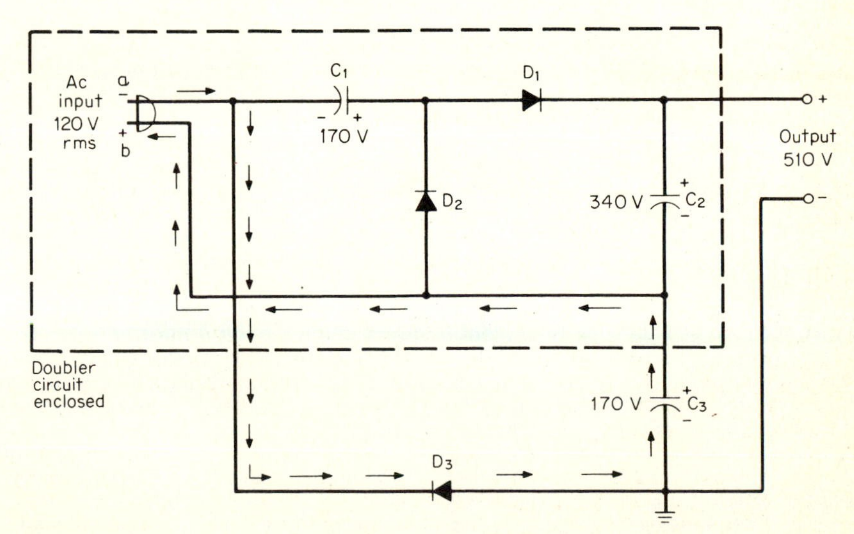

Fig. 16.10 A voltage tripler.

the *full-wave doubler* are examples. Other diode circuits that multiply the input voltage are triplers, quadruplers, quintuplers, etc.) The circuit of Fig. 16.10 is a voltage tripler. The portion of the circuit enclosed in dotted lines is exactly the same as for the half-wave voltage doubler previously discussed, and illustrated in Fig. 16.8. Capacitor C_2 is charged to twice the peak input line voltage in this part of the circuit. Diode D_3 and capacitor C_3 have been added to the doubler circuit to convert it into a tripler.

On one-half cycle, point *a* on the plug will be negative with respect to point *b*, and this places a negative voltage on the cathode of D_3. The electron current flow through D_3 starts at point *a*, flows through the diode up into capacitor C_3, and then out of the capacitor and back to point *b*. This electron current path is indicated by solid arrows on Fig. 16.10. During the same half-cycle when C_3 is charged, D_2 is also conducting and charging capacitor C_1.

On the next half-cycle point, *a* is positive with respect to point *b*, and the diode D_3 is cut off. Diode D_1 conducts, and charges capacitor C_2 to twice the peak line voltage, as discussed previously for the doubler circuit.

The overall result is that the voltage across C_2 is twice the peak line voltage, and the voltage across C_3 is equal to the peak line voltage. Assume, as an example, that the line voltage is 120 V rms. This will result in a voltage across C_2 of 340 V, and a voltage across C_3 of 170 V. The total voltage across the two capacitors (in series) is 510 V—which is three times the peak input line voltage. This voltage appears across the output terminals of the supply. The output voltage of the tripler is high, but the circuit cannot sustain a heavy load current.

If a negative output voltage is required, the diodes in any of the circuits can all be reversed. If this is done, the connections for the electrolytic capacitors must also be reversed because these capacitors are *polarized*—that is, they can only be connected into a circuit with the correct polarity.

LOAD VERSUS LOAD RESISTANCE The terms load and load resistance have been used interchangeably in discussions up to now, but these terms actually mean different things. The load of a power supply always refers to the amount of *current* that the power supply must deliver. Therefore, when it is written that a power supply is under a *heavy load*, it means that the power supply must deliver a heavy current to the load. The load resistance, on the other hand, is the value of load resistance across the power-supply output terminals. The higher the load resistance value, the lower the amount of current that must be delivered by the power supply. Therefore, a high load resistance means a low power-supply load.

COMBINATION POSITIVE AND NEGATIVE SUPPLIES In some circuits it is desirable to obtain both a positive and a negative output voltage from the same power supply. For example, an amplifier may use both pnp and npn transistors, and a vacuum-tube amplifier requires a positive plate voltage and a negative grid voltage. Transformer-type power supplies can be readily adapted to provide both polarities from the same circuit. Such a circuit is shown in Fig. 16.11. Instead of using a common tie point at the center tap of the transformer secondary winding, this point is connected to one side of a series resistance branch comprised of R_1 and R_2. The junction of the two resistors is tied to the common point of the circuit.

Since this is a full-wave rectifier circuit, the current will flow through the resistors on both half-cycles of input signal. The arrows show the paths of current flow. The polarities of the resulting voltage drops across the resistors are marked on the diagram. Since the current is flowing from point *b* toward ground, point *b* must be a negative voltage with respect to ground. At the same time current is flowing from the ground point toward point *a*, making point *a* positive with respect to ground. Both voltage polarities are obtained from the same power-supply circuit. The output voltages are usually filtered by two different filter circuits—one for each polarity of voltage.

If the circuit to which the power-supply voltage is being delivered requires a highly positive voltage for operation of the plates of a tube and a relatively low negative voltage for operating the grid, a separate transformer may be used to obtain the negative voltage. For transformerless power supplies, the positive voltage may be obtained with a half-wave rectifier or other transformerless supply, and the negative dc voltage

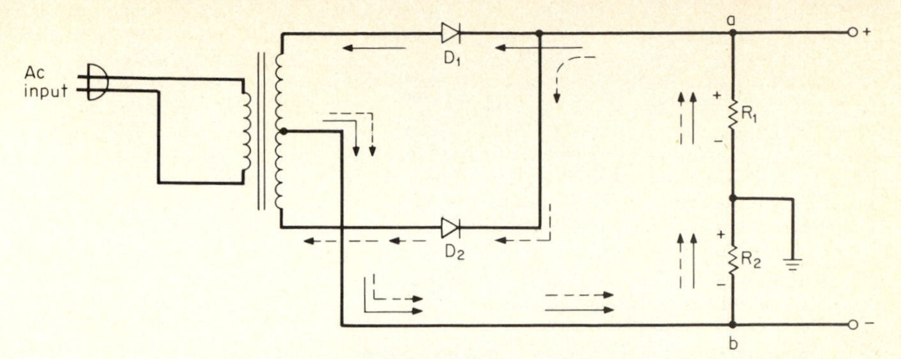

Fig. 16.11 This circuit will supply a positive and a negative output voltage.

may be obtained by a separate transformer supply. The advantage of using two separate supplies over that of the circuit shown in Fig. 16.11 is that the negative supply can be more efficiently designed. Normally the grid circuits of vacuum-tube amplifiers do not draw current, and therefore the load on such a supply is small. A small, inexpensive transformer may be used for the negative grid circuit supply. The identical power-supply principles are equally applicable to power supplies for solid-state circuits.

16.3 REGULATED POWER SUPPLIES

An ideal regulated dc power supply is one that maintains a constant dc output, regardless of changes in external load requirements or changes in input voltage. The degree to which the performance of an actual power supply approaches this ideal is defined by its specifications, which will be discussed in depth later in this chapter. Figure 16.12 shows a general-purpose regulated power supply used in laboratories.

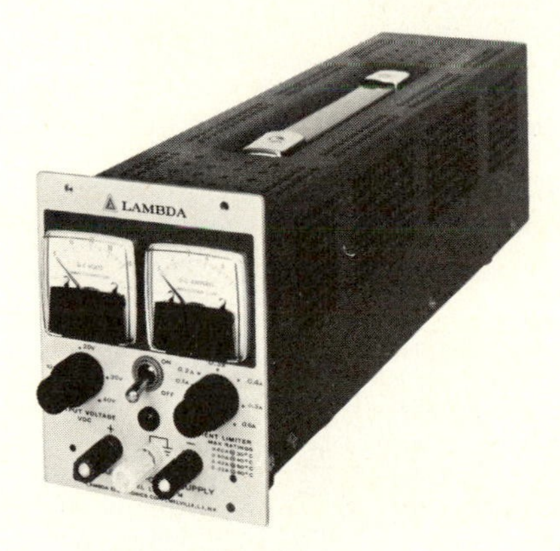

Fig. 16.12 A general-purpose lab power supply. (*Courtesy Lambda Electronics Corp.*)

In general, dc power supplies may be divided into two functional categories: constant-current and constant-voltage types. Constant-current supplies maintain a constant current through a load resistor, regardless of the value of the load resistance, within a certain voltage range. Constant-voltage supplies maintain a fixed voltage across the load resistor as long as the current through the load resistor stays within a specified range. Equations (16.1 and 16.2) give the maximum and minimum values of load resistance for a constant current power supply.

$$R_{max} = \frac{E_{max}}{I} \quad (16.1)$$

$$R_{min} = \frac{E_{min}}{I} \quad (16.2)$$

where R_{max} = maximum value of load resistance, Ω
R_{min} = minimum value of load resistance, Ω
E_{max} = maximum value of output voltage, V
E_{min} = minimum value of output voltage, V
I = output current, A

example 16.1 A constant-current power supply has an output current of one ampere and a voltage range (called the *compliance*) of 0 to 50 V. What are the maximum and minimum values of load resistance?

solution Using Eq. (16.1), we may find the maximum load resistance:

$$R_{max} = \frac{E_{max}}{I} = \frac{50 \text{ V}}{1 \text{ A}} = 50 \ \Omega \quad (16.1)$$

Using Eq. (16.2), we may find the minimum load resistance:

$$R_{min} = \frac{E_{min}}{I} = \frac{0 \text{ V}}{1 \text{ A}} = 0 \ \Omega \quad (16.2)$$

In this case, the minimum load resistance is zero ohms, which is a short circuit.

Equations (16.3) and (16.4) give the maximum and minimum values of load resistance for a constant-voltage power supply.

$$R_{max} = \frac{E}{I_{min}} \quad (16.3)$$

$$R_{min} = \frac{E}{I_{max}} \quad (16.4)$$

where R_{max}, R_{min} = same values as stated above
E = output voltage, V
I_{min} = minimum value of output current, A
I_{max} = maximum value of output current, A

In most supplies, $I_{min} = 0$; the maximum load resistance for a constant-voltage supply is then infinity, or an open circuit.

example 16.2 A constant-voltage power supply has an output voltage of 50 V and an output current range from 50 mA to one ampere. What are the maximum and minimum values of load resistance?

solution Using Eq. (16.3) to find the maximum load resistance,

$$R_{max} = \frac{E}{I_{min}} = \frac{50 \text{ V}}{50 \text{ mA}} = 1000 \ \Omega$$

Using Eq. (16.4) to find the minimum load resistance,

$$R_{min} = \frac{E}{I_{max}} = \frac{50 \text{ V}}{1 \text{ A}} = 50 \ \Omega$$

If this supply had been capable of going to a no-load condition, zero amperes output current, the maximum load resistance would be infinite or an open circuit.

Many power supplies are designed to operate in either a constant-voltage mode or in a constant-current mode. They are usually equipped with *automatic crossover*, a feature that allows the supply to go directly from one mode of operation to the other as the load resistance changes.

Many supplies are designed only for constant-voltage operation over a narrow range. These are frequently referred to as *slot-range* supplies. They generally are equipped with a *foldback* current-limiter circuit, which functions to decrease the output voltage and current simultaneously when the load resistance is too low; i.e., the output current rating of the supply would be exceeded if the output voltage remained at its normal value.

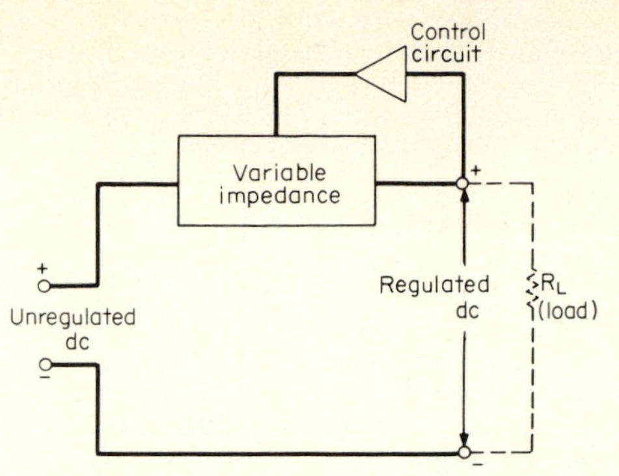

Fig. 16.13 Block diagram of a series-regulator power supply.

BASIC TYPES Regulated dc power supplies may also be classified by the type of input supply they utilize: ac or dc. The ac type usually employs a transformer to isolate the input from the output and to either raise or lower the input voltage to the level required. The ac output of the transformer is then rectified and filtered. The filter output is an unregulated dc voltage. From this point on, the regulator which follows may be considered a "dc-to-dc regulator." Thus, once voltage changing, rectification, and filtering are accomplished, the ac-to-dc and the dc-to-dc regulators may be treated alike.

There are three basic methods of regulating an unregulated dc source. Many power systems utilize a combination of two or in some cases all three of these.

Series type The most common system employed is the series regulator, which is shown in block form in Fig. 16.13. The series regulator is essentially a variable impedance placed between the unregulated input and the regulated output. By varying this impedance, the output voltage (or current) is maintained constant. The variable impedance is an active device, usually a power transistor in modern equipment, but vacuum tubes are utilized in older supplies.

Shunt type Another frequently encountered regulation system is the shunt regulator, shown in block form in Fig. 16.14. Here a fixed-source impedance is placed between the unregulated source and the regulated output. A variable impedance, called a "shunt" or "parallel" impedance, is placed across the output. The shunt impedance changes to maintain the output voltage (or current) constant. As in the case of the series regulator, the variable impedance is an active device, either a transistor or a vacuum tube.

Switching type In either of these two above approaches, all energy not delivered to the load must be dissipated in the regulating elements. More efficient systems involve switching circuits which utilize some form of *duty-cycle* control. The details

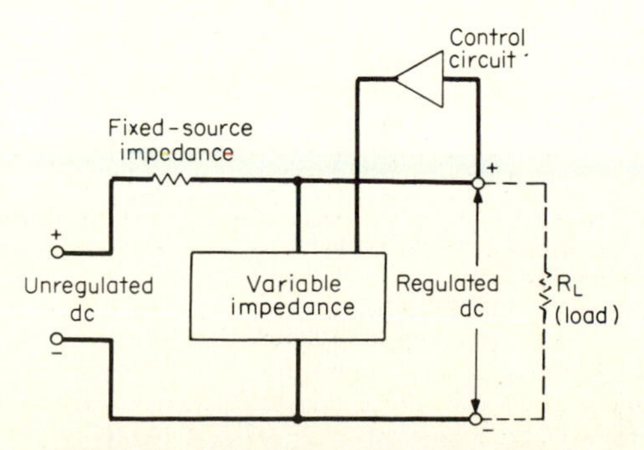

Fig. 16.14 Block diagram of a shunt regulator.

of this *nondissipative* type of regulator will be discussed in more detail later in this chapter. Some systems use only a nondissipative circuit, while others employ this circuitry as a coarse preregulator followed by a fine regulator of either the shunt or series type. This produces the finer regulation of the dissipative control, while having the efficiency advantage of a "nondissipative" circuit.

16.4 POWER-SUPPLY REGULATOR CIRCUITS

Figure 16.15 shows a simple power-supply regulator consisting of a series resistor R_s and a zener diode D_1. A gaseous regulator tube would serve the same purpose as the zener diode in this circuit, but it is not available with low-voltage ratings. It usually takes more than 70 V to "fire" a gaseous regulator tube, while the zener has the advantage that it is available in a wide selection of voltage values. Another advantage of the zener diode over the gaseous regulator is that it is available in a much wider range of current-handling abilities.

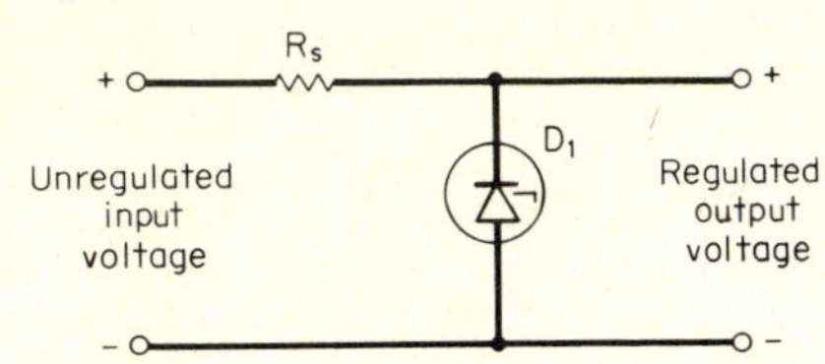

Fig. 16.15 A simple power-supply regulator.

The value of resistance for the series resistor in the simple shunt regulator of Fig. 16.15 can be calculated from the equation

$$R_s = \frac{E_i - E_o}{I_m + I_z} \tag{16.5}$$

where E_i = unregulated input voltage, V
E_o = regulated output voltage, V
I_m = maximum load current, A
I_z = current through the zener diode, A

As a general rule, the zener diode current I_z should be made equal to about 10 percent of the value of the maximum load current. Mathematically,

$$I_z = 0.1 I_m \tag{16.6}$$

where I_m and I_z have the same meaning as for Eq. (16.5).

The power rating of R_s can be determined from the equation

$$P_s = 2(I_m + I_z)^2 R_s \tag{16.7}$$

Once the zener current rating is obtained from Eq. (16.7), then its power rating can be determined from the equation

$$P_z = E_o I_z \tag{16.8}$$

Manufacturer's specifications can be consulted in order to choose the zener that meets the particular requirements of a power supply. A sample problem will show how the regulated zener supply can be designed.

example 16.3 Design a 14-V regulated supply that will provide a maximum load current of 100 mA from an unregulated power supply that delivers 25 V.

solution A 14-V zener diode will be needed.

The zener current rating:

$$\begin{aligned} I_z &= 0.1 I_m \\ &= 0.1 \times 100 \text{ mA} = 10 \text{ mA} \end{aligned} \tag{16.6}$$

Resistance value of R_s from Eq. (16.5):

$$R_s = \frac{E_i - E_o}{I_m + I_z} = \frac{25 - 14}{0.1 + 0.01} = 100\ \Omega$$

Power rating of R_s from Eq. (16.9):

$$P_s = 2(I_m + I_z)^2 R_s = 2(0.1 + 0.01)^2 \times 100 = 2.42 \text{ W}$$

$$P_z = E_o I_z = 14 \times 0.01 = 0.14 \text{ W}$$

The series regulator will have a resistance of 100 Ω and a power rating of over 2.4 W. The 14-V zener will have a current rating of 10 mA and a power rating of over 0.28 W.

Figure 16.16 shows a closed-loop power-supply regulator circuit. At the unregulated input side of the regulator, there is a series circuit comprised of R_1, Q_1, and zener diode D_1. At the regulated output terminals there is a series resistor circuit comprised of R_2, R_3, and R_4.

Resistor R_2 is a variable resistor that sets the base voltage of Q_1. The series resistor string is sometimes called the *sense circuit* because it delivers any change in the output voltage to the base of transistor Q_1.

In order to understand the operation of the power-supply regulator shown in Fig. 16.16, it will be assumed that the regulated output voltage begins to rise for some reason (such as a decrease in load current). The emitter voltage of Q_1 is a fixed value set by the zener diode. Therefore, when the output voltage rises, it causes the base of Q_1 to become more positive, causing Q_1 to conduct harder through resistor R_1. The base voltage for control amplifier transistor Q_2 is obtained from the drop across R_1. Increasing the current through R_1 causes the base of Q_2 to become less positive. This decreases the current through Q_2, which controls the base current of Q_3. Decreasing the base current of Q_3, the power regulator transistor, has the same effect as increasing the resistance between the emitter and collector of the transistor.

To summarize, a rise in the voltage at the output terminals is accompanied by an increase in resistance at Q_2, causing an increased drop across Q_3 and bringing the voltage at the output terminals back to its regulated value.

A decrease in the output voltage causes the base at Q_1 to become less positive. Therefore, it conducts less through resistor R_1. That causes the base of Q_2 to go in a positive direction, increasing the current through Q_2 and also through Q_3. This is equivalent to decreasing the resistance of Q_3 and decreasing the voltage drop across it.

To summarize, a decrease in voltage at the output terminals is accompanied by a decrease in drop across Q_3, and the output voltage remains constant. In this closed-loop feedback regulator circuit, Q_3 serves as a variable resistor which is automatically adjusted to compensate for changes in load voltage. The voltage across Q_3 increases or decreases, as required, in order to maintain the voltage across the output terminals at a constant.

The closed-loop regulator of Fig. 16.16 will also regulate against changes in input voltage. Suppose, for example, that the input voltage—that is, the unregulated voltage

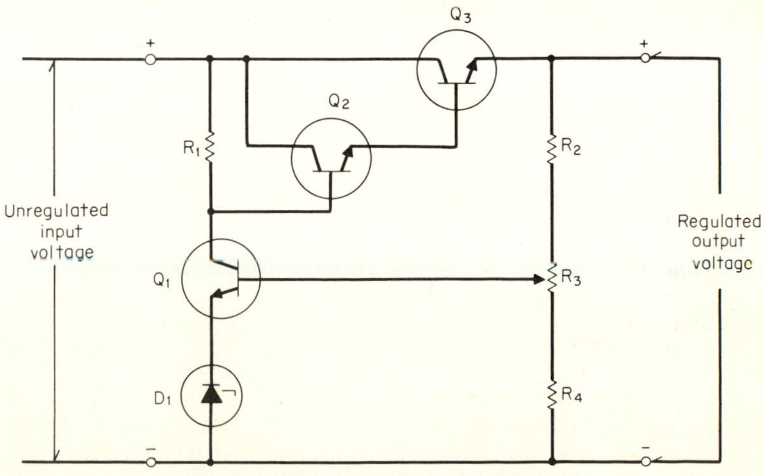

Fig. 16.16 A closed-loop regulator.

– increases. This makes the collector voltage of Q_1 greater and causes it to conduct harder. The base of transistor Q_2 becomes less positive, decreasing the current through that transistor and hence decreasing the current through Q_3. This is the same as increasing the emitter-to-collector resistance of Q_3 and increasing the voltage drop across it. Thus, an increase in the input voltage on the unregulated side is accompanied by an increase in the voltage drop across Q_3, and the output voltage at the regulated side does not change.

A similar chain of events occurs when the output voltage decreases. The result is that when the output voltage decreases, the feedback loop sets the base of Q_3 so that it conducts harder. This is the same as lowering the resistance of Q_3 and decreasing the voltage drop across it. Thus, when the input voltage decreases, the decrease also occurs across Q_3, and the voltage remains unchanged.

16.5 CHARACTERISTICS OF FILTERING CIRCUITS FOR UNREGULATED SUPPLIES

Regulated power supplies often have only a simple capacitive filter between the unregulated output and the regulator circuitry. The regulator itself serves to filter out the ripple, and may be referred to as an *electronic filter*. For unregulated supplies the use of filters is common practice.

The subject of filters was covered in Chap. 11, but the relationship between the filter and the power supply will be taken up here. The three most commonly encountered power-supply filter configurations are shown in Fig. 16.17. In practice, there may be several identical filter sections, but the basic principle is the same. They are *low-pass filters*, which means that they will pass direct current but reject ripple frequencies. The capacitors bypass the ripple frequencies to ground and the inductors (or resistors) oppose the passage of ripple frequencies.

Ideally the filter would pass only direct current – the lowest possible frequency – and reject all ripple frequencies. However, when using lumped components as shown in

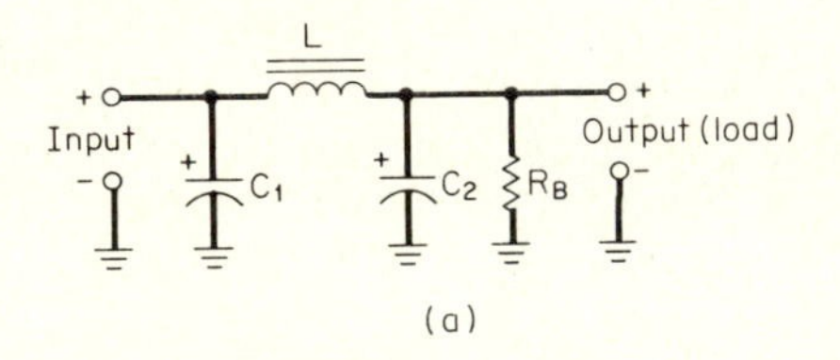

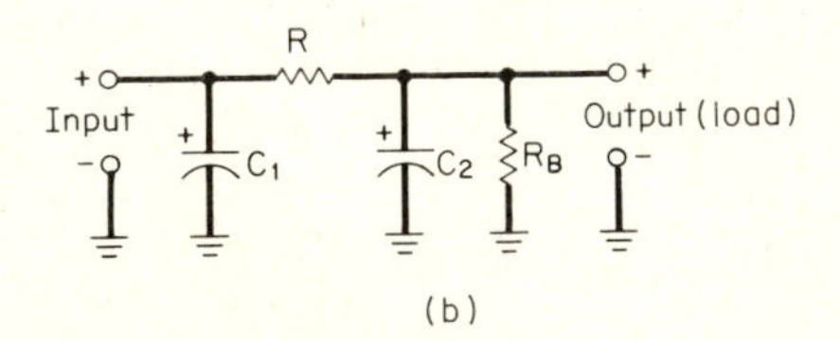

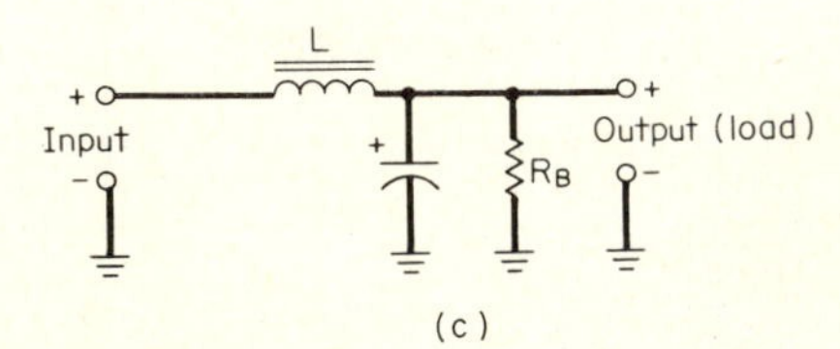

Fig. 16.17 Popular power-supply filters: (*a*) Capacitive input filter. (*b*) *RC* filter. (*c*) Choke-input filter.

Fig. 16.17, it is not possible to achieve these ideal conditions, although they can be approached. For a half-wave rectifier on a 60-Hz power line the ripple frequency is 60 Hz, and the cutoff frequency of the filter must be below this value. For a full-wave rectifier on a 60-Hz power line the ripple frequency is 120 Hz, and it is easier to achieve a cutoff frequency below this than for a half-wave supply.

Figure 16.17*a* shows a *capacitor input filter*, so designated because the first component following the supply is C_1. The filter configurations of Fig. 16.17*a* and *b* are also called *pi filters* because of their resemblance to the Greek letter π. Capacitor C_1 will charge to the peak of the ripple frequency, and an important characteristic of capacitive-input filters is that they provide a higher voltage at the output terminals when compared to choke-input filters. However, this type of filter must never be used with a gaseous rectifier because the initial input capacitor charging current could damage the tube.

The "peak-inverse" voltage across the rectifier may be twice the peak-input voltage. Figure 16.18 shows why. The input capacitor has charged to the peak voltage, as

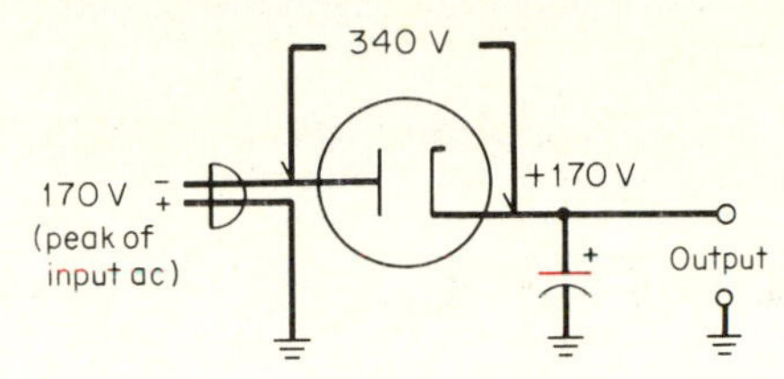

Fig. 16.18 This illustration shows why the peak-inverse voltage across a diode is larger than applied peak voltage, when a capacitor-input filter is used.

shown in Fig. 16.18. (The drop across the rectifier has been disregarded in this illustration.) On the half-cycle that puts the rectifier plate at a potential of −170 V, the inverse voltage across the diode is clearly seen to be 340 V. This may be excessive for some rectifiers. Both tube and semiconductor devices are rated by the maximum peak-inverse voltage (p.i.v.) they can withstand, and this must be taken into consideration when making a replacement during servicing.

The percent regulation of a power supply is given as the equation

$$\text{Percent regulation} = 100\,\frac{E_1 - E_2}{E_2} \tag{16.9}$$

where E_1 = power-supply output voltage when no load is connected
E_2 = power-supply output voltage when delivering the maximum load current

Supply regulation is affected by the filter since it has resistance and it is between the supply and the load. This is more readily apparent from the pi filter of Fig. 16.17*b*. Any change in load current will cause a change in the voltage drop across the filter resistor R and will, therefore, directly affect the output voltage. The advantage of using a choke instead of a resistor is that the supply regulation will be better. The disadvantage is that a choke costs considerably more than a power resistor. The advantage of better regulation does not justify the cost of the power supply if used in a system where the load is relatively constant. For example, in a radio the load consists of a fixed number of tubes or transistors in operation, and regulation of the supply is not an important factor.

The filter resistor (Fig. 16.17*b*) must dissipate power in the form of heat. If the power amplifiers of a system conduct through the filter resistor, its power rating must be quite high. However, power amplifiers are relatively insensitive to small changes in voltage on their anodes, so they can be connected to the power-supply side—that is, to point *a* in Fig. 16.17*b*.

The choke-input filter of Fig. 16.17*c* provides better regulation than a capacitor-input filter, but it does not have as high an output voltage for a given input. Choke-input filters are used with gaseous rectifiers.

For choke-input filters the inductance of the choke must be equal to or greater than a certain minimum value called the *critical inductance*. This inductance is defined by the equation

$$\text{Critical inductance (in henrys)} = \frac{E \text{ (in volts)}}{I \text{ (in milliamperes)}} \tag{16.10}$$

where E = output voltage of the supply
I = current flow through the choke

All the filters of Fig. 16.17 show a bleeder resistor R_B across the output terminals. An important function of this resistor is to discharge the filter capacitors when the supply is turned off. If properly designed, regulation is improved by use of a bleeder.

16.6 SPECIFICATIONS

Each power-supply manufacturer provides his customers with performance specifications for his product. Although the actual details of the specifications vary from one manufacturer to another, a basic understanding of the parameters will enable one to interpret most "spec" sheets.

REGULATION The basic purpose of a regulated power supply is to provide a constant output (either voltage or current, depending on the type of supply), independent of changes in either the specified input line or output load. The specifications will enable the user to determine the maximum deviation of the output voltage or current that will occur owing to changes in line or load.

Line regulation is defined as the change in output voltage, or current, for a given line change, with the load and the ambient temperature held constant. It is usually denoted as a fixed change plus a percentage change for any line-voltage change within the operating range of the supply: for example, 0.01% + 1 mV for a constant-voltage supply, or 0.05% + 2 mA for a constant-current supply. Equations (16.11*a*) and (16.11*b*) describe the line regulation for a constant-voltage and constant-current power supply, respectively.

$$\Delta E_o = (\text{percentage regulation})E_o + \text{fixed error} \tag{16.11a}$$

where ΔE_o = change in output voltage due to an input line change
E_o = value of output voltage before the line is changed

$$\Delta I_o = (\text{percentage regulation})I_o + \text{fixed error} \tag{16.11b}$$

where ΔI_o = change in output current due to an input line change
I_o = value of the output current before the line is changed

Load regulation is defined as the change in output voltage, or output current, for any given load change within the specified limits with the input line voltage and ambient temperature held constant. In a constant-voltage supply, this means that the load impedance may be changed in such a manner that the output load current remains within the current range of the supply. In a constant-current supply, this means that the load impedance may change between the limits defined by Eqs. (16.1) and (16.2). As with line regulation, load regulation is usually specified as a percentage of the output plus a fixed change, for example, 0.01% + 1 mV or 0.02% + 2 mA.

Equations (16.11*a*) and (16.11*b*) are valid for both line and load regulation.

TEMPERATURE COEFFICIENT The temperature coefficient of a power supply is the change in output voltage (or output current) due to changes in the external ambient temperature, with the input line and output load held constant. This change is due to the temperature coefficient of the reference element, usually a zener diode, and of the amplifier offset and gain. The temperature coefficient is most often specified as percentage of the output plus a fixed change per degree Celsius, e.g., (0.01% + 1 mV)/°C or (0.05% + 1 mA)/°C. Equations (16.5*c*) and (16.5*d*) may be used to determine the change in output for a given temperature change.

$$\Delta E_o = \Delta T_x \text{ (percentage temperature coefficient } E_o + \text{fixed change)/°C} \tag{16.11c}$$

$$\Delta I_o = \Delta T \text{ (percentage temperature coefficient } I_o + \text{fixed change)/°C} \tag{16.11d}$$

where ΔE_o = change in output voltage of a constant-voltage supply for a change in ambient temperature ΔT

ΔI_o = change in output current of a constant-current supply for a change in ambient temperature ΔT

ΔT = change in ambient temperature in degrees Celsius

STABILITY Line regulation, load regulation, and temperature coefficient describe the changes in the power-supply output due to external variations. They imply that if the line, load, and ambient temperature are held constant, the output will not vary. In general, this is not true. There are output changes due to internal component changes caused by temperature changes within the supply. These temperature shifts are the result of localized internal heating and external air currents which are rarely completely eliminated.

If all the effects of temperature gradients are eliminated, the output will still change somewhat because of component aging changes. These changes affect the "stability" of a power supply. Stability is defined as the change in output, either voltage or current, with line, load, and ambient temperature held constant. The measurement is taken over a fixed time period, for example, over 30 days. The time period is usually either eight hours or thirty days.

Most power supplies do not specify stability. In order to obtain a guaranteed degree of stability, it is necessary to mount sensitive components in such a way that they will not be affected by normal air flows and, therefore, will not be subject to the resulting temperature variations. In addition to these precautions, it is necessary to use components whose stability is controlled. This control is usually obtained through an aging process during which all unstable parts are culled out. This is an expensive and time-consuming procedure. These additional expenses make these techniques uneconomical in all but the most expensive laboratory supplies where knowledge of the supply stability is essential. In units of this type, the reference and parts of the amplifier are usually placed in a constant temperature oven in order to reduce the effects of external ambient changes.

TRANSIENT RESPONSE Line and load regulation define the steady-state limits of the output of the supply for line or load changes. It is often of interest to know what the output does during the transition from one state to the other. This transition period is known as the transient response.

Methods of specifying transient response vary. In most cases, load transient response is more important than line transient response. The load transient response is specified for a fixed step load change, usually from 10 percent of full load to full load, or from full load to 10 percent of load. There are two important factors: the maximum amplitude of the transient response, and the time required for the output to return to some specified point. The specified point is usually within 50 mV of the final value.

Figure 16.19 shows the transient which occurs when a constant-voltage power supply

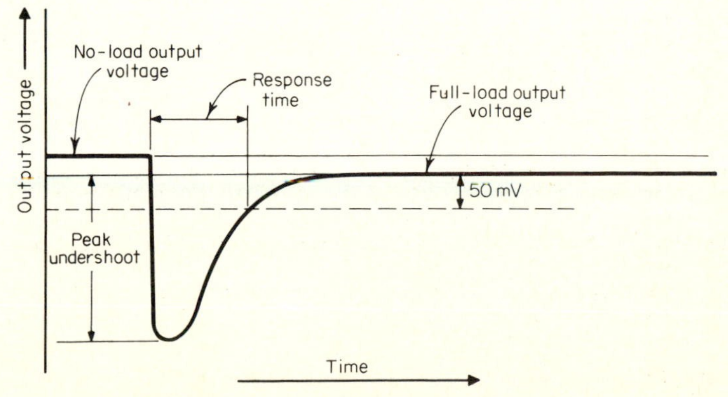

Fig. 16.19 Transient response due to a change in load.

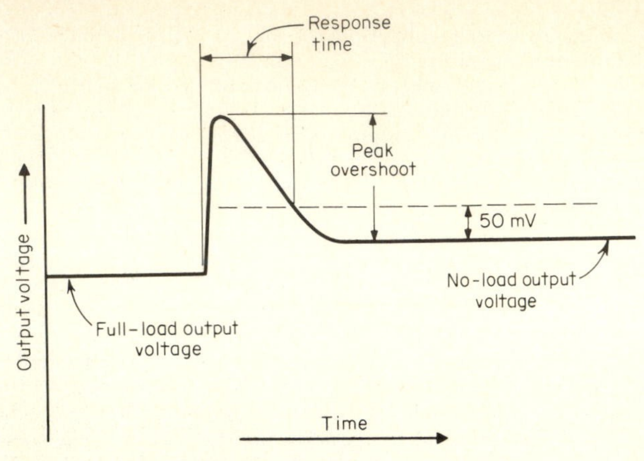

Fig. 16.20 Load off transient response.

goes from a light-load to a full-load condition. The rapid drop in output voltage represents the discharge of the output capacitor of the supply. The rate of discharge is limited only by the wiring resistance and inductance. During this time the control circuitry has not reacted, and the supply is still providing the original load current. Until the control circuitry reacts, the output capacitor is supplying the current to the load. As the control circuit turns on, the rate of decay of the output voltage decreases. When the power circuit is completely on, the power supply then provides load current and recharges the output capacitor. The output voltage then increases to its new value within the regulation band.

There is also a transient that occurs when the load is suddenly removed. This overshoot is shown in Fig. 16.20. The main source of overshoot is that it is impossible to locate the output capacitor directly at the point of regulation. Therefore, there is a voltage drop due to wiring resistance from the output terminal of the supply. When the load current is decreased, this drop decreases. The voltage on the capacitor cannot change instantaneously; the voltage difference must appear at the output terminals. This is illustrated in Fig. 16.21. For a one-ampere supply, 50 mΩ of resistance will provide 50 mV of overshoot.

RIPPLE The final parameter that is usually specified is output ripple. This is the ac component which is superimposed on the dc output. The ripple is generally specified as an rms (root mean square) value and a peak-to-peak value. A typical specification might read:

Output ripple and noise, 0.250 mV rms, 1 mV p-p

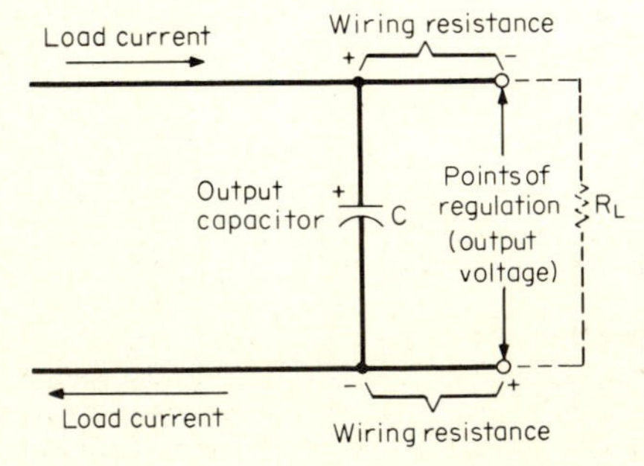

Fig. 16.21 Relation of output capacitor location and internal wiring resistance to transient response.

The amount of ripple that may be tolerated varies from user to user. High ripple, particularly any switching spikes that might be present, would be unacceptable in logic circuits where the power-supply noise could appear as logic inputs and cause an error. In transmitting and receiving equipment, it is usually necessary to have a fairly low ripple to avoid hum, especially in the audio sections.

Constant-current power supplies also have ripple specifications which are similar to those of constant-voltage supplies except that they are in terms of current instead of voltage. For example, a supply may be rated as 0.5 mA rms, 2 mA p-p.

16.7 SPECIAL DESIGN FEATURES

Four important power-supply design features will now be discussed. They are found as either standard or optional features on most high-quality power supplies.

REMOTE SENSING Modern power supplies are capable of load or line regulation in the order of tens or hundreds of microvolts. If this regulation is to be utilized, it is necessary to have the load at the point of regulation.

Figure 16.22 shows the output terminals of a 1-A, 10-V power supply with the load connected by 20 ft of no. 24 wire which has a resistance of 0.026 Ω/ft. If the load is changed by one ampere, the voltage change at the output of the supply is 2 mV, but the change at the load is 2 mV + 1 A × 0.52 Ω, the total wiring resistance.

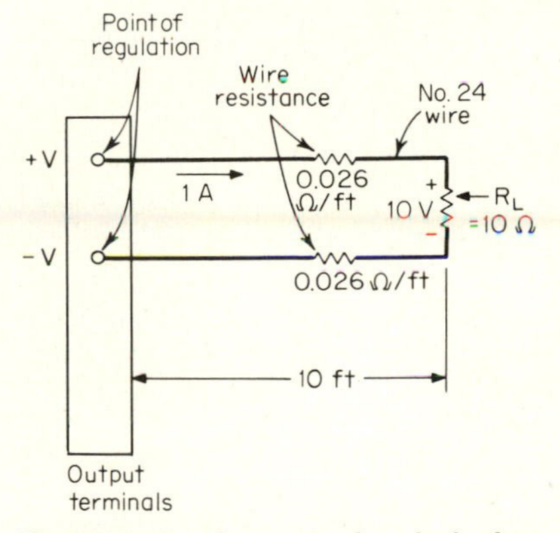

Fig. 16.22 Local sensing when the load is away from the power-supply output terminals.

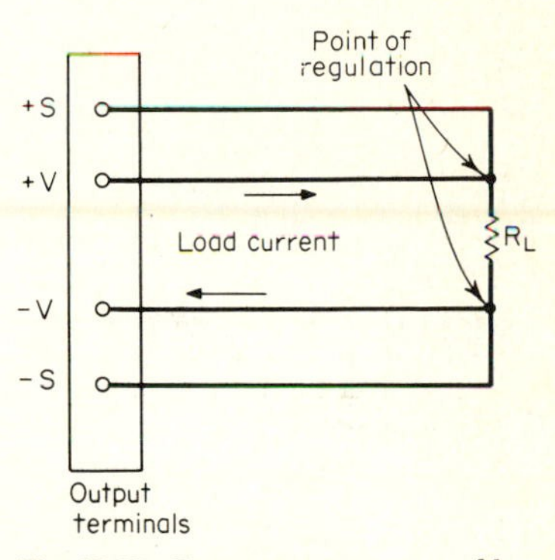

Fig. 16.23 Remote sensing to avoid loss of load regulation due to power-line voltage drops.

To eliminate this degeneration of load regulation, most modern power supplies are equipped with two additional terminals, the sensing terminals, $+S$ and $-S$, for remote sensing. Wires may be run from these terminals to the point where regulation is desired. Since no load current flows through these "sensing" leads, the effects of the line drops in the power leads are compensated for. It is generally advisable to twist the leads from $+S$ and $-S$ together to eliminate the pickup of any stray fields. In applications where the load is relatively constant, it is possible to connect the $+V$ terminal to the $+S$ terminal and the $-S$ terminal to the $-V$ terminal at the power-supply barrier strip. Figure 16.23 shows a power supply hooked up in a remote sensing mode.

REMOTE PROGRAMMING Most power supplies come equipped with an internal potentiometer which may be used to control the output voltage. In many applications it is desirable to have the output voltage control at some point remote from the power supply itself. The ability to have such remote programming is found in most modern power supplies.

Remote programming is usually specified in terms of resistive programming and voltage programming. The former specifies a certain number of ohms per volt which

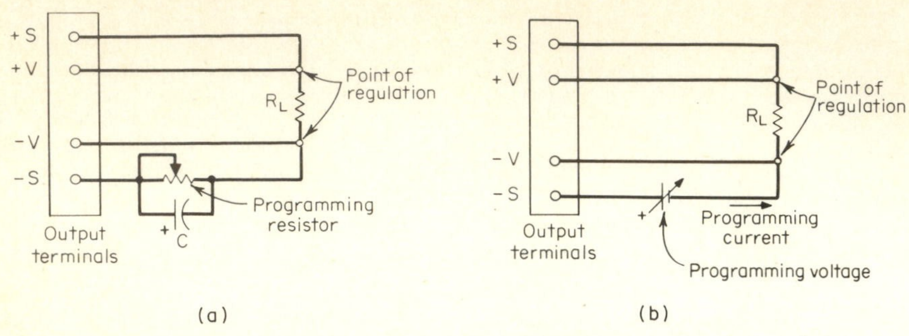

Fig. 16.24 Remote-programming systems: (*a*) Resistive remote programming. (*b*) Voltage remote programming.

gives the required external programming resistance change for a one-volt change in output voltage. The latter specifies a certain number of volts per volt, usually one, which gives the required external programming voltage change for a one-volt change in output voltage.

Figure 16.24 shows a voltage remote-programming system. Figure 16.24*a* shows a typical hookup for remote resistance programming. In this case, the internal control is set to zero and all control is maintained by the external resistor. The programming resistance is usually between 200 and 1000 Ω/V. The capacitor across the programming resistor is used to keep the output ripple to a minimum.

In Fig. 16.24*b* a power supply is shown being programmed by an external voltage source. It is necessary that the external voltage source be able to draw a reverse current equal to the programming current of the power supply, usually between 1 and 5 mA. If the programming voltage cannot do this, it should be shunted by a fixed resistor which will draw the required programming current at the operating voltage of the programming supply.

OVERVOLTAGE PROTECTION In many applications, particularly those involving integrated circuits, an abnormally high voltage at the output of the power supply could be catastrophic. In applications such as these, it is usually recommended that some type of overvoltage protection be employed.

There are two basic types of overvoltage protectors, or "O.V.'s" as they are frequently called. The first compares the output voltage of the supply to a preset level. When the output voltage exceeds this level, the O.V. *crowbars* (or short-circuits) the output by firing an SCR (silicon-controlled rectifier), which is directly across the output terminals of the supply. If the overvoltage condition has been caused by a failure in the power supply, either a fuse will open or a breaker will trip to disable the set. If the overvoltage condition was caused by an operator inadvertently adjusting the output voltage beyond the set point of the O.V., then the supply will go into current limiting. It is then necessary to turn the supply off and lower the voltage adjust setting to reset the overvoltage protector.

Figure 16.25 shows two block diagrams used for protection purposes. Figure 16.25*a* is a block diagram of an overvoltage protector. The comparator compares the output voltage of the power supply to a reference voltage. If the output voltage exceeds the reference, the comparator output actuates a triggering circuit, which in turn fires the SCR which crowbars the output of the supply. The output voltage is reduced to the forward conduction voltage of the SCR which is usually about one volt.

The other basic type is described as a *tracking O.V.* This type compares the reference voltage of the power supply to the output voltage. It will fire only when the supply has malfunctioned and there is too large a difference between the output voltage and the reference voltage. This type will not protect the external circuitry from operator error.

The overvoltage protector is a safety device placed between the supply and the load. As such, its circuitry should be independent of the power-supply control circuitry. This is not true of the tracking overvoltage protector.

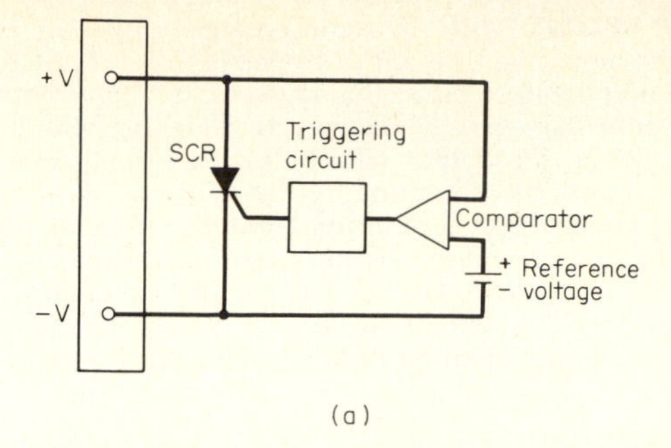

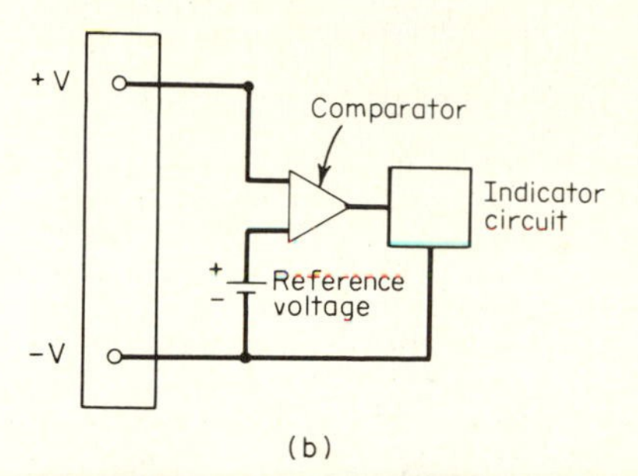

Fig. 16.25 Two block diagrams of circuits used for protection purposes: (*a*) Block diagram of an overvoltage protection circuit. (*b*) Block diagram of an undervoltage warning circuit.

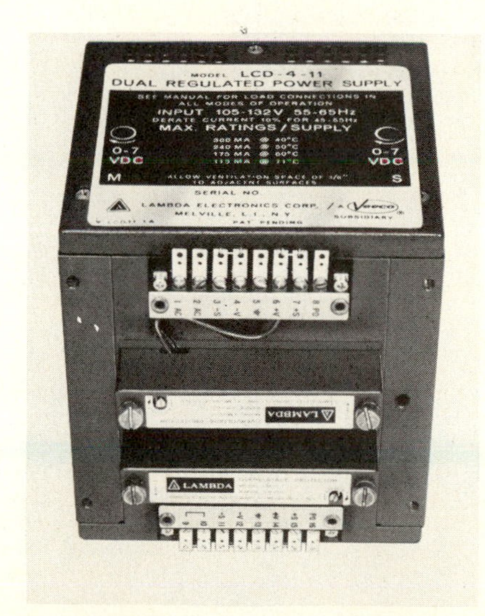

Fig. 16.26 Example of a dual-output power supply, capable of remote sensing, remote programming, and series and parallel operation. (*Courtesy Lambda Electronics Corp.*)

UNDERVOLTAGE PROTECTION In some circuits, especially memory circuits, it is undesirable for the output voltage of the power supply to fall below a certain level. Then some form of undervoltage protection is used.

The simplest undervoltage protection merely compares the supply voltage to a preselected value. When the output falls below this value, some type of signal is actuated to alert the user to this condition.

Figure 16.25*b* is a block diagram of an undervoltage protection circuit. The comparator compares the output voltage of the power supply to the reference voltage. When the output falls below the reference voltage, the comparator output actuates the indicator circuit. The indicator circuit may be a light, a bell, or any other type of signal that would call the operator's attention to this condition.

A more sophisticated approach is for the undervoltage protector to switch in an auxiliary power source.

Undervoltage protection is generally not found as a standard feature on most supplies. One reason is that most undervoltage conditions are due to power-line (input ac) failure. In this case, the power source for any indicating system or an auxiliary power supply is also disabled. Thus, an undervoltage protection system would have to be tailored to a particular installation.

Figure 16.26 illustrates a Lambda LCD-2-55 dual-output power supply. This type of supply is used primarily in systems applications. The supply is capable of remote sensing, remote programming, and series and parallel operation. The two appendages are overvoltage protectors.

Chapter 17

Batteries

17.1 INTRODUCTION

The use of chemical reactions to produce electricity dates from March 1800 when the Italian physicist, Alessandro Volta (1745–1827), described two batteries to the Royal Society of London. The first of these was called a "Crown of Cups" since it consisted of a group of cups arranged in a circular pattern. Each cup was filled with salt solution (the electrolyte) and contained two strips (electrodes) of dissimilar metals (silver and zinc), with each of the two strips externally connected to its adjoining mate. These, of course, were series-connected and were the forerunner of our various modern cylindrical cell systems. In the other design that Volta presented, the dissimilar metals were stacked in the form of a "pile," very much like coins in a dispenser, with a piece of paper soaked in salt separating alternate layers of metal. This was the forerunner of our modern-day wafer or flat-plate cells.

For the most part, batteries representing variations of the Crown of Cups concept were used until about 1860 when Gaston Plante described a battery using lead and lead oxide electrodes and an electrolyte of sulfuric acid to the French Academy of Science. This, of course, ultimately developed into our familiar automobile battery, and has by now evolved into a myriad of sizes and engineering types to serve a very large number of applications. Plante's contribution was particularly noteworthy in that it achieved high-power levels.

Working at about the same time, another Frenchman, Georges LeClanche, conceived the idea of having all the battery's electrolyte absorbed by its electrodes so as to essentially function as though it were dry. In 1868 LeClanche received a patent for such a "dry" cell and undertook its manufacture. As one of the electrodes in his cell, LeClanche used manganese dioxide and a carbon-rod current collector. A cup of zinc not only formed the cell container, but also served as the other electrode. LeClanche's cell, therefore, was the direct lineal forebear of our familiar flashlight dry cells, and was a most significant development since it provided a truly mobile source of electric energy, thereby triggering the growth of the battery industry and a host of battery-using devices.

Today we have a broad spectrum of devices that yield electric energy from a controlled chemical reaction. In size, these range from tiny mercury "button" cells such as are used to power hearing aids, to the large lead-acid cells used to provide all power requirements, including propulsion of submerged submarines. A small button cell (i.e., the M5 size) will be less than $^5/_{16}$ in. in diameter by a little more than $^1/_8$ in high and will weigh about 0.02 oz. A perfectly conventional submarine cell will be about 18 in wide, 18 in long, $5^1/_2$ ft high, and will weigh over one ton. The little M5 has a capacity of 36 mAh whereas its giant submarine cell cousin has a capacity measured in kiloampere-hours.*

* The abbreviations "h" for "hour" and "s" for "second" follow the SI system recommended by the IEEE.

Not only is there an almost continuum of sizes between these two extremes, but there are now, as regularly available commercial products, a number of different cell systems, each having a combination of properties against which specific needs can be matched. In addition to Plante's cell and LeClanche's cell, we can consider nickel-cadmium batteries for extended cycle life, nickel-iron cells for ruggedness, silver-zinc cells for very high power, mercury cells for very high capacities, alkaline cells for outstanding versatility, magnesium dry cells for stability in hot, humid climates, and so on. In addition to sizes and types, additional variety of operating properties is obtained from different basic physical forms. Cells are available in cylindrical designs, rectilinear forms (using flat electrodes or tubular electrodes). Other forms include wafer cells (both round or rectangular) for easy series stacking for high voltages, and now even in pile form for still easier stacking to still higher voltages. In spite of this variety, the underlying principles whereby chemical reactions are used to provide electric energy—or the process is reversed so that electric energy can be used to bring about chemical reactions—can be readily understood.

GENERAL CONSIDERATIONS Of the various kinds of chemical reactions that can take place, one type generally called a redox reaction (i.e., reduction-oxidation) is of particular interest to us. A redox reaction not only involves a chemical reaction between two substances, but also a transfer of electrons from one of the substances (the reducing agent) to other substances (the oxidizing agent). As might be expected, if a redox reaction can be set up in such a manner as to cause the transfer of electrons to take place through a wire, we would in effect have a means of generating an electric current. Just such conditions are set up in a cell or battery which thereby acts as a source of electricity. (A battery is a group of cells connected in series and parallel to provide the desired voltage and power. By common usage, single cells are also commonly referred to as batteries.)

There are three essential components in a battery: an anode, a cathode, and an electrolyte. The anode and cathode together are referred to as the electrodes and are made up of the chemicals which will ultimately react to cause the electron transfer. In addition, the electrodes must provide a means for physically supporting the reacting chemicals, and a suitable conductive path for facilitating the flow of the electrons. This latter structure is referred to as the grid of the electrode. The anode is the electrode which will release electrons (and become oxidized) during the discharge of the battery. Similarly the cathode is the electrode which will absorb electrons (and be reduced) during discharge of the battery. The electrolyte is a liquid, usually an aqueous solution made conductive by the dissolution of an acid, a base, or a salt. The electrolyte completes the conductive path between the electrodes of opposite polarity. We can summarize and illustrate this with a conventional primary dry cell.

PRIMARY BATTERY In the common dry cell (also referred to as a carbon-zinc or LeClanche cell), manganese dioxide (MnO_2) is used as a cathode and zinc (Zn) as an anode. The electrolyte is a solution of ammonium chloride and/or zinc chloride. When the anode and cathode are connected through a wire so as to permit electrons to flow from anode to cathode, the reaction between MnO_2 and Zn can take place, and an electric current is set up in the external circuit and the battery is discharging, as illustrated in Fig. 17.1.

In operation, zinc is converted to zinc oxide (ZnO), releasing two electrons which then flow through the external circuit to become available to the MnO_2. These electrons are then absorbed by the MnO_2, thereby changing the latter to a lower oxide of manganese (such as Mn_2O_3). It should be noted that hydroxyl ions (OH) and water also enter into the reactions, but their net participation is canceled out. (Hydroxyl ions are particles in solution made up of hydrogen and oxygen atoms and carry a single negative charge.) As can be seen, the battery can continue to set up an electron flow until one of the other chemical reactants is exhausted, in which case the battery is said to be discharged.

In summary, a battery stores chemicals (not electricity) and can bring about a reaction between these chemicals in such a manner as to cause an electron flow through an external circuit.

The cell we have just described is classified as a primary battery since the chemical

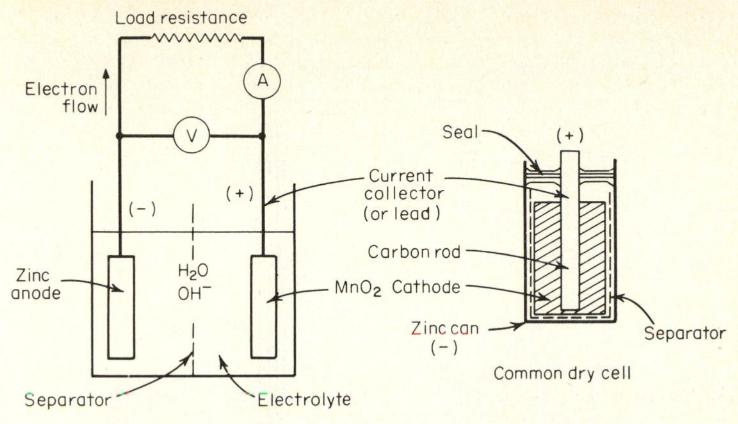

Fig. 17.1 Fundamental operation of a dry (primary) battery.

Anode reaction:

$$Zn + 2OH^- \rightarrow ZnO + H_2O + 2e$$

Cathode reaction:

$$2MnO_2 + H_2O + 2e \rightarrow Mn_2O_3 + 2OH^-$$

Overall reaction (obtained by adding anode and cathode reactions):

$$Zn + 2MnO_2 \rightarrow ZnO + Mn_2O_3$$

reactions that gave rise to the electric current are not readily reversible. By convention, the electric current flow is opposite to that of the electron flow, and the cathode is therefore designated as the positive electrode, and the anode as the negative.

SECONDARY BATTERY Once the active materials are used up, a primary battery is not capable of yielding any more electric energy and is then discarded. Other battery systems use materials that can be restored to their original chemical state by reversing the current flow, i.e., by providing electric energy to the cell from some external source. This process is known as charging, and a battery that is capable of undergoing a number of discharge-charge cycles is called a secondary battery. An example of a secondary battery is the lead-acid system, illustrated in Fig. 17.2. By definition, the anode on discharge becomes the cathode on charge. Similarly, the cathode on discharge becomes the anode on charge. However, the terminal designations of positive and negative do not change since the current flow reverses when the roles of the electrodes of secondary batteries are most often referred to as positives and negatives, and while this designation can also be used for primaries, the electrodes of the latter are more usually referred to as anodes and cathodes.

In their most common forms both primary and secondary batteries are entirely operative when they are made. Because of this, slow deteriorations can occur owing to a small amount of spontaneous reaction or "local action" that goes on all the time in activated batteries.

RESERVE BATTERIES Where extra long shelf life (i.e., ability to yield its electric energy after long storage) is desired, the electrolyte is withheld from the battery, and steps are taken to exclude moisture since water, in some form, is usually required for the local action reactions. Such batteries are extremely stable in storage, and are activated by the addition of the appropriate electrolyte. Since the power of such batteries is thereby kept in reserve, they are frequently referred to as "reserve batteries." A special type of reserve battery is sometimes also referred to as a "dry charged" battery. In this case the electrodes of the battery are brought to a state of full charge at the time they are made, and then deactivated by the removal of electrolyte. The restoration of electrolyte at the time and place the battery needed results in a virtually

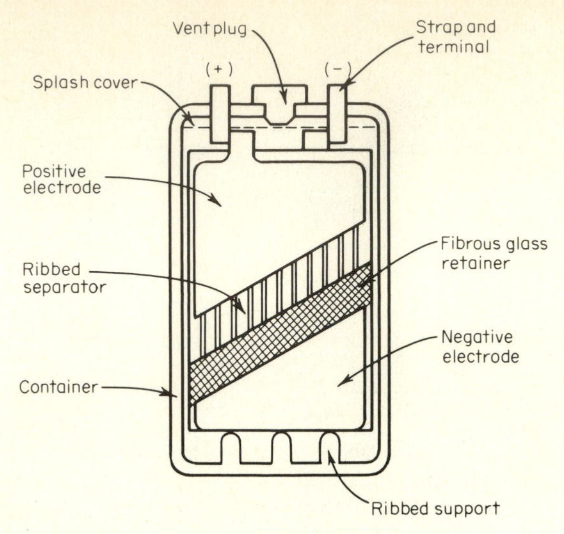

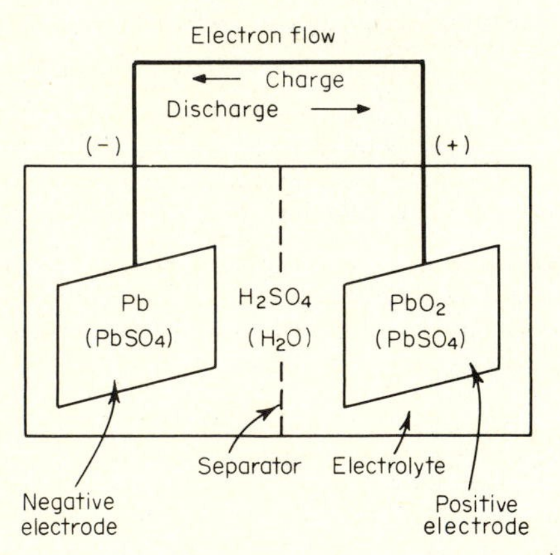

Fig. 17.2 Fundamental operation of a lead-acid (secondary) battery.

Discharge:

$$Pb + SO_4^{--} \rightarrow PbSO_4 + 2e$$
$$PbO_2 + SO_4^{--} + 4H^+ + 2e \rightarrow PbSO_4 + 2H_2O$$

Charge:

$$PbSO_4 + 2e \rightarrow Pb + SO_4^{--}$$
$$PbSO_4 + 2H_2O \rightarrow PbO_2 + SO_4^{--} + 4H^+ + 2e$$

factory-fresh battery. A great many automobile batteries are now shipped in the dry charged condition with the station attendant adding the (sulfuric acid) electrolyte at the time of installation.

There is a special class of reserve batteries that use nonaqueous or molten salt electrolytes. Such electrolytes contain no water and utilize organic liquids, or salts that become conductive in the molten condition. This latter kind of electrolyte must be liquefied before it becomes functional, and the battery can therefore be activated by heat. This is usually accomplished by a heat squib, which melts the electrolyte to activate the battery. Such batteries are known as thermal-type reserve batteries, heat-activated reserve batteries, or simply thermal batteries. A typical thermal battery will be in the form of a small button cell with a calcium or magnesium anode, a silver chromate cathode, and an electrolyte consisting of a mixture of potassium, sodium, and lithium chlorides. Such batteries have an indefinite storage capability, but discharge times of only a few minutes at best. However, they have found special uses such as power for proximity fuses. Further descriptions of reserve batteries will be considered later in this chapter along with other special types.

17.2 CHARACTERISTICS OF IMPORTANT COMMERCIAL PRIMARY BATTERIES

A large number of different primary-cell types are now being produced commercially, but only relatively few have reached the point where they are in volume production and widely distributed. Actually, each cell system has physical and electrical properties different from all the others, so that the user now has an extensive choice from which to select the type most adaptable to his particular application. Table 17.1 shows the comparative properties of the different primary-cell systems that are most widely distributed and of commercial importance.

TABLE 17.1 Characteristics of Commercially Important Primary Cells

Type	Open-circuit voltage, V	Average operating voltage, V	Wh/lb	Wh/in³
LeClanche	1.50–1.65	1.25	5–40	0.5–3
Alkaline	1.52	1.20	20–50	2.0–3.5
Mercury	1.35–1.40	1.25	25–55	5–8
Magnesium dry	1.90–1.95	1.50	15–50	1–4

Of these, the most important is the LeClanche system, also known as carbon-zinc cells or simply "dry cells." This latter designation has come into widespread use although the cells are certainly not dry. As a matter of fact, they contain electrolyte just as any other battery does, but this electrolyte is absorbed into the cathode active material so that the cell appears to be dry.

BASIC TYPES The dry-cell industry has standardized on three basic engineering types: round cells, flat cells, and wafer cells. The industry has further standardized dimensions for 14 round cell sizes, 4 flat cell sizes, and 5 wafer sizes. Virtually all batteries are made up from these standard sizes, detailed in Table 17.2. Many manufacturers, however, issue their own catalog information showing the exact dimensions of their finished cells and batteries. Dry cells are called upon to provide a wide variety of electrical needs in terms of capacity, voltage, and current. For example, a flashlight may require 0.25 A at 3 V, while a small transistor radio may require only 5 mA at 9 V. To satisfy an almost endless variety of capacity-voltage-current requirements, the industry has developed these standard sizes and types, which are then combined in series-parallel arrangements to satisfy almost any desired power requirement.

POLARIZATION OF DRY CELLS During the discharge of a dry cell, the electrolyte actively enters into the electrode reactions. In the vicinity of the actual reaction sites, the electrolyte at one electrode becomes depleted of its reactants, while the electrolyte at the other electrode becomes concentrated with the products of the reaction. This condition, known as polarization, slows down (or may even block) any further reaction. This has the effect of lowering the voltage at which the battery is actually discharging, or restricting the magnitude of the current that can be taken from the battery. When the battery is allowed to stand on open circuit, the natural diffusion of the components of the electrolyte permits the reactants and products to redistribute themselves more uniformly, thereby reducing (or even entirely overcoming) the effects of polarization. After this occurs, the battery can continue its discharge. Thus, dry cells can yield

TABLE 17.2 Standard Commercial LeClanche (Dry) Cell Sizes

Cell designation		Nominal dimensions°, in		
American National Standards Institute	International Electrochemical Commission	Diameter		Height of can
		Round Cells		
N	R1	27/64		1 1/64
N	R1	7/16		1 3/32
R	R4	17/52		1 3/8
AAA	R03	25/64		1 11/16
AA	R6	17/32		1 7/8
A	R8	5/8		1 7/8
B	R12	3/4		2 1/8
C	R14	15/16		1 13/16
CD	R18	1		3 3/16
D	R20	1 1/4		2 1/4
E	R22	1 1/4		2 7/8
F	R25	1 1/4		3 7/16
G	R26	1 1/4		3 13/16
6	R40	2 1/2		6
		Length	Width	Thickness
		Flat Cells		
F40	F40	1 1/4	27/32	7/32
F70	F70	2	1 1/4	7/32
F90	F90	1 5/8	1 9/16	13/16
F100	F100	2 13/32	1 13/16	1 27/32
		Wafer Cells		
F15	F15	9/16	9/16	1/8
F20	F20	15/16	17/32	1/8
F30	F30	1 1/4	27/32	1/8
F25	F25	15/16	15/16	7/32
F40	F40	1 1/4	27/32	9/32

° Bare cells, i.e., less manufacturer's finishing and labeling.

more energy on an intermittent discharge routine than on a continuous one. As can be expected from the foregoing, a normal discharge of a dry cell shows a continuously decreasing voltage until the desired cutoff voltage (usually 0.9 V) is reached. Since diffusion is working against polarization, cells discharging at low rates (where diffusion can keep pace with polarization) show very slowly decreasing voltages, dropping off rapidly, however, as the battery is exhausted. Cells discharging at high rates (where polarization effects dominate) show very rapidly decreasing voltages during discharge, but good recovery after a "rest" period. This is illustrated in Fig. 17.3.

PERFORMANCE DATA Because of the wide variety of circumstances under which dry cells are discharged, performance data are difficult to summarize. Nevertheless, some fairly good methods of correlating the anticipated performance with the use conditions have been developed. One fairly widely used procedure is to correlate the capacity, expressed as operating time, with the initial drain rate at the start of the discharge, i.e., after initial stabilization of battery voltage. Such discharges are based on the load having a constant resistance. A graphical presentation usually gives the service life in hours as the ordinate, against an abscissa of initial drain rate in milliamperes. For convenience, a log-log plot is used. As indicated earlier, minimum serviceable voltage also has an effect, and such plots usually show a family of curves for several end of discharge voltages. Figures 17.4 and 17.5 present representative rating curves for a round cell and a wafer cell, operating continuously and intermittently.

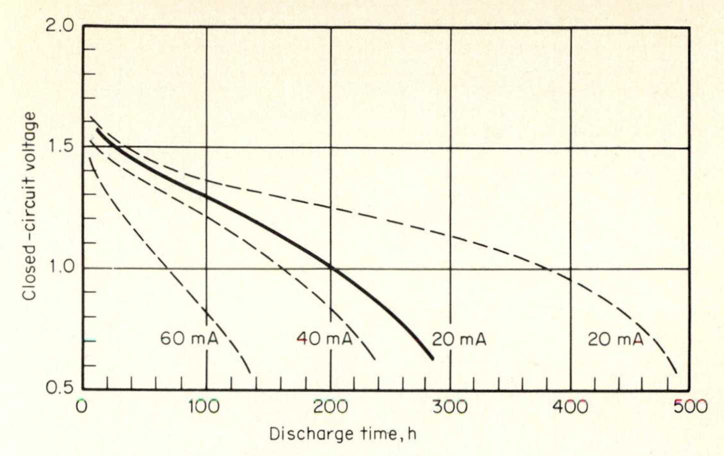

Fig. 17.3 Discharge characteristics of LeClanche dry cells; D-size cells showing continuous discharge (solid line) and high and low rates of intermittent discharges for 4 h/day (dashed lines).

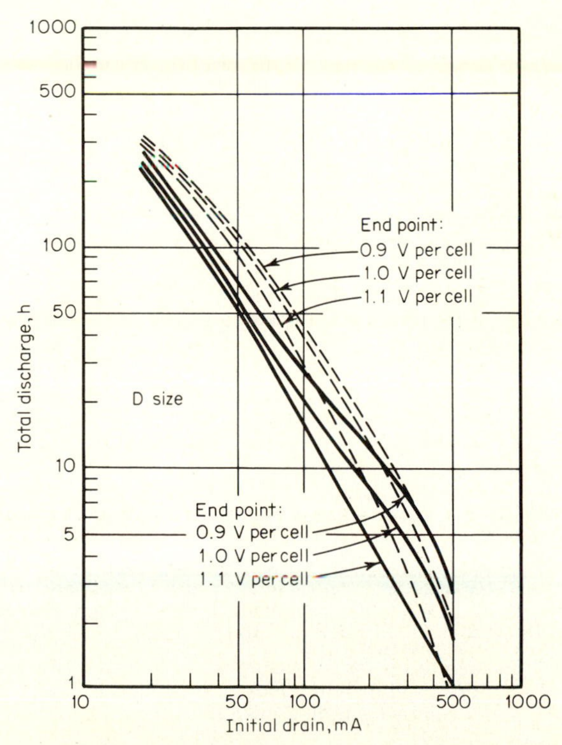

Fig. 17.4 Rating curves for typical round LeClanche cell. Solid line shows continuous discharge; dashed line intermittent discharge (4 h/day).

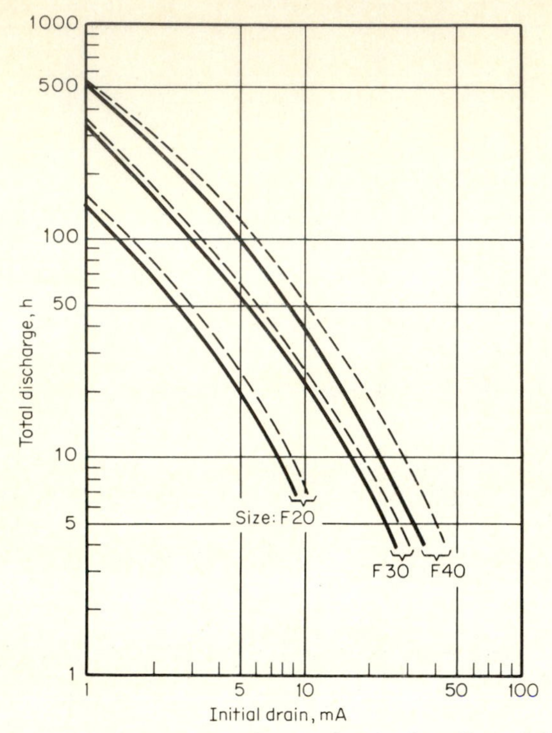

Fig. 17.5 Rating curves for typical wafer dry cells. Solid lines show continuous discharge; dashed lines show intermittent discharge (4 h/day).

example 17.1 How long can a D cell be expected to continuously operate a resistive load of 25 Ω to an end point of 0.9 V?

To approximate the initial drain, note that the initial stable cell voltage will be about 1.5 V. (See Fig. 17.3.)

$$I = \frac{E}{R} = \frac{1.5}{25} = 0.06 \text{ A}$$
$$= 60 \text{ mA}$$

On Fig. 17.4, locate an initial drain of 60 mA on the abscissa (horizontal axis). Follow this point vertically upward until it intersects the topmost solid curve (which represents a 0.9-V end point) and read 53 h as the ordinate (vertical value) of that point.

example 17.2 How long can the cell of Example 17.1 be expected to operate if used intermittently for four hours per day to the same end point?

For this problem, follow the 60-mA initial drain line until it intersects the topmost dashed line, and read its ordinate as 90 h.

The manufacturers of dry batteries have considerable control over the performance of their batteries through control of the mix formulations. Those batteries intended for essentially low drain use, but requiring high capacity (as for example in powering transistor radios) use mixes designed to give these properties; others needing only low capacity, but at high drain rates (as for example in photoflash applications) use mix formulations that give this combination. This is illustrated in Fig. 17.6 showing how discharge characteristics can be controlled even in a single physical size.

example 17.3 How much longer will a flashlight having an initial drain of 200 mA operate on a heavy-duty D than on a general-service flashlight battery if both are operated to a 1.0-V/cell end point?

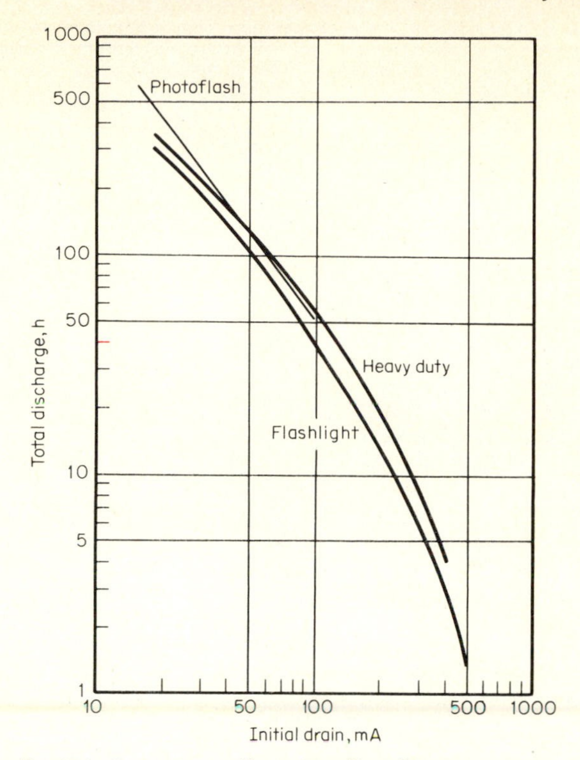

Fig. 17.6 Rating curves illustrating effect of mix composition; D-size cells showing intermittent discharge (4 h/day), low voltage percent.

Using Fig. 17.6, find the initial drain of 200 mA on the abscissa, and follow it vertically upward until it intersects the flashlight battery line and read 15 h as the ordinate. Then continue vertically until the 200-mA line intersects the heavy-duty battery line and read 20 h as the ordinate. The heavy-duty battery will therefore yield $20 - 5 = 5$ h of additional service.

Another means of specifying capacity is in terms of the amount of electricity that can be drained under the conditions of discharge. This is usually stated in ampere-hours, milliampere-hours, or ampere-minutes. Capacity expressed in this way, however, is sometimes difficult to arrive at since with a constant load the current will vary with time as a result of the battery voltage varying with time. Under such conditions, the ampere-hour capacity can be taken as the product of the average current and the discharge time.

AMERICAN NATIONAL STANDARDS INSTITUTE REQUIREMENTS For many purposes, the American National Standards Institute has developed test procedures and minimum specified requirements. They are described in detail in National Bureau of Standards Handbook 71, but some of the more important test procedures and specifications are abstracted in Table 17.3.

example 17.4 A radio A battery uses 3F-size dry cells in series. How shall it be tested for compliance with ANSI Standards? Does this battery meet ANSI Standards if it runs 41 days on this test?

Using Table 17.3, the test shown for radio A service is set up to operate the battery daily for one four-hour period each day. The load will be $3 \times 25 = 75\ \Omega$ to an end point of $3 \times 1.0 = 3.0$ V. If the battery runs 46 days, it will have yielded $41 \times 4 = 164$ h. Since the ANSI requirement is 140 h, this battery is well over the minimum requirement.

TABLE 17.3 ANSI Test Procedures and Minimum Performance Specifications

Test name	Daily discharge schedule	Resistance per $1\frac{1}{2}$-V unit, Ω	End point per $1\frac{1}{2}$-V unit, Ω	Cell size	ANSI initial requirement
General-purpose flashlight to represent 0.5-A lamp	1.5-min period	2.25	0.65	D general purpose	400 min
General-purpose flashlight to represent 0.3-A lamp	1.5-min period	4	0.75	C AA	325 min 80 min
General-purpose flashlight to represent 0.22-A lamp	1.5-min period	5	0.75	AAA	50 min
Heavy industrial flashlight to represent 0.3-A lamp	32 4-min periods	4	0.9	D industrial	800 min
Light industrial flashlight to represent 0.3-A lamp	8 4-min periods	4	0.9	D general purpose D industrial	600 min 950 min
Railroad lantern to represent 0.15-A lamp	8 $\frac{1}{2}$-h periods	8	0.9	F	45 h
Photoflash test	60 1-s periods (1 discharge/min for 1 h/day)	0.15	0.5 D size, 0.25 C & AA sizes	D photoflash C photoflash AA photoflash	800 s 700 s 150 s
Radio A	1 4-h period	25	1.0	F G	140 h 170 h
Radio B	1 4-h period	$166\frac{2}{3}$	1.0	F40 A F90	30 h 130 h 225 h
Heavy intermittent	2 1-h periods	$\frac{2}{3}$	0.85	F (4 cells in parallel) #6 general purpose #6 industrial	70 h

SHELF LIFE Among the special characteristics of dry batteries, a very important one is shelf life. As previously discussed, batteries produce their electric energy as a result of reactions that take place simultaneously at two otherwise independent electrodes. Under certain circumstances it is possible for each electrode reaction to take place independently of the other, thus not providing useful electric energy to an external circuit. Such reactions are referred to as "local action" phenomena, and have the effect of using up the materials without providing useful electric energy. In a battery, local action predominately affects the anode or negative electrode. Usually one of the by-products of local action is the generation of hydrogen gas which can result in a buildup of internal pressure, and some provision must therefore be made to allow hydrogen to escape. This is usually done by making the central carbon electrode porous so that hydrogen can slowly diffuse safely from the cell. It should be noted that this represents a design difficulty, since water which must be kept in the cell in order for the cell to operate properly, can also escape by this route (particularly at elevated temperatures).

Through developmental efforts over the years, the shelf life of quality dry batteries in casual storage (i.e., at normal temperatures and humidities) has reached the 18 to 21 months level, at which time the battery can still give up to two-thirds of its original service life.

In considering the storage of LeClanche cells, low temperatures are very beneficial. For example, a dry cell stored at 32°F will retain virtually all its original capacity for as long as two years. It should be noted, however, that the battery should be allowed to warm up to room temperature before it can yield its full capacity. This should be done slowly to prevent condensed moisture from physically disrupting the battery assembly.

The LeClanche system is also capable of designs that produce high voltages in small spaces. For such batteries, the flat-type cell or wafer-type cell is most frequently used. These batteries find application where high capacities are not required, but high voltages are. Since the flat cell can be readily stacked in rectangular spaces, the efficiency of space utilization for such batteries is high. In addition, the series connection of a large number of cells to produce the required voltages can be efficiently handled. Flat cells or wafer-cell batteries have therefore found extensive use in instrumentation, as radio B batteries and even radio A batteries (particularly where the power requirements are not excessive).

The special electrical properties and variety of available designs and shapes have resulted in the very wide use of LeClanche batteries in flashlights, radio receivers and transmitters, photographic equipment, portable meters, and other (generally low-power) devices.

ALKALINE BATTERIES At first appearance, alkaline-MnO_2 primary cells, sometimes called alkaline dry cells, or simply alkaline cells, seem to be very closely related to LeClanche cells. In point of fact, there are a great many similarities such as the electrochemically active materials which are the same (MnO_2 cathode, zinc anode) as illustrated in Fig. 17.7. In addition, the similarity is continued in available physical

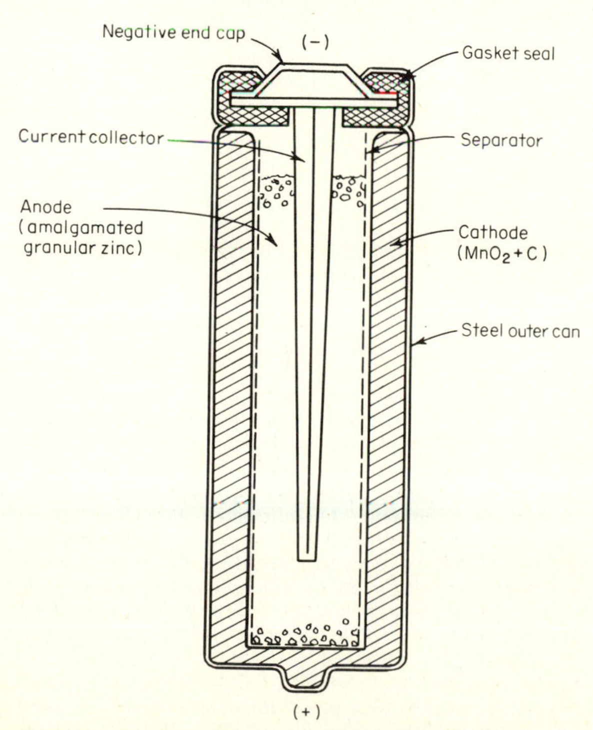

Fig. 17.7 Schematic diagram of a typical alkaline primary cell.

sizes and shapes. However, in the alkaline cell, a solution of a strong alkali, potassium hydroxide, is used as the electrolyte in place of the $ZnCl_2$ and NH_4Cl of the conventional LeClanche system. In addition, the zinc exists in granular form so that it presents a very extensive surface to the electrolyte. These factors greatly alter the electrical and other characteristics of the battery. Polarization effects in alkaline cells, for example, are much less pronounced than in conventional dry cells (under equivalent discharge conditions), and much of the energy can be withdrawn at higher power levels. Along related lines, it is also found that their discharge characteristics are more easily defined and catalogued since the voltage variation during discharge is likely to be somewhat less than in a LeClanche cell. Alkaline dry batteries are, therefore, rated according to their capacity, usually cited at a particular drain rate. For convenience the initial drain rate is used, and the capacity is usually expressed in milliampere-hours or ampere-hours at this rate. Table 17.4 summarizes the sizes and electrical characteristics of the commercially important alkaline cell sizes.

TABLE 17.4 Sizes and Characteristics of Alkaline Primary Cells

	Capacity			
Size designation	mAh	At initial drain of mA	Diameter,* in	Height,* in
N	580	18	0.47	1.12
AAA	750	18	0.41	1.75
AA	1,800	60	0.55	1.94
C	5,000	100	1.31	1.93
D	10,000	300	1.31	2.37

* Finished cells; i.e., including manufacturer's labels, etc.

example 17.5 How long can an alkaline AA cell be expected to operate at a drain of (*a*) 60 mA, (*b*) 40 mA, (*c*) 100 mA?

solution From Table 17.4, the capacity of an AA alkaline cell is 1800 mAh at a drain of 60 mA. The operating time therefore will be 1800/60 = 30 h. At 40 mA the capacity will be slightly higher, but for estimating purposes would still be taken as 1800 mAh: 1800/40 = 45 h. At 100 mA the capacity will be slightly lower, but for estimating purposes can still be taken as 1800 mAh: 1800/100 = 18 h. (At higher current drains, the capacity should be determined for the specific use and circumstances, the capacity given in the table being used only for approximations.)

Like LeClanche cells, alkaline cells have generally good shelf-life characteristics. As a matter of fact, alkaline cells are actually sealed since local action reactions in healthy cells are not serious. As a consequence, hydrogen production is very much slower than in LeClanche cells, and in fact by proper care during manufacture can be made negligible. Since alkaline cells are sealed, there is also no loss of moisture during storage, and alkaline cells therefore accept prolonged storage very well. As a matter of fact, a shelf life of two to three years (to 80 percent of original capacity) is common. Also, because they are sealed, alkaline cells can be stored at higher temperatures than LeClanche cells. Storage up to 130°F for extended periods of time is entirely feasible. However, since elevated temperatures accelerate local action reactions, it is not recommended that they be exposed to temperatures above 130°F for any length of time. In the other direction, as the temperature of the cells is reduced, there is only a slow falloff in capacity for alkaline cells, making them well-suited for operation in moderately low-temperature ambients. Quantitatively the effects of temperature are summarized in Table 17.5.

Although the alkaline dry cell is generally not troubled with the evolution of hydrogen gas, under adverse conditions such as overdischarging or excessive temperatures some gassing can occur, and most alkaline cells are therefore provided with a safety blowoff which permits hydrogen to be vented at a certain pressure so that the hydrogen can be harmlessly relieved.

Because of their basic and special properties and ability to provide relatively high power levels, alkaline dry batteries are now extensively used to power portable radios (particularly where high power levels are required), in photographic equipment, and

TABLE 17.5 Effect of Ambient Temperature on Capacities of Alkaline Primary Cells

Discharge temperature, °F	Approximate Service, %*		
	Light drain	Medium drain	Heavy drain
113	100	100	100
70	100	100	25
30	70	40	25
−4	25	20	5

Note. When discharge times (at normal ambient temperatures) exceed approximately 30 h, drains are considered light; when less than approximately 15 h, drains are considered heavy.
* Cells cooled to ambient temperature before discharge; service at 70°F to 0.8-V end point taken as 100%.

for driving fractional horsepower electric motors, such as are used in electric shavers, hobbycraft, and the like.

MERCURY DRY BATTERIES Another cell system using an alkaline electrolyte that has achieved a significant commercial position is the mercury dry cell. The outstanding characteristics of mercury cells is that they represent the highest available capacities for a given volume of any of the primary batteries. As can be seen in Fig. 17.8, one type of mercury battery is very similar in construction to the alkaline battery. Physically the major difference is that mercuric oxide (HgO) is substituted for the manganese dioxide, but this brings about some profound changes in cell characteristics. In addition to having very high capacities, mercury cells also have an outstanding shelf life since mercury or mercury compounds have a strong effect in reducing local action. As a matter of fact, the shelf-life problem as far as mercury cells are concerned is the ability to thoroughly seal the cell rather than any chemical action that might take place.

Because of this, the practical shelf life seems to be limited to about two or three years on larger sizes, and perhaps to one or two years on the smaller sizes. Mercury cells are marketed in essentially the same sizes as alkaline and LeClanche dry cells, but in addition are available in extremely small sizes called button cells, schematically illustrated in Fig. 17.8. Table 17.6 gives the general sizes and weights of the more important commercial sizes.

The temperature characteristics of mercury cells are very similar to those of alkaline cells, except that at low temperatures capacities fall off a little more rapidly, so that the lowest temperature suitable for mercury-cell operation would be about 30°F, as shown in Table 17.7.

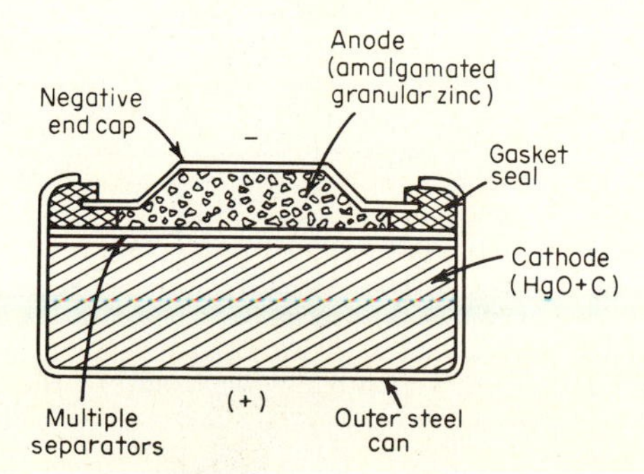

Fig. 17.8 Schematic diagram of typical mercury button cell. Note that cylindrical mercury cells closely resemble alkaline primary cells, shown in Fig. 17.7, differing principally in the composition of the cathode mix.

TABLE 17.6 Sizes of Commercially Important Mercury Cells

Size designation, ANSI	Rated capacity		Nominal dimensions, in*	
	mAh	At drain of mA	Diameter	Height
M5	36	2	0.31	0.14
M8	100	3	0.46	0.13
M15	160	5	0.46	0.21
M25	350	3	0.46	0.57
M40	1000	35	0.63	0.65
M55(AA)	2400	50	0.55	1.97
M100(D)	14000	250	1.28	2.39

* Bare cells, i.e., less manufacturer's finishing and labeling.

TABLE 17.7 Temperature Characteristics of Mercury Cells

Discharge temperature, °F	Approximate service, %*	
	Light drain	Heavy drain
113	100	100
70	100	100
40	95	7
30	6	2

Note. When discharge times (at normal ambient temperatures) exceed approximately 25 h, drains are considered light; when less than approximately 25 h, drains are considered heavy.
* Cells cooled to ambient temperature before discharge; service at 70°F to 0.9-V end point taken as 100%.

example 17.6 A mercury battery can operate a particular radio receiver for 160 h at normal ambients. How long can it be expected to operate the receiver in an ambient of 35°F.?
Table 17.7 indicates that the battery would operate the radio (i.e., a light drain) for 0.95(160 = 152 h at 40°F, and only 0.06(160) = 9.6 h at 30°F. To estimate performance at 35°F, it is necessary to extrapolate between these values: 0.5(152 − 9.6) + 9.6 = 80.8 h.

In addition to their extremely good shelf life and high capacity, mercury cells have another outstanding characteristic: namely, uniformity of discharge voltage. As can be seen from Fig. 17.9, mercury cells have an almost uniform discharge voltage, except for a short time at the very beginning and of course at the very end of the discharge.

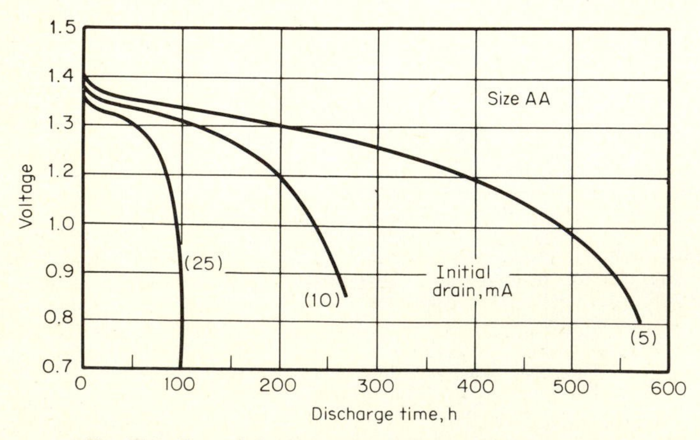

Fig. 17.9 Typical discharge characteristics of mercury cells.

This permits designers of instrumentation and electronic equipment to use circuits that need not accommodate voltage fluctuations. Not only is the voltage of mercury cells stable, but it is also highly reproduceable. Mercury cells, therefore, find wide application in instrumentation as stable voltage working cells (to be periodically matched against standard cells for very high precision and long service). As a consequence of their high capacity, availability in very small sizes, good shelf life, and stable discharge voltage, mercury cells have found a wide range of applications, as, for example, in portable voice recorders, radio transceivers, and the like.

MAGNESIUM DRY BATTERIES No discussion of primary batteries would be complete without significant mention of the new magnesium dry cell. Fundamentally, the magnesium dry cell is an exact counterpart of the LeClanche system in which the outer can is made of magnesium alloy instead of zinc.

This rather "small" change is accompanied by a number of more subtle changes that result in an entirely new system with a set of characteristics all its own. The outstanding properties of the magnesium dry battery are its ability to withstand high-temperature, high-humidity storage, and extremely long shelf life under casual storage. In addition, the magnesium dry battery operates at voltage levels considerably higher than conventional LeClanche or mercury dry cells. This is due to the fact that magnesium is a far more electropositive (i.e., active) metal than zinc. However, because of its reactivity, one would ordinarily expect a magnesium cell to be unstable, but what actually happens is that magnesium forms a very thin but very impervious oxide coating, thereby protecting the underlying metal from local action. Although special electrolytes must be used (generally solutions of magnesium bromide or magnesium perchlorate) and the cell must be closed by very special techniques in order to retain moisture and still permit hydrogen to escape (since hydrogen is liberated as a by-product of the discharge), it is really the protective film which forms on the magnesium that has made the magnesium dry cell possible. At their present stage of development, these batteries can be successfully stored at 160°F for 30 days without significant loss of capacity (i.e., less than 10 percent).

As yet, the total production from all suppliers is being funneled into military applications, but it is anticipated that units will soon be available for normal commercial applications. Cell sizes have not as yet been standardized, and only cursory performance data can be presented at this time. What general information is available is presented in Table 17.8 and Figure 17.10.

TABLE 17.8 Sizes of Typical Magnesium Cells

Cell designation, ANSI	Nominal dimensions, in*	
	Diameter	Height
N	0.42	1.09
R	0.53	1.38
A	0.63	1.88
CD	1.00	3.19
D	1.25	2.25

* Bare cells, i.e., less manufacturer's finishing and labeling.

As can be seen from Table 17.1, magnesium dry batteries are fully in a position to compete with mercury and alkaline batteries in terms of power capability per unit weight and capacity per unit volume. Ultimately, magnesium dry cells should also be more favorably priced than their mercury or alkaline counterparts. At least for the immediate future, however, on the basis of costs, magnesium dry cells are not expected to make serious inroads into conventional dry battery areas, except for those applications where extremely long shelf life (even up to five years) is required, or where the batteries are to be used in hot, humid areas where the logistics of getting fresh LeClanche cells to the user are poor. Magnesium batteries, since they can withstand extremely adverse storage conditions for extended periods of time, can be delivered

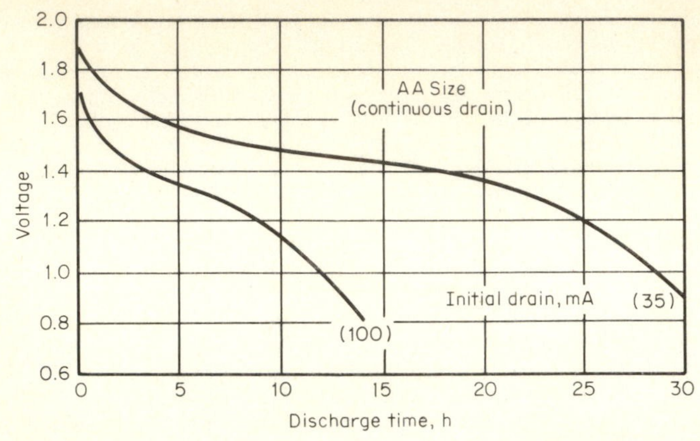

Fig. 17.10 Typical discharge characteristics of magnesium dry cells.

to the ultimate user in tropical and semitropical countries at full or almost full capacity. The outstanding shelf life of the system (as compared to that of conventional dry cells) is illustrated by Fig. 17.11. Here both the carbon-zinc and magnesium types show very little capacity loss after three months' storage at room temperature, but for storage at elevated temperatures the superiority of the magnesium system quickly asserts itself. At really high storage temperatures, the carbon-zinc type fails rather quickly, while the magnesium type retains most of its original capability.

There are other primary batteries that have achieved a degree of commercial value, such as silver-zinc, air-depolarized, and others, but these are not yet in sufficiently general use to warrant discussion at this time. However, it should be mentioned that the silver-zinc primary does have an outstanding high rate capability and a capacity as much as 80 percent greater than its secondary counterpart, described later in this chapter. Physically it is very similar to the dry charged form of the secondary version (shown in Fig. 17.15) except that it usually does not have an ion-restructive diaphragm. The higher capacity is due to the fact that the silver electrode can be oxidized to a higher level before cell assembly than can conveniently be done after cell assembly and activation (as would be the case for secondary operation).

Before going on to rechargeable systems, Fig. 17.12 is included to illustrate the actual physical forms of some of the primary systems we have considered.

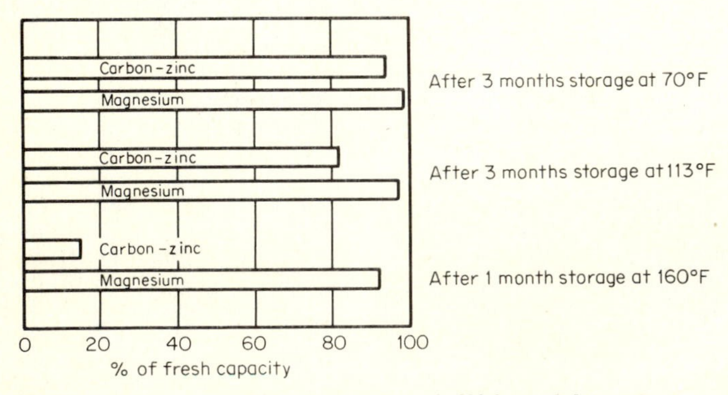

Fig. 17.11 Comparison of high-temperature shelf-life capabilities of carbon-zinc and magnesium dry batteries.

Fig. 17.12 Sample display of primary batteries. *(Courtesy Burgess Battery Company)*

17.3 CHARACTERISTICS OF IMPORTANT COMMERCIAL SECONDARY TYPES

Although there are a large number of types among the secondary, or rechargeable batteries, only three have achieved any really significant commercial importance. These are the lead-acid, nickel-cadmium, and silver-zinc systems.

LEAD-ACID BATTERY Of these, the lead-acid system is the most important from a commercial point of view and is widely distributed. It is available in a very large variety of sizes and shapes to satisfy applications requiring a broad range of capacities and power levels. A summary of commercially important types is given in Table 17.9. It should be noted that while the battery sizes are arranged in order of increasing capacity, some are for 6-V (three-cell), 12-V (six-cell), or 2-V (single-cell) units so that the table covers a much broader range than at first appearance.

An interesting point should be noted concerning the 50- and 100-Ah units which are approximately the same size and weight in spite of the twofold capacity difference. This is explained by the fact that they both have the same energy content, 600 Wh (obtained by multiplying the capacity in ampere-hours by the battery voltage, as indicated above). It is important to realize that the capacity and power capability of a rechargeable battery depend in very large measure on its prior history. A typical situation, particularly in lead-acid batteries, is that the capacity builds up during early cycle life,

TABLE 17.9 Representative Sizes and Weights of Lead-Acid Batteries

Nominal size Ah capacity @ 20-h rate	Number of cells	Nominal dimensions, in*			Approximate weight
		Length	Width	Height	
4	3	2¾	1⅞	4	1¾
8	3	5½	2	4¾	4
30	6	8	5⅜	9⅛	27
50	6	11½	7⅛	9⅛	50
100	3	10⅜	7⅛	9⅛	46
200	6	21⅝	11⅛	10¼	147
280	1	5⅛	6¼	15¼	60
420	1	5⅛	6¼	21⅜	84

* Overall dimensions, i.e., including terminals and vent plugs.

then maintains a substantially uniform value for most of the battery's useful life, and finally falls off (more and more rapidly) toward the end of life. Unless some accidental life-terminating event occurs, such as a short circuit or loss of electrolyte through a cracked case, etc., the end of life of the battery will be reached when its capacity falls below that which is useful for the application. For a great many applications, this will be at about 60 percent of the original capacity.

As for the rating of new batteries, different manufacturers hold to somewhat different views. Some, for example, might rate a battery on what it will do on the very first discharge cycle in the user's hands, whereas others may take advantage of the fact that the capacity normally builds up during early cycle life, and will rate a battery after it has been "conditioned" in use. Where the capacity of the battery is critical, the user should determine which of these systems is the basis for rating the battery.

The conditions of discharge also have a very profound effect on the capacity and power capability of rechargeable batteries. As a general rule, a battery will display its greatest capacity when discharged at very low rates. For lead-acid batteries this rate is generally chosen to be the 20-h rate. A great deal of confusion has arisen as a result of the industry method of expressing the discharge rate of a battery. The phrase "the 20-h rate" means that the battery is being discharged at such a current as will exhaust the capacity of the battery in 20 h. The confusion results from the fact that it is necessary to know what the capacity of the battery is at the particular current drain in order to determine the current at which it should be discharged.

To simplify this situation, the discharge current is now generally related to a nominal capacity of the battery. For example, if the nominal capacity of a battery is 10 Ah, a discharge rate of 0.5 A would be called the "20-h rate." Most leading producers now cite capacity at the 20-h rate and at normal (78°F) ambient temperatures. For lead-acid batteries, there is usually very little difference in capacity as the rate of discharge is increased from the 20-h rate to perhaps the 10-h rate. But thereafter, as the discharge rate is further increased, there is an increasingly rapid falloff of capacity. Except for very special designs, the half-hour rate is generally considered to be the maximum sustained rate at which lead-acid batteries can be used with any degree of efficiency.

The lead-acid system has a number of characteristics that are particularly noteworthy. Although these batteries do not withstand high-temperature storage very well (anything above 130°F being somewhat deleterious), they do withstand low-temperature storage extremely well and are capable of yielding considerable power and operation at low temperatures. As a typical example, the capacity of a lead-acid battery is likely to drop off only about 30 percent as the temperature is decreased from room temperature to 0°F. Additional information is given in Table 17.10. In determining the capacity of lead-acid batteries from Table 17.10, it should be noted that the discharge rate has considerable significance. In using the table, the column calling for the discharge rate at which the battery is to be operated as measured at room temperature must first be found, and this column is then followed downward to the appropriate temperature. The figure cited is the capacity (expressed as a percent of the 20-h rate capacity at room temperature) at the same current drain.

TABLE 17.10 Effect of Temperature on Capacity of Lead-Acid Batteries

Temperature, °F	Running time, hours*					
	1	5	10	50	100	500
	Drain current, amperes					
100	118	21	10	1.4	0.6	0.06
80	114	20	9.0	1.3	0.5	0.05
60	106	19	8.3	1.1	0.5	0.05
40	97	17	7.5	1.0	0.4	0.04
20	90	15	6.7	0.9	0.3	0.03
10	82	13	5.6	0.8	0.2	0.02

* Capacity at 80°F and 20-h rate = 100 Ah.

example 17.7 A battery has a capacity of 100 Ah at the 20-h rate at room temperature. What is its capacity at the 10-h rate at room temperature and at 20°F?

solution For a 100-Ah battery the current drain at the 20-h rate is 5A. From Table 17.10, its capacity at room temperature at the 10-h rate will be 91 percent of 100 Ah, or 91 Ah. This is the nominal capacity of the battery at room temperature and at the 10-h rate. For this battery, therefore, the discharge current at room temperature and the 10-h rate is 9.1 A. Following the 10-h rate column downward to 20°F, the capacity at a drain of 9.1 A will be 66 percent of what it was at room temperature, or 60 Ah. This table may seem a little difficult to use, but it is a very serviceable one in that it contains a great deal of information since interpolations for other rates and temperatures give entirely acceptable results.

In a rechargeable battery, charge retention, i.e., the ability to maintain capacity after charge, corresponds to shelf life in a primary battery. In a conventional lead-acid battery (sometimes referred to as an antimonial alloy lead battery), the charge retention varies considerably from the early stages of life to the latter stages, falling off slowly but continuously as the battery is cycled. When new, the battery is very likely to be capable of losses of less than 10 percent per month in (casual) storage, but this figure could easily approach 100 percent for an old unit nearing the end of its useful life. A newer type of lead-acid battery is now available, however (i.e., the calcium alloy lead battery), which has a remarkable degree of charge retention that does not fall off to any significant extent as the battery is used. In a calcium alloy type, the charge retention is likely to be better than 85 percent after as long as one year in casual storage, and this will be fairly uniform throughout its life. Generally speaking, this latter type is used primarily for standby power since it has only a limited cycle life capability.

Because of the variety of properties that can be developed by design and/or electrical treatment, lead-acid batteries have achieved an amazing variety of applications. Apart from the popular ones of providing starting, lighting, and ignition power for automobiles, operating electric fork-lift trucks and similar commercial applications, they have now found significant applications in standby power (for telephones, industrial purposes, hospitals, emergency lighting, etc.), as well as in a variety of special applications such as portable TV, electric power tools, and ordnance guidance systems, to name but a few of the more important ones.

NICKEL-CADMIUM BATTERIES A second type of rechargeable battery of great and continuously growing commercial importance is the nickel-cadmium system. Within the basic system, there are several different engineering types and designs, each design reflecting the development of specific properties for a specific field of application. The pocket-type nickel-cadmium battery, for example, is an extremely long-life unit capable of thousands of very deep discharges. Because it can also moderate rates of discharge, it is used principally where motive power is desired, such as in electric wagons and the like. Of more recent development, the sintered-plate-type nickel-cadmium battery is available in both the sealed and vented forms. The sintered-plate system is by far the fastest growing type in the nickel-cadmium field and will be discussed in more detail.

Basically, the sintered-plate type utilizes a very porous nickel to support the active materials and serve as a current path. This matrix is made by sintering finely divided particles of carbonyl nickel powder (a special grade of "macelike" particles). The resulting plaques can be as much as 80 percent porous and are impregnated with the active materials, nickelic hydroxide in the positive electrode, and metallic cadmium in the negative. As is common in all nickel-cadmium types, the sintered-plate nickel-cadmium battery utilizes a potassium hydroxide electrolyte. A schematic diagram of a sintered-plate nickel-cadmium battery is shown in Fig. 17.13, and common commercially available sizes are given in Table 17.11.

Vented cells are usually encased in plastic containers, while sealed cells are usually in thin-walled steel containers. Both are extremely rugged, and the battery can take a good deal of physical abuse, as well as electrical abuse, without permanently deleterious effects. Nickel-cadmium batteries are capable of very high rates of discharge, and as a matter of fact many applications will discharge such batteries at the one, and even the one-half hour rate at good efficiencies. The high rate capability of these batteries also makes them useful for "pulsed" discharges (i.e., very high rate, very short duration discharges).

In addition to their fine physical characteristics, nickel-cadmium batteries perform

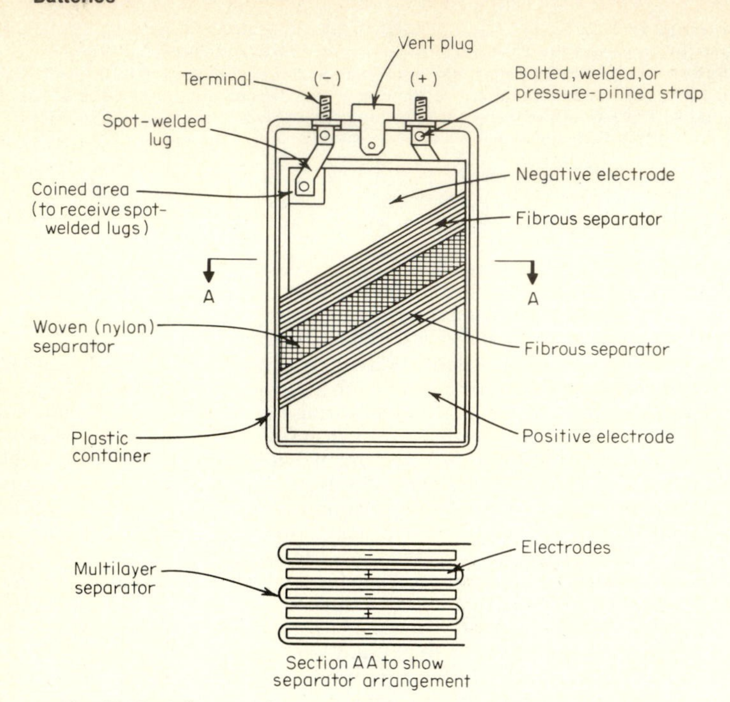

Fig. 17.13 Schematic diagram of sintered-plate nickel-cadmium cell.

TABLE 17.11 Sizes of Common Nickel-Cadmium Cells

ANSI size designation	Nominal capacity, mAh	Nominal capacity, At drain of mA	Size, in: Diameter	Size, in: Height
		Sealed, Round		
	50	5	0.61	0.24
	150	15	0.98	0.28
	450	45	1.70	0.30
AA	450	45	0.56	1.95
C	1200	120	0.88	1.67
D	4000	400	1.30	2.43

Capacity: Nominal Ah	Capacity: At drain of amperes	Size, in: Length	Size, in: Width	Size, in: Height
		Sealed, Prismatic		
1.5	0.15	1.36	0.75	3.36
4.5	0.45	1.93	1.67	3.13
7.5	0.75	1.93	1.67	4.25
15	1.50	3.56	1.34	4.90
		Vented, Prismatic		
6.5	0.65	0.96	2.18	4.06
24	2.4	1.08	3.16	8.28
40	4.0	1.39	3.16	9.40
120	12.0	2.17	6.81	8.73
230	23.0	3.17	8.14	9.40

well at both high and low temperatures, but capacity does fall off if the temperature is raised or lowered from ambient room temperatures. The temperature characteristics of typical nickel-cadmium cells are shown in Table 17.12. As can be seen from Table 17.12, the falloff in capacity as the temperature is lowered is about the same as for lead-acid types for the temperature range covered. However, as the temperature is lowered beyond the range covered in Tables 17.10 and 17.12, the performance of the nickel-cadmium battery falls off more slowly than its lead-acid counterpart. Here only qualitative data can be cited (since we are in an area where specific designs play a most important role), but a nickel-cadmium battery can be quite useful down to −40°F (i.e., yielding about 5 to 10 percent of its normal capacity), whereas a lead-acid battery would probably have a practical limit of −20°F. There is another very important factor to consider as regards high rate capability at low temperatures, in that the current capability per unit weight or per unit volume, for short-duration discharges, is very favorable for nickel-cadmium sintered-plate batteries. For these reasons this type of battery is favored for low-temperature engine-starting applications.

TABLE 17.12 Temperature Characteristics of Nickel-Cadmium Cells

Discharge temperature, °F	% of room-temperature capacity at 10-h rate
115	93
78	100
40	93
32	90
0	60
−5	50

The charge retention characteristics of a nickel-cadmium battery are not too far different from those of a lead-acid battery, but in at least one respect the nickel-cadmium battery does have an outstanding characteristic: it can withstand extremely long periods of idle storage (up to 7 to 10 years being not at all uncommon) without permanent, deleterious effects. In fact, when it is desired to store a battery, it should be completely discharged all the way down to zero volts (by even short-circuiting the battery at low rates when its voltage has dropped to a point where this can be done safely). By contrast, a lead-acid battery must be electrically maintained during idle storage (by a monthly low-rate trickle, or some other routine-freshening charge). A nickel-cadmium battery needs no such electrical maintenance, and when placed into service after long idle storage will, after a few cycles, pretty much return to its proper capacity.

Perhaps one of the most outstanding characteristics of the nickel-cadmium battery is its ability to accept a recharge very quickly. Since the back emf of the battery rises relatively slowly during charge, as indicated on Fig. 17.14, most of the charge can be completed at high rates before the back emf begins to seriously limit the charge acceptance. However, since the battery is capable of accepting charges at very high rates, it is not overly important to control the charge to a great degree, particularly at the early stages. This is true for vented cells to a much greater degree than for sealed cells since the latter can also accept high overcharge rates. During overcharge of vented cells, however, water is electrolyzed into hydrogen and oxygen, using up water and electricity, but otherwise having no further deleterious effect on the battery.

The nickel-cadmium system is thus a sophisticated combination of virtues and limitations. In spite of its high initial cost, when used under proper circumstances it can be quite economical since it provides the closest approach the battery industry has yet made toward the goal of long-life low (or no)-maintenance batteries. These batteries have therefore found extensive use in jet aircraft starting, radio and telephone communications, subcutaneous low-power medical applications, cold weather starting, low power starting (as in lawnmowers, chain saws, etc.), and, of course, in cordless appliances (such as toothbrushes, shavers, etc).

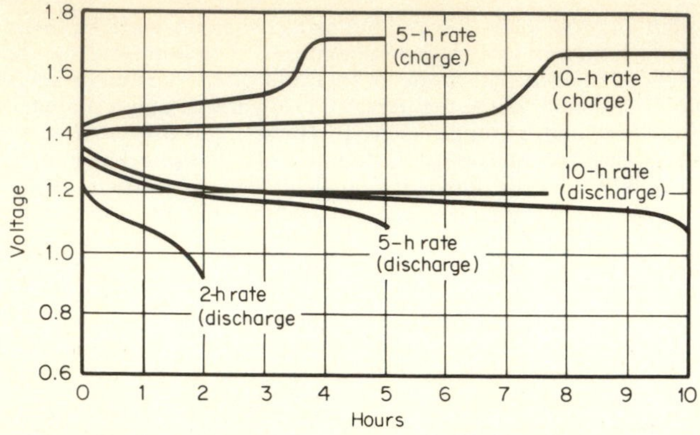

Fig. 17.14 Constant-current charge-discharge characteristics of nickel-cadmium batteries.

SILVER-ZINC BATTERIES In the spectrum of battery power sources, the silver-zinc system has achieved a firmly entrenched position principally because of its power capability. Whereas nickel-cadmium batteries do quite well at discharge rates as high as the one-hour rate, silver-zinc primaries and secondaries easily go to the ten-minute rate, and for pulse power even to the one-second rate with relatively good efficiencies. The general construction of silver-zinc secondaries is shown in Fig. 17.15. Essentially the primary version is made in about the same way, but the ion-restrictive diaphragm need not be used, nonwoven fabrics being used instead (to provide lower resistance and more rapid activation). Apart from stability in the dry condition (making it suitable for reserve battery applications), one of the principal virtues of the primary version is that it can take advantage of higher degrees of oxidation of the silver (since the positive electrode is "formed" before cell assembly under conditions favorable to silver oxidation). This manifests itself as a greatly increased discharge capacity.

Typical primary and secondary characteristics are shown in Fig. 17.16. This figure shows the voltage-time curves for a charge and discharge cycle of a silver-zinc secondary, and for the discharge of a primary. It is interesting to note that Fig. 17.16 makes

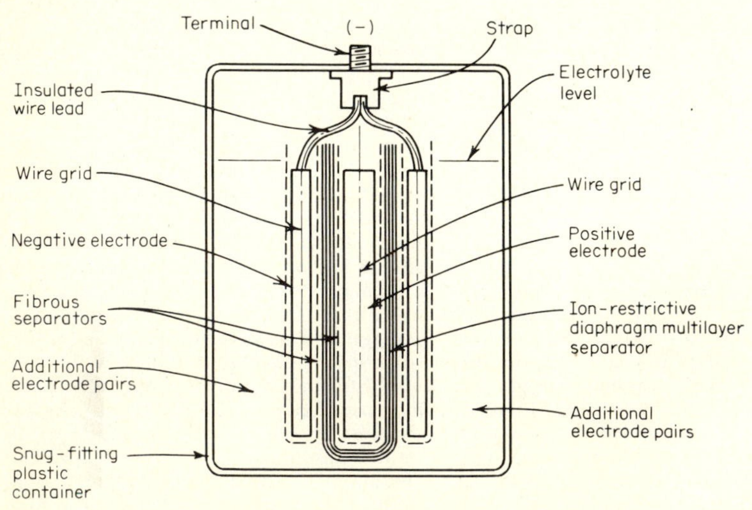

Fig. 17.15 Schematic diagram of typical silver-zinc secondary cell (end view to show separator configuration).

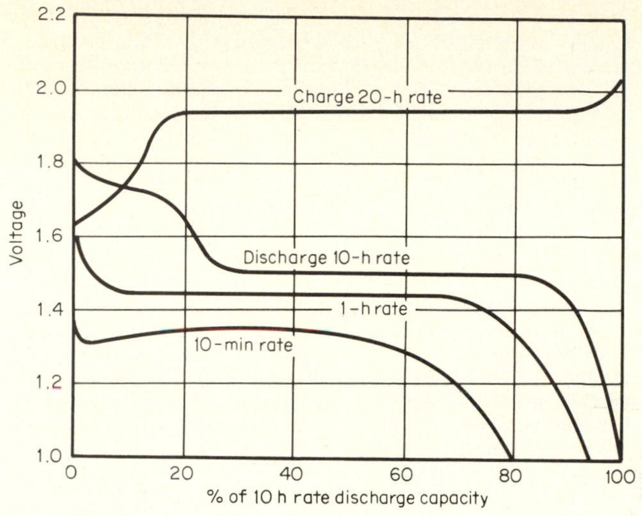

Fig. 17.16 Charge-discharge characteristics of typical silver-zinc secondary cell.

no reference to size (i.e., capacity) since the general shape of the curves fit all sizes with remarkable uniformity when each is operated at its appropriate current value. It is also interesting to note that while secondary batteries are usually furnished dry and uncharged, they can be furnished dry charged, in which case the first discharge will have the characteristics of a typical primary, reverting back to normal secondary characteristics for subsequent cycles. Of course, because of the presence of an ion-restrictive barrier, which usually is slow wetting, the activation time of a dry charged secondary will be as long as for any other silver-zinc secondary.

There are, of course, too many physical forms that have been developed for secondary batteries, but a representative sampling is shown in Fig. 17.17.

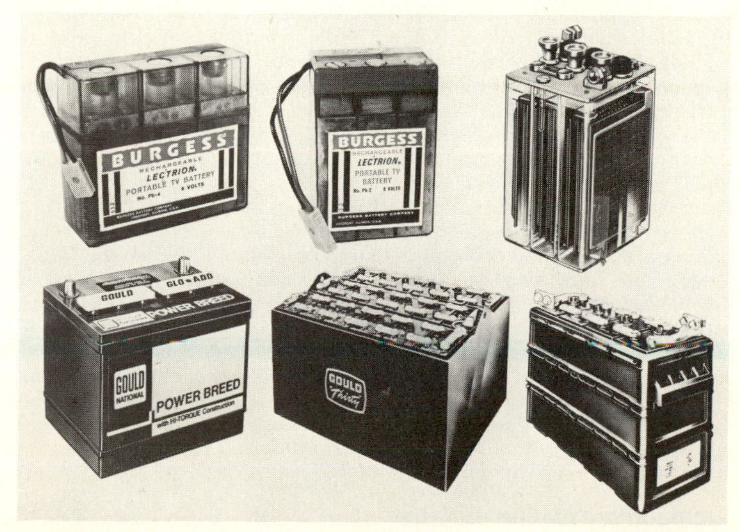

Fig. 17.17 Sample display of secondary batteries. (*Courtesy Burgess Battery Company*)

Referring now to the secondary version, the silver-zinc system shows a marked buildup in capacity during early (perhaps 20 percent of) life, followed by a long period (perhaps 70 percent of life) of reasonably stable capacity. Toward the end of life there is a rapid decline in capacity to a normal end point, usually about 60 percent of rated capacity. In operation the cycle life is highly variable, depending on use circumstances, but generally the range is from 20 for deep, rapidly repeated cycles, to perhaps 200 cycles of moderate (50 percent) depth, provided that sufficient time is allowed between discharge-charge-discharge to avoid temperature buildup.

High temperatures can be particularly harmful since the solubility of the negative electrode in the electrolyte is greatly increased, and deleterious degradation of the diaphragm also occurs. Overcharge also has seriously deleterious effects, but as in calcium-alloy batteries, a pronounced increase in back emf at the approach of the full-charge condition permits convenient control of the charge so that serious overcharging can be readily avoided.

Since silver oxide is one of the active materials, the cost of silver-zinc batteries is obviously high, limiting its applications to date to essentially military gear and special communication devices. Nevertheless it has an extremely good low-temperature capability, a modest charge retention (although this has been and is being improved), and a reasonable cycle life. For those applications requiring very high power levels, or extremely high bursts of energy (pulse loads) silver-zinc batteries indeed provide a convenient, reliable power source in small, lightweight packages.

17.4 MAINTENANCE

Unlike primary batteries which require no special attention, the maintenance of secondary batteries is of extreme importance, particularly where maximum performance and life are to be obtained. Charging has a profound effect in this regard, and several good methods have been developed. The constant-current method simply returns the capacity that has been discharged at a fixed current for a fixed period of time, or until the back emf (i.e., the countervoltage) of the battery rises to some predetermined value. It is therefore sometimes called the time-controlled or voltage-controlled method of charging. This is perhaps the simplest method in terms of the required control equipment, but does not account for, or take advantage of the fact that the battery has the ability to accept a charge very rapidly at the early stages of recharge (particularly after a deep discharge) and that this situation is reversed as the battery approaches the fully charged condition.

As the battery approaches full charge, the efficiency with which the battery accepts the charge falls off very rapidly (reaching perhaps only 4 to 5 percent at about 95 percent of full charge). This results in an undesirable condition in constant-current charging, since it is too slow at the beginning of the recharge (when the battery can accept charge efficiently) and too fast toward the end of recharge (when the battery can accept charge only slowly). A wasteful recharge not only manifests itself as an undesirable cost, but also as excessive positive grid corrosion (which will result in short life) and excessive watering frequency.

To improve this situation, we can take advantage of the fact that the back emf of the battery rises as the state of charge increases. Many manufacturers now market charging devices which use a modified constant-potential procedure in which the initial stages of the recharge are carried out at a high, but constant current, the value of which is essentially dependent on the power capability of the charging source. As the back emf of the battery reaches a predetermined point, which is toward the latter part of the recharge, automatic switching circuits change the system to a fixed potential recharge. This potential is set so that it is a little higher than the end of charge back emf of the battery. As the back emf of the battery approaches the potential of the charging circuit, the charging current falls off as the ability of the battery to accept the charge falls off. This turns out to be efficient, not only in terms of time and cost, but also in the avoidance of overcharge and in reducing watering frequency.

LEAD-ACID BATTERY MAINTENANCE Apart from the selection of a suitable charging procedure, the electrolyte in lead-acid batteries must be maintained at the proper concentration and level. Negative electrodes exposed to the atmosphere as a result of low electrolyte level are likely to oxidize very rapidly, a process that is only partially reversible by charging, and results in varying degrees of permanent damage to the electrode. When the concentration of sulfuric acid is allowed to fall, the capacity of the battery also suffers rather quickly since sulfate ions are a participant in the discharge reactions. In addition, the negative plates tend to sulfate (i.e., form an irreducible lead sulfate) in low-gravity acid.

On the other hand, if the electrolyte concentrates in the battery, the positive grids are likely to corrode more rapidly, especially during overcharge periods, and again the total life of the battery will be impaired. Since sulfuric acid is not lost from the battery as a result of the direct electrolytic reactions, only water must be replaced periodically to maintain the electrolyte composition and level. During overcharge, gases are evolved as a result of the decomposition of water, and this not only represents a loss of water, but also the gases produce an electrolyte spray above the upper portions of the cell. This electrolyte spray is extremely fine, and so quickly finds its way out of the cell as entrainment with the gases being evolved.

If this should occur, the top of the battery is very likely to get wet with sulfuric acid, which, being a very good electric conductor, permits stray currents to be set up, not only from cell to cell, but also from each cell to ground. With poor top-of-battery cleanliness, these currents can indeed reach very significant values, and the battery can be rather quickly discharged or even destroyed. To avoid this, many manufacturers have developed ingenious spray traps and have incorporated these in their vent plugs. However, careful and proper charging can go a long way toward eliminating this difficulty. Where electrolyte has come out of the cell, it is necessary to neutralize it (boric acid or bicarbonate of soda being recommended) followed by a thorough washing and drying of the top of the battery. This procedure will minimize the destructive effects of corrosion on external battery parts, terminals, and holders.

The loss of sulfuric acid as a result of spray entrainment will, of course, also impair the capacity of the battery. However, this sulfuric acid can be replaced, using chemically pure sulfuric acid, to bring the concentration (or specific gravity) of the electrolyte up to the manufacturer's recommended value.

In doing this, it is best to remove as much of the free electrolyte to which the additional acid is to be mixed to an external vessel. The new acid mixture is properly adjusted for concentration and can then be returned to the battery. If fresh acid is simply added to the battery, the dense, fresh acid will simply settle to the bottom and will probably attack the separators on the way down. The addition of fresh sulfuric acid also generates quite a bit of heat, and, of course, concentrated sulfuric acid is quite dangerous to handle. This operation should, therefore, be conducted only with proper protective clothing and facilities, and under qualified supervision.

SEALED UNITS To minimize maintenance, a number of brands of lead-acid batteries have been put on the market as "sealed" or "closed" units. Such batteries are generally made using the calcium alloy system which has an extremely high back emf (of the order of 2.8 to 3.0 V/cell when approaching the full charge condition). For this reason inexpensive automatic charge control equipment can be used to limit the amount of overcharge. This, coupled with the fact that local-action shelf losses are extremely small for calcium alloy batteries, has resulted in a condition in which not a great deal of water is used up during the normal use of the battery, nor is there any serious gassing on standby. Under these conditions the cell can essentially be "closed," since the need to replace water will have been reduced to the point where the cell can go without rewatering for its useful life. The fact that only very little gas is produced and even this will be principally hydrogen (a gas that is very difficult to contain) permits the gas to slowly but surely leak out via even very minute flaws in the seal. Since such batteries can, therefore, go for very extended periods without any maintenance, their use, particularly in small power applications and automobiles, is increasing.

Since many small power applications require a degree of portability (as, for example, power tools, etc.) it is also desirable to immobilize the electrolyte. This is usually done by forming a "gel" in the electrolyte by the addition of silica gel (finely divided SiO_2) to the electrolyte. This in effect absorbs the electrolyte, permitting the battery to be turned even upside down without reorientation of the electrolyte.

However, in immobilized electrolyte-sealed types, it is very desirable to avoid disruption of the gel, such as would occur with excessive gassing. Consequently such batteries must be recharged with perhaps a little more care than those having free electrolyte, although the usual rules still apply.

CHARGING NICKEL-CADMIUM BATTERIES The ability to be recharged quickly, and here we are talking of within an hour or so from a full discharge, is also true for the sealed version. However, in the sealed version, the danger of disrupting the battery physically during or even approaching the overcharge period is greater than in the vented type. The sealed nickel-cadmium battery is made possible by the fact that the negative electrode is capable of recombining the oxygen liberated by the positive electrode during the overcharge period, while at the same time it is possible to arrange the relative capacities of the electrodes so that the liberation of hydrogen at the negative electrode during the overcharge period is entirely avoided.

In addition, local-action losses during storage are such as to avoid the production of hydrogen, still further enhancing the conditions that permit the sealing of such cells. Thus, while a sealed cell can also be recharged rapidly, greater precautions (such as exercise of greater electric control) must be taken to reduce the rate as the cell approaches gassing conditions. Generally speaking, a sealed nickel-cadmium cell will be operated at somewhere in the vicinity of 20 to 30 percent lower available capacity than a similarly sized vented cell. However, the convenience of no physical maintenance, and the cleanliness of external components (i.e., freedom from spray and spillage) have made the sealed nickel-cadmium cell very popular in small power devices where extended cycle life is particularly important.

Although nickel-cadmium batteries are indeed rugged, there is an electrical condition that can quickly lead to the demise of either vented or sealed types. This, known as "runaway," occurs when the battery is charged in such a manner as to create a great deal of internal heat. This can result from excessive charge rates, or from having the battery in an environment where the heat of the charge (particularly during the overcharge period) cannot be readily dissipated. As the temperature of the battery rises, its internal resistance drops very sharply. When the charge is being carried out from a "system-controlled" power source (i.e., essentially a constant-potential charge), a lowering of the internal resistance will result in an increase in the charging current. This increased current in turn results in a lowering of the efficiency of the charge acceptance, still further increasing the heat generation, still further raising the temperature of the cell. The cell can thus very quickly be brought to a condition where the heat becomes physically disruptive. This runaway condition is prevented by providing for better electric controls and adequate ventilation of the cells and the compartment in which it is housed.

An interesting characteristic of nickel-cadmium batteries is the so-called "memory effect" in which the battery seems to "remember" the manner in which it has been cycled. The memory effect shows up only in a highly repetitive cycling routine, and manifests itself as an apparently permanent loss of reserve capacity. Thus, if a 10-Ah battery is cycled on a routine procedure to a depth of only 4 Ah, with the intention to keep the remaining 6 Ah in reserve against emergencies or to enhance long cycle life, the actual available capacity of the battery falls to 4 Ah after a fairly short time on this routine. To avoid this, batteries in such service are periodically given conditioning cycles in which the depth of charge and discharge, and/or the rates at which these are performed, are varied from the routine. This treatment completely avoids the memory effect, and in fact can restore the capacity of a battery so affected, although the latter is more difficult and requires a greater number and variety of conditioning cycles.

CHARGING SILVER-ZINC BATTERIES All the general rules for charging secondary batteries apply to silver-zinc types as well, only perhaps more so. Since such batteries are frequently used in higher-voltage systems, in fact up to 150 to 200 V, top cleanliness to avoid leakage currents and stray currents to ground becomes most important. The deleterious effects of excessive overcharge, and particularly high end-of-charge rates, are also magnified in this kind of battery. In overcharge resulting in gassing, the loss of electrolyte can be quite serious since (unlike lead-acid batteries, for example) there is very little free electrolyte (i.e., that which is not absorbed by the separators and electrodes). Any loss of free electrolyte causes drastic changes in the electrolyte level, and its replacement is difficult, especially if the distribution of the electrolyte among the various electrochemical components has been disturbed. Then too, since the zinc electrode is partially soluble in the electrolyte, a loss of the latter also actually means a loss of electrode material.

A still more serious effect of overcharging can occur where heat is also allowed to build up inside the cell. This is rather easy to do since silver-zinc batteries are almost always housed in plastic cell containers which are poor thermal conductors. A number of close-packed cells, charged and overcharged at excessive rates, will rapidly result in high temperatures, particularly at the center of the battery. If we remember that silver-zinc cells have organic separators and diaphragms under contact pressure with a strong oxidant such as the silver oxide of the positive electrode (see Fig. 17.15), there is little doubt that high temperatures result in the failure of these components.

In charging silver-zinc secondaries, therefore, sufficient time should be allowed between the end of discharge and the start of charge to allow the battery to cool down. A similar rest would also be very desirable between the end of charge and subsequent discharge. Since very little heat gets out around the sides of the battery, a good practice is to blow air gently over the terminals and intercell connectors. These, being of silver or silver-plated copper, can conduct a good deal of heat from inside the battery to where it can be conveniently dissipated.

While voltage-actuated control systems can be used, Fig. 17.16 indicates that the largest voltage rise as a silver-zinc cell is charged occurs when the battery is only about two-thirds charged, and that the rise at full charge is not very pronounced. Also, voltage is not a clear-cut means of establishing the advent of overcharge. Although not shown in Fig. 17.16, the battery voltage at the end of charge is not at all fixed, slowly increasing from about 1.88 V for a fresh battery to even as high as 2.1 V for one that is nearing end of life. For these reasons the most common practice is to recharge silver-zinc batteries at constant current for a fixed period of time, and to do so for such a time as to return 105 to 125 percent of the ampere-hours removed during the previous discharge. A slightly more sophisticated system takes advantage of the fact that silver-zinc batteries will accept a charge at very high rates at the start of a charge (after a deep discharge), and in fact will do so up to the first sharp voltage rise at about two-thirds full charge. This charging system simply uses a high constant-current charge period with a time control to switch to a lower constant-current level for completion of the charge. This two-step system has a great deal of merit, and many manufacturers now recommend such a charging procedure.

17.5 SPECIAL TYPES

A considerable amount of research and development is going on in the battery field today relating to new electrode combinations. Since practically any redox reaction can be conducted electrolytically in a cell from which electric energy can be withdrawn, a great many couples (electrode combinations) have been proposed. A few of these have found specific applications. Among the more important are:

1. Magnesium dry cells
2. Organic depolarized cells
3. Cuprous chloride–magnesium cells
4. Silver chloride–magnesium cells
5. Air-depolarized cells
6. Air-zinc cells

The general properties of these batteries are outlined in Table 17.13.

Each of these has at least one and sometimes even several special properties that has been a key factor in its growth. Magnesium dry cells are capable of storage under high-temperature high-humidity conditions; organic depolarized dry cells are potentially capable of very high capacities with light weights; cuprous chloride–magnesium types are capable of operation in very low pressure and low temperature ambients; silver chloride–magnesium batteries are quickly activated by dilute saline solutions; air-depolarized cells require practically no maintenance; air-zinc cells are capable of high energy yields per unit weight; and so on. It is important, therefore, to consider these properties in greater detail, and also in combination with other properties.

TABLE 17.13 Characteristics of Special Cell Systems

Type	Open-circuit voltage, V	Average discharge voltage, V	Capacity, Wh	
			Per lb	Per in³
Magnesium dry	1.95	1.60	45	3
Organic depolarizer (magnesium anode)	1.65	1.20	64	3
Cuprous chloride–magnesium	1.5	1.15	31	2
Silver chloride–magnesium	1.9	1.5	75	8
Air-depolarized	1.45	1.05	71	2
Air-zinc	1.45	1.2	52	6

MAGNESIUM BATTERIES At the head of the list, in terms of commercial production, is the magnesium dry cell which we have already introduced under primary batteries. This particular battery very much resembles a conventional LeClanche dry cell in which a magnesium can has simply been substituted for the conventional zinc can. This accomplishes several things including raising the voltage of the cell, increasing its power capability, and (as a result of other design modifications) also its capacity.

However, it is the nature of the magnesium anode that it very quickly forms a very protective film when in contact with the battery electrolyte, and it is this film that gives the magnesium dry cell its outstanding property. This film protects the underlying metal and permits the battery to have an extremely long shelf life. It also helps it to withstand very adverse storage conditions such as high temperatures and high humidities. Not a great deal of reliable data are yet pertinent to the shelf life of magnesium dry cells, but there are the indications that they can easily be stored for as long as five to seven years with only modest losses (i.e., 10 to 20 percent for cells of current designs). In terms of adverse storage conditions, magnesium dry batteries can be stored continuously for 30 days (and many for 60 and even 90 days) in ambients of 160°F and 90 to 95 percent humidity with losses of only about 10 percent in capacity; and even this is largely due to loss of moisture instead of anode deterioration. This latter feature makes the magnesium dry cell extremely attractive for use in those areas of the world which are subject to hot, humid climates, and where distribution logistics require a fairly long time interval between manufacturer and user.

ORGANIC DEPOLARIZED BATTERIES The organic depolarized cell is physically similar to the magnesium dry cell and, in fact, may even employ the same magnesium anode can, but in this case reducible organic material takes the place of the usual MnO_2 depolarizer used in magnesium or LeClanche dry cells. Organic depolarizers, in addition to being synthetically produced materials (and, therefore, of more uniform and continuing supply than conventional depolarizers), have the outstanding characteristic of being able to absorb a large number of electrons during the course of their reduction. This produces batteries of very high capacity for a given space and weight. However, because the organic depolarizers are completely nonconductive electrically, the battery must have a substantial grid structure (usually a carbon "chain" as in other dry cells), and this tends to reduce its space advantage. Even so, organic depolarizer cells show a fairly high internal resistance, but where the drain rates are low, as in radio receivers, it warrants, and is receiving, a great deal of attention.

CUPROUS CHLORIDE–MAGNESIUM BATTERIES The cuprous chloride-magnesium (CuCl-Mg) cell is again a very specialized type which has found considerable use in radiosonde work. Since it is stored in a dry condition, the CuCl-Mg battery is classified as a reserve type, activated by immersion in water (either fresh or saline). For radiosonde work, the CuCl-Mg battery uses a separator which is highly absorbent so that the electrolyte remains within the cell after it is withdrawn from the activating liquid. In this condition the cell can be sent to very high altitudes where pressures and temperatures are extremely low. The special characteristics of the cuprous chloride–magnesium battery which permits this to be done is the fact that as soon as

Fig. 17.18 Cuprous chloride–magnesium radiosonde battery. (*Courtesy Clevite Corp.*)

the battery starts to discharge, copper is transferred from the cathode to the anode where it catalyzes the local action reactions. The resulting high local action rate produces a great deal of heat which keeps the battery at a suitable operating temperature, in spite of the fact that it may be operating in extremely low ambients.

Figure 17.18 shows a typical cuprous chloride–magnesium battery that powers a radiosonde for high-altitude weather observations. This unit supplies both A and B power, the latter being provided by 78 series-connected cells in three parallel stacks. The peculiar spatial arrangement provides the proper balance between the heat generated and the heat lost to the outside necessary to keep the battery at a proper operating temperature.

SILVER CHLORIDE–MAGNESIUM BATTERIES The silver chloride–magnesium (AgCl-Mg) battery is, in a general way, similar to the cuprous chloride–magnesium battery, and in fact has been used for radiosonde work where higher power requirements are needed than can be supplied by a CuCl-Mg battery. The principal application for silver chloride–magnesium batteries, however, is to power sonobuoys and torpedoes. The AgCl-Mg system has an outstanding characteristic in being extremely stable in storage without any special preparation or care, and units that have been stored for more than ten years have given full performance when activated and discharged.

In sonobuoy or torpedo batteries, the saline water of the ocean itself forms the electrolyte. Because this type (dilute salt) of electrolyte is a fairly poor conductor, the electrodes must be spaced very closely together, thus achieving the low internal resistance which makes possible the high power levels at which this type of battery is generally used. Here again local action at the magnesium anode produces a good deal of heat, so that means are usually provided for continuously flushing the battery with sea water. This not only keeps the internal water temperature down, but also keeps the electrodes free of sludge resulting from the discharge. The silver chloride–magnesium battery, the characteristics of which are given in Fig. 17.19, is capable of an extremely wide range of rates, performing well even at very high power levels.

AIR-DEPOLARIZED BATTERIES Since the air-depolarized cell actually uses atmospheric oxygen as its depolarizer, it can have very high capacities within a given weight. However, it does not have very high or even medium rate capabilities, and although it has been on the market for a number of years, it has found only limited appli-

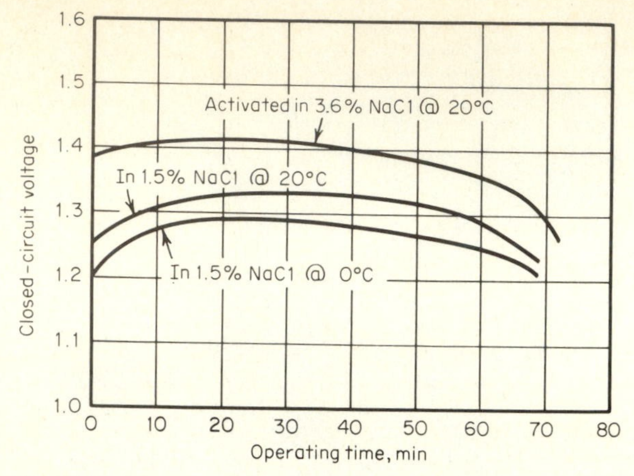

Fig. 17.19 Characteristics of (sonobuoy type) silver chloride–magnesium cell.

cation in low-power devices at inaccessible locations. The air-depolarized cell uses a cast block of zinc as its anode, and a porous carbon electrode open to the air as its cathode. The electrolyte is potassium hydroxide, and is usually present in large excess of operating requirements so as to practically eliminate watering and other maintenance procedures.

AIR-ZINC BATTERIES Recently more sophisticated versions of the air-depolarized cell have been under intensive development, and such cells (renamed air-zinc cells to differentiate them from the earlier versions) are now made in a variety of sizes. These are beginning to find one application in powering radio transceivers. Still further applications are envisioned in the form of motive power (and even automobile) systems, since the cells receive their depolarizer from the air as needed.

Fig. 17.20 Air-zinc battery with replaceable anodes. (*Courtesy Clevite Corp.*)

In the modern air-zinc cell, special low-cost cathode catalysts are employed to achieve respectably high rates of discharge. However, no really standardized sizes have yet been presented for general commercial use, since most batteries developed to date have been against highly specific military uses. In a broad sense, air-zinc cells can also be considered to be reserve batteries since they are packaged in dry form and are activated by addition of water. They carry solid potassium hydroxide (usually in the pores of the anode), which by combining with the water forms the electrolyte of the cell. After activation, the shelf life of the battery is fairly poor, and pretty nearly all its capacity will be dissipated in a matter of weeks. However, some considerable work is now being done to improve this situation, principally through amalgamation of the zinc with mercury, a process that greatly reduces the rate at which the zinc will spontaneously react with the electrolyte.

In addition to improvement of the primary version, work is also being directed toward recharging the zinc electrodes in air-zinc cells in a more or less conventional manner, thus producing a secondary battery. At the present time such batteries have not reached any really commercial performance, but an alternate development, the so-called mechanically rechargeable battery, has become available. In this type of battery, illustrated in Fig. 17.20, the zinc anodes are designed in such a way as to be easily removed from the battery upon completion of their discharge, and new ones are slipped into their place. During this process the electrolyte is also replaced, and the result is a battery virtually capable of brand-new performance. The positive, or air electrode, has been developed to a current life expectancy equivalent to about 50 anodes, making this a fairly economical battery to operate.

17.6 FUEL CELLS

The discussion of air-zinc cells, where one of the reactants (oxygen) enters the cell as needed, leads directly into a group of electrochemical generators usually referred to as fuel cells. As indicated earlier, many electrochemical reactions are reversible. From this the concept was evolved that if water can be broken down electrochemically into hydrogen and oxygen, it should be possible to recombine hydrogen and oxygen in an electrochemical cell (i.e., on suitable electrode surfaces) so as to yield electric energy. This indeed proved to be true, and cells were developed as early as the 1920s which accomplished the pertinent reactions at reasonably good efficiencies. In a great many respects, such cells had the characteristics of a prime mover–generator combination since the electrochemical materials (hydrogen and oxygen) were gases, and could be stored outside the cell to be metered into the cell as needed. Within limits it was found that the greater the rate at which the gases were fed, the greater was the electric power generated by the cell. Furthermore these gases acted very much as they might have as a fuel and an oxidant, and hence the popular term fuel cell.

In more recent times, a great variety of "fuels" have been found to be suitable, as have a fairly large number of oxidants, so that by now there is a variety of fuel-cell systems. Also, the concept has been enlarged to embrace other than gaseous fuels, and typically, in addition to hydrogen, such materials as hydrocarbons, hydrazine $(N_2H)_2$, and molten metals (such as sodium, lithium, potassium, and other reactive easily fused metals) have been used. Oxidants such as air, chlorine, bromine, and others, have also provided interesting cell characteristics. As a matter of fact, the various fuel-cell combinations of oxidants and reductants that have been investigated have resulted in fuel-cell systems with highly specialized characteristics for very specific applications.

Basically, however, pretty nearly all fuel cells are variations of the fundamental design shown schematically in Fig. 17.21. This figure shows a hydrogen-oxygen cell and the auxiliary equipment needed to support it. This auxiliary equipment turns out to be extremely important for the extended operation of a fuel cell. In the cell illustrated, the overall reaction involves the oxidation of hydrogen to water. This water enters the electrolyte and, thereby, dilutes the latter. Some system must be provided, therefore, to remove the water in order to maintain the electrolyte concentration at appropriate levels. If the hydrogen or oxygen contains any impurities, these will enter the cell and ultimately build up to the point where they will seriously impair the performance of the cell, and the fuel and oxidant must therefore be brought to a high level of purity before entering the cell. These turn out to be very difficult engineering problems, and a good

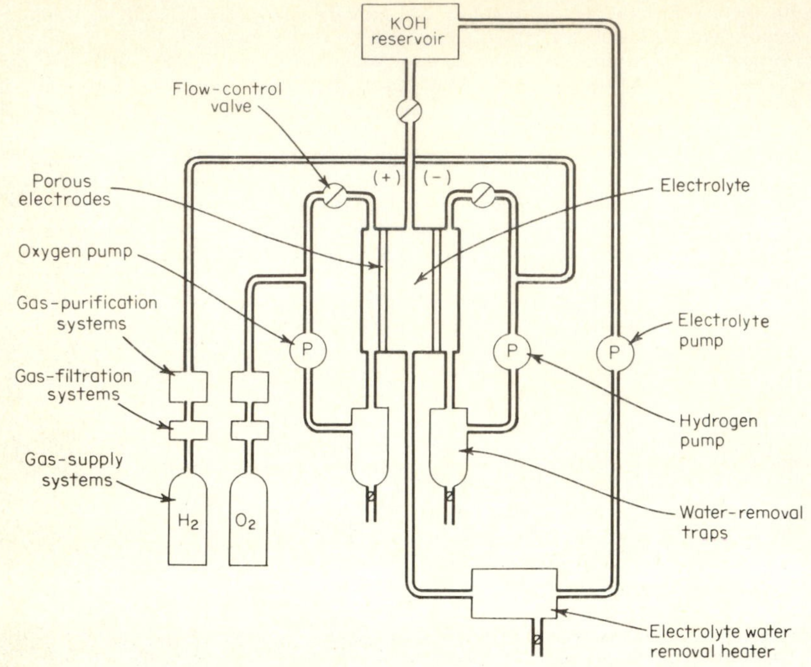

Fig. 17.21 Schematic diagram of hydrogen-oxygen fuel-cell system.

Fig. 17.22 Hydrogen-oxygen fuel-cell power system. (*Courtesy Union Carbide Corp.*)

TABLE 17.14 Characteristics of Typical Fuel-Cell Systems

Power capability, W	Application	Fuel	Oxidant	Unattended life, h	Power capability Kw/lb	Power capability Kw/ft^3
200	Radio communication	Hydrogen	Oxygen	1500	0.006	0.4
500	General power	Hydrazine	Air	1000	0.005	0.2
10 000	Space power	Alcohol	Air	2500	0.000	0.3
25 000	Marine	Hydrocarbon	Air	7500	0.01	1.3

deal of the progress that has been made has resulted from careful development of the auxiliary equipment. A photograph of an actual fuel-cell system is shown in Fig. 17.22, with general operating characteristics of a number of fuel-cell systems detailed in Table 17.14.

Fuel cells are, therefore, no longer simple electrochemical power generators in that the movement of fuel and oxidant into the cell and the products of the oxidation out of the cell have become a fairly complicated materials movement problem, and we find that modern fuel cells have a great deal of pumping, filtering, throttling, and control equipment associated with the electrochemical generator itself. In many instances, as for the H_2O_2 cell, it has been necessary to go to very high pressures and temperatures (to achieve high power outputs) further complicating the equipment. Others, such as the lithium-fluorine system, require extremely high temperatures in order to operate properly.

Many fuel cells are now available with fairly large power capabilities and have found use as the power source for manned space vehicles, as for example in the Apollo program. Fuel cells using alcohol have been produced in smaller sizes and have been promoted for military communication. To date, fuel-cell designs are developed against very specific purposes, and there are no generalized sizes, or standardization of components, so that tabularized information of this nature is not currently available. Nevertheless the wide variety of properties that can be achieved, coupled with long-life catalysts which greatly increase the rate of reaction possible at the electrode surfaces (and thereby increase the power capability of the cell), have accounted for the tremendous current interest being shown in these devices. In addition, a great many organizations, both government and private, are supporting major projects to increase still further the efficiency and the ease of the operation of these units, and it is very likely that standard forms will be evolved over the next few years.

At the moment, the indications are that these forms will be directed particularly toward motive power applications (small automobiles and the like), stationary generators (particularly for standby service), and radio communication. Fuel cells have indeed shown themselves to be capable of producing usable quantities of energy directly from conventional fuel sources and have a long life with relatively little maintenance. As a matter of fact, it is now entirely feasible to have completely unattended remote fuel-cell power plants, or to utilize waste fuels (such as hydrogen from caustic production) to produce "convenient" power. At the present time, however, the cost of fuel cells and their operation are high, but there is every expectation that as their use becomes more widespread and improvements continue, and as some degree of standardization is achieved, fuel cells will occupy an important position as electrochemical power sources.

Chapter 18
Vacuum Tubes

18.1 INTRODUCTION

Regardless of the type of amplifying device used in a circuit (vacuum tube, transistor, field-effect transistor, etc.), the circuit operation starts with some method of supplying electrons and controlling them as they move from one point to another. This is, essentially, the meaning of *electronics:* supplying and controlling electrons. In a vacuum tube, the electrons are supplied by some type of *emission*—that is, they are released from the surface of a material, and caused to move through a vacuum. The motion of electrons is controlled by either an electric or a magnetic field which exists somewhere between the *cathode* (their point of emission) and the *plate* (their destination).

This study of vacuum-tube theory starts with electron emission. After this, the various methods of controlling the motion of electrons in a vacuum will be discussed. Finally, the various types of tubes made and the methods of predicting their behavior in circuits will be discussed.

18.2 SOME BASIC PROPERTIES OF ELECTRONS

Physicists know that the electron has both wave and particle properties. In other words, under certain conditions it behaves like a wave, and under certain other conditions it behaves like a particle. Fortunately, as far as the study of electronics is concerned it can be considered to be a very small, spinning, negative particle. This simplifies the description of circuit operation since the wave properties require a considerable amount of mathematical analysis. The particle electron can be interpreted in terms of the *Bohr atom* which pictures the atom as a tiny solar system with a nucleus comprised of protons and neutrons, and with electrons moving about the nucleus in orbits (or shells).

In addition to moving around the nucleus, the electrons are considered to be spinning on their own axis. This is somewhat like the Earth spinning on its axis as it moves around the sun. There is one important difference that should be noted. The Earth moves around the sun in a plane (as far as we know) but the fast-moving electron path covers the nucleus like a shell. One of the main problems in electronics involves getting this electron away from its atom.

The electron is the smallest unit of negative electric charge. The negative charge on a single electron is 1.6×10^{-19} C, and the mass of an electron is 9.1×10^{-31} kg. These extremely small values can be better appreciated when written without powers of ten:*

* The abbreviations "C" for "coulomb" and "eV" for "electronvolt" follow the SI system recommended by the IEEE.

Charge of one electron:

0.00 000 000 000 000 000 016 C!

Mass of one electron:

0.00 000 000 000 000 000 000 000 000 000 091 kg!

So small is the electron that when one ampere of current flows, there are 6.25×10^{18} electrons (that is, 6250 000 000 000 000 000.0 electrons) flowing past a point in the circuit every second. It is interesting to reflect on the tremendous impact that such a small particle has had on civilization.

Energy may be defined as the capacity to do work. An electron possesses energy by virtue of its motion (called *kinetic energy*) and energy by virtue of its position in the atom (called *potential energy*). One of the peculiar properties of an electron is that it can exist only at certain discrete amounts of total energy—called *energy levels*. Figure 18.1 shows how these energy levels are usually depicted for an electron in an atom for one kind of material. An electron can exist only at the levels shown by lines but never in the forbidden bands. For an atom of another kind of material, the lines would be in different positions.

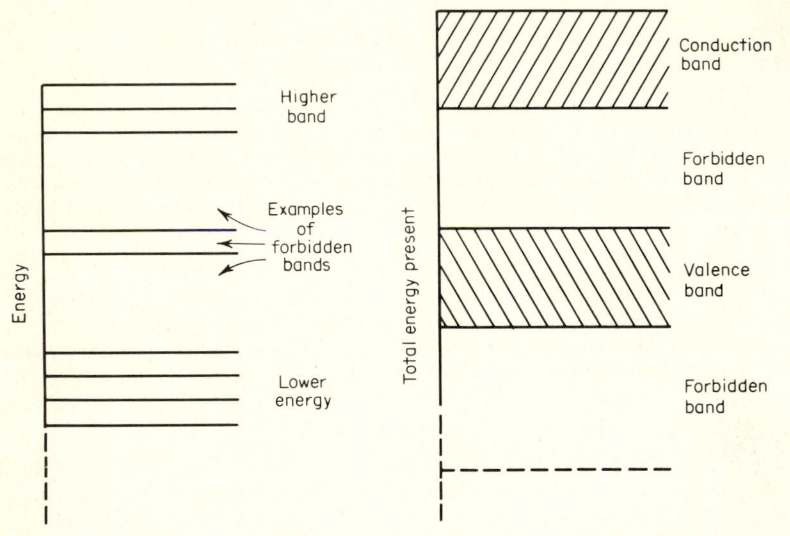

Fig. 18.1 Energy levels of an electron in an atom. The possible levels of existence are shown by horizontal lines. An electron in this atom cannot exist at any level between the lines.

Fig. 18.2 Energy "bands" occur when a large number of atoms are closely packed together.

The energy levels shown in Fig. 18.1 are only for a single electron in an isolated atom. If atoms are close together, their energy levels overlap somewhat, and it is a more common practice to refer to them as energy bands.

In order to get an electron from one energy level to the next higher one, it is necessary to supply the amount of energy required for it to get completely across the forbidden band. If you supply energy to an electron, and it is not enough to get it to move completely across the forbidden band, then it will stay where it is.

Only the electrons in the two outer bands of energy levels influence the chemical or electrical behavior of an atom. These bands are called the *conduction band* and the *valence band*. They are shown in Fig. 18.2. (The valence band establishes the chemical properties of the material.) Electrons in the conduction band have the highest energy level. The actual magnitude of the energy level depends largely on the temperature of the material containing the electron.

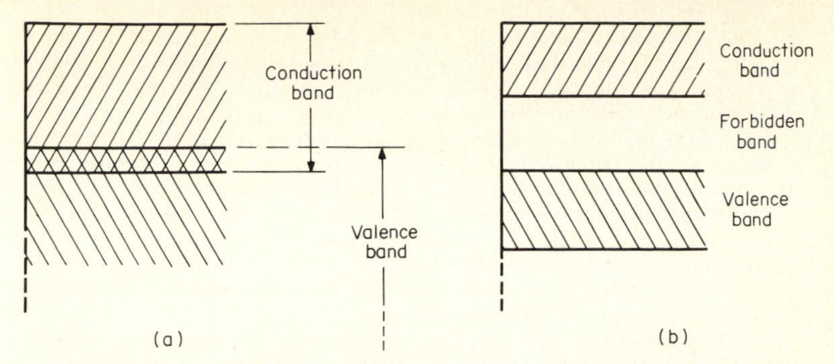

Fig. 18.3 Comparison of energy bands for conductors and insulators. (*a*) The energy levels of the valence and conduction bands overlap in materials that are electric conductors. There is no forbidden band between these levels. (*b*) For materials that are electric insulators, the conduction and valence bands are widely spaced.

Figure 18.3 shows how the energy bands of conductors compare with the energy levels of insulating material. The outer energy levels of metallic conductors overlap, as shown in Fig. 18.3*a*. Therefore it is relatively easy to raise an electron to the conduction band from the valence band. Once in the conduction band, the electron can be removed from the atom with very little effort. This explains why it is possible to get an electron current flow by placing a small voltage across an electric conductor. The wide forbidden band between the conduction and valence bands in an insulator, as shown in Fig. 18.3*b*, must be crossed by an electron before it can get to a point where it will contribute to current flow. Since not many electrons make it across this forbidden band, it is very difficult to get any appreciable electron current flow in an insulator.

Although there are free electrons moving around in a conductor, it is quite unlikely that they would gain a sufficient amount of energy to escape from the surface. Therefore, to obtain electron emission, the free electrons will have to receive an additional amount of energy from some source. The problem of obtaining electron emission, then, is one of adding a sufficient amount of energy to free electrons to allow them to escape from the surface of the material.

18.3 TYPES OF ELECTRON EMISSION

Four different methods are used to obtain electron emission in tubes: (1) thermionic emission, (2) field emission, (3) photoemission, and (4) secondary emission.

In each method, a sufficient amount of energy is added to free electrons to permit them to escape from the surface of a material. We will discuss these methods briefly at this time, and will take them one by one in a more detailed discussion later in this chapter.

Thermionic emission is obtained by heating the material containing the free electrons. The heat energy is imparted to the electrons to allow them to escape. The number of electrons leaving each square centimeter of cathode surface per second is a function of the square of the temperature of the cathode as measured in degrees Kelvin, that is, degrees above absolute zero. (Emission is also related to the *work function* which will be discussed later.)

Since very little thermionic emission takes place at temperatures below 1000 K, only certain materials can be used as cathodes, assuming that it is desired to have the cathode operate over a period of time. Furthermore, not all the metals that can withstand such temperatures are efficient emitters of electrons, and this further limits the number of materials that are suitable for use as cathodes.

Field emission is obtained by placing a high positive electric field near the surface of the material and literally pulling electrons out of the material. Since unlike charges attract, the positive field attracts the negative electrons with a force sufficient to allow it to escape from the surface.

Certain materials emit electrons when a light falls on their surface. The process of obtaining electron emission with light energy is called *photoemission*. The materials are able to emit electrons when electromagnetic energy at a certain frequency—or band of frequencies—falls on their surface.

Regardless of the fact that electrons are extremely small physically, they can acquire a sufficient amount of energy when accelerated to high speeds to knock other electrons off the surface of a material. This process is called *secondary emission*. As with other types of emission, certain materials will give up electrons by this method more readily than others. Of course, the speed of the electrons bombarding the surface (called *primary electrons*) is also a determining factor as to the number of secondary electrons leaving the surface. Also, the angle at which the primary electrons strike the emitting surface affects the number of secondary electrons emitted.

Unlike other types of emission, secondary emission can be obtained from the surface of insulating materials, but this is not necessarily an advantage of the method. In one case history, the electron beam of a picture tube in a television receiver was striking the neck of the tube. The constant bombardment of electrons drove secondary electrons off the glass surface. Since glass is an insulator, other electrons could not move in to take the place of those emitted; so the glass became more and more positively charged. Ultimately a voltage breakdown occurred between the positively charged glass and a nearby ground point. This breakdown punctured the glass and destroyed the tube.

In some special types of tubes, such as television camera tubes, several types of electron emission are used within the same device. The majority of small receiving tubes used in home entertainment systems operate by thermionic emission.

18.4 WORK FUNCTION

At the very instant when an electron leaves the surface of a material, that material becomes positively charged, and it exerts an attraction for the electron. The electron must not only attain a sufficient amount of energy to leave the surface, but it must also have a sufficient amount of energy to get away from the positive potential resulting from its loss. The amount of kinetic energy needed by an electron to escape from the surface of a material is called the work function of that material. It is harder to obtain electron emission from the surface of a material with a high work function than for a material with a low work function.

Work function is measured in electronvolts. An electronvolt is the amount of energy acquired by one electron as it falls through a potential of one volt. Thus, work function is a unit of energy. The work function of a material depends upon so many different factors, such as the amount of impurities in the material and the smoothness of the surface material, that it is difficult to obtain consistent measurements. If you would look in several different sources for the work function of a material, you might get several different values. The values given in Table 18.1 were obtained by averaging values from more than one source. The table is given to show which are the best emitters of electrons. For example, platinum is a poor emitter compared to oxide-coated nickel. There are some applications where a high work function is desirable. Grid wires in tubes should definitely *not* emit electrons. All the electrons emitted in a tube should come from the cathode. In order to prevent emission from grid structures, they are sometimes coated with a material, such as gold, that has a high work function.

TABLE 18.1 Work Functions of Materials

Material	Work function, eV
Oxide-coated nickel	1–2
Cesium	1.81
Thorium on tungsten	2.63
Calcium	2.7
Nickel	3.9
Tantalum	4.07
Copper	4.2
Platinum	6.0

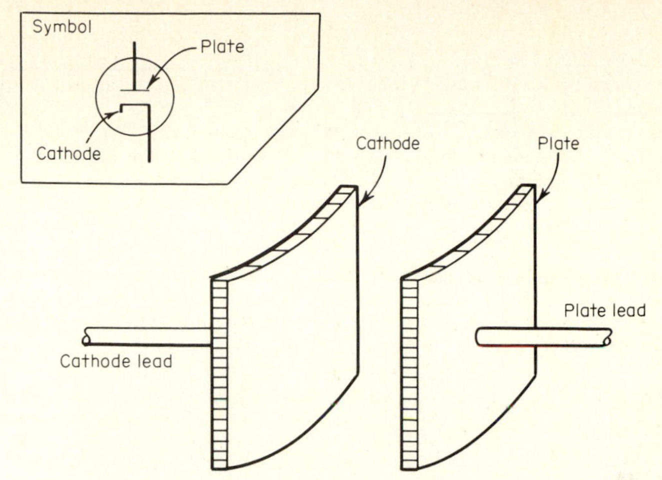

Fig. 18.4 Cathode and plate of a theoretical diode. The schematic symbol for a diode is shown in the inset.

18.5 THERMIONIC EMISSION IN A DIODE

For the purpose of discussing thermionic emission, we will consider a simple diode as shown in Fig. 18.4. It consists of two electrodes (or *elements* as they are often called): a *cathode* for emitting electrons, and a *plate* for collecting the electrons. These elements are usually enclosed in a vacuum. As will be shown later, diodes are not actually constructed with flat plates and cathodes as shown in Fig. 18.4, but this illustration may be considered to be a section of a typical diode found in practice. The schematic symbol of a diode is shown in the inset of Fig. 18.4.

When the voltage on the plate is positive with respect to the heated cathode, an electron current flows through the tube. This is sometimes called the *Edison effect*. The negative electrons emitted from the cathode are attracted by the positive plate voltage (unlike charges attract). An ammeter—or milliammeter—in series with the plate lead will indicate a current flow. The electron current through a diode from the heated cathode to the positive plate is often referred to as the forward plate current or forward cathode current.

When the plate is negative with respect to the cathode, it repels electrons (like charges repel), and no electron current flows through the tube. A meter in the plate circuit will indicate zero current under this condition. When a diode has a voltage across it so that an electron current flows through it, it is said to be *forward-biased*. When the diode has a voltage across it so that no electron current flows through it, it is said to be *reverse-biased*. These two connections are depicted in Fig. 18.5. Figure

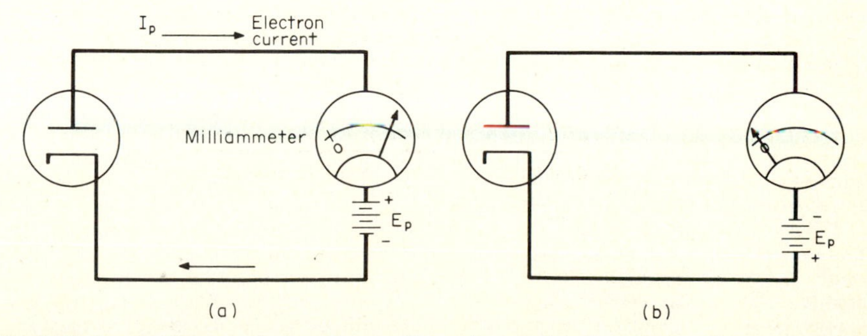

Fig. 18.5 Current flow in a diode with a positive and a negative plate voltage. It is assumed that the cathode is heated and emitting electrons.

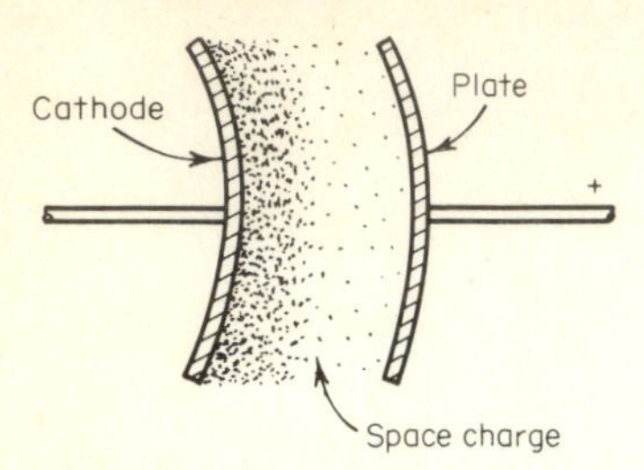

Fig. 18.6 Location of space charge between a heated cathode and plate. Dots represent electrons: (*a*) When the battery is connected in this way, current flows. (*b*) When the battery polarity is reversed, no current flows.

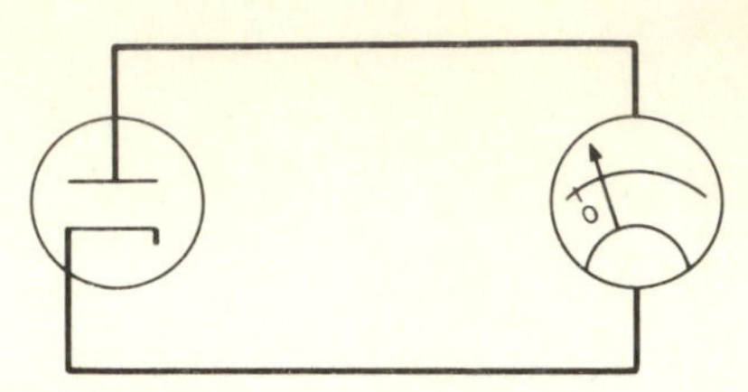

Fig. 18.7 Circuit for producing Edison effect current. The cathode is presumed to be hot and emitting electrons.

18.5*a* shows a battery connected so as to produce a forward current flow through the tube. The cathode is presumed to be heated, even though no heater circuit appears in the illustration. Figure 18.5*b* shows that current stops when the battery polarity is reversed—that is, when the diode is reverse-biased.

When the voltage on the plate is zero with respect to the cathode, electrons emitted from the cathode surface accumulate in a *cloud* around the cathode. This cloud of electrons is called the *space charge*. Once the cloud is formed, it prevents further electrons from being emitted from the cathode. The space charge is negative, and it repels the negative electrons back to the cathode.

Figure 18.6 illustrates the space charge. It would seem that under this condition no plate current could flow since the plate voltage is zero and the space charge tends to prevent electrons from leaving the cathode. However, a very small amount of current *does* flow because a few electrons reach a sufficient amount of velocity to arrive at the plate. This small amount of current flowing in the absence of a positive voltage on the plate is sometimes called contact current or the Edison effect current. Figure 18.7 shows the circuit for producing an Edison effect current. Although quite small, this current is important in certain vacuum-tube applications.

18.6 CHARACTERISTIC CURVES

The amount of plate current in a tube depends on a number of things, the most important being the amount of *plate voltage* and the *tube geometry*. Plate voltage is the amount of positive voltage on the plate with respect to the cathode. It is important to remember that tube voltages are always measured with respect to the cathode. For example, a plate voltage of +100 V means that the plate is 100 V positive *with respect to the cathode*. The tube geometry includes such things as the distance between the plate and cathode, and the total area of the plate and the cathode surfaces.

Characteristic curves are graphs that describe the tube operation under certain conditions. If the temperature of the cathode is held constant, then the plate current may be graphed as a function of the plate voltage. Figure 18.8 illustrates some typical characteristic curves. Figure 18.8*a* shows a typical characteristic curve for a diode with several points of interest marked for reference. When the plate voltage is 0 volts, there is a plate current flowing as indicated by point *b*. This is the Edison effect current. In order to reduce the plate current to 0 mA, it is necessary to make the plate voltage negative by an amount marked *a* on the curve. In some characteristic curves the small amount of Edison effect current is not shown, and the curve is presumed to start at 0, as shown by the dotted line in Fig. 18.8*a*.

It is evident from the characteristic curve that the plate current does not immediately increase to its maximum positive value when the plate voltage is made positive. Instead, it increases gradually along the curve from point *b* to point *c*. In this region, the positive plate voltage attracts electrons out of the space charge. For each electron that leaves the space charge, one is emitted from the cathode to take its place. As shown in Fig. 18.6, the concentration of electrons in the space charge is near the cathode. Since

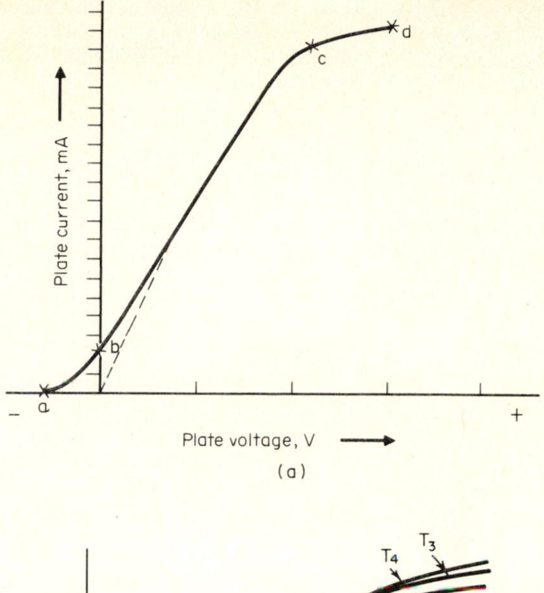

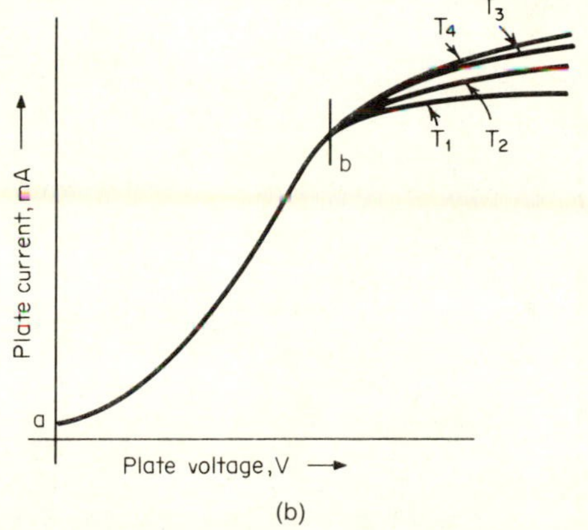

Fig. 18.8 Some examples of characteristic curves: (*a*) A typical diode characteristic curve for a given cathode temperature. (*b*) A family of characteristic curves.

the space charge limits the number of electrons that can leave the cathode, and therefore the number arriving at the plate, the current in the region of the curve bounded by points *b* and *c* of Fig. 18.8*a* is said to be *space-charge-limited.*

When the tube current is space-charge-limited, the positive plate does not attract electrons directly from the cathode, but, rather, obtains them from the space charge. Since the space charge is supplying the electrons, it is sometimes referred to as a *virtual cathode.*

Point *c* on the curve of Fig. 18.8*a* is called the knee of the curve, or the *saturation point.* At this point, all the electrons leaving the cathode go directly to the plate, and the space charge has been eliminated. Since the number of electrons arriving at the plate under this condition depends on the number being emitted (which, in turn, depends on temperature), the plate current is said to be *emission-limited,* or *temperature-limited.*

It would seem that a further increase in positive plate voltage beyond the saturation point could not produce an increase in plate current since all the electrons leaving the cathode are arriving at the plate. However, the presence of the highly positive plate voltage reduces the work function of the cathode and produces a small increase. This is shown in the region from point *c* to point *d* on the curve of Fig. 18.8*a*. The presence of a positive voltage near the surface of a metal reduces the work function; and if the voltage is sufficiently positive, the work function is reduced to the point that electrons are emitted from the surface. When this happens, *field emission* takes place.

The curve of Fig. 18.8*a* is drawn for a fixed cathode temperature. If the cathode temperature is increased or decreased, the shape of the curve will change slightly. If several curves are drawn on the same graph, one for each different cathode temperature, the result is called a *family of curves*. Figure 18.8*b* shows a family of characteristic curves that was obtained by operating the cathode at four different temperatures: T_1, T_2, T_3, and T_4. In this group, T_1 is the lowest temperature and T_4 the highest.

An important feature of the family of curves is that the space-charge-limited current—between points *a* and *b* of Fig. 18.8*b*—is identical for all four curves. This is usually stated in a different way: The space-charge-limited current of a tube is independent of the cathode temperature.

When diode cathodes are coated with thorium, they have a very low work function. As shown by the characteristic curve of Fig. 18.9, there is no saturation point on the characteristic curve of such a diode. The rate of increase in plate current slows somewhat after the voltage goes beyond point *a*, but not sufficient to produce a knee on the curve. If the plate voltage is increased beyond the manufacturer's recommendation, the cathode emission will ultimately become so great that it will destroy the cathode.

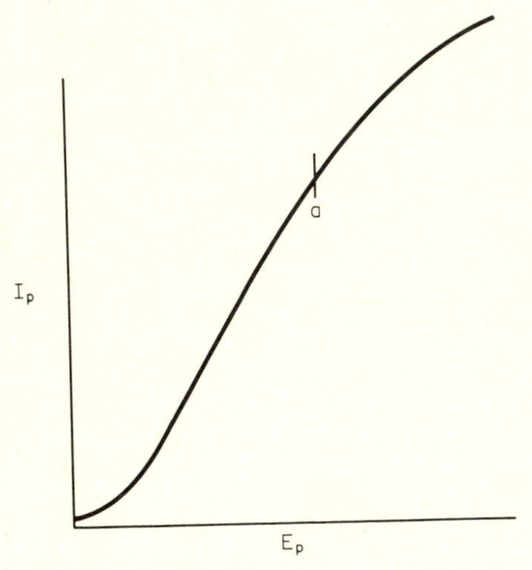

Fig. 18.9 The characteristic curve of a tube with an oxide-coated cathode does not have a saturation point.

Some tubes are made with oxide-coated cathodes that emit electrons at a very low temperature. The filament current is so low in these tubes that they operate—for all practical purposes—cold. Such tubes are said to have *dark heaters*.

Vacuum tubes are usually operated in the space-charge-limited current region—that is, in the region below saturation—in order to assure a long tube life.

18.7 THREE-HALVES POWER LAW

Provided that the tube is operated in the space-charge-limited current region, then the amount of plate current is dependent only on the value of plate voltage and the geometry of the tube. The mathematical relationship between the plate current and the plate voltage is called the three-halves power law. It is also known as the *Langmuir-*

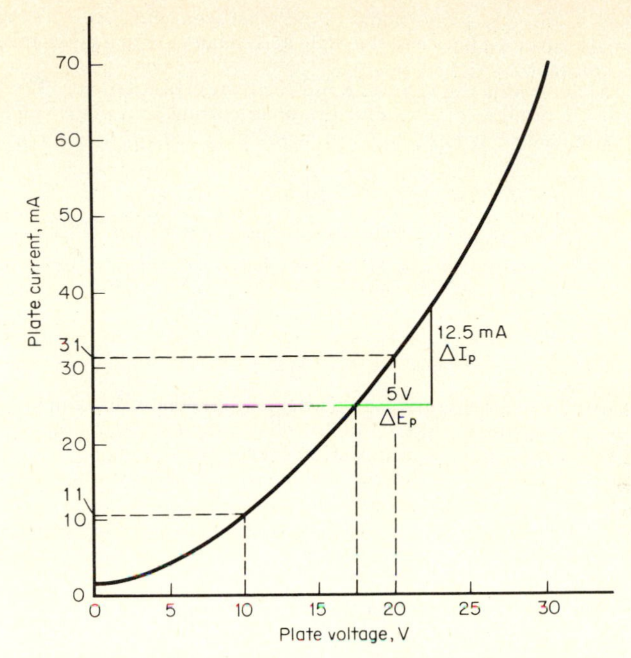

Fig. 18.10 Diode characteristic curve. Dotted lines are used for examples in text.

Child's law or simply as *Child's law*. Mathematically, the three-halves power law is given as

$$I_p = KE_p^{3/2} \tag{18.1}$$

where I_p = plate current, A
E_p = plate voltage, V
K = a constant related to the geometry of the tube

Note: To raise a number to the $\frac{3}{2}$ power, first cube it, and then take the square root of the result.

example 18.1 Assuming that a tube's current is related to its plate voltage by the three-halves power law, what increase in current can be expected when the plate voltage is doubled? (Check the answer graphically, using the characteristic curve of Fig. 18.10.)

solution We are looking for the ratio between two currents (I_{p1} and I_{p2}) when the voltage is equal to E_p and when it is equal to $2E_p$.

When the plate voltage is E_p:

$$I_{p1} = K\sqrt{(E_p)^3}$$

When the plate voltage is $2E_p$:

$$I_{p2} = K\sqrt{(2E_p)^3}$$

The ratio of I_{p2} to I_{p1} is

$$\frac{I_{p2}}{I_{p1}} = \frac{\not{K}\sqrt{(2E_p)^3}}{\not{K}\sqrt{E_p^3}} = \frac{\sqrt{8E_p^3}}{\sqrt{E_p^3}} = \frac{2.83\sqrt{\not{E}_p^3}}{\sqrt{\not{E}_p^3}} = \frac{I_{p2}}{I_{p1}} = \frac{2.83}{1}$$

or $\quad I_{p2} = 2.83I_{p1}$ *Answer*

This means that doubling the plate voltage should increase the plate current by about 2.8 times. On the curve of Fig. 18.10, a plate voltage of *10 V* (E_p) corresponds to a plate current of 11 mA (I_{p1}); and a plate voltage of *20 V* corresponds to a plate current of 31 mA. This gives a ratio of

$$\frac{I_{p2}}{I_{p1}} = \frac{31}{11} = 2.8$$ *Answer*

This corresponds very closely to the ratio of 2.83 obtained mathematically.

The value of K is directly proportional to the plate area, and inversely proportional to the square of the distance between the plate and cathode. For a given tube type, the value of K is a constant.

It is possible to determine the value of K indirectly by measurement. The procedure is to measure (or determine graphically) the plate current and plate voltage. Then Eq. (18.1) is rearranged so as to make K a function of I_p and $E_p^{3/2}$:

$$K = \frac{I_p}{E_p^{3/2}} \tag{18.2}$$

where K, I_p, and E_p have the same meanings as described for Eq. (18.1).

example 18.2 The graph of Fig. 18.10 shows that a plate voltage of 10 V will produce a plate current of 11 mA. What is the value of K for Eq. (18.2)?

solution

$$K = \frac{I_p}{E_p^{3/2}} = \frac{0.011}{10^{3/2}} = \frac{0.011}{31.6} = 349 \times 10^{-6}$$

Once the value of K is determined for a particular tube, that value can be substituted into Eq. (18.1). Then, for any value of plate voltage, the corresponding value of plate current could be found mathematically. However, since the exponential value of E_p may not be exactly $3/2$, a graphical determination of I_p (for a given value of E_p) is usually obtained from the characteristic curves of the tube.

example 18.3 Using the value of K that was determined in Example 18.2, determine the current through the tube when the plate voltage is 10 V. Check the answer with the graph of Fig. 18.10.

solution The value of K is 349×10^{-6}, and the plate voltage E_p is 20 V

$$I_p = KE_p^{3/2} \tag{18.1}$$
$$= 349 \times 10^{-6} \times 20^{3/2}$$
$$= 0.03 \times 10^{-3} = 31 \text{ mA} \qquad \textit{Answer}$$

The graph of Fig. 18.10 shows that a current of 31 mA will flow when the plate voltage is 20 V.

18.8 DIODE TUBE CONSTRUCTION

Vacuum tubes have either directly heated or indirectly heated cathodes. Figure 18.11 shows the two types. *Directly heated cathodes* are filament wires similar to those used in incandescent light bulbs. The filament wires are heated by passing a current through them, and electron emission takes place from the filament surface. Figure 18.11*a* shows two examples of directly heated cathodes and the schematic symbol used to represent them. When the filament emitter is coated with an alkaline-earth oxide, it has a very low work function. This has two advantages: *First,* a low current will produce a sufficient amount of emission for tube operation; and *second,* since emission takes place at a very low temperature, the tube can be placed into operation in a very short time.

The first advantage makes directly heated cathodes useful in tubes designed for use

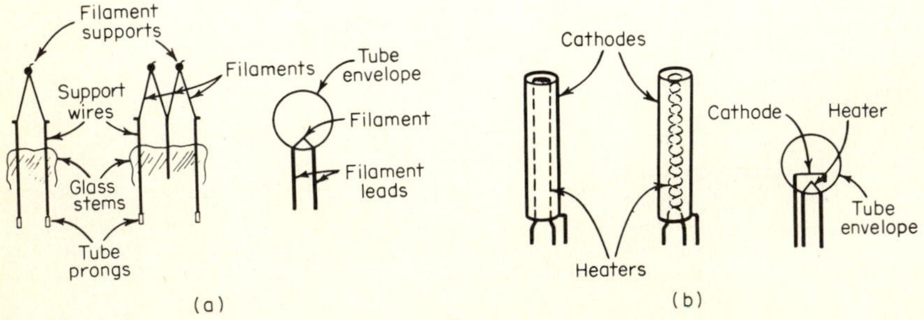

Fig. 18.11 Two types of cathodes used in vacuum tubes: (*a*) Directly heated cathode and schematic symbol. (*b*) Indirectly heated cathode and schematic symbol.

in battery-operated equipment. The low-filament current requirement results in prolonged battery life. The second advantage makes directly heated cathodes useful in tubes designed for equipment that is used intermittently. An example is a low-power transmitter designed for occasional operation. It is desirable to be able to energize such a transmitter in a few seconds, but to maintain it in the deenergized state when not in use.

Instead of emission being obtained directly from the filament, an *indirectly heated cathode* may be employed. Figure 18.11*b* shows two examples of indirectly heated cathodes and the schematic symbol used to represent them. The cathode here is a metal sleeve with a filament in the center. The filament heats the cathode, and electron emission takes place at the cathode surface. The symbol used for tubes with indirectly heated cathodes often does not show the filament, and in such cases the filament is understood to be present. (See Fig. 18.5 for an example.)

When an alternating current is used to heat a filament, there is a fluctuation in temperature with each half-cycle of current. This fluctuation causes variations in emission from moment to moment which, in turn, causes undesirable variations in plate current. With indirectly heated cathodes, the ac fluctuations in filament current do not produce noticeable changes in cathode temperature or plate current, and this is its most important advantage. Most of the television receiver tubes, including the picture tube, are constructed in this manner.

The plate of the tube surrounds the cathode in a typical diode. Figure 18.12 shows the construction of a tube with a cathode type of emitter and also one with a filament type of emitter. In both there are two diodes in the same tube *envelope*—the name given for the container in which the tube is constructed. The name *duodiode* is used in reference to tubes having two diodes in the same envelope.

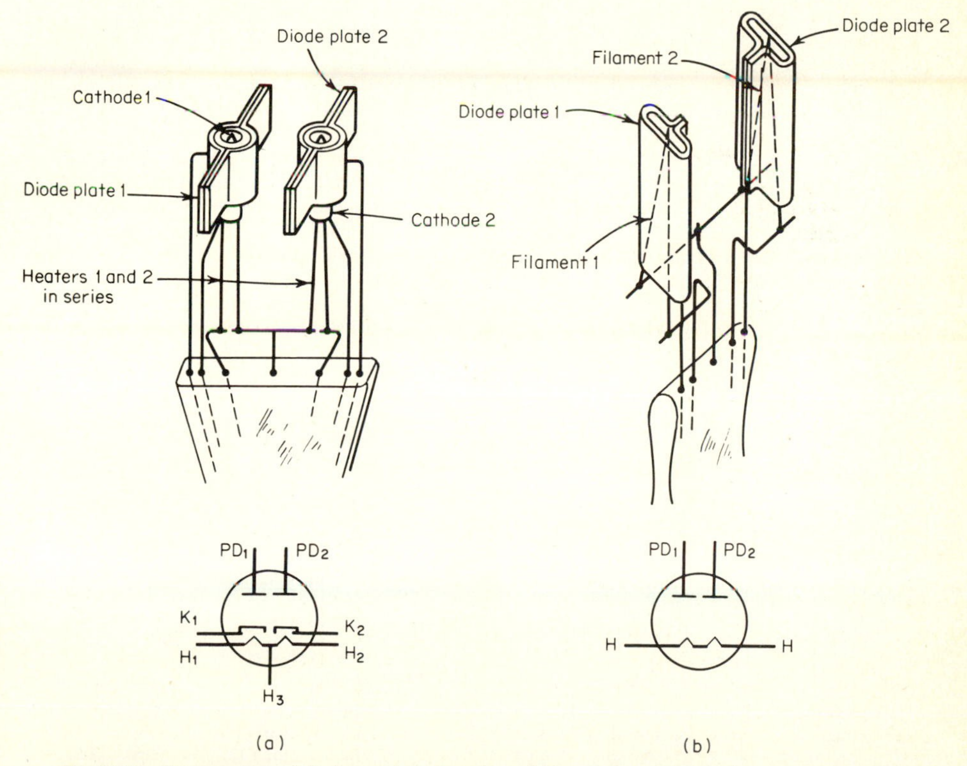

Fig. 18.12 Construction of diodes with indirectly and directly heated cathodes. (*a*) Diodes with indirectly heated cathodes. (*b*) Diodes with directly heated cathodes.

18.9 PLATE RESISTANCE

When a plate voltage is applied to a diode tube, a certain amount of plate current flows provided, of course, that the cathode is sufficiently hot to emit electrons. Figure 18.5*a* shows a simple diode circuit with the plate voltage E_p and plate current I_p marked. (The symbols E_b and I_b are sometimes used instead of E_p and I_p.) The only limit to current flow in the circuit is the space charge in the tube. The resistance that a tube offers to the flow of direct current is called the dc plate resistance. Since the measurements are made under unvarying conditions, which are often referred to as *static conditions*, this resistance is also known as the *static plate resistance*.

The dc plate resistance of a diode may be obtained by measuring E_p and I_p, and then using Ohm's law in the equation

$$R_p = \frac{E_p}{I_p} \tag{18.3}$$

where R_p = dc plate resistance, Ω
E_p = dc plate voltage, V
I_p = dc plate current, A

The values of E_p and I_p can also be determined from the tube characteristic curves.

example 18.4 Figure 18.10 shows a typical response curve. The dotted lines on this curve show that, for the particular points where they meet the curve, the plate voltage is 17.5 V and the plate current is 25 mA (0.025 A). What is the plate resistance?

solution The plate resistance is calculated as follows:

$$R_p = \frac{E_p}{I_p} = \frac{17.5}{0.025} = 700\ \Omega \qquad \textit{Answer}$$

If a different point had been used on the characteristic curve for determining E_p and R_p, the value of dc plate resistance would have been different, because the diode is a nonlinear device. It is necessary, then, that the value of dc plate resistance obtained from a characteristic curve must be calculated for the actual plate voltage and corresponding plate current at which the tube is to be operated.

Instead of determining the plate resistance from a single point on the characteristic curve, the value can be determined over a range of values on the curve. When obtained in this manner it is called the ac plate resistance or *dynamic plate resistance* and is defined as the opposition that a tube offers to the flow of alternating current. Although the present discussion is concerned with diode tubes, the definitions of dc and ac plate resistance are the same for other tube types also.

The ac plate resistance r_p is obtained from the equation

$$r_p = \frac{\Delta E_p}{\Delta I_p} \tag{18.4}$$

where r_p = ac plate resistance, Ω
ΔE_p = a small change in plate voltage
ΔI_p = change in plate current that results from change in plate voltage

example 18.5 Suppose that the plate voltage of a certain diode is raised from 10 to 25 V. The change in voltage, ΔE_p, is 15 V. If this increase in plate voltage results in an increase in plate current from 20 to 50 mA, then the change in plate current, ΔI_p, is 30 mA (0.03 A). What is the ac plate resistance of the tube?

solution By substituting the values of ΔE_p and ΔI_p into Eq. (18.4) the value of r_p for that tube is obtained:

$$r_p = \frac{\Delta E_p}{\Delta I_p} = \frac{15\text{ V}}{0.03\text{ A}} = 500\ \Omega \qquad \textit{Answer}$$

The dynamic plate resistance can also be obtained graphically.

example 18.6 Returning to the characteristic curve of Fig. 18.10, a change in plate voltage, ΔE_p, and a corresponding change in plate current, ΔI_p, are marked on the graph. The value of ΔE_p is 5 V because the voltage rises from 17.5 to 22.5 V along the line marked ΔE_p. This produces a corresponding increase in plate current from 25 to 37.5 mA—an increase of 12.5 mA which is the value of ΔI_p. What is the ac plate resistance?

solution Using Eq. (18.4),

$$r_p = \frac{\Delta E_p}{\Delta I_p} = \frac{5\text{ V}}{0.0125\text{ A}} = 4.0\ \Omega \qquad \textit{Answer}$$

As with static plate resistance, the value of r_p depends on the points where the value of ΔE_p and the corresponding value of ΔI_p are taken on the curve. The points should be taken as closely as possible to the actual point of operation. The values of ΔE_p and ΔI_p taken for the example problem just worked are larger than those generally used for calculation of r_p.

Usually the value of ac plate resistance will be lower than the value of dc plate resistance. When the term "plate resistance" is used without designating whether dc or ac value is meant, it is generally correct to assume that it is the ac plate resistance.

The plate resistance of a diode, or for any type of tube, is an important factor in determining how the tube will operate in a circuit. More will be said about plate resistance later in this chapter when additional tube parameters are discussed.

18.10 DIODE RATINGS

There are certain limitations (placed by the manufacturer) on how each type of diode tube is to be used. Many of the methods employed for rating diodes are also applicable to other types of tubes. Although a safety factor is incorporated into these ratings so that they can withstand an occasional abuse, for prolonged tube life the tube should be operated well within the limits stated. The following ratings can be obtained from a tube manual for any specific tube type.

FILAMENT VOLTAGE Up to a certain point (where the filament burns out), increasing the voltage across the filament will increase the filament current. This, in turn, increases the cathode temperature and causes a greater electronic emission. However, the increase in emission is at the expense of filament life, so tubes should always be operated at their rated filament voltage.

The first number in the tube designation often indicates the required filament voltage. Thus a 5U4 is a diode operated with a filament voltage of 5 V; and a 6AL3 is a diode operated with a filament voltage of 6.3 V. (The first number only *indicates* the voltage; it is not necessarily the exact value of voltage. Thus the 6AL3 has a 6.3-V filament, and a 12AL5 has a 12.6-V filament.)

FILAMENT CURRENT In addition to the filament voltage, the manufacturer's specifications list the filament current. This is the amount of current that will flow in the filament when the proper filament voltage is applied.

The reason for specifying both the voltage and the current is that several tube filaments may be connected in series, parallel, or series parallel. Figure 18.13 shows series and parallel connections. Whenever filaments are connected in series, they must have the same filament current ratings! This follows from the fact that the current must be the same in all parts of a series circuit. The voltage ratings of series filaments must add to equal the applied voltage for the circuit. Figure 18.13*a* shows how two filaments are connected in series across a source.

Filaments of different current ratings may be connected in parallel, but they must have identical voltage ratings for this type of connection. This follows from the fact that the voltage across all parts of a parallel circuit is the same. The filaments may have any number of different current ratings when connected in parallel, provided the sum of the currents is not greater than the amount of current that the filament supply can provide. Figure 18.13*b* shows a parallel connection for filaments.

The filament power supply—which is a transformer in Fig. 18.13—must be able to deliver both the required voltage and the total current necessary for operating the filaments at their full rated value. For example, the transformer of Fig. 18.13*a* must deliver 18.9 V at 0.15 A, and the one in Fig. 18.13*b* must deliver 6.3 V at 0.45 A.

PEAK HEATER-TO-CATHODE VOLTAGE This rating is applicable to tubes with indirectly heated cathodes. In many applications the cathode and filament of a tube may be operated at different potentials. Since they are so closely spaced, an arc discharge can occur at a relatively low voltage difference between the two electrodes.

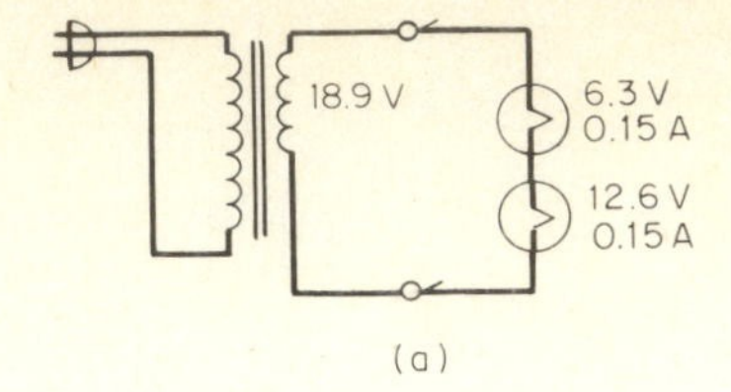

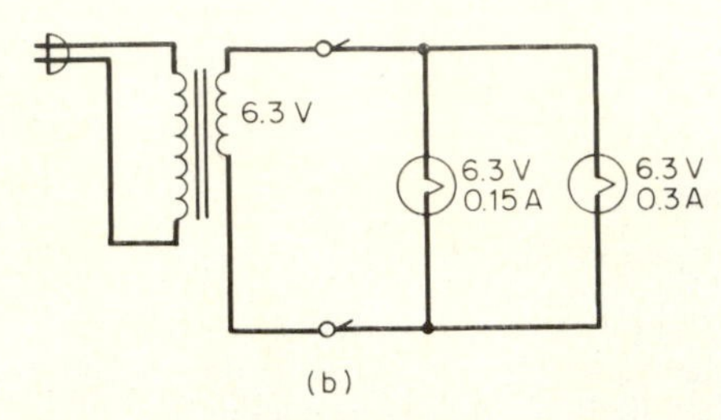

Fig. 18.13 Series and parallel connections of filaments: (*a*) Series-connected filaments. (*b*) Parallel-connected filaments.

The manufacturer normally gives the maximum allowable voltage difference in the rating marked "Peak Heater-to-Cathode Voltage." As with other tube ratings, this voltage difference should not be exceeded.

PEAK INVERSE VOLTAGE The plate voltage of a diode may be negative with respect to the cathode during certain periods of operation. During these periods the polarity of the plate voltage is as illustrated in Fig. 18.5*b*. Since electrons cannot be emitted by the plate, no current flows in the tube under this condition. If, however, the plate voltage is made sufficiently negative, an arc discharge will take place within the tube. This occurs because there is no such thing as a perfect insulator. Not even a vacuum can prevent current from flowing if the voltage is sufficiently large. Given a sufficient amount of voltage, *any* medium will conduct electricity. In a vacuum tube with a highly negative plate voltage, the discharge takes place in the form of an arc discharge (like a tiny bolt of lightning) between the plate and the cathode.

Since an arc discharge will usually destroy the tube, manufacturers give a peak inverse voltage rating—usually marked PIV—which is the maximum negative voltage that can be placed on the plate without the risk of an arc discharge. This rating should never be exceeded.

MAXIMUM PLATE CURRENT There is a limit to the amount of electron current that a tube can handle. This limit is fixed by the maximum temperature that the tube can dissipate, just as the maximum allowable current through a resistor depends on how well it can radiate heat. If the current rating is exceeded, the plate will become red hot, and the glass envelope may become soft from the heat. Also, the plate and cathode leads may open because of overheating.

In some applications the maximum allowable plate current for a tube may not be sufficient. In such cases it is possible to operate the tubes in parallel. Two tubes connected in parallel as shown in Fig. 18.14 can handle twice the current of a single tube.

PLATE DISSIPATION In order to be sure that a resistor will not burn out during normal use, the maximum allowable current through it could be stated. It is more of a common practice, however, to give the limits of operation in terms of a *power rating*. A 10-kΩ 5-W (10 000 Ω, 5 W) resistor can carry more current, and also dissipate more heat, than a 10-kΩ 1-W resistor. As with resistors, vacuum tubes may be rated by power

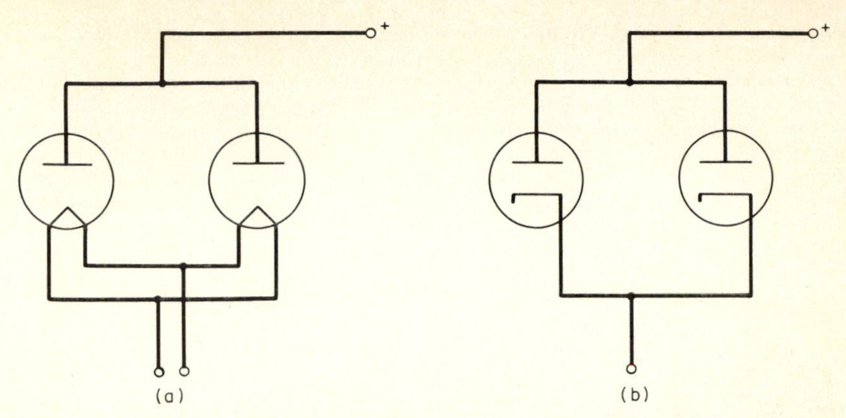

Fig. 18.14 Parallel connections of diode tubes for greater current flow: (*a*) Parallel tubes with filament emitters. (*b*) Parallel tubes with cathode emitters.

dissipation rather than maximum current. The plate dissipation of a tube is the total power delivered to the plate minus the power delivered to the load. This can be stated in equation form:

$$\text{Plate dissipation} = \text{total power} - \text{load power} \qquad (18.5)$$

where plate dissipation = power lost in the form of heat at the plate of the tube, measured in watts
total power = power delivered to the tube circuit and load combined; total power measured in watts
load power = power dissipated by the tube load, measured in watts

Assuming that a diode is connected into a circuit that has no phase angle between the voltage and current; its plate dissipation can also be calculated by multiplying the plate voltage and plate current:

$$P = E_p \times I_p \qquad (18.6)$$

where P = plate dissipation, W
E_p = plate voltage, V
I_p = plate current, A

example 18.7 In the example given for Fig. 18.10, the plate voltage was 17.5 V, and the corresponding plate current was 0.025 A. Find the plate dissipation.

solution For this point of operation, the plate dissipation is found as follows:

$$P = E_p \times I_p = 17.5 \times 0.025 = 0.4375 \text{ W} \qquad \textit{Answer}$$

18.11 TRIODE TUBES

The year 1907 is generally regarded as being the time of birth for the electronics industry. It was in that year that Lee De Forest invented the triode tube, which he called an *audion*. Up to that time vacuum-tube diodes had been in use in a limited number of applications. While a vacuum-tube diode can *modify* the waveform of an incoming signal, it cannot *amplify*. This important feature—amplification—was introduced with the triode tube.

Basically, a triode is a diode with a screen inserted between the cathode and the plate. The purpose of the screen—which is called a *control grid*—is to regulate the number of electrons that pass through it, and therefore the number that arrive at the plate. The voltage applied to the control grid determines the number of electrons that arrive at the plate at any given instant.

Figure 18.15 indicates the schematic symbols of two types of triodes. In Fig. 18.15*a* a triode with a direct (filament-type) emitter is shown, while the triode in Fig. 18.15*b* employs a cathode-type emitter. The construction of the cathodes and the plates of these tubes is similar to that described for diodes.

The control grid of the triode is located physically close to the cathode, and is constructed in such a way that a majority of the electrons can pass through it on their way to the plate whenever the grid voltage is zero. When a highly negative voltage is applied to the grid, the electrons are repelled by the negative voltage and cannot pass through to the plate. On the other hand, if the grid is made only slightly negative, some electrons can pass through. *An important feature of the triode is that the number of electrons that arrive at the plate is a direct function of the amount of negative voltage on the grid!*

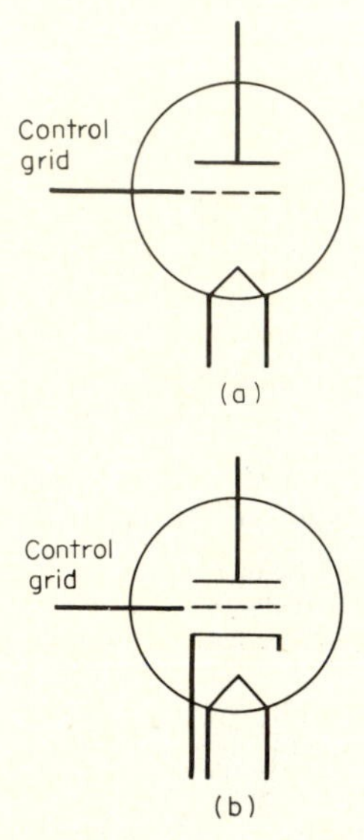

Fig. 18.15 Two types of triodes: (*a*) Triode with filament-type emitter. (*b*) Triode with cathode-type emitter.

Figure 18.16 shows the construction details of a number of different control grids that are used in triode tubes. Each type of grid construction has its own particular advantage and disadvantage. This illustration does not represent the limit of possible types of grids in use.

With the help of the circuit in Fig. 18.17 we can investigate the important characteristics of the triode tube. In this illustration, the voltage applied to the plate of the tube is marked E_B, and the plate-supply voltage is called E_{BB}. The control grid voltage is identified as E_C, and the grid-supply voltage is E_{CC}. At one time, vacuum-tube triode circuits were operated by batteries, and the batteries were identified by letter symbols. Batteries that supplied the filament current were known as *A batteries*, plate-supply batteries were known as *B batteries*, and grid-control supply batteries were known as *C batteries*. The subscripts in the identifications of Fig. 18.17 are carried over from battery-operated tube circuits of earlier times. Today the plate voltage may be called either E_B or E_P, and the grid voltage may be referred to as E_C or E_g—depending on the author's preference.

Variable resistor R in the circuit of Fig. 18.17 permits the amount of negative voltage on the grid to be selected. The term *bias voltage* is used to refer to the amount of negative voltage on the control grid with respect to the cathode. For example, if the grid is 10 V negative with respect to cathode, the grid tube is said to have a −10-V bias. It is important to note that all voltages around the tube circuit are always measured *with*

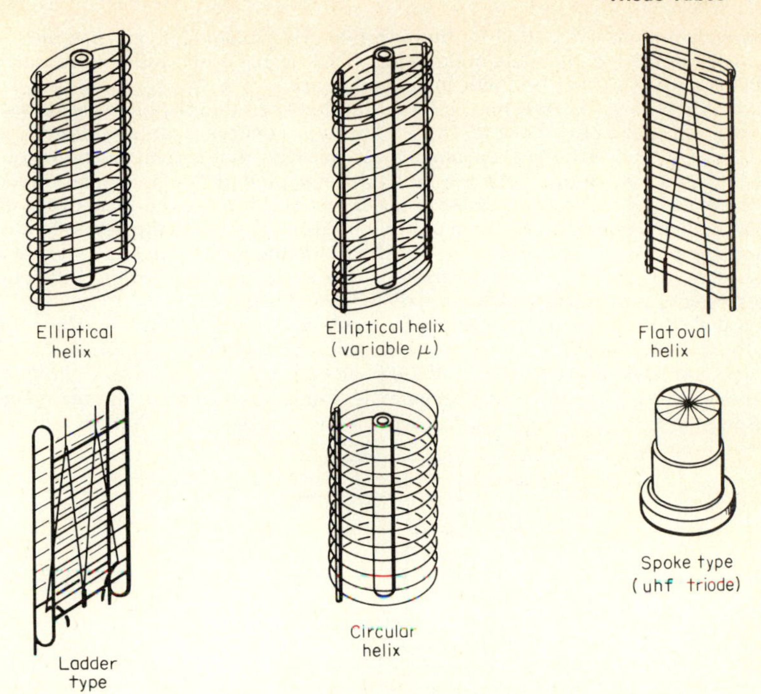

Fig. 18.16 Typical grid construction.

respect to the cathode as far as tube operation is concerned. However, as a matter of convenience the voltages with respect to ground may be given in troubleshooting literature.

Assume that the arm of variable resistor R is adjusted to point a. In this case, the grid bias voltage will be zero volts because the grid is essentially connected directly to the cathode. Under this condition, the plate current will be quite high. This is especially true if the cathode is of the oxide-coated variety. As a matter of fact, most triode tubes designed for receiver and home entertainment equipment are made in such a way that the grid voltage should *never* be made zero volts with respect to the cathode because of the excessive plate current that would result.

As the arm of the variable resistor R is moved toward point b, the grid voltage becomes more and more negative with respect to the cathode. When this happens, more

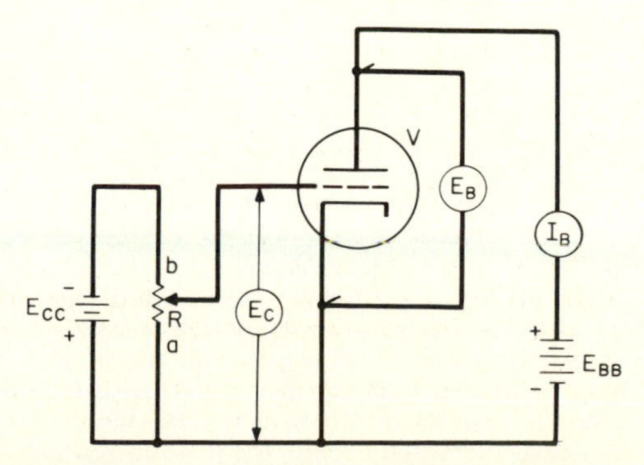

Fig. 18.17 Circuit for obtaining the static characteristics of a triode tube.

and more electrons are repelled by the negative grid voltage and are thus prevented from passing through to the plate of the tube. Thus, as the grid is made more and more negative, the plate current becomes lower and lower.

Figure 18.18 shows the graphical relationship between the plate current I_B and the grid voltage E_C in the circuit of Fig. 18.17. This type of curve is often referred to as an E_gI_p *characteristic curve.* Tube manufacturers provide such curves for their triodes. It will be noted that on the curve the grid voltage is plotted in a horizontal direction, while the plate current is plotted in a vertical direction. It is standard practice to graph the independent variable along the x axis (the horizontal axis) and the dependent variable along the y axis (the vertical axis). For the triode the grid voltage is varied at will, and therefore it is the *independent variable.* The value of plate current is dependent on the setting of grid voltage, and so the plate current is the dependent variable.

When the grid voltage (for the particular triode for which the curve of Fig. 18.18 is drawn) is at −6 V, the amount of plate-current flow is negligible. The point at which the negative voltage on the grid cuts off the plate current is called the *tube cutoff point.* Note that as the grid-bias voltage becomes less and less negative, the plate current increases.

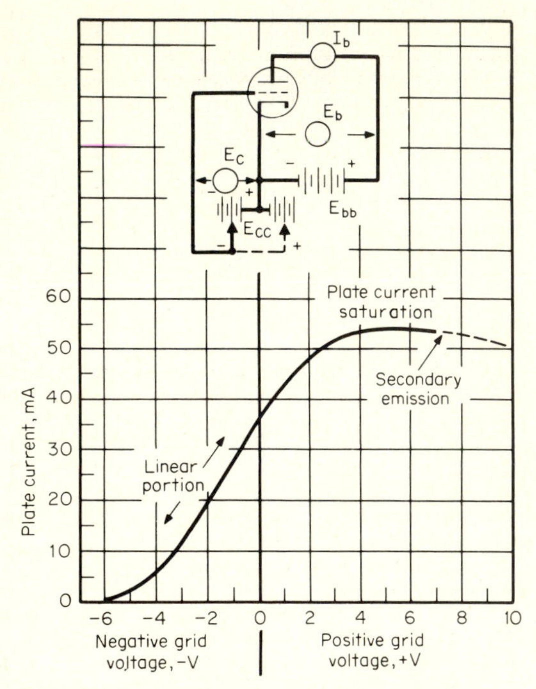

Fig. 18.18 A typical plate-current grid-voltage characteristic curve.

When the grid voltage is near the cutoff region, the increase in plate current is *not* directly proportional to the increase in grid voltage. (When we say *increase in grid voltage,* we are referring to a change in grid voltage in the *positive direction.*) The nonlinear portion of the curve near the cutoff point is referred to as the *lower knee of the curve.* As the grid-bias voltage is further increased—that is, made more positive—a point will be reached on the curve which is called the *linear portion.* Here the plate-current increase is directly related to the grid-voltage increase. This linear portion is important for operating a tube as an amplifier, as will be shown later.

When the grid voltage has increased to zero, the plate current has reached a value of approximately 36 mA. Moving the grid-bias voltage into the positive region results in a further increase in the plate current. The reason for this is that the positive grid voltage attracts electrons away from the space charge surrounding the cathode, and therefore reduces the internal resistance of the tube.

When the grid voltage is positive, not only does the plate current increase, but there is also a grid-current flow. During the portion of the E_gI_p curve in which the voltage is negative, no electrons are attracted to the negative grid. This is because the grid and the electrons are both negative, and like charges repel. Triodes used for home entertainment equipment are designed in such a way that they must always be operated with a negative grid voltage, and therefore no grid current is presumed to flow. Even a small amount of grid current will immediately destroy such a tube.

Returning to the characteristic curve of Fig. 18.18, it is seen that after the positive grid voltage reaches a value of about four volts, the plate current no longer takes on higher values. This is called the *saturation point of the triode.* A further increase in positive grid voltage will ultimately result in a *lowering* of the plate current, because high-velocity electrons striking the plate knock other electrons loose. This process is called *secondary emission.* The secondary electrons will actually flow to the positive grid, thus decreasing the total number of electrons that flow in the plate circuit.

Characteristic curves for a triode are very important because they allow the designer to predict the amount of current flow in a circuit when the plate voltage and the grid-bias voltage are known. The characteristic curve of Fig. 18.18 has been plotted for a varying value of plate voltage. If the plate voltage is increased, and the curve replotted, the shape of the curve will change. Since the tube manufacturer has no way of knowing what value of plate voltage is to be used in a particular circuit—although he may prescribe the *limits* of such voltages—he will ordinarily supply a *family of curves:* a group of curves plotted with each curve being different from the others because one of the parameters is changed. For example, in an E_gI_p family of curves each curve is plotted for a different value of plate voltage.

Figure 18.19 shows a family of curves. Here the grid voltage versus plate current is plotted for five different values of plate voltage: 300, 250, 200, 150, and 100 V. The curves are not permitted to go into the positive grid region which is a normal operation for a triode.

Referring again to the circuit of Fig. 18.17, we see that there are actually three different parameters important to the operation of the tube: the grid voltage, the plate voltage, and the plate current. Since it is not possible to show all three parameters on a two-dimensional graph, the characteristic curves of triodes will always imply that one of the parameters is held constant. For each curve in Fig. 18.19 the value of plate voltage is held constant for that particular curve. Another important curve used in predicting the operation of the tube in the circuit is shown in Fig. 18.20: the E_pI_p curve—that is, the plate voltage–plate current characteristic curve. In this case the grid voltage is held constant for each of the curves drawn.

You should study both sets of characteristic curves (Figs. 18.19 and 18.20) and satisfy yourself that the following general rules are correct:

1. *The more negative the grid voltage, the lower the amount of plate current—that flows—provided the plate voltage is held constant.*
2. *The more positive the plate voltage, the greater the amount of plate current—provided the grid voltage is held constant.*
3. *If the plate current increases, either or both of the following must have occurred—the grid voltage has been made less negative, or the plate voltage has been made more positive.*
4. *If the grid is made sufficiently negative, no plate current will flow.*

The characteristic curves that have been discussed are referred to as *static curves* because either the plate voltage or the grid voltage is held constant for each curve. Static curves are necessary for determining the dc operating characteristics of a tube in a circuit when no signal is applied. Ohm's law cannot be used because the tube is a nonlinear device. During normal operation of the triode, however, the grid voltage is allowed to vary by a certain amount, and a corresponding change in plate current and plate voltage is utilized in an amplifier circuit. Since a knowledge of varying characteristics is essential for an understanding of amplifiers, it will be necessary to define some tube constants that are based on changing values.

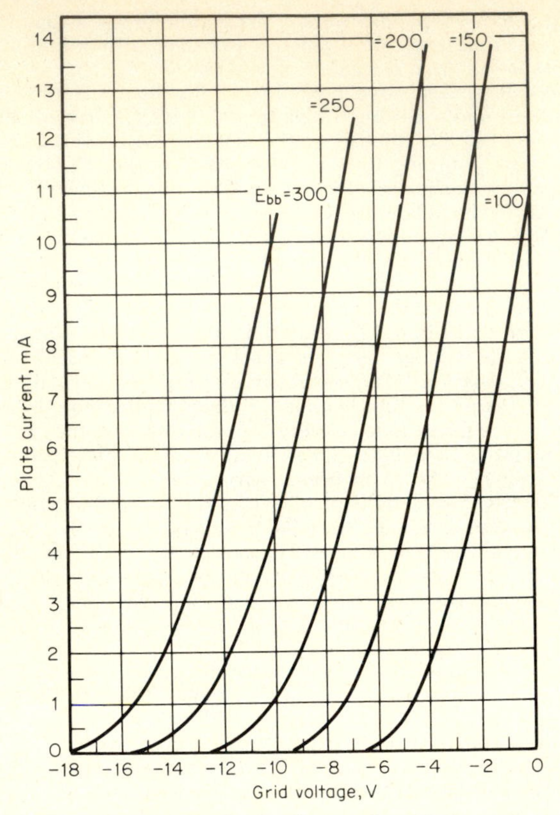

Fig. 18.19 A family of curves for a triode.

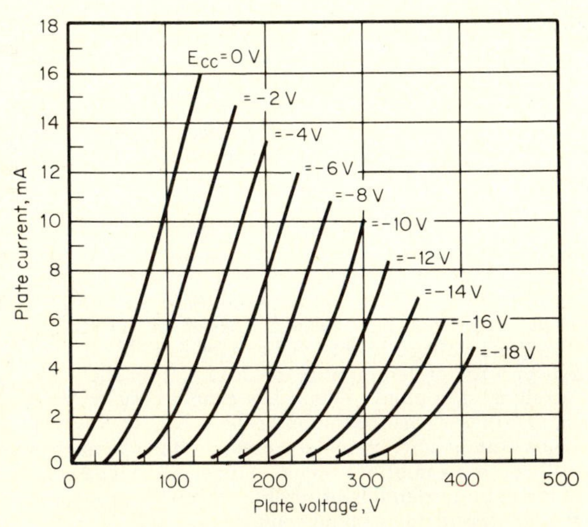

Fig. 18.20 Another type of characteristic curve.

18.12 TUBE CONSTANTS BASED ON CHANGING VALUES

The static curves cannot be used to directly determine a tube's value as an amplifier, although the required information can be ascertained with the help of the curves. The important dynamic characteristics (as opposed to static characteristics) are amplification factor, transconductance, and dynamic plate resistance.

AMPLIFICATION FACTOR For a triode to serve as a satisfactory amplifier, it is necessary that the control grid exercise a strong influence over the amount of plate current flowing at all times. Specifically, a small change in *control grid* voltage must produce a relatively large change in *plate current.* A change in plate voltage will also produce a change in plate current. However, if the tube is a good amplifier, a larger change in the plate voltage is necessary to produce the same amount of change in plate current than can be produced by a small change in grid voltage.

The amplification factor (represented by the symbol μ) is a measure of the effectiveness of the control grid in regulating plate current in comparison to the effectiveness of the plate voltage in regulating the current. Mathematically, the amplification factor is expressed as

$$\text{Amplification factor} = \frac{\begin{pmatrix}\text{small change in plate voltage that produces a} \\ \text{change in plate current}\end{pmatrix}}{\begin{pmatrix}\text{small change in grid voltage required to} \\ \text{produce the same in plate current}\end{pmatrix}} \qquad (18.7)$$

It is not convenient to write the equation in this form. The symbol Δ (pronounced delta) is used to mean a small change. It is often called an increment. Thus, ΔE_p means a small change in plate voltage, and ΔE_g means a small change in grid voltage. Using symbols, the equation becomes

$$\mu = \frac{\Delta E_p}{\Delta E_g} \qquad (18.7a)$$

The amplification factor of a tube can be determined directly from its $E_g I_p$ curve. The actual value changes according to the point on the curve where the measurements are made. Generally, the measurements are made with regard to the actual plate voltage and grid voltage that is to be used on the tube when it is placed in a particular amplifier circuit. This normally implies that the curve will be linear (that is, nearly a straight line) at the points where the measurements are made.

example 18.8 A vacuum tube with the characteristic curve shown in Fig. 18.21 is to be used as an amplifier. The grid-bias voltage is to be −8 V, and the plate voltage 250 V. Determine the amplification factor of the tube relative to this point.

solution Point *A* on the curve is the operating point described in the problem. The amplification factor must be determined by using points near this point on the curve.

Point *B* is chosen next. It should be near point *A*, but if it is too near, no solution can be obtained. (This will be clear after the construction lines are drawn.) A horizontal line, *B–C*, is drawn to the 200-V curve. Thus, for ΔE_p we are taking a value of 250 − 200 = 50 V.

Projecting down to the *x* axis, it is seen that line *B–C* also represents a change in grid voltage ΔE_g of 9.6 − 7 = 2.6 V.

Both changes—in plate voltage and in grid voltage—produce the same amount of plate-current change. The actual plate-current change ΔI_p is not needed for calculation of the amplification factor, but the value can be determined by projecting to the left from points *B* and *D* on the curve. Doing this, we find that ΔI_p is 11.7 − 5.4, or 6.3 mA.

It is very important to note that when the grid-bias voltage is maintained at a constant value of −7 V, the plate voltage must be changed from point *D* to point *C* to produce the current change ΔI_p of 6.3 mA. At the same time, if the plate voltage is held at a constant value of 250 V, then the grid voltage must be changed from point *B* to point *C* to obtain the same amount of plate-current change ΔI_p.

You can now readily see that if point *B* was chosen much closer to point *A*, then point *D* would have been off scale.

Using the values, $\Delta E_g = 2.6$ V and $\Delta E_p = 50$ V,

$$\mu = \frac{\Delta E_p}{\Delta E_g} = \frac{50}{2.6} = 19.2 \qquad \textit{Answer}$$

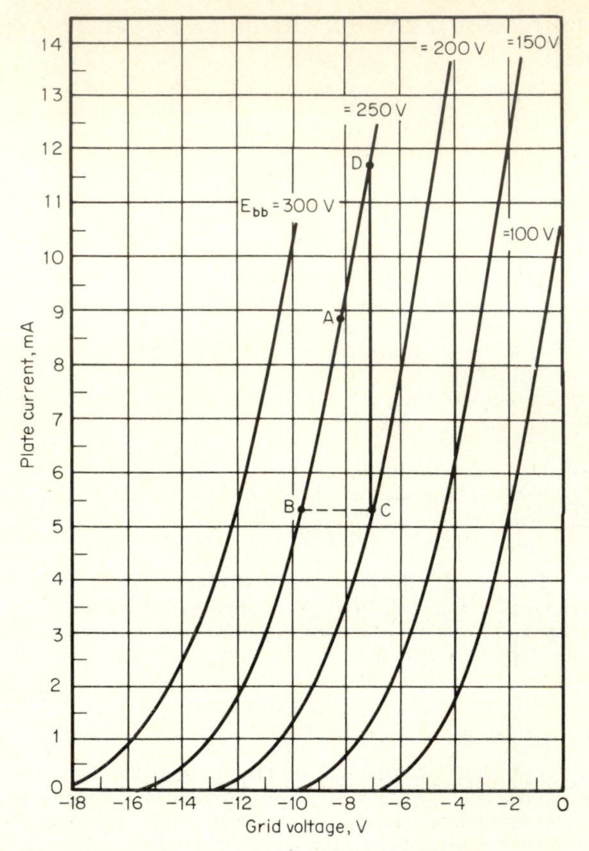

Fig. 18.21 Grid family of curves used to determine amplification factor.

A little practice is needed for obtaining the best possible results by choosing the optimum operating point and the reference points on the curve. Note that the portion of the curve from point *B* to point *D* is relatively linear. This is necessary for the best results. If the points that you choose cause you to work in nonlinear portions of the curve, it is best to start over with new points.

PLATE RESISTANCE When a vacuum-tube triode is connected into a circuit as shown in Fig. 18.17, it behaves very much like a variable resistor. In other words, the direct current is limited to a value I_b by the presence of the tube itself.

In order to calculate the size of a required power supply for a given tube (or for a number of tubes in a circuit), it is necessary to know how much current the tube will *draw*—that is, how much resistance the tube offers to the flow of direct current. The dc plate resistance of a triode is defined as the opposition that a tube offers to the flow of direct current when the tube is placed across a dc source of supply. It is represented by the symbol R_p. The dc plate resistance of a triode tube can be easily calculated by Ohm's law.

$$R_p = \frac{E_{bb}}{I_b} \tag{18.8}$$

where R_p = dc plate resistance, Ω
E_{bb} = dc plate voltage, V
I_b = dc plate current, V

Thus, the dc plate resistance (which is also known as the static resistance) of the triode is found in the same way as for a diode. [See Eq. (18.3).]

The dc plate resistance of a triode can be determined directly from the tube's characteristic curve. It is only necessary to find the plate current that corresponds to the value of plate voltage and the value of grid voltage to be used. Dividing the dc plate current into the dc plate voltage gives the dc plate resistance for the particular value of plate voltage and grid voltage used. The more negative the grid voltage, the lower the amount of plate current. It follows then that the dc plate resistance of a tube is increased by making the grid more negative.

Not only will the tube oppose the flow of direct current, but it will also oppose the flow of alternating current. The ac plate resistance (also called the dynamic plate resistance or plate impedance) of the tube is given by the equation

$$r_p = \frac{\text{change in plate voltage}}{\text{corresponding change in plate current}} \tag{18.9}$$

This equation is valid if the grid voltage is held constant while the plate voltage and plate current are changed. Equation (18.9) is more easily written as

$$r_p = \frac{\Delta e_b}{\Delta i_b} \tag{18.9a}$$

where r_p = dynamic plate resistance (also called plate impedance), Ω
Δe_b = change in plate voltage, V
Δi_b = change in plate current, A, resulting from change in plate voltage (provided grid voltage is held constant)

The dynamic plate resistance of a triode can be readily obtained from its E_pI_p characteristic curve (like the one shown in Fig. 18.20).

example 18.9 The grid voltage of a certain triode tube, which has the characteristic curve shown in Fig. 18.20, is to be maintained at a constant value of −6 V. Calculate the dynamic plate resistance corresponding to a plate-voltage change from 175 to 200 V.

solution When the plate voltage is 175 V and the grid voltage is −6 V, the graph shows that the plate current is 5 mA (0.005 A). When the plate voltage is 200 V, the graph shows that the plate current (corresponding to the grid voltage of −6 V is 6.75 mA. The change in plate voltage Δe_b is 200 − 175 = 25 V; and the change in plate current is 6.75 − 5 = 1.75 mA (0.00175 A). Substituting these values into Eq. (18.9*a*), we get

$$r_p = \frac{\Delta e_b}{\Delta i_b} = \frac{25}{0.001\,75} = 14\,285\ \Omega \qquad \textit{Answer}$$

This answer can be rounded off to a value of 14 300 Ω.

As with dc plate resistance, the more negative the grid of a triode, the higher its dynamic plate resistance. Furthermore, increasing the value of positive plate voltage will reduce the plate resistance. Whenever the term "plate resistance" is used in technical literature, the dynamic plate resistance is usually implied. If dc plate resistance is meant, it will be specifically identified as such.

TRANSCONDUCTANCE The plate current in a triode can be changed in either of two ways: by changing the grid voltage or the plate voltage. The transconductance of a tube is a measure of how much the plate current of the tube will change when the grid voltage is changed (and at the same time the plate voltage is held constant). Transconductance is represented by the symbol g_m, and is given by the equation

$$g_m = \frac{\Delta i_b}{\Delta e_c} \tag{18.10}$$

where g_m = transconductance of tube, mhos
Δe_c = change in grid voltage, V
Δi_b = change in plate current resulting in grid-voltage change (provided plate voltage is held constant), A

The transconductance of a tube is readily found from the E_pI_p characteristic curve. An example will show how this is done.

example 18.10 For the tube having the characteristic curve shown in Fig. 18.20, a steady plate voltage of 200 V is applied. Find the transconductance by assuming that the grid voltage is changed from −6 to −8 V.

solution For a value of 200 V on the plate, when the grid voltage is −6 V, the plate current is 7.5 mA; and when the grid voltage is −8 V, the plate current is 3.5 mA. The change in plate current (Δi_b) is $7.5 - 3.5 = 4$ mA, or 0.004 A. The change in grid voltage Δe_c that produced this amount of plate-current change is $(-6) - (-8) = 2$ V. Substituting $\Delta i_b = 0.004$ A and $\Delta e_c = 2$ into Eq. (18.10) gives

$$g_m = \frac{\Delta i_b}{\Delta e_c} = \frac{0.004}{2} = 0.002 \text{ mho}$$

For convenience, transconductance is usually expressed in micromhos. The answer to the problem can be converted from mhos to micromhos by moving the decimal point to the right six places. Thus,

$$g_m = 0.002 \text{ mho} = 2000 \text{ micromhos}$$ *Answer*

The value of transconductance is considered to be one of the most important parameters of a triode tube. It determines how much influence the grid voltage has in changing the plate current. In general, a transconductance of 2000 in one tube compared to 1000 in another tube means that the first one will be a better amplifier. The transconductance of a tube is so important that most dynamic tube checkers measure this value.

RELATIONSHIP AMONG PARAMETERS The amplification factor, dynamic plate resistance, and transconductance of a tube are called its parameters. By definition, a parameter is one of the constants entering into a functional equation and corresponding to some characteristic property, dimension, or degree of freedom.

A little mathematical manipulation can be used to show that there is a direct relationship among the three tube parameters:

$$g_m = \frac{\Delta i_b}{\Delta e_c} \quad (18.10) \qquad r_p = \frac{\Delta e_b}{\Delta i_b} \quad (18.9a)$$

Multiplying g_m by r_p [Eq. (18.10) by Eq. (18.9a)], we get

$$g_m \times r_p = \frac{\Delta i_b}{\Delta e_c} \times \frac{\Delta e_b}{\Delta i_b}$$

or

$$g_m \times r_p = \frac{\Delta e_b}{\Delta e_c}$$

But $\Delta e_b / \Delta e_c$ is the amplification factor μ. Thus,

$$g_m \times r_p = \mu \quad (18.11)$$

where g_m = transconductance, mhos
r_p = dynamic plate resistance, Ω
μ = amplification factor (as a number)

If any two parameters for a tube are known, the third can be found from Eq. (18.11).

example 18.11 A certain tube has an amplification factor of 100 and a transconductance of 1000 micromhos. What is the plate resistance of the tube?

solution In order to use Eq. (18.11) for the solution to this problem, the value of transconductance must be expressed in mhos. To convert 2000 micromhos to mhos, move the decimal point to the *left* six places:

$$2000 \text{ micromhos} = 0.002 \text{ mho}$$

Substituting $\mu = 100$ and $g_m = 0.002$ mhos into Eq. (18.11),

$$g_m \times r_p = \mu$$
$$0.002 r_p = 100$$
$$r_p = 50\,000\ \Omega$$ *Answer*

18.13 VOLTAGE GAIN OF TRIODE CIRCUITS

Tube parameters are needed for calculating a tube's performance in a circuit. As an example, suppose it is desired to find the gain of the amplifier stage shown in Fig. 18.22.

The output signal of the circuit is developed across a load resistor R_L. The voltage amplification of the stage A_e is the ratio of the output signal voltage e_o across the load resistor divided by the input signal to the grid e_s. In other words,

$$A_e = \frac{e_o}{e_s} \tag{18.12}$$

where A_e = voltage gain of the stage
e_o = output signal, V
e_s = input signal, V
(Voltage gain is a number.)

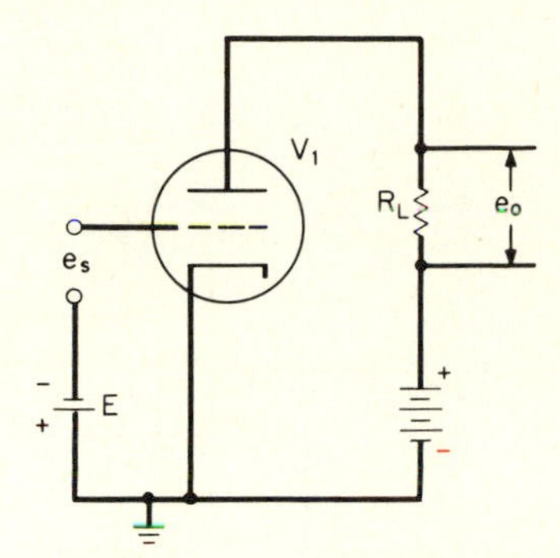

Fig. 18.22 A simple voltage amplifier circuit.

example 18.12 In the circuit of Fig. 18.22 an input signal voltage of ½ V results in an output signal voltage of 10 V. What is the voltage gain of the amplifier?

solution Using Eq. (18.12),

$$A_e = \frac{e_o}{e_s} = \frac{10}{0.5} = 20 \qquad \textit{Answer}$$

Calculation of the voltage gain from Eq. (18.12) requires that the input and output signal voltages be known. This means that the circuit must be constructed in order to find out how much gain it will have. A better method is to calculate the gain from the value of R_L and the tube parameters. Mathematically, it can be shown that, for the circuit in Fig. 18.22, the voltage gain is

$$A_e = \frac{\mu}{1 + r_p/R_L} \tag{18.13}$$

where A_e = voltage gain (a number)
μ = amplification factor of triode (a number)
r_p = dynamic plate resistance of tube, Ω
R_L = resistance of load, Ω

example 18.13 In the circuit of Fig. 18.22, the following things are known: $g_m = 2000$ micromhos, $r_p = 40\,000$ Ω, $R_L = 10\,000$ Ω. What is the voltage gain of the stage?

solution In order to apply Eq. (18.13), the amplification factor must be known. Since g_m and r_p are given in the problem, the amplification factor μ can be determined by using Eq. (18.11). Substituting $g_m = 0.002$ mhos and $r_p = 40\,000$ Ω into the equation

$$\mu = g_m \times r_p = 0.002 \times 40\,000 = 80$$

Now Eq. (18.13) can be used to find the gain of the stage for the circuit.

$$A_e = \frac{\mu}{1 + r_p/R_L} = \frac{80}{1 + 40\,000/10\,000} = \frac{80}{5} = 16 \qquad \textit{Answer}$$

An important thing about Eq. (18.13) is that it shows that the tube parameters make it possible to predict tube circuit behavior.

18.14 POWER AMPLIFICATION AND CIRCUIT EFFICIENCY

It has already been noted that the voltage amplification A_e of a voltage amplifier stage is equal to the output signal voltage e_o divided by the input signal voltage e_s. It might seem that the power amplification of a stage, then, would be found by simply dividing the output power by the input power. However, it will be noted that most tubes operate with no grid current flow, and if no current flows in the grid circuit no power will be dissipated in it. Thus the power amplification in such an amplifier could not be determined. Instead, a rating known as the *power sensitivity* is used. This is simply the output power divided by the square of the input signal voltage. The equations for power amplification and for power sensitivity are as follows:

$$\text{Power amplification} = \frac{\text{output power}}{\text{input power}} \tag{18.14}$$

where the output and input power values are expressed in watts. (Power amplification is a number.)

$$\text{Power sensitivity} = \frac{P_o}{(E_s)^2} \tag{18.15}$$

where P_o = output power of signal, in watts
$(E_s)^2$ = rms value of signal voltage input (Power sensitivity is a number.)

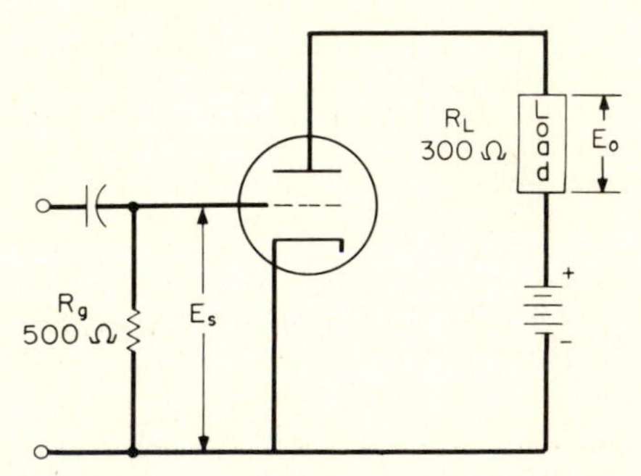

Fig. 18.23 A simple power amplifier circuit used for determining power gain and power sensitivity.

example 18.14 In the circuit of Fig. 18.23, the following things are known: Input rms signal voltage E_s is 1.2 V: output signal voltage E_o is 12 V. Calculate the power amplification and the power sensitivity of the stage.

solution

$$\text{Power amplification} = \frac{\text{output power}}{\text{input power}} \tag{18.14}$$

Since power = E^2/R, this equation can be rewritten as

$$\text{Power amplification} = \frac{E_o^2/R_L}{E_s^2/R_g} = \frac{(12)^2/300}{(1.2)^2/500}$$

$$= \frac{144/300}{1.44/500} = 167$$

$$\text{Power sensitivity} = \frac{P_o}{(E_s)^2} = \frac{(12)^2/300}{(1.2)^2}$$

$$= \frac{144/300}{1.44} = 0.33$$

It should be clearly understood that power gain and power sensitivity are not the same thing.

The efficiency of any device is defined as the output power divided by the input power. Expressed as a percentage,

$$\% \text{ efficiency} = \frac{\text{output power}}{\text{input power}} \times 100 \tag{18.16}$$

where the input and output powers are expressed in watts.

The input power of any device is always greater than the output power because of losses incurred, and therefore the efficiency is always less than 100 percent. In a vacuum-tube amplifier the output power is the power of the output *signal.* This is the only usable power as far as the amplifier is concerned. The input power is actually the *dc power* required to operate the amplifier stage. Therefore the efficiency of the power amplifier stage—which is known as the plate efficiency—is given by the equation

$$\text{Plate efficiency} = \frac{\text{ac signal output power}}{\text{dc input power}} \times 100 \tag{18.17}$$

The plate efficiency is also known as the *tube efficiency.* As a general rule, the greater the amplification factor of a tube, the *lower* its plate efficiency. You will remember that the amplification factor is a measure of how well the grid voltage is capable of producing a change in the plate current, compared to how well the plate voltage is capable of producing the same change in plate current. In general, the higher the amplification factor, the greater the voltage gain of an amplifier, and this implies that the grid must be spaced closely to the cathode. Conversely, the higher the amplification factor, the lower the plate efficiency. The reason is that the signal voltage amplitude cannot be as great in tubes with a high amplification factor, since a small voltage will completely cut the tube off. This is an important fact because it explains why power amplifiers are very poor voltage amplifiers.

In addition to the fact that the plate efficiency decreases with amplification factor, there is also another point for consideration. Power amplifier tubes have a relatively high current flow, which means that their cathodes must be quite hot to obtain the electron emission needed for the tube operation. Locating the grid very close to such a hot cathode would, in turn, mean problems with grid emission, and this is another reason why the power amplifier grid is located some distance from the cathode in comparison to the grid-to-cathode spacing in voltage amplifier tubes. Locating the grid farther away from the cathode reduces the amplification factor and at the same time increases the plate efficiency of the stage.

18.15 EVALUATION OF TRIODE TUBES

The more closely the grid is spaced to the cathode, the greater effect it has on the flow of electrons between the cathode and the plate. When electrons first leave the cathode, they have a relatively low velocity. Under this condition, a very small negative voltage on the grid can exercise a great influence over the electron motion. When the grid is located farther away from the cathode, by the time the electrons reach the grid they have obtained a high velocity, and a much greater negative voltage on the grid, which is required in order to influence their motion.

At first it would seem that the best tubes would be those that have their grids located closest to the cathode. However, additional problems are encountered by close grid-to-cathode spacing. In the first place, the cathode is hot. When the grid is located close to the cathode, the grid wires become heated. Most metal materials will emit electrons when heated, and the grid is no exception.

Grid emission is highly undesirable. In an ideal amplifier stage, all the electrons that arrive at the plate should be a direct function of the number that left the cathode, and also of the amount of negative voltage on the grid. If a greater number of electrons arrive at the plate than left the cathode, the effect will be an increase in plate current—but *not* an increase in plate signal.

One way to solve the problem of grid emission is to coat the grid with a material that has a low work function. An example is gold. Gold does not emit electrons even when heated to a relatively high temperature. Thus, in some very high gain amplifier tubes, gold-plated grids are used.

In the normal operation of a triode, the grid is always maintained at a negative voltage. If the grid should become positive, it will attract electrons from the cathode. Again, it means that the number of electrons arriving at the plate is not a direct function of the number leaving the cathode and the grid-bias voltage, and this is undesirable. Even more undesirable is the fact that the grid itself must dissipate power when grid current is flowing. The very small wires of an ordinary receiving-type tube cannot withstand the resulting temperature rise—that is, they cannot dissipate any appreciable amount of heat because of grid-current flow. In such tubes even a small amount of current can destroy the grid. In high-power triodes, the grid wires are often made in such a way that they can dissipate relatively large amounts of heat.

18.16 HIGH-FREQUENCY LIMITATIONS OF TRIODE TUBES

The electrodes of a triode include the plate, grid, and cathode. These are areas of metal separated by a vacuum, which is a *dielectric*. Therefore there is capacitance between the electrodes. Figure 18.24 shows a triode tube with the capacitance values between tube electrodes represented as external capacitors. As far as the circuit is concerned, we could consider the triode to be a tube that has no internal capacitance, but is always connected into the circuit with the external capacitors shown in Fig. 18.24.

The tube input capacitance—that is, the capacitance that the signal sees as it looks into the grid of a tube—is the grid-to-plate capacitance C_{GP} and the grid-to-cathode capacitance C_{GK}. The output signal—that is, the signal taken from the plate—is shunted by the plate-to-cathode capacitance C_{PK}. At low frequencies these interelectrode capacitance values can be ignored. Usually they have values of a few picofarads, and the capacitive reactance at low frequencies is so high that the capacitors appear to be open circuits to the signal. However, as the frequency increases, both the input and output capacitances affect both the input and output signals.

One way to reduce the effect of the interelectrode capacitance is to reduce the size of the tube electrodes. The capacitance of a capacitor decreases as the area of the capacitor plates decreases. This, among many other reasons, is why tubes designed for very high frequency operation are quite small compared to those used for low-frequency operation. Of course, if we decrease the physical size of the electrodes, we also decrease the amount of power that they can dissipate. The input capacitance of a triode presents a serious deterrent to amplification at high frequencies. If the input signal is shunted to ground by the interelectrode capacitance, it cannot be amplified. Likewise, if the output signal is shunted around the load by the plate-to-cathode capacitance, it will not appear in the load.

The problem of tube capacitance is even more complicated because of a phenomenon known as the *Miller effect*. The Miller effect states that the input capacitance of a tube increases with the gain of the stage. Mathematically the input capacitance is given by the equation

$$C_i = C_{GK} + C_{GP}\,(A_e + 1) \tag{18.18}$$

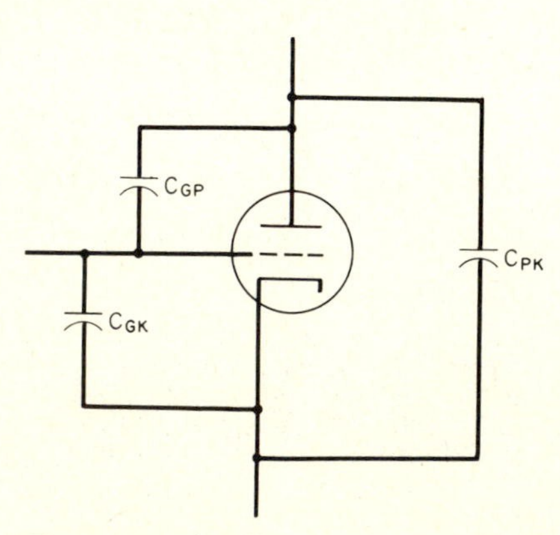

Fig. 18.24 A triode with its interelectrode capacitances.

where C_i = input capacitance of the tube as "seen" by the input signal, pF
C_{GK} = grid-to-cathode capacitance, pF
C_{GP} = grid-to-plate capacitance, pF
A_e = voltage gain of the stage

Equation (18.18) is for the input capacitance of voltage amplifiers having a purely resistive load. Note that the grid-to-plate capacitance is multiplied by the value of voltage gain of the stage, A_e plus one. In other words, the greater the gain, the greater the input capacitances of the stage. The equation also shows why it is often more desirable to have many stages of amplification at low gain instead of one single stage of amplification with very high gain when dealing with r-f signals.

example 18.15 The following tube capacitance values are given for one-half of a 6CS7 twin triode tube. (A twin triode is two triodes located within the same tube envelope.) C_{GK} = 1.8 pF; C_{PK} = 0.5 pF; C_{GP} = 2.6 pF. Typical voltage gain = 12. What is the input capacitance for the tube?

solution

$$C_1 = C_{GK} + C_{GP}(A_e + 1) = 1.8 + 2.6(12 + 1)$$
$$= 1.8 + 2.6(13) = 35.6 \text{ pF}$$
Answer

The voltage gain of an amplifier stage is dependent on a number of factors such as the amplification factor of the tube. Obviously tubes with high amplification factors will normally exhibit more input capacitance. Furthermore, the gain of the stage is a function of the load resistance. Since the input capacitance must be reduced in order to overcome the effect on the r-f signal, some method must be used to reduce the gain of the stage, and to reduce the tube's grid-to-plate capacitance. The grid-to-plate capacitance can be reduced to a negligible value by inserting electrodes between the grid and the plate that serve as a Faraday shield. This is the advantage of the so-called *multigrid vacuum tubes* which will be discussed in the next section. Another way to minimize the problem of input capacitance, besides decreasing the gain, is to add inductors into the circuit which tune to the interelectrode capacitance values. This procedure is called peaking compensation, and the inductors used are called *peaking coils.*

18.17 MULTIGRID TUBES

There are certain disadvantages to triodes which have led to the development of other tubes such as the *tetrode* and *pentode*. Because these tubes employ more than one grid, they are called *multigrid tubes*. It will be helpful to summarize the disadvantages of the triode which led to the development of multigrid tubes.

Because of the Miller effect, grid-to-plate capacitance is a very serious problem because it is multiplied by the gain of the stage. This is shown in the equation for input capacitance to the tube [Eq. (18.18)]. From this equation it is apparent that a decrease in grid-to-plate capacitance will have a greater effect on decreasing the *input capacitance* of the tube than a reduction in the grid-to-cathode capacitance would have.

There are a number of ways in which the grid-to-plate capacitance of a tube can be minimized. One is by *neutralization*, as shown in the circuit of Fig. 18.25. This circuit presents a simple triode amplifier in which the output signal is transformer-coupled to the next stage from L_1 to L_2. The signal developed at the plate of this tube will be fed back to the grid by way of the grid-to-plate capacitance C_{GP}, and it will be combined with the incoming signal in the grid circuits. This plate-to-grid feedback is often undesirable.

For circuits having tuned primary and secondary input and output stages, the feedback is *regenerative*. (Regenerative feedback occurs when the feedback signal is *in phase* with the input signal. The two signals add. If the feedback signal is small, the result is an increase in amplification. If the feedback signal is large, the result is oscillation.) In *RC*-coupled amplifiers the feedback is *degenerative*. (With degenerative feedback, the signal is fed from the plate to the grid so that it is 180° out of phase with the grid signal. This causes the two signals to almost cancel, or at least reduce, the amplitude of the grid signal.)

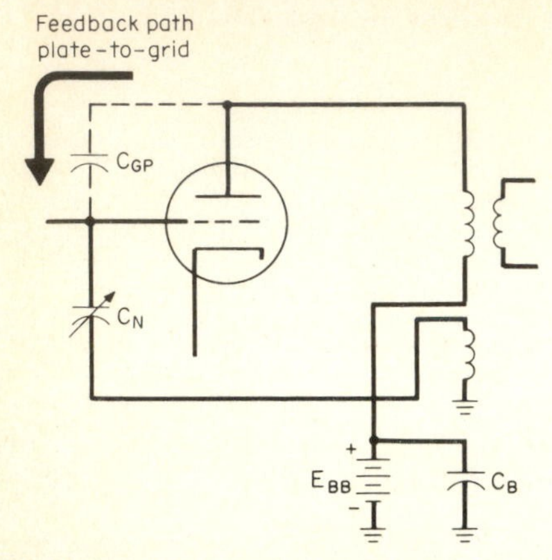

Fig. 18.25 One method of obtaining neutralization is to place an extra winding in the plate transformer. The neutralizing signal is fed to the grid via C_N. Capacitor C prevents the signal from entering the power supply.

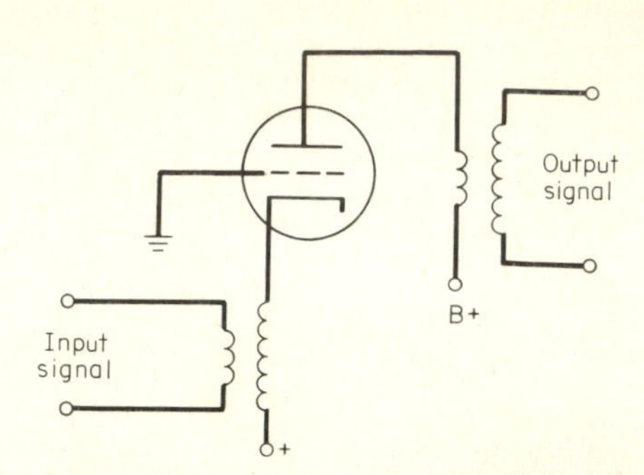

Fig. 18.26 In the grounded-grid circuit the input signal is applied to the cathode and the output signal is taken from the plate. Although the cathode is shown returned to a positive point in this illustration, it may be returned to chassis or ground.

In either case—regenerative or degenerative feedback—an unwanted plate-to-grid feedback signal follows the path shown by an arrow in Fig. 18.25. This feedback signal is eliminated by using another winding on the transformer, L_3, which provides a feedback signal equal in amplitude and opposite in phase to the grid-to-plate feedback signal flowing through C_{GP}. The feedback signal is coupled to the grid of the amplifier stage by way of a *neutralizing capacitor* C_N, which is made variable in order to make it possible to adjust the amount of feedback to neutralize the stage. During the neutralizing procedure the tube is not conducting, and C_N is adjusted until the plate-to-grid feedback signal is exactly canceled.

There are other methods of obtaining the neutralizing signal for feedback. That shown in Fig. 18.25 is only one representative example. In other methods the feedback signal may be taken from the secondary of the coupling transformer, or from a special voltage-dividing network in the primary circuit. In the latter case, the neutralizing signal is obtained by providing a tap on the primary winding. In any event, the purpose of neutralizing is always the same: to obtain a feedback signal equal in amplitude and opposite in phase to the undesired feedback.

The method of neutralization just described eliminates the effective grid-to-plate capacitance. However, it is not without problems. One disadvantage is that the feedback circuit is frequency-selective. In other words, it will feed back only a narrow range of frequencies for the purpose of neutralization. If it is necessary to pass a wide band of frequencies, it is almost impossible to obtain the necessary feedback over a wide range. Another disadvantage of the neutralization circuit is that at very high frequencies the circuit becomes quite difficult to adjust, and any small change in stray capacitance or tube capacitance will necessitate the readjustment of C_N.

Another method of decreasing the effect of grid-to-plate capacitance is to operate the tube in a *grounded-grid configuration* like the one shown in Fig. 18.26. Here the grid is operated at ground potential, and the cathode is operated at a slightly positive voltage with respect to ground. Thus, the cathode is positive with respect to the grid, or the grid is negative with respect to the cathode, as required for the operation of the triode tube. The input signal is delivered to the cathode, and the output signal is taken from the plate. The grid located between the cathode and the plate serves as a *Faraday shield*—that is, an electrostatic shield which prevents capacitive coupling between the output signal at the plate and the input signal at the cathode. With this type of circuit configuration, the input signals and the output signals are isolated from each other, and the effective feedback capacitance is minimized.

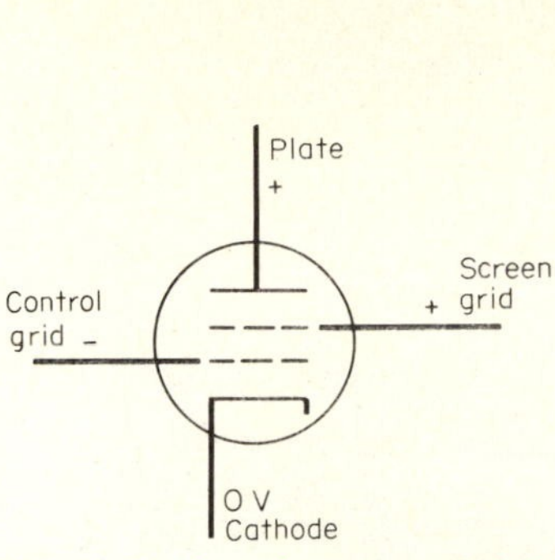

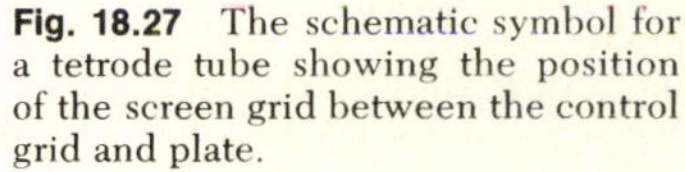

Fig. 18.27 The schematic symbol for a tetrode tube showing the position of the screen grid between the control grid and plate.

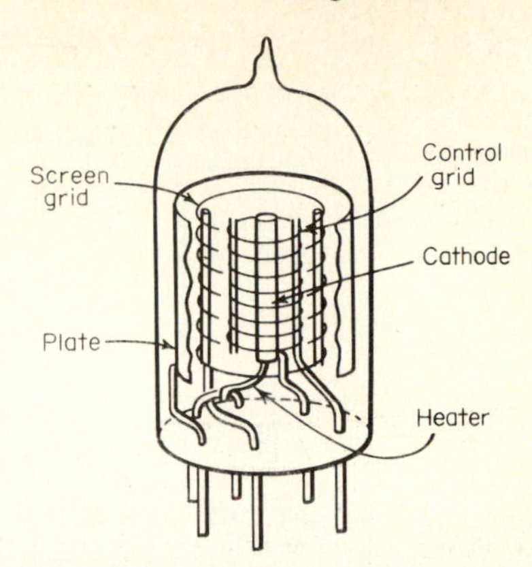

Fig. 18.28 Construction details of a tetrode tube.

A disadvantage of the grounded-grid circuit is that the gain is somewhat lower than can be obtained from a conventional (grounded-cathode) configuration.

In the earlier days of radio a considerable amount of difficulty was experienced in obtaining sufficient gain in each stage with a triode circuit. This was an important factor because tubes were expensive. In order to get the required amount of signal increase, it was necessary to use a number of stages. The development of multigrid tubes caused a high increase in gain per stage, and therefore reduced the number of tubes required. This advantage is no longer so important because of the very high gain per stage available in modern triode circuits. However, it is mentioned here because it did lead to the early development of multigrid tubes.

THE TETRODE One of the first methods for reducing plate-to-grid capacitance was to insert an additional grid between the control grid and the plate. This additional electrode was called the *screen grid*. (See Fig. 18.27.) The relative voltage polarities on the screen-grid tube are shown in this illustration.

All voltages around a vacuum tube are taken with respect to the cathode. In other words, the cathode is considered to be zero volts, and all other dc operating voltages are either positive or negative with respect to the cathode. Note that the control grid is *negative* with respect to the cathode as in the triode. The plate is *positive* with respect to the cathode, and so is the screen grid. This is an important point: The screen grid is maintained at *signal ground* potential in order that it may act as a Faraday shield between the plate and the control grid, but as far as the dc operating voltage is concerned, it is operated at a positive potential. It is important in all tube circuits to make a distinction between *signal ground* and *dc ground*, and the screen grid tube demonstrates one of the reasons for this.

Figure 18.28 shows the construction for a simple tetrode tube. This is sometimes referred to as *coaxial construction* because the cathode is at the center and the control grid, screen grid, and plate are all surrounding the cathode.

Figure 18.29 shows the circuit connection of a typical battery-operated tetrode amplifier circuit. The input signal to this amplifier circuit is developed across grid resistor R_g. Negative bias for the control grid is obtained from a C battery E_C. The screen grid obtains its positive dc operating voltage from the same battery that is used for the plate. However, in normal operation of the tetrode, the screen is often made slightly less positive than the plate, and therefore a screen-dropping resistor R_S is included in the circuit. A capacitor C_S connecting the screen to ground maintains the screen at *signal ground potential*. The plate is connected to the power supply through load resistor R_L, and a bypass capacitor C_B prevents any signal-current variations (which flow through the internal resistance of the battery) from introducing voltage variations

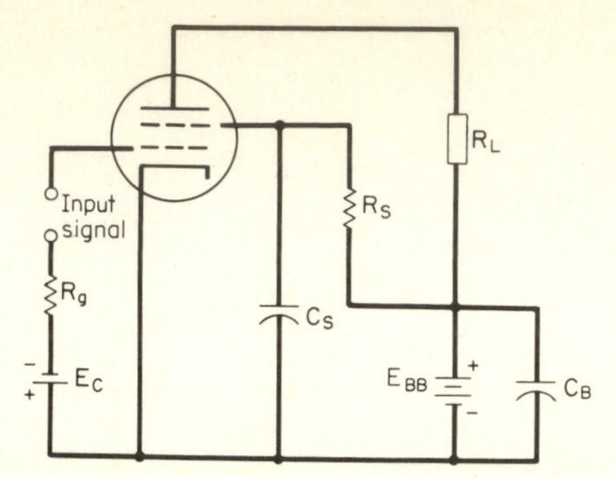

Fig. 18.29 An amplifier circuit using a tetrode tube.

across the battery terminals. If the screen—along with R_S and C_S—are removed in the circuit of Fig. 18.29, it is a simple triode amplifier.

The tetrode tube solved the problem of interelectrode capacitance between the grid and the plate, but introduced problems of its own which were (in some cases) worse than the ones that it was supposed to solve. In a triode, not much attention needs to be given to the problem of secondary emission. The only highly positive electrode in the triode is the plate. Therefore any secondary electrons that are produced are immediately attracted back to the plate, and they do not appreciably affect the operation of the tube.

In the tetrode there are two highly positive electrodes: the *plate* and the *screen*. Secondary electrons can easily leave the region of influence of the positive screen potential. It is this problem that causes the so-called "negative resistances" shown between points x and y on the curve of Fig. 18.30. It is a characteristic of tetrodes, and makes the tube unstable for use with relatively low plate voltages.

To overcome the problem of negative resistance in the tetrode, one approach is to use a very high positive plate voltage. This makes it more difficult for the electrons to get out of the region of influence of the plate, and reduces the amount of negative-resistance "kink" in a characteristic curve. The effect of the high positive voltage can be seen in the characteristic curve of Fig. 18.31. It will be noted that the low positive plate voltage—that is, voltage below 100 V—causes the tube to operate in the unstable region, but at plate voltages above 125 V, the characteristic curve is linear. The fact

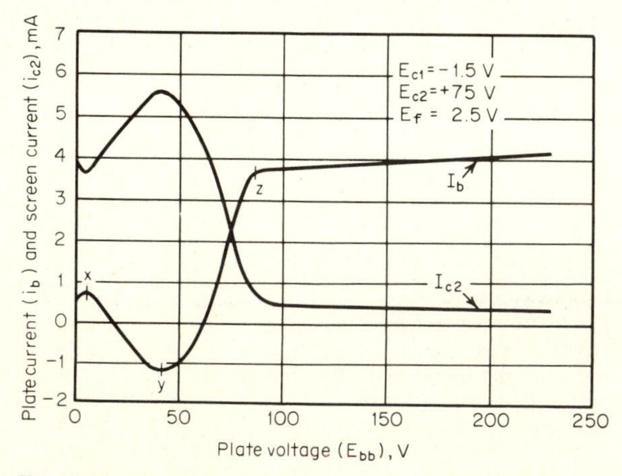

Fig. 18.30 Characteristic curves showing plate and screen currents versus applied voltage.

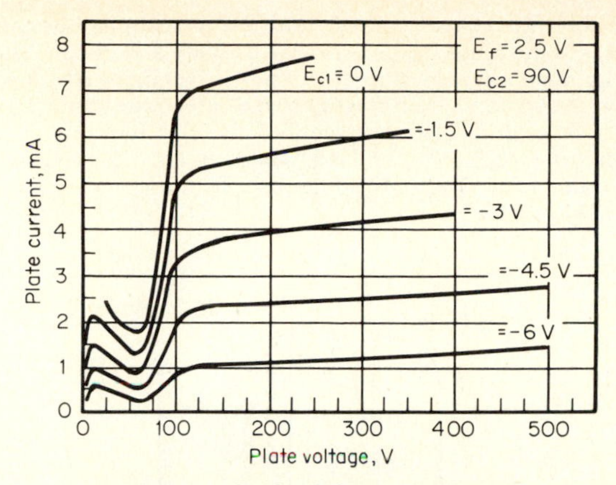

Fig. 18.31 A family of characteristic curves.

that tetrodes must be operated with high positive plate voltages can be considered a disadvantage for this type of tube.

Another method of reducing the problem of secondary emission from a plate (which causes the problem of negative-resistance tetrodes) is to chemically coat or treat the surface of the plate with materials that do not emit electrons easily. This method has been applied with some success in tetrodes, but it is not possible to completely remove the negative-resistance "kink" by this manner.

THE PENTODE The problem of secondary emission is minimized considerably by placing a negatively charged grid between the screen and the plate. The purpose of the grid—called the *suppressor grid*—is to repel the secondary electrons back to the plate, and thus prevent them from reaching the screen grid. This is the theory of operation of pentode tubes. Figure 18.32 shows the symbols used for a pentode, and the relative voltages on the electrodes of the pentode with respect to the cathode. The difference between the tube pentode symbols is that in one the suppressor grid is brought to an external connection and in the other the suppressor grid is connected internally—that is, inside the tube envelope—to the cathode. In either case the suppressor grid is operated at zero volts or a slightly negative voltage. Note that if the suppressor grid is at zero volts and the plate and screen are both positive, then the suppressor is negative with respect to both the screen and the plate. This is necessary in order for the suppressor to be capable of repelling negative secondary electrons back to the plate.

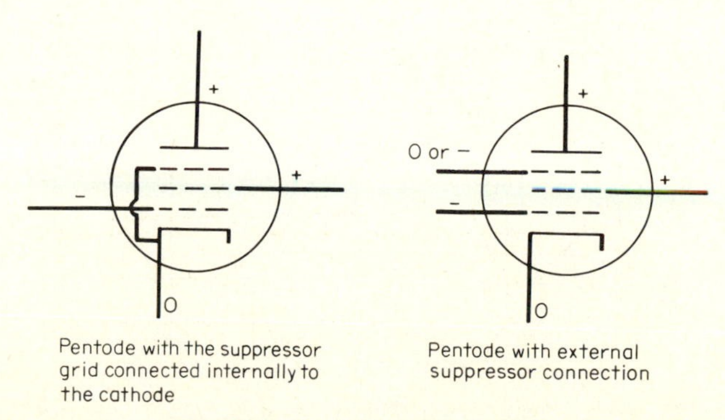

Fig. 18.32 Symbol for pentode tubes.

Figure 18.33 shows the coaxial construction of a typical pentode tube with the cathode at the center and the plate surrounding the cathode and grids. The electrodes in a pentode serve the same purpose as in the tetrode. The input signal is usually applied to the control grid, and the screen prevents interelectrode capacitance coupling between the plate and the grid. The plate collects the electrons that are emitted from the cathode and delivers a current to the load, which varies in accordance with the input signal voltage on the grid. The only additional electrode in comparison with the tetrode is the suppressor grid.

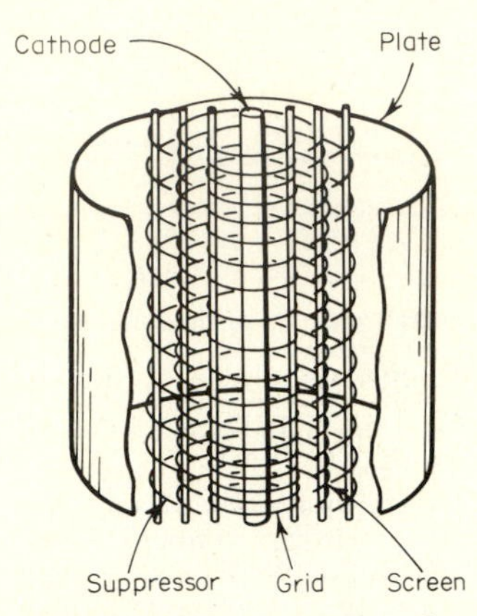

Fig. 18.33 Cutaway view of a pentode showing relative positions of the electrodes.

In a typical pentode amplifier, the suppressor grid is operated at either ground or cathode potential, and it does not in any way influence the primary flow of electrons between cathode and plate. In order to accomplish this, the suppressor-grid windings are quite far apart so that they do not intercept electrons on their path from cathode to plate. (In a few special applications of pentodes, the suppressor grid is used for influencing electron current flow, but these are not general applications. In these, the pentodes with suppressor connections delivered to an external tube pin must be used.)

Figure 18.34 shows a family of curves for a pentode tube. These are similar to the curve for tetrode tubes, except that they do not have the negative-resistance "kinks" in the low plate-voltage region. In order to make these curves, the suppressor-grid and the screen-grid voltages are held at a constant dc value (0 V for the suppressor grid and 100 V for the screen grid). In addition to the fact that there is no negative-resistance

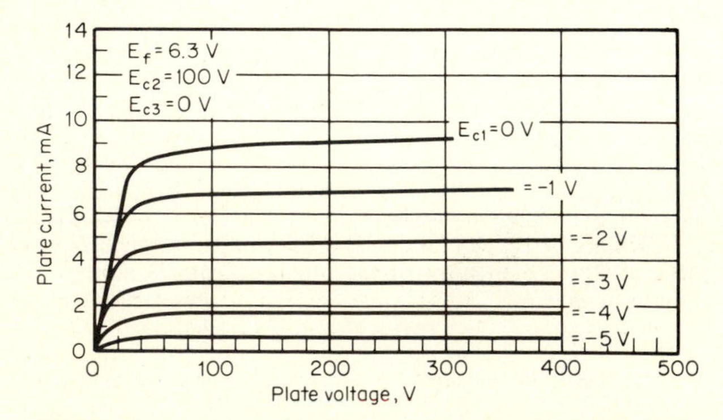

Fig. 18.34 Family of characteristic curves for a typical pentode tube.

region in the characteristic curve, it will also be noted that plate current rises to its maximum value much more rapidly than for the tetrode.

BEAM-POWER TUBES Although tetrode and pentode tubes are relatively free from plate-to-grid capacitance, they introduce their own problems when used for amplifying high frequencies. These must be taken into consideration for a complete evaluation of a tube's performance. One problem is that electrons in passing from the cathode to the plate often strike the additional electrodes in the tetrode and pentode tubes. Every time an electron strikes an electrode, it gives up energy and produces electric noise within the tube. This noise appears in the form of a hissing sound in the speaker of receivers designed to receive signals at those frequencies.

Another problem with both tetrodes and pentodes is that the screen conducts a relatively large amount of current, and this takes current away from the plate circuit. Ideally all the output power of a tube would be delivered to the plate circuit since that is where the useful load is connected. Therefore any current taken away by the screen can be considered to be a reduction of current for producing power in the plate.

The beam-power tube, which may be either a tetrode or a pentode, is specially designed to eliminate these problems. Figure 18.35 shows the construction of a beam-power tube and the flow of electrons between the cathode and the plate within the tube. The cross section in this illustration is of a tetrode beam-power tube. In some types of beam-power tubes an additional grid structure replaces the *beam-confining electrode* in which case it is considered to be a pentode beam-power tube.

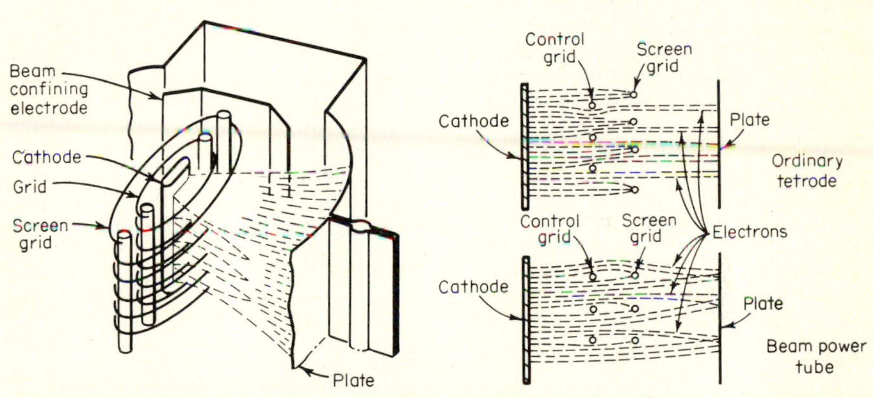

Fig. 18.35 Internal construction of a beam-power tube, showing concentration of electrons flowing between cathode and plate. (*Courtesy of RCA.*)

Fig. 18.36 Electron flow in the beam-power tube.

The cathode control grid, screen grid, and plate serve the same purposes as they do in an ordinary tetrode. (See Fig. 18.36.) However, in the beam-power tube the screen grid is maintained at a higher potential than for normal operation of tetrodes. This means that the electrons are actually slowed down after they leave the screen. This is true because the plate is negative with respect to the screen. The reduced velocity of electrons between the screen and plate regions results in a concentration of electrons called a *virtual cathode* or *space cloud.* The high negative density of electrons produces a negative charge which prevents secondary electrons from leaving the plate and returning to the screen. Thus the beam-power tube has the advantages of the pentode tube which uses a suppressor grid for preventing secondary electrons from reaching the screen. It has the added advantage that there is no electrode in the region between the screen grid and plate, and this reduces the amount of noise generated in the tube.

The *beam-confining electrodes* (also called *beam-forming plates*) serve two purposes: They keep the electrons concentrated in a sheet or beam so that the high electron den-

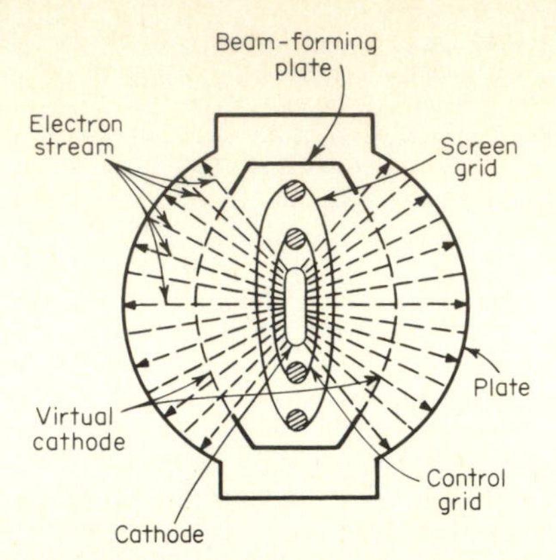

Fig. 18.37 Top view of tetrode type of beam-power tube.

sity can be formed in the plate region, and they prevent any secondary electrons from moving around to the screen from the sides.

Another feature of the beam-power tube is that the screen is positioned so that the screen conductors are hidden behind the grid conductors (as viewed from the cathode). Figure 18.36 shows the relationship of grid and screen construction. In the ordinary tetrode the screen wires are positioned between the control-grid wires. This allows them, because of their positive voltage, to intercept a relatively large number of electrons moving between the cathode and the plate. In the beam-power tube the screen is hidden behind the grid wires. Therefore, there will be a much lower screen-grid current. Obviously, if the screen-grid current is low, the electrons that are not intercepted by the screen will be able to pass on to the plate, which will result in a larger plate current. The advantage in terms of output power has already been discussed.

Figure 18.37 shows a top view of a tetrode-type beam-power tube illustrating the virtual cathode and the relative positions of the cathode, control grid, screen grid, and beam-forming electrodes and plate. The schematic symbols for beam-power tubes are shown in Fig. 18.38. In Fig. 18.38*a*, a tube with beam-forming plates is illustrated. Note that there are only four electrode connections (not counting heaters), and therefore the tube is joined into a circuit in the same way as a tetrode or pentode. Figure 18.38*b* shows a pentode-type beam-power-tube schematic diagram, and again there are only four external connections which makes this tube very similar to a tetrode or pentode in its circuit connections.

One difference between the beam-power-tube and the tetrode or pentode circuits is that in the beam-power tube the screen is normally operated at a higher positive voltage than the plate. This is one useful way of distinguishing between a beam-power tube and pentodes in a circuit.

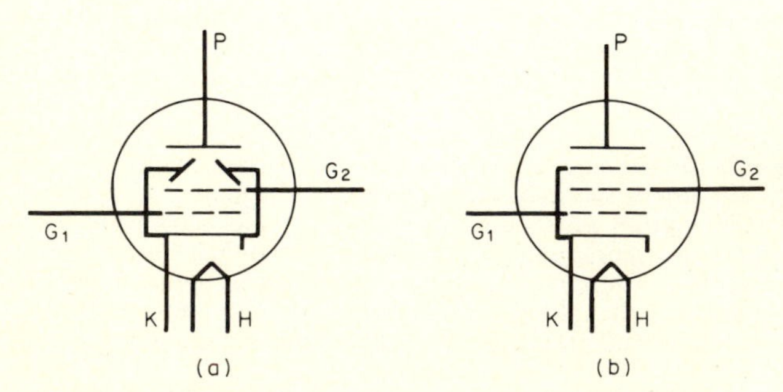

Fig. 18.38 Symbols for beam-power tubes: (*a*) Tetrode type of beam-power tube with beam-forming plates. (*b*) Pentode type of beam-power tube with a grid serving as the virtual cathode.

In discussing tetrodes and pentodes, it was noted that the amount of plate voltage swing allowable is limited to the regions beyond the knee of the characteristic curve, assuming that distortion is to be minimized. Figure 18.39 shows a comparison of pentode and beam-power-tube characteristic curves. Note that the knee of the curve—that is, the saturation point—occurs much sooner for the beam-power tube. Another way of saying this is that the plate of the beam-power tube becomes independent of the

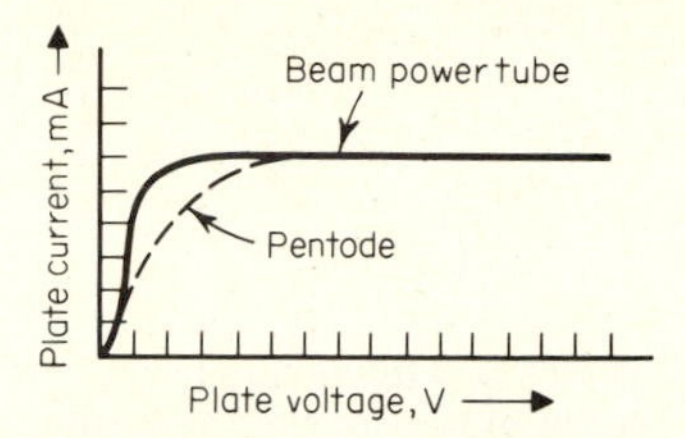

Fig. 18.39 Comparison of pentode and beam-power tube characteristic curves.

plate voltage at a lower positive value of plate voltage. The obvious advantage is that the plate voltage can be allowed to swing through a greater range of values before the knee of the curve is reached. Thus, the beam-power tube can handle much higher signal voltages than either a pentode or a tetrode.

PENTAGRID TUBES For certain special applications tubes with more than three grids have been designed. One example is the pentagrid tube (Fig. 18.40) with five grids. Actually two of the grids serve as a plate for delivering current. Figure 18.40*a*

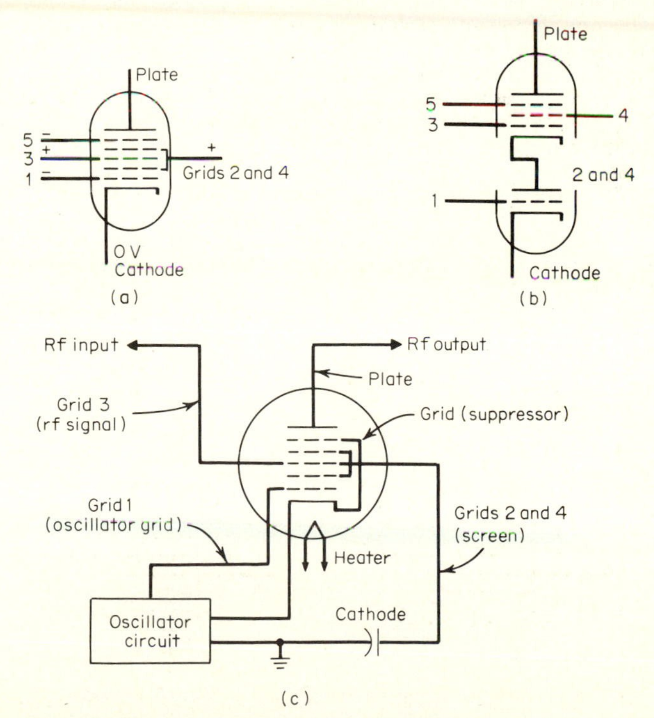

Fig. 18.40 The pentagrid tube: (*a*) Symbol for a pentagrid tube. (*b*) The pentagrid tube acts as though it were two tubes connected in series. (*c*) An example of how the pentagrid tube is used.

shows the schematic symbol of a pentagrid tube and the polarities of the voltages on the grids with respect to the cathode.

Grid 1 serves as a control grid. Any signal on this grid will exercise control over the electron current flowing between the cathode and plate. Grids 2 and 4 serve as a plate which attracts some (but not all) electrons out of the stream. The combination of the cathode and grids 1, 2, and 4 serves as a triode.

Grid 3 is another control grid. As in grid 1, any signal of this control grid will exercise control over the electron current flowing between the cathode and plate. Grid 5 serves as a suppressor, and is usually tied to the cathode within the tube envelope. The combination of the cathode, grids 3, 4, and 5, and the plate may be thought of as being a pentode.

Since the triode and the pentode elements within the tube both serve to exercise control over the electron stream, it is sometimes convenient to imagine them as being two separate tubes in series, as shown in Fig. 18.40*b*. The numbers on the series-connected tubes correspond to the grid numbers of the pentagrid tube.

An important application of the pentagrid tube is in the converter stage of a radio. Figure 18.40*c* shows this application. The converter stage consists of an oscillator (the triode) and a r-f amplifier (the pentode). Both signals combine in the tube to produce the following: r-f signal, oscillator signal, oscillator signal plus r-f signal, and oscillator signal minus r-f signal. The plate circuit is tuned so that it is receptive only to the difference frequency—that is, the oscillator signal minus the r-f signal.

18.18 MULTIUNIT TUBES

From the standpoint of compactness and initial circuit manufacturing costs, many tubes have been constructed with more than one type of tube in the same envelope. The savings in circuit manufacturing costs is that only one tube socket needs to be wired to accommodate such a tube. An example of such a *dual-purpose tube* is the dual diode used in low-voltage power supplies. A full-wave rectifier requires two diodes for its operation. One way of making such a circuit would be to employ separate diodes, separate tube sockets, and the attending separate connections. However, it is much simpler to include both the diodes in the same envelope, with only one tube socket needed. This leads to construction of a dual-diode tube, sometimes called a duodiode.

Figure 18.41 shows the symbols for a number of dual-purpose tubes. There is a limit to the number of tubes that can be placed in the same envelope because of the limited

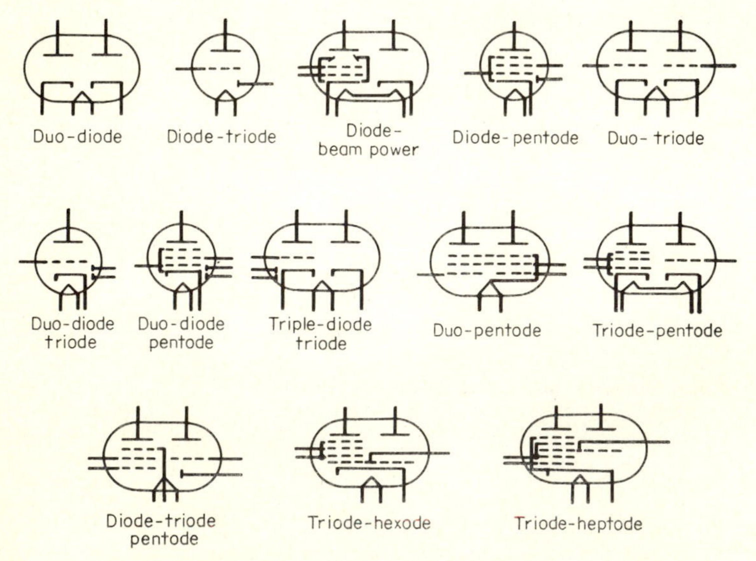

Fig. 18.41 A number of multiunit tubes.

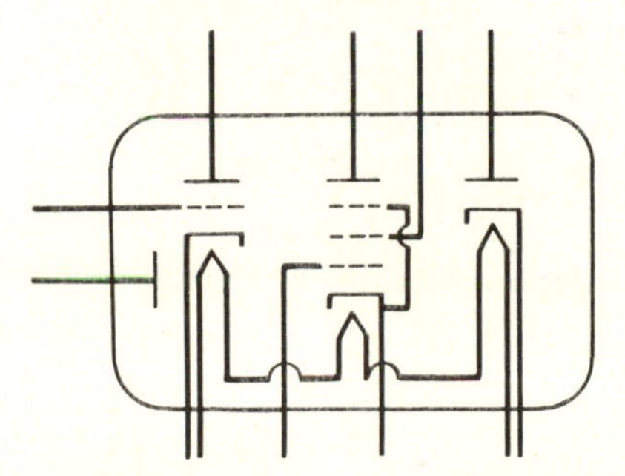

Fig. 18.42 Compactron tubes: (*a*) A grouping of compactron tubes. There are many different types in wide use in TV receivers. (*Courtesy of General Electric Company.*) (*b*) Symbol for a typical compactron tube: two diodes, one triode, and one pentode in the same envelope.

number of pins that are connected to a standard tube base. Miniature tubes have seven or nine pin connections, while some of the older tubes (called *octal tubes*) have eight connections. None of the symbols shown in Fig. 18.41 require more than nine pins for external connections in the tube base.

Manufacturers solved the problem of the limited number of tube base connections by designing a special tube specifically designed for mounting several different tubes within the same envelope. These tubes are called *compactrons*. Figure 18.42*a* shows the appearance of compactron tubes. They have 12 pin connections in the base. Figure 18.42*b* shows a typical schematic symbol for a compactron tube.

The initial cost of wiring the chassis to accommodate the tube is less for multiunit tubes. However, if one tube within the envelope burns out, all the tubes have to be replaced. Therefore, the replacement cost is somewhat higher, and this is a disadvantage.

18.19 NUVISTOR TUBES

An important limitation of an ordinary vacuum tube, when used at high frequencies, is the amount of time that it takes the electron to travel from the cathode to the plate within

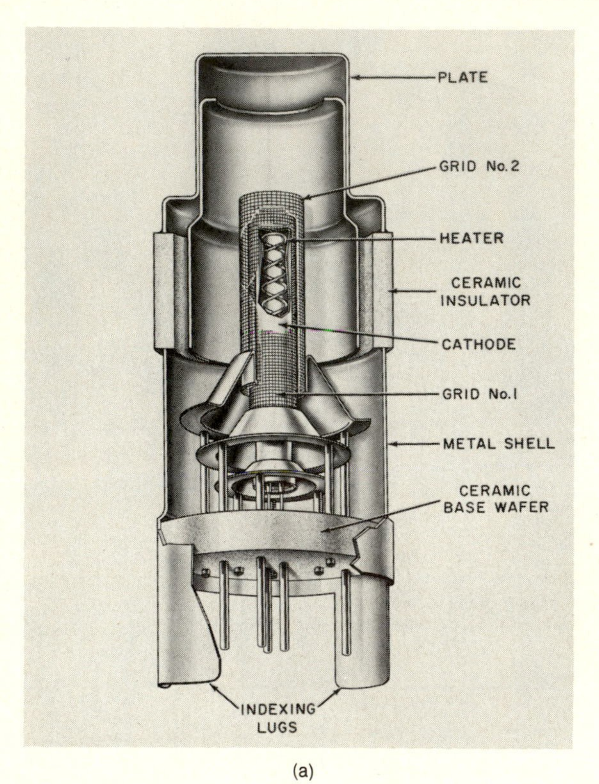

(a)

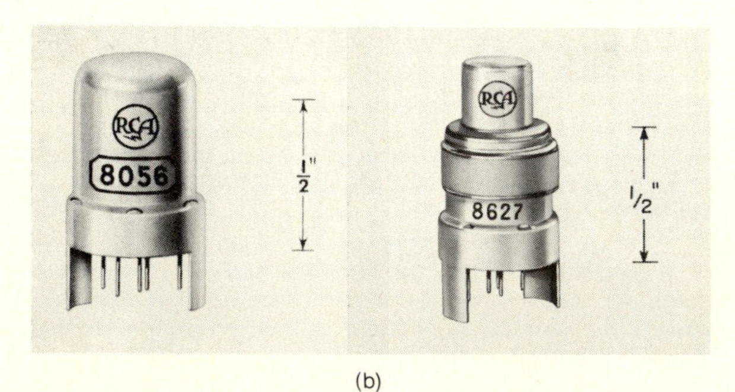

(b)

Fig. 18.43 The Nuvistor tube is a five-pin ultraminiature high-frequency tube no larger than some transistors: (*a*) Cutaway view of a double-ended power tetrode. (*b*) At left is a single-ended triode; at right is a double-ended triode. (*Courtesy of RCA.*)

the tube. This time, called *transit time,* is virtually instantaneous in dealing with lower frequencies. However, when the frequency being amplified is very high, the amount of transit time can be appreciable when compared to the time for one complete cycle of input signal. As a general rule, the transit time should never be greater than the time required for one-fourth cycle of input signals, although one-tenth of a cycle period is much more desirable. It can be shown that the efficiency of the tube decreases appreciably as the transit time (in comparison with the cycle time) increases.

One way to reduce the transit-time effect is to use very close spacing between the cathode and the plate. This is one feature of the *Nuvistor* tube, which is a five-pin ultraminiature type no larger than some transistors. In addition to the reduced transit time, this tube also has a relatively low interelectrode capacitance due to the very small physical size of the electrodes. (The interelectrode capacitance of a tube decreases as the sizes of the electrodes decrease). Additional features of the Nuvistor tube are its very low power consumption and its resistance to shock and vibration.

Figure 18.43 shows the Nuvistor tube. A cutaway view of the tube is presented in Fig. 18.43*a*, and several versions of the tube are depicted in Fig. 18.43*b*. Normally these tubes are used in very-high frequency and ultrahigh-frequency applications.

Appendix

A. The Greek Alphabet

Name	Large	Small
Alpha	A	α
Beta	B	β
Gamma	Γ	γ
Delta	Δ	δ
Epsilon	E	ϵ
Zeta	Z	ζ
Eta	H	η
Theta	Θ	θ
Iota	I	ι
Kappa	K	κ
Lambda	Λ	λ
Mu	M	μ
Nu	N	ν
Xi	Ξ	ξ
Omicron	O	o
Pi	Π	π
Rho	P	ρ
Sigma	Σ	σ
Tau	T	τ
Upsilon	Υ	υ
Phi	Φ	ϕ
Chi	X	χ
Psi	Ψ	ψ
Omega	Ω	ω

B. Probable Values of General Physical Constants*

Constant	Symbol	Value
Electronic charge	q	1.602×10^{-19} C
Electronic mass	m	9.109×10^{-31} kg
Ratio of charge to mass of an electron	q/m	1.759×10^{11} C/kg
Mass of atom of unit atomic weight (hypothetical)	. . . . .	1.660×10^{-27} kg
Mass of proton	m_p	1.673×10^{-27} kg
Ratio of proton to electron mass	m_p/m	1.837×10^{3}
Planck's constant	h	6.626×10^{-34} J-s
Boltzmann constant	$\bar{k}$	1.381×10^{-23} J/K
	k	8.620×10^{-5} eV/K
Stefan-Boltzmann constant	σ	5.670×10^{-8} W/(m^2)(K^4)
Avogadro's number	N_A	6.023×10^{23} molecules/mol
Gas constant	R	8.314 J/(deg)(mol)
Velocity of light	c	2.998×10^{8} m/s
Faraday's constant	F	9.649×10^{3} C/mol
Volume per mole	V_o	2.241×10^{-2} m^3
Acceleration of gravity	g	9.807 m/s^2
Permeability of free space	μ_o	1.257×10^{-6} H/m
Permittivity of free space	ϵ_o	8.849×10^{-12} F/m

* E. A. Mechtly, "The International System of Units: Physical Constants and Conversion Factors," National Aeronautics and Space Administration, NASA SP-7012, Washington, D.C., 1964.
From J. Millman and C. Halkias, "Integrated Electronics: Analog Digital Circuits and Systems," McGraw-Hill Book Company, 1972.

C. Conversion Factors and Prefixes*

1 ampere (A)	= 1 C/s	1 lumen per square foot	= 1 ft-candle (fc)
1 angstrom unit (Å)	= 10^{-10} m	mega (M)	= × 10^{6}
	= 10^{-4} μm	1 meter (m)	= 39.37 in
1 atmosphere pressure	= 760 mm Hg	micro (μ)	= × 10^{-6}
1 coulomb (C)	= 1 A-s	1 micron	= 10^{-6} m
1 electronvolt (eV)	= 1.60 × 10^{-19} J		= 1 μm
1 farad (F)	= 1 C/V	1 mil	= 10^{-3} in
1 foot (ft)	= 0.305 m		= 25 μm
1 gram-calorie	= 4.185 J	1 mile	= 5,280 ft
giga (G)	= × 10^{9}		= 1.609 km
1 henry (H)	= 1 V-s/A	milli (m)	= × 10^{-3}
1 hertz (Hz)	= 1 cycle/s	nano (n)	= × 10^{-9}
1 inch (in)	= 2.54 cm	1 newton (N)	= 1 kg-m/s^2
1 joule (J)	= 10^{7} ergs	pico (p)	= × 10^{-12}
	= 1 W-s	1 pound (lb)	= 453.6 g
	= 6.25 × 10^{18} eV	1 tesla (T)	= 1 Wb/m^2
	= 1 N-m	1 ton	= 2,000 lb
	= 1 C-V	1 volt (V)	= 1 W/A
kilo (k)	= × 10^{3}	1 watt (W)	= 1 J/s
1 kilogram (kg)	= 2.205 lb	1 weber (Wb)	= 1 V-s
1 kilometer (km)	= 0.622 mi	1 weber per square meter (Wb/m^2)	= 10^{4} gauss
1 lumen	= 0.0016 W (at 0.55 μm)		

From J. Millman and C. Halkias, "Integrated Electronics: Analog Digital Circuits and Systems," McGraw-Hill Book Company, 1972.

D. Decibel Table

Db	Current and voltage ratio		Power ratio		Db	Current and voltage ratio		Power ratio	
	Gain	Loss	Gain	Loss		Gain	Loss	Gain	Loss
0.1	1.01	0.989	1.02	0.977	8.0	2.51	0.398	6.31	0.158
0.2	1.02	0.977	1.05	0.955	8.5	2.66	0.376	7.08	0.141
0.3	1.03	0.966	1.07	0.933	9.0	2.82	0.355	7.94	0.126
0.4	1.05	0.955	1.10	0.912	9.5	2.98	0.335	8.91	0.112
0.5	1.06	0.944	1.12	0.891	10.0	3.16	0.316	10.00	0.100
0.6	1.07	0.933	1.15	0.871	11.0	3.55	0.282	12.6	0.079
0.7	1.08	0.923	1.17	0.851	12.0	3.98	0.251	15.8	0.063
0.8	1.10	0.912	1.20	0.832	13.0	4.47	0.224	19.9	0.050
0.9	1.11	0.902	1.23	0.813	14.0	5.01	0.199	25.1	0.040
1.0	1.12	0.891	1.26	0.794	15.0	5.62	0.178	31.6	0.032
1.1	1.13	0.881	1.29	0.776	16.0	6.31	0.158	39.8	0.025
1.2	1.15	0.871	1.32	0.759	17.0	7.08	0.141	50.1	0.020
1.3	1.16	0.861	1.35	0.741	18.0	7.94	0.126	63.1	0.016
1.4	1.17	0.851	1.38	0.724	19.0	8.91	0.112	79.4	0.013
1.5	1.19	0.841	1.41	0.708	20.0	10.00	0.100	100.0	0.010
1.6	1.20	0.832	1.44	0.692	25.0	17.8	0.056	3.16×10^{2}	3.16×10^{-3}
1.7	1.22	0.822	1.48	0.676	30.0	31.6	0.032	10^{3}	10^{-3}
1.8	1.23	0.813	1.51	0.661	35.0	56.2	0.018	3.16×10^{3}	3.16×10^{-4}
1.9	1.24	0.803	1.55	0.646	40.0	100.0	0.010	10^{4}	10^{-4}
2.0	1.26	0.794	1.58	0.631	45.0	177.8	0.006	3.16×10^{4}	3.16×10^{-5}
2.2	1.29	0.776	1.66	0.603	50.0	316	0.003	10^{5}	10^{-5}
2.4	1.32	0.759	1.74	0.575	55.0	562	0.002	3.16×10^{5}	3.16×10^{-6}
2.6	1.35	0.741	1.82	0.550	60.0	1,000	0.001	10^{6}	10^{-6}
2.8	1.38	0.724	1.90	0.525	65.0	1,770	0.0006	3.16×10^{6}	3.16×10^{-7}
3.0	1.41	0.708	1.99	0.501	70.0	3,160	0.0003	10^{7}	10^{-7}
3.2	1.44	0.692	2.09	0.479	75.0	5,620	0.0002	3.16×10^{7}	3.16×10^{-8}
3.4	1.48	0.676	2.19	0.457	80.0	10,000	0.0001	10^{8}	10^{-8}
3.6	1.51	0.661	2.29	0.436	85.0	17,800	0.00006	3.16×10^{8}	3.16×10^{-9}
3.8	1.55	0.646	2.40	0.417	90.0	31,600	0.00003	10^{9}	10^{-9}
4.0	1.58	0.631	2.51	0.398	95.0	56,200	0.00002	3.16×10^{9}	3.16×10^{-10}
4.2	1.62	0.617	2.63	0.380	100.0	100,000	0.00001	10^{10}	10^{-10}
4.4	1.66	0.603	2.75	0.363	105.0	178,000	0.000006	3.16×10^{10}	3.16×10^{-11}
4.6	1.70	0.589	2.88	0.347	110.0	316,000	0.000003	10^{11}	10^{-11}
4.8	1.74	0.575	3.02	0.331	115.0	562,000	0.000002	3.16×10^{11}	3.16×10^{-12}
5.0	1.78	0.562	3.16	0.316	120.0	1,000,000	0.000001	10^{12}	10^{-12}
5.5	1.88	0.531	3.55	0.282	130.0	3.16×10^{6}	3.16×10^{-7}	10^{13}	10^{-13}
6.0	1.99	0.501	3.98	0.251	140.0	10^{7}	10^{-7}	10^{14}	10^{-14}
6.5	2.11	0.473	4.47	0.224	150.0	3.16×10^{7}	3.16×10^{-8}	10^{15}	10^{-15}
7.0	2.24	0.447	5.01	0.199	160.0	10^{8}	10^{-8}	10^{16}	10^{-16}
7.5	2.37	0.422	5.62	0.178	170.0	3.16×10^{8}	3.16×10^{-9}	10^{17}	10^{-17}

From F. E. Terman, "Radio Engineers' Handbook," McGraw-Hill Book Company, 1943.

E. Natural Sines and Cosines

NOTE.—For cosines use right-hand column of degrees and lower line of tenths.

Deg	°0.0	°0.1	°0.2	°0.3	°0.4	°0.5	°0.6	°0.7	°0.8	°0.9	
0°	0.0000	0.0017	0.0035	0.0052	0.0070	0.0087	0.0105	0.0122	0.0140	0.0157	89
1	0.0175	0.0192	0.0209	0.0227	0.0244	0.0262	0.0279	0.0297	0.0314	0.0332	88
2	0.0349	0.0366	0.0384	0.0401	0.0419	0.0436	0.0454	0.0471	0.0488	0.0506	87
3	0.0523	0.0541	0.0558	0.0576	0.0593	0.0610	0.0628	0.0645	0.0663	0.0680	86
4	0.0698	0.0715	0.0732	0.0750	0.0767	0.0785	0.0802	0.0819	0.0837	0.0854	85
5	0.0872	0.0889	0.0906	0.0924	0.0941	0.0958	0.0976	0.0993	0.1011	0.1028	84
6	0.1045	0.1063	0.1080	0.1097	0.1115	0.1132	0.1149	0.1167	0.1184	0.1201	83
7	0.1219	0.1236	0.1253	0.1271	0.1288	0.1305	0.1323	0.1340	0.1357	0.1374	82
8	0.1392	0.1409	0.1426	0.1444	0.1461	0.1478	0.1495	0.1513	0.1530	0.1547	81
9	0.1564	0.1582	0.1599	0.1616	0.1633	0.1650	0.1668	0.1685	0.1702	0.1719	80°
10°	0.1736	0.1754	0.1771	0.1788	0.1805	0.1822	0.1840	0.1857	0.1874	0.1891	79
11	0.1908	0.1925	0.1942	0.1959	0.1977	0.1994	0.2011	0.2028	0.2045	0.2062	78
12	0.2079	0.2096	0.2113	0.2130	0.2147	0.2164	0.2181	0.2198	0.2215	0.2232	77
13	0.2250	0.2267	0.2284	0.2300	0.2317	0.2334	0.2351	0.2368	0.2385	0.2402	76
14	0.2419	0.2436	0.2453	0.2470	0.2487	0.2504	0.2521	0.2538	0.2554	0.2571	75
15	0.2588	0.2605	0.2622	0.2639	0.2656	0.2672	0.2689	0.2706	0.2723	0.2740	74
16	0.2756	0.2773	0.2790	0.2807	0.2823	0.2840	0.2857	0.2874	0.2890	0.2907	73
17	0.2924	0.2940	0.2957	0.2974	0.2990	0.3007	0.3024	0.3040	0.3057	0.3074	72
18	0.3090	0.3107	0.3123	0.3140	0.3156	0.3173	0.3190	0.3206	0.3223	0.3239	71
19	0.3256	0.3272	0.3289	0.3305	0.3322	0.3338	0.3355	0.3371	0.3387	0.3404	70°
20°	0.3420	0.3437	0.3453	0.3469	0.3486	0.3502	0.3518	0.3535	0.3551	0.3567	69
21	0.3584	0.3600	0.3616	0.3633	0.3649	0.3665	0.3681	0.3697	0.3714	0.3730	68
22	0.3746	0.3762	0.3778	0.3795	0.3811	0.3827	0.3843	0.3859	0.3875	0.3891	67
23	0.3907	0.3923	0.3939	0.3955	0.3971	0.3987	0.4003	0.4019	0.4035	0.4051	66
24	0.4067	0.4083	0.4099	0.4115	0.4131	0.4147	0.4163	0.4179	0.4195	0.4210	65
25	0.4226	0.4242	0.4258	0.4274	0.4289	0.4305	0.4321	0.4337	0.4352	0.4368	64
26	0.4384	0.4399	0.4415	0.4431	0.4446	0.4462	0.4478	0.4493	0.4509	0.4524	63
27	0.4540	0.4555	0.4571	0.4586	0.4602	0.4617	0.4633	0.4648	0.4664	0.4679	62
28	0.4695	0.4710	0.4726	0.4741	0.4756	0.4772	0.4787	0.4802	0.4818	0.4833	61
29	0.4848	0.4863	0.4879	0.4894	0.4909	0.4924	0.4939	0.4955	0.4970	0.4985	60°
30°	0.5000	0.5015	0.5030	0.5045	0.5060	0.5075	0.5900	0.5105	0.5120	0.5135	59
31	0.5150	0.5165	0.5180	0.5195	0.5210	0.5225	0.5240	0.5255	0.5270	0.5284	58
32	0.5299	0.5314	0.5329	0.5344	0.5358	0.5373	0.5388	0.5402	0.5417	0.5432	57
33	0.5446	0.5461	0.5476	0.5490	0.5505	0.5519	0.5534	0.5548	0.5563	0.5577	56
34	0.5592	0.5606	0.5621	0.5635	0.5650	0.5664	0.5678	0.5693	0.5707	0.5721	55
35	0.5736	0.5750	0.5764	0.5779	0.5793	0.5807	0.5821	0.5835	0.5850	0.5864	54
36	0.5878	0.5892	0.5906	0.5920	0.5934	0.5948	0.5962	0.5976	0.5990	0.6004	53
37	0.6018	0.6032	0.6046	0.6060	0.6074	0.6088	0.6101	0.6115	0.6129	0.6143	52
38	0.6157	0.6170	0.6184	0.6198	0.6211	0.6225	0.6239	0.6252	0.6266	0.6280	51
39	0.6293	0.6307	0.6320	0.6334	0.6347	0.6361	0.6374	0.6388	0.6401	0.6414	50°
40°	0.6428	0.6441	0.6455	0.6468	0.6481	0.6494	0.6508	0.6521	0.6534	0.6547	49
41	0.6561	0.6574	0.6587	0.6600	0.6613	0.6626	0.6639	0.6652	0.6665	0.6678	48
42	0.6691	0.6704	0.6717	0.6730	0.6743	0.6756	0.6769	0.6782	0.6794	0.6807	47
43	0.6820	0.6833	0.6845	0.6858	0.6871	0.6884	0.6896	0.6909	0.6921	0.6934	46
44	0.6947	0.6959	0.6972	0.6984	0.6997	0.7009	0.7022	0.7034	0.7046	0.7059	45
	°1.0	°0.9	°0.8	°0.7	°0.6	°0.5	°0.4	°0.3	°0.2	°0.1	Deg

From "Standard Handbook for Electrical Engineers," 7th ed.

E. Natural Sines and Cosines (*Concluded*)

Deg	°0.0	°0.1	°0.2	°0.3	°0.4	°0.5	°0.6	°0.7	°0.8	°0.9	
45	0.7071	0.7083	0.7096	0.7108	0.7120	0.7133	0.7145	0.7157	0.7169	0.7181	44
46	0.7193	0.7206	0.7218	0.7230	0.7242	0.7254	0.7266	0.7278	0.7290	0.7302	43
47	0.7314	0.7325	0.7337	0.7349	0.7361	0.7373	0.7385	0.7396	0.7408	0.7420	42
48	0.7431	0.7443	0.7455	0.7466	0.7478	0.7490	0.7501	0.7513	0.7524	0.7536	41
49	0.7547	0.7559	0.7570	0.7581	0.7593	0.7604	0.7615	0.7627	0.7638	0.7649	40°
50°	0.7660	0.7672	0.7683	0.7694	0.7705	0.7716	0.7727	0.7738	0.7749	0.7760	39
51	0.7771	0.7782	0.7793	0.7804	0.7815	0.7826	0.7837	0.7848	0.7859	0.7869	38
52	0.7880	0.7891	0.7902	0.7912	0.7923	0.7934	0.7944	0.7955	0.7965	0.7976	37
53	0.7986	0.7997	0.8007	0.8018	0.8028	0.8039	0.8049	0.8059	0.8070	0.8080	36
54	0.8090	0.8100	0.8111	0.8121	0.8131	0.8141	0.8151	0.8161	0.8171	0.8181	35
55	0.8192	0.8202	0.8211	0.8221	0.8231	0.8241	0.8251	0.8261	0.8271	0.8281	34
56	0.8290	0.8300	0.8310	0.8320	0.8329	0.8339	0.8348	0.8358	0.8368	0.8377	33
57	0.8387	0.8396	0.8406	0.8415	0.8425	0.8434	0.8443	0.8453	0.8462	0.8471	32
58	0.8480	0.8490	0.8499	0.8508	0.8517	0.8526	0.8536	0.8545	0.8554	0.8563	31
59	0.8572	0.8581	0.8590	0.8599	0.8607	0.8616	0.8625	0.8634	0.8643	0.8652	30°
60°	0.8660	0.8669	0.8678	0.8686	0.8695	0.8704	0.8712	0.8721	0.8729	0.8738	29
61	0.8746	0.8755	0.8763	0.8771	0.8780	0.8788	0.8796	0.8805	0.8813	0.8821	28
62	0.8829	0.8838	0.8846	0.8854	0.8862	0.8870	0.8878	0.8886	0.8894	0.8902	27
63	0.8910	0.8918	0.8926	0.8934	0.8942	0.8949	0.8957	0.8965	0.8973	0.8980	26
64	0.8988	0.8996	0.9003	0.9011	0.9018	0.9026	0.9033	0.9041	0.9048	0.9056	25
65	0.9063	0.9070	0.9078	0.9085	0.9092	0.9100	0.9107	0.9114	0.9121	0.9128	24
66	0.9135	0.9143	0.9150	0.9157	0.9164	0.9171	0.9178	0.9184	0.9191	0.9198	23
67	0.9205	0.9212	0.9219	0.9225	0.9232	0.9239	0.9245	0.9252	0.9259	0.9265	22
68	0.9272	0.9278	0.9285	0.9291	0.9298	0.9304	0.9311	0.9317	0.9323	0.9330	21
69	0.9336	0.9342	0.9348	0.9354	0.9361	0.9367	0.9373	0.9379	0.9385	0.9391	20°
70°	0.9397	0.9403	0.9409	0.9415	0.9421	0.9426	0.9432	0.9438	0.9444	0.9449	19
71	0.9455	0.9461	0.9466	0.9472	0.9478	0.9483	0.9489	0.9494	0.9500	0.9505	18
72	0.9511	0.9516	0.9521	0.9527	0.9532	0.9537	0.9542	0.9548	0.9553	0.9558	17
73	0.9563	0.9568	0.9573	0.9578	0.9583	0.9588	0.9593	0.9598	0.9603	0.9608	16
74	0.9613	0.9617	0.9622	0.9627	0.9632	0.9636	0.9641	0.9646	0.9650	0.9655	15
75	0.9659	0.9664	0.9668	0.9673	0.9677	0.9681	0.9686	0.9690	0.9694	0.9699	14
76	0.9703	0.9707	0.9711	0.9715	0.9720	0.9724	0.9728	0.9732	0.9736	0.9740	13
77	0.9744	0.9748	0.9751	0.9755	0.9759	0.9763	0.9767	0.9770	0.9774	0.9778	12
78	0.9781	0.9785	0.9789	0.9792	0.9796	0.9799	0.9803	0.9806	0.9810	0.9813	11
79	0.9816	0.9820	0.9823	0.9826	0.9829	0.9833	0.9836	0.9839	0.9842	0.9845	10°
80°	0.9848	0.9851	0.9854	0.9857	0.9860	0.9863	0.9866	0.9869	0.9871	0.9874	9
81	0.9877	0.9880	0.9882	0.9885	0.9888	0.9890	0.9893	0.9895	0.9898	0.9900	8
82	0.9903	0.9905	0.9907	0.9910	0.9912	0.9914	0.9917	0.9919	0.9921	0.9923	7
83	0.9925	0.9928	0.9930	0.9932	0.9934	0.9936	0.9938	0.9940	0.9942	0.9943	6
84	0.9945	0.9947	0.9949	0.9951	0.9952	0.9954	0.9956	0.9957	0.9959	0.9960	5
85	0.9962	0.9963	0.9965	0.9966	0.9968	0.9969	0.9971	0.9972	0.9973	0.9974	4
86	0.9976	0.9977	0.9978	0.9979	0.9980	0.9981	0.9982	0.9983	0.9984	0.9985	3
87	0.9986	0.9987	0.9988	0.9989	0.9990	0.9990	0.9991	0.9992	0.9993	0.9993	2
88	0.9994	0.9995	0.9995	0.9996	0.9996	0.9997	0.9997	0.9997	0.9998	0.9998	1
89	0.9998	0.9999	0.9999	0.9999	0.9999	1.000	1.000	1.000	1.000	1.000	0°
	°1.0	°0.9	°0.8	°0.7	°0.6	°0.5	°0.4	°0.3	°0.2	°0.1	Deg

From "Standard Handbook for Electrical Engineers," 7th ed.

F. Natural Tangents and Cotangents

NOTE.—For cotangents use right-hand column of degrees and lower line of tenths.

Deg	°0.0	°0.1	°0.2	°0.3	°0.4	°0.5	°0.6	°0.7	°0.8	°0.9	
0°	0.0000	0.0017	0.0035	0.0052	0.0070	0.0087	0.0105	0.0122	0.0140	0.0157	89
1	0.0175	0.0192	0.0209	0.0227	0.0244	0.0262	0.0279	0.0297	0.0314	0.0332	88
2	0.0349	0.0367	0.0384	0.0402	0.0419	0.0437	0.0454	0.0472	0.0489	0.0507	87
3	0.0524	0.0542	0.0559	0.0577	0.0594	0.0612	0.0629	0.0647	0.0664	0.0682	86
4	0.0699	0.0717	0.0734	0.0752	0.0769	0.0787	0.0805	0.0822	0.0840	0.0857	85
5	0.0875	0.0892	0.0910	0.0928	0.0945	0.0963	0.0981	0.0998	0.1016	0.1033	84
6	0.1051	0.1069	0.1086	0.1104	0.1122	0.1139	0.1157	0.1175	0.1192	0.1210	83
7	0.1228	0.1246	0.1263	0.1281	0.1299	0.1317	0.1334	0.1352	0.1370	0.1388	82
8	0.1405	0.1423	0.1441	0.1459	0.1477	0.1495	0.1512	0.1530	0.1548	0.1566	81
9	0.1584	0.1602	0.1620	0.1638	0.1655	0.1673	0.1691	0.1709	0.1727	0.1745	80°
10°	0.1763	0.1781	0.1799	0.1817	0.1835	0.1853	0.1871	0.1890	0.1908	0.1926	79
11	0.1944	0.1962	0.1980	0.1998	0.2016	0.2035	0.2053	0.2071	0.2089	0.2107	78
12	0.2126	0.2144	0.2162	0.2180	0.2199	0.2217	0.2235	0.2254	0.2272	0.2290	77
13	0.2309	0.2327	0.2345	0.2364	0.2382	0.2401	0.2419	0.2438	0.2456	0.2475	76
14	0.2493	0.2512	0.2530	0.2549	0.2568	0.2586	0.2605	0.2623	0.2643	0.2661	75
15	0.2679	0.2698	0.2717	0.2736	0.2754	0.2773	0.2792	0.2811	0.2830	0.2849	74
16	0.2867	0.2886	0.2905	0.2924	0.2943	0.2962	0.2981	0.3000	0.3019	0.3038	73
17	0.3057	0.3076	0.3096	0.3115	0.3134	0.3153	0.3172	0.3191	0.3211	0.3230	72
18	0.3249	0.3269	0.3288	0.3307	0.3327	0.3346	0.3365	0.3385	0.3404	0.3424	71
19	0.3443	0.3463	0.3482	0.3502	0.3522	0.3541	0.3561	0.3581	0.3600	0.3620	70°
20°	0.3640	0.3659	0.3679	0.3699	0.3719	0.3739	0.3759	0.3779	0.3799	0.3819	69
21	0.3339	0.3859	0.3879	0.3899	0.3919	0.3939	0.3959	0.3979	0.4000	0.4020	68
22	0.4040	0.4061	0.4081	0.4101	0.4122	0.4142	0.4163	0.4183	0.4204	0.4224	67
23	0.4245	0.4265	0.4286	0.4307	0.4327	0.4348	0.4369	0.4390	0.4411	0.4431	66
24	0.4452	0.4473	0.4494	0.4515	0.4536	0.4557	0.4578	0.4599	0.4621	0.4642	65
25	0.4663	0.4684	0.4706	0.4727	0.4748	0.4770	0.4791	0.4813	0.4834	0.4856	64
26	0.4877	0.4899	0.4921	0.4942	0.4964	0.4986	0.5008	0.5029	0.5051	0.5073	63
27	0.5095	0.5117	0.5139	0.5161	0.5184	0.5206	0.5228	0.5250	0.5272	0.5295	62
28	0.5317	0.5340	0.5362	0.5384	0.5407	0.5430	0.5452	0.5475	0.5498	0.5520	61
29	0.5543	0.5566	0.5589	0.5612	0.5635	0.5658	0.5681	0.5704	0.5727	0.5750	60°
30°	0.5774	0.5797	0.5820	0.5844	0.5867	0.5890	0.5914	0.5938	0.5961	0.5985	59
31	0.6009	0.6032	0.6056	0.6080	0.6104	0.6128	0.6152	0.6176	0.6200	0.6224	58
32	0.6249	0.6273	0.6297	0.6322	0.6346	0.6371	0.6395	0.6420	0.6445	0.6469	57
33	0.6494	0.6519	0.6544	0.6569	0.6594	0.6619	0.6644	0.6669	0.6694	0.6720	56
34	0.6745	0.6771	0.6796	0.6822	0.6847	0.6873	0.6899	0.6924	0.6950	0.6976	55
35	0.7002	0.7028	0.7054	0.7080	0.7107	0.7133	0.7159	0.7186	0.7212	0.7239	54
36	0.7265	0.7292	0.7319	0.7346	0.7373	0.7400	0.7427	0.7454	0.7481	0.7508	53
37	0.7536	0.7563	0.7590	0.7618	0.7646	0.7673	0.7701	0.7729	0.7757	0.7785	52
38	0.7813	0.7841	0.7869	0.7898	0.7926	0.7954	0.7983	0.8012	0.8040	0.8069	51
39	0.8098	0.8127	0.8156	0.8185	0.8214	0.8243	0.8273	0.8302	0.8332	0.8361	50°
40°	0.8391	0.8421	0.8451	0.8481	0.8511	0.8541	0.8571	0.8601	0.8632	0.8662	49
41	0.8693	0.8724	0.8754	0.8785	0.8816	0.8847	0.8878	0.8910	0.8941	0.8972	48
42	0.9004	0.9036	0.9067	0.9099	0.9131	0.9163	0.9195	0.9228	0.9260	0.9293	47
43	0.9325	0.9358	0.9391	0.9424	0.9457	0.9490	0.9523	0.9556	0.9590	0.9623	46
44	0.9657	0.9691	0.9725	0.9759	0.9793	0.9827	0.9861	0.9896	0.9930	0.9965	45
	°1.0	°0.9	°0.8	°0.7	°0.6	°0.5	°0.4	°0.3	°0.2	°0.1	Deg

From "Standard Handbook for Electrical Engineers," 7th ed.

F. Natural Tangents and Cotangents (*Concluded*)

Deg	°0.0	°0.1	°0.2	°0.3	°0.4	°0.5	°0.6	°0.7	°0.8	°0.9	
45	1.0000	1.0035	1.0070	1.0105	1.0141	1.0176	1.0212	1.0247	1.0283	1.0319	44
46	1.0355	1.0392	1.0428	1.0464	1.0501	1.0538	1.0575	1.0612	1.0649	1.0686	43
47	1.0724	1.0761	1.0799	1.0837	1.0875	1.0913	1.0951	1.0990	1.1028	1.1067	42
48	1.1106	1.1145	1.1184	1.1224	1.1263	1.1303	1.1343	1.1383	1.1423	1.1463	41
49	1.1504	1.1544	1.1585	1.1626	1.1667	1.1708	1.1750	1.1792	1.1833	1.1875	40°
50°	1.1918	1.1960	1.2002	1.2045	1.2088	1.2131	1.2174	1.2218	1.2261	1.2305	39
51	1.2349	1.2393	1.2437	1.2482	1.2527	1.2572	1.2617	1.2662	1.2708	1.2753	38
52	1.2799	1.2846	1.2892	1.2938	1.2985	1.3032	1.3079	1.3127	1.3175	1.3222	37
53	1.3270	1.3319	1.3367	1.3416	1.3465	1.3514	1.3564	1.3613	1.3663	1.3713	36
54	1.3764	1.3814	1.3865	1.3916	1.3968	1.4019	1.4071	1.4124	1.4176	1.4229	35
55	1.4281	1.4335	1.4388	1.4442	1.4496	1.4550	1.4605	1.4659	1.4715	1.4770	34
56	1.4826	1.4882	1.4938	1.4994	1.5051	1.5108	1.5166	1.5224	1.5282	1.5340	33
57	1.5399	1.5458	1.5517	1.5577	1.5637	1.5697	1.5757	1.5818	1.5880	1.5941	32
58	1.6003	1.6066	1.6128	1.6191	1.6255	1.6319	1.6383	1.6447	1.6512	1.6577	31
59	1.6643	1.6709	1.6775	1.6842	1.6909	1.6977	1.7045	1.7113	1.7182	1.7251	30°
60°	1.7321	1.7391	1.7461	1.7532	1.7603	1.7675	1.7747	1.7820	1.7893	1.7966	29
61	1.8040	1.8115	1.8190	1.8265	1.8341	1.8418	1.8495	1.8572	1.8650	1.8728	28
62	1.8807	1.8887	1.8967	1.9047	1.9128	1.9210	1.9292	1.9375	1.9458	1.9542	27
63	1.9626	1.9711	1.9797	1.9883	1.9970	2.0057	2.0145	2.0233	2.0323	2.0413	26
64	2.0503	2.0594	2.0686	2.0778	2.0872	2.0965	2.1060	2.1155	2.1251	2.1348	25
65	2.1445	2.1543	2.1642	2.1742	2.1842	2.1943	2.2045	2.2148	2.2251	2.2355	24
66	2.2460	2.2566	2.2673	2.2781	2.2889	2.2998	2.3109	2.3220	2.3332	2.3445	23
67	2.3599	2.3673	2.3789	2.3906	2.4023	2.4142	2.4262	2.4383	2.4504	2.4627	22
68	2.4751	2.4876	2.5002	2.5129	2.5257	2.5386	2.5517	2.5649	2.5782	2.5916	21
69	2.6051	2.6187	2.6325	2.6464	2.6605	2.6746	2.6889	2.7034	2.7179	2.7326	20°
70°	2.7475	2.7625	2.7776	2.7929	2.8083	2.8239	2.8397	2.8556	2.8716	2.8878	19
71	2.9042	2.9208	2.9375	2.9544	2.9714	2.9887	3.0061	3.0237	3.0415	3.0595	18
72	3.0777	3.0961	3.1146	3.1334	3.1524	3.1716	3.1910	3.2106	3.2305	2.2506	17
73	3.2709	3.2914	3.3122	3.3332	3.3544	3.3759	3.3977	3.4197	3.4420	3.4646	16
74	3.4874	3.5105	3.5339	3.5576	3.5816	3.6059	3.6305	3.6554	3.6806	3.7062	15
75	3.7321	3.7583	3.7848	3.8118	3.8391	3.8667	3.8947	3.9232	3.9520	3.9812	14
76	4.0108	4.0408	4.0713	4.1022	4.1335	4.1653	4.1976	4.2303	4.2635	4.2972	13
77	4.3315	4.3662	4.4015	4.4374	4.4737	4.5107	4.5483	4.5864	4.6252	4.6646	12
78	4.7046	4.7453	4.7867	4.8288	4.8716	4.9152	4.9594	5.0045	5.0504	5.0970	11
79	5.1446	5.1929	5.2422	5.2924	5.3435	5.3955	5.4486	5.5026	5.5578	5.6140	10°
80°	5.6713	5.7297	5.7894	5.8502	5.9124	5.9758	6.0405	6.1066	6.1742	6.2433	9
81	6.3138	6.3859	6.4596	6.5350	6.6122	6.6912	6.7720	6.8548	6.9395	7.0264	8
82	7.1154	7.2066	7.3002	7.3962	7.4947	7.5958	7.6996	7.8062	7.9158	8.0285	7
83	8.1443	8.2636	8.3863	8.5126	8.6427	8.7769	8.9152	9.0579	9.2052	9.3572	6
84	9.5144	9.677	9.845	10.02	10.20	10.39	10.58	10.78	10.99	11.20	5
85	11.43	11.66	11.91	12.16	12.43	12.71	13.00	13.30	13.62	13.95	4
86	14.30	14.67	15.06	15.46	15.89	16.35	16.83	17.34	17.89	18.46	3
87	19.08	19.74	20.45	21.20	22.02	22.90	23.86	24.90	26.03	27.27	2
88	28.64	30.14	31.82	33.69	35.80	38.19	40.92	44.07	47.74	52.08	1
89	57.29	63.66	71.62	81.85	94.49	114.6	143.2	191.0	286.5	573.0	0°
	°1.0	°0.9	°0.8	°0.7	°0.6	°0.5	°0.4	°0.3	°0.2	°0.1	Deg

From "Standard Handbook for Electrical Engineers," 7th ed.

G. Common Logarithms of Numbers

N	0	1	2	3	4	5	6	7	8	9
10	0000	0043	0086	0128	0170	0212	0253	0294	0334	0374
11	0414	0453	0492	0531	0569	0607	0645	0682	0719	0755
12	0792	0828	0864	0899	0934	0969	1004	1038	1072	1106
13	1139	1173	1206	1239	1271	1303	1335	1367	1399	1430
14	1461	1492	1523	1553	1584	1614	1644	1673	1703	1732
15	1761	1790	1818	1847	1875	1903	1931	1959	1987	2014
16	2041	2068	2095	2122	2148	2175	2201	2227	2253	2279
17	2304	2330	2355	2380	2405	2430	2455	2480	2504	2529
18	2553	2577	2601	2625	2648	2672	2695	2718	2742	2765
19	2788	2810	2833	2856	2878	2900	2923	2945	2967	2989
20	3010	3032	3054	3075	3096	3118	3139	3160	3181	3201
21	3222	3243	3263	3284	3304	3324	3345	3365	3385	3404
22	3424	3444	3464	3483	3502	3522	3541	3560	3579	3598
23	3617	3636	3655	3674	3692	3711	3729	3747	3766	3784
24	3802	3820	3838	3856	3874	3892	3909	3927	3945	3962
25	3979	3997	4014	4031	4048	4065	4082	4099	4116	4133
26	4150	4166	4183	4200	4216	4232	4249	4265	4281	4298
27	4314	4330	4346	4362	4378	4393	4409	4425	4440	4456
28	4472	4487	4502	4518	4533	4548	4564	4579	4594	4609
29	4624	4639	4654	4669	4683	4698	4713	4728	4742	4757
30	4771	4786	4800	4814	4829	4843	4857	4871	4886	4900
31	4914	4928	4942	4955	4969	4983	4997	5011	5024	5038
32	5051	5065	5079	5092	5105	5119	5132	5145	5159	5172
33	5185	5198	5211	5224	5237	5250	5263	5276	5289	5302
34	5315	5328	5340	5353	5366	5378	5391	5403	5416	5428
35	5441	5453	5465	5478	5490	5502	5514	5527	5539	5551
36	5563	5575	5587	5599	5611	5623	5635	5647	5658	5670
37	5682	5694	5705	5717	5729	5740	5752	5763	5775	5786
38	5798	5809	5821	5832	5843	5855	5866	5877	5888	5899
39	5911	5922	5933	5944	5955	5966	5977	5988	5999	6010
40	6021	6031	6042	6053	6064	6075	6085	6096	6107	6117
41	6128	6138	6149	6160	6170	6180	6191	6201	6212	6222
42	6232	6243	6253	6263	6274	6284	6294	6304	6314	6325
43	6335	6345	6355	6365	6375	6385	6395	6405	6415	6425
44	6435	6444	6454	6464	6474	6484	6493	6503	6513	6522
45	6532	6542	6551	6561	6571	6580	6590	6599	6609	6618
46	6628	6637	6646	6656	6665	6675	6684	6693	6702	6712
47	6721	6730	6739	6749	6758	6767	6776	6785	6794	6803
48	6812	6821	6830	6839	6848	6857	6866	6875	6884	6893
49	6902	6911	6920	6928	6937	6946	6955	6964	6972	6981
50	6990	6998	7007	7016	7024	7033	7042	7050	7059	7067
51	7076	7084	7093	7101	7110	7118	7126	7135	7143	7152
52	7160	7168	7177	7185	7193	7202	7210	7218	7226	7235
53	7243	7251	7259	7267	7275	7284	7292	7300	7308	7316
54	7324	7332	7340	7348	7356	7364	7372	7380	7388	7396

From "Standard Handbook for Electrical Engineers," 7th ed.

G. Common Logarithms of Numbers (*Concluded*)

N	0	1	2	3	4	5	6	7	8	9
55	7404	7412	7419	7427	7435	7443	7451	7459	7466	7474
56	7482	7490	7497	7505	7513	7520	7528	7536	7543	7551
57	7559	7566	7574	7582	7589	7597	7604	7612	7619	7627
58	7634	7642	7649	7657	7664	7672	7679	7686	7694	7701
59	7709	7716	7723	7731	7738	7745	7752	7760	7767	7774
60	7782	7789	7796	7803	7810	7818	7825	7832	7839	7846
61	7853	7860	7868	7875	7882	7889	7896	7903	7910	7917
62	7924	7931	7938	7945	7952	7959	7966	7973	7980	7987
63	7993	8000	8007	8014	8021	8028	8035	8041	8048	8055
64	8062	8069	8075	8082	8089	8096	8102	8109	8116	8122
65	8129	8136	8142	8149	8156	8162	8169	8176	8182	8189
66	8195	8202	8209	8215	8222	8228	8235	8241	8248	8254
67	8261	8267	8274	8280	8287	8293	8299	8306	8312	8319
68	8325	8331	8338	8344	8351	8357	8363	8370	8376	8382
69	8388	8395	8401	8407	8414	8420	8426	8432	8439	8445
70	8451	8457	8463	8470	8476	8482	8488	8494	8500	8506
71	8513	8519	8525	8531	8537	8543	8549	8555	8561	8567
72	8573	8579	8585	8591	8597	8603	8609	8615	8621	8627
73	8633	8639	8645	8651	8657	8663	8669	8675	8681	8686
74	8692	8698	8704	8710	8716	8722	8727	8733	8739	8745
75	8751	8756	8762	8768	8774	8779	8785	8791	8797	8802
76	8808	8814	8820	8825	8831	8837	8842	8848	8854	8859
77	8865	8871	8876	8882	8887	8893	8899	8904	8910	8915
78	8921	8927	8932	8938	8943	8949	8954	8960	8965	8971
79	8976	8982	8987	8993	8998	9004	9009	9015	9020	9025
80	9031	9036	9042	9047	9053	9058	9063	9069	9074	9079
81	9085	9090	9096	9101	9106	9112	9117	9122	9128	9133
82	9138	9143	9149	9154	9159	9165	9170	9175	9180	9186
83	9191	9196	9201	9206	9212	9217	9222	9227	9232	9238
84	9243	9248	9253	9258	9263	9269	9274	9279	9284	9289
85	9294	9299	9304	9309	9315	9320	9325	9330	9335	9340
86	9345	9350	9355	9360	9365	9370	9375	9380	9385	9390
87	9395	9400	9405	9410	9415	9420	9425	9430	9435	9440
88	9445	9450	9455	9460	9465	9469	9474	9479	9484	9489
89	9494	9499	9504	9509	9513	9518	9523	9528	9533	9538
90	9542	9547	9552	9557	9562	9566	9571	9576	9581	9586
91	9590	9595	9600	9605	9609	9614	9619	9624	9628	9633
92	9638	9643	9647	9652	9657	9661	9666	9671	9675	9680
93	9685	9689	9694	9699	9703	9708	9713	9717	9722	9727
94	9731	9736	9741	9745	9750	9754	9759	9763	9768	9773
95	9777	9782	9786	9791	9795	9800	9805	9809	9814	9818
96	9823	9827	9832	9836	9841	9845	9850	9854	9859	9863
97	9868	9872	9877	9881	9886	9890	9894	9899	9903	9908
98	9912	9917	9921	9926	9930	9934	9939	9943	9948	9952
99	9956	9961	9965	9969	9974	9978	9983	9987	9991	9996

From "Standard Handbook for Electrical Engineers," 7th ed.

H. Natural, Napierian, or Hyperbolic Logarithms

N	0	1	2	3	4	5	6	7	8	9
0	— ∞	0.0000	0.6931	1.0986	1.3863	1.6094	1.7918	1.9459	2.0794	2.1972
10	2.3026	2.3979	2.4849	2.5649	2.6391	2.7081	2.7726	2.8332	2.8904	2.9444
20	2.9957	3.0445	3.0910	3.1355	3.1781	3.2189	3.2581	3.2958	3.3322	3.3673
30	3.4012	3.4340	3.4657	3.4965	3.5264	3.5553	3.5835	3.6109	3.6376	3.6636
40	3.6889	3.7136	3.7377	3.7612	3.7842	3.8067	3.8286	3.8501	3.8712	3.8918
50	3.9120	3.9318	3.9512	3.9703	3.9890	4.0073	4.0254	4.0431	4.0604	4.0775
60	4.0943	4.1109	4.1271	4.1431	4.1589	4.1744	4.1897	4.2047	4.2195	4.2341
70	4.2485	4.2627	4.2767	4.2905	4.3041	4.3175	4.3307	4.3438	4.3567	4.3694
80	4.3820	4.3944	4.4067	4.4188	4.4308	4.4427	4.4543	4.4659	4.4773	4.4886
90	4.4998	4.5109	4.5218	4.5326	4.5433	4.5539	4.5643	4.5747	4.5850	4.5951
100	4.6052	4.6151	4.6250	4.6347	4.6444	4.6540	4.6634	4.6728	4.6821	4.6913
110	4.7005	4.7095	4.7185	4.7274	4.7362	4.7449	4.7536	4.7622	4.7707	4.7791
120	4.7875	4.7958	4.8040	4.8122	4.8203	4.8283	4.8363	4.8442	4.8520	4.8598
130	4.8675	4.8752	4.8828	4.8903	4.8978	4.9053	4.9127	4.9200	4.9273	4.9345
140	4.9416	4.9488	4.9558	4.9628	4.9698	4.9767	4.9836	4.9904	4.9972	5.0039
150	5.0106	5.0173	5.0239	5.0304	5.0370	5.0434	5.0499	5.0562	5.0626	5.0689
160	5.0752	5.0814	5.0876	5.0938	5.0999	5.1059	5.1120	5.1180	5.1240	5.1299
170	5.1358	5.1417	5.1475	5.1533	5.1591	5.1648	5.1705	5.1761	5.1818	5.1874
180	5.1930	5.1985	5.2040	5.2095	5.2149	5.2204	5.2257	5.2311	5.2364	5.2417
190	5.2470	5.2523	5.2575	5.2627	5.2679	5.2730	5.2781	5.2832	5.2883	5.2933
200	5.2983	5.3033	5.3083	5.3132	5.3181	5.3230	5.3279	5.3327	5.3375	5.3423
210	5.3471	5.3519	5.3566	5.3613	5.3660	5.3706	5.3753	5.3799	5.3845	5.3891
220	5.3936	5.3982	5.4027	5.4072	5.4116	5.4161	5.4205	5.4250	5.4293	5.4337
230	5.4381	5.4424	5.4467	5.4510	5.4553	5.4596	5.4638	5.4681	5.4723	5.4765
240	5.4806	5.4848	5.4889	5.4931	5.4972	5.5013	5.5053	5.5094	5.5134	5.5175
250	5.5215	5.5255	5.5294	5.5334	5.5373	5.5413	5.5452	5.5491	5.5530	5.5568
260	5.5607	5.5645	5.5683	5.5722	5.5759	5.5797	5.5835	5.5872	5.5910	5.5947
270	5.5984	5.6021	5.6058	5.6095	5.6131	5.6168	5.6204	5.6240	5.6276	5.6312
280	5.6348	5.6384	5.6419	5.6454	5.6490	5.6525	5.6560	5.6595	5.6630	5.6664
290	5.6699	5.6733	5.6768	5.6802	5.6836	5.6870	5.6904	5.6937	5.6971	5.7004
300	5.7038	5.7071	5.7104	5.7137	5.7170	5.7203	5.7236	5.7268	5.7301	5.7333
310	5.7366	5.7398	5.7430	5.7462	5.7494	5.7526	5.7557	5.7589	5.7621	5.7652
320	5.7683	5.7714	5.7746	5.7777	5.7807	5.7838	5.7869	5.7900	5.7930	5.7961
330	5.7991	5.8021	5.8051	5.8081	5.8111	5.8141	5.8171	5.8201	5.8230	5.8260
340	5.8289	5.8319	5.8348	5.8377	5.8406	5.8435	5.8464	5.8493	5.8522	5.8551
350	5.8579	5.8608	5.8636	5.8665	5.8693	5.8721	5.8749	5.8777	5.8805	5.8833
360	5.8861	5.8889	5.8916	5.8944	5.8972	5.8999	5.9026	5.9054	5.9081	5.9108
370	5.9135	5.9162	5.9189	5.9216	5.9243	5.9269	5.9296	5.9322	5.9349	5.9375
380	5.9402	5.9428	5.9454	5.9480	5.9506	5.9532	5.9558	5.9584	5.9610	5.9636
390	5.9661	5.9687	5.9713	5.9738	5.9764	5.9789	5.9814	5.9839	5.9865	5.9890
400	5.9915	5.9940	5.9965	5.9989	6.0014	6.0039	6.0064	6.0088	6.0113	6.0137
410	6.0162	6.0186	6.0210	6.0234	6.0259	6.0283	6.0307	6.0331	6.0355	6.0379
420	6.0403	6.0426	6.0450	6.0474	6.0497	6.0521	6.0544	6.0568	6.0591	6.0615
430	6.0638	6.0661	6.0684	6.0707	6.0730	6.0753	6.0776	6.0799	6.0822	6.0845
440	6.0868	6.0890	6.0913	6.0936	6.0958	0.0981	6.1003	6.1026	6.1048	6.1070
450	6.1092	6.1115	6.1137	6.1159	6.1181	6.1203	6.1225	6.1247	6.1269	6.1291
460	6.1312	6.1334	6.1356	6.1377	6.1399	6.1420	6.1442	6.1463	6.1485	6.1506
470	6.1527	6.1549	6.1570	6.1591	6.1612	6.1633	6.1654	6.1675	6.1696	6.1717
480	6.1738	6.1759	6.1779	6.1800	6.1821	6.1841	6.1862	6.1883	6.1903	6.1924
490	6.1944	6.1964	6.1985	6.2005	6.2025	6.2046	6.2066	6.2086	6.2106	6.2126

From "Standard Handbook for Electrical Engineers," 7th ed.

NOTE 1: Moving the decimal point n places to the right (or left) in the number is equivalent to adding (or subtracting) n times 2.3026.

NOTE 2:

$$\log_e x = 2.3026 \log_{10} x$$
$$\log_{10} x = 0.4343 \log_e x$$
$$\log_e 10 = 2.3026$$
$$\log_{10} e = 0.4343$$

n	$n \times 2.3026$
1	2.3026 = 0.6974–3
2	4.6052 = 0.3948–5
3	6.9078 = 0.0922–7
4	9.2103 = 0.7897–10
5	11.5129 = 0.4871–12
6	13.8155 = 0.1845–14
7	16.1181 = 0.8819–17
8	18.4207 = 0.5793–19
9	20.7233 = 0.2767–21

H. Natural, Napierian, or Hyperbolic Logarithms (*Concluded*)

N	0	1	2	3	4	5	6	7	8	9
500	6.2146	6.2166	6.2186	6.2206	6.2226	6.2246	6.2265	6.2285	6.2305	6.2324
510	6.2344	6.2364	6.2383	6.2403	6.2422	6.2442	6.2461	6.2480	6.2500	6.2519
520	6.2538	6.2558	6.2577	6.2596	6.2615	6.2634	6.2653	6.2672	6.2691	6.2710
530	6.2729	6.2748	6.2766	6.2785	6.2804	6.2823	6.2841	6.2860	6.2879	6.2897
540	6.2916	6.2934	6.2953	6.2971	6.2989	6.3008	6.3026	6.3044	6.3063	6.3081
550	6.3099	6.3117	6.3135	6.3154	6.3172	6.3190	6.3208	6.3226	6.3244	6.3261
560	6.3279	6.3297	6.3315	6.3333	6.3351	6.3368	6.3386	6.3404	6.3421	6.3439
570	6.3456	6.3474	6.3491	6.3509	6.3256	6.3544	6.3561	6.3578	6.3596	6.3613
580	6.3630	6.3648	6.3665	6.3682	6.3699	6.3716	6.3733	6.3750	6.3767	6.3784
590	6.3801	6.3818	6.3835	6.3852	6.3869	6.3886	6.3902	6.3919	6.3936	6.3953
600	6.3969	6.3986	6.4003	6.4019	6.4036	6.4052	6.4069	6.4085	6.4102	6.4118
610	6.4135	6.4151	6.4167	6.4184	6.4200	6.4216	6.4232	6.4249	6.4265	6.4281
620	6.4297	6.4313	6.4329	6.4345	6.4362	6.4378	6.4394	6.4409	6.4425	6.4441
630	6.4457	6.4473	6.4489	6.4505	6.4520	6.4536	6.4552	6.4568	6.4583	6.4599
640	6.4615	6.4630	6.4646	6.4661	6.4677	6.4693	6.4708	6.4723	6.4739	6.4754
650	6.4770	6.4785	6.4800	6.4816	6.4831	6.4846	6.4862	6.4877	6.4892	6.4907
660	6.4922	6.4938	6.4953	6.4968	6.4983	6.4998	6.5013	6.5028	6.5043	6.5058
670	6.5073	6.5088	6.5103	6.5117	6.5132	6.5147	6.5162	6.5177	6.5191	6.5206
680	6.5221	6.5236	6.5250	6.5265	6.5280	6.5294	6.5309	6.5323	6.5338	6.5352
690	6.5367	6.5381	6.5396	6.5410	6.5425	6.5439	6.5453	6.5468	6.5482	6.5497
700	6.5511	6.5525	6.5539	6.5554	6.5568	6.5582	6.5596	6.5610	6.5624	6.5639
710	6.5653	6.5667	6.5681	6.5695	6.5709	6.5723	6.5737	6.5751	6.5765	6.5779
720	6.5793	6.5806	6.5820	6.5834	6.5848	6.5862	6.5876	6.5889	6.5903	6.5917
730	6.5930	6.5944	6.5958	6.5971	6.5985	6.5999	6.6012	6.6026	6.6039	6.6053
740	6.6067	6.6080	6.6093	6.6107	6.6120	6.6134	6.6147	6.6161	6.6174	6.6187
750	6.6201	6.6214	6.6227	6.6241	6.6254	6.6267	6.6280	6.6294	6.6307	6.6320
760	6.6333	6.6346	6.6359	6.6373	6.6386	6.6399	6.6412	6.6425	6.6438	6.6451
770	6.6464	6.6477	6.6490	6.6503	6.6516	6.6529	6.6542	6.6554	6.6567	6.6580
780	6.6593	6.6606	6.6619	6.6631	6.6644	6.6657	6.6670	6.6682	6.6695	6.6708
790	6.6720	6.6733	6.6746	6.6758	6.6771	6.6783	6.6796	6.6809	6.6821	6.6834
800	6.6846	6.6859	6.6871	6.6884	6.6896	6.6908	6.6921	6.6933	6.6946	6.6958
810	6.6970	6.6983	6.6995	6.7007	6.7020	6.7032	6.7044	6.7056	6.7069	6.7081
820	6.7093	6.7105	6.7117	6.7130	6.7142	6.7154	6.7166	6.7178	6.7190	6.7202
830	6.7214	6.7226	6.7238	6.7250	6.7262	6.7274	6.7286	6.7298	6.7310	6.7322
840	6.7334	6.7346	6.7358	6.7370	6.7382	6.7393	6.7405	6.7417	6.6429	6.7441
850	6.7452	6.7464	6.7476	6.7488	6.7499	6.7511	6.7523	6.7534	6.7546	6.7558
860	6.7569	6.7581	6.7593	6.7604	6.7616	6.7627	6.7639	6.7650	6.7662	6.7673
870	6.7685	6.7696	6.7708	6.7719	6.7731	6.7742	6.7754	6.7765	6.7776	6.7788
880	6.7799	6.7811	6.7822	6.7833	6.7845	6.7856	6.7867	5.7878	6.7890	6.7901
890	6.7912	6.7923	6.7935	6.7946	6.7957	6.7968	6.7979	6.7991	6.8002	6.8013
900	6.8024	6.8035	6.8046	6.8057	6.8068	6.8079	6.8090	6.8101	6.8112	6.8123
910	6.8134	6.8145	6.8156	6.8167	6.8178	6.8189	6.8200	6.8211	6.8222	6.8233
920	6.8244	6.8255	6.8265	6.8276	6.8287	6.8298	6.8309	6.8320	6.8330	6.8341
930	6.8352	6.8363	6.8373	6.8384	6.8395	6.8405	6.8416	6.8427	6.8437	6.8448
940	6.8459	6.8469	6.8480	6.8491	6.8501	6.8512	6.8522	6.8533	6.8544	8.8554
950	6.8565	6.8575	6.8586	6.8596	6.8607	6.8617	6.8628	6.8638	6.8648	6.8659
960	6.8669	6.8680	6.8690	6.8701	6.8711	6.8721	6.8732	6.8742	6.8752	6.8763
970	6.8773	6.8783	6.8794	6.8804	6.8814	6.8824	6.8835	6.8845	6.8855	6.8865
980	6.8876	6.8886	6.8896	6.8906	6.8916	6.8926	6.8937	6.8947	6.8957	6.8967
990	6.8977	6.8987	6.8997	6.9007	6.9017	6.9027	6.9037	6.9047	6.9057	6.9068

From "Standard Handbook for Electrical Engineers," 7th ed.

I. Degrees and Minutes Expressed in Radians

Degrees						Hundredths				Minutes	
1°	0.0175	61°	1.0647	121°	2.1118	0°.01	0.0002	0°.51	0.0089	1′	0.0003
2	0.0349	62	1.0821	122	2.1293	0 .02	0.0003	0 .52	0.0091	2′	0.0006
3	0.0524	63	1.0996	123	2.1468	0 .03	0.0005	0 .53	0.0093	3′	0.0009
4	0.0698	64	1.1170	124	2.1642	0 .04	0.0007	0 .54	0.0094	4′	0.0012
5°	0.0873	65°	1.1345	125°	2.1817	0 .05	0.0009	0 .55	0.0096	5′	0.0015
6	0.1047	66	1.1519	126	2.1991	0 .06	0.0010	0 .56	0.0098	6′	0.0017
7	0.1222	67	1.1694	127	2.2166	0 .07	0.0012	0 .57	0.0099	7′	0.0020
8	0.1396	68	1.1868	128	2.2340	0 .08	0.0014	0 .58	0.0101	8′	0.0023
9	0.1571	69	1.2043	129	2.2515	0 .09	0.0016	0 .59	0.0103	9′	0.0026
10°	0.1745	70°	1.2217	130°	2.2689	0°.10	0.0017	0°.60	0.0105	10′	0.0029
11	0.1920	71	1.2392	131	2.2864	0 .11	0.0019	0 .61	0.0106	11′	0.0032
12	0.2094	72	1.2566	132	2.3038	0 .12	0.0021	0 .62	0.0108	12′	0.0035
13	0.2269	73	1.2741	133	2.3213	0 .13	0.0023	0 .63	0.0110	13′	0.0038
14	0.2443	74	1.2915	134	2.3387	0 .14	0.0024	0 .64	0.0112	14′	0.0041
15°	0.2618	75°	1.3090	135°	2.3562	0 .15	0.0026	0 .65	0.0113	15′	0.0044
16	0.2793	76	1.3265	136	2.3736	0 .16	0.0028	0 .66	0.0115	16′	0.0047
17	0.2967	77	1.3439	137	2.3911	0 .17	0.0030	0 .67	0.0117	17′	0.0049
18	0.3142	78	1.3614	138	2.4086	0 .18	0.0031	0 .68	0.0119	18′	0.0052
19	0.3316	79	1.3788	139	2.4260	0 .19	0.0033	0 .69	0.0120	19′	0.0055
20°	0.3491	80°	1.3963	140°	2.4435	0°.20	0.0035	0°.70	0.0122	20′	0.0058
21	0.3665	81	1.4137	141	2.4609	0 .21	0.0037	0 .71	0.0124	21′	0.0061
22	0.3840	82	1.4312	142	2.4784	0 .22	0.0038	0 .72	0.0126	22′	0.0064
23	0.4014	83	1.4486	143	2.4958	0 .23	0.0040	0 .73	0.0127	23′	0.0067
24	0.4189	84	1.4661	144	2.5133	0 .24	0.0042	0 .74	0.0129	24′	0.0070
25°	0.4363	85°	1.4835	145°	2.5307	0 .25	0.0044	0 .75	0.0131	25′	0.0073
26	0.4538	86	1.5010	146	2.5482	0 .26	0.0045	0 .76	0.0133	26′	0.0076
27	0.4712	87	1.5184	147	2.5656	0 .27	0.0047	0 .77	0.0134	27′	0.0079
28	0.4887	88	1.5359	148	2.5831	0 .28	0.0049	0 .78	0.0136	28′	0.0081
29	0.5061	89	1.5533	149	2.6005	0 .29	0.0051	0 .79	0.0138	29′	0.0084
30°	0.5236	90°	1.5708	150°	2.6180	0°.30	0.0052	0°.80	0.0140	30′	0.0087
31	0.5411	91	1.5882	151	2.6354	0 .31	0.0054	0 .81	0.0141	31′	0.0090
32	0.5585	92	1.6057	152	2.6529	0 .32	0.0056	0 .82	0.0143	32′	0.0093
33	0.5760	93	1.6232	153	2.6704	0 .33	0.0058	0 .83	0.0145	33′	0.0096
34	0.5934	94	1.6406	154	2.6878	0 .34	0.0059	0 .84	0.0147	34′	0.0099
35°	0.6109	95°	1.6581	155°	2.7053	0 .35	0.0061	0 .85	0.0148	35′	0.0102
36	0.6283	96	1.6755	156	2.7227	0 .36	0.0063	0 .86	0.0150	36′	0.0105
37	0.6458	97	1.6930	157	2.7402	0 .37	0.0065	0 .87	0.0152	37′	0.0108
38	0.6632	98	1.7104	158	2.7576	0 .38	0.0066	0 .88	0.0154	38′	0.0111
39	0.6807	99	1.7279	159	2.7751	0 .39	0.0068	0 .89	0.0155	39′	0.0113
40°	0.6981	100°	1.7453	160°	2.7925	0°.40	0.0070	0°.90	0.0157	40′	0.0116
41	0.7156	101	1.7628	161	2.8100	0 .41	0.0072	0 .91	0.0159	41′	0.0119
42	0.7330	102	1.7802	162	2.8274	0 .42	0.0073	0 .92	0.0161	42′	0.0122
43	0.7505	103	1.7977	163	2.8449	0 .43	0.0075	0 .93	0.0162	43′	0.0125
44	0.7679	104	1.8151	164	2.8623	0 .44	0.0077	0 .94	0.0164	44′	0.0128
45°	0.7854	105°	1.8326	165°	2.8798	0 .45	0.0079	0 .95	0.0166	45′	0.0131
46	0.8029	106	1.8500	166	2.8972	0 .46	0.0080	0 .96	0.0168	46′	0.0134
47	0.8203	107	1.8675	167	2.9147	0 .47	0.0082	0 .97	0.0169	47′	0.0137
48	0.8378	108	1.8850	168	2.9322	0 .48	0.0084	0 .98	0.0171	48′	0.0140
49	0.8552	109	1.9024	169	2.9496	0 .49	0.0086	0 .99	0.0173	49′	0.0143
50°	0.8727	110°	1.9199	170°	2.9671	0°.50	0.0087	1°.00	0.0175	50′	0.0145
51	0.8901	111	1.9373	171	2.9845	....	......	....	......	51′	0.0148
52	0.9076	112	1.9548	172	3.0020	....	......	....	......	52′	0.0151
53	0.9250	113	1.9722	173	3.0194	....	......	....	......	53′	0.0154
54	0.9425	114	1.9897	174	3.0369	....	......	....	......	54′	0.0157
55°	0.9599	115°	2.0071	175°	3.0543	....	......	....	......	55′	0.0160
56	0.9774	116	2.0246	176	3.0718	....	......	....	......	56′	0.0163
57	0.9948	117	2.0420	177	3.0892	....	......	....	......	57′	0.0166
58	1.0123	118	2.0595	178	3.1067	....	......	....	......	58′	0.0169
59	1.0297	119	2.0769	179	3.1241	....	......	....	......	59′	0.0172
60°	1.0472	120°	2.0944	180°	3.1416	....	......	....	......	60′	0.0175

Arc 1° = 0.01745. Arc 1′ = 0.0002909. Arc 1″ = 0.000004848.
1 radian = 57°.296 = 57° 17′.75 = 57° 17′ 44″.81.
From Lionel S. Marks, "Mechanical Engineers' Handbook."

J. Radians Expressed in Degrees

Rad	Deg	Rad	Deg	Rad	Deg	Rad	Deg	Rad	Deg
0.01	0°.57	0.64	36°.67	1.27	72°.77	1.90	108°.86	2.53	144°.96
0.02	1°.15	0.65	37°.24	1.28	73°.34	1.91	109°.43	2.54	145°.53
0.03	1°.72	0.66	37°.82	1.29	73°.91	1.92	110°.01	2.55	146°.10
0.04	2°.29	0.67	38°.39	1.30	74°.48	1.93	110°.58	2.56	146°.68
0.05	2°.86	0.68	38°.96	1.31	75°.06	1.94	111°.15	2.57	147°.25
0.06	3°.44	0.69	39°.53	1.32	75°.63	1.95	111°.73	2.58	147°.82
0.07	4°.01	0.70	40°.11	1.33	76°.20	1.96	112°.30	2.59	148°.40
0.08	4°.58	0.71	40°.68	1.34	76°.78	1.97	112°.87	2.60	148°.97
0.09	5°.16	0.72	41°.25	1.35	77°.35	1.98	113°.45	2.61	149°.54
0.10	5°.73	0.73	41°.83	1.36	77°.92	1.99	114°.02	2.62	150°.11
0.11	6°.30	0.74	42°.40	1.37	78°.50	2.00	114°.59	2.63	150°.69
0.12	6°.88	0.75	42°.97	1.38	79°.07	2.01	115°.16	2.64	151°.26
0.13	7°.45	0.76	43°.54	1.39	79°.64	2.02	115°.74	2.65	151°.83
0.14	8°.02	0.77	44°.12	1.40	80°.21	2.03	116°.31	2.66	152°.41
0.15	8°.59	0.78	44°.69	1.41	80°.79	2.04	116°.88	2.67	152°.98
0.16	9°.17	0.79	45°.26	1.42	81°.36	2.05	117°.46	2.68	153°.55
0.17	9°.74	0.80	45°.84	1.43	81°.93	2.06	118°.03	2.69	154°.13
0.18	10°.31	0.81	46°.41	1.44	82°.51	2.07	118°.60	2.70	154°.70
0.19	10°.89	0.82	46°.98	1.45	83°.08	2.08	119°.18	2.71	155°.27
0.20	11°.46	0.83	47°.56	1.46	83°.65	2.09	119°.75	2.72	155°.84
0.21	12°.03	0.84	48°.13	1.47	84°.22	2.10	120°.32	2.73	156°.42
0.22	12°.61	0.85	48°.70	1.48	84°.80	2.11	120°.89	2.74	156°.99
0.23	13°.18	0.86	49°.27	1.49	85°.37	2.12	121°.47	2.75	157°.56
0.24	13°.75	0.87	49°.85	1.50	85°.94	2.13	122°.04	2.76	158°.14
0.25	14°.32	0.88	50°.42	1.51	86°.52	2.14	122°.61	2.77	158°.71
0.26	14°.90	0.89	50°.99	1.52	87°.09	2.15	123°.19	2.78	159°.28
0.27	15°.47	0.90	51°.27	1.53	87°.66	2.16	123°.76	2.79	159°.86
0.28	16°.04	0.91	52°.14	1.54	88°.24	2.17	124°.33	2.80	160°.43
0.29	16°.62	0.92	52°.71	1.55	88°.81	2.18	124°.90	2.81	161°.00
0.30	17°.19	0.93	53°.29	1.56	89°.38	2.19	125°.48	2.82	161°.57
0.31	17°.76	0.94	53°.86	1.57	89°.95	2.20	126°.05	2.83	162°.15
0.32	18°.33	0.95	54°.43	1.58	90°.53	2.21	126°.62	2.84	162°.72
0.33	18°.91	0.96	55°.00	1.59	91°.10	2.22	127°.20	2.85	163°.29
0.34	19°.48	0.97	55°.58	1.60	91°.67	2.23	127°.77	2.86	163°.87
0.35	20°.05	0.98	56°.15	1.61	92°.25	2.24	128°.34	2.87	164°.44
0.36	20°.63	0.99	56°.72	1.62	92°.82	2.25	128°.92	2.88	165°.01
0.37	21°.20	1.00	57°.30	1.63	93°.39	2.26	129°.49	2.89	165°.58
0.38	21°.77	1.01	57°.87	1.64	93°.97	2.27	130°.06	2.90	166°.16
0.39	22°.35	1.02	58°.44	1.65	94°.54	2.28	130°.63	2.91	166°.73
0.40	22°.92	1.03	59°.01	1.66	95°.11	2.29	131°.21	2.92	167°.30
0.41	23°.49	1.04	59°.59	1.67	95°.68	2.30	131°.78	2.93	167°.88
0.42	24°.06	1.05	60°.16	1.68	96°.26	2.31	132°.35	2.94	168°.45
0.43	24°.64	1.06	60°.73	1.69	96°.83	2.32	132°.93	2.95	169°.02
0.44	25°.21	1.07	61°.31	1.70	97°.40	2.33	133°.50	2.96	169°.60
0.45	25°.78	1.08	61°.88	1.71	97°.98	2.34	134°.07	2.97	170°.17
0.46	26°.36	1.09	62°.45	1.72	98°.55	2.35	134°.65	2.98	170°.74
0.47	26°.93	1.10	63°.03	1.73	99°.12	2.36	135°.22	2.99	171°.31
0.48	27°.50	1.11	63°.60	1.74	99°.69	2.37	135°.79	3.00	171°.89
0.49	28°.07	1.12	64°.17	1.75	100°.27	2.38	136°.36	3.01	172°.46
0.50	28°.65	1.13	64°.74	1.76	100°.84	2.39	136°.94	3.02	173°.03
0.51	29°.22	1.14	65°.32	1.77	101°.41	2.40	137°.51	3.03	173°.61
0.52	29°.79	1.15	65°.89	1.78	101°.99	2.41	138°.08	3.04	174°.18
0.53	30°.37	1.16	66°.46	1.79	102°.56	2.42	138°.66	3.05	174°.75
0.54	30°.94	1.17	67°.04	1.80	103°.13	2.43	139°.23	3.06	175°.33
0.55	31°.51	1.18	67°.61	1.81	103°.71	2.44	139°.80	3.07	175°.90
0.56	32°.09	1.19	68°.18	1.82	104°.28	2.45	140°.37	3.08	176°.47
0.57	32°.66	1.20	68°.75	1.83	104°.85	2.46	140°.95	3.09	177°.04
0.58	33°.23	1.21	69°.33	1.84	105°.42	2.47	141°.52	3.10	177°.62
0.59	33°.80	1.22	69°.90	1.85	106°.00	2.48	142°.09	3.11	178°.19
0.60	34°.38	1.23	70°.47	1.86	106°.57	2.49	142°.67	3.12	178°.76
0.61	34°.95	1.24	71°.05	1.87	107°.14	2.50	143°.24	3.13	179°.34
0.62	35°.52	1.25	71°.62	1.88	107°.72	2.51	143°.81	3.14	179°.91
0.63	36°.10	1.26	72°.19	1.89	108°.29	2.52	144°.39	3.15	180°.48

Interpolation	
0.0002	0°.01
0.0004	0°.02
0.0006	0°.03
0.0008	0°.05
0.0010	0°.06
0.0012	0°.07
0.0014	0°.08
0.0016	0°.09
0.0018	0°.10
0.0020	0°.11
0.0022	0°.13
0.0024	0°.14
0.0026	0°.15
0.0028	0°.16
0.0030	0°.17
0.0032	0°.18
0.0034	0°.19
0.0036	0°.21
0.0038	0°.22
0.0040	0°.23
0.0042	0°.24
0.0044	0°.25
0.0046	0°.26
0.0048	0°.28
0.0050	0°.29
0.0052	0°.30
0.0054	0°.31
0.0056	0°.32
0.0058	0°.33
0.0060	0°.34
0.0062	0°.36
0.0064	0°.37
0.0066	0°.38
0.0068	0°.39
0.0070	0°.40
0.0072	0°.41
0.0074	0°.42
0.0076	0°.44
0.0078	0°.45
0.0080	0°.46
0.0082	0°.47
0.0084	0°.48
0.0086	0°.49
0.0088	0°.50
0.0090	0°.52
0.0092	0°.53
0.0094	0°.54
0.0096	0°.55
0.0098	0°.56

Multiples of π		
1	3.1416	180°
2	6.2832	360°
3	9.4248	540°
4	12.5664	720°
5	15.7080	900°
6	18.8496	1080°
7	21.9911	1260°
8	25.1327	1440°
9	28.2743	1620°
10	31.4159	1800°

From Lionel S. Marks, "Mechanical Engineers' Handbook."

K. Exponentials

(e^n and e^{-n})

n	e^n	Diff.	n	e^n	Diff.	n	e^n	n	e^{-n}	Diff.	n	e^{-n}	n	e^{-n}
0.00	1.000	10	0.50	1.649	16	1.0	2.718†	0.00	1.000	−10	0.50	0.607	1.0	0.368
0.01	1.010	10	0.51	1.665	17	1.1	3.004	0.01	0.990	−10	0.51	0.600	1.1	0.333
0.02	1.020	10	0.52	1.682	17	1.2	3.320	0.02	0.980	−10	0.52	0.595	1.2	0.301
0.03	1.030	11	0.53	1.699	17	1.3	3.669	0.03	0.970	− 9	0.53	0.589	1.3	0.273
0.04	1.041	10	0.54	1.716	17	1.4	4.055	0.04	0.961	−10	0.54	0.583	1.4	0.247
0.05	1.051	11	0.55	1.733	18	1.5	4.482	0.05	0.951	− 9	0.55	0.577	1.5	0.223
0.06	1.062	11	0.56	1.751	17	1.6	4.953	0.06	0.942	−10	0.56	0.571	1.6	0.202
0.07	1.073	10	0.57	1.768	18	1.7	5.474	0.07	0.932	− 9	0.57	0.566	1.7	0.183
0.08	1.083	11	0.58	1.786	18	1.8	6.050	0.08	0.923	− 9	0.58	0.560	1.8	0.165
0.09	1.094	11	0.59	1.804	18	1.9	6.686	0.09	0.914	− 9	0.59	0.554	1.9	0.150
0.10	1.105	11	0.60	1.822	18	2.0	7.389	0.10	0.905	− 9	0.60	0.549	2.0	0.135
0.11	1.116	11	0.61	1.840	19	2.1	8.166	0.11	0.896	− 9	0.61	0.543	2.1	0.122
0.12	1.127	12	0.62	1.859	19	2.2	9.025	0.12	0.887	− 9	0.62	0.538	2.2	0.111
0.13	1.139	11	0.63	1.878	18	2.3	9.974	0.13	0.878	− 9	0.63	0.533	2.3	0.100
0.14	1.150	12	0.64	1.896	20	2.4	11.02	0.14	0.869	− 8	0.64	0.527	2.4	0.0907
0.15	1.162	12	0.65	1.916	19	2.5	12.18	0.15	0.861	− 9	0.65	0.522	2.5	0.0821
0.16	1.174	11	0.66	1.935	19	2.6	13.46	0.16	0.852	− 8	0.66	0.517	2.6	0.0743
0.17	1.185	12	0.67	1.954	20	2.7	14.88	0.17	0.844	− 9	0.67	0.512	2.7	0.0672
0.18	1.197	12	0.68	1.974	20	2.8	16.44	0.18	0.835	− 8	0.68	0.507	2.8	0.0608
0.19	1.209	12	0.69	1.994	20	2.9	18.17	0.19	0.827	− 8	0.69	0.502	2.9	0.0550
0.20	1.221	13	0.70	2.014	20	3.0	20.09	0.20	0.819	− 8	0.70	0.497	3.0	0.0498
0.21	1.234	12	0.71	2.034	20	3.1	22.20	0.21	0.811	− 8	0.71	0.492	3.1	0.0450
0.22	1.246	13	0.72	2.054	21	3.2	24.53	0.22	0.803	− 8	0.72	0.487	3.2	0.0408
0.23	1.259	12	0.73	2.075	21	3.3	27.11	0.23	0.795	− 8	0.73	0.482	3.3	0.0369
0.24	1.271	13	0.74	2.096	21	3.4	29.96	0.24	0.787	− 8	0.74	0.477	3.4	0.0334
0.25	1.284	13	0.75	2.117	21	3.5	33.12	0.25	0.779	− 8	0.75	0.472	3.5	0.0302
0.26	1.297	13	0.76	2.138	22	3.6	36.60	0.26	0.771	− 8	0.76	0.468	3.6	0.0273
0.27	1.310	13	0.77	2.160	21	3.7	40.45	0.27	0.763	− 7	0.77	0.463	3.7	0.0247
0.28	1.323	13	0.78	2.181	22	3.8	44.70	0.28	0.756	− 8	0.78	0.458	3.8	0.0224
0.29	1.336	14	0.79	2.203	23	3.9	49.40	0.29	0.748	− 7	0.79	0.454	3.9	0.0202
0.30	1.350	13	0.80	2.226	22	4.0	54.60	0.30	0.741	− 8	0.80	0.449	4.0	0.0183
0.31	1.363	14	0.81	2.248	22	4.1	60.34	0.31	0.733	− 7	0.81	0.445	4.1	0.0166
0.32	1.377	14	0.82	2.270	23	4.2	66.69	0.32	0.726	− 7	0.82	0.440	4.2	0.0150
0.33	1.391	14	0.83	2.293	23	4.3	73.70	0.33	0.719	− 7	0.83	0.436	4.3	0.0136
0.34	1.405	14	0.84	2.316	24	4.4	81.45	0.34	0.712	− 7	0.84	0.432	4.4	0.0123
0.35	1.419	14	0.85	2.340	23	4.5	90.02	0.35	0.705	− 7	0.85	0.427	4.5	0.0111
0.36	1.433	15	0.86	2.363	24			0.36	0.698	− 7	0.86	0.423		
0.37	1.448	14	0.87	2.387	24	5.0	148.4	0.37	0.691	− 7	0.87	0.419	5.0	0.00674
0.38	1.462	15	0.88	2.411	24	6.0	403.4	0.38	0.684	− 7	0.88	0.415	6.0	0.00248
0.39	1.477	15	0.89	2.435	25	7.0	1097.	0.39	0.677	− 7	0.89	0.411	7.0	0.000912
0.40	1.492	15	0.90	2.460	24	8.0	2981.	0.40	0.670	− 6	0.90	0.407	8.0	0.000335
0.41	1.507	15	0.91	2.484	25	9.0	8103.	0.41	0.664	− 7	0.91	0.403	9.0	0.000123
0.42	1.522	15	0.92	2.509	26	10.0	22026.	0.42	0.657	− 6	0.92	0.399	10.0	0.000045
0.43	1.537	16	0.93	2.535	25	$\pi/2$	4.810	0.43	0.651	− 7	0.93	0.395		
0.44	1.553	15	0.94	2.560	26	$2\pi/2$	23.14	0.44	0.644	− 6	0.94	0.391		
0.45	1.568	16	0.95	2.586	26	$3\pi/2$	111.3	0.45	0.638	− 7	0.95	0.387	$\pi/2$	0.208
0.46	1.584	16	0.96	2.612	26	$4\pi/2$	535.5	0.46	0.631	− 6	0.96	0.383	$2\pi/2$	0.0432
0.47	1.600	16	0.97	2.638	26	$5\pi/2$	2576	0.47	0.625	− 6	0.97	0.379	$3\pi/2$	0.00898
0.48	1.616	16	0.98	2.664	27	$6\pi/2$	12392	0.48	0.619	− 6	0.98	0.375	$4\pi/2$	0.00187
0.49	1.632	17	0.99	2.691	27	$7\pi/2$	59610	0.49	0.613	− 6	0.99	0.372	$5\pi/2$	0.000388
0.50	1.649		1.00	2.718		$8\pi/2$	286751	0.50	0.607		1.00	0.368	$6\pi/2$	0.000081

$e = 2.71828$. $1/e = 0.367879$. $\log_{10} e = 0.4343$. $1/(0.4343) = 2.3026$.
From Lionel S. Marks, "Mechanical Engineers' Handbook."
† NOTE: Do not interpolate in this column.

L. Trigonometric Relations

$$\sin x = \frac{A}{B}$$

$$\cos x = \frac{C}{B}$$

$$\tan x = \frac{A}{C} = \frac{\sin x}{\cos x}$$

$$\sec x = \frac{B}{C} = \frac{1}{\cos x}$$

$$\operatorname{cosec} x = \frac{B}{A} = \frac{1}{\sin x}$$

$$\cot x = \frac{C}{A} = \frac{1}{\tan x} = \frac{\cos x}{\sin x}$$

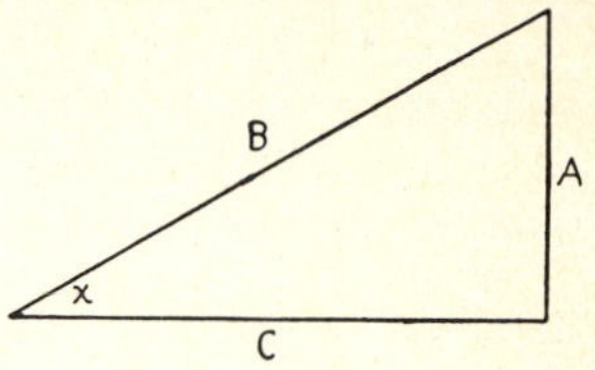

Fig. 1.

$$\sin(-x) = -\sin x$$
$$\cos(-x) = \cos x$$
$$\tan(-x) = -\tan x$$

$$\sin\left(\frac{\pi}{2} - x\right) = \cos x \qquad \sin\left(\frac{\pi}{2} + x\right) = \cos x$$

$$\cos\left(\frac{\pi}{2} - x\right) = \sin x \qquad \cos\left(\frac{\pi}{2} + x\right) = -\sin x$$

$$\tan\left(\frac{\pi}{2} - x\right) = \cot x \qquad \tan\left(\frac{\pi}{2} + x\right) = -\cot x$$

$$\sin(\pi - x) = \sin x \qquad \sin(\pi + x) = -\sin x$$
$$\cos(\pi - x) = -\cos x \qquad \cos(\pi + x) = -\cos x$$
$$\tan(\pi - x) = -\tan x \qquad \tan(\pi + x) = \tan x$$

$$\left.\begin{aligned} \sin(x + 2\pi n) &= \sin x \\ \cos(x + 2\pi n) &= \cos x \\ \tan(x + 2\pi n) &= \tan x \end{aligned}\right\} \text{(n a positive or negative integer)}$$

$$\sin(x + y) = \sin x \cos y + \cos x \sin y$$
$$\sin(x - y) = \sin x \cos y - \cos x \sin y$$
$$\cos(x + y) = \cos x \cos y - \sin x \sin y$$
$$\cos(x - y) = \cos x \cos y + \sin x \sin y$$

$$\tan(x + y) = \frac{\tan x + \tan y}{1 - \tan x \tan y}$$

$$\tan(x - y) = \frac{\tan x - \tan y}{1 + \tan x \tan y}$$

$$\sin 2x = 2 \sin x \cos x \qquad \cos 2x = 2\cos^2 x - 1 = 1 - 2\sin^2 x$$

$$\tan 2x = \frac{2 \tan x}{1 - \tan^2 x} \qquad \cot 2x = \frac{(\cot^2 x - 1)}{2 \cot x}$$

$$\sin 3x = 3 \sin x - 4 \sin^3 x \qquad \cos 3x = 4\cos^3 x - 3\cos x$$

$$\sin \frac{1}{2} x = \left[\frac{1}{2}(1 - \cos x)\right]^{1/2} = \frac{1}{2}(1 + \sin x)^{1/2} - \frac{1}{2}(1 - \sin x)^{1/2}$$

$$\cos \frac{1}{2} x = \left[\frac{1}{2}(1 + \cos x)\right]^{1/2} = \frac{1}{2}(1 + \sin x)^{1/2} + \frac{1}{2}(1 - \sin x)^{1/2}$$

$$\tan \frac{1}{2} x = \frac{(1 - \cos x)^{1/2}}{(1 + \cos x)^{1/2}} = \frac{1 - \cos x}{\sin x} = \frac{\sin x}{1 + \cos x}$$

$$\sin^2 x = \frac{1}{2}(1 - \cos 2x) \qquad \cos^2 x = \frac{1}{2}(1 + \cos 2x)$$

$$\sin^3 x = \frac{1}{4}(3 \sin x - \sin 3x) \qquad \cos^3 x = \frac{1}{4}(\cos 3x + 3 \cos x)$$

L. Trigonometric Relations (*Concluded*)

$$\sin x \sin y = \frac{1}{2}\cos(x-y) - \frac{1}{2}\cos(x+y)$$

$$\cos x \cos y = \frac{1}{2}\cos(x-y) + \frac{1}{2}\cos(x+y)$$

$$\sin x \cos y = \frac{1}{2}\sin(x-y) + \frac{1}{2}\sin(x+y)$$

$$\sin x + \sin y = 2\sin\frac{1}{2}(x+y)\cos\frac{1}{2}(x-y)$$

$$\sin x - \sin y = 2\cos\frac{1}{2}(x+y)\sin\frac{1}{2}(x-y)$$

$$\cos x + \cos y = 2\cos\frac{1}{2}(x+y)\cos\frac{1}{2}(x-y)$$

$$\cos x - \cos y = -2\sin\frac{1}{2}(x+y)\sin\frac{1}{2}(x-y)$$

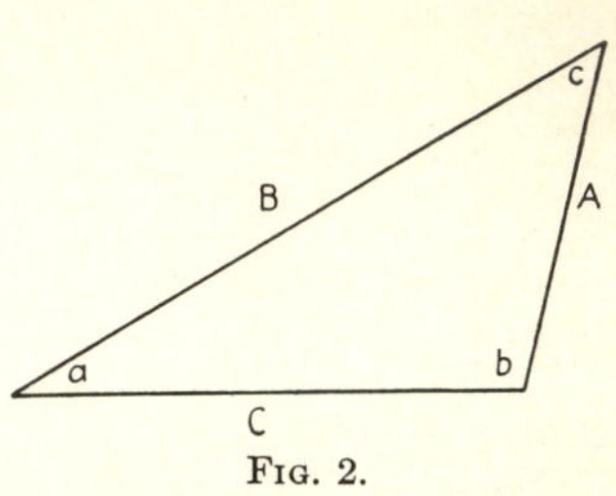

FIG. 2.

$$\tan x + \tan y = \frac{\sin(x+y)}{\cos x \cos y} \qquad \tan x - \tan y = \frac{\sin(x-y)}{\cos x \cos y}$$

$$\sin^2 x - \sin^2 y = \sin(x+y)\sin(x-y)$$
$$\cos^2 x - \cos^2 y = -\sin(x+y)\sin(x-y)$$
$$\cos^2 x - \sin^2 y = \cos(x+y)\cos(x-y)$$
$$\sin^2 x + \cos^2 x = 1 \qquad \sec^2 x - \tan^2 x = 1$$

In any triangle (Fig. 2):

$$\frac{A}{\sin a} = \frac{B}{\sin b} = \frac{C}{\sin c} \qquad \text{(law of sines)}$$

$$A^2 = B^2 + C^2 - 2BC\cos a \qquad \text{(law of cosines)}$$

$$\frac{A+B}{A-B} = \frac{\tan\frac{1}{2}(a+b)}{\tan\frac{1}{2}(a-b)} = \frac{\sin a + \sin b}{\sin a - \sin b} \qquad \text{(law of tangents)}$$

$$a + b + c = 180^\circ \qquad A = B\cos c + C\cos b$$
$$B = C\cos a + A\cos c \qquad C = A\cos b + B\cos a$$

M. Properties of Hyperbolic Functions

$$\sinh x = \frac{\epsilon^x - \epsilon^{-x}}{2} = x + \frac{x^3}{3!} + \frac{x^5}{5!} + \cdots$$

$$\cosh x = \frac{\epsilon^x + \epsilon^{-x}}{2} = 1 + \frac{x^2}{2!} + \frac{x^4}{4!} + \cdots$$

$$\tanh x = \frac{\sinh x}{\cosh x}$$

$$\sinh(-x) = -\sinh x$$

$$\cosh(-x) = \cosh x$$

$$\cosh^2 x = 1 + \sinh^2 x$$

$$\frac{d(\sinh x)}{dx} = \cosh x$$

$$\frac{d(\cosh x)}{dx} = \sinh x$$

$$\int \sinh x \cdot dx = \cosh x$$

$$\int \cosh x \cdot dx = \sinh x$$

$$\sinh(x \pm j\pi) = -\sinh x$$

$$\cosh(x \pm j\pi) = -\cosh x$$

$$\sinh(x \pm j2\pi) = \sinh x$$

$$\cosh(x \pm j2\pi) = \cosh x$$

FIG. 1.—REACTANCE CHART*

Always obtain approximate value from Fig. 2 before using Fig. 1

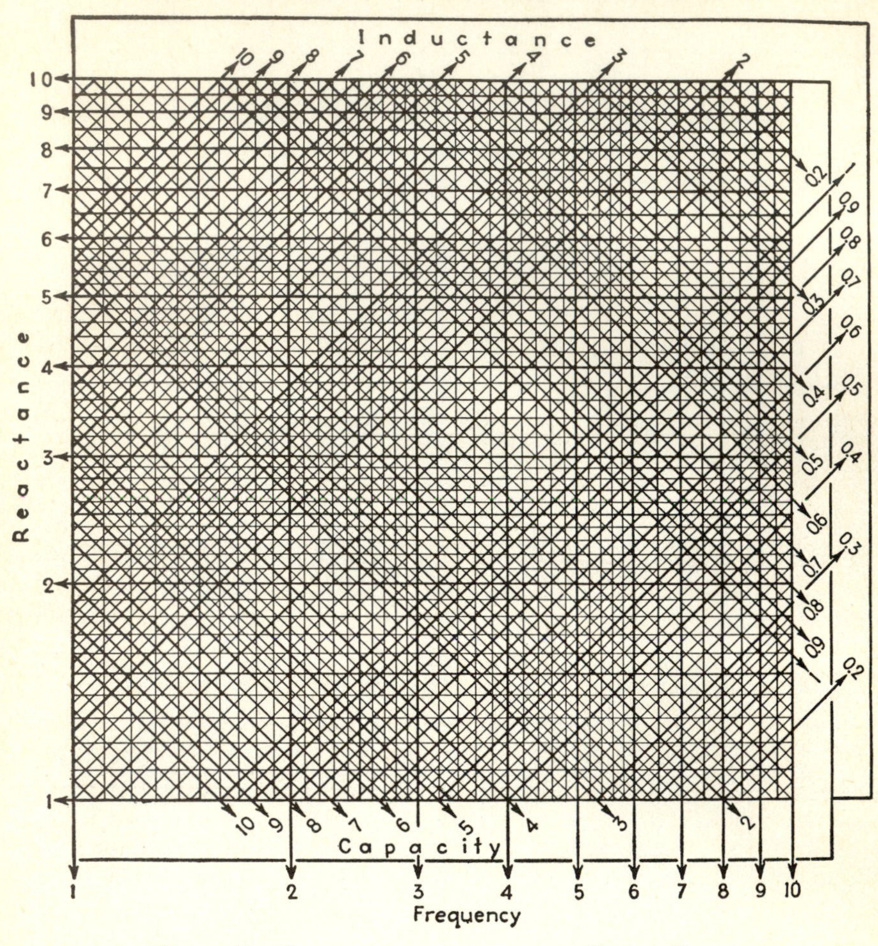

Use of Fig. 1

Figure 1 is used to obtain additional precision of reading but does not place the decimal point which must be located from a preliminary entry on Fig. 2. Since the chart necessarily requires two logarithmic decades for inductance and capacitance for every single decade of frequency and reactance, unless the correct decade for L and C is chosen, erroneous results are obtained.

Typical Results

1. Find reactance of inductance of 0.00012 henry at 960 kc(kHz).

Answer: 720 ohms.

2. What capacity will have 265 ohms reactance at 7000 kc?

Answer: 86 $\mu\mu$f(pF).

3. What is the resonant frequency with $L = 21\ \mu$h and $C = 45\mu\mu$f?

Answer: 5.18 mc(MHz).

* Figures 1 and 2 are from General Radio Company Reactance Chart.

FIG. 2.—REACTANCE CHART*

Always use corresponding scales

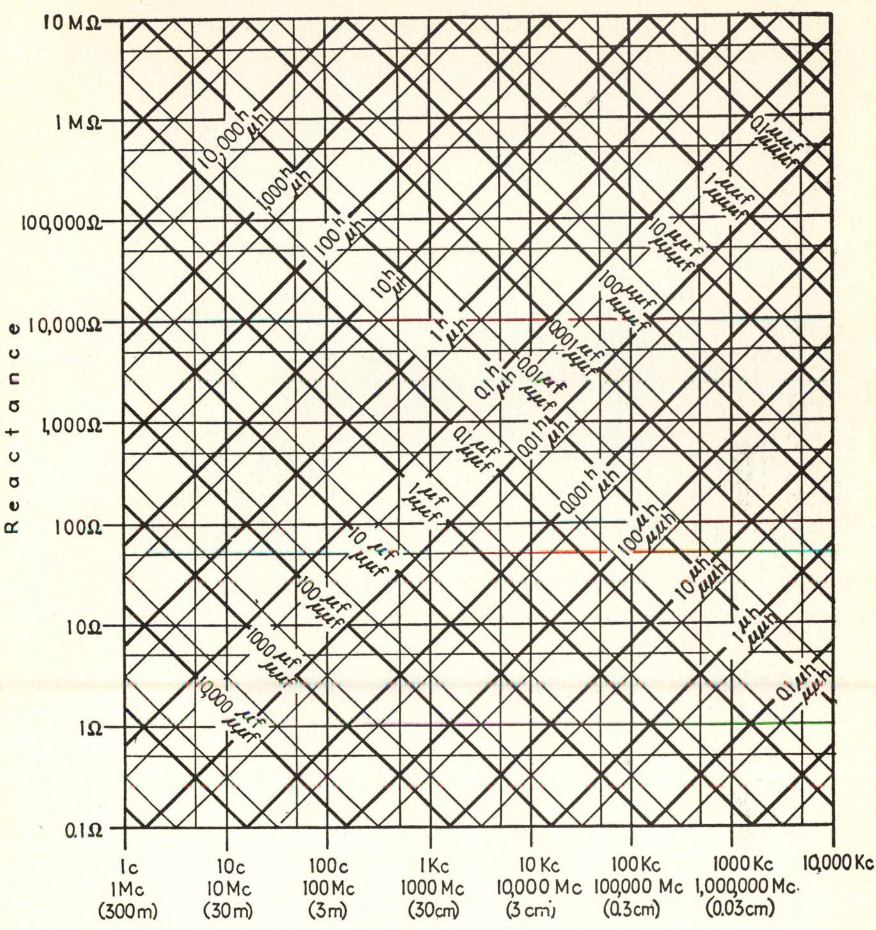

The accompanying chart may be used to find

(1) The reactance of a given inductance at a given frequency.

(2) The reactance of a given capacity at a given frequency.

(3) The resonant frequency of a given inductance and capacitance.

In order to facilitate the determination of magnitude of the quantities involved to two or three significant figures the chart is divided into two parts. Figure 2 is the complete chart to be used for rough calculations. Figure 1, which is a single decade of Fig. 2 enlarged approximately 7 times, is to be used where the significant two or three figures are to be determined.

To Find Reactance

Enter the charts vertically from the bottom (frequency) and along the lines slanting upward to the left (inductance) or to the right (capacity). Corresponding scales (upper or lower) must be used throughout. Project horizontally to the left from the intersection and read reactance.

To Find Resonant Frequency

Enter the slanting lines for the given inductance and capacity. Project downward from their intersection and read resonant frequency from the bottom scale. Corresponding scales (upper or lower) must be used throughout.

* Figures 1 and 2 are from General Radio Company Reactance Chart.